Switching Power Supplies A–Z

Switching Power Supplies A–Z

Sanjaya Maniktala

AMSTERDAM • BOSTON • HEIDELBERG • LONDON
NEW YORK • OXFORD • PARIS • SAN DIEGO
SAN FRANCISCO • SINGAPORE • SYDNEY • TOKYO

Newnes is an imprint of Elsevier

ELSEVIER

Newnes

Newnes is an imprint of Elsevier
The Boulevard, Langford Lane, Kidlington, Oxford OX5 1GB, UK
225 Wyman Street, Waltham, MA 02451, USA

Notice
No responsibility is assumed by the publisher for any injury and/or damage to persons or property as a matter of products liability, negligence or otherwise, or from any use or operation of any methods, products, instructions or ideas contained in the material herein. Because of rapid advances in the medical sciences, in particular, independent verification of diagnoses and drug dosages should be made

British Library Cataloguing-in-Publication Data
A catalogue record for this book is available from the British Library

Library of Congress Cataloging-in-Publication Data
A catalog record for this book is available from the Library of Congress

ISBN: 978-0-12-386533-5

For information on all Newnes publications
visit our web site at www.newnespress.com

Typeset by MPS Limited, a Macmillan Company, Chennai, India
www.macmillansolutions.com

Printed in Great Britain

11 12 13 14 15   10 9 8 7 6 5 4

Working together to grow
libraries in developing countries

www.elsevier.com | www.bookaid.org | www.sabre.org

ELSEVIER   BOOK AID International   Sabre Foundation

Contents

Preface

I am delighted to be presenting to you the new edition of my A to Z book, also known as "the red book." It is a book that *almost never was*. I will explain why, and also share the reason it did finally make it into your hands today.

After my most recent writing effort in 2007, I was quite convinced I would never write a single word more on the subject. I stopped writing switching-power app notes, EE Times guest column articles, and so on, things I had revelled in the past. I became a wannabe Cat Stevens of sorts in the field of power conversion. Professionally, I moved on to quite a different ball-game, Power-over-Ethernet (which may even be the subject of my next book). Looking back however, perhaps the real reason for my self-imposed reclusiveness was that I was merely trying to avoid being crushed under the wheels of an oncoming freight train we all call "writer's block" (sounds more like chopping block to me). Now I can finally confess, that in vague moments of nervous bravado past, I even secretly contemplated writing a novel instead, something original, perhaps like J.K. Rowling. I would actually make some money for a change (big money oh yeah, not chump change). But then I fell right out of bed. So coming back to the future, 4 years later, I am actually pleased to discover I have managed to add on almost another entire book on top of the existing A to Z book. Really, I couldn't stop even after 9 months of rather grueling nonstop late-night and weekend writing sprees. I dreamt up vividly etched Mathcad programming loops emblazoned with Greek symbols in jaw-dropping *high-definition* quality (just a few years ago they were all in Technicolor, whatever happened?). Every now and then I had to be re-introduced to my wife and daughter, they tell me. One by the other, successively.

Meanwhile six, scratch that, eight new chapters were added. Keep in mind that the seven EMI chapters of the first edition are now four. But they are not condensed, just reorganized. These have actually been enhanced with better graphics and tables. In addition, several detailed design examples have now been thrown in, and old typos/errors hopefully removed.

Coming to magnetics, you will see that all the original equations on flyback core sizing that I had virtually pulled out of a hat in the first edition, yes, those very equations that had seemed so suspiciously simple, or perhaps just too good to be true (especially considering you couldn't find them in any other book), are now all derived in *Chapter 5*. They were accurate all along, and that's as much of a relief to me, the author, as perhaps to you. But there's much more in *Chapter 5*, too, starting with the unique micro-joule-by-micro-joule energy transfer diagrams of the three topologies. I am pretty sure the process of power conversion has never been explained in related literature in so elemental, or fundamental, a manner. Yes, I really do like to cut to the chase. All the new material should go a long way in appeasing the magnetics skeptics in particular. But to be on the safe side, I have also presented all the equations and the original Mathcad worksheet, behind the AC resistance/proximity analysis charts I had previously published in *Chapter 3* (though this time there is no accompanying Mathcad CD, only Mathcad-in-text). Air-gapped cores are also covered in a very simple manner in *Chapter 5*, using my unique "*z*-factor" treatment. I admit that particular idea came to my head years ago while talking to my mentor (and all-time hero of course), Doctor G.T. Murthy (now retired). As you may remember, I worked under him for 5 years in Mumbai (the first edition of that was called Bombay incidentally); and yes, Mumbai was also the undeclared city that formed the underlying inspiration for my well-known power conversion "train terminus analogy" on Page 1 of this book. Incidentally, I realized *Chapter 1* had been so obviously liked by most readers that I did not have the heart to change it one bit, except perhaps to fix some graphics and minor typos. *Chapters 2–4* were improved significantly, as were the old chapters titled "Conduction and Switching Losses" and "PCB Layout." They were actually quite solid (and guaranteed to confuse if the need arose). But the perfectionist got hold of me here.

I have always felt that magnetics need not be as scary and challenging as some people make it out to be. But though I do want to simplify difficult subjects, I do not want to cut corners either and somehow *pretend* it is "oh-so-simple." Because it really isn't — nothing about power conversion is. We have all learned rather painfully over the years, never to judge a book by its cover, or a converter by its size (or its component count). Magnetics, in particular, can be so counterintuitive at times, we need to be on our guard always. The same holds true for loop stability: that is not so "easy" either, but with the right guidance, it can get much easier than you may have initially thought. I certainly don't want young engineers thinking, as I did once, that the only way to understand this rather tricky area of power is to go out frantically geocaching in search of *that exclusive invitation*, the one that grants entry into a privileged club of experienced (and rather snooty) designers, who to the outsider, seem to be leading enviable lives in some mystical imaginary plane strewn with glittering poles and zeros. So, to make things even simpler than I did in the last edition, I rewrote, or rather reorganized, the entire chapter on loop stability. Then I fortified it with a detailed solved example contained in *Chapter 19*. I also added a new section on subharmonic instability and slope compensation, thereby acknowledging a rather glowing but constructive web review I had received on my previous book (you can still find that review on "Analogzone," now "En-genius.net").

In this process of restructuring the entire book to make all these substantial additions as seamless as possible ("no patchwork," I had promised myself), and yet be thorough and simple at the same time (paradoxical I know), I did finally run into an obvious quandary. I wanted to maintain the oft-remarked readability of my previous book by *not* interposing very heavy equations and derivations throughout — I did not want to turn it into a scary textbook guaranteed to drive away all but the most captive EE students. I have always wanted my books to be used for real products, not just for good grades. On the other hand, I really *did* want to include all this extra new "heavy" material, for the more seasoned and demanding practitioners and professionals. So, I opted for a notable compromise: in the form of some rather busy-looking "wall-charts." Initial impressions aside, and I will give you all the time to catch your breath here, the idea is actually quite simple: you, as a reader, can and *should* bypass these charts at first sight, assuming you don't want to get into that much detail so early on in the game. The conclusions from these charts are well summarized in the accompanying text anyway. Later, when you are ready (or experienced enough in power) to get down and dirty, so to say, you can return to these very charts. And at that moment you will likely find all the additional information you need — available at your fingertips, in rather cramped but clearly demarcated pages that I call "wall-charts." You can also use these pages as a quick reference, or cheat-sheet, going forward. In other words, I am hoping you will, not immediately, but *eventually*, love the idea — not only for what it did to the book, but also for what it *didn't* do.

There are also some advanced and contemporary (emerging) topics included this time around: coupled inductors for example (*Chapter 13*). Unfortunately, these topics do need a good amount of mathematics and not much intuition can be brought to bear on them. So, feel free to skip them for now, but I think they are useful and very comprehensive going forward (I do not know of any power conversion book that distills these rather tricky topics for the average reader). I have based all my analyses on some scattered but excellent articles available on the IEEE Xplore website (IEL). Those sources have been duly acknowledged in the updated references of this edition. One time-consuming lesson for me while writing this particular chapter was to realize I shouldn't accept at face value every single article or paper I came across on the subject, and I advise you to do the same to avoid needless confusion. I saw some papers/articles claiming *everything* on Planet Earth improved as a result of inductor coupling. But I know for sure that power conversion is, if anything, all about trade-offs and design compromises. As engineers, we almost instinctively *expect* to see *both* pros *and* cons listed out logically; two sides of the same coin. If not, I admit I tend to get worried that some potentially good engineering presentations from some fine corporations otherwise, were run past marketing first.

One long chapter, "Discovering New Topologies" (*Chapter 9*), has some really unique ideas, but I will let you judge its eventual worth. Is it a game-changer or is it not? Let me know. Same for the front-end of AC−DC power supplies (*Chapter 14*). See if you like that too. Authors always need to allow their readers to pass eventual judgment, even though a lot of work may have gone into a particular chapter. All I will ask of you is: these are tricky chapters, please don't form an opinion after skimming through them just once.

Finally, time for some off-the-cuff observations: for example, you may have noticed that quite coincidentally, this edition too is largely red in color. I can therefore continue to refer to it in the same old-fashioned way. At my age, people quickly realize it is never a good idea to try and change habits; proposed changes habitually end up dead much faster by the roadside than habits. Ask my wife of two decades! So, I am just plain happy I can stick to calling this the "red one" and staying well within my comfort zone. But names aside, something has obviously changed drastically this time around: not just under the hood, but all around it, in fact maybe everything *except* the hood itself. Why? Well, this does happen to be the *next* edition of a *previous* book. And I believe getting to this point implies the earlier edition was a great success. So, for that milestone and achievement alone, besides my ever-supportive publishers, I really need to warmly thank my fantastic readers over the years. Especially those who, as complete strangers, took precious time off their busy schedules, either to write to me personally or post some really nice and obviously heart-felt 4- or 5-star reviews on the web (I am disregarding a couple of obvious trolls out there, for reasons *known*). However, I do need to apologize to some of you out there since I did not always manage to reply to your rather encouraging e-mails. But please be very clear about one thing: the only reason for the book you hold in your hands today is *you. You* made it happen. I got my writing spirit back largely because of *you.* Therefore, this book, in any shape, form, or color, would always remain my way of saying: *thank you so much for your support and wishes in the past, and I hope you like this even more than the previous one.*

–Sanjaya Maniktala
(*Fremont, California*)

Acknowledgments

Many people went into the creation of this book in one way or another. I would like to thank them all as follows.

(a) My most engaged and competent technical reviewer this time around was Sheshagiri Haniyur. He read several of these chapters very carefully indeed, made great comments, and very appropriately even asked me on one occasion to completely rewrite the culprit (*Chapter 14*), which I promptly did.

(b) I had absolutely great technical feedback from two very smart readers of mine, who contacted me by e-mail over the years, and then proceeded to literally microscopically examine almost every equation in every chapter of my previous edition. They really helped me improve important sections of this edition. I must thank Chee How from Malaysia and Meroni Silvano from Italy, listed in the order they came into the picture. Thanks guys or gals, I have never met you but I loved your attitude, technical skills, and sheer diligence. You caught my mistakes quite a few times too, and I have kept all you told me in mind for this edition. So let me know if I got it right and if it is easy to understand too.

(c) A favorite reviewer of mine in the last edition helped me out again on several chapters this time. He is Harry Holt. Thanks to upright and brutally honest Harry as always. You make it better every time. Some new, and very engaged reviewers plus supporters this time include an ex-colleague of mine, Dipak Patel. Many thanks guys.

(d) Other key readers/reviewers who made this book possible are Ken Coffman, Gautam (Tom) Nath, and Inder Dhingra.

(e) Not to forget the best PR person I ever knew: Mike He, who literally turned me into a known author years ago at National, and is still a great friend of mine.

(f) Elsevier folks are always great people to work with, as I have discovered. I have not forgotten my first commissioning editor, the wonderful Chuck Glaser. This time around, the commissioning editor was the equally likeable and encouraging Tim Pitts. You should understand that Tim is the primary reason for the existence of this second edition. He also gave me extraordinary support through his wonderful staff, namely Miss Charlotte ("Charlie") Kent and Sally Mortimore.

(g)   My wife said it is a cliché to thank her every time, and that this time I shouldn't bother. But it is really true. For 9 months I was only *seemingly* present in her life. But she chugged along stoically, being extremely understanding and supportive. As did my wonderful daughter Aartika. And as for my two furry little four-legged friends, Munchi and Cookie, they were the two Maltipoos who kept me thoughtfully hydrated (from the outside), as I toiled away almost unaware.

(h)   We should not forget this book is an extension of Doctor G.T. Murthy's work and undying influence on many of us across the world today. He was not only my mentor but also gave me a life and career when I needed it the most. He was certainly one of the most progressive, ethical, and also "human" managers I ever knew. He could, and often would, use his considerable influence to fight for his men and also for what he strongly believed in. He was not the one who strived merely to sound politically correct, make the right sounds, and appear "good" in an unfortunate modern, stereotypical corporate sense. I also remember he always seemed to uncannily understand *me* far better than my very own blood. So I am still very proud to know him, and wish him health and a very long life.

(i)   A lot of readers write in to me with sincere thanks and encouragement. They have taken the extra, genuine effort to make authors like me feel it was worth it, and are the real reason why we return with a new edition. So, I must thank some of them equally sincerely here: Gheorghe (Gigi) Plaesoianu, Mark MarKell, Chris Themelis, Roberto Zanzottera, Mirza Kolakovic, Rajan Darekar, Sanjay Agrawal, Gene Krzywinski, Ramesh Tirumala, Stephen Blake, Xiaohong Zhu, Charles Potter, Robert Haugum, Xia Heng, Eric Wen, Roc Zhu, Cyril Aloysius Quinto, Sridhar Gurram, Alex Byrley, Meng (Mark) Jianhui, Bingbing Song, Michael Chang, Wei Guan, Ronald Moradkhan, Amalendu Iyer, Georg Glock, besides several more that I may have forgotten to mention. They are all part of the very reason for this book.

(j)   I also sincerely wanted to thank Debbie Clark, Production supervisor of Elsevier, for her tremendous patience getting this book right eventually, and looking so presentable in my opinion.

CHAPTER 1

The Principles of Switching Power Conversion

Introduction

Imagine we are at some busy "metro" terminus one evening at peak hour. Almost instantly, thousands of commuters swarm the station trying to make their way home. Of course, there is no train big enough to carry all of them *simultaneously*. So, what do we do? Simple! We *split* this sea of humanity into several *trainloads* — and move them out in rapid succession. Many of these outbound passengers will later transfer to alternative forms of transport. So, for example, trainloads may turn into bus-loads or taxi-loads, and so on. But eventually, all these "packets" will merge once again, and a throng will be seen, exiting at the destination.

Switching power conversion is remarkably similar to a mass transit system. The difference is that instead of people, it is *energy* that gets transferred from one level to another. So we draw energy continuously from an "input source," *chop* this incoming stream into packets by means of a "switch" (a transistor), and then transfer it with the help of components (inductors and capacitors) that are able to *accommodate* these energy packets and exchange them among themselves as required. Finally, we make all these packets merge again and thereby get a smooth and steady flow of energy into the output.

So, in either of the cases above (energy or people), from the viewpoint of an *observer*, a stream will be seen entering and a similar one exiting. But at an *intermediate* stage, the transference is accomplished by breaking up this stream into more *manageable packets*.

Looking more closely at the train station analogy, we also realize that to be able to transfer a given number of passengers *in a given time* (note that in electrical engineering, energy transferred in unit time is "power") — either we need bigger trains with departure times spaced relatively far apart OR several *smaller* trains leaving in *rapid* succession. Therefore, it should come as no surprise that in switching power conversion, we always try to *switch*

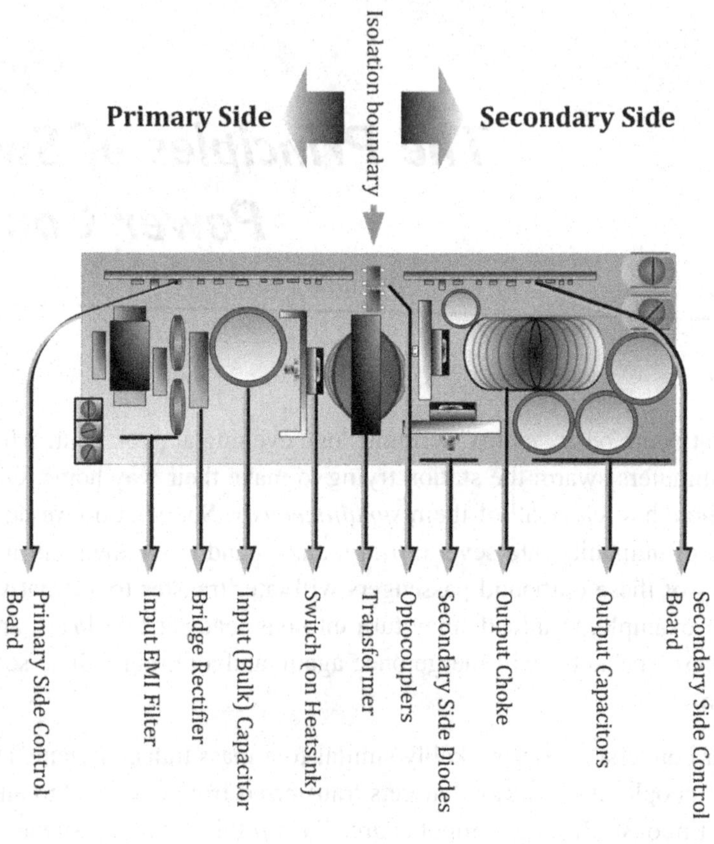

Isolation boundary

Primary Side   **Secondary Side**

Primary Side Control Board

Input EMI Filter

Bridge Rectifier

Input (Bulk) Capacitor

Switch (on Heatsink)

Transformer

Optocouplers

Secondary Side Diodes

Output Choke

Output Capacitors

Secondary Side Control Board

Figure 1.1: Typical off-line power supply.

at high frequencies. The primary purpose for that is to *reduce the size* of the energy packets and thereby also the *size of the components* required to store and transport them.

Power supplies that use this principle are called "switching power supplies" or "switching power converters."

"DC–DC converters" are the basic building blocks of modern high-frequency switching power supplies. As their name suggests, they "convert" an available *DC* (direct current) input voltage rail "V_{IN}" to another more *desirable or usable* DC output voltage level "V_O." "AC–DC converters" (see Figure 1.1), also called "off-line power supplies," typically run off the mains input (or "line input"). But they first *rectify* the incoming sinusoidal AC (alternating current) voltage "V_{AC}" to a DC voltage level (often called the "HVDC rail" or "high voltage DC rail") — which then gets applied at the input of what is essentially just *another DC–DC converter stage* (or derivative thereof). We thus see that power conversion is, in essence, almost always a *DC–DC voltage conversion process*.

But it is also equally important to create a *steady* DC output voltage level, from what can often be a widely *varying* and different DC input voltage level. Therefore, a "control circuit" is used in all power converters to constantly monitor and compare the output voltage against an internal "reference voltage." Corrective action is taken if the output drifts from its set value. This process is called "output regulation" or simply "regulation." Hence, the generic term "voltage regulator" for supplies which can achieve this function, switching, or otherwise.

In a practical implementation, "application conditions" are considered to be the applied input voltage V_{IN} (also called the "line voltage"), the current being drawn at the output, that is, I_O (the "load current"), and the set output voltage V_O. Temperature is also an application condition, but we will ignore it for now, since its effect on the system is usually not so dramatic. Therefore, for a given output voltage, there are *two* specific application conditions whose variations can cause the output voltage to be immediately impacted (were it not for the control circuit). Maintaining the output voltage steady when V_{IN} varies over its stated operating range V_{INMIN} to V_{INMAX} (minimum input to maximum input) is called "line regulation," whereas maintaining regulation when I_O varies over its operating range I_{OMIN} to I_{OMAX} (minimum-to-maximum load) is referred to as "load regulation." Of course, nothing is ever "perfect," so nor is the regulation. Therefore, despite the correction, there is a small but measurable change in the output voltage, which we call "ΔV_O" here. Note that mathematically, line regulation is expressed as "$\Delta V_O/V_O \times 100\%$ (implicitly implying it is over V_{INMIN} to V_{INMAX})." Load regulation is similarly "$\Delta V_O/V_O \times 100\%$" (from I_{OMIN} to I_{OMAX}).

However, the *rate* at which the output can be corrected by the power supply (under *sudden* changes in line and load) is also important — since no physical process is "instantaneous." So, the property of any converter to provide quick regulation (correction) under external disturbances is referred to as its "loop response" or "AC response." Clearly, the loop response is in general, a combination of "step-load response" and "line transient response."

As we move on, we will first introduce the reader to some of the most basic terminology of power conversion and its key concerns. Later, we will progress toward understanding the behavior of the most vital component of power conversion — the *inductor*. It is this component that even some relatively experienced power designers still have trouble with! Clearly, real progress in any area cannot occur without a clear understanding of the components and the basic concepts involved. Therefore, only after understanding the *inductor* well enough, can we go on to demonstrate the fact that switching converters are not all that mysterious either — in fact they evolve quite naturally out of a keen understanding of the inductor.

Overview and Basic Terminology

Efficiency

Any regulator carries out the process of power conversion with an "efficiency," defined as

$$\eta = \frac{P_O}{P_{IN}}$$

where P_O is the "output power," equal to

$$P_O = V_O \times I_O$$

and P_{IN} is the "input power," equal to

$$P_{IN} = V_{IN} \times I_{IN}$$

Here, I_{IN} is the *average* or DC current being drawn from the source.

Ideally we want $\eta = 1$, and that would represent a "perfect" conversion efficiency of 100%. But in a *real* converter, that is, with $\eta < 1$, the *difference* "$P_{IN} - P_O$" is simply the wasted power "P_{loss}" or "dissipation" (occurring within the converter itself). By simple manipulation, we get

$$P_{loss} = P_{IN} - P_O$$

$$P_{loss} = \frac{P_O}{\eta} - P_O$$

$$P_{loss} = P_O \times \left(\frac{1 - \eta}{\eta}\right)$$

This is the loss expressed in terms of the output power. In terms of the input power, we would similarly get

$$P_{loss} = P_{IN} \times (1 - \eta)$$

The loss manifests itself as *heat* in the converter, which in turn causes a certain measurable "temperature rise" ΔT over the surrounding "room temperature" (or "ambient temperature"). Note that high temperatures affect the *reliability* of all systems — the rule of thumb being that every 10°C rise causes the failure rate to double. Therefore, part of our skill as designers is to reduce this temperature rise and also achieve higher efficiencies.

Coming to the *input current* (drawn by the converter), for the hypothetical case of 100% efficiency, we get

$$I_{IN\_ideal} = I_O \times \left(\frac{V_O}{V_{IN}}\right)$$

So, in a real converter, the input current increases from its "ideal" value by the factor $1/\eta$.

$$I_{\text{IN\_measured}} = \frac{1}{\eta} \times I_{\text{IN\_ideal}}$$

Therefore, if we can achieve high efficiency, the current drawn from the input (keeping application conditions unchanged) will decrease — *but only up to a point*. The input current clearly cannot fall below the "brickwall," that is, "$I_{\text{IN\_ideal}}$," because this current is equal to P_O/V_{IN} — that is, related only to the "useful power" P_O, delivered by the power supply, which we are assuming has not changed.

Further, since

$$V_O \times I_O = V_{\text{IN}} \times I_{\text{IN\_ideal}}$$

by simple algebra, the dissipation in the power supply (energy lost per second as heat) can also be written as

$$P_{\text{loss}} = V_{\text{IN}} \times (I_{\text{IN\_measured}} - I_{\text{IN\_ideal}})$$

This form of the dissipation equation indicates a little more explicitly how *additional* energy (more input current for a given input voltage) is pushed into the input terminals of the power supply by the applied DC source — to compensate for the wasted energy inside the power supply — even as the converter continues to provide the useful energy P_O being constantly demanded by the load.

A modern switching power supply's efficiency can typically range from 65% to 95% — that figure being considered attractive enough to have taken switchers to the level of interest they arouse today and their consequent wide application. *Traditional regulators* (like the "linear regulator") provide much poorer efficiencies — and that is the main reason why they are slowly but surely getting replaced by switching regulators.

Linear Regulators

"Linear regulators," equivalently called "series–pass regulators," or simply "series regulators," also produce a regulated DC output rail from an input rail. But they do this by placing a transistor in series between the input and the output. Further, this "series–pass transistor" (or "pass transistor") is operated in the *linear* region of its voltage–current characteristics — thus acting like a variable *resistance* of sorts. As shown in the uppermost schematic of Figure 1.2, this transistor is made to literally "drop" (abandon) the unwanted or "excess" voltage across itself.

The excess voltage is clearly just the difference "$V_{\text{IN}} - V_O$" — and this term is commonly called the "headroom" of the linear regulator. We can see that the headroom needs to be a

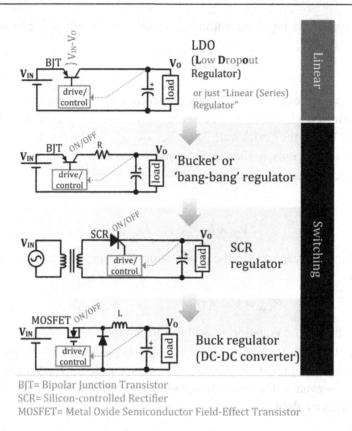

BJT= Bipolar Junction Transistor
SCR= Silicon-controlled Rectifier
MOSFET= Metal Oxide Semiconductor Field-Effect Transistor

Figure 1.2: Basic types of linear and switching regulators.

positive number always, thus implying $V_O < V_{IN}$. Therefore, linear regulators are, in principle, always "step-down" in nature — that being their most obvious limitation.

In some applications (e.g., battery-powered portable electronic equipment), we may want the output rail to remain well regulated even if the input voltage dips very low — say down to within *0.6 V or less* of the set output level V_O. In such cases, the *minimum possible headroom* (or "dropout") achievable by the linear regulator stage may become an issue.

No switch is perfect, and even if held fully conducting, it does have some voltage drop across it. So the dropout is simply the minimum achievable "forward-drop" across the switch. Regulators which can continue to work (i.e., regulate their output), with V_{IN} barely exceeding V_O, are called "*low*-dropout" regulators or "LDOs." But note that there is really no precise voltage drop at which a linear regulator "officially" becomes an LDO. So the term is sometimes applied rather loosely to linear regulators in general. However, the rule of thumb is that a dropout of about 200 mV or lower qualifies as an LDO, whereas older devices (conventional linear regulators) have a typical dropout voltage of around 2 V.

There is also an intermediate category called "quasi-LDOs" that have a dropout of about 1 V, that is, somewhere in between the two.

Besides being step-down in principle, linear regulators have another limitation — poor efficiency. Let us understand why that is so. The instantaneous power dissipated in any device is by definition the cross-product $V \times I$, where V is the instantaneous voltage drop across it and I the instantaneous current through it. In the case of the series−pass transistor, under steady application conditions, both V and I are actually constant with respect to time — V in this case being the headroom $V_{IN} - V_O$ and I the load current I_O (since the transistor is always in *series* with the load). So we see that the $V \times I$ dissipation term for linear regulators can, under certain conditions, become a significant proportion of the useful output power P_O. And that simply spells *"poor efficiency"*! Further, if we stare hard at the equations, we will realize there is also nothing we can do about it — how can we possibly argue against something as basic as $V \times I$? For example, if the input is 12 V, and the output is 5 V, then at a load current of 100 mA, the dissipation in the regulator is necessarily $\Delta V \times I_O = (12-5) \text{ V} \times 100 \text{ mA} = 700$ mW. The useful (output) power is, however, $V_O \times I_O = 5 \text{ V} \times 100 \text{ mA} = 500$ mW. Therefore, the efficiency is $P_O/P_{IN} = 500/(700+500) = 41.6\%$. What can we do about that? Blame Georg Ohm?

On the positive side, linear regulators are very "quiet" — exhibiting none of the noise and electromagnetic interference (EMI) that have unfortunately become a "signature" or "trademark" of modern switching regulators. Switching regulators need *filters* — usually both at the input and at the output, to quell some of this noise, which can interfere with other gadgets in the vicinity, possibly causing them to malfunction. Note that sometimes the usual input/output capacitors of the converter may themselves serve the purpose, especially when we are dealing with "low-power" (and "low-voltage") applications. But in general, we may require filter stages containing *both* inductors and capacitors. Sometimes these stages may need to be cascaded to provide even greater noise attenuation.

Achieving High Efficiency through Switching

Why are switchers so much more efficient than "linears"?

As their name indicates, in a *switching* regulator, the series transistor is not held in a perpetual *partially conducting* (and therefore dissipative) mode — but is instead *switched* repetitively. So there are only two *states* possible — either the switch is held "ON" (fully conducting) or it is "OFF" (fully nonconducting) — there is no "middle ground" (at least not in principle). When the transistor is ON, there is (ideally) zero *voltage* across it ($V=0$), and when it is OFF, we have zero *current* through it ($I=0$). So, it is clear that the cross-product "$V \times I$" is also zero for *either* of the two states. And that simply implies zero "switch dissipation" at all times. Of course, this too represents an impractical or "ideal" case. Real switches do dissipate. One reason for that is they are never either *fully* ON nor *fully* OFF. Even when they are supposedly ON, they have a small voltage drop across them, and when they are supposedly "OFF," a small current still flows through them. Further, no device switches "instantly" either — there is always a definable period in which the device is *transiting between states*. During this interval too, $V \times I$ is not zero and some additional dissipation occurs.

We may have noticed that in most introductory texts on switching power conversion, the switch is shown as a *mechanical* device — with contacts that simply open ("switch OFF") or close ("switch ON"). So, a mechanical device comes very close to our definition of a "perfect switch" — and that is the reason why it is often the vehicle of choice to present the most basic principles of power conversion. But one obvious problem with actually *using* a mechanical switch in any practical converter is that such switches can wear out and fail over a relatively short period of time. So in practice, we always prefer to use a *semiconductor device* (e.g., a transistor) as the switching element. As expected, that greatly enhances the life and reliability of the converter. But the most important advantage is that since a semiconductor switch has none of the mechanical "inertia" associated with a mechanical device, it gives us the ability to switch *repetitively* between the ON and OFF states — and does so *very fast*. We have already realized from the metro terminus analogy on page 1 that that will lead to *smaller* components in general.

We should be clear that the phrase "switching fast," or "high switching speed," has slightly varying connotations, even within the area of switching power conversion. When it is applied to the overall *circuit*, it refers to the frequency at which we are repeatedly switching — ON OFF, ON OFF, and so on. This is the converter's basic *switching frequency* "*f*" (in Hz). But when the same term is applied specifically to the *switching element* or device, it refers to the *time* spent transiting *between* its two states (i.e., from ON to OFF and OFF to ON) and is typically expressed in "ns" (nanoseconds). This transition interval is then rather *implicitly* and intuitively being compared to the total "time period" T (where $T = 1/f$) and therefore to the switching frequency — though we should be clear there is no *direct* relationship between the transition time and the switching frequency.

We will learn shortly that the ability to *crossover* (i.e., transit) quickly between switching states is in fact rather crucial. Yes, up to a point, the switching speed is almost completely determined by how "strong" and effective we can make our external "drive circuit." But ultimately, the speed becomes limited purely by the device and its technology — an "inertia" of sorts at an electrical level.

Basic Types of Semiconductor Switches

Historically, most power supplies used the bipolar junction transistor (BJT) shown in Figure 1.2. It is admittedly a rather *slow* device by modern standards. But it is still relatively cheap! In fact its "NPN" version is even cheaper and therefore more popular than its "PNP" version. Modern switching supplies prefer to use a "MOSFET" (metal oxide semiconductor field effect transistor), often simply called a "FET" (see Figure 1.2 again). This modern high-speed switching device also comes in several "flavors" — the most commonly used ones being the *N-channel* and *P-channel* types (both usually being the "enhancement mode" variety). The *N-channel MOSFET* happens to be the favorite in terms of cost effectiveness and performance for most applications. However, sometimes, P-channel devices may be preferred for various reasons — mainly because they usually require simpler drive circuits.

Despite the steady course of history in favor of MOSFETs in general, there still remain some arguments for continuing to prefer BJTs in certain applications. Some points to consider and debate here are:

a. It is often said that it is *easier to drive a MOSFET than a BJT*. In a BJT, we do need a large drive current (injected into its "base" terminal) — to turn it ON. We also need to *keep* injecting base current to *keep it* in that state. On the other hand, a MOSFET is considered easier to drive. In theory, we just have to apply a certain voltage at its "Gate" terminal to turn it ON and also keep it that way. Therefore, a MOSFET is called a "voltage-controlled" device, whereas a BJT is considered a "current-controlled" device. However, in reality, a modern MOSFET needs a certain amount of Gate current *during* the time it is *in transit* (ON to OFF and OFF to ON). Further, to make it change state *fast*, we may in fact need to push in (or pull out) a *lot* of current (typically 1 to 2A).

b. The *drive requirements of a BJT may actually turn out easier to implement* in many cases. The reason for that is, to turn an NPN BJT ON for example, its Gate has to be taken only about 0.8 V above its Emitter (and can even be tied directly to its Collector on occasion), whereas in an N-channel MOSFET, its Gate has to be taken *several volts* higher than its Source. Therefore, in certain types of DC−DC converters, when using an N-channel MOSFET, it can be shown that we need a "drive rail" that is significantly *higher* than the (available) input rail V_{IN}. And how else can we hope to have such a rail except by a circuit that can somehow manage to "push" or "pump" the input voltage to a higher level? When thus implemented, such a rail is called the "bootstrap" rail.

> *Note: The most obvious implementation of a "bootstrap circuit" may just consist of a small capacitor that gets charged by the input source (through a small signal diode) whenever the switch turns OFF. Thereafter, when the switch turns ON, we know that a certain voltage node in the power supply suddenly "flips" whenever the switch changes state. But since the "bootstrap capacitor", one end of which is connected to this (switching) node, continues to hold on to its acquired voltage (and charge), its other end, which forms the bootstrap rail, gets pushed up to a level higher than the input rail as desired. This rail then helps drive the MOSFET properly under all conditions.*

c. The main advantage of BJTs is that they are known to generate *significantly less EMI and "noise and ripple"* than MOSFETs. That ironically is a positive outcome of their *slower* switching speed!

d. BJTs are also often better suited for *high-current* applications — because their "forward drop" *(on-state voltage drop)* is *relatively constant*, even for very high switch currents. This leads to significantly lower "switch dissipation," more so when the switching frequencies are not too high. On the contrary, in a MOSFET, the forward drop is almost proportional to the current passing through it — so its dissipation can become significant at high loads. Luckily, since it also switches faster (lower transition times), it usually more than makes up for that loss term, and so in fact becomes much better in terms of the *overall* loss — more so when compared at very high switching frequencies.

> *Note: In an effort to combine the "best of both worlds," a "combo" device called the "IGBT" (insulated Gate bipolar transistor is also often used nowadays. It is driven like a MOSFET (voltage-controlled) but behaves like a BJT in other ways (the forward drop and switching speed). It too is therefore suited mainly for low-frequency and high-current applications but is considered easier to drive than a BJT.*

Semiconductor Switches Are Not "Perfect"

We mentioned that all semiconductor switches suffer losses. Despite their advantages, they are certainly not the perfect or ideal switches we may have imagined them to be at first sight.

So, for example, unlike a mechanical switch, in the case of a semiconductor device, we may have to account for the small but measurable "leakage current" flowing through it when it is considered "fully OFF" (i.e., nonconducting). This gives us a dissipation term called the "leakage loss." This term is usually not very significant and can be ignored. However, there is a small but *significant* voltage drop ("forward drop") across the semiconductor when it is considered "fully ON" (i.e., conducting) — and that gives us a significant "conduction loss" term. In addition, there is also a brief moment as we transition *between* the two switching states, when the current and voltage in the switch need to *slew up or down almost simultaneously* to their new respective levels. So, during this "transition time" or "crossover time," we *have neither* $V = 0$ *nor* $I = 0$ instantaneously, and therefore nor is $V \times I = 0$. This therefore leads to some additional dissipation and is called the "crossover loss" (or sometimes just "switching loss"). Eventually, we need to learn to minimize all such loss terms if we want to improve the efficiency of our power supply.

However, we must remember that power supply design is by its very nature full of *design tradeoffs* and subtle compromises. For example, if we look around for a transistor with a very low forward voltage drop, possibly with the intent of minimizing the conduction loss, we usually end up with a device that also happens to transition more slowly — thus leading to a higher crossover loss. There is also an overriding concern for cost that needs to be constantly looked into, particularly in the commercial power supply arena. So, we should not underestimate the importance of having an astute and seasoned engineer at the helm of affairs, one who can really grapple with the finer details of power supply design. As a corollary, neither can we probably ever hope to replace him or her (at least not entirely), by some smart automatic test system, nor by any "expert design software" that the upper management may have been dreaming of.

Achieving High Efficiency through the Use of Reactive Components

We have seen that one reason why switching regulators have such a high efficiency is because they use a *switch* (rather than a transistor that "thinks" it is a resistor, as in an LDO). Another root cause of the high efficiency of modern switching power supplies is their effective use of *both capacitors and inductors*. Capacitors and inductors are categorized as "reactive" components because they have the unique ability of being able to *store energy*. However, that is also why they cannot ever be made to *dissipate* energy either (at least not *within* themselves) — they just store all the energy "thrown at them"! On the other hand, we know that "resistive" components dissipate energy but, unfortunately, can't store any!

A capacitor's stored energy is called *electrostatic*, equal to $1/2 \times C \times V^2$, where C is the "capacitance" (in Farads) and V the voltage across the capacitor. Whereas an inductor's stored energy is called *magnetic*, equal to $1/2 \times L \times I^2$, with L being the "inductance" (in Henrys) and I the current passing through it (at any given moment).

But we may well ask — despite the obvious efficiency concerns, do we really *need* reactive components *in principle*! For example, we may have realized we don't really need an input or output capacitor for implementing a *linear regulator* — because the series–pass element (the BJT) is all that is required to block any excess voltage. For switching regulators, however, the reasoning is rather different. This leads us to the general *"logic of switching power conversion"* summarized below.

- A transistor is needed to establish control on the output voltage and thereby bring it into regulation. The reason we *switch* it is as follows — dissipation in this control element is related to the product of the voltage across the control device and the current through it, that is $V \times I$. So, if we make either V or I zero (or very small), we will get zero (or very small) dissipation. By switching *constantly* between ON and OFF states, we can keep the switch dissipation down, but at the same time, by controlling the *ratio* of the ON and OFF intervals, we can *regulate* the output, based on average energy flow considerations.

- But whenever we switch the transistor, we effectively *disconnect the input from the output* (during either the ON or the OFF state). However, the output (load) always demands a *continuous* flow of energy. Therefore, we need to introduce *energy-storage* elements somewhere inside the converter. In particular, we use output *capacitors* to "hold" the voltage steady across the load during the above-mentioned input-to-output "disconnect" interval.
- But as soon as we put in a capacitor, we now also need to limit the *inrush current* into it — all capacitors connected directly across a DC source will exhibit an uncontrolled inrush — and that can't be good either for noise, for EMI, or for efficiency. Of course, we could simply opt for a *resistor* to subdue this inrush, and that in fact was the approach behind the early "bucket regulators" (Figure 1.2).
- But unfortunately a resistor always *dissipates* — so what we may have saved in terms of transistor dissipation may ultimately end up in the resistor! To maximize the overall efficiency, we therefore need to *use only reactive* elements in the conversion process. Reactive elements can *store* energy but do not dissipate any (in principle). Therefore, an *inductor becomes our final choice* (along with the capacitor), based on its ability to *non-dissipatively limit the (rate of rise of) current*, as is desired for the purpose of limiting the capacitor inrush current.

Some of the finer points in this summary will become clearer as we go on. We will also learn that *once the inductor has stored some energy, we just can't **wish** this stored energy away at the drop of a hat.* We need to do something about it! And that in fact gives us an actual working converter down the road.

Early RC-Based Switching Regulators

As indicated above, a possible way out of the "input-to-output disconnect" problem is to use *only an output capacitor*. This can store some extra energy when the switch connects the load to the input, and then provide this energy to the load when the switch disconnects the load.

But we still need to limit the capacitor *charging current* ("inrush current"). And as indicated, we could use a resistor. That was in fact the basic principle behind some early linear-to-switcher "crossover products" like the bucket regulator shown in Figure 1.2.

The bucket regulator uses a transistor driven like a *switch* (as in modern switching regulators), a small *series resistor* to limit the current (not entirely unlike a linear regulator), and an *output capacitor* (the "bucket") to store and then provide energy when the switch is OFF. Whenever the output voltage falls below a certain threshold, the switch turns ON, "tops up" the bucket, and then turns OFF. Another version of the bucket regulator uses a cheap low-frequency switch called an SCR ("semiconductor controlled rectifier") that works off the Secondary windings of a step-down transformer connected to an AC mains supply, as also shown in Figure 1.2. Note that in this case, the resistance of the windings (usually) serves as the (only) effective limiting resistance.

Note also that in either of these RC-based bucket regulator implementations, the switch ultimately ends up being *toggled repetitively* at a certain rate — and in the process, a rather crudely regulated stepped-down output DC rail is created. By definition, that makes these regulators *switching regulators* too!

But we realize that the very use of a resistor in any power conversion process always bodes ill for efficiency. So, we may have just succeeded in *shifting* the dissipation away from the transistor — into the resistor! If we really want to maximize overall efficiency, we need to do away with *any intervening resistance* altogether.

So, we attempt to use an inductor instead of a resistor for the purpose — we don't really have many other component choices left in our bag! In fact, if we manage to do that, we get our first modern *LC-based switching regulator* — the "Buck regulator" (i.e., step-down converter), as also presented in Figure 1.2.

LC-Based Switching Regulators

Though the detailed functioning of the modern Buck regulator of Figure 1.2 will be explained a little later, we note that besides the obvious replacement of *R* with an *L*, it looks very similar to the bucket regulator — *except for a "mysterious" diode*. The basic principles of power conversion will in fact become clear only *when we realize the purpose of this diode*. This component goes by several names — "catch diode," "freewheeling diode," "commutation diode," and "output diode," to name but a few! But its basic purpose is always the same — a purpose we will soon learn is intricately related to the behavior of the *inductor* itself.

Aside from the Buck regulator, there are *two* other ways to implement the basic goal of switching power conversion (using *both inductors and capacitors*). Each of these leads to a distinct *"topology."* So besides the *Buck* (step-down), we also have the *"Boost"* (step-up), and the *"Buck-Boost"* (step-up or step-down). We will see that though all these are based on the same *underlying* principles, they are set up to look and behave quite differently. As a prospective power supply designer, we really do need to learn and master each of them almost on an *individual basis*. We must also keep in mind that in the process, our *mental picture will usually need a drastic change as we go from one topology to another*.

> **Note:** *There are some other capacitor-based possibilities — in particular "charge pumps" — also called "inductor-less switching regulators." These are usually restricted*

to rather low powers and produce output rails that are rather crudely regulated multiples of the input rail. In this book, we are going to ignore these types altogether. Then there are also some other types of LC-based possibilities — in particular the "resonant topologies." Like conventional DC−DC converters, these also use both types of reactive components (L and C) along with a switch. However, their basic principle of operation is very different. Without getting into their actual details, we note that these topologies do not maintain a constant switching frequency, which is something we usually rather strongly desire. From a practical standpoint, any switching topology with a variable switching frequency can lead to an unpredictable *and varying EMI spectrum and noise signature. To mitigate these effects, we may require rather complicated filters. For such reasons, resonant topologies have not really found widespread acceptance in commercial designs, and so we too will largely ignore them from this point on.*

The Role of Parasitics

In using conventional LC-based switching regulators, we may have noticed that their constituent inductors and capacitors do get fairly *hot* in most applications. But if, as we said, these components are *reactive*, why at all are they getting hot? We need to know why, because any source of heat impacts the overall efficiency! And *efficiency* is what modern switching regulators are all about!

The heat arising from *real-world* reactive components can invariably be traced back to dissipation occurring within the small "parasitic" *resistive* elements, which always accompany any such (reactive) component.

For example, a real inductor has the basic property of inductance L, but it also has a certain non-zero *DC resistance* ("DCR") term, mainly associated with the copper windings used. Similarly, any real capacitor has capacitance C, but it also has a small *equivalent series resistance* ("ESR"). Each of these terms produces "ohmic" losses — that can all add up and become fairly significant.

As indicated previously, a real-world semiconductor switch can also be considered as having a parasitic resistance "strapped" *across* it. This parallel resistor in effect "models" the leakage current path and thus the "leakage loss" term. Similarly, the forward drop across the device can also, in a sense, be thought of as a *series* parasitic resistance — leading to a conduction loss term.

But any real-world component also comes with various *reactive* parasitics. For example, an inductor can have a significant parasitic capacitance across its terminals — associated with electrostatic effects between the layers of its windings. A capacitor can also have an *equivalent series inductance* ("ESL") — coming from the small inductances associated with its leads, foil, and terminations. Similarly, a MOSFET also has various parasitics — for example, the "unseen" capacitances present *between* each of its terminals (within the package). In fact, these MOSFET parasitics play a major part in determining the limits of its switching speed (transition times).

In terms of dissipation, we understand that reactive parasitics certainly cannot dissipate heat — at least not *within the parasitic element itself*. But more often than not, these reactive parasitics do manage to "dump" their stored energy (at specific moments during the switching cycle) into a nearby *resistive* element — thus increasing the overall losses indirectly.

Therefore, we see that to improve efficiency, we generally need to go about minimizing all such parasitics — *resistive or reactive*. We should not forget they are the very reason we are not getting 100% efficiency from our converter in the first place. Of course, we have to learn to be able to do this optimization to within *reasonable* and *cost-effective* bounds, as dictated by market compulsions and similar constraints.

But we should also bear in mind that *nothing is straightforward in power*! So these parasitic elements should not be considered entirely "useless" either. In fact, they do play a rather helpful and stabilizing role on occasion.

- For example, if we short the outputs of a DC–DC converter, we know it is unable to regulate, however hard it tries. In this "fault condition" ("open-loop"), the momentary "overload current" within the circuit can be "tamed" (or mitigated) a great deal by the very presence of certain identifiably "friendly" parasitics.
- We will also learn that the so-called "voltage-mode control" switching regulators may actually *rely on the ESR* of the output capacitor for ensuring "loop stability" — even under normal operation. As indicated previously, loop stability refers to the ability of a power supply to regulate its output quickly, when faced with sudden changes in line and load, without undue oscillations or ringing.

Certain other parasitics, however, may just prove to be a nuisance and some others a sheer bane. But their actual *roles too may keep shifting* depending upon the prevailing conditions in the converter. For example

- A certain parasitic *inductance* may be quite *helpful* during the *turn-on* transition of the switch — by acting to limit any current spike trying to pass through the switch. But it can be *harmful* due to the high *voltage spike* it creates across the switch at *turn-off* (as it tries to release its stored magnetic energy).
- On the other hand, a parasitic *capacitance* present across the switch, for example, can be *helpful* at *turn-off* — but *unhelpful* at *turn-on*, as it tries to dump its stored electrostatic energy inside the switch.

  Note: We will find that during turn-off, the parasitic capacitance mentioned above helps limit or "clamp" any potentially destructive voltage spikes appearing across the switch by absorbing *the energy residing in that spike. It also helps* decrease the crossover loss *by slowing down the rising ramp of voltage and thereby reducing the V–I "overlap" (between the transiting V and I waveforms of the switch). However at* turn-on, *the same parasitic capacitance now has to discharge whatever energy it acquired during the preceding turn-off transition — and that leads to a* current spike *inside the switch. Note that this spike is externally "invisible" — apparent only by the higher-than-expected switch dissipation and the resulting higher-than-expected temperature.*

Therefore, generally speaking, *all parasitics constitute a somewhat "double-edged sword,"* one that we just can't afford to overlook for very long in *practical* power supply design. However, as we too will do in some of our discussions that follow, sometimes we can consciously and *selectively* decide to ignore some of these second-order influences initially, just to build up *basic concepts* in power first. Because the truth is if we don't do that, we just run the risk of feeling quite overwhelmed, too early in the game!

Switching at High Frequencies

In attempting to generally reduce parasitics and their associated losses, we may notice that these are often dependent on various external factors — *temperature* for one. Some losses increase with temperature — for example, the conduction loss in a MOSFET. And some may decrease — for example, the conduction loss in a BJT *(when operated with low currents)*. Another example of the latter type is the ESR-related loss of a typical aluminum electrolytic capacitor, which also decreases with temperature. On the other hand, some losses may have rather "strange" shapes. For example, we could have an inverted "bell-shaped" curve — representing an optimum operating point somewhere *between the two extremes*. This is what the "core loss" term of many modern "ferrite" materials (used for inductor cores) looks like — it is at its minimum at around 80–90 °C, increasing on either side.

From an overall perspective, it is hard to predict how all these variations with respect to temperature add up — and how the efficiency of the power supply is thereby affected by changes in temperature.

Coming to the dependency of parasitics and related loss terms on *frequency*, we do find a somewhat clearer trend. In fact, it is rather rare to find any loss term that *decreases* at higher frequencies (though a notable exception to this is the loss in an aluminum electrolytic capacitor — because its ESR decreases with frequency). Some of the loss terms are virtually *independent* of frequency (e.g., conduction loss). And the remaining losses actually *increase* almost *proportionally* to the switching frequency — for example, the crossover loss. So, in general, we realize that *lowering, not increasing, the switching frequency would almost invariably help improve efficiency.*

There are other frequency-related issues too besides efficiency. For example, we know that switching power supplies are inherently noisy and generate a lot of EMI. By going to higher switching frequencies, we may just be making matters worse. We can mentally visualize that even the small connecting wires and "printed circuit board" (PCB) traces become very effective antennas at high frequencies and will likely spew out *radiated EMI* in every direction.

This therefore begs the question: ***why at all are we face to face with a modern trend of ever-increasing switching frequencies?*** Why should we not *decrease* the switching frequency?

The first motivation toward higher switching frequencies was to simply take "the action" beyond audible human hearing range. Reactive components are prone to creating sound pressure waves for various reasons. So, the early LC-based switching power supplies switched at around 15−20 kHz and were therefore barely audible, if at all.

The next impetus toward even higher switching frequencies came with the realization that the bulkiest component of a power supply, that is, *the inductor, could be almost proportionately reduced in size if the switching frequency was increased* (everybody does seem to want smaller products after all!). Therefore, successive generations of power converters moved upward in almost arbitrary steps, typically 20 kHz, 50 kHz, 70 kHz, 100 kHz, 150 kHz, 250 kHz, 300 kHz, 500 kHz, 1 MHz, 2 MHz, and often even higher today. This actually helped simultaneously *reduce* the size of the *conducted EMI* and input/ output filtering components — including the capacitors! High switching frequencies can also almost proportionately enhance the *loop response* of a power supply.

Therefore, we realize that the only thing holding us back at any moment of time from going to even higher frequencies are the *"switching losses."* This term is in fact rather broad — encompassing all the losses that occur *at the moment* when we actually switch the transistor (i.e., from ON to OFF and/or OFF to ON). Clearly, the *crossover loss* mentioned earlier is just one of several possible switching loss terms. Note that it is easy to visualize why such losses are (usually) exactly proportional to the switching frequency — since energy is lost *only whenever we actually switch* (transition) — therefore, the greater the number of times we do that (in a second), the more energy is lost (dissipation).

Finally, we also do need to learn how to *manage* whatever dissipation is still remaining in the power supply. This is called "thermal management," and that is one of the most important goals in any good power supply design. Let us look at that now.

Reliability, Life, and Thermal Management

Thermal management basically just means trying to get the heat out from the power supply and into the surroundings — thereby lowering the local temperatures at various points inside it. The most basic and obvious reason for doing this is to keep all the components to within their maximum rated operating temperatures. But in fact, that is rarely enough.

We always strive to reduce the temperatures even further, and every couple of *degrees Celsius* (°C) may well be worth fighting for.

The reliability "*R*" of a power supply at any given moment of time is defined as $R(t) = e^{-\lambda t}$. So at time $t = 0$ (start of operational life), the reliability is considered to be at its maximum value of 1. Thereafter it decreases exponentially as time elapses, "λ" is the failure rate of a power supply, that is, the number of supplies failing over a specified period of time. Another commonly used term is "MTBF" or *mean time between failures*. This is the reciprocal of the overall failure rate, that is, $\lambda = 1/\text{MTBF}$. A typical commercial power supply will have an MTBF of between 100,000 h and 500,000 h — *assuming it is being operated at a fairly typical and benign "ambient temperature" of around 25°C.*

Looking now at the variation of failure rate with respect to temperature, we come across the well-known rule of thumb — *failure rate doubles every 10 °C rise in temperature.* If we apply this admittedly loose rule of thumb to each and every component used in the power supply, we see it must also hold for the entire power supply too — since the overall failure rate of the power supply is simply the sum of the failure rates of each component comprising it ($\lambda = \lambda_1 + \lambda_2 + \lambda_3 + \cdots$). All this clearly gives us a good reason to try to reduce temperatures of *all* the components even further.

But aside from failure rate, which clearly applies to *every* component used in a power supply, there are also certain "lifetime" considerations that apply to *specific* components. The "life" of a component is stated to be the duration it can work for continuously without *degrading* beyond certain specified limits. At the end of this "useful life," it is considered to have become a "wearout failure" — or simply put — it is "worn-out." Note that this need not imply the component has failed "catastrophically" — more often than not, it may be just "out of spec." The latter phrase simply means the component no longer provides the expected performance — as specified by the limits published in the electrical tables of its datasheet.

> *Note: Of course, a datasheet can always be "massaged" to make the part look good in one way or another — and that is the origin of a rather shady but widespread industry practice called "specmanship." A good designer will therefore keep in mind that not all vendors' datasheets are equal — even for what may seem to be the same or equivalent part number at first sight.*

As designers, it is important that we not only do our best to extend the "useful life" of any such component, but also account *upfront* for its slow degradation over time. In effect, that implies that the power supply may *initially* perform better than its minimum specifications. Ultimately, however, the worn-out component, especially if it is present at a critical location, could cause the entire power supply to "go out of spec" and even fail catastrophically.

Luckily, most of the components used in a power supply have no meaningful or definable lifetime — at least not within the usual 5–10 years of useful life expected from most electronic products. We therefore usually don't, for example, talk in terms of an inductor or transistor "degrading" (over a period of time) — though of course either of these components can certainly *fail* at any given moment, even under normal operation, as evidenced by their non-zero failure rates.

> *Note: Lifetime issues related to the* materials *used in the construction of a component can affect the life of the component indirectly. For example, if a semiconductor device is operated* well beyond its usual maximum rating of 150 °C, *its* plastic package *can exhibit wearout or degradation — even though nothing happens to the semiconductor itself up to a much higher temperature. Subsequently, over a period of time, this degraded package can cause the junction to get severely affected by environmental factors, causing the device to fail catastrophically — usually taking the power supply (and system) with it too! In a similar manner, inductors made of a "powdered iron" type of core material are also known to degrade under extended periods of high temperatures — and this can produce not only a failed inductor, but a failed power supply too.*

A common example of lifetime considerations in a commercial power supply design comes from its use of aluminum electrolytic capacitors. Despite their great affordability and respectable performance in many applications, such capacitors are a victim of wearout due to the steady evaporation of their enclosed electrolyte over time. Extensive calculations are needed to predict their *internal* temperature ("core temperature") and thereby estimate the true rate of evaporation and thereby extend the capacitor's useful life. The rule recommended for doing this life calculation is — *the useful life of an aluminum electrolytic capacitor halves every 10 °C rise in temperature.* We can see that this relatively hard-and-fast rule is uncannily similar to the rule of thumb for failure rate. But that again is just a coincidence, since life and failure rate are really two different issues altogether as discussed in Chapter 6.

In either case, we can now clearly see that the way to extend life *and* improve reliability is to *lower the temperatures of all the components in a power supply* and also the *ambient temperature inside the enclosure of the power supply.* This may also call out for a better-ventilated enclosure (more air vents), more exposed copper on the PCB, or say, even a built-in fan to push the hot air out. Though in the latter case, we now have to start worrying about both the failure rate and life of the fan itself!

Stress Derating

Temperature can ultimately be viewed as a "thermal stress" — one that causes an increase in failure rate (and life if applicable). But how severe a stress really is, must naturally be judged *relative to the "ratings" of the device.* For example, most semiconductors are rated for a "maximum junction temperature" of 150 °C. Therefore, keeping the junction no higher than 105 °C in a given application represents a *stress reduction factor*, or alternately — a "temperature derating" factor equal to 105/150 = 70%.

In general, "stress derating" is the established technique used by good designers to diminish internal stresses and thereby reduce the failure rate. Besides temperature, the failure rate (and life) of any component can also depend on the applied *electrical* stresses — voltage and current. For example, a typical "voltage derating" of 80% as applied to semiconductors means that the worst-case operating voltage across the component never exceeds 80% of the maximum specified voltage rating of the device. Similarly, we can usually apply a typical "current derating" of 70–80% to most semiconductors.

The practice of derating also implies that we need to select our components judiciously *during the design phase itself* — with well-considered and built-in operating *margins*. And though, as we know, some loss terms decrease with temperature, contemplating raising the temperatures just to achieve better efficiency or performance is clearly not the preferred direction, because of the obvious impact on system reliability.

A good designer eventually learns to *weigh reliability and life concerns against cost, performance, size, and so on.*

Advances in Technology

But despite the best efforts of many a good power supply designer, certain sought-after improvements may still have remained merely on our annual Christmas wish list! Luckily, there have been significant accompanying advances in the technology of the *components* available to help enact our goals. For example, the burning desire to reduce resistive losses and simultaneously make designs suitable for high-frequency operation has ushered in significant improvements in terms of a whole new generation of high-frequency, low-ESR ceramic, and other specialty capacitors. We also have diodes with very low forward voltage drops and "ultrafast recovery," much faster switches like the MOSFET, and several new low-loss ferrite material types for making the transformers and inductors.

> **Note:** *"Recovery" refers to the ability of a diode to quickly change from a conducting state to a nonconducting (i.e., "blocking") state as soon as the voltage across it reverses. Diodes which do this well are called "ultrafast diodes." Note that the "Schottky diode" is preferred in certain applications because of its low forward drop (∼0.5 V). In principle, it is also supposed to have zero recovery time. But unfortunately, it also has a comparatively higher parasitic "body capacitance" (across itself) that in some ways tends to mimic conventional recovery phenomena. Note that it also has a higher leakage current and is typically limited to blocking voltages of less than 100 V.*

However, we observe that the actual *topologies* used in power conversion have not really changed significantly over the years. We still have just *three* basic topologies: the Buck, the Boost, and the Buck-Boost. Admittedly, there have been significant improvements like zero voltage switching ("ZVS"), "current-fed converters," and "composite topologies" like the "Cuk converter" and the single-ended primary inductance converter ("SEPIC"), but all these are perhaps best viewed as icing on a three-layer cake. The basic building blocks (or topologies) of power conversion have themselves proven to be quite fundamental. And that is borne out by the fact that they have stood the test of time and remained virtually unchallenged to date.

So, *finally, we can get on with the task of really getting to understand these topologies well. We will soon realize that the best way to do so is via the route that takes us past that rather enigmatic component — **the inductor**. And that's where we begin our journey now.*

Understanding the Inductor

Capacitors/Inductors and Voltage/Current

In power conversion, we may have noticed that we always talk rather instinctively of *voltage* rails. That is why we also have DC−DC *voltage* converters forming the subject of this book. But why not *current* rails or *current* converters, for example?

We should realize that the world we live in, keenly interact with, and are thus comfortable with, is ultimately one of voltage, not current. So, for example, every electrical gadget or appliance we use runs off a specified *voltage* source, the currents drawn from which being largely ours to determine. So, for example, we may have 110-VAC or 115-VAC "mains input" in many countries. Many other places may have 220-VAC or 240-VAC mains input. So, if for example, an electric room heater is connected to the "mains outlet," it would draw a very large *current* (∼ 10−20 A), but the line voltage itself would hardly change in the process. Similarly, a clock radio would typically draw only a few hundred milliamperes of current, the line voltage again remaining fixed. That is by definition a voltage source. On the other hand, imagine for a moment that we had a 20-A *current source* outlet available in our wall. By definition, this would try to push out 20 A, *come what may* — even adjusting the voltage if necessary to bring that about. So, if we don't connect any appliance to it, it would even attempt to arc over, just to keep 20 A flowing. No wonder we hate current sources!

We may have also observed that *capacitors* have a rather more direct relationship with voltage, rather than current. So $C = Q/V$, where C is the capacitance, Q is the charge on either plate of the capacitor, and V is the voltage across it. This gives capacitors a somewhat imperceptible, but natural, association with our more "comfortable" world of voltages. It's perhaps no wonder we tend to understand their behavior so readily.

Unfortunately, capacitors are not the only power-handling component in a switching power supply! Let us now take a closer look at the main circuit blocks and components of a typical off-line power supply as shown in Figure 1.1. Knowing what we now know about capacitors and their natural relationship to voltage, we are not surprised to find there are capacitors present at both the input and the output ends of the supply. But we also find an *inductor* (or "choke") — in fact a rather bulky one at that too! We will learn that this behaves like a *current source*, and therefore, quite naturally, we don't relate too well to it! However, to gain mastery in the field of power conversion, we need to understand *both* the key components involved in the process: capacitors *and inductors*.

Coming in from a more seemingly natural world of voltages and capacitances, it may require a certain degree of mental readjustment to understand inductors well enough. Sure, most power supply engineers, novice or experienced, are able to faithfully reproduce the Buck converter *duty cycle equation*, for example (i.e., the relationship between input and output voltage). Perhaps they can even derive it too on a good day! But scratch the surface, and we can surprisingly often find a noticeable lack of "feel" for inductors. We would do well to recognize this early on and remedy it. With that intention, we are going to start at the very basics.

The Inductor and Capacitor Charging/Discharging Circuits

Let's start by a simple question, one that is sometimes asked of a prospective power supply hire (read "nervous interviewee"). This is shown in Figure 1.3.

Note that here we are using a mechanical switch for the sake of simplicity, thus also assuming it has none of the parasitics we talked about earlier. At time $t = 0$, we close the switch (ON) and thus apply the DC voltage supply (V_{IN}) across the capacitor (C) through the small series limiting resistor (R). What happens?

Most people get this right. The capacitor *voltage* increases according to the well-known exponential curve $V_{IN} \times (1 - e^{-t/\tau})$, with a "time constant" of $\tau = RC$. The capacitor *current*, on the other hand, starts from a high initial value of V_{IN}/R and then decays exponentially according to $(V_{IN}/R) \times e^{-t/\tau}$. Yes, if we wait "a very long time," the capacitor would get charged up almost fully to the applied voltage V_{IN}, and the current would correspondingly fall (almost) to zero. Let us now *open* the switch (OFF), though not necessarily having waited a very long time. In doing so, we are essentially attempting to force the *current* to zero (that is what a *series* switch is always supposed to do). What happens? The capacitor remains charged to whatever voltage it had already reached, and its current goes down immediately to zero (if not already there).

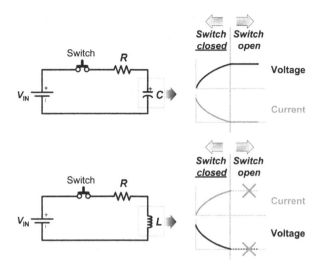

Figure 1.3: Basic charging/discharging circuits for capacitor and inductor.

Now let us repeat the same experiment, but *with the capacitor replaced by an inductor* (L), as also shown in Figure 1.3. Interviewees usually get the "charging" part (switch-closed phase) of this question right too. They are quick to point out that the current in the inductor behaves just as the voltage across the capacitor did during the charging phase. And the voltage across the inductor decays exponentially, just as the capacitor current did. They also seem to know that the time constant here is $\tau = L/R$, not RC.

This is actually quite encouraging, as it seems we have, after all, heard of the *"duality principle."* In simple terms this principle says that *a capacitor can be considered as an inverse (or "mirror") of an inductor because the voltage–current equations of the two devices can be transformed into one another by exchanging the voltage and current terms.* So, in essence, capacitors are analogous to inductors, and voltage to current.

But wait! Why are we even interested in this exotic-sounding new principle? Don't we have enough on our hands already? Well, it so happens, that by using the duality principle we can often *derive a lot of clues about any L-based circuit from a C-based circuit, and vice versa* — right off the bat — *without* having to plunge headlong into a web of hopelessly non-intuitive equations. So, in fact, we would do well to try to use the duality principle to our advantage if possible.

With the duality principle in mind, let us attempt to *open* the switch in the inductor circuit and try to predict the outcome. What happens? No! Unfortunately, things don't remain almost "unchanged" as they did for a capacitor. In fact, the behavior of the inductor during the off-phase is really no replica of the off-phase of the capacitor circuit.

So does that mean we need to jettison our precious duality principle altogether? Actually we don't. The problem here is that the two circuits in Figure 1.3, despite being deceptively similar, are *really not duals of each other*. And for that reason, we really can't use them to derive any clues either. A little later, we will construct *proper* dual circuits. But for now we may have already started to suspect that we really don't understand inductors as well as we thought, nor in fact the duality principle we were perhaps counting on to do so.

The Law of Conservation of Energy

If a nervous interviewee hazards the guess that the current in the inductor simply "goes to zero immediately" on opening the switch, a gentle reminder of what we all learnt in high school is probably due. The stored energy in a capacitor is $CV^2/2$, and so there is really no problem opening the switch in the capacitor circuit — the capacitor just continues to hold its stored energy (and voltage). But in an inductor, the stored energy is $LI^2/2$. Therefore, if we speculate that the current in the inductor circuit is at some finite value before the switch is opened and zero *immediately* afterward, the question arises: to *where did all the stored inductor energy suddenly disappear? Hint:* we have all heard of the *law of conservation of energy* — energy can change its form, but it just cannot be wished away!

Yes, sometimes a particularly intrepid interviewee will suggest that the inductor current *"decays exponentially to zero"* on opening the switch. So the question arises — where is the current in the inductor flowing to and from? We always need a *closed* galvanic path for current to flow (from Kirchhoff's laws)!

But, wait! Do we even fully understand the *charging phase* of the inductor well enough? Now this is getting really troubling! Let's find out for ourselves!

The Charging Phase and the Concept of Induced Voltage

From an intuitive viewpoint, most engineers are quite comfortable with the mental picture they have acquired over time *of a capacitor being charged* — the accumulated charge keeps trying to repel any charge trying to climb aboard the capacitor plates till finally a balance is reached and the incoming charge (current) gets reduced to near-zero. This picture is also intuitively reassuring because at the back of our minds, we realize it corresponds closely with our understanding of *real-life* situations — like that of an overcrowded bus during rush hour, where the number of commuters that manage to get on board at a stop depends on the capacity of the bus (double-decker or otherwise) and also on the sheer desperation of the commuters (the applied voltage).

But coming to the inductor charging circuit (i.e., switch closed), we can't seem to connect this too readily to any of our immediate real-life experiences. Our basic question here is — why does the charging *current* in the inductor circuit actually *increase* with time?

Or equivalently, what prevents the current from being high to start with? We know there is no mutually repelling "charge" here, as in the case of the capacitor. So why?

We can also ask an even more basic question — why is there *any voltage* even present across the inductor? We always accept a voltage across a *resistor* without argument — because we know *Ohm's law* ($V = I \times R$) all too well. But an inductor has (almost) *no resistance* — it is basically just a length of solid conducting copper wire (wound on a certain core). So how does it manage to "hold-off" any voltage across it? In fact, we are comfortable about the fact that a capacitor can hold voltage across it. But for the inductor — we are not very clear! Further, if what we have learnt in school is true — that electric field by definition is the *voltage gradient* dV/dx ("x" being the distance), we are now faced with having to explain a mysterious *electric field* somewhere inside the inductor! Where did that come from?

It turns out that, according to Lenz and/or Faraday, the current takes time to build up in an inductor only because of *"induced voltage."* This voltage, by definition, opposes any external effort to *change* the existing flux (or current) in an inductor. So, if the current is fixed, yes, there is no voltage present across the inductor — it then behaves just as a piece of conducting wire. But the moment we try to *change* the current, we get an induced voltage across it. By definition, the *voltage measured across an inductor at any moment* (whether the switch is open or closed, as in Figure 1.3) is the "induced voltage."

> *Note: We also observe that the analogy between a capacitor/inductor and voltage/current, as invoked by the duality principle, doesn't stop right there! For example, it was considered equally puzzling at some point in history, how at all any current was apparently managing to flow through a capacitor — when the applied voltage across it was changed. Keeping in mind that a capacitor is basically two metal plates with an interposing (nonconducting) insulator, it seemed contrary to the very understanding of what an "insulator" was supposed to be. This phenomenon was ultimately explained in terms of a "displacement current" that flows (or rather seems to flow) through the plates of the capacitor when the voltage changes. In fact, this current is completely analogous to the concept of "induced voltage" — to explain the fact that a voltage was being observed across an inductor when the current through it was changing.*

So let us now try to figure out exactly how the induced voltage behaves when the switch is closed. Looking at the inductor charging phase in Figure 1.3, the inductor current is initially zero. Thereafter, by closing the switch, we are attempting to cause a sudden change in the current. The induced voltage now steps in to try to keep the current down to its initial value (zero). So we apply "Kirchhoff's voltage law" to the closed loop in question. Therefore, at the moment the switch closes, the induced voltage must be exactly equal to the applied voltage, since the voltage drop across the series resistance R is initially zero (by Ohm's law).

As time progresses, we can think intuitively in terms of the applied voltage "winning." This causes the current to rise up progressively. But that also causes the voltage drop across R to increase, and so the induced voltage must fall *by the same amount* (to remain faithful to Kirchhoff's voltage law). That tells us exactly what the *induced voltage* (voltage across inductor) is during the entire switch-closed phase.

Why does the applied voltage "win"? For a moment, let's suppose it didn't. That would mean the applied voltage and the induced voltage have managed to completely counterbalance each other — and the current would then *remain* at zero (or constant). However, that cannot be because zero rate of change in current implies no induced voltage either! In other words, the very existence of induced voltage depends on the fact that current changes, and it *must* change.

We also observe rather thankfully that all the laws of nature bear each other out. There is no contradiction whichever way we look at the situation. For example, even though the current in the inductor is subsequently higher, its *rate of change* is less, and therefore, so is the induced voltage (on the basis of Faraday's/Lenz's law). And this "allows" for the additional drop appearing across the resistor, as per Kirchhoff's voltage law!

But we still don't know how the induced voltage behaves when the switch turns OFF! To unravel this part of the puzzle, we actually need some more analysis.

The Effect of the Series Resistance on the Time Constant

Let us ask — what are the *final* levels at the end of the charging phase in Figure 1.3 — that is, of the current in the inductor and the voltage across the capacitor. This requires us to focus on the exact role being played by R. Intuitively, we expect that for the capacitor circuit, increasing the R will increase the charging time constant τ. This is borne out by the equation $\tau = RC$ too, and is what happens in reality too. But for the inductor charging circuit, we are again up against another seemingly counter-intuitive behavior — *increasing R actually decreases the charging time constant*. That is in fact indicated by $\tau = L/R$ too.

Let us attempt to explain all this. Looking at Figure 1.4 which shows the inductor charging current, we can see that the $R = 1\ \Omega$ current curve does, indeed, rise faster than the $R = 2\ \Omega$ curve (as intuitively expected). But the *final value* of the $R = 1\ \Omega$ curve is *twice* as high. Since by definition, the time constant is "the time to get to 63% of the *final value*," the $R = 1\ \Omega$ curve has a *larger* time constant, despite the fact that it did rise much faster from the get-go. So, that explains the inductor *current* waveforms.

But looking at the inductor *voltage* waveforms in Figure 1.5, we see there is still some explaining to do. Note that for a decaying exponential curve, the time constant is defined as the time it takes to get to 37% of the *initial* value. So, in this case, we see that though the

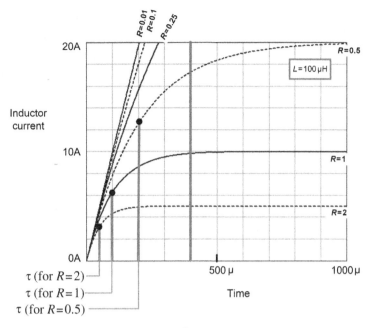

Figure 1.4: Inductor current during charging phase for different *R* (in ohms), for an applied input of 10 V.

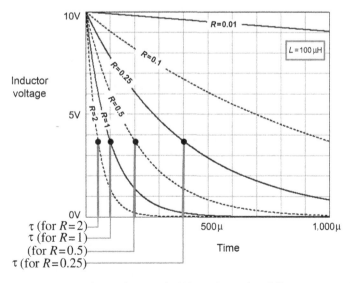

Figure 1.5: Inductor voltage during charging phase for different *R* (in ohms), for an applied input of 10 V.

initial values of all the curves are the same, *yet, for example, the R = 1 Ω curve has a slower decay (larger time constant) than the R = 2 Ω curve!* There is actually no mystery involved here, since we already know what the current is doing during this time (Figure 1.4), and therefore the voltage curves follow automatically from Kirchhoff's laws.

The conclusion is that if, in general, we ever make the mistake of looking *only* at an inductor *voltage* waveform, we may find ourselves continually baffled by an inductor! *For an inductor, we should always try to see what the **current** in it is trying to do.* That is why, as we just found out, the voltage during the off-time is determined entirely by the current. The voltage just follows the dictates of the current, not the other way around. In fact, in *Chapter 8*, we will see how this particular behavioral aspect of an inductor determines the exact shape of the voltage and current waveforms during a switch transition and thereby determines the crossover (transition) loss too.

The Inductor Charging Circuit with R = 0 and the "Inductor Equation"

What happens if *R* is made to decrease to zero?

From Figure 1.5, we can correctly guess that the only reason that the voltage across the inductor during the on-time changes at all from its initial value V_{IN} is the presence of *R*! So if *R* is 0, we can expect that the voltage across the inductor *never changes* during the on-time! The induced voltage must then be equal to the applied DC voltage. That is not strange at all — if we look at it from the point of view of Kirchhoff's voltage law, there is *no* voltage drop present across the resistor — simply because there is no resistor! So in this case, all the applied voltage appears across the inductor. And we know it can "hold-off" this applied voltage, provided the current in it is changing. Alternatively, *if any voltage is present across an inductor, the current through it **must** be changing!*

So now, as suggested by the low-*R* curves of Figures 1.4 and 1.5, we expect that the *inductor current will keep ramping up with a constant slope during the on-time.* Eventually, it will reach an *infinite* value (in theory). In fact, this can be mathematically proven by differentiating the inductor charging current equation with respect to time, and then putting *R* = 0 as follows:

$$I(t) = \frac{V_{IN}}{R}(1 - e^{-tR/L})$$

$$\frac{dI(t)}{dt} = \frac{V_{IN}}{R}\left(\frac{R}{L}e^{-tR/L}\right)$$

$$\left.\frac{dI(t)}{dt}\right|_{R=0} = \frac{V_{IN}}{L}$$

So, we see that when the inductor is connected directly across a voltage source V_{IN}, the slope of the line representing the inductor current is constant and equal to V_{IN}/L (the current *rising* constantly).

Note that in the above derivation, the voltage across the inductor happened to be equal to V_{IN} because R was 0. But in general, if we call "V" the voltage actually present *across* the inductor (at any given moment), I being the current through it, we get the general "inductor equation"

$$\frac{dI}{dt} = \frac{V}{L} \quad (inductor\ equation)$$

This equation applies to an ideal inductor ($R = 0$), in any circuit, under any condition. For example, it not only applies to the "charging" phase of the inductor, but also applies to its "discharging" phase!

> *Note: When working with the inductor equation, for simplicity, we usually plug in only the* magnitudes *of all the quantities involved (though we do mentally keep track of what is really happening, i.e., current rising or falling).*

The Duality Principle

We now know how the voltage and current (rather its rate of change) are mutually related in an inductor during both the charging and discharging phases. Let us use this information, along with a more complete statement of the *duality principle*, to finally understand what really happens when we try to interrupt the current in an inductor.

The principle of duality concerns the transformation between two apparently different circuits, which have similar properties when current and voltage are interchanged. Duality transformations are applicable to planar circuits only and involve a topological conversion: capacitor and inductor interchange, resistance and conductance interchange, and **voltage source and current source** *interchange.*

We can thus spot our "mistakes" in Figure 1.3. First, we were using an input *voltage* source applied to *both* circuits — whereas we should have used a *current* source for the "other" circuit. Second, we used a *series* switch in both the circuits. We note that the primary function of a series switch is only to interrupt the flow of *current* — not to change the voltage (though that may happen as a result). So, if we really want to create proper mirror (dual) circuits, then forcing the current to zero in the inductor is the dual of forcing the *voltage* across the capacitor to zero. And to implement that, we obviously need to place a switch in *parallel* to the capacitor (not in series with it). With these changes in mind, we have finally created *true* dual circuits as shown in Figure 1.6 (*both* are actually *equally* impractical in reality!).

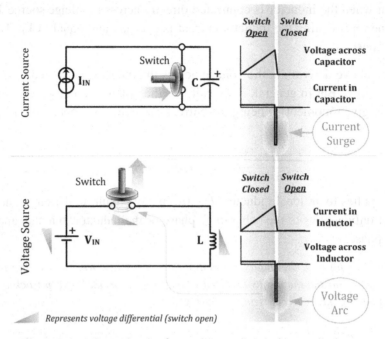

Figure 1.6: Mirror circuits for understanding inductor discharge.

The "Capacitor Equation"

To analyze what happens in Figure 1.6, we must first learn the "capacitor equation" — analogous to the "inductor equation" derived previously. If the duality principle is correct, both the following equations must be valid:

$$V = L\frac{\mathrm{d}I}{\mathrm{d}t} \quad \text{(inductor equation)}$$

$$I = C\frac{\mathrm{d}V}{\mathrm{d}t} \quad \text{(capacitor equation)}$$

Further, if we are dealing with "straight-line segments" (constant V for an inductor and constant I for a capacitor), we can write the above equations in terms of the corresponding *increments* or *decrements* during the given time segment.

$$V = L\frac{\Delta I}{\Delta t} \quad \text{(inductor equation for constant applied voltage)}$$

$$I = C\frac{\Delta V}{\Delta t} \quad \text{(capacitor equation for constant applied current)}$$

It is interesting to observe that the duality principle is actually helping us understand how the *capacitor* behaves when being charged (or discharged) by a *current source*. We can guess that the *voltage* across the capacitor will then ramp up in a straight line — to near infinite values — just as the inductor *current* does with an applied *voltage* source. And in both cases, the final values reached (of the voltage across the capacitor and the current through the inductor) *are dictated only by various parasitics* that we have not considered here — mainly the ESR of the capacitor and the DCR of the inductor respectively.

The Inductor Discharge Phase

We now analyze the mirror circuits of Figure 1.6 in more detail.

We know intuitively (and also from the capacitor equation) what happens to a *capacitor* when we attempt to suddenly discharge it (by means of the parallel switch). Therefore, we can now easily guess what happens when we suddenly try to "discharge" the inductor (i.e., force its current to zero by means of the series switch).

We know that if a "short" is applied across any capacitor terminals, we get an extremely high-current surge for a brief moment — during which time the capacitor discharges and the voltage across it ramps down steeply to zero. So, we can correctly infer that if we try to interrupt the current through an inductor, we will get *a very high voltage across it — with the current simultaneously ramping down steeply to zero*. So the mystery of the inductor "discharge" phase is solved — *with the help of the duality principle!*

But we still don't know *exactly* what the actual *magnitude* of the voltage spike appearing across the switch/inductor is. That is simple — as we said previously, during the off-time, the voltage will take on any value to force *current continuity*. So, a brief arc will appear across the contacts as we try to pull them apart (see Figure 1.6). If the contacts are separated by a greater distance, the voltage will increase automatically to maintain the spark. And during this time, the current will ramp down steeply. The arcing will last for *as long* as there is *any* remaining inductor stored energy — that is, till the current *completely* ramps down to zero. The rate of fall of current is simply V/L, from the inductor equation. So eventually, *all the stored energy in the inductor is completely dissipated* in the resulting flash of heat and light, and the current returns to zero simultaneously. At this moment, the induced voltage collapses to zero too, its purpose complete. This is in fact the basic principle behind the automotive spark plug, and the camera flash too (occurring in a more controlled fashion).

But wait — we have stated above that the rate of fall of current in the inductor circuit was "V/L." What is V? V is the voltage *across the inductor, not* the voltage across the contacts. In the following sections, we will learn that the voltage *across an inductor* (almost always) reverses when we try to interrupt its current. If that is true, then by Kirchhoff's voltage law, since the algebraic sum of all the voltage drops in any closed circuit must add up to zero, the voltage *across the contacts* will be equal to the sum of the magnitudes of the induced voltage and the applied DC rail — however, the sign of the voltage across the contacts (i.e., its direction) will necessarily be *opposite* to the other voltages (see the gray triangles in the lower schematic of Figure 1.6). Therefore, we conclude that the magnitude of the voltage spike across the inductor is equal to the magnitude of the voltage across the contacts minus the magnitude of the input DC voltage.

Finally, we know everything about the puzzling inductor discharge phase!

Flyback Energy and Freewheeling Current

The energy that "must get out" of the inductor when we try to open the switch is called the "flyback" energy. The current that continues to force its way through is called the "freewheeling" current. Note that this not only sounds, but, in fact, is very similar to another real-world situation — that of a *mechanical spinning wheel* or of a "flywheel." In fact, understanding the flywheel can help greatly in gaining an intuitive insight into the behavior of an inductor.

Just as the inductor has stored energy related to the current flowing through it, the flywheel stores energy related to its spinning action. And neither of these energy terms can be wished away in an instant. In the case of the flywheel, we can apply "brakes" to dissipate its rotational energy (as heat in the brake linings) — and we know this will produce a *progressive* reduction in the spinning. Further, if the brakes are applied more emphatically, the time that will elapse till the spinning stops entirely gets proportionately decreased. That is very similar to an inductor — with the *induced voltage (during the off-time) playing the part of the "brakes" and the current being akin to the spinning.* So, the induced voltage causes a *progressive* reduction in the current. If we have a higher induced voltage, this will cause a steeper fall in the current. In fact, that is also indicated by the inductor equation $V = L dI/dt$!

However, we have also learned something more fundamental about the behavior of an inductor, as described next.

Current Must Be Continuous, Its Slope Need Not Be

The keyword in the previous section was *progressive*. From a completely mathematical/geometrical point of view now, we should understand that any curve representing inductor *current* cannot be *discontinuous* (no *sudden* jumps allowed) — because that will in effect cause *energy* to be discontinuous, which we know is impossible. But we can certainly cause the *slope* of the current (i.e., its dI/dt) to have "jumps." So we can, for example, change the slope of current (dI/dt) *in an instant* — from one representing a rising ramp (increasing stored energy) to one representing a falling ramp (opposite sign, i.e., decreasing energy).

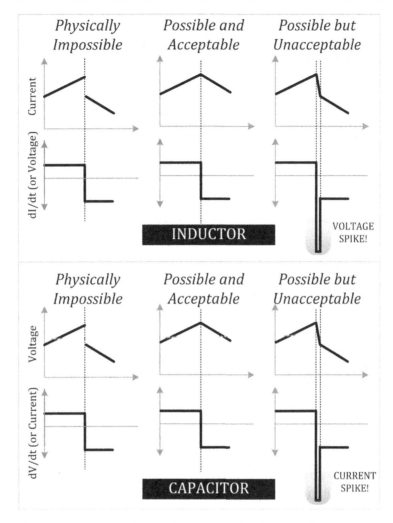

Figure 1.7: Inductor current must be continuous, but its slope need not be. Capacitor voltage must be continuous, but its slope need not be.

However, the current itself must always be continuous. This is shown in Figure 1.7 under the choices marked "possible."

Note that there are *two* options in the figure that are "possible." Both are so, simply because they do not violate any known physical laws. However, one of these choices is considered "unacceptable" because of the *huge spike* — which we know can damage the switch. The other choice, marked "acceptable," is in fact what really happens in any switching converter topology, as we will soon see.

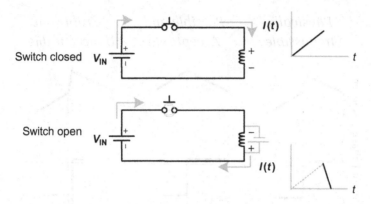

Figure 1.8: How the voltage reverses on attempted interruption of inductor current.

The Voltage Reversal Phenomenon

We mentioned there is a *voltage reversal* across the switch when we try to interrupt its current. Let us try to understand that better now.

An intuitive (but not necessarily rigorous) way to visualize it is shown in Figure 1.8. Here we note that when the switch is closed (upper schematic), the *current is shown leaving the positive terminal of the applied DC voltage source* — that being the normal convention for describing the direction of current flow. During this on-time, the *upper* end of the inductor gets set to a higher voltage than its lower end. Subsequently, when the switch opens, the input DC source gets disconnected from the inductor. But we have just learnt that the current demands to keep flowing (at least for a while) — *in the same direction as previously flowing.* So during the switch off-time, we can mentally visualize the inductor as becoming a sort of "voltage source," forcing the current to keep flowing. For that reason, we have placed an imaginary (gray) voltage source (battery symbol) across the inductor in the lower half of the figure — its polarity in accordance with the convention that current must leave by the positive terminal of any voltage source. Thus, we can see that this causes the lower end of the inductor to now be at a higher voltage than its upper end. Clearly, voltage reversal has occurred — simply by the need to maintain current continuity.

The phenomena of voltage reversal can be traced back to the fact that induced voltage always opposes any change (in current). However, in fact, voltage reversal does *not always* occur. For example, voltage reversal does *not* occur during the *initial* startup ("power-up") phase of a Boost converter. That is because the primary requirement is only that the inductor *current* somehow needs to keep flowing — voltage takes a backseat. So hypothetically, if a circuit is wired in a certain way, and the conditions are "right," it is certainly possible that voltage reversal won't occur, so long as current continuity can still be maintained.

However, we must be clear that *if and when a converter reaches a "steady state," voltage reversal will necessarily occur at every switch transition.*

For that we now have to understand what a "steady state" is.

A Steady State in Power Conversion and the Different Operating Modes

A *steady state* is, as the name indicates, *stable*. So, it is in essence the opposite of a runaway or unstable condition. But we can easily visualize that we will in fact get an unstable condition if at the end of every cycle, we **don't** *return to the current we started the cycle with* — because then, every successive cycle, we will accumulate a net increase or decrease of current, and the situation will keep changing forever (in theory).

From $V = L\Delta I/\Delta t$, it is clear that if the current is ramping *up* for a positive (i.e., applied) voltage, the current must ramp *down* if the voltage reverses. So the following equations must apply (magnitudes only):

$$V_{\text{ON}} = L\frac{\Delta I_{\text{ON}}}{\Delta t_{\text{ON}}}$$

$$V_{\text{OFF}} = L\frac{\Delta I_{\text{OFF}}}{\Delta t_{\text{OFF}}}$$

Here the subscript "ON" refers to the switch being closed and "OFF" refers to the switch being open. V_{ON} and V_{OFF} are the respective voltages *across the inductor* during the durations Δt_{ON} and Δt_{OFF}. Note that very often, Δt_{ON} is written simply as "t_{ON}," the switch on-time. And similarly, Δt_{OFF} is simply "t_{OFF}," the switch off-time.

Now suppose we are able to create a circuit in which the *amount* the current ramps *up* by in the on-time (ΔI_{ON}) is *exactly equal* to the amount the current ramps *down* by during the off-time (ΔI_{OFF}). If that happens, we would have reached a steady state. Now we could repeat the same sequence an innumerable amount of times, and get the *same* result each and every time. In other words, *every "switching cycle" would then be an exact replica of the previous cycle.* Further, we could also perhaps get our circuit to deliver a steady stream of (identical) energy packets continuously to an output capacitor and load. If we could do that, by definition, we would have created a *power converter*!

Achieving a steady state is luckily not as hard as it may sound. Nature *automatically* tries to help every natural process move toward a stable state (without "user intervention").

So, in our case, all we need to do on our part is to *provide* a circuit that *allows* these conditions to develop *naturally* (over several cycles). And if we have created the right conditions, a steady state *will* ultimately result. Further, this would be self-sustaining thereafter. Such a circuit would then be called a switching *"topology"*!

Conversely, any valid topology must be able to reach a state described by the following key equation $\Delta I_{ON} = \Delta I_{OFF} \equiv \Delta I$. If it can't get this to happen, it is *not* a topology. Therefore, this simple current increment/decrement equation forms the litmus test for validating any new switching topology.

Note that the inductor equation, and thereby the definition of "steady state," refers only to the *increase/decrease* in current — it says nothing about the actual (absolute) value of the current at the start (and end) of every cycle. So, there are in fact several possibilities. We could have a steady state in which the current returns to *zero* every cycle, and this is called a "discontinuous conduction mode" (DCM). However, if the current stays pegged at some non-zero value throughout, we will have "continuous conduction mode" (CCM). The latter mode is the most common mode of operation encountered in power conversion. In Figure 1.9, we have graphically shown these operating modes (all in steady state). We also have some other modes that we will talk about very soon. Note that in the figure, the "square" waveform is the voltage across the inductor and the slowly ramping waveform is the inductor current. Let us make some related observations:

a.   We see that the voltage across the inductor *always* reverses at every switching event (as expected in steady state).

b.   We note that since the inductor equation relates voltage to the *slope of the current*, not to the actual current, therefore, for a given V_{ON} and V_{OFF}, several current waveforms are possible (all having the *same dI/dt* for corresponding segments). Each of these possibilities has a name — CCM, DCM, BCM (boundary conduction mode, also called critical conduction mode), and so on. Which of these operating modes actually occurs depends on the specific circuit (i.e., the topology) and also the application conditions (how much output power we are demanding and what the input and output voltages are).

c.   The inductor voltages, V_{ON} and V_{OFF} shown in the figure, are related to the application conditions V_{IN} and/or V_O. Their exact relationship will become known a little later, and we will also learn that it depends on the specific topology.

d.   A key question is — what is the exact relationship between the average *inductor* current and the *load* current? We will soon see that that too depends on the specific *topology*. However, in all cases, the average inductor current ("I_{AVG}" or "I_L") is proportional to the load current ("I_O"). So if, for example, I_O is 2 A and I_{AVG} is 10 A, then if I_O is decreased to 1 A, I_{AVG} will fall to 5 A. Therefore on decreasing the load current, we can get I_{AVG} to decrease, as indicated in Figure 1.9.

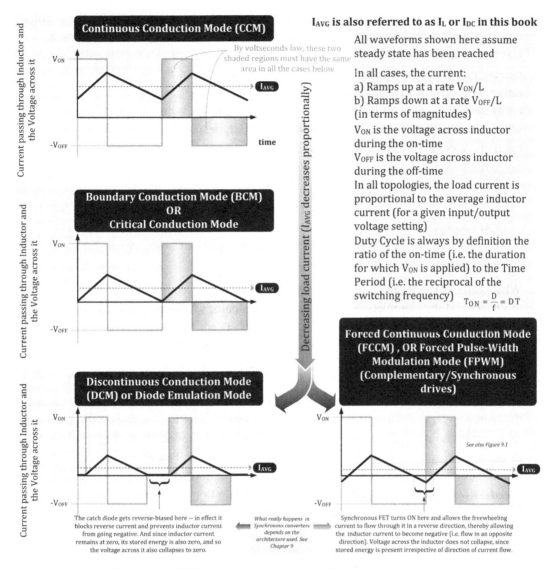

Figure 1.9: Different operating modes of switching regulators.

e. Typically, we transit *automatically* from CCM to DCM, just by reducing the load current of the converter. But note that we will necessarily have to pass through BCM along the way.

f. "BCM" is just that — a *"boundary* conduction mode" — situated exactly between CCM and DCM. It is therefore a purely philosophical question to ask whether BCM should be viewed as CCM or DCM (at their respective extremes) — it really doesn't matter.

g. Note that in all the cases shown in Figure 1.9, with the exception of DCM, the average inductor current I_{AVG} is just the *geometrical center of the ramp* part of the current waveform. In DCM, however, we have an additional interval in which there is *no current* passing for a while. So, to find the average value of the inductor current, a rather more detailed calculation is required. In fact, that is the primary reason why *DCM equations turn out looking so complicated* — to the point that many engineers seem to rather instinctively ignore DCM altogether, despite some advantages of operating a converter in DCM instead of CCM.

> *Note: Expectedly, all DCM equations lead to exactly the same* numerical *results as the CCM equations* — when the converter is in BCM. *Practically speaking, we can freely pick and choose whether to use the CCM equations, or the more formidable-looking DCM equations, for evaluating a converter in BCM. Of course, there is no reason why we would ever want to struggle through complicated equations, when we can use much simpler equations to get the same results!*

h. What really is the average inductor current "I_{AVG}" as shown in Figure 1.9? A nice way to understand this parameter is through the "car analogy." Suppose we press the gas pedal of a car. The car responds by increasing its speed. In an analogous fashion, when we apply a voltage across an inductor (the on-time voltage "V_{ON}"), the current ramps up. Subsequently, suppose we press on the brakes of the car. The car will then respond by decreasing its speed. Similarly, when the applied voltage is removed from the inductor, voltage reversal occurs, and an induced voltage (the "brakes") appears across the inductor "V_{OFF}." Since it is in the opposite direction as V_{ON}, it causes the current to ramp down. So now, if we press the gas pedal (V_{ON}), followed by the brakes (V_{OFF}), in *quick succession*, and with the right timing, we could still make the car continue to move forward despite the constant lurching. It would then have a certain *average speed* — depending on the ratio of the gas pedal duration and the subsequent braking duration. In power conversion, this "lurching" is analogous to the "current ripple" $\Delta I = \Delta I_{ON} = \Delta I_{OFF}$. And quite similarly, we have an "average inductor current" I_{AVG} too, as shown in Figure 1.9. However, we do understand that in a power converter, the output capacitor eventually absorbs (or smoothens) this "lurching" and thus manages to deliver a steady DC current to the load as desired.

i. Some control ICs manage to *maintain* the converter in BCM mode under all application conditions. Examples of these are certain types of "hysteretic controllers" and self-oscillating types called "ringing choke converters" (RCCs). However, we know that the current ramps down at a rate V/L. And since V depends on the input/output voltages, the time to get to zero current depends on the specific application conditions. Therefore, in any BCM implementation, we always lose the advantage of fixed switching frequency operation.

j. Most conventional topologies are nowadays labeled "non-synchronous" — to distinguish them from more recent "synchronous" topologies. In the former, a diode is always present (the catch diode) that *prevents the inductor current from reversing direction* any time during the switching cycle. That is why, on reducing output power and/or increasing input voltage, we automatically transit from CCM to DCM. However, in synchronous topologies, the catch diode is either replaced completely by, or supplemented with, a low-drop MOSFET placed across it. So, whenever the diode is supposed to conduct, we force this extra MOSFET into conduction for that duration. Since the drop across this MOSFET is much lower than across a diode, not only do we manage to significantly reduce the conduction loss in the freewheeling path, but we can also now allow *reverse inductor current* — that is, current moving instantaneously *away* from the load. However, note that the average inductor current could still be positive — see Figure 1.9. Further, with negative currents now being "allowed," we no longer get DCM on reducing output power, but rather enter FPWM/FCCM (forced continuous conduction mode) as described in the figure.

> *Note: It is fortunate that almost all the standard CCM design equations (for non-synchronous topologies) apply equally to FCCM. So from the viewpoint of a harried designer, one of the "advantages" of using synchronous topologies is that the complicated DCM equations are a thing of the past! However, there are some new complications and nuances of synchronous topologies that we need to understand eventually.*

The Voltseconds Law, Inductor Reset and Converter Duty Cycle

There is another way to describe a *steady state*, by bringing in the inductor equation $V = L\Delta I/\Delta t$.

We know that during a steady state $\Delta I_{\text{ON}} = \Delta I_{\text{OFF}} \equiv \Delta I$. So what we are also saying is that

in steady state, the product of the voltage applied across the inductor, multiplied by the duration we apply it for (i.e., the on-time), must be equal to the voltage that appears across the inductor during the off-time, multiplied by the duration that it lasts for.

Therefore, we get

$$V_{\text{ON}} \times t_{\text{ON}} = V_{\text{OFF}} \times t_{\text{OFF}}$$

The product of the voltage and the time for which it appears across the inductor is called the "voltseconds" across the inductor. So equivalently, what we are also saying is that

if we have an inductor in a steady state, the voltseconds present across it during the on-time (i.e., current ramp-up phase) must be exactly equal in magnitude, though opposite in sign, to the voltseconds present across it during the off-time (i.e., during the current ramp-down phase).

That also means that if we plot the inductor voltage versus time, *the area under the voltage curve during the on-time must be equal to the area under the voltage curve during the off-time.* But we also know that *voltage reversal* always occurs in steady state. So clearly, these two areas must have the opposite signs. See the shaded segments in Figure 1.9.

Therefore, we can also say that *the **net** area under the voltage curve of an inductor must be equal to zero (in any switching cycle under steady state operation).*

Note that since the typical times involved in modern switching power conversion are so small, "voltseconds" turns out to be a very small number. Therefore, to make numbers more *manageable*, we usually prefer to talk in terms of "*Et*" or the "*voltµseconds.*" *Et* is clearly just the voltage applied across the inductor multiplied by the time in *microseconds* (not seconds). Further, we know that typical inductance values used in power conversion are also better expressed in terms of "µH" (microhenrys), not H. So, from $V = LdI/dt$ we can write

$$\Delta I_{ON} = \frac{V_{ON} \times t_{ON}}{L} = \frac{V_{ON} \times t_{ON\_\mu H}}{L_{\mu H}} = \frac{Et}{L_{\mu H}}$$

or simply

$$\Delta I = \frac{Et}{L} \quad (steady\ state, L\ in\ \mu H)$$

Note: If in any given equation Et *and* L *appear together, it should be generally assumed that* L *is in µH. Similarly, if we are using voltseconds, that would usually imply* L *is in H (unless otherwise indicated).*

Another term often used in power, one that tells us that we have managed to return to the same inductor current (and energy) that we started off with, is called inductor "*reset.*" Reset occurs at the very moment when the equality $\Delta I_{OFF} = \Delta I_{ON}$ is established. Of course, we could also have a *nonrepetitive* (or "single-shot") event, where the current starts at *zero* and then returns to *zero* — and that too would be inductor "reset."

The corollary is that in a repetitive switching scenario (steady state), *an inductor must be able to reset every cycle.* Reversing the argument — *any circuit configuration that makes inductor reset an impossibility* — is not a viable switching topology.

When we switch repetitively at a switching frequency "*f*," the "time period" (*T*) is equal to $1/f$. We can also define the "duty cycle" (*D*) of a power converter as the ratio of the on-time of the switch to the time period. So,

$$D = \frac{t_{ON}}{T} \quad (duty\ cycle\ definition)$$

Note that we can also write this as

$$D = \frac{t_{ON}}{t_{ON} + (T - t_{ON})} \quad \text{(duty cycle definition)}$$

At this point, we should be very clear how we are defining "t_{OFF}" in particular. While applying the voltseconds law, we had implicitly assumed that t_{OFF} was *the time for which the induced voltage V_{OFF} lasts*, not necessarily the time *for which the switch is OFF* (i.e., $T - t_{ON}$). In DCM, they are **not** the same (see Figure 1.9)! Only in CCM do we get

$$t_{OFF} = T - t_{ON} \quad \text{(duty cycle in CCM)}$$

and therefore

$$D = \frac{t_{ON}}{t_{ON} + t_{OFF}} \quad \text{(duty cycle in CCM)}$$

If working in DCM, we should stick to the more general definition of duty cycle given initially.

Using and Protecting Semiconductor Switches

We realize that all topologies exist only because they can achieve a steady state. In an "experimental" topology in which we can't make $\Delta I_{ON} = \Delta I_{OFF}$ happen, the inductor may see a *net increase of current every cycle*, and this can eventually escalate to a very large, almost uncontrolled value of current in just a few cycles. The name given to this progressive ramping-up (or down) of current (or inductor energy), one that is ultimately limited only by parasitics like the ESR and DCR, is called "staircasing." The switch will also turn ON into the same current and can thus be destroyed — that is if the induced voltage spike hasn't already done so (which can happen, if the situation is anything similar to the "unacceptable" plots shown in Figure 1.7!).

> *Note: The very use of the inductor equation $V = LdI/dt$ actually implies we are ignoring its parasitic resistance, DCR. The inductor equation is an idealization, applying only to a "perfect" inductor. That is why we had to put $R = 0$ when we derived it previously.*

In an actual power supply, the "mechanical switch" is replaced with a modern semiconductor device (like the MOSFET) — largely because then the switching action can be implemented reliably and also at a *very high repetition rate*. But semiconductor devices have certain electrical *ratings* that we need to be well aware of.

Every semiconductor device has an "absolute maximum *voltage* rating" that, unlike any typical mechanical relay, cannot be exceeded *even momentarily* — without possibly causing its *immediate* destruction. So, most MOSFETs do *not* allow any "latitude" whatsoever in terms of their *voltage* ratings.

Note: There are some "avalanche-rated" MOSFETs available, which can internally "clamp" the excess voltage appearing across them to some extent. In doing so, they are basically dissipating the excess energy associated with the voltage spike, within their internal clamp. Therefore, they can survive a certain amount of excess voltage (and energy), but only for a short duration (since the device heats up quickly).

There is also a maximum semiconductor device *"current* rating," but that is usually more *long term* in nature, dictated by the comparatively slower process of internal heat build-up inside the device. So hypothetically speaking, we could perhaps exceed the current rating somewhat, though only for a *short* time. Of course, we don't want to run a device constantly in this excess-current condition. However, under "abnormal conditions," like an "overload" on the output of the converter (or the extreme case of a shorted output), we may *judiciously* allow for a certain amount of "temporary/transient abuse" with regard to the current rating — but certainly not with the absolute maximum voltage rating!

In a practical implementation, we have to design the converter, select the switch, and then lay it all out on a PCB with great care — to ensure in particular that there is no *voltage* spike that can "kill" the switch (or other semiconductor devices present on the board). Occasionally, we may therefore need to add an external "snubber" or "clamp" across the switch so as to truncate any remnant spikes to within the voltage ratings of the switch.

To protect the switch (and converter) from excess currents, a "current limit" is usually required. In this case, the current in the inductor, or in the switch, is sensed, and then compared against a set threshold. If and when that is momentarily exceeded, the control circuitry forces the switch to turn OFF *immediately for the remainder of the switching cycle* so as to protect itself. In the next cycle, no "memory" is usually retained of what may have happened in the preceding cycle. Therefore, every switching cycle is started "afresh," with the current being continuously monitored to ensure it is at a "safe" level. If not, protective action is again initiated and can be repeated every cycle for several cycles if necessary, until the "overcurrent" condition ceases.

Note: One of the best known examples of the perils of "previous-cycle memory" in implementing current limit occurs in the popular "Simple Switcher®" family of parts (at www.national.com). In the "third-generation" LM267x family, the control circuit rather surprisingly reduces the duty cycle to about 45% for several cycles after any single current limit event. It then tries to progressively allow the duty cycle to increase over several successive cycles back to its required value. But this causes severe output "foldback" and consequent inability to regulate up to full rated load, particularly in applications that require a duty cycle greater than 50%. This condition is further exacerbated with large-output capacitances because the higher currents required to charge the output capacitor after the removal of an abnormal condition (e.g., output short) can lead to another current limit event (and consequent foldback for several cycles again) — before the duty cycle has been able to return to its desired value. *In effect, the converter goes into a continuous "motorboating" condition on removal of the output short, and so the output never recovers. This is rather obliquely "revealed" only deep within the product datasheets under the intriguing heading "Additional Applications Information."*

With the introduction to power conversion now complete, we turn our attention to how switching topologies develop naturally out of the behavior of an inductor.

Evolution of Switching Topologies

Controlling the Induced Voltage Spike by Diversion through a Diode

We realize that our "problem" with using an inductor is *two-fold*: either we are going to end up with *near-infinite induced voltage* spikes, as shown in Figures 1.6 and 1.7, or if we do somehow manage to control the induced voltage to some finite level, the equation $V = LdI/dt$ tells us we could very well end up with *near-infinite currents* (staircasing).

And further, coming to think of it, our basic purpose is still not close to being fulfilled — we still don't know how to derive any *useful power* from our circuit!

Luckily, *all* the above-described problems can be solved in one stroke! And in doing so, we will arrive at our very first "switching topology." Let's now see how that comes about.

We recollect from Figure 1.6 that the spike of induced voltage at switch turn-off occurs only because the current (previously flowing in the inductor) was still *demanding* a path along which to flow — *and somehow unknowingly, we had failed to provide any.* Therefore, nature, in search of the "weakest link," *found this in the switch itself* — and produced an arc across it to try to move the current across anyway.

But suppose we consciously provide a "diversionary path." Then there would be no problem turning the switch OFF and stopping the inductor current flowing through the switch — because it could continue to flow via this alternate route. The inductor would then no longer "complain" in the form of a dangerous voltage spike. Thereafter, perhaps we can even reroute the current back into the switch when it turns ON again. Finally, we can perhaps even repeat the ON−OFF−ON−OFF process indefinitely at a certain switching frequency.

In Figure 1.10, we have created such an *alternate path*. We will see that the way the diode is pointed, this path can come into play automatically, and *only* when the switch turns OFF.

Just to make things clearer, we have used some sample numbers in Figure 1.10. We have taken the applied input voltage to be 12 V and assumed a typical Schottky diode forward

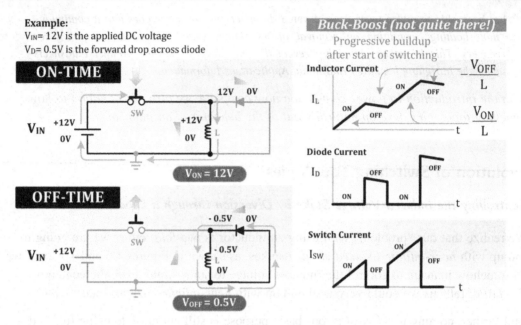

Figure 1.10: Providing a "diversion" for the inductor current through a diode.

drop of 0.5 V. Note that we are assuming a "perfect" switch here (no forward drop), for the sake of simplicity. We make the following observations:

- When the switch is ON (closed), the voltage at the upper end of the inductor L is at 12 V and the lower end is at 0 V ("ground"). So the diode is reverse-biased and does not conduct. Energy is then being built up in the inductor by the applied DC voltage source.

    The magnitude of the voltage applied across the inductor during the on-time of the switch (i.e., "V_{ON}") is equal to 12 V.

- When the switch turns OFF (open), an alternate path is *available* for the inductor current to flow — through the diode. And we can be sure that "nature" (in our case the "induced voltage") will attempt to exploit this path — by forcing the diode to conduct. But for that, the diode must get *"forward-biased,"* that is, its anode must get to a voltage 0.5 V higher than the cathode. But the anode is being held at ground (0-V rail). Therefore, the cathode must fall to −0.5 V.

    The magnitude of the voltage applied across the inductor during the off-time of the switch (i.e., "V_{OFF}") is equal to 0.5 V.

- Note that the induced voltage during the switch off-time has had its polarity reversed.
- The rate of rise of the current (in the inductor and switch) during the on-time is equal to V_{ON}/L. And during the off-time, the current ramps down (much more slowly), at a rate of V_{OFF}/L (in the inductor and diode).

• Yes, if we *wait long enough*, the inductor current will finally ramp down to zero (inductor "reset"). But if we don't wait and turn the switch back ON again, the current will again start to ramp up (staircasing), as shown in Figure 1.10.

• Note that *both* the switch and the diode currents have a "choppy" waveform — since one takes over where the other left off. This is in fact always true for any switching power converter (or topology).

Summarizing: We see that having provided a diversionary path for the current, the inductor isn't "complaining" anymore and there is no uncontrolled induced voltage spike anymore. But we certainly have now ended up with a possible problem of escalating currents. And come to think of it, neither do we have a *useful* output rail yet, which is what we are basically looking to do finally. In fact, *all that we are accomplishing in* Figure 1.10 *is dissipating some of the stored energy built-up in the inductor during the on-time, within the diode during the off-time.*

Achieving a Steady State and Deriving Useful Energy

We realize that to prevent staircasing, we need to somehow induce *voltseconds balance*. Yes, as mentioned, we could perhaps wait long enough before turning the switch ON again. But that still won't give us a useful output rail.

To finally solve all our problems in one go, let us take a hint from our "natural world of voltages." Since we realize we are looking for an *output DC voltage rail*, isn't it natural to use a *capacitor* somewhere in the circuit of Figure 1.10? Let us therefore now interpose a capacitor in series with the diode, as shown in Figure 1.11. If we do that, the diode (freewheeling) current would charge the capacitor up — and hopefully the capacitor voltage would eventually reach a steady level "V_O"! Further, *since that would increase the voltage drop appearing across the inductor during the off-time* (V_{OFF}), *it would increase the rate at which the inductor current can ramp down* — which we recognize was the basic problem with the circuit in Figure 1.10. So, we are finally seeing light at the end of the tunnel — by making V_{OFF} comparable to V_{ON}, we are hoping to achieve *voltseconds balance* expressed by $V_{ON} \times t_{ON} = V_{OFF} \times t_{OFF}$.

In Figure 1.11, the current escalates initially, but then after several cycles, it *automatically* levels out in what is clearly a steady state. That is because every cycle the capacitor charges up, it progressively increases the slope of the down-ramp, eventually allowing the converter to settle down *naturally* into the basic condition $\Delta I_{ON} = \Delta I_{OFF} \equiv \Delta I$. And once that is achieved, it is self-sustaining!

We also have a *useful* rail now — available across the output capacitor, from which we can draw some stored energy. So, we have shown a DC current passing through to the load by the dashed arrows in Figure 1.11.

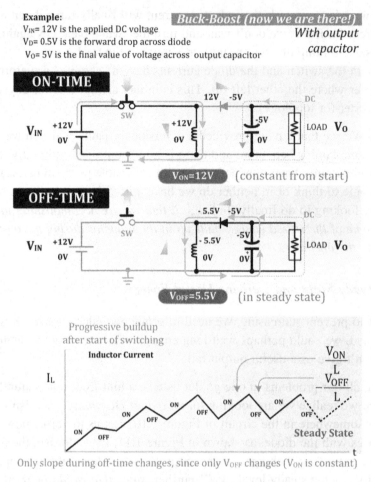

Example:
V_{IN}= 12V is the applied DC voltage
V_D= 0.5V is the forward drop across diode
V_O= 5V is the final value of voltage across output capacitor

Buck-Boost (now we are there!)

With output capacitor

Only slope during off-time changes, since only V_{OFF} changes (V_{ON} is constant)

Figure 1.11: Evolution of the Buck-Boost topology.

In fact, *this is our very first switching topology — the **Buck-Boost topology**.*

> *Note: Under the abnormal condition of an output short for example, Figure 1.11 effectively reduces to Figure 1.10! Therefore, to protect the converter under such conditions, a* current limit *is required.*

The Buck-Boost Converter

To understand Figure 1.11 better, we are actually going to work *backward* from here. So let us *assume* we have achieved a steady state — and therefore the output capacitor too has reached a steady value of say 5 V. Let us now find the *conditions* needed to make that a reality.

In Figure 1.11, the slope of the rising ramp is unchanged every cycle, being equal to V_{IN}/L. The slope of the falling ramp is initially V_D/L, where "V_D" is the drop across the diode. So from the inductor equation, initially $\Delta I_{ON} > \Delta I_{OFF}$. Thus, the current starts to staircase. But the magnitude of the slope of the falling ramp, and therefore ΔI_{OFF}, keeps getting larger and larger as the capacitor charges up. Eventually, we will reach a steady state defined by $\Delta I_{OFF} = \Delta I_{ON}$. At that moment, the voltseconds law applies.

$$V_{ON} \times t_{ON} = V_{OFF} \times t_{OFF}$$

Using the numbers of the example, we get

$$12 \times t_{ON} = 5.5 \times t_{OFF}$$

We see that a 5-V output is possible only if we have been switching with a *constant ratio* between the switch ON- and switch OFF-time, as given by

$$\frac{t_{OFF}}{t_{ON}} = \frac{12}{5.5} = 2.18$$

So to get the voltseconds to balance out for this case (5-V output and a 12-V input), we have to make the *off-time* 2.18 times larger than the on-time. Why so? Simply because the voltage during the *on-time* (across the inductor) is larger by *exactly the same proportion*: 12 V during the on-time as compared to 5.5 V during the off-time. Check: 12/5.5 = 2.18.

The duty cycle (assuming CCM) is therefore equal to

$$D = \frac{t_{ON}}{t_{ON} + t_{OFF}} = \frac{1}{1 + (t_{OFF}/t_{ON})} = \frac{1}{1 + 2.18} = 0.314$$

Now, had we taken a semiconductor switch instead of a mechanical one, we would have had a non-zero forward voltage drop of say "V_{SW}." This forward drop effectively just subtracts from the applied DC input during the on-time. So, had we done the above calculations symbolically, we would get

$$V_{ON} = V_{IN} - V_{SW} \quad (Buck\text{-}Boost)$$

and

$$V_{OFF} = V_O + V_D \quad (Buck\text{-}Boost)$$

Then, from the voltseconds law

$$\frac{t_{OFF}}{t_{ON}} = \frac{V_{IN} - V_{SW}}{V_O + V_D} \quad (Buck\text{-}Boost)$$

We thus get the duty cycle

$$D = \frac{V_O + V_D}{V_{IN} - V_{SW} + V_O + V_D} \quad (Buck\text{-}Boost)$$

If the switch and diode drops are small as compared to the input and output rails, we can simply write

$$D \approx \frac{V_O}{V_{IN} + V_O} \quad (Buck\text{-}Boost)$$

We can also write the relationship between the input and the output as follows:

$$V_O = V_{IN} \times \frac{D}{1 - D} \quad (Buck\text{-}Boost)$$

Note that some other easily derivable and convenient relationships to remember are

$$\frac{t_{ON}}{t_{OFF}} = \frac{D}{1 - D} \quad (any\ topology)$$

$$t_{ON} = \frac{D}{f} \quad (any\ topology)$$

$$t_{OFF} = \frac{1 - D}{f} \equiv \frac{D'}{f} \quad (any\ topology)$$

where we have defined $D' = 1 - D$ as the "duty cycle of the diode," since the diode is conducting for the remainder of the switching cycle duration (in CCM).

Ground-Referencing Our Circuits

We need to clearly establish what is referred to as the "ground" rail in any DC–DC switching topology. We know that there are two rails by which we apply the DC input voltage (current goes in from one and returns from the other). Similarly, there are also two output rails. But all practical topologies generally have *one rail that is common to both the input and the output*. It is this *common* rail that, by convention, is called the system "ground" in DC–DC converter applications.

However, there is yet another convention in place — the ground is also considered to be "0 V" (zero volts).

The Buck-Boost Configurations

In Figure 1.12, the common (ground) rails have been highlighted in bold gray background.

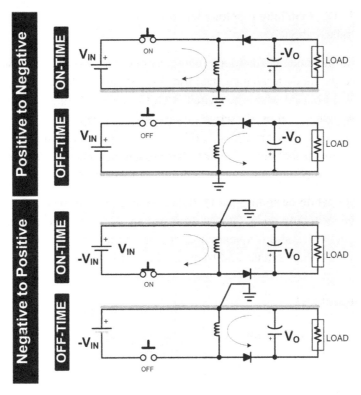

Figure 1.12: The two configurations of the Buck-Boost (inverting) topology.

We now realize that the Buck-Boost we presented in Figure 1.11 is actually a "positive (input)-to-negative (output)" Buck-Boost. There is another possibility, as shown in the lower half of Figure 1.12. We have relabeled its ground in accordance with the normal convention. Therefore, this is a "negative-to-positive Buck-Boost."

For either configuration, we see that whatever polarity is present at the input, it gets reversed at the output. Therefore, *the Buck-Boost is often simply called an "inverting" topology* (though we should keep in mind that that allows for *two different configurations*).

The Switching Node

Very simply put — the "point of detour" for the inductor current, that is, between the switch and the diode, is called the "switching node." Current coming into this node from the inductor can go either into the diode or into the switch, depending upon the state of the switch. Every DC–DC switching topology has this node (without it we would get the huge voltage spike we talked about!).

Since the current at this node needs to alternate between the diode and the switch, it needs to alternately force the diode to change state too (i.e., be reverse-biased when the switch turns ON and forward-biased when the switch is OFF). So, the voltage at this node must necessarily be "swinging." An oscilloscope probe connected here (with its ground clip connected to the power supply ground, i.e., 0 V) will always see a voltage waveform with "square edges." This is in fact very similar to the voltage across the inductor, except that it is DC level-shifted by a certain amount, depending on the topology.

On a practical level, while designing the PCB, we have to be cautious in not putting too much copper at the switching node. Otherwise it becomes an effective electric-field antenna, spewing radiated radio frequency interference all around. The output cables can thereafter pick up the radiated noise and transmit it directly to the load.

Analyzing the Buck-Boost

In Figure 1.13, we have drawn a line "I_L" through the *geometric center* of the ramp portion of the steady-state inductor current waveform. This is defined as the *average inductor current*. The switch current also has an average value of I_L, but only during the interval t_{ON}. Similarly, the average of the diode current is also I_L during t_{OFF}. However, the switch and diode currents when *averaged over the entire cycle* (i.e., over both the ON and OFF durations) are by simple mathematics their respective *weighted* averages.

$$I_{SW\_AVG} = I_L \times \frac{t_{ON}}{T} = I_L \times D \quad (Buck\text{-}Boost)$$

$$I_{D\_AVG} = I_L \times \frac{t_{OFF}}{T} = I_L \times D' = I_L \times (1 - D) \quad (Buck\text{-}Boost)$$

where D' is the duty cycle of the diode, that is, $1 - D$. It is also easy to visualize that for this particular topology, the *average input current is equal to the average switch current*. Further, as we will see in the following sections, the *average diode current* is equal to the *load current*. This is what makes the Buck-Boost topology quite different from the Buck topology.

Properties of the Buck-Boost

We now make some observations based on Figures 1.11–1.13:

- For example, a "positive-to-negative" Buck-Boost can convert 12 V to −5 V (step-down) or 12 V to −15 V (step-up). A "negative-to-positive" Buck-Boost can convert

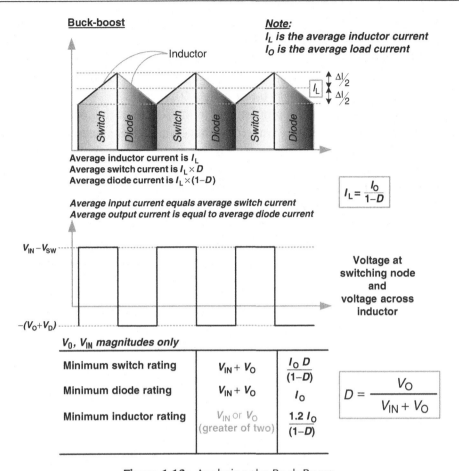

Figure 1.13: Analyzing the Buck-Boost.

say -12 V to 5 V, or 5 V to 15 V, and so on. The *magnitude* of the output voltage can thus be either smaller or larger (or equal to) the *magnitude* of the input voltage.

- When the switch is ON, energy is delivered *only* into the inductor by the input DC source (via the *switch*), and *none* of it passes through to the output.
- When the switch is OFF, *only* the stored energy of the inductor is pushed into the output (through the *diode*), and *none* comes directly from the input DC source.
- The above two observations make the Buck-Boost topology *the only "pure flyback" topology around, in the sense that **all** the energy transferred from the input to the output must have been previously stored in the inductor.* No other topology shares this unique property.
- The current coming from the *input capacitor* (DC source) is "choppy," that is, pulsating. That is because this current, combined with the steady DC current (I_{IN})

coming in from the DC source, basically forms the *switch* current waveform (which we know is always choppy for any topology) (see Figure 1.9).

- Similarly, the current into the *output capacitor* is also choppy because combined with the steady DC current into the load (I_{OUT}), it forms the diode current (which we know is always choppy for any topology) (see Figure 1.9).

- We know that heat dissipation is proportional to the square of the root mean square (RMS) current. And since choppy waveforms have high RMS values, the efficiency of a Buck-Boost is not very good. Also, there is generally a relatively high level of noise and ripple across the board. Therefore, the Buck-Boost may also demand much better filtering at its input, and often at its output too.

- Though current enters the output capacitor to charge it up when the switch turns ON and leaves it to go into the load when the switch is OFF, the *average capacitor current is always zero*. In fact, any capacitor in "steady state" must, by definition, have *zero average current* passing through it — otherwise it would keep charging or discharging until it too reaches a steady state, just like the inductor current.

Since the average current from the output capacitor is zero, therefore, for *the Buck-Boost, the average diode current must be equal to the load current* (where else can the current come from?). Therefore,

$$I_{D\_AVG} = I_O = I_L \times (1 - D)$$

So,

$$I_L = \frac{I_O}{1 - D} \quad (Buck\text{-}Boost)$$

This is the relationship between the average inductor current and the load current. Note that in Figure 1.13, in the embedded table, we have asked for an inductor rated for $1.2 \times I_O/(1 - D)$. The factor "1.2" comes from the fact, that by typical design criteria, the peak of the inductor current waveform is about 20% higher than its average. So we need to look for an inductor rated at least for a current of $1.2 \times I_L$.

Why Three Basic Topologies Only?

There are certainly several ways to set up circuits using an inductor, which provide a "freewheeling path" too, for the inductor current. But some of these are usually disqualified simply because the input and output do not share a common rail, and thus there is no proper *ground reference* available for the converter and the rest of the system. Two examples of such "working-but-unacceptable" converters are the Buck-Boost configurations shown in Figure 1.14. Compare these with Figure 1.12 to see what the problem is! However, note

Buck–boost configurations with no proper ground reference

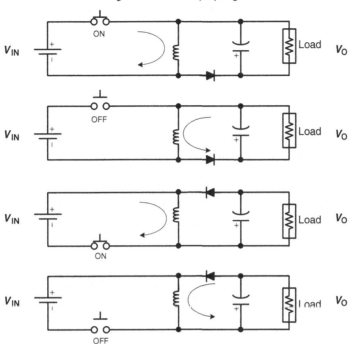

Figure 1.14: Improperly referenced Buck-Boost configurations.

that if these were "front-end converters," the system ground could be established starting at the output of this converter itself and may thus be acceptable.

Of the remaining ways, several are just "configurations" of a *basic topology* (like the two configurations in Figure 1.12). Among the basic topologies, we actually have just three — the Buck, the Boost, and the Buck-Boost. Why only three? That is because of *the way the inductor is connected*. Note that with proper ground-referencing in place, there are only three distinct rails possible — the input, the output, and the (common) ground. So if one end of the inductor is connected to the ground, it becomes a Buck-Boost! On the other hand, if it is connected to the input, it becomes a Boost. And if connected to the output, it becomes a Buck (see Figure 1.15).

The Boost Topology

In Figure 1.16, *we* have presented the schematic of the Boost topology. The direct and the freewheeling paths are indicated therein. In Figure 1.17, we have the corresponding analysis, including the key waveforms.

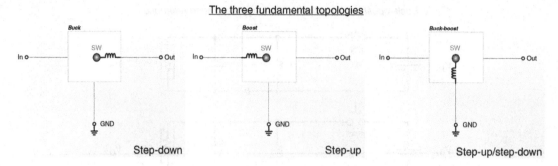

The three fundamental topologies

In all cases, one end of the inductor is tied to one of the three available DC rails, (IN, OUT, or GND). That determines the "topology". The other end of the inductor in all cases gets alternately connected via the switch to the input source (energy pulled in) and then via the catch diode to the output (energy delivered). Therefore, the volage on this "other" end is constantly switching — it is therefore called the switching node ("SW" above). The inductor voltage reversal (flipping of SW node) is with respect to the steady (DC) end of the inductor. This voltage reversal is what indirectly leads to the observed input voltage step-down, input voltage step in or input voltage step-up/step-down behavior of the concerned topology.

Figure 1.15: Three basic topologies possible only.

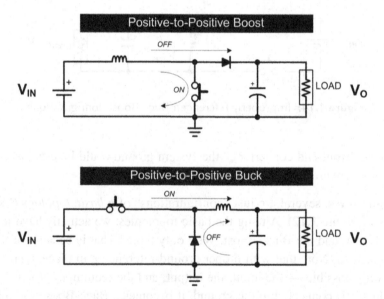

Figure 1.16: The (positive) Boost and Buck topologies.

We now make some observations:

- For example, a "positive-to-positive" Boost can convert 12 V to 50 V. A "negative-to-negative" Boost would be able to convert say -12 V to -50 V. The *magnitude* of the output voltage must therefore always be larger than the *magnitude* of the input voltage. So a Boost converter only steps-up and also does not change the polarity.

- In the Boost, when the switch is ON, energy is delivered *only* into the inductor by the input DC source (via the *switch*), and *none* of it passes through to the output.

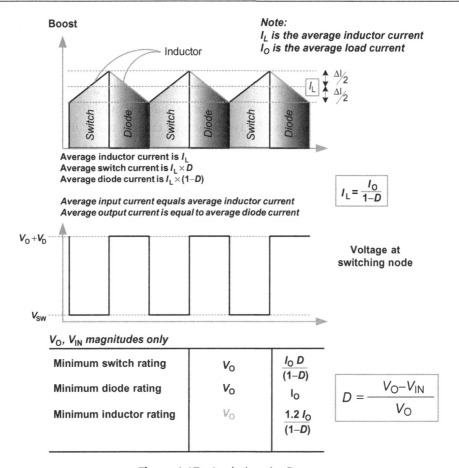

Figure 1.17: Analyzing the Boost.

- When the switch is OFF, the stored energy of the inductor is pushed into the output (through the *diode*). ***But some of it also comes from the input DC source.***
- The current coming from the *input capacitor* (DC source) is "smooth," since it is in series with the inductor (which prevents sudden jumps in current).
- However, the current into the *output capacitor* is "choppy" because combined with the steady DC current into the load (I_{OUT}), it forms the diode current (which we know is always choppy for any topology) (see Figure 1.9).
- Since the average current from the output capacitor is zero, therefore, for *the Boost, the average diode current must be equal to the load current* (where else can the current come from?). Therefore,

$$I_{D\_AVG} = I_O = I_L \times (1 - D)$$

So,

$$I_L = \frac{I_O}{1 - D} \quad (Boost)$$

This is the relationship between the average inductor current and the load current. Note that in Figure 1.17, in the embedded table, we have asked for an inductor rated for $1.2 \times I_O/(1 - D)$. The factor "1.2" comes from the fact that by typical design criteria, the peak of the inductor current waveform is about 20% higher than its average. So, we need to look for an inductor rated at least for a current of $1.2 \times I_L$.

Let us analyze the Boost topology in terms of the voltseconds in steady state. We have

$$V_{ON} = V_{IN} - V_{SW} \quad (Boost)$$

and

$$V_{OFF} = V_O + V_D - V_{IN} \quad (Boost)$$

So, from the voltseconds law

$$\frac{t_{OFF}}{t_{ON}} = \frac{V_{IN} - V_{SW}}{V_O + V_D - V_{IN}} \quad (Boost)$$

Performing some algebra on this to eliminate t_{OFF},

$$\frac{t_{OFF}}{t_{ON}} + 1 = \frac{V_{IN} - V_{SW}}{V_O + V_D - V_{IN}} + 1$$

$$\frac{t_{OFF} + t_{ON}}{t_{ON}} = \frac{V_{IN} - V_{SW} + V_O + V_D - V_{IN}}{V_O + V_D - V_{IN}}$$

Finally, the "duty cycle" of the converter D, which is defined as

$$D = \frac{t_{ON}}{T} \quad (any\ topology)$$

is the reciprocal of the preceding equation. So,

$$D = \frac{V_O + V_D - V_{IN}}{V_O + V_D - V_{SW}} \quad (Boost)$$

We have thus derived the classical DC transfer function of a *Boost converter*.

If the switch and diode drops are small as compared to the input and output rails, we can just write

$$D \approx \frac{V_O - V_{IN}}{V_O} \quad (Boost)$$

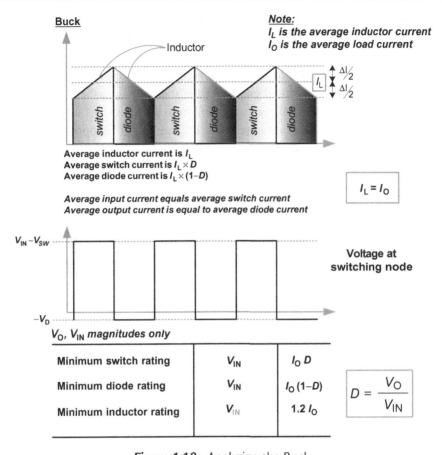

Figure 1.18: Analyzing the Buck.

We can also write the relationship between the input and output as follows:

$$V_O = V_{IN} \times \frac{1}{1-D} \quad (Boost)$$

The Buck Topology

In Figure 1.16, *we* had also presented the schematic of the Buck topology. The direct and the freewheeling paths are indicated therein. In Figure 1.18, we have the corresponding analysis, including the key waveforms.

We now make some observations:

•   For example, a "positive-to-positive" Buck can convert 12 V to 5 V. A "negative-to-negative" Buck would be able to convert say −12 V to −5 V. The *magnitude* of the output voltage must therefore always be smaller than the *magnitude* of the input voltage. So, a Buck converter only steps-down and also does not change the polarity.

• When the switch is ON, energy is delivered to the inductor by the input DC source (via the *switch*). **But some of it also passes through to the output**.

• When the switch is OFF, the stored energy of the inductor is pushed into the output (through the *diode*). And *none* of it now comes from the input DC source.

• The current coming from the *input capacitor* (DC source) is "choppy." That is because this current, combined with the steady DC current (I_{IN}) coming in from the DC source, basically forms the *switch* current waveform (which we know is always choppy for any topology).

• However, the current into the *output capacitor* is "smooth" because it is in series with the inductor (which prevents sudden jumps in current).

• Since the average current from the output capacitor is zero, for *the Buck, the average inductor current must be equal to the load current* (where else can the current come from?). Therefore,

$$I_L = I_O \quad (Buck)$$

This is the relationship between the average inductor current and the load current. Note that in Figure 1.18, in the embedded table, we have asked for an inductor rated for $1.2 \times I_O$. The factor "1.2" comes from the fact that by typical design criteria, the peak of the inductor current waveform is about 20% higher than its average. So, we need to look for an inductor rated at least for a current of $1.2 \times I_L$.

Let us analyze the Buck topology in terms of the voltseconds in steady state. We have

$$V_{ON} = V_{IN} - V_{SW} - V_O \quad (Buck)$$

and

$$V_{OFF} = V_O - (-V_D) = V_O + V_D \quad (Buck)$$

As before, using the voltseconds law and simplifying, we get the "duty cycle" of the converter:

$$D = \frac{V_O + V_D}{V_{IN} + V_D - V_{SW}} \quad (Buck)$$

We have thus derived the classical DC transfer function of a *Buck converter*. If the switch and diode drops are small as compared to the input and output rails, we can just write

$$D \approx \frac{V_O}{V_{IN}} \quad (Buck)$$

We can also write the relationship between the input and the output as follows:

$$V_O = V_{IN} \times D \quad (Buck)$$

Advanced Converter Design

This should serve as an introduction to understanding and designing switching power converters. More details and worked examples can be found in *Chapter 2*. The reader can also at this point briefly scan *Chapter 4* for some finer nuances of design. A full design table is also available in the Appendix for future reference.

DC−DC Converter Design and Magnetics

The reader is strongly advised to read *Chapter 1* before attempting this chapter.

The magnetic components of any switching power supply are an integral part of its topology. The design and/or selection of the magnetics can affect the selection and cost of all the other associated power components, besides dictating the overall performance and size of the converter itself. Therefore, we really should not try to design a converter, without looking closely at its magnetics, and vice versa. With that in mind, in this chapter, we will be introducing the basic concepts of magnetics — in parallel with a formal DC−DC converter design procedure.

Note that in the area of *DC−DC converters*, we have only a single magnetic component to consider — its inductor. Further, in this particular area of power conversion, it is customary to just pick an *off-the-shelf* inductor for most applications. Of course there cannot possibly be enough "standard" inductors going around to cover all possible application scenarios. But the good news is that, given a certain inductor, and knowing its performance under a *stated set of conditions*, we can easily calculate how it will perform under our specific *application conditions*. And thereby, we can either validate or invalidate our initial selection. It may take more than one iteration or attempt, but moving in this direction, we can almost always find a standard inductor that fits our application.

In the next chapter we will introduce *"off-line"* power supply design. Such converters usually work off an AC (mains) input that ranges from 90 V to 270 V. To protect users from the high voltage, these converters almost invariably use an isolating *transformer* — in addition to, or in place of, the inductor. But though these topologies are really just derivatives of standard DC−DC topologies, in terms of *magnetics*, they are quite *different*. For example, we encounter significant (non-negligible) *high-frequency* effects within the transformer — like skin depth and proximity effects — the analysis of which can be quite challenging. In addition, we find that there are definitely not enough general-purpose (off-the-shelf) parts going around, that can meet all possible permutations and combinations of requirements, as can arise in off-line applications. So, in these applications, we usually always end up having to *custom-design* the magnetics. And as mentioned, this is not a mean task. But by trying to first understand *DC−DC converter* design, and the selection of *off-the-shelf* inductors, we are in a much better position to tackle off-line power supplies.

We can thereby build up basic concepts and skills, while garnering a much-needed "feel" for magnetics.

Off-line converters and DC−DC converters are also relatively quite different in terms of some rather implicit (often completely unstated) differences in basic design strategy — like the issue relating to the size of the magnetics vis-à-vis the current limit of the converter, as we will soon learn. With regard to their *similarities*, we should remember that both can have a *wide-input* voltage range, *not a single-value input voltage*, as is often assumed in related literature. Having a wide input raises the following question — what voltage point within the prescribed input range is the "worst-case" (or maximum) for a given stress parameter? Note that in selecting a power component we often need to consider the *worst-case* stress it is going to endure in our application. And then, provided that the particular stress parameter happens to be a relevant and decisive factor in its selection, we usually add an additional amount of safety margin, for the sake of reliability. However, the problem is that *different stress parameters do not attain their worst-case values at the same input voltage point*. We, therefore, realize that the design of a wide-input converter is necessarily going to be "tricky." For sure, designing a *functional* switching converter may be considered "easy," but designing it *well* certainly isn't.

Toward the end of this chapter, we will present a detailed DC−DC converter design procedure. But to account for a wide-input range, we will proceed in two distinct steps:

- A *"general inductor design procedure,"* for choosing and validating an off-the-shelf inductor for our application. We will see that depending on the topology at hand, this is to be carried out at a certain, specified voltage end — one that we will identify as being the "worst-case" *from the viewpoint of the inductor.*
- Then we will consider the other power components. We will point out which particular stress parameters are important in each case, and also the input voltage at which they reach their maximum, and how to ultimately select the component.

Note that, although the design procedure may be seen to specifically address only the Buck topology, the accompanying annotations clearly indicate how a particular step or equation may need to change if the procedure were being carried out for a Boost or a Buck-Boost topology.

DC Transfer Functions

When the switch turns ON, the current ramps up in the inductor according to the inductor equation $V_{ON} = L \times \Delta I_{ON}/t_{ON}$. The current *increment* during the on-time is $\Delta I_{ON} = (V_{ON} \times t_{ON})/L$. When the switch turns OFF, the inductor equation $V_{OFF} = L \times \Delta I_{ON}/t_{OFF}$ leads to a current *decrement* $\Delta I_{OFF} = (V_{OFF} \times t_{OFF})/L$.

Table 2.1: Derivation of DC Transfer Functions of the Three Topologies.

	Applying Voltseconds Law and $D = t_{ON}/(t_{ON} + t_{OFF})$		
Steps	$V_{ON} \times t_{ON} = V_{OFF} \times t_{OFF}$ $\dfrac{t_{ON}}{t_{OFF}} = \dfrac{V_{OFF}}{V_{ON}}$ $\dfrac{t_{ON}}{t_{ON} + t_{OFF}} = \dfrac{V_{OFF}}{V_{OFF} + V_{ON}}$ Therefore, $\boxed{D = \dfrac{V_{OFF}}{V_{ON} + V_{OFF}}}$ *(duty cycle equation for all topologies)*		
	Buck	**Boost**	**Buck-Boost**
V_{ON}	$V_{IN} - V_O$	V_{IN}	V_{IN}
V_{OFF}	V_O	$V_O - V_{IN}$	V_O
DC Transfer Functions	$\boxed{D = \dfrac{V_O}{V_{IN}}}$	$\boxed{D = \dfrac{V_O - V_{IN}}{V_O}}$	$\boxed{D = \dfrac{V_O}{V_{IN} + V_O}}$

The current increment ΔI_{ON} must be equal to the decrement ΔI_{OFF}, so that the current at the end of the switching cycle returns to the *exact* value it had at the start of the cycle — otherwise we wouldn't be in a repeatable (steady) state. Using this argument, we can derive the input–output (DC) transfer functions of the three topologies, as shown in Table 2.1. It is interesting to note that the reason why the transfer functions turn out different in each of the three cases can be traced back to the fact that the *expressions for V_{ON} and V_{OFF} are different*. Other than that, the derivation and its underlying principles remain the same *for all topologies*.

The DC Level and the "Swing" of the Inductor Current Waveform

From $V = L \, dI/dt$, we get $\Delta I = V \, \Delta t/L$. So, the *"swinging"* component of the inductor current "ΔI" is *completely* determined by the applied voltseconds and the inductance. Voltseconds is the *applied voltage multiplied by the time it is applied for*. To calculate it, we can either use V_{ON} times t_{ON} (where $t_{ON} = D/f$), or V_{OFF} times t_{OFF} (where $t_{OFF} = (1 - D)/f$) — and we will get the same result (for that is how D gets defined in the first place!). But note also, that if we apply 10 V across a given inductor for 2 μs, we will get the same current swing ΔI, if we apply say, 20 V for 1 μs, or 5 V for 4 μs, and so on. So, *for a given inductor*, talking about either the voltseconds or the ΔI is effectively one and the same thing.

What does the voltseconds depend on? It depends on the input/output voltages (i.e. duty cycle) and time, via the switching frequency. ***Therefore, only by changing L, f, or D can we affect ΔI***. Nothing else! See Table 2.2. In particular, *changing the load current I_O does*

Table 2.2:   **How Varying the Inductance, Frequency, Load Current, and Duty Cycle Influence ΔI and I_{DC}.**

		Action											
		$L\uparrow$ (increasing)			$I_O\uparrow$ (increasing)			$D\uparrow$ (increasing)			$f\uparrow$ (increasing)		
		Buck	Boost	Buck-Boost	Buck	Boost	Buck-Boost	Buck	Boost	Buck-Boost	Buck	Boost	Buck-Boost
Response	$\Delta I = ?$	$\downarrow$	$\downarrow$	$\downarrow$	$\times$	$\times$	$\times$	$\downarrow$	$\uparrow\downarrow^a$	$\downarrow$	$\downarrow$	$\downarrow$	$\downarrow$
	$I_{DC} = ?$	$\times$	$\times$	$\times$	$\uparrow(=)$	$\uparrow$	$\uparrow$	$\times$	$\uparrow$	$\uparrow$	$\times$	$\times$	$\times$

($\uparrow\downarrow$) indicates it increases and decreases over the range; ($\times$) indicates no change; ($\uparrow(=)$) indicates I_{DC} is increasing and is equal to I_O.
[a]Maximum at $D = 0.5$.

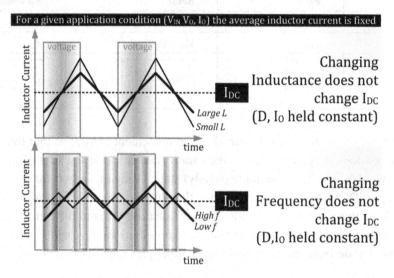

Figure 2.1: If D and I_O are fixed, I_{DC} cannot change.

nothing to ΔI. I_O is therefore in effect, an altogether *independent* influence on the inductor current waveform. But what part of the inductor current does it specifically influence/determine? We will see that I_O *is proportional to the average inductor current.*

The inductor current waveform is considered to have another (independent) component besides its swing ΔI — it is the *DC* (average) level "I_{DC}," defined as the level *around* which the swing ΔI takes place symmetrically — that is, $\Delta I/2$ above it, and $\Delta I/2$ below it. See Figure 2.1. Geometrically speaking, this is the "center of the ramp." It is sometimes also called the "platform" or "pedestal" of the inductor current. The important point to note is that I_{DC} is based only on *energy flow requirements* — that is, the need to maintain an *average* rate of energy flow consistent with the input/output voltages and desired output power. So, if the "application conditions," that is, the output power and the input/output voltages, do not change, there is in fact *nothing* we can do to alter this DC level — in that sense, I_{DC} is rather "stubborn" (see Figure 2.1). In particular

- Changing the inductance L doesn't affect I_{DC}.
- Changing the frequency f doesn't affect I_{DC}.
- Changing the duty cycle D *does* affect I_{DC} — for the Boost and Buck-Boost.

To understand the last bullet above, we should note the following equations that we will derive a little later

$$I_{DC} = I_O \quad (Buck)$$

$$I_{DC} = \frac{I_O}{1-D} \quad (Boost \text{ and } Buck\text{-}Boost)$$

The intuitive reason why the above relations are different is that in a Buck, the output is in series with the inductor (from the standpoint of the DC currents, the output capacitor contributes nothing to the DC current distribution), and therefore the average inductor current must at all times be equal to the load current. Whereas, in a Boost and Buck-Boost, the output is in series with the diode, and so the *average diode current* must equal the load current.

Therefore, if we keep the load current constant, and change only the input/output voltages (duty cycle), we can affect I_{DC} — in all cases **except** for the Buck. In fact, the *only* way to change the DC inductor current level for a Buck is to change the load current. Nothing else will work!

In the Buck, I_{DC} and I_O are equal. But in the Boost and Buck-Boost, I_{DC} depends also on the duty cycle. That makes the design/selection of magnetics for these two topologies rather different from a Buck. For example, if the duty cycle is 0.5, the average inductor current is twice the load current. Therefore, using a 5 A inductor for a 5 A load current may be a recipe for disaster. But for a Buck it is OK except for high-voltage applications (discussed later).

One thing we can be sure of is that in the Boost and Buck-Boost, I_{DC} is *always greater* than the load current. We may be able to cause this DC level to fall and even approach the load current value if we reduce the *duty cycle* close to 0 (i.e., a very small *difference between the input and output voltages*). But then, on increasing the duty cycle toward 1, the DC level of the inductor current will climb steeply. It is important we recognize this clearly and early on.

Another thing we can conclude with certainty is that in *all* the topologies, the DC level of the inductor current is *proportional* to the load current. So, doubling the load current for example (keeping everything else the same), doubles the DC level of the inductor current (whatever it was to start with). So, in a Boost with a duty cycle of 0.5 for example, if we have a 5 A load, then the I_{DC} is 10 A. And if I_O is increased to 10 A, I_{DC} will become 20 A.

Changing the input/output voltages (i.e. duty cycle) does affect the DC level of the inductor current for the Boost and the Buck-Boost. Changing D also affects the *swing ΔI in all three topologies*, because it changes the duration of the applied voltage and thereby changes the voltseconds. Summarizing:

- Changing the duty cycle affects I_{DC} for the Boost and the Buck-Boost.
- Changing the duty cycle affects ΔI for all topologies.

> *Note: The off-line Forward converter transformer is probably the only known exception to the above logic. We will learn that if we, for example, double the duty cycle (i.e., double t_{ON}), then almost coincidentally, V_{ON} halves, and therefore the voltseconds does not change (and nor does ΔI). In effect, ΔI is then independent of duty cycle.*

Based on the discussions above, and also the detailed design equations, we have summarized these "variations" in Table 2.2. This table should hopefully help the reader eventually develop a more intuitive and analytical "feel" for converter and magnetics design, one which can come in handy at a later stage. We will continue to discuss certain aspects of this table, in more detail, a little later.

Defining the AC, DC, and Peak Currents

In Figure 2.2, we see how the AC, DC, peak-to-peak, and peak values of the inductor current waveform are defined. In particular, we note that the AC value of the current waveform is defined as

$$I_{AC} = \frac{\Delta I}{2}$$

We should also note from Figure 2.2 that $I_L \equiv I_{DC}$. **Therefore, sometimes in our discussions that follow, we may refer to the DC level of the inductor current as "I_{DC}" and sometimes as the average inductor current "I_L" but they are actually synonymous.** In particular, we should not get confused by the subscript "L" in "I_L." The "L" stands for *inductor*, not *load*. The load current is always designated as "I_O." Of course, we do realize that $I_L = I_O$ for a *Buck*, but that is just happenstance.

In Figure 2.2 we have also defined another key parameter called "r," or the "current ripple ratio." This *connects the two independent current components I_{DC} and ΔI*. We will explore this particular parameter in much greater detail a little later. Here, it suffices to mention that r needs to be set to an "optimum" value in any converter — usually approximately 0.3–0.5, *irrespective of the specific application conditions, the switching frequency, and even the topology itself*. This, therefore, becomes a universal design *rule of thumb*. We will also learn that the choice of r affects the current stresses and dissipation in all the power

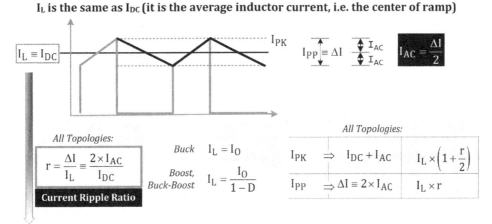

I_L is the same as I_{DC} (it is the average inductor current, i.e. the center of ramp)

Example:

<u>Buck:</u> If load current is 1A, I_L is 1A.
So if $r=0.4$, peak-to-peak current ('ΔI') is 0.4A, and peak current is 1+0.2=1.2A.

<u>Boost/Buck-Boost:</u> If load current is 1A and D=0.5, then I_L is 2A.
So if $r=0.4$, the peak-to-peak current ('ΔI') is 0.8A, and the peak current is 2+0.4=2.4A.

Figure 2.2: The AC, DC, peak, and peak-to-peak currents, and the current ripple ratio "*r*" defined.

components, and thereby impacts their selection. Therefore, **setting *r* should be the *first* step when commencing any power converter design**.

The DC level of the inductor current (largely) determines the I^2R losses in the copper windings ('copper loss'). However, the final temperature of the inductor is also affected by another term — the "*core loss*" — that occurs *inside* the magnetic material (core) of the inductor. Core loss is, to a first approximation, determined *only* by the AC (swinging) component of the inductor current (ΔI), and is therefore virtually independent of the DC level (I_{DC} or "DC bias").

We must pay the closest attention to the *peak current*. Note that in any converter, the terms "peak inductor current," "peak switch current," and "peak diode" current are all *synonymous*. Therefore, in general, we just refer to all of them as simply the "peak current" I_{PK} where

$$I_{PK} = I_{DC} + I_{AC}$$

The peak current is in fact the *most critical current component* of all — because it is not just a source of *long-term* heat buildup and consequent temperature rise, but a potential cause of *immediate* destruction of the switch. We will show later that the *inductor current is instantaneously proportional to the magnetic field* inside the core. So, at the exact moment when the current reaches its peak value, so does this field. We also know that real-world inductors can "saturate" (start losing their inductance) if the field inside them exceeds a certain "safe" level — that value being dependent on the actual *material* used for the core (not on the geometry, or number of turns or even the air-gap, for example). Once saturation occurs, we may get an almost *uncontrolled* surge of current passing through the switch — because, the ability to limit current (which is one of the reasons the inductor is used in switching power supplies in the first place), depends on the inductor *behaving* like one. Therefore, losing inductance is certainly not going to help! In fact, we *usually* cannot afford to allow the inductor to "saturate" *even momentarily*. And for this reason, *we need to monitor the peak current closely* (usually on a cycle-by-cycle basis). As indicated, the peak is the likeliest point of the inductor current waveform where saturation can start to occur.

> **Note:** *A slight amount of core saturation* may turn out to be acceptable *on occasion, especially if it occurs only under temporary conditions, like power-up for example. This will be discussed in more detail later.*

Understanding the AC, DC, and Peak Currents

We have seen that the AC component ($I_{AC} = \Delta I/2$) is derivable from the voltseconds law. From the basic inductor equation $V = L \, dI/dt$, we get

$$2 \times I_{AC} = \Delta I = \frac{\text{voltseconds}}{\text{inductance}}$$

So, the current swing $I_{PP} \equiv \Delta I$, can be *intuitively visualized* as "*voltseconds per unit inductance*." If the applied voltseconds doubles, so does the current swing (and AC component). And if the inductance doubles, the swing (and AC component) is halved.

Let us now consider the DC level again. Note that *any capacitor has zero average (DC) current through it in steady state*, so all capacitors can be considered to be "missing altogether" when calculating DC current distributions. Therefore, for a Buck, since energy flows into the output during *both* the on-time and off-time, and *via the inductor*, the average *inductor* current must always be equal to the load current. So,

$$I_L = I_O \quad (Buck)$$

On the other hand, in both the Boost and the Buck-Boost, energy flows into the output only during the off-time, and *via the diode*. Therefore, in this case, the average *diode* current must be equal to the load current. Note that the diode current has an average value equal to I_L *when it is conducting* (see the dashed line passing through the center of the down-ramp in the upper half of Figure 2.3). But if we calculate the average of this diode current over

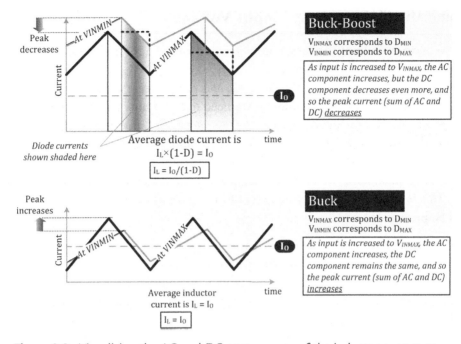

Figure 2.3: Visualizing the AC and DC components of the inductor current as input voltage varies.

the *entire switching cycle*, we need to weight it by *its* duty cycle, that is, $1 - D$. Therefore, calling "I_D" the average diode current, we get

$$I_D = I_L \times (1 - D) = I_O$$

solving

$$I_L = \frac{I_O}{1 - D} \quad (\textit{Boost and Buck-Boost})$$

Note also, that ***for any topology, a high-duty cycle corresponds to a low-input voltage, and a low-duty cycle is equivalent to a high input***. So, increasing D amounts to decreasing the input voltage (its magnitude) in all cases. Therefore, *in a Boost or Buck-Boost, if the difference between the input and output voltages is large, we get the highest DC inductor current.*

Finally, with the DC and AC components known, we can calculate the peak current using

$$I_{PK} = I_{AC} + I_{DC} \equiv \frac{\Delta I}{2} + I_L$$

Defining the "Worst-Case" Input Voltage

So far, we have been implicitly assuming a *fixed* input voltage. In reality, in most practical applications, the input voltage is a certain *range*, say from "V_{INMIN}" to "V_{INMAX}." We therefore also need to know how the AC, DC, and peak current components *change as we vary the input voltage*. Most importantly, we need to know at what specific voltage within this range we get the maximum *peak* current. As mentioned, **the peak is critical from the standpoint of ensuring there is no inductor saturation**. Therefore, defining the "worst-case" voltage (for inductor design) as the point of the input voltage range where the peak current is at its maximum, we need to design/select our inductor *at this particular point always*. This is in fact the underlying basis of the *"general inductor design procedure"* that we will be presenting soon.

We will now try to understand where and *why* we get the highest peak currents for each topology. In Figure 2.3, we have drawn various inductor current waveforms to help us better visualize what really happens as the input is varied. We have chosen two topologies here, the Buck and the Buck-Boost, for which we display two waveforms each, corresponding to two different input voltages. Finally, in Figure 2.4 we have plotted out the AC, DC, and peak values. Note that these plots are based on the actual design equations, which are also presented within the same figure. While interpreting the plots, we

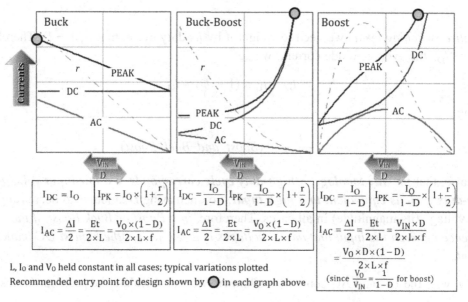

Figure 2.4: Plotting how the AC, DC, and peak currents change with duty cycle.

should again keep in mind that for all topologies, a high D corresponds to a low input. The following analysis will also explain certain cells of the previously provided Table 2.2, where the variations of ΔI and I_{DC}, with respect to D, were summarized.

a) *For the Buck*, the situation can be analyzed as follows:
   - As the input *increases*, the duty cycle decreases in an effort to maintain regulation. But *the slope of the down-ramp $\Delta I/t_{OFF}$ cannot change*, because it is equal to V_{OFF}/L, that is, V_O/L, and we are assuming V_O is fixed. But now, since t_{OFF} has increased, but the slope $\Delta I/t_{OFF}$ has not changed, the only possibility is that ΔI must have increased (proportionally). So, we conclude that the *AC component* of the Buck inductor current actually *increases as the input increases* (even though the duty cycle decreased in the process).
   - On the other hand, the center of the ramp I_L is fixed at Io, so we know the *DC level does not change.*
   - Finally, since the peak current is the *sum* of the AC and DC components, it *increases at high-input voltages* (see relevant plot in Figure 2.4).
       Therefore, *for a Buck, it is always preferable to start the inductor design at V_{INMAX} (i.e., at D_{MIN}).*

b) **For the Buck-Boost**, the situation can be analyzed as follows:
   - As the input *increases*, the duty cycle decreases. But *the slope of the down-ramp $\Delta I/t_{OFF}$ cannot change*, because it is equal to V_{OFF}/L, that is, V_O/L, and V_O is fixed (same situation as for the Buck). But since t_{OFF} has increased, ΔI must also increase to keep the slope $\Delta I/t_{OFF}$ unchanged. So, we see that the *AC component* ($\Delta I/2$) *increases as the input increases* (duty cycle decreasing). Note that up till this point, the analysis is the same as for the Buck — traced back to the fact that in both these topologies $V_{OFF} = V_O$.
   - But now coming to the DC level I_L of the Buck-Boost, we will find it *must change* for this topology (it remained fixed for the Buck). Note that the shaded portion of the waveform in the upper half of Figure 2.3 represents the diode current. The average value of this *during the off-time* is the square dashed line passing through its center, that is, I_L. So, the average diode current, calculated *over the entire switching cycle*, is $I_L \times (1 - D)$. And we know this must equal the load current I_O. Therefore, as the input *increases* and duty cycle decreases, the term $(1 - D)$ increases. So, the only way $I_L \times (1 - D)$ can remain constant at the value I_O is if I_L *decreases as D decreases*. We therefore realize that the *DC level decreases as the input increases* (duty cycle decreasing).
   - Since the peak current is the sum of the AC and DC components, it also *decreases at high-input voltages* (see relevant plot in Figure 2.4).
       Therefore, for a *Buck-Boost, we should always start the inductor design at V_{INMIN} (i.e., at D_{MAX}).*

c) *For the Boost*, the situation is a little *trickier* to understand. On the face of it, it is quite similar to the Buck-Boost, *but there is a notable difference* — and that is why we did not even try to include it in Figure 2.3.

- Once again, as the input *increases*, the duty cycle decreases. But the difference here is that **the slope of the down-ramp $\Delta I/t_{OFF}$ must** *decrease* — because it is equal to V_{OFF}/L, that is, $(V_O - V_{IN})/L$ (magnitudes only) — and we know that $V_O - V_{IN}$ *is decreasing*. Further, the required decrease in the slope $\Delta I/t_{OFF}$ can come about in *two* ways — either from an increase in t_{OFF} (which is already occurring as the duty cycle decreases), *or* from a decrease in ΔI. In fact, ΔI is allowed to increase or decrease (as we increase the input). For example, if t_{OFF} increases faster than ΔI — then $\Delta I/t_{OFF}$ will still decrease as required. And in practice, that is what actually does happen in the case of the Boost. With some detailed math, we can show that ΔI increases as *D approaches 0.5*, but decreases after that (see Table 2.2 and Figure 2.4).
- However, the increase/decrease in the AC level *does not dominate* in a Boost, and therefore, the peak current ends up being dictated *only by the DC component*. But we already know that the DC level of a Boost changes in exactly the same way as for the Buck-Boost (discussed above) — *it decreases as the input increases* (duty cycle decreasing).
- We conclude that the peak current for the Boost *decreases at high-input voltages* (see relevant plot in Figure 2.4).

    Therefore, for a *Boost*, we should always start the inductor design at V_{INMIN} (i.e., at D_{MAX}).

The Current Ripple Ratio "*r*"

In Figure 2.2, we first introduced the most basic, yet far-reaching design parameter of the power supply itself — its *current ripple ratio "r."* This is a geometrical ratio that compares and connects the AC value of the inductor current to its associated DC value. So,

$$r = \frac{\Delta I}{I_L} \equiv 2 \times \frac{I_{AC}}{I_{DC}}$$

Here, we have used $\Delta I = 2 \times I_{AC}$. ***Once r is set by the designer (at maximum load current and worst-case input), almost everything else is pre-ordained*** — like the currents in the input and output capacitors, the "RMS" (root mean square) current in the switch, and so on. Therefore, *the choice of* r *affects component selection and cost, and it must be understood clearly, and picked carefully.*

Note that the ratio *r* is defined for CCM (*continuous conduction mode*) operation only. Its valid range is from 0 to 2. When *r* is 0, ΔI must be 0, and the inductor equation then

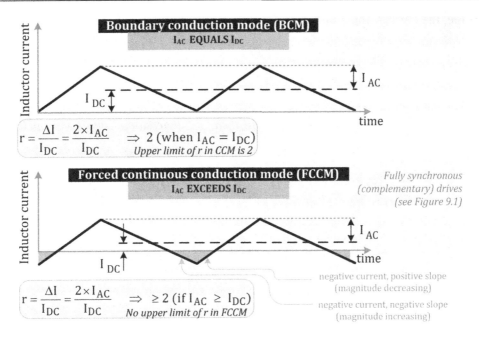

Figure 2.5: BCM and forced CCM operating modes.

implies a very large (infinite) inductance. Clearly, $r = 0$ is not a practical value! If r equals 2, the converter is operating at the *boundary* of continuous and discontinuous conduction modes (boundary conduction mode or "BCM") (see Figure 1.9 and Figure 2.5). In this so-called boundary (or "critical") conduction mode, $I_{AC} = I_{DC}$ by definition. *Note that readers can refer back to* **Chapter 1,** *in which CCM, DCM (discontinuous conduction mode), and BCM were all initially introduced and explained*.

Note that an exception to the "valid" range of r from 0 to 2 occurs in "forced CCM" mode, discussed in more detail later.

Relating r to the Inductance

We know that current swing is given as voltseconds per unit inductance. So, we can also write

$$\Delta I = \frac{Et}{L_{\mu H}} \quad \text{(any topology)}$$

Here "Et" is defined as the (magnitude of the) *voltmicroseconds* across the inductor (either during the on-time or off-time — both being necessarily equal in steady state), and $L_{\mu H}$ is the inductance in μH. The reason for defining Et is that this number is simply easier to manipulate than voltseconds because of the very small time intervals involved in modern power conversion.

Therefore, the current ripple ratio is

$$r = \frac{\Delta I}{I_L} = \frac{Et}{L_{\mu H} I_L} \quad \textit{(any topology)}$$

Note also that from now on, *whenever L is paired up with Et in any given equation, we will drop the subscript of L, that is, "μH." It will then be "understood" that L is in μH.*

Finally, we have the following key relationships between *r* and *L*:

$$r = \frac{Et}{(L \times I_L)} \equiv \frac{V_{ON} \times D}{(L \times I_L) \times f} \equiv \frac{V_{OFF} \times (1 - D)}{(L \times I_L) \times f} \quad \textit{(any topology)}$$

Incidentally, the preceding equation, that is, the one involving V_{OFF}, assumes CCM, because it assumes that t_{OFF} (the time for which V_{OFF} is applied) is equal to the full available off-time $(1 - D)/f$.

Conversely, *L* as a function of *r* is

$$L = \frac{V_{ON} \times D}{r \times I_L \times f} \quad \textit{(any topology)}$$

In subsequent sections we will often use the following *easy-to-remember* form of the previous equations. We are going to nickname this the *"L × I" equation* (or rule)

$$L \times I_L = \frac{Et}{r} \quad \textit{(any topology)}$$

But perhaps we are still wondering — why do we even need to talk in terms of *r* — why not talk *directly* in terms of *L*? We do realize from the above equations that *L* and *r* are related. However, the "desirable" value of inductance depends on the *specific application conditions, the switching frequency, and even the topology.* So, it is just not possible to give a general design rule for picking *L*. But there is in fact such a general design rule of thumb for selecting *r* — one that applies almost universally. We mentioned that it should be approximately 0.3–0.5 in all cases. And that is why **it makes sense to calculate L by first setting the value of r**. Of course, once we pick *r*, *L* gets automatically determined *for a given set of application conditions and switching frequency.*

The Optimum Value of *r*

It can be shown that, in terms of overall stresses in a converter and size, $r \approx 0.4$ represents an "optimum" of sorts. We will now try to understand why this is so, *and later we will try to point out exceptions to this reasoning.*

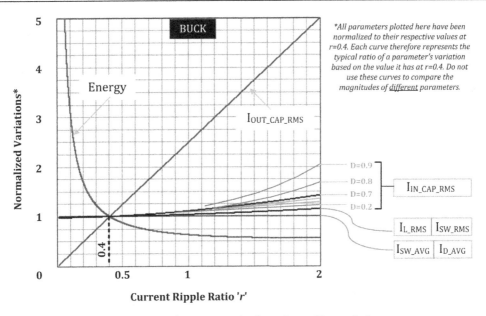

Figure 2.6: How varying the current ripple ratio r affects all the components.

The size of an inductor can be thought of as being virtually proportional to its *energy-handling capability* (the effect of air-gap on size will be studied later). So, for example, we probably already know intuitively that we need bigger cores to handle higher powers. The energy-handling capability of the selected core must, at a bare minimum, match the energy we need to store in it in our application — that is, $(1/2) \times L \times I_{PK}^2$. Otherwise, the inductor will saturate. (Later, read *Chapter 5* to understand the related topology-dependency aspect.)

In Figure 2.6, we have plotted the energy, $E = (1/2) \times L \times I_{PK}^2$, as a function of r. We see that it has a "knee" at around 0.4. This tells us that if we try to reduce r much lower than 0.4, we will certainly need a *very large inductor*. On the other hand, if we increase r, there isn't much greater reduction in the size of the inductor. In fact, we will see that beyond $r \sim 0.4$, we enter a *region of diminishing returns*.

In Figure 2.6, we have also plotted the *capacitor* RMS currents for a Buck converter. We see that if r is increased beyond 0.4, the currents will increase significantly. This will lead to increased heat generation inside the capacitors (and other related components too). Eventually, we may be forced to pick a capacitor with a lower ESR and/or lower case-to-air thermal resistance (more expensive/bigger).

Note: The RMS value of the current through any component is the current component responsible for the heat developed in it — via the equation $P = I_{RMS}^2 \times R$, where P is the dissipation, and R is the series resistance term associated with the particular component *(e.g., the DC resistance (DCR) of an inductor, or the ESR of a capacitor). However, it can be shown that the switch, diode, and inductor RMS current values are not very "shape-dependent." Therefore, the heat developed in them does not depend much on* r, *but mainly on the average value of the current. On the other hand, the RMS of the capacitor current waveforms can increase significantly, if* r *is increased. So,* capacitor currents are very "shape-dependent," *and therefore depend strongly on* r. *The reason for that is fairly obvious — any capacitor in a steady state has* zero average *(DC) current through it. Since a capacitor effectively subtracts out the DC level of the accompanying current waveform, we are left with a capacitor current waveform that has a large "ramp portion" built-in into it. Therefore, changing* r *changes this ramp portion, thereby impacting the capacitor current greatly.*

Note that in Figure 2.6, though we have used the Buck topology as an example, the *energy curve in particular is exactly the same for any topology.* The capacitor current curves though, may not be identical to those of the Buck, but are *similar*, and so the conclusions above still apply.

Therefore, in general, **a current ripple ratio of around 0.4 is a good design target for any topology, any application, and any switching frequency**.

Later, we will discuss some reasons/considerations for *not* adhering to this $r \sim 0.4$ rule of thumb (under certain conditions).

Do We Mean Inductor? or Inductance?

Note that in the previous section, we said nothing explicitly about what the *inductance* was — we just talked about the *size* of the *inductor*. We know that in theory, we can put almost any number of turns on a given core, and get almost any inductance. So, *inductance* and *size* of inductor are not *necessarily* related. However, we will now see that in power conversion they often do turn out to be so, though rather indirectly.

Looking at Figure 2.6, we can see that a smaller *r* will require a higher energy-handling capability, and thus a *larger* inductor. Let us now formally go through *all* the possible ways of reducing *r*.

Since *we are assuming our application conditions are fixed*, the load current and input/output voltages are also fixed. Therefore, I_{DC} is fixed too. The only way we can cause *r* to decrease under these circumstances is to make ΔI smaller. However, ΔI is

$$\Delta I = \frac{\text{voltseconds}}{\text{inductance}} \quad \text{(V-s/H)}$$

But we know the applied voltseconds is fixed too (input and output voltages being fixed). So, the *only way to decrease r (for a given set of application conditions) is to increase the inductance*. We can therefore conclude that if we choose a *high inductance*, we *will invariably require a bigger inductor*. It is therefore no surprise that when power supply designers instinctively ask for a "large inductance," they might well mean a "large inductor." Therefore, *the designer is cautioned against being too "ripple-phobic" in their designs. A certain amount of ripple is certainly "healthy."*

However, we must not forget that if, for example, we *increase the load current* (i.e., a change in application conditions), and we will clearly need to move to a *larger inductor* (with greater energy-handling capability). *But simultaneously, we will need to **decrease** the inductance*. That's because I_{DC} will increase, and so to keep to the "optimum" value of r, we will need to *increase* ΔI in the same proportion as the increase in I_{DC}. And to do this, we have to decrease, rather than increase, L.

This highlights the importance of thinking in terms of r to ease power supply design.

How Inductance and Inductor Size Depend on Frequency

The following discussion applies to *all* the topologies.

If keeping everything else fixed (including D) we *double the frequency*, the voltseconds will halve, because the durations t_{ON} and t_{OFF} have halved. But since ΔI is "voltseconds per unit inductance," it too will halve. Further, since I_{DC} has not changed, $r = \Delta I/I_{DC}$ will also halve. So, if we started off with $r = 0.4$, we now have $r = 0.2$.

If we want to return the converter to the optimum value of $r = 0.4$, we will now need to somehow double the ΔI (that we were left with at the end of the last step). The way to do that is to *halve the inductance*.

- Therefore, we can generally state that **inductance is inversely proportional to frequency**. Finally, having restored r to 0.4, the peak will still be 20% higher than the DC level. But the DC level has not changed. So, the peak value is also unchanged (since r hasn't changed either, eventually). However, the energy-handling requirement (size of inductor) is $(1/2) \times L \times I_{PK}^2$. Now, since L has halved, and I_{PK} is unchanged, the required size of the inductor has halved.
- Therefore, we can generally state that the **size of the inductor is inversely proportional to frequency**.
- Note also that the required current rating of the inductor is independent of the frequency (since peak is unchanged).

How Inductance and Inductor Size Depend on Load Current

For all topologies, if we *double the load current* (keeping input/output voltages and D fixed), r will tend to halve since ΔI has not changed but I_{DC} has doubled. Therefore, to restore r to its optimum value of 0.4, we need to get ΔI to double too. But we know that ΔI is simply "voltseconds per unit inductance," and in this case the voltseconds has not changed. So, the only way to get ΔI to double is to *halve the inductance*.

- Therefore, we can generally state that **inductance is inversely proportional to the load current**.

  What about the size? Since we doubled the load current, but still kept r at 0.4, the peak current $I_{DC}(1 + r/2)$ has also doubled. But the inductance has halved. So, the energy-handling requirement (size of inductor), $(1/2) \times L \times I_{PK}^2$, will double.

- Therefore, we can generally state that the **size of the inductor is proportional to the load current**.

How Vendors Specify the Current Rating of an Off-the-shelf Inductor and How to Select It

The "energy-handling capability" of an inductor, $1/2 \times LI^2$, is one way of picking the size of the inductor. But most vendors do not provide this number upfront. However, they do provide one or more "current ratings" for us to decipher. And if we interpret these current rating(s) correctly, that serves the purpose.

The current rating may be expressed by the vendor either as a maximum rated I_{DC}, or a maximum rated I_{RMS}, or/and a maximum I_{SAT}. The first two are usually considered synonymous, since the RMS and DC values of a typical inductor current waveform are almost equal (we had indicated previously that the RMS of the inductor current is not very "shape-dependent"). So, the DC/RMS rating of an inductor is by definition basically the direct current we can pass through it, such that we get a specified temperature rise (typically 40–55 °C depending on the vendor). The last rating, that is, the I_{SAT}, is the maximum current we can pass, just before the *core starts saturating*. At that point, the inductor is considered close to the useful limit of its energy-storing capability.

We will also find that many, if not most, vendors have chosen the wire gauge in such a manner that the I_{DC} and I_{SAT} ratings of any inductor are also virtually the same. And by doing this, they can publish one (*single*) current rating — for example, "the inductor is rated for 5 A." Basically, having determined the I_{SAT} of the inductor, the vendor has then consciously tweaked the wire gauge (at the saturation current level), so as to also get the specified temperature rise.

The rationale for wanting to set $I_{DC} = I_{SAT}$ is as follows — suppose the inductor had a DC rating of 3 A and an I_{SAT} of 5 A. The 5-A rating is then likely to be *superfluous*, because users would probably never select this inductor for an application that required more than 3 A anyway. Therefore, the excessive I_{SAT} rating in this case essentially amounts to an *unnecessarily over-sized core.* Of course, if we do find an inductor with different I_{DC} and I_{SAT} ratings, it is also possible the vendor may have (unsuccessfully) tried to exploit the larger size of the chosen core (by increasing the wire thickness), but the stumbling block was that the selected core geometry was somehow not conducive to doing so — maybe it just did not have enough *window space* for accommodating the thicker windings.

In general, an inductor with a "single" current rating is usually the most optimum/cost-effective too.

However, in some rare off-the-shelf inductors, we may even find I_{SAT} stated to be less than I_{DC}. But what use is that? We can't operate beyond I_{SAT} in any case! So, the only advantage, if any, that can be gleaned from such an inductor is that the temperature rise in a real application will be less than the maximum specified. Automotive applications?

*In general, for most practical purposes, the current rating of the inductor that we need to consider is the **lowest** rating of all the published current ratings. We can usually simply ignore all the rest.*

There are some subtle considerations and exceptions to the argument for always preferring an inductor with $I_{DC} \approx I_{SAT}$. For example, under transient/temporary conditions, the *momentary* current may exceed the *normal steady operating current by a wide margin.* So, for example, suppose we are using a switcher with an internally *fixed* current limit "I_{CLIM}" or "I_{LIM}" of 5 A — in *a 3-A application*. Then under startup (or sudden line/load steps), the current is very likely to hit the limiting value of 5 A for several cycles in succession as the control circuitry struggles to bring up the output rail into regulation. We will discuss this issue in greater detail below — in particular, *whether this is even a concern to start with*! However, assuming for now that it is, it then seems that it may actually make sense to use an inductor rated for 3 A continuous current, with an I_{SAT} rating of 5 A (provided such an inductor is freely available, and cheap). Of course, alternatively, we could just pick a standard "5 A inductor" (for the 3-A application), and thereby we would certainly avoid inductor saturation under all conditions (and the consequent likelihood of switch destruction). But we realize that in doing so, our inductor may be considered slightly *over-designed* from the viewpoint of its copper/temperature rise — the wire being unnecessarily thicker. However, we should keep in mind that larger cores certainly affect cost, but a little more copper rarely does!

What Is the Inductor Current Rating We Need to Consider for a Given Application?

Whenever we start-up, or subject the converter to sudden line/load transients, the current no longer stays at the steady value it has under *normal* operation (i.e., when delivering the required maximum rated load current). For example, if we suddenly short the output the control circuitry in an effort to regulate the output may momentarily expand the duty cycle to the *highest permissible value* (as set by the controller). We then are no longer in steady state, and so under the increased on-time voltseconds, the current ramps up progressively, and can reach the set current limit.

But then, the inductor would probably be saturating! For example, if we are using a 5-A fixed current limit *Buck* switcher IC for a 3-A application, we have probably picked an inductor rated for only around 3 A. But when we short the output, the current momentarily hits the current limit (which may be set typically around 5.3 A for a "5-A Buck switcher").

So the question is — *should we select an inductor with a rating based on the current limit threshold (that it may encounter under severe transients), or simply on the basis of the maximum continuous normal operating current (under steady state operation in our application)*? In fact, this question is not as philosophical as it may seem — it virtually separates standard industry off-line design procedures from those of DC–DC converters. To answer it effectively, a lot of factors may need to be considered, often on an individual or case-by-case basis. Let us address some of these concerns next.

Luckily, in most *low-voltage applications*, a certain amount of core saturation doesn't cause any problem. The reason for this is that if in the above example, the switch is rated for 5 A, and the current-limiting circuit in the IC is known to act *fast enough* to prevent the current from ever rising beyond 5 A, then even if the inductor has started saturating as it gets to 5 A, there is no cause for concern — *after all, if the switch doesn't break, we don't have a problem! And since the current doesn't exceed 5 A, the switch cannot break*. Hence, in this case, we could certainly pick a cost-effective "3-A inductor" for our application, knowing well in advance that it would saturate somewhat under various non-steady conditions. Of course we don't want to operate a switching converter constantly (under its rated maximum load conditions) with a saturating inductor — we just "allow" it to do so under abnormal and temporary conditions, so long as we are sure that the switch can never be damaged.

However, the above logic begs another key question to be answered — *what exactly constitutes "fast enough"* — that is, which factors affect our ability to turn the switch OFF fast enough to protect it from the consequences of a saturating inductor? Since this consideration may eventually end up dictating the size and cost of the inductor, it is important to understand this *response-time issue* well.

(a)   All current limit circuitry takes some finite time to respond. There are inherent (internal) "propagation delays" as we move the overcurrent signal through the internal comparators of the IC, its op-amps, level-shifters, driver, and so on to the IC pin driving the switch.

(b) If we are using a controller IC (as opposed to an "integrated switcher," i.e., with an internal switch), the switch will necessarily be at a certain physical distance from its driver (which is usually inside the IC). In that case, the parasitic inductances of the intervening PCB traces (roughly *20 nH/in. of trace*) will resist any sudden change in current, thereby creating an additional delay before the turn-off command issued by the IC actually reaches the Gate/Base of the switch.

(c) Theoretically speaking, even if the current-limiting circuitry had responded *immediately* to the overcurrent condition, *and* if the intervening traces had truly negligible inductance, the switch may still take a little time before it really turns itself OFF. During this delay, if the inductor is saturating, it will not be able to effectively prevent or limit the current spike that can get pushed through the transistor by the applied input DC source. The current could go well beyond the "safe" current limit threshold.

Bipolar junction transistors (BJTs) are inherently slow, as compared to more modern devices like MOSFETs. But large MOSFETs (e.g., high-current, high-voltage devices) also produce delays because of their higher internal parasitic Gate resistance and inductance and significant *inter-electrode parasitic capacitances* (that demand to be either discharged or charged as the case may be, before they allow the switch to change its state). Matters can get worse if we parallel several such MOSFETs together, as say for a very high-current application.

(d) Many controllers and ICs incorporate an internal "blanking time" — during which they deliberately "do not look" at the current waveform. The basic purpose is to avoid false triggering of the current limit circuitry by the noise generated at the turn-on transition. But this delay time could prove fatal to the switch, especially if the inductor has already started saturating, because the current limit circuitry won't even "know" if there is any overcurrent condition during this blanking interval. Further, in current-mode control ICs, the ramp to the PWM (pulse-width modulator) comparator stage is usually derived from the (noisy) switch current. So, the blanking time is typically set even higher — typically about 100 ns for low-voltage applications and up to 300 ns for off-line applications.

(e) Integrated high-frequency switchers (i.e., with the MOSFET or BJT switch contained within the *same package* as the control and driver) are usually the best-protected and most reliable, because the intervening inductances are minimized. Also, the blanking times can be set more accurately and optimally, since there is not going to be much variation in terms of different switches with widely varying characteristics. Therefore, integrated switchers can usually survive momentarily saturating inductors with almost no problem — *unless the input voltage is very high* (typically above 40–60 V), and in addition, the inductor is sized "very small."

(f) If the input voltage is high, the *rate of rise of the saturating inductor current ramp* can become very large ("steep"). This follows from the basic equation $V = L \, dI/dt$.

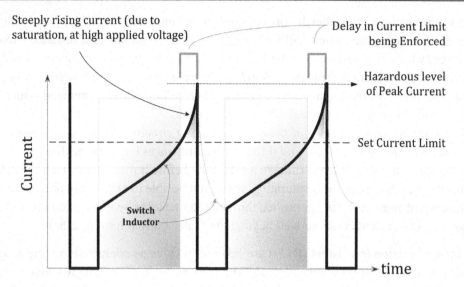

Steeply rising current (due to
saturation, at high applied voltage)

Delay in Current Limit
being Enforced

Hazardous level
of Peak Current

Set Current Limit

Current

Switch
Inductor

time

Figure 2.7: How higher voltages combined with inherent response-time delays can cause
overstress in the switch when the inductor starts saturating.

Here, if $L \to 0$, since V is fixed, the dI/dt must increase dramatically (see Figure 2.7).
So now, even a small delay can prove fatal because a *large* ΔI can take place during a
very small interval. The current can therefore overshoot the set current limit threshold
by a very large amount, thereby endangering the switch. That is why, especially when
we come to off-line applications, it is actually *customary* to select a core *large enough
to avoid saturation at the current limit threshold.* And that usually gives enough time
for the current limit circuitry to act — *before* the slope of the current has gone
completely out of control.

Note however, that the copper windings still only need to be proportioned to handle
the continuous current (i.e., based on the maximum operating load).

In effect, what we are always implicitly doing in off-line applications is *designing the
transformer such that its* I_{SAT} *is higher than its* I_{DC} *rating.* That is clearly not what we
usually do in low-voltage DC–DC converter design, where we tend to equate the two.

(g) Generally speaking, in most *low-voltage applications* (i.e., V_{IN} typically less than
about 40 V), the inductors are selected based only *on the maximum operating load
current.* The set current limit is therefore, in effect, virtually ignored! This is the usual
industry practice for DC–DC converter design, though it is probably not clearly
spelled out in this way most of the time. Luckily, it seems to have worked!

The Spread and Tolerance of the Current Limit

Any specification, including the current limit, either set by the user or fixed internally in the IC, will have a certain inherent *tolerance band* — that includes spreads over process variations and over temperature. All these variations are combined together inside the electrical tables of the datasheet of the device, under its "MIN" and "MAX" limits. In a practical converter design, a good designer learns to pay heed to such spreads.

But let us first summarize the general procedure for selecting the *inductance* for a switching power converter. Then we will look at the practical issues concerning spreads/tolerance.

The normal procedure is to determine the inductance by requiring that the current ripple ratio is about 0.4 — because we know that it represents an optimum of sorts for the entire converter. But there may be another possible limitation when dealing with switcher ICs, especially those with internally set (fixed) current limits — *if our normal operating peak currents are close to the set current limit of the device* (i.e., we are operating close to the maximum current capability of the switcher IC), *we need to ensure that the inductance is large enough not to cause the calculated operating peak current (within any given cycle) to exceed the current limit* — otherwise foldback (reduction in output voltage) will occur at the current limit threshold, and so the maximum output power cannot be guaranteed.

For example, if we have a "5-A Buck switcher IC," being operated at 5-A load, with an r of 0.4, then the normal operating peak current is $5 \times (1 + 0.4/2) = 5 \times 1.2 = 6$ A. So *ideally*, we would want the current limit of the device to be at least 6 A. *Unfortunately, when we come to such integrated switchers, that much "margin" is rarely available* — manufacturers always like to "bolster" the advertised ratings of their parts to be close to the maximum stress limits. So yes, if this particular part was declared to be a "4-A IC" instead of a "5-A IC," we would have been just fine. But as things stand, manufacturers usually pay scant regard as to what may constitute an *optimum* rating for the device, in relationship to its associated components and the overall design strategy. Therefore, for example, a certain commercial "5-A switcher IC" may have a published (set) current limit of only 5.3 A. But on analysis, we see that it allows only 0.3 A above, and 0.3 A below, the average level of 5 A. Therefore, the maximum allowed ΔI at 5 A load is only 0.6 A. And the maximum r is $0.6/5 = 0.12$ (when operated at a load current of 5 A). We can see that it is clearly much less than the optimum r of 0.4. And no doubt, this lowered r will adversely impact the size of the inductor (and converter). So is it truly a 5-A IC?

Now we take up the issue of the *spread* in current limit. I_{CLIM} actually has two limits — I_{CLIM_MIN} and I_{CLIM_MAX} (i.e., the MIN and MAX of the current limit, respectively). The question is — *which of these limits should we consider for designing the inductor?*

- To *guarantee output power*, we need to look at the *MIN of the current limit only*. In most low-voltage DC–DC converter applications, the MIN limit is the only threshold that really counts — we can usually completely ignore the MAX (and of course the TYP value). The basic criterion for guaranteeing output power is — we must ensure that the calculated normal operating peak current in our application is always *less than the MIN* value of the current limit. Of course, if we are not operating close to the current limit of the device, this condition will be met without any struggle, and so we can then just focus on setting r to about 0.4.

- But like all components, inductors also have a typical tolerance — usually about ±10%. So, if we are operating very close to the limits of the device, and thereby r is being effectively dictated by the MIN of the current limit (rather than by its optimum or desirable value), then the (nominal) value of inductance we ultimately choose should be about *10% higher than the calculated value*. That will guarantee output power unconditionally — under all possible variations in current limit *and inductance*.

- Note that ideally, we would also like to leave at least 20% additional margin (*headroom*) between the peak current of our application and the MIN of the current limit. This is usually necessary for getting a quick *response (correction) to a sudden increase in load*. So in general, if we somehow manage to curtail the ability of the converter to respond quickly (e.g., by not providing sufficient headroom in the current limit and/or maximum duty cycle), the inductor will not be able to ramp up current quickly enough to meet the sudden increase in energy demand. Therefore, the output will droop rather severely for several cycles, before it eventually recovers.

But unfortunately, when dealing with fixed current limit (integrated) switchers, we will find that this "nice-to-have transient headroom" may be a *luxury we just can't afford* — because in most cases, the MIN current limit is set only slightly higher than the declared "rating" of the device. First, a 20% headroom may not be available! Under these conditions, to try to forcibly create some "headroom for good transient response" by increasing the inductance, becomes *counterproductive* — a large inductance takes even more time for its current to ramp up, and thereby effectively slows down the transient (loop) response — opposite to what we were hoping to achieve here! Therefore, we almost always end up ignoring this desirable 20% or so *step-response headroom/margin*, especially when dealing with integrated switcher ICs.

As for the MAX of the current limit, *whenever we deem that inductor saturation is of real concern to us* (as in high-voltage applications), *we must look at the MAX of the current limit* to decide upon the size of the inductor — that being the worst-case in terms of peak current under overloads, inductor energy storage, and its possible saturation.

Therefore, in general, *in high-voltage DC–DC (or in off-line) applications, the MIN of the current limit may sometimes need to be considered when selecting inductance (as when operating close to current limit), but the MAX of the current limit will certainly always be used to determine the size of the inductor.*

As a corollary, manufacturers of (low-voltage) *DC–DC converter ICs* actually need *not* (and probably justifiably do not) struggle too hard to minimize the spreads and tolerances of current limit (provided of course the MIN of the current limit is at least set high enough not to intrude on the declared power-handling capability of the IC). And for *low-voltage DC–DC converter applications*, the current limit is typically ignored by users — the final selection of inductor current rating (and size) is simply based on the cycle-by-cycle peak inductor current under *normal* (steady) operation (i.e., the maximum load of the application, at the worst-case input voltage end).

On the other hand, manufacturers of *off-line* switcher ICs *do* need to maintain a *tight tolerance* on the current limit. In their case, the maximum power-handling capability of their particular device is in effect dependent only on the 'MIN' (minimum limit) of the current limit specification, whereas, the transformer size is determined entirely by the "MAX" of the current limit specification. So, in this case, a "loose" current limit specification effectively amounts to requiring *bigger components* (transformer) for the same maximum power-handling capability.

> *Note: Some makers of off-line integrated switcher ICs (e.g., the "TOPSWITCH®" from Power Integrations) often tout their "precise" current limit — thus suggesting that we get the best power-to-size ratio (i.e., converter power density) when using their products. However, we should remember that in most cases, their product families have a discrete set of fixed current limits. And that is a problem! For example, we may have devices available with current limits in steps of 2 A, 3 A, 4 A, and so on. So yes, we may indeed get a higher power density when operating at the maximum rated output power of a particular IC. But when operating at a power level between available current limits, we are not going to get an optimum solution. For example, in an application where the peak current is 2.2 A, then we would need to select the 3 A current limit part, and we will need to design our magnetics to avoid core saturation at 3 A. So in effect, we have a very imprecise current limit now! The best solution is to look for a part (integrated switcher or controller plus MOSFET solution) where we can precisely set the current limit externally, depending upon our application.*

With all these subtle considerations in mind, a designer can hopefully pick a more appropriate inductor current rating for his or her application. Clearly, there are no hard and fast rules. Engineering judgment needs to be applied as usual, and perhaps some further bench-testing may also be needed to validate the final choice of inductor.

In the worked examples that follow, the general approach and design procedure will become clearer.

Worked Example (1)

A Boost converter has an input range of 12−15 V, a regulated output of 24 V, and a maximum load current of 2 A. What would be a reasonable goal for its inductance, if the switching frequency is (a) 100 kHz, (b) 200 kHz, and (c) 1 MHz? What is the peak current in each case? And what is the energy-handling requirement?

The first thing we have to remember is that for this topology (as for the Buck-Boost), the *worst-case* is the *lowest end* of the input range, since that corresponds to the highest duty cycle and thus the highest average current $I_L = I_O/(1 - D)$. So, for all practical purposes, we can completely disregard V_{INMAX} here — in fact it was a red herring to start with, for this particular analysis!

From Table 2.1, the duty cycle is

$$D = \frac{V_O - V_{IN}}{V_O} = \frac{24 - 12}{24} = 0.5$$

Therefore,

$$I_L = \frac{I_O}{1 - D} = \frac{2}{1 - 0.5} = 4 \text{ A}$$

Let us target a current ripple ratio of 0.4. So,

$$I_{PK} = I_L\left(1 + \frac{r}{2}\right) = 4 \times \left(1 + \frac{0.4}{2}\right) = 4.8 \text{ A}$$

- We should remember that $r = 0.4$ *always implies that the peak is 20% higher than the average.* So, we realize that in effect, the peak current does *not* depend on the frequency. The inductor must be able to handle the above peak current without saturating. So, in this example, we would be fine just picking an inductor rated for 4.8 A (or more), *irrespective of frequency*. In fact we had previously learned that *the required current rating of an inductor is independent of the frequency (since the peak is unchanged)*. However, the size does change with frequency, because size is $(1/2) \times L \times I_{PK}^2$, and L changes as follows.

To calculate the inductance corresponding to the chosen value of r, we can use the following equation (presented previously). We also note from Table 2.1 that $V_{ON} = V_{IN}$ for the Boost. Therefore, for $f = 100$ kHz

$$L = \frac{V_{ON} \times D}{r \times I_L \times f} = \frac{12 \times 0.5}{0.4 \times 4 \times 100 \times 10^3} \Rightarrow 37.5 \text{ } \mu\text{H}$$

For $f = 200$ kHz, we would get *half* of this, that is, 18.75 μH. And for $f = 1$ MHz, we get 3.75 μH. We clearly see that *high frequencies lead to smaller inductances*.

We have previously observed that *for a given application*, small inductances invariably lead to small inductors. Therefore, we conclude that *on increasing the switching frequency, we will get smaller-sized inductors too*. And that is the basic reason for hiking up switching frequencies in general.

The energy-handling requirement, if desired, can be explicitly calculated in each case, by using $E = (1/2) \times L \times I_{PK}^2$.

So far, we have been generally targeting r *= 0.4 as an optimum value. Let us now understand all the reasons why this may* **not** *be a good choice on occasion.*

Current Limit Considerations in Setting r

We had indicated previously that the current limit may be too low to allow *r* from being set to its optimum. Now, we will also include the impact of *spread* on the current limit.

So for example, in Table 2.3 we have the published specifications for the current limit of an integrated "5-A" switcher, the LM2679. To be able to guarantee the specified power output (or load current in this case) unconditionally, we need to guarantee that the peak current in our application never reaches the *lower limit* ("MIN") of the published current limit specification. So in fact, in Table 2.3, we need to disregard all the numbers except for the "MIN" value — given as *5.3 A*.

Now, if we are trying to get 5 A out of our converter with an *r* of 0.4, the estimated peak current will be $1.2 \times 5 = 6$ A. Clearly, as mentioned earlier, we are *not going to get there* with the LM2679! Unless we *lower* the value of *r (increase inductance)*. Maximum value of *r* is

$$I_{PK} = I_O \times \left(1 + \frac{r}{2}\right) \leq I_{CLIM\_MIN}$$

Solving, with $I_O = 5$ A, and $I_{CLIM\_MIN} = 5.3$ A, we get

$$r \leq 2\left(\frac{I_{CLIM\_MIN}}{I_O} - 1\right) = 2\left(\frac{5.3}{2} - 1\right) = 0.12$$

We can see from Figure 2.6, *that this calls for an energy-handling capability (size of inductor) almost 3× the optimum!*

Table 2.3: Published Current Limit Specs for the LM2679.

	Conditions		TYP	MIN	MAX	Unit
Current limit "I_{CLIM}"	$R_{CLIM} = 5.6$ kΩ	Room temperature	6.3	5.5	7.6	A
		Full operating temperature range		5.3	8.1	

Actually, it turns out that this part is just *specified* inappropriately. It in reality is the one with an *adjustable* current limit. And so, we could have probably adjusted the current limit adjust resistor quoted in the electrical tables to allow for a "better" value of current limit, and thereby a better value of *r* (at maximum rated load). But unfortunately, that is not clarified in the tables.

We should always remember that the minimum and maximum limits of the electrical tables are the only parts of a datasheet really guaranteed by any vendor (certainly not the *typical* values!). So, as a matter of fact, *any* other information in a datasheet just amounts to general design "guidance" — and that includes any "typical performance curves" provided. A prudent designer would never second-guess the vendor — in this case as to whether the current limit resistor can indeed be adjusted to give us a smaller inductor, or not. Therefore, as it stands, if we are using the LM2679 for a 5-A load current application, we *do need an inductor three times larger than the optimum*. Note that if the current limit could indeed be adjusted higher, the vendor should have picked the appropriate value for the current limit adjusted for a resistor in the "conditions" column of the electrical table (and declared the limits accordingly).

Note also that when we talk of a "5-A Buck IC," it implies the part is supposed to deliver 5-A *load current*. The current limit of course needs to be set (and stated) correctly for the rated load, as discussed above. However, we should be very clear that **when we are talking of Boost or Buck-Boost switcher ICs, a "5 A" part for example, does *not* give us a 5-A load current**. That is because the DC inductor current is not equal to I_O, but $I_O/(1 - D)$ for these topologies. So, a "5-A" rating in this case only refers to the *current limit* of the device. What load current we can derive from a "non-Buck" IC depends on our specific application — in particular on the D_{MAX} (duty cycle at V_{INMIN}). For example, if the desired load current is 5 A, and the (maximum) duty cycle in our application is 0.5, then the average inductor current is actually $I_O/(1 - D) = 10$ A. Further, with an r of 0.4, the peak would be 20% higher, that is, $1.2 \times 10 = 12$ A. So, for an optimum case, we would need to actually look for a device whose *minimum* current limit is 12 A or more in this case. At the bare minimum, we need a device with a current limit higher than 10 A, just to guarantee output power.

Continuous Conduction Mode Considerations in Fixing r

As discussed previously, under various conditions, we may enter discontinuous conduction mode (DCM). From Figure 2.5 we can see that just as DCM starts to occur, the current ripple ratio is 2. However, we can pose the question in the following manner — what if we have set the current ripple ratio to a certain value r' (i.e., the current ripple ratio *at the maximum load current*, I_{O_MAX}). And then we decrease the load current slowly — at what load does the converter enter DCM?

By simple geometry it can be shown that the transition to DCM will occur at r′/2 times the maximum load. For example suppose we set r' to 0.4 at 3 A load, the converter will transition into DCM at $(0.4/2) \times 3 = 0.6$ A.

But designers know that when DCM is entered, a lot of things within the converter change suddenly! The duty cycle, for one, will now start pinching off toward zero as we decrease the load current further. In addition, the loop response of the converter (its ability to correct quickly for disturbances in line and load) also usually gets degraded in DCM. The noise and EMI profile can change suddenly too, and so on. Of course there are some advantages of operating in DCM too, but let us for now assume that for various reasons, the designer wishes to avoid DCM altogether, if possible.

We see that maintaining the converter in CCM, down to the minimum load of our application, enforces a certain *maximum* value for r'. For example, if the minimum load is $I_{O\_MIN} = 0.5$ A, then to maintain the converter in CCM at 0.5 A, the set current ripple ratio (r' at 3 A) needs to be lowered. Back calculating, we get the required condition for this

$$I_O \times \frac{r'}{2} = I_{O\_MIN}$$

So,

$$r' = \frac{2 \times I_{O\_MIN}}{I_{O\_MAX}}$$

In our case we get

$$r' = \frac{2 \times 0.5}{3} = 0.333$$

We therefore need to set the current ripple ratio to less than 0.333 at maximum load, to ensure CCM at I_{O_MIN}. This was the traditional design criterion for inductors, before Figure 2.6 was published and understood.

Note that generally speaking, we can make the converter operate in boundary conduction mode (BCM), or in full DCM, in three ways — (a) by decreasing the load, (b) choosing a small inductance, or (c) *increasing the input voltage*.

We realize that decreasing the load will proportionally decrease I_{DC} to virtually any value, and so the condition $r \geq 2$ (BCM to DCM) will certainly occur sooner or later — below a certain load current. Similarly, decreasing L will necessarily increase ΔI, and so at some point we can expect the ratio $\Delta I/I_{DC}$ (i.e., r) to try to become greater than 2 (implying DCM).

However, as far as the *third* method of entering DCM mentioned above is concerned, we should realize that **solely** *increasing the input voltage just might not do the trick! DCM or*

BCM can only happen under an input (line) variation, provided the load current is simultaneously below a certain value to start with (the value being dependent on L).

It is instructive to study the three topologies separately in this regard. Note that the general equation for r is

$$r = \frac{V_{ON} \times D}{I_L \times L \times f} \quad \text{(any topology, any mode)}$$

Applying the voltseconds law in CCM (or BCM), we also get

$$r = \frac{V_{OFF} \times (1-D)}{I_L \times L \times f} \quad \text{(any topology, CCM or BCM only)}$$

(a)   From the plots of r in Figure 2.4, we see that both the **Buck and the Buck-Boost** have the highest value of r when D approaches zero, i.e., at maximum input voltage. For these topologies, the equation for r (derivable from the more general equation for r just given immediately above) is

$$r = \frac{V_O}{I_O \times L \times f}(1-D) \quad \text{(Buck)}$$

$$r = \frac{V_O}{I_O \times L \times f}(1-D) \quad \text{(Buck-Boost)}$$

So, putting $r = 2$ and $D = 0$ (i.e., highest input voltage plus BCM), we get the limiting condition

$$I_O = \frac{1}{2} \times \frac{V_O}{L \times f} \quad \text{(Buck and Buck-Boost)}$$

Therefore, for these two topologies, if I_O is *greater* than the above limiting value, we will *always remain in CCM, no matter how high we increase the input voltage.*

(b)   Coming to the **Boost**, the situation is not so obvious. From Figure 2.4, we see that r peaks at $D = 0.33$ (corresponding to the input being exactly two-thirds of the output). *So, the Boost is most likely to enter DCM at $D = 0.33$* — not say, at $D = 0$ or $D = 1$. We can derive the following (exact) equation for r

$$r = \frac{V_O}{I_O \times L \times f} D \times (1-D)^2 \quad \text{(Boost)}$$

So, putting $D = 0.33$, and $r = 2$ in this equation, we get the following limiting condition

$$I_O = \frac{2}{27} \times \frac{V_O}{Lf} \quad \text{(Boost)}$$

Therefore, for the Boost topology, if I_O is *greater* than this value, we will *always remain in CCM*, no matter how high we increase the input voltage.

Note that, if we do manage to enter DCM, the most likely input point for this to happen is an *input of 0.67 times the output*. In other words, if we are not in DCM at this particular input voltage, we can be sure we will be in CCM throughout the entire input range (whatever it may be).

Setting r to Values Higher than 0.4 when Using Low-ESR Capacitors

Nowadays, with improvements in capacitor technology, we are seeing a new generation of very "low-ESR" capacitors — like monolithic multilayer ceramic capacitors ("MLCs" or "MLCCs"), polymer capacitors, and so on. Due to their extremely low ESRs, these capacitors usually have very high ripple (RMS) current ratings. Therefore, the required size of such capacitors in any application *is no longer dictated by their ripple current handling capability*. In addition, these capacitors also have almost no *ageing* characteristics (or lifetime issues) that we need to account for beforehand in the design (as we customarily do for electrolytic capacitors — which "dry out" over time). Further, due to their very high dielectric constant, these new capacitors have also become very *small in size*. So in fact nowadays, *increasing* r *may not necessarily cause a noticeable increase in the space occupied by the capacitors (or size of converter)*. On the other hand, increasing *r* may still lead to a relatively significant reduction in size of the inductor.

Summing up, with modern capacitors to the rescue, it may start making perfect sense to increase *r* from its traditional "optimum" of 0.4, to say approximately 0.6–1 on occasion (provided other considerations do not restrict this). If we do so, Figure 2.6 tells us, we can still get an additional 30–50% reduction in the size of the inductor. And that is certainly not insignificant, provided of course that the advantage is not offset by having to use larger capacitors in the bargain (due to higher filtering capacitances required)!

Setting r to Avoid Device "Eccentricities"

Surprisingly, device *eccentricities* may on occasion play a part in defining the limits of *r* too. For example in Figure 2.8 we have presented the current limit plot of an integrated high-voltage flyback switcher IC called the "TOPSWITCH®." On it we have superimposed a typical switch current waveform, just to make things a little clearer.

We see that surprisingly, the current limit of this device is *time-dependent* for about 1.5 μs after the turn-on transition — something we don't intuitively ever expect. This "initial current limit" of the device occurs just as its internal current limit comparator starts to come out of its (valid) "leading edge blanking" time. As mentioned, during this blanking time the IC is just "not looking" at the current at all to avoid spurious triggering on the

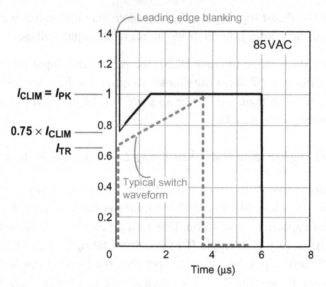

Figure 2.8: The "initial current limit" of the TOPSWITCH®.

noise edge of the turn-on transition. But the problem is that once the current limit circuit gets down to monitoring the switch current again, it takes a *certain time* for the current limit threshold to settle down — and during this time it can be triggered at only about 75% of the supposed current limit!

Looking at the switch (or inductor) current waveform, we know that the current at the moment the switch turns ON is always *less* than the average value by the amount $\Delta I/2$. In other words, this *trough (valley)* current "I_{TR}" is related to r according to the equation

$$I_{TR} = I_L \times \left(1 - \frac{r}{2}\right)$$

We realize that to avoid hitting the initial current limit of the device, we need to ensure that the trough falls below $0.75 \times I_{CLIM}$. So,

$$I_{TR} = I_L \times \left(1 - \frac{r}{2}\right) \leq 0.75 \times I_{CLIM}$$

Now, we are assuming the power supply is at maximum load in this analysis. Therefore, the peak current is set equal to the current limit I_{CLIM}

$$I_{PK} = I_L \times \left(1 - \frac{r}{2}\right) = I_{CLIM}$$

Equating the above two equations, we get the limiting condition for r

$$\left(1 - \frac{r}{2}\right) \leq 0.75 \times \left(1 + \frac{r}{2}\right)$$

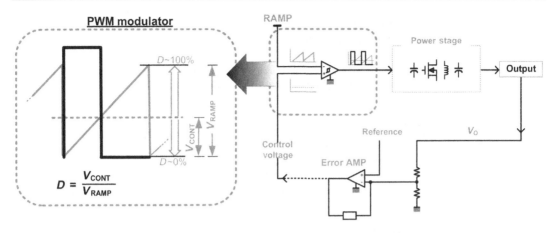

Figure 2.9: The pulse width modulator section of a power converter.

or

$$r \geq 0.286$$

Since r in any case is typically set to about 0.4, we should normally have no trouble with this "initial current limit" issue. However, note that on finer examination of the electrical tables of the datasheet, this $0.75 \times$ factor is specified *only at* 25 °C. Unfortunately, very few power devices stay at 25 °C for long! So, the bottom line is that, we, as designers, *do not really know* the value of the current limit as the device *heats up*. Yes, we can certainly make an educated guess, possibly leave an additional safety margin when fixing r, and certainly, we may face no problem whatsoever. But the truth is we are *on our own now* — the vendor has *not* provided the requisite data (in the form of *guaranteed* limits within the electrical tables).

Setting r to Avoid Subharmonic Oscillations

Looking at Figure 2.9, we see that in any converter, the output voltage is first compared against an internal reference voltage. Then, the difference between the two (the "error") is filtered, amplified, and inverted by an "error amplifier," the output of which (the "*control voltage*") is fed to one of the two inputs of a "pulse width modulator" (PWM) comparator. On the other input of this PWM comparator, a *ramp* is applied, and this produces the switching pulses. So, for example, if the error at the output increases, the control voltage will decrease, and the duty cycle will thus decrease in an effort to reduce the output voltage. That is how regulation usually works.

In voltage-mode control, the ramp applied to the PWM comparator is derived from an internal (fixed) clock. However, in current-mode control, it is derived from the *inductor current (or switch current)*. And the latter leads to a rather odd situation where even a

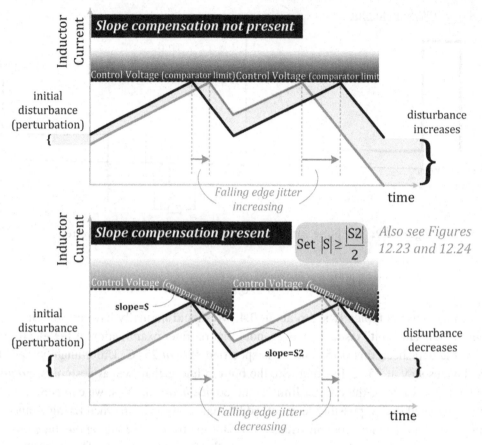

Figure 2.10: Subharmonic instability in current-mode control, and avoiding it by slope compensation.

slight disturbance in the inductor current waveform can become *worse in the next cycle* (see upper half of Figure 2.10).

Eventually, the converter may lapse into a strange "one pulse wide, one pulse narrow" switching waveform. This represents an operating mode that is definitely not "legitimate" or desirable for several reasons — in particular, the output voltage ripple is now much higher, and the loop response is severely degraded.

To get the disturbance to *decrease* every cycle and eventually die out, it can be shown that we need to do one of two things. Actually, both methods effectively amount to mixing a *little voltage-mode control into current-mode control*. So,

(a)   either we add a small fixed (clock-derived) voltage ramp to the sensed voltage ramp (derived from the inductor/switch)

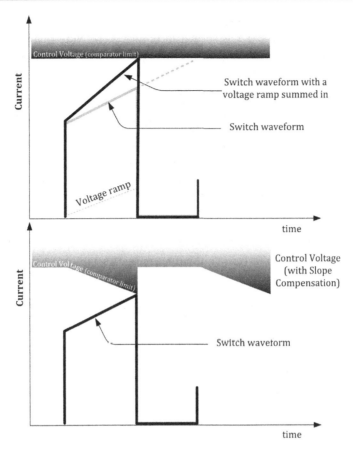

Figure 2.11: Adding a fixed ramp to the sensed signal, or modifying the control voltage, are equivalent methods of slope compensation in current-mode control.

(b)   or we subtract the same fixed voltage ramp from the control voltage (output of error amplifier)

As we can see from Figure 2.11, both are equivalent. That is in fact not surprising at all, considering that both the ramp and the control voltage go to the pins of a *comparator*. So, if we compare a signal $A + B$ with a signal C, that is exactly equivalent to comparing A to $C - B$. And in both cases, equality at the input pins is established when $A + B = C$.

This technique is called "slope compensation," and is the most recognized way of quenching the alternate wide and narrow pulsing (or "subharmonic instability") associated with current-mode control (see lower half of Figure 2.10). See also *Chapter 12*.

It can be shown that to avoid subharmonic instability, we need to ensure that the amount of slope compensation (expressed in A/s) is equal to *half* the slope of the falling inductor current ramp, *or more*. Note that in principle, subharmonic instability can occur only if D is (close to or) greater than 50%. So, slope compensation can be applied either over the full duty cycle range, or just for $D \geq 0.5$ as shown in Figure 2.10. Note that subharmonic instability can occur only if we are operating in continuous conduction mode (CCM). However, one way of avoiding this instability altogether is to operate in DCM.

If the amount of slope compensation is *fixed* by the controller, then as designers, we need to personally ensure that the slope of the falling inductor current ramp is equal to *twice* the slope compensation — or *less* (note that we are talking in terms of the magnitudes of the slopes only). This will in effect dictate a certain *minimum value of inductance*. And in terms of *r*, this tells us that we could have a situation where we may need to set *r* to *less* than the optimum of 0.4 — for example, if the control IC has an inadequate amount of built-in slope compensation.

As a result of more detailed modeling of current-mode control, optimum relationships for the minimum inductance required (to avoid subharmonic instability) have been generated as follows:

$$L \geq \frac{D - 0.34}{slopecomp} \times V_{IN} \ \mu H \quad (Buck)$$

$$L \geq \frac{D - 0.34}{slopecomp} \times V_O \ \mu H \quad (Boost)$$

$$L \geq \frac{D - 0.34}{slopecomp} \times (V_{IN} + V_O) \ \mu H \quad (Buck\text{-}Boost)$$

where the slope compensation is in A/μs.

Note that for all these topologies, we have to do the preceding calculation at the *maximum input voltage point at which the duty cycle is greater than 50%, AND we are also simultaneously in CCM.*

More details on subharmonic instability and slope compensation can be found in *Chapter 12.*

Quick Selection of Inductors Using "L × I" and "Load Scaling" Rules

Finally, having decided upon the value of *r* based on all the considerations outlined so far, we first present a *quick* method of picking an inductor for a given application. After that we will proceed to a more detailed analysis and worked example.

As mentioned previously, from the inductor equation $V = L \ dI/dt$, we can derive another useful relationship that we have named the "$L \times I$" equation ("el-ex-eye" equation)

$$(L \times I_L) = \frac{Et}{r} \quad (any \ topology)$$

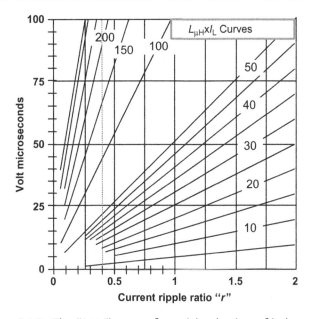

Figure 2.12: The "$L \times I$" curves for quick selection of inductance.

Symbolically,

$$L \times I = \frac{\text{voltseconds}}{\text{current ripple ratio}} \quad (\textit{any topology})$$

If we know the voltseconds (from our application conditions), and have a target value for r, we can calculate "$L \times I$." Then by knowing I, we can calculate L.

Note that $L \times I$ *can be visualized as a sort of* **inductance "per" Ampere** — except that the relationship is *inverse* — that is, if we *increase* the current, we need to *decrease* the inductance (by the same amount). So, for example, if we get an inductance of 100 μH for a 2-A application, then for a 1-A application, the inductance must be 200 μH, and for a 4-A application, the inductance would be 50 μH, and so on.

Note that because the $L \times I$ equation doesn't depend on topology, switching frequency, or on the specific input/output voltages, we can graph it out universally, as in Figure 2.12. That helps quickly pick an inductance for any application. Let us now exemplify the $L \times I$ graphical selection method for each topology.

Worked Examples (2, 3, and 4)

Buck: *Suppose we have an input of 15–20 V, an output of 5 V, and a maximum load current of 5 A. What is the recommended inductance if the switching frequency is 200 kHz?*

(a)   We need to start the inductor design at V_{INMAX} (20 V) for a Buck.

(b)   The duty cycle from Table 2.1 is $V_O/V_{IN} = 5/20 = 0.25$.

(c)   The time period is $1/f = 1/200$ kHz $= 5$ μs.

(d)   The off-time t_{OFF} is $(1 - D) \times T = (1 - 0.25) \times 5 = 3.75$ μs.

(e)   The voltseconds (calculated using the *off-time*) is $V_O \times t_{OFF} = 5 \times 3.75 = 18.75$ μs.

(f)   From Figure 2.12, with $r = 0.4$, and $Et = 18.75$ μs, we get $L \times I = 45$ μH A.

(g)   For a 5 A load, $I_L = I_O = 5$ A.

(h)   Therefore, we need $L = 45/5 = 9$ μH.

(i)   The inductor must be rated for at least $(1 + r/2) \times I_L = 1.2 \times 5 = 6$ A.

Summarizing, we need a 9 μH/6 A inductor (or closest available).

Boost: *Suppose we have an input of 5–10 V, an output of 25 V, and a maximum load current of 2 A. What is the recommended inductance if the switching frequency is 200 kHz?*

(a)   We need to start the inductor design at V_{INMIN} (5 V) for a Boost.

(b)   The duty cycle from Table 2.1 is $(V_O - V_{IN})/V_O = (25 - 5)/25 = 0.8$.

(c)   The time period is $1/f = 1/200$ kHz $= 5$ μs.

(d)   The on-time t_{ON} is $D \times T = 0.833 \times 5 = $ μs.

(e)   The voltseconds (calculated using the *on-time*) is $V_{IN} \times t_{ON} = 5 \times 4 = 20$ μs.

(f)   From Figure 2.12, with $r = 0.4$, and $Et = 20$ μs, we get $L \times I = 47$ μH A.

(g)   For a 2 A load, $I_L = I_O/(1 - D) = 2/(1 - 0.8) = 10$ A.

(h)   Therefore, we need $L = 47/10 = 4.7$ μH.

(i)   The inductor must be rated for at least $(1 + r/2) \times I_L = 1.2 \times 10 = 12$ A.

Summarizing, we need a 4.7 μH/12 A inductor (or closest available).

Buck-Boost: *Suppose we have an input of 5–10 V, an output of −25 V output, and a maximum load current of 2 A. What is the recommended inductance if the switching frequency is 200 kHz?*

(a)   We need to start the inductor design at V_{INMIN} (5 V) for a Buck-Boost.

(b)   The duty cycle from Table 2.1 is $V_O/(V_{IN} + V_O) = 25/(5 + 25) = 0.833$.

(c)   The time period is $1/f = 1/200$ kHz $= 5$ μs.

(d)   The on-time t_{ON} is $D \times T = 0.833 \times 5 = 4.17$ μs.

(e)   The voltseconds (calculated using the *on-time*) is $V_{IN} \times I_{ON} = 5 \times 4.17 = 20.83$ μs.

(f)   From Figure 2.12, with $r = 0.4$, and $Et = 20.83$ μs, we get $L \times I = 52$ μH A.

(g)   For a 2 A load, $I_L = I_O/(1 - D) = 2/(1 - 0.833) = 12$ A.

(h)   Therefore, we need $L = 52/12 = 4.3$ μH.

(i)   The inductor must be rated for at least $(1 + r/2) \times I_L = 1.2 \times 12 = 14.4$ A.

Summarizing, we need a 4.3 μH/14.4 A inductor (or closest available).

The Current Ripple Ratio r in Forced Continuous Conduction Mode ("FCCM")

Finally, before we move on to magnetic fields, we make some closing remarks on designing with forced continuous conduction mode ("FCCM"). More on this in *Chapter 9*.

As discussed previously, r is defined only for CCM, and therefore cannot exceed 2 (since that marks the boundary between CCM and DCM). However, in *synchronous regulators* (with diode replaced or supplemented by a low-drop MOSFET across it), we actually never enter DCM (unless the IC is deliberately designed to go into "diode emulation mode"). So now, on decreasing the load, we actually continue to remain in CCM. That is because for DCM to ever occur, the inductor current must be forced to *stay* at least for some part of the switching cycle at zero. And to get that to happen, we need to have a reverse-biased diode that prevents the inductor current from "going the other way." But in synchronous regulators, the MOSFET across the diode allows reverse-conduction even if the diode is reverse-biased, so we do not get DCM.

The CCM-type mode that replaces the DCM mode in synchronous regulators is distinguished from the usual (normal) CCM mode, by calling it the "forced continuous conduction mode" (FCCM). The main switch is usually identified as the *top* (or "high-side") MOSFET, whereas the MOSFET across the diode is called the *bottom* (or "low-side") MOSFET. Further, in FCCM, r *is legitimately allowed to exceed 2* (see Figure 2.5).

We can visualize FCCM as starting to occur when the load current is decreased sufficiently to cause part of the inductor current waveform to become "submerged" below "sea-level" — that is, with parts of it having a negative value (inductor current flowing momentarily away from the load). But note that as long as we are still drawing some load current *out* of the output terminals of the converter, the *average* value of the waveform, I_{DC} (center of ramp), is still positive — that is, going toward the load — *on an average*. Further, because I_{DC} is always proportional to the load current, it can be made to decrease all the way *down to zero* while still maintaining CCM. Since the *swing* in current, ΔI, depends *only* on the input and output voltages, which we have assumed have not changed, the ratio $r = \Delta I/I_L$ not only exceeds 2 but can in fact become extremely large.

All the basic design equations we can write for the RMS, DC, AC, or peak currents in the input/output capacitors and the switch, when operating in conventional CCM, *apply to the converter in FCCM too* (though there may be some *additional* losses, as for example when the current flows through the body diode of the top MOSFET). This, despite the fact that r can now exceed 2. In other words, the CCM equations do *not* get invalidated in FCCM. However, a specific computational problem can arise in some cases, because if r is infinite (zero load current), we can get a singularity — a "0" in the denominator. At first sight, this seems to make the CCM equations (presented the way we have been doing) unusable. But, one trick that can be employed to avoid the singularity is to *assume* a few milliamperes of minimum load, however small. Alternatively, we can substitute $r = \Delta I / I_{DC}$ back into the equations, and we will then see that I_{DC} cancels out (does not appear in the denominator anywhere). Either way, the equations of CCM (see the Appendix), apply to FCCM too.

Basic Magnetic Definitions

Having understood basic concepts like voltseconds, current components, worst-case voltage, and also how to do an initial (quick) selection of an off-the-shelf inductor, we will now try to go *inside* the magnetic component, so as to learn what happens in terms of the *magnetic fields* present inside its core. We will then use this information to do a more complete validation of a selected off-the-shelf inductor. Then we will find the remaining (worst-case) stresses of the converter.

At the outset, we should note that in magnetics, there are several different *systems of units* in use. This can become very confusing, since even the basic equations look different depending on the system in use. It is therefore a wise policy to stick to one system of units all the way through — converting to a different system, if required, *only at the very end*, that is, only at the level of *numerical* results (not at the level of *equations*).

Further, unless otherwise stated, the reader can safely assume we are using the *meter–kilogram–seconds* system of units — that is, "MKS" system, also called the "SI" system (for System International).

The basic definitions are as follows:

*   *H-field*: Also called "field strength," "field intensity," "magnetizing force," "applied field," and so on. Its units are A/m.
*   *B-field*: Also called "flux density" or "magnetic induction." Units of B are Tesla ("T") or Webers per square meter (Wb/m^2).
*   *Flux*: This is the integral of B over a given surface area expressed as Webers (Wb). It is

$$\phi = \int_S B \, dS \text{ Webers}$$

where dS is a differential of surface area. If B is constant over the surface, we get the more common form $\phi = BA$, where A is the area of the surface through which the flux in question is flowing.

> *Note: The integral of* B *over a closed surface is zero since flux lines do not start or end at any given point but are continuous.*

* B is related to H *at any given point* by the equation $B = \mu H$, where μ is the *permeability* of the material. Note that later we will use the symbol μ for "relative

permeability," that is, the *ratio* of the permeability of the material to that of air. So, in MKS units we should actually preferably write $B = \mu_c H$, where μ_c is the permeability of the core (magnetic material). By definition, $\mu_c = \mu \mu_o$.

- The permeability of air, denoted by μ_o, is equal to $4\pi \times 10^{-7}$ Henrys/m in MKS units. In CGS units it is equal to 1. That is why in CGS units $\mu_c = \mu$, where μ is also automatically the relative permeability of the material (though units are different).
- Faraday's law of induction (also called Lenz's law) relates the induced voltage V that is developed across the ends of a coil (N turns), to the (time varying) B-field passing through it. So,

$$V = N\frac{d\phi}{dt} = NA\frac{dB}{dt}$$

- The "inertia" of a coil to a change in flux through it due to a time-varying current through it is its "inductance" L, defined as

$$L = \frac{N\phi}{I} \text{ Henrys}$$

- Since it can be shown that the flux is proportional to the number of turns N, the inductance L is proportional to the *square* of the number of turns. This proportionality constant is called the "inductance index" and is denoted by "A_L." It is usually expressed as nH/turns^2 (though sometimes it is considered to be mH/1,000 turns^2, both being *numerically* the same). So,

$$L = A_L \times N^2 \times 10^{-9} \text{ Henrys}$$

- When H is integrated over a *closed loop*, we get the current enclosed by the loop

$$\oint H \, dl = I \text{ Amperes}$$

where the integration symbol above reflects the fact that it is being performed over a closed loop. This is also called "Ampere's circuital law."

- Combining Lenz's law with the inductor equation $V = L \, dI/dt$, we get

$$V = N\frac{d\phi}{dt} = NA\frac{dB}{dt} = L\frac{dI}{dt}$$

- From this we get the *two key equations* used in power conversion

$$\Delta B = \frac{L \, \Delta I}{NA} \quad (\text{"voltage–independent equation"})$$

$$\Delta B = \frac{V \, \Delta t}{NA} \quad (\text{"voltage–dependent equation"})$$

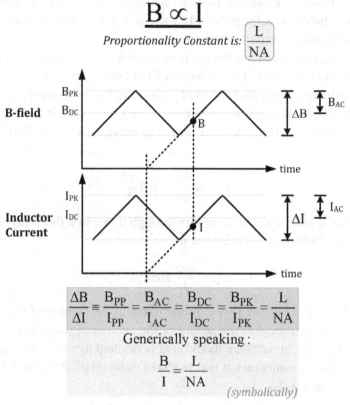

Further, since at any given point, B=μH, H is also proportional to I

Figure 2.13: *B* and *I* can be usually considered to be proportional to each other.

The first equation can be written symbolically as

$$B = \frac{LI}{NA} \quad (voltage-independent\ equation)$$

And the latter equation can be written in a more "power-conversion-friendly" form as follows

$$B_{AC} = \frac{V_{ON}\ D}{2 \times NAf} \quad (voltage-dependent\ equation)$$

For most inductors used in power conversion, if we reduce the current to zero, the field inside the core also goes to zero (not a permanent magnet). An implicit assumption of complete linearity is also usually made — that is, *B and I are considered proportional to each other* as shown in Figure 2.13 (unless of course the core starts saturating, at which point, all bets are off!). The *voltage-independent equation* can then be expressed as any of the equations shown in the figure — in other words, *this proportionality applies to the peak values of current and field, their average values, their AC values, their DC values, and so on.* The constant of proportionality is equal to

$$\frac{L}{NA} \quad (\text{proportionality constant linking B and I})$$

where N is the number of turns and A the actual geometrical cross-sectional area of the core (its center limb usually, or simply the "effective area" A_e given in the datasheet of the core). See *Chapter 5* too.

Worked Example (5) — When Not to Increase the Number of Turns

Note that the voltage-independent equation is useful if, for example, we want to do a *quick check to see if our core may be saturating.* Suppose we are custom-designing our inductor. We have wound 40 turns on a core with an area of $A = 2$ cm^2. Its *measured* inductance is 200 µH, and the peak inductor current in our given application is 10 A. Then the peak flux density can be calculated as follows:

$$B_{PK} = \frac{L}{NA}I_{PK} = \frac{200 \times 10^{-6}}{40 \times (2/10^4)} = 0.25 \text{ T}$$

Note that we have converted the area to m^2 in the above equation, because we are using the MKS version of the equation.

For most ferrites, an operating flux density of 0.25 T is acceptable, since the saturation flux density is typically around 0.3 T.

Based on the B and I linearity, we can also *linearly extrapolate* and thus conclude that the peak current in our application should under no condition be allowed to exceed $(0.3/0.25) \times 10 = 12$ A, because at 12 A, the field will be 0.3 T, and the core will then start to saturate.

But note that *the number of turns should not be increased any further* (at 12 A). Looking at the B_{PK} equation above, it seems at first sight that increasing the number of turns will reduce the B-field. However, *inductance increases as N^2* (from the A_L equation given previously); so, the numerator will increase much faster than the denominator. Therefore, in reality, the B-field will *increase*, rather than decrease if we increase the number of turns, and we know we can't afford to exceed 0.3 T.

In other words — we usually tend to instinctively rely on the current-limiting properties of an inductor. And in general, increasing the inductance will certainly help increase the inductance, and therefore help limit the current. However, *if we are already close to the energy-storage limits of the material of the core, we have to be very careful* — a few extra turns could take us "over the edge" (saturation), and then in fact, the inductance will start collapsing rather than increasing.

We should also not forget our basic premise of inductors in power conversion — for a given application, *a large inductance does usually end up requiring a large inductor!* So, increasing the number of turns, without increasing the size, may naturally turn out to be a recipe for disaster.

The "Field Ripple Ratio"

Since I and B are proportional to each other, and r happens to be a ratio, we realize that r must apply equally to the field components as it does to the current components. So, in that sense, r can be looked at as a "field ripple ratio" too. We can therefore extend the definition of r as follows

$$r = 2\frac{I_{AC}}{I_{DC}} = 2\frac{B_{AC}}{B_{DC}}$$

Therefore, r can also be used to relate the peak, AC, and DC values of both the current and field according to the equations

$$B_{DC} = \frac{2 \times B_{PK}}{r+2} \quad \text{or} \quad I_{DC} = \frac{2 \times I_{PK}}{r+2}$$

$$B_{AC} = \frac{r \times B_{PK}}{r+2} \quad \text{or} \quad I_{AC} = \frac{r \times I_{PK}}{r+2}$$

We can relate the peak to the swing too as follows:

$$B_{PK} = \frac{r+2}{2 \times r} \times \Delta B \quad \text{or} \quad I_{PK} = \frac{r+2}{2 \times r} \times \Delta I$$

The latter form will in fact be used later by us in a worked example that follows.

The Voltage-Dependent Equation in Terms of Voltseconds (MKS Units)

When discussing the current swing ΔI, we related it to the voltseconds. Now we can do the same for the B-field

$$\Delta B = \frac{L \times \Delta I}{N \times A} = \frac{Et}{N \times A} \text{ Tesla}$$

So as for current, the voltseconds in our application also determines the *swing* of the magnetic field — though not its DC level.

Table 2.4:   Magnetic Systems of Units and Their Conversions.

	CGS Units	MKS Units	Conversions
Magnetic flux	Line (or Maxwell)	Weber	1 Weber = 10^6 Lines
Flux density (B)	Gauss	Tesla (or Wb/m^2)	1 Tesla = 10^4 Gauss
Magnetomotive force	Gilbert	Ampere-turn	1 Gilbert = 0.796 Ampere-turn
Magnetizing force field (H)	Oersted	Ampere-turn/meter	1 Oersted = $1{,}000/4\pi$ = 79.577 Ampere/meter
Permeability	Gauss/Oersted	Weber/m-Ampere-turn	$\mu_{\text{MKS}} = \mu_{\text{CGS}} \times (4\pi \times 10^{-7})$

CGS Units

We may personally prefer to use the more broadly accepted MKS units, but we have to deal with the ground reality of the situation — that certain vendors (especially North American ones) still use "CGS" (centimeter-gram-seconds) units. Since we would certainly be evaluating and looking at their datasheets too, we will need to use the conversions in Table 2.4.

In particular, we should remember that the saturation flux density B_{SAT}, which is around 0.3 T (300 mT) for most ferrites, is 3,000 Gauss ("G") in CGS units. Also note that permeability of a material in MKS units needs to be divided by $4\pi \times 10^{-7}$ to get the permeability in CGS units. The reason for that is that permeability of air is set to 1 in CGS units, but in MKS units it is (numerically) equal to $4\pi \times 10^{-7}$.

The Voltage-Dependent Equation in Terms of Voltseconds (CGS Units)

It is also therefore helpful to know how to write the voltage-dependent equation (expressed in terms of *Et*) in CGS units instead.

So, converting A in m^2 to A in cm^2, we get from the previous equation

$$\Delta B = \frac{100 \times Et}{N \times A} \text{Gauss} \quad (A \text{ in cm}^2)$$

Core Loss

The core loss depends on various factors — the flux swing ΔB, the (switching) frequency f, and the temperature (though we usually ignore this latter dependency for most estimates). Note however, that **when vendors of magnetic materials express the dependency of core loss on a certain "B," what they are really talking about is $\Delta B/2$, that is, B_{AC}.** This happens to be the usual industry convention, but it is often quite confusing to power supply designers. In fact, there is more confusion caused by the fact that "B" may be expressed by the vendor, either in terms of Gauss or in Tesla. In fact, the dissipation also (due to the core loss) may be expressed either as mW or as W.

Table 2.5:   The Different Systems in Use for Describing Core Loss (and Their Conversions).

	Constant	Exponent of B	Exponent of f	B	f	V_e	Units
System A	$Cc = \dfrac{C \times 10^{4 \times P}}{10^3}$	$Cb = p$	$Cf = d$	Tesla	Hz	cm^3	W/cm^3
System B	$C = \dfrac{Cc \times 10^3}{10^{4 \times Cb}}$	$p = Cb$	$d = Cf$	Gauss	Hz	cm^3	mW/cm^3
System C	$Kp = \dfrac{C}{10^3}$	$n = p$	$m = d$	Gauss	Hz	cm^3	W/cm^3

First, let us look at the general form of core loss.

$$\text{Core loss} = (\text{core toss per unit volume}) \times \text{volume}$$

where "core loss per unit volume" is expressed generally as

$$\text{constant } 1 \times B^{\text{constant } 2} \times f^{\text{constant } 3}$$

In Table 2.5 we have indicated the three main systems of units in use for describing the core loss per unit volume, and also provided the rules for converting between them. Note that we are using "V_e" (effective volume) here — this usually can be considered as the actual physical volume of the core, or we can just look it up in the datasheet of the core.

In Table 2.6, we have provided values for the constants in the core loss equation in one of these systems of units, besides some other operating limits. The reader is however advised to confirm these values from the respective vendors.

Worked Example (6) — Characterizing an Off-the-Shelf Inductor in a Specific Application

Now we will present the *"general inductor design procedure"* we have been talking about. We will be considering a *wide-input* voltage range here. The procedure is to be carried out at the *"worst-case input voltage end"* **with respect to the peak current**. The basic purpose is to ensure that we are avoiding inductor saturation under normal operation. So for the **Buck**, we will work at V_{INMAX}, because that is the point at which the peak current is at its maximum. For a **Boost or a Buck-Boost**, we need to conduct this procedure at V_{INMIN}, not V_{INMAX}, since that is the worst-case input voltage end with regard to the peak current, for these topologies.

The procedure will be illustrated by means of a step-by-step worked example. Though it is carried out for a Buck, throughout the calculation, we will indicate precisely how the

Table 2.6:   Typical Core Loss Coefficients of Common Materials.

Material (Vendor)	Grade	c	p (B^p)	d (f^d)	μ	$\approx B_{SAT}$ (Gauss)	f_{MAX} (MHz)
Powdered iron (Micrometals)	8	4.3E−10	2.41	1.13	35	12,500	100
	18	6.4E−10	2.27	1.18	55	10,300	10
	26	7E−10	2.03	1.36	75	13,800	0.5
	52	9.1E−10	2.11	1.26	75	14,000	1
Ferrite (Magnetics Inc.)	F	1.8E−14	2.57	1.62	3,000	3,000	1.3
	K	2.2E−18	3.1	2	1,500	3,000	2
	P	2.9E−17	2.7	2.06	2,500	3,000	1.2
	R	1.1 E−16	2.63	1.98	2,300	3,000	1.5
Ferrite (Ferroxcube)	3C81	6.8E−14	2.5	1.6	2,700	3,600	0.2
	3F3	1.3E−16	2.5	2	2,000	3,700	0.5
	3F4	1.4E−14	2.7	1.5	900	3,500	2
Ferrite (TDK)	PC40	4.5E−14	2.5	1.55	2,300	3,900	1
	PC50	1.2E−17	3.1	1.9	1,400	3,800	2
Ferrite (Fair Rite)	77	1.7E−12	2.3	1.5	2,000	3,700	1

Note: (a)E−(b) is $(a) \times 10^{-(b)}$

procedure and equations may need to change, were this a Boost or a Buck-Boost. So for example, to the right of any equation presented below, we have indicated in brackets, which topology it is valid for.

A Buck converter has an input of 18–24 V, an output of 12 V, and a maximum load of 1 A. We desire a current ripple ratio of 0.3 (at maximum load). We assume $V_{SW} = 1.5$ V, $V_D = 0.5$ V, and f = 150,000 Hz. An off-the-shelf inductor is to be selected and characterized for this application.

As mentioned, all the steps involved in the "general inductor design procedure" below are being carried out at a certain "V_{IN}" — which is the *maximum input voltage* for a **Buck**, and *minimum input voltage* for a **Boost or a Buck-Boost**.

Estimating Requirements

For a Buck regulator, the duty cycle is (now including the switch and diode forward drops)

$$D = \frac{V_O + V_D}{V_{IN} - V_{SW} + V_D} \quad (Buck)$$

So,

$$D = \frac{12 + 0.5}{24 - 1.5 + 0.5} = 0.543$$

(For Boost, use $D = (V_O - V_{IN} + V_D)/(V_O - V_{SW} + V_D)$, for Buck-Boost, use $D = (V_O + V_D)/(V_{IN} + V_O - V_{SW} + V_D)$.)

The switch on-time is therefore

$$t_{ON} = \frac{D}{f} \Rightarrow \frac{0.543}{150,000} \Rightarrow 3.62 \ \mu s \quad (\textit{any topology})$$

$$t_{ON} = 3.62 \ \mu s$$

The voltage across the inductor when the switch is ON is

$$V_{ON} = V_{IN} - V_{SW} - V_O = 24 - 1.5 - 12 = 10.5 \ V \quad (\textit{Buck})$$

(For Boost and Buck-Boost, use $V_{ON} = V_{IN} - V_{SW}$.)

So, the voltsmicroseconds is

$$Et = V_{ON} \times t_{ON} = 10.5 \times 3.62 = 38.0 \ V \ \mu s \quad (\textit{any topology})$$

Using the "L × I" equation

$$(L \times I_L) = \frac{Et}{r} \quad (\textit{any topology})$$

we get

$$(L \times I_L) = \frac{38}{0.3} = 127 \ \mu H \ A$$

But the average inductor current is

$$I_L = I_O \quad (\textit{Buck})$$

(For a Boost and Buck-Boost, use $I_L = (I_O)/(1 - D)$.)

Therefore,

$$L = \frac{(L \times I_L)}{I_L} \equiv \frac{(L \times I_O)}{I_O} = \frac{127}{1} = 127 \ \mu H \quad (\textit{any topology})$$

The peak current will be 15% higher than I_L for $r = 0.3$. This follows from

$$I_{PK} = \left(1 + \frac{r}{2}\right) \times I_L = 1.15 \times 1 = 1.15 \ A \quad (\textit{any topology})$$

Table 2.7:   Specifications of a Selected Inductor (the PO150).

I_{DC} (A)	L_{DC} (μH)	Et (V μs)	DCR (mΩ)	Et_{100} (V μs)
0.99	137	59.4	387	10.12

- The inductor is such that 380 mW dissipation corresponds to 50 °C rise in temperature.
- The core loss equation for the core is $6.11 \times 10^{-18} \times B^{2.7} \times f^{2.04}$ mW, where f is in Hz and B is in Gauss.
- Et_{100} is the voltmicroseconds at which "B" is 100 G.
- "B" is B_{AC}, i.e., $\Delta B/2$.
- Rated frequency of operation is 250 kHz.

We now pick a promising off-the-shelf inductor — the PO150 from Pulse Electronics. Its inductance is 137 μH, which is close to our requirement of 127 μH, and it is rated for a continuous DC of 0.99 A, which is very close to our requirement of 1 A. Its datasheet is reproduced in Table 2.7. Note that the other conditions mentioned by the vendor *do not match our application* (but that is not unexpected — what are the chances of an off-the-shelf inductor that precisely matches a given application?). Nevertheless, we can perform a full analysis, and thus either validate or invalidate our choice of component.

Current Ripple Ratio

We use the "$L \times I$" rule

$$(L \times I_{L}) = \frac{Et}{r} \quad (any \ topology)$$

So,

$$r = \frac{Et}{L \times I_{L}} \quad (any \ topology)$$

The inductor has been designed by its vendor, for an r of

$$r = \frac{59.4}{137 \times 0.99} = 0.438$$

In our application we will get

$$r = \frac{38}{137 \times 1} = 0.277$$

This is very close to (and less than) our target of $r = 0.3$, and is therefore acceptable.

Peak Current

The inductor has been designed for a peak current of

$$I_{PK} = \left(1 + \frac{r}{2}\right) \times I_L = \left(1 + \frac{0.438}{2}\right) \times 0.99 = 1.21 \ \text{A} \quad (\textit{any topology})$$

In our application we will get

$$I_{PK} = \left(1 + \frac{r}{2}\right) \times I_L = \left(1 + \frac{0.277}{2}\right) \times 1 = 1.14 \ \text{A} \quad (\textit{any topology})$$

The peak current in our application is considered "safe," being *less* than what the inductor was originally designed for. Therefore, we can safely assume that the peak B-field of our application must also be within the design limits of the inductor. However, it is instructive to confirm that directly, as we do next.

Note that the frequency has not even entered the picture directly so far, since voltseconds is all that really matters to an inductor. **Different applications, with the same DC level of current, and the same voltseconds, are essentially the same application from the viewpoint of the inductor**. It just "doesn't care," for example, what topology it is, or what is the duty cycle. It doesn't even care about the frequency directly (though the exception to this is the core loss term, because that depends not only on the voltseconds, i.e., the current swing, but on the frequency too). However, we will also see that the core loss term is much smaller anyway, compared to the copper loss. So, for all practical purposes, **if the rated voltseconds of a given inductor (current swing), and its DC current rating correspond to the applied voltseconds and DC current of our application, we are almost certainly going to be fine right off-the-bat. However, even if the rated voltseconds and DC level are quite different, as long as the applied peak flux density is close to or less than the rated value, we are OK from the saturation point of view. That's a good start, and we can then proceed to do a full validation analysis** — of the temperature rise and so on, under our specific application conditions.

Flux Density

The vendor provides the following information (see Table 2.7):

$$Et_{100} = 10.12 \ \text{V} \ \mu s$$

This means that the voltµseconds that produces a B_{AC} of 100 G is 10.12. Since $B_{AC} = \Delta B/2$, the corresponding ΔB is 200 G (for every 10.12 V µs). Note that G stands for Gauss.

We had previously presented the following relationship between ΔB and Et:

$$\Delta B = \frac{100 \times Et}{N \times A} \quad \text{Gauss} \ (\textit{any topology})$$

Since ΔB and Et are proportional to each other (for a given inductor), we can conclude that the inductor has been designed for a flux density swing of

$$\Delta B = \frac{Et}{Et_{100}} \times 200 = \frac{59.4}{10.12} \times 200 = 1174 \text{ G} \quad (any \ topology)$$

and a peak flux density of

$$B_{\text{PK}} = \frac{r+2}{2 \times r} \times \Delta B = \frac{0.438 + 2}{2 \times 0.438} \times 1174 = 3267 \text{ G} \quad (any \ topology)$$

In our application this will give us a swing of

$$\Delta B = \frac{Et}{Et_{100}} \times 200 = \frac{38}{10.12} \times 200 = 751 \text{ G} \quad (any \ topology)$$

and a peak of

$$B_{\text{PK}} = \frac{r+2}{2 \times r} \times \Delta B = \frac{0.277 + 2}{2 \times 0.277} \times 751 = 3087 \text{ G} \quad (any \ topology)$$

We see that the peak field in our application is within the design limits of the inductor, as expected, so we need not worry about core saturation. This is a basic qualification the inductor must pass before we can proceed with the rest of the analysis.

Note that the proportionality constant connecting B and I (for this inductor) is

$$\frac{L}{NA} = \frac{B_{\text{PK}}}{I_{\text{PK}}} = \frac{3,087}{1.14} = 2,708 \text{ G/A} \quad (any \ topology)$$

> *Note: If we break open the inductor and measure the number of turns, and also estimate/ measure the cross-sectional area of the central limb of its core, we can verify the number 2,708 above.*

Copper Loss

From the equations contained in Figure 2.14, we can calculate the RMS of the inductor current waveform. The inductor was designed for an RMS *squared* of

$$I_{\text{RMS}}^2 = \frac{\Delta I^2}{12} + I_{\text{DC}}^2 = I_{\text{DC}}^2 \left(1 + \frac{r^2}{12} \right) = 0.99^2 \left(1 + \frac{0.438^2}{12} \right) = 0.996 \text{ A}^2 \quad (any \ topology)$$

and a copper loss of

$$P_{\text{CU}} = I_{\text{RMS}}^2 \times \text{DCR} = 0.996 \times 387 = 385 \text{ mW} \quad (any \ topology)$$

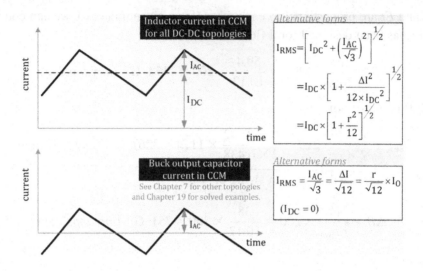

Figure 2.14: RMS value of the inductor current waveform.

Whereas in our application we will get

$$I_{RMS}^2 = I_L^2 \left(1 + \frac{r^2}{12}\right) = 1^2 \left(1 + \frac{0.277^2}{12}\right) = 1.006 \ \text{A}^2 \quad (any \ topology)$$

and a copper loss of

$$P_{CU} = I_{RMS}^2 \times \text{DCR} = 1.006 \times 387 = 389 \ \text{mW} \quad (any \ topology)$$

Core Loss

Note that the vendor has already factored in the *volume of the core* and thus provided the following overall equation for the core loss of the inductor:

$$P_{CORE} = 6.11 \times 10^{-18} \times B^{2.7} \times f^{2.04} \ \text{mW} \quad (any \ topology)$$

where f is in Hz and B is in Gauss. Note that "B" is $\Delta B/2$ here as per convention. So, the core loss that the inductor was originally designed for is

$$P_{CORE} = 6.11 \times 10^{-18} \times \left(\frac{1,174}{2}\right)^{2.7} \times (250 \times 10^3)^{2.04} = 18.8 \ \text{mW}$$

Whereas in our application

$$P_{CORE} = 6.11 \times 10^{-18} \times \left(\frac{751}{2}\right)^{2.7} \times (150 \times 10^3)^{2.04} = 2 \ \text{mW}$$

DC–DC Converter Design and Magnetics

In general, we will find that in most ferrite-based off-the-shelf inductors, the designed core loss is only 5–10% of the total inductor loss (copper-plus-core loss). However, if the inductor uses a "powdered iron" core, this number may rise to about 20–30%.

> *Note: Powdered iron cores tend to saturate more "softly" than ferrites, and that usually enhances their ability to withstand severe abnormal currents without leading to immediate switch destruction. On the other hand, powdered iron cores may have "lifetime" issues caused by slow degradation of the organic binder that holds their iron particles together. The vendor must be consulted about this possibility, and the steps necessary to avoid a premature end to our converter!*

Temperature Rise

The vendor has stated that the inductor is such that 380 mW dissipation corresponds to 50 °C rise in temperature. In effect this tells us that the thermal resistance of the core "Rth" is

$$\text{Rth} = \frac{\Delta T}{\text{W}} = \frac{50}{0.38} = 131.6 \, ^\circ C/W \quad (any \ topology)$$

The inductor was originally designed for a total loss of

$$P = P_{\text{CORE}} + P_{\text{CU}} = 385 + 18.8 = 403.8 \, \text{mW} \quad (any \ topology)$$

This would have given a temperature rise of

$$\Delta T = \text{Rth} \times P = 131.6 \times 0.404 = 53 \, ^\circ C \quad (any \ topology)$$

In our application

$$P = P_{\text{CORE}} + P_{\text{CU}} = 389 + 2 = 391 \, \text{mW}$$

This will give a temperature rise of

$$\Delta T = \text{Rth} \times P = 131.6 \times 0.391 = 51 \, ^\circ C$$

Provided we accept this temperature rise in our application (that will depend on our maximum operating ambient temperature), we can validate the chosen inductor. We have already confirmed it does not saturate in our application, and further, the current ripple ratio it provides is acceptable too.

This completes the general inductor design procedure.

Calculating "Other" Worst-case Stresses and their Selection Criteria

*Having validated our choice of inductor, we can look a little more closely at the important issue of how the wide-input range impacts the **other** key parameters and stresses in our proposed converter. This also helps in correctly selecting the other power components.*

Worst-case Core Loss

In the above so-called "general inductor design procedure," we have actually been working at V_{INMAX} for a Buck, and at V_{INMIN} for a Boost or Buck-Boost. The reason was that the *inductor* sees the highest peak current at this voltage end, so we have to "ensure" the magnetics design at these particular points (extremes). But this point *may not be the worst-case for the other stresses in the power supply*, and we need to start understanding that clearly now.

Let us first focus on the inductor. The point at which we are doing the inductor design usually gives us the worst-case temperature rise too. *But that is because the I_{DC} component of the inductor current is usually the dominant term.* If for any reason, we are interested in knowing what the maximum *core loss* component of the total loss is, we should realize, looking back at Figure 2.4, that though the DC level may be going up as input voltage falls, the AC component (on which the core loss term depends) may actually be decreasing (or having an odd-shaped profile, as for the Boost).

From Figure 2.4, we see that I_{AC} increases at high-input voltages for both the Buck and the Buck-Boost, but not necessarily for the Boost. For a **Buck**, the general inductor design calculation above was carried out at V_{INMAX} and that just happens to be the point at which the core loss is a maximum too. Therefore, calculating the core loss at V_{INMAX} as we did in the previous example does coincidentally also give us the worst-case core loss.

However, if we were doing the calculation for a **Buck-Boost**, our general inductor design calculation starts at V_{INMIN}. But the core loss is a maximum at V_{INMAX}. Similarly, for a **Boost**, we would also start the general inductor design calculation at V_{INMIN}. But the worst-case core loss for this topology actually occurs at $D = 0.5$ (see the I_{AC} curve for Boost in Figure 2.4). Note that from the duty cycle equation of Boost, $D = 0.5$ corresponds to an input voltage equal to half the output.

> *Note: If for the Boost, the input range of the given application does* not *include the* $D = 0.5$ *point, we need to identify which voltage end of the range provides a duty cycle closest to* $D = 0.5$*. And we need to then do the worst-case core loss calculation at that end. That would be the input voltage extreme with a* D *closest to* D = 0.5*.*

Generally, the core loss term, being such a small component of the total loss, is of no great concern to us, so we won't even bother to do a numerical calculation here. But the general procedure to handle such cases will become apparent as we study the other worst-case loss terms of the converter below.

First let us start *annotating (or subscripting)* some of the terms derived so far, just for gaining clarity in the discussion to follow.

- **For a Buck**: The general inductor design procedure was carried out at V_{INMAX}, that is, D_{MIN}. So for example, the r we have set to 0.3–0.4 (and possibly re-calculated with the selected inductor) is now referred to as "r_{DMIN}." Similarly, the voltseconds, Et, we have calculated so far is actually "Et_{DMIN}."
- **For a Boost and Buck-Boost**: If a similar general inductor design procedure were carried out for these topologies, it would be done at V_{INMIN} that is, D_{MAX}. So for example, the r we would have set to 0.3–0.4 (and possibly re-calculated with the selected inductor) would actually be called "r_{DMIN}." Similarly, the voltseconds, Et, we would have calculated so far is "Et_{DMIN}."

We need to keep these distinctions in mind, otherwise the following discussion can become confusing no end!

Worst-case Diode Dissipation

The general equation for the average diode current is

$$I_{\text{D}} = I_{\text{L}} \times (1 - D) \quad (any \; topology)$$

or equivalently

$$I_{\text{D}} = I_{\text{O}} \times (1 - D) \quad (Buck)$$
$$I_{\text{D}} = I_{\text{O}} \quad (Boost \; and \; Buck\text{-}Boost)$$

This leads to a diode dissipation of

$$P_{\text{D}} = V_{\text{D}} \times I_{\text{D}} = V_{\text{D}} \times I_{\text{O}} \times (1 - D) \quad (Buck)$$
$$P_{\text{D}} = V_{\text{D}} \times I_{\text{D}} = V_{\text{D}} \times I_{\text{O}} \quad (Boost \; and \; Buck\text{-}Boost)$$

For the Buck, as the input voltage is raised, the duty cycle falls, and because the average inductor current I_{L} remains fixed at I_{O}, the average diode current *increases*. That means we get the worst-case diode current (and dissipation) at V_{INMAX} for a Buck. *Therefore, we can just use the numbers already derived while carrying out the general inductor design procedure* (at V_{INMAX}).

For the Boost and the Buck-Boost, as the input is raised, D decreases, but the average inductor current also falls, thereby keeping I_{D} fixed at I_{O}. (We should remember that the Boost and the Buck-Boost are unique in the sense that *all* the output current must pass through the diode when it conducts, so I_D must necessarily equal I_O at all times.) That means the diode dissipation is independent of input voltage for these topologies. So we can, if we want, just *use the numbers already derived while carrying out the general inductor design procedure* (at V_{INMIN}).

Finally, for the ongoing Buck converter design example, the calculation is as follows:

$$P_D = V_D \times I_O \times (1 - D_{MIN}) = 0.5 \times 1 \times (1 - 0.543) = 0.23 \text{ W} \quad (Buck)$$

Note that the General Diode Selection Procedure is as Follows

The rule of thumb is to pick a diode with a current rating at least equal to, but preferably at least *twice* the worst-case average diode current given below (for low losses, since the diode forward drop decreases substantially if its current rating is increased):

- For a **Buck** — maximum diode current is $I_O \times (1 - D_{MIN})$.
- For a **Boost** — maximum diode current is I_O.
- For a **Buck-Boost** — maximum diode current is I_O.

Its voltage rating is usually picked to be at least 20% higher ("~80% derating" — i.e., safety margin) than the worst-case diode voltage given below:

- For a **Buck** — maximum diode voltage is V_{INMAX}.
- For a **Boost** — maximum diode voltage is V_O.
- For a **Buck-Boost** — maximum diode voltage is $V_O + V_{INMAX}$.

Worst-case Switch Dissipation

For all topologies the average input current (and therefore switch current) must increase as the input voltage decreases, so as to continue to satisfy the basic power requirement expressed by $P_{IN} = I_{IN} \times V_{IN} = P_O/\eta$ (where η is the efficiency, assumed fixed). Therefore, the switch RMS current is a maximum at V_{INMIN} (i.e., D_{MAX}) *for all topologies.*

For the Boost and Buck-Boost, the general inductor design procedure is at D_{MAX} in any case. So, *we can directly use the numbers derived from that* to find the switch RMS current using the equation below:

$$I_{RMS\_SW} = I_{L\_DMAX} \times \sqrt{D_{MAX} \times \left(1 + \frac{r_{DMAX}{}^2}{12}\right)} \quad (any \ topology)$$

where I_{L_DMAX} and r_{DMAX} are, respectively, the average inductor current and current ripple ratio at D_{MAX} (i.e., at V_{INMIN}). D_{MAX} can be calculated using

$$D_{\text{MAX}} = \frac{V_O - V_{\text{INMIN}} + V_D}{V_O - V_{\text{SW}} + V_D} \quad (Boost)$$

$$D_{\text{MAX}} = \frac{V_O + V_D}{V_{\text{INMIN}} + V_O - V_{\text{SW}} + V_D} \quad (Buck\text{-}Boost)$$

and we should remember that

$$I_{\text{L\_DMAX}} = \frac{I_O}{1 - D_{\text{MAX}}} \quad (Boost \ and \ Buck\text{-}Boost)$$

For the Buck, the general inductor design procedure is at D_{MIN}. So, *we cannot directly use the numbers derived* from that to find the switch RMS current (by the previously given equation). We need to calculate r_{DMAX}, but we only know t_{DMIN} so far. Let us proceed with the required steps.

$$r_{\text{DMAX}} = \frac{Et_{\text{DMAX}}}{L \times I_{\text{L}}} \quad (any \ topology)$$

In other words, if we know the voltseconds at V_{INMIN}, we will know the corresponding current ripple ratio r_{DMAX} for the chosen inductor. But first we have to calculate D_{MAX}

$$D_{\text{MAX}} = \frac{V_O + V_D}{V_{\text{INMIN}} - V_{\text{SW}} + V_D} = \frac{12 + 0.5}{18 - 1.5 + 0.5} = 0.735 \quad (Buck)$$

The switch on-time is therefore

$$t_{\text{ON\_DMAX}} = \frac{D_{\text{MAX}}}{f} \Rightarrow \frac{0.735 \times 10^6}{150,000} = 4.9 \ \mu s \quad (any \ topology)$$

The voltage across the inductor when the switch is ON is

$$V_{\text{ON\_DMAX}} = V_{\text{INMIN}} - V_{\text{SW}} - V_O = 18 - 1.5 - 12 = 4.5 \text{ V} \quad (Buck)$$

So, the voltmicroseconds is

$$Et_{\text{DMAX}} = V_{\text{ON\_DMAX}} \times t_{\text{ON\_DMAX}} = 4.5 \times 4.9 = 22 \text{ V} \ \mu s \quad (any \ topology)$$

Therefore,

$$r_{\text{DMAX}} = \frac{Et_{\text{DMAX}}}{L \times I_O} = \frac{22}{137 \times 1} = 0.16 \quad (Buck)$$

Finally, we are in a position to calculate the switch dissipation

$$I_{\text{RMS\_SW}} = I_O \times \sqrt{D_{\text{MAX}} \times \left(1 + \frac{r_{\text{DMAX}}^2}{12}\right)} = 1 \times \sqrt{0.735 \times \left(1 + \frac{0.16^2}{12}\right)} = 0.86 \text{ A} \quad (Buck)$$

If, for example, the Drain-to-Source resistance is 0.5 Ω, the dissipation in the MOSFET is

$$P_{SW} = I_{RMS\_SW}{}^2 \times R_{DS} = 0.86^2 \times 0.5 = 0.37 \text{ W} \quad (any \; topology)$$

Note that the General Switch Selection Procedure is as Follows

The rule of thumb is to pick a switch with a current rating at least equal to, but preferably at least *twice* the worst-case RMS switch current calculated above (for low losses, since the switch forward drop will decrease substantially if its current rating is increased).

Its voltage rating is usually picked to be at least 20% higher ("~80% derating" — i.e., safety margin) than the worst-case switch voltage given below:

- For a **Buck** — maximum switch voltage is V_{INMAX}.
- For a **Boost** — maximum switch voltage is V_O.
- For a **Buck-Boost** — maximum switch voltage is $V_O + V_{INMAX}$.

Worst-case Output Capacitor Dissipation

Coincidentally, *the worst-case output capacitor RMS current for all three topologies occurs at the same point at which the general inductor design procedure for each of them is carried out.* In other words, this point is V_{INMAX} for the Buck, and V_{INMIN} for the Boost and Buck-Boost. So, we should have no trouble, *directly using the numbers derived from the general inductor design procedure*, to find the worst-case RMS current of the output capacitor, using the equations below:

For the Buck, we get

$$I_{RMS\_OUT} = I_O \times \frac{r_{DMIN}}{\sqrt{12}} = 1 \times \frac{0.277}{\sqrt{12}} = 0.08 \text{ A} \quad (Buck)$$

So for example, if the ESR of the output capacitor is 10 Ω, we get the dissipation

$$P_{SW} = I_{RMS\_OUT}{}^2 \times \text{ESR} = 0.08^2 \times 10 = 0.064 \text{ W} \quad (any \; topology)$$

For the Boost and the Buck-Boost, we need to use

$$I_{RMS\_OUT} = I_O \times \sqrt{\frac{D_{MAX} + (r_{DMAX}{}^2/12)}{1 - D_{MAX}}} \quad (Boost \; and \; Buck\text{-}Boost)$$

Note that the General Output Capacitor Selection Procedure is as Follows

The rule of thumb is to pick an output capacitor with a ripple current rating equal to or greater than the worst-case RMS capacitor current calculated above. Its voltage rating is usually picked to be at least 20–50% higher than what it will see in the application (i.e., V_O for all topologies). The output voltage ripple of the converter is also usually a concern. The total peak-to-peak output voltage ripple produced by the output capacitor is equal to its ESR multiplied by the worst-case peak-to-peak output current given below (ignoring the ESL of the capacitor):

- For a **Buck** — peak-to-peak capacitor current is $I_O \times r_{DMIN}$. This is the same point at which the general inductor design procedure would have been carried out, and so r_{DMIN} is already known.
- For a **Boost** — peak-to-peak capacitor current is $I_O \times (1 + r_{DMAX}/2)/(1 - D_{MAX})$. This is the same point at which the general inductor design procedure would have been carried out for this topology, so r_{DMAX} and D_{MAX} are already known.
- For a **Buck-Boost** — peak-to-peak capacitor current is $I_O \times (1 + r_{DMAX}/2)/(1 - D_{MAX})$. This is the same point at which the general inductor design procedure would have been carried out for this topology, so r_{DMAX} and D_{MAX} are already known.

Worst-case Input Capacitor Dissipation

For the Buck-Boost, things are much simpler, since the worst-case input capacitor RMS current occurs at D_{MAX}, which is also the point at which we carry out the general inductor design procedure. So, all the numbers available from that procedure can be used directly in the equation below:

$$I_{RMS\_IN} = I_{L\_DMAX} \times \sqrt{D_{MAX} \times \left(1 - D_{MAX} + \frac{r_{DMAX}^2}{12}\right)} \quad (Buck\text{-}Boost)$$

For the Buck and the Boost, the worst-case input RMS capacitor current occurs at $D = 0.5$. Therefore, we have to calculate "r_{50}," that is, the *current ripple ratio at $D = 50\%$* (or whatever voltage within the specified input range of our application range is closest to this point).

Let us do the numerical calculation for the Buck, and the procedure will become clearer.

The input voltage at which $D = 50\%$ occurs for the Buck is

$$V_{IN\_50} = 2 \times V_O + V_{SW} + V_D = 2 \times 12 + 1.5 + 0.5 = 26 \text{ V} \quad (Buck)$$

and for the Boost

$$V_{IN\_50} = \frac{V_O + V_{SW} + V_D}{2} \approx \frac{V_O}{2} \quad (Boost)$$

We see that our input range does not include this point. But the closest to it is V_{INMAX}. However, coincidentally, this is already the point at which the general inductor design procedure was carried out. So, we can use all the numbers derived from that procedure to calculate the input capacitor RMS current, using the equation below:

$$I_{RMS\_IN} = I_O \times \sqrt{D \times \left(1 - D + \frac{r^2}{12}\right)} = 1 \times \sqrt{0.543 \times \left(1 - 0.543 + \frac{0.277^2}{12}\right)} \quad (Buck)$$

and for the Boost

$$I_{RMS\_IN} = \frac{I_O}{1 - D} \times \frac{r}{\sqrt{12}} \quad (Boost)$$

So finally

$$I_{RMS\_IN} = 0.502 \text{ A}$$

> *Note: If for our worked Buck example, the input range was not 18–24 V but say 30–45 V, then the general inductor design procedure would clearly be carried out at 45 V. However, the input capacitor current would be a maximum at 30 V. So, we can use the above equation for the RMS current, but we would now need to use r_{DMIN} and D_{MAX}. Therefore, knowing only r_{DMAX} so far we would need to calculate r_{DMIN} by the same procedure presented earlier — that is, by recalculating the voltseconds, and so on.*

A full solved example is available in *Chapter 19*. See also *Chapters 6* and *7* for stress derivations and sample calculations.

Note that the General Input Capacitor Selection Procedure is as Follows

The rule of thumb is to pick an input capacitor with a ripple current rating equal to or greater than the worst-case RMS capacitor current calculated above. Its voltage rating is usually picked to be at least 20–50% higher than what it will see in the application (i.e., V_{IN_MAX} for all topologies). The input voltage ripple of the converter is also usually a concern because a small part of it does get transmitted to the output. There can also be EMI considerations involved. In addition, every control IC has a certain (usually unspecified) amount of input noise and ripple rejection, and it may misbehave if the ripple is too much. Typically, the input ripple needs to be kept down to less than ±5% to ±10% of the input voltage. The total peak-to-peak input voltage ripple produced by the input capacitor is equal to its ESR multiplied by the worst-case peak-to-peak input current given below (ignoring the ESL of the capacitor):

- For a *Buck* — peak-to-peak capacitor current is $I_O \times (1 + r_{DMIN}/2)$. This is the same point at which the general inductor design procedure would have been carried out, and so r_{DMIN} is already known.

- For a *Buck-Boost* — peak-to-peak capacitor current is $I_O \times (1 + r_{DMAX}/2)/(1 - D_{MAX})$. This is the same point at which the general inductor design procedure would have been carried out for this topology, so r_{DMAX} and D_{MAX} are already known.
- For a *Boost* — peak-to-peak capacitor current at the worst-case point for this parameter (i.e., $D = 0.5$) is equal to $2 \times I_O \times r_{50}$ where

$$r_{50} = \frac{V_{IN\_50}}{4 \times f \times L \times I_O} \quad \text{and} \quad V_{IN\_50} = \frac{V_O + V_{SW} + V_D}{2} \approx \frac{V_O}{2}$$

Note that if the input range does not include the $D = 0.5$ point, we need to look for the input voltage end closest to $D = 0.5$. Then we can use the general equation for the peak-to-peak input capacitor current

$$I_{PK\_PK} = \frac{I_O \times r}{1 - D}$$

where r and D correspond to this particular worst-case input voltage end. To find r we can use

$$r = \frac{V_O - V_{SW} + V_D}{I_O \times L \times f} \times D \times (1 - D)^2$$

where L is in H, and f is in Hz.

That completes the converter and magnetics design procedure. Next we will move on to off-line converters.

For further clarification on the techniques introduced in this admittedly tricky chapter, please refer to *Chapters 6, 7* and *19* later.

CHAPTER 3

Off-Line Converter Design and Magnetics

Off-line converters are *derivatives* of standard DC—DC converter topologies. For example, the flyback topology, popular for low-power applications (typically <100 W), is really a Buck-Boost, with its usual single-winding inductor replaced by an inductor with multiple windings. Similarly, the Forward converter, popular for medium-to-high powers, is a Buck-derived topology, with the usual inductor ("choke") supplemented by a transformer. The flyback inductor actually behaves both as an inductor and as a transformer. It stores magnetic energy as any inductor would, but it also provides "mains isolation" (mandated for safety reasons), just like any transformer would. In the Forward converter, the energy-storage function is fulfilled by the choke, whereas its transformer provides the necessary mains isolation.

Because of the similarities between DC—DC converters and off-line converters, most of the spadework for this chapter is in fact contained in *Chapter 2*. The basic magnetic definitions have also been presented therein. Therefore, the reader should read that chapter before attempting this one. More information is available in *Chapter 5* too.

Note that in both the flyback and the Forward converters, the transformer, besides providing the necessary mains isolation, also provides another very important function — that of a *fixed-ratio down-conversion step*, determined by the "turns ratio" of the transformer. The turns ratio is the number of turns of the *input* ("Primary") winding, divided by the number of turns of the *output* ("Secondary") winding. The question arises — why do we even feel the need for a transformer-based step-down-conversion stage, when in principle, a switching converter should by itself have been able to up-convert or down-convert at will? The reason will become obvious if we carry out a sample calculation — we will then find that without any additional "help," the converter would require *impractically low values of duty cycle* — to down-convert from such a high-input voltage to such a low-output voltage. Note that the worst-case AC mains input can be as high as 270 V in certain countries. So, when this AC voltage is rectified by a conventional bridge-rectifier stage, it becomes a DC rail of almost $\sqrt{2} \times 270 = 382$ V, which is fed to the input of the switching converter stage that follows. But the corresponding output voltage can be very low (5 V, 3.3 V, 1.8 V, and so on), so the required DC transfer ratio (*conversion ratio*) is extremely hard to meet, given the minimum on-time limitations of any typical converter, especially when switching at high frequencies. Therefore, in both the flyback and Forward converters, we can intuitively think of the transformer as performing a rather coarse fixed-ratio step-down of the input to a more amenable (lower) value, from which point onward the converter does the rest (including the regulation function).

Flyback Converter Magnetics

Polarity of Windings in a Transformer

In Figure 3.1, the *turns ratio* is $n = N_P/N_S$, where N_P is the number of turns of the Primary winding, and N_S is the number of turns of the Secondary winding.

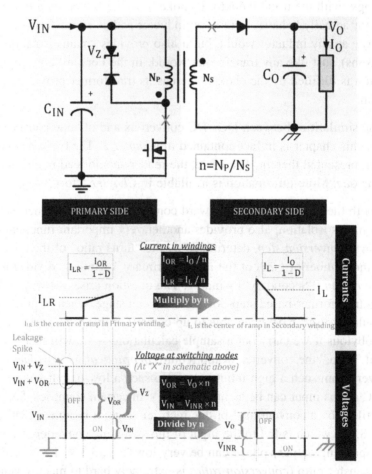

Figure 3.1: Voltage and currents in a flyback.

We have also placed a *dot* on one end of *each* of the windings. All dotted ends of a transformer are considered to be mutually "equivalent." All non-dotted ends are also obviously *mutually* equivalent. This means that when the voltage on a given dotted end goes "high" (to whatever value), so does the voltage on the dotted ends of all other windings. That happens because all windings share the *same magnetic core*, despite the fact that they are not physically (galvanically) connected to each other. Similarly, all the dotted ends also go "low" at the same time. Clearly, the dots are only an indication of *relative* polarity. Therefore, in any given schematic, we can always swap all the dotted and non-dotted ends of the transformer simultaneously, without changing the schematic in any way.

In a flyback, the relative polarity of the windings is deliberately arranged such that when the Primary winding conducts, the Secondary winding is *not* allowed to do so. So, when the switch conducts, the dotted end at the Drain of the MOSFET in Figure 3.1 goes low. And therefore, the anode of the output diode also goes low, thereby reverse-biasing the diode. We should recall that the basic purpose of a Buck-Boost (which this in fact also is) is to allow incoming energy from the source during the switch on-time to build up in the inductor (*only*), and then later, during the off-time, to "collect" all this energy (and no more) at the output. Note that this is the unique property that distinguishes the Buck-Boost (and the flyback) from the Buck and the Boost. For example, in a Buck, energy from the input source gets delivered to the inductor *and* the output (during the on-time). Whereas, in a Boost, stored energy from the inductor *and* the input source gets delivered to the output (during the off-time). Only in a Buck-Boost do we have *complete separation* between the energy-storage and the collection process, during the on-time and the off-time. So, now we start to understand why the flyback is considered to be just a Buck-Boost derivative. More on this in *Chapter 5*.

We know that every DC−DC topology has a so-called "switching node." This node represents the point of *diversion* of the inductor current — from its *main* path (i.e., by which the inductor *receives* energy from the input) to its *freewheeling* path (i.e., by which the inductor *provides* stored energy to the output). So, clearly, the switching node is necessarily the node *common* to the switch, the inductor, and the diode. We thus find that the voltage at this node is always "swinging" — because that is what is required to get the diode to alternately forward and reverse-bias, as the switch toggles. But looking at Figure 3.1, we see that with a transformer replacing the traditional DC−DC inductor, there are now, in effect, *two* "switching nodes" — one on each side of the transformer, as indicated by the "X" markings in Figure 3.1 — one "X" is at the Drain of the MOSFET, and the other "X" is at the anode of the output diode. These two nodes are clearly "equivalent" because of the dots, as explained above. And since at both these nodes, the voltage is swinging, both are considered to be "switching nodes" (of the transformer-based topology). Note that if we had, say, three windings (e.g., an additional output winding), we would have had three switching nodes.

Transformer Action in a Flyback and Its Duty Cycle

Classic "transformer action" implies that the voltages *across* the windings of the transformer, and the currents *through* each of them, *scale* according to the turns ratio, as described in Figure 3.1. But it is perhaps not immediately apparent why the flyback inductor exhibits transformer action since the windings do not conduct at the same time.

When the switch turns ON, a voltage V_{IN} (the rectified AC input) gets impressed across the Primary winding of the transformer. And at the same time, a voltage equal to $V_{INR} = V_{IN}/n$ ("R" stands for reflected) gets impressed across the Secondary winding (in a direction that causes the output diode to get reverse-biased). Therefore, there is no current in the Secondary winding when the Primary winding is conducting.

Let us calculate what V_{INR} is. The voltage translation across the isolation boundary follows from the induced voltage equation applied to each winding:

$$V_P = -N_P \frac{d\phi}{dt} \quad \text{and} \quad V_S = -N_S \frac{d\phi}{dt}$$

Note that both windings enclose the same magnetic core, so the flux ϕ is the same for both, and so is the rate of change of flux $d\phi/dt$ for each winding. Therefore,

$$V_S = -N_S \times \left(\frac{V_P}{-N_P} \right)$$

or

$$V_S = N_S \times \left(\frac{V_{IN}}{N_P} \right) = \frac{V_{IN}}{n} \equiv V_{INR}$$

Also,

$$\frac{V_P}{N_P} = \frac{V_S}{N_S}$$

$$\frac{V_P}{V_S} = n$$

This above equation represents classic "transformer action" with respect to the voltages involved. But we also learn from the preceding equation that the *Volts/turn* for *any winding* (at *any given instant*) is the same for all the windings present on a given magnetic core — and this is what eventually leads to the observed *voltage scaling*.

Note also that voltage scaling in any transformer occurs *irrespective of whether a given winding is passing current or not*. That is because, whether a given winding is contributing to the net flux ϕ present in the core or not, each winding encloses this entire flux, and so the basic equation $V = -N \times d\phi/dt$ applies to all windings, and so does voltage scaling.

We know that energy is built up in the transformer during the on-time. When the switch turns OFF, this stored energy (and its associated current) needs to flyback/freewheel. We also know that the voltages will automatically try to adjust themselves in any possible way, so as to make that happen. So, we can safely assume the diode will somehow conduct during the switch off-time. Now, assuming we have reached a "steady state," the voltage on the output capacitor has stabilized at some fixed value V_O. Therefore, the voltage at the Secondary-side switching node gets clamped at V_O (ignoring the diode drop). Further, since one end of the Secondary winding is tied to ground, the voltage *across* this winding is now equal to V_O. By transformer action, this reflects a voltage *across* the Primary winding, equal to $V_{OR} = V_O \times n$. But the switch is OFF during this time. Therefore, under normal circumstances, the voltage at the Primary-side switching node would have settled at V_{IN}. However, now this reflected output voltage V_{OR}, coming through the transformer, adds to that. Therefore, the voltage at the Primary-side switching node eventually goes up to $V_{IN} + V_{OR}$ (for now, we are ignoring the leakage spike encircled in Figure 3.1).

> *Note: During the on-time, the Primary side is the one determining the voltages across* all *the windings. And during the off-time, it is the Secondary winding that gets to "call the shots"!*

We can calculate duty cycle from the most basic equation (from voltseconds law):

$$D = \frac{V_{OFF}}{V_{OFF} + V_{IN}}$$

We have the option of performing this calculation, either on the Primary winding, or on the Secondary winding. Either way, we get the same result, as shown in Table 3.1.

We should be always very clear that transformer action applies only to the voltages *across* windings. And "voltage *across*" is not necessarily "voltage *at*"! To describe the voltage *at* a given point, we have to consider what the reference level (i.e., "ground" by definition) is,

Table 3.1: Derivation of DC Transfer Function of Flyback.

	Primary-side	*Secondary-side*
V_{ON}	V_{IN}	$V_{INR} \equiv V_{IN} / n$
V_{OFF}	$V_{OR} \equiv V_O \times n$	V_O
Duty Cycle	$D = \dfrac{V_{OFF}}{V_{ON} + V_{OFF}}$	
	$D = \dfrac{V_{OR}}{V_{IN} + V_{OR}}$	$D = \dfrac{V_O}{V_{INR} + V_O}$
	$D = \dfrac{nV_O}{V_{IN} + nV_O}$	

with respect to which its voltage needs to be measured, or stated. In fact, the reference level (i.e., by definition, "ground") is called the "Primary ground" on the Primary side and the "Secondary ground" on the Secondary side. Note that these are indicated by different ground symbols in Figure 3.1.

To find out the (absolute) voltage *at* the swinging end of any winding, we can use the following level-shifting rule:

To get the value of the voltage at the swinging end of any winding, we must add the voltage across the winding to the DC voltage present at its "nonswinging" end.

So, for example, to get the voltage at the Drain of the MOSFET (swinging end of Primary winding), we need to add V_{IN} (voltage at other end of winding) to the voltage waveform that represents the voltage across the Primary winding. That is how we got the voltage waveforms shown in Figure 3.1.

Coming to the question of how currents actually reflect from one side of the transformer to the other, it must be pointed out that even though the final current-scaling equations of a flyback transformer are exactly the same as in the case of an actual transformer, this is not strictly "classic transformer action." The difference from a conventional transformer is that in the flyback, the Primary and Secondary windings do not conduct at the same time. So, in fact, it seems a mystery why their currents are related to each other at all!

The current scaling that occurs in a flyback actually follows from *energy considerations*. The energy in a core is in general written as

$$E = \frac{1}{2} L I^2$$

We know the windings of our flyback conduct at different times, so the energy associated with each of them must be equal to the energy in the core and must therefore be equal to each other (we are ignoring the ramp portion of the current here for simplicity). Therefore,

$$E = \frac{1}{2} L_P I_P^2 = \frac{1}{2} L_S I_S^2$$

where L_p is the inductance measured across the Primary winding with the Secondary winding floating (no current), and L_S the inductance measured across the Secondary winding with the Primary winding floating. But we know that

$$L = N^2 \times A_L \times 10^{-9} \, \text{H}$$

where A_L is the *inductance index*, defined previously. Therefore, in our case we get

$$L_P = N_P^2 \times A_L \times 10^{-9}$$
$$L_S = N_S^2 \times A_L \times 10^{-9}$$

Substituting in the energy equation, we get the well-known current-scaling equations:

$$N_P I_P = N_S I_S$$

or

$$\frac{I_P}{I_S} = \frac{1}{n}$$

We see that analogous to the Volts/turns rule, the *Ampere-turns* also need to be preserved at all times. In fact, the core itself doesn't really "care" which particular winding is passing current at any given moment, so long as there is no sudden change in the *net* Ampere-turns of the transformer. This becomes the "transformer-version" of the basic rule we learned in *Chapter 1* — that the current through an inductor cannot change discontinuously. Now we see that *the net Ampere-turns of a transformer cannot change discontinuously.*

Summarizing, transformer action works as follows — when reflecting a voltage from Primary side to Secondary side, we need to divide by the turns ratio. When going from the Secondary side to the Primary side, we need to multiply by the turns ratio. The rule reverses for currents — so we multiply by the turns ratio when going from Primary to Secondary and divide in the opposite direction.

The Equivalent Buck-Boost Models

Because of the many similarities, and also because of the way voltages scale in the transformer, it becomes very convenient (most of the time) to study the flyback as an equivalent DC–DC (inductor-based) Buck-Boost. In other words, we separate out the coarse fixed-ratio step-down ratio and incorporate it into equivalent (reflected) voltages and currents. *We thereby manage to reduce the flyback transformer into a simple energy-storage medium, just like any conventional DC–DC Buck-Boost inductor.* In other words, for most practical purposes, the transformer goes "out of the picture." The advantage is that almost all the equations and design procedures we can write for a conventional Buck-Boost now apply to this *equivalent* Buck-Boost model. One exception to this is the leakage inductance issue (and everything related to it — the clamp, the loss in efficiency due to it, the turn-off voltage spike on the switch, and so on). We will discuss this exception later. But other than that, all other parameters — such as the capacitor, diode, and switch currents — can be more readily visualized and calculated if we use this DC–DC model approach.

The *equivalent DC–DC model* is created essentially by reflecting the voltages and currents across the isolation boundary of the transformer to one side. But again, as in the case of the duty cycle calculation (see Table 3.1), we have two options here — we can reflect everything either to the Primary side *or* to the Secondary side. We thus get the *two* equivalent Buck-Boost models as shown in Figure 3.2. We can use the Primary-side

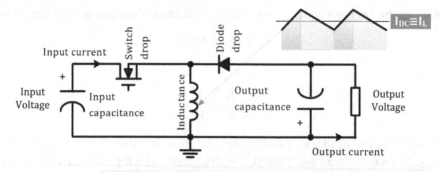

	Primary-side equivalent buck-boost	Secondary-side equivalent buck-boost
Input Voltage	V_{IN}	$V_{INR}=V_{IN}/n$
Input current	I_{IN}	$I_{INR}=I_{IN} \times n$
Input capacitance	C_{IN}	$n^2 \times C_{IN}$
Inductance	L_P	$L_S=L_P/n^2$
Switch drop	V_{SW}	V_{SW}/n
Output Voltage	$V_{OR}=V_O \times n$	V_O
Output current	$I_{OR}=I_O/n$	I_O
Inductor Current (I_{DC})	$I_{OR}/(1-D)=I_O/[n \times (1-D)]$	$I_O/(1-D)$
Output capacitance	C_O/n^2	C_O
Diode drop	$V_D \times n$	V_D
Duty Cycle	D	D
Current Ripple Ratio	r	r

Figure 3.2: The equivalent Buck-Boost models of the flyback.

equivalent model to calculate all the voltages and currents on the Primary side of the original flyback and the Secondary-side equivalent model for calculating all the currents and voltages on the Secondary side of the original flyback.

We know that voltages and currents reflect across the boundary by getting either multiplied or divided by the turns ratio. In fact, the "reflected output voltage" V_{OR} is one of the most important parameters of a flyback. As the name indicates, ***V_{OR} is effectively the output voltage as seen by the Primary side***. In fact, if we compare the switch waveform of the flyback in Figure 3.1 with that of a Buck-Boost, we will realize that to the switch, it seems as if the *output voltage is really V_{OR}*. See Figure 3.2.

As an example, suppose we have a 50-W converter with an output of 5 V at 10 A and a turns ratio of 20. The V_{OR} is therefore $5 \times 20 = 100$ V. Now, if we change the set output to say 10 V and reduce the turns ratio to 10, the V_{OR} is still 100 V. We will find that none of the Primary-side *voltage* waveforms change in the process (assuming efficiency doesn't change). Further, if we have also kept the output *power* constant in the process, that is, by changing the load to 5 A for an output at 10 V, all the *currents* on the Primary side will also be unaffected. Therefore, *the switch will never "know the difference"*. In other words, the switch virtually "thinks" that it is a simple DC–DC Buck-Boost — delivering an output voltage of V_{OR} with a load current of I_{OR}.

As mentioned, the only difference between a transformer-based flyback that "thinks" it is providing an output of V_{OR} with I_{OR} and an inductor-based version that *really* is providing an output of V_{OR} with I_{OR} is the "leakage inductance" of the flyback transformer. This is that part of the Primary side inductance that is *not coupled* to the Secondary side and therefore cannot partake in the transfer of useful energy from the input to the output. We can confirm from Figure 3.1 that the only portion of the Primary-side (switch) voltage waveform that "doesn't make it" to the Secondary side is the *spike* occurring just after the turn-off transition. This spike comes from the uncoupled leakage inductance, as we will soon see.

Note that in the equivalent Buck-Boost models, the reactive component values also get reflected — though as the *square* of the turns ratio. We can understand this fact easily from *energy considerations*. For example, the output capacitor C_O in the original flyback was charged up to a value of V_O. So, its stored energy was $1/2 C_O V_O^2$. In the Primary-side Buck-Boost model, the output of the converter is V_{OR}, that is, $V_O \times n$. Therefore, to keep the energy stored in this capacitor invariant (in the DC–DC model, as in the flyback), the output capacitance must get reflected to the Primary side according to C_O/n^2. Notice also from Figure 3.2 how the inductance reflects. This is consistent with the fact that $L \propto N^2$.

The Current Ripple Ratio for the Flyback

Looking at the equivalent Buck-Boost models in Figure 3.2, the *center of the ramp* on the Secondary side (average inductor current, "I_L" or I_{DC}) must be equal to $I_O/(1-D)$, as for a Buck-Boost (because the average diode current must equal the load current). This Secondary-side "inductor" current gets reflected to the Primary side, and so the center of the Primary-side inductor current ramp is "I_{LR}," where $I_{LR} = I_L/n$. Equivalently, it is equal to $I_{OR}/(1-D)$, where I_{OR} is the *reflected load current*, that is, $I_{OR} = I_O/n$. Similarly, the current swings on the Primary and Secondary sides are also related via scaling (turns ratio n). Therefore, we see that the *ratio* of the swing to the center of the ramp is *identical* on both sides (Primary- and Secondary-side DC–DC models). We are thus in a position to define a *current ripple ratio r* for the flyback topology too — just as we did for a DC–DC converter. We just need to visualize r in a slightly different manner this time — ***in terms of the center*** *of the ramp (switch or diode) rather than the DC inductor level* (because there is no inductor present really). And as for DC–DC converters, we should try to set it to around 0.4 in most cases.

The value of *r* for a flyback is the same for both the Primary and the Secondary DC–DC equivalent models.

The Leakage Inductance

The leakage inductance can be thought of as a *parasitic inductance* in series with the Primary-side inductance of the transformer. So, just at the moment the switch turns OFF, the current flowing through both these inductances is "I_{PKP}," that is, the peak current on the Primary side. However, when the switch turns OFF, the energy in the Primary inductance has an available freewheeling path (through the output diode), but the leakage inductance energy has nowhere to go. So, it expectedly "complains" in the form of a huge voltage spike (see Figure 3.1). This spike (or a scaled version of it) is *not* seen on the Secondary side, simply because this is not a coupled inductance, like the Primary inductance.

If we don't make any effort to collect this leakage energy, the spike can be very large, causing switch destruction. Since we certainly can't get this energy to transfer to the Secondary side, we have just two options — either we can try to *recover* it and cycle it back into the input capacitor, or we burn it (dissipation). The latter approach is usually preferred for the sake of simplicity. It is commonly accomplished by means of a straightforward "zener diode clamp," as shown in Figure 3.1. Of course, the zener voltage must be chosen according to the maximum voltage the switch can tolerate. Note that for several reasons, in particular efficiency, it is preferable to connect this zener *across* the Primary winding, as shown (via a blocking diode in series with it). An alternative method is to connect it from the switching node to Primary ground.

We can ask — where does the leakage inductance reside? *Most* of it is inside the Primary winding of the transformer, though some of it lies in the printed circuit board (PCB) trace sections and transformer terminations, especially with those associated with the *Secondary* winding, as we will see further.

Zener Clamp Dissipation

If we intend to burn the energy in the leakage, it is important to know how this affects the efficiency. It is sometimes intuitively felt the energy dissipated every cycle is $1/2 \times L_{LKP}I_{PK}^2$, where I_{PK} is the peak switch current and L_{LKP} is the Primary-side leakage. That certainly is the energy residing *in the leakage inductance* (at the moment the switch turns OFF), but it is *not* the *entire* energy that gets dissipated in the zener clamp on account of the leakage.

The Primary winding is in series with the leakage, so during the small interval that the leakage inductance is trying, in effect, to attain *reset* by freewheeling into the zener, the Primary winding *is forced to follow suit* and continues to provide this in-series current. Though the Primary winding is certainly trying (and managing partially) to freewheel into the Secondary side, a part of its energy also gets diverted into the zener clamp — continuing till the leakage inductance achieves full reset (zero clamp current). In other words, some energy from the Primary inductance gets virtually plucked out by the series leakage inductance, and this energy also finds its way into the zener, along with the energy residing in the leakage itself. A detailed calculation (see Figure 7.10) reveals that the zener dissipation actually is

$$P_Z = \frac{1}{2} \times L_{LK} \times I_{PK}^2 \times \frac{V_Z}{V_Z - V_{OR}}\, W$$

So, the energy in the leakage $1/2 \times L_{LK} I_{PK}^2$ gets *multiplied* by the term $V_z/(V_z - V_{OR})$ (this additional term is from the Primary inductance).

Note that if the zener voltage is too close to the chosen V_{OR}, the dissipation in the clamp goes up steeply. *V_{OR} therefore always needs to be picked with great care*. That simply means that *the turns ratio has to be chosen carefully*!

Secondary-Side Leakages also Affect the Primary Side

Why did we use the symbol "L_{LK}" in the dissipation equation above? Why didn't we identify it as the *Primary-side* leakage ("L_{LKP}")? The reason is that L_{LK} represents the *overall* leakage inductance as seen by the switch. So, it is partly L_{LKP} — but it also is influenced by the *Secondary-side* leakage inductance. This is a little hard to visualize, since by definition, the Secondary-side leakage inductance is not supposed to be coupled to the Primary side (and vice versa). So, how could it be affecting anything on the Primary side?

The reason is that just as the Primary-side leakage *prevents* the Primary-side current from freewheeling into the output immediately following the turn-off transition (thereby causes an increase in the zener dissipation), any Secondary-side inductance also prevents the freewheeling path from *becoming available* immediately. Basically, the Secondary-side inductance *insists* that we ("politely" and) slowly build up the current through it — respecting the fact that it is an inductance after all! However, until the current in the bona fide freewheeling path can build up to the required level, the Primary-side current still needs to freewheel *somewhere*! The path the inductor current therefore seeks out is the one containing the zener clamp (that being the only path available). The zener can therefore see significant dissipation, even assuming *zero* Primary-side leakage.

In brief, the Secondary-side leakage has created much the same effect as a Primary-side leakage.

When both Primary- and Secondary-side leakages are present, we can calculate the effective Primary-side leakage (as seen by the switch and zener clamp) as

$$L_{LK} = L_{LKP} + n^2 L_{LKS}$$

So, like any other reactive element, the Secondary-side leakage also reflects onto the Primary side according to the *square of the turns ratio*, where it adds up in series with any Primary-side leakage present (see Figure 3.2).

For a given V_{OR}, if the output voltage is "low" (e.g., 5 V or 3.3 V), the turns ratio is much greater. Therefore, *if the chosen V_{OR} is very high, the reflected Secondary-side leakage can become even greater than any Primary-side leakage*. This can become quite devastating from the efficiency standpoint.

Measuring the Effective Primary-side Leakage Inductance

The best way to know what L_{LK} really is, is by *measuring* it! Commonly, a leakage inductance measurement is done by shorting the Secondary winding pins and then measuring the inductance across the ends of the (open) Primary winding. By shorting, we virtually cancel out all coupled inductance. And so what we measure is just the Primary-side leakage inductance in this case.

However, the best method to measure leakage is actually an *in-circuit* measurement so that we include the Secondary-side PCB traces in the measurement. The recommended procedure is as follows.

On the given application board, a thick piece of copper foil (or a thick section of braided copper strands), with as short a length as possible, is placed directly across the diode solder pads on the PCB. A similar piece of conductor is placed across the output capacitor solder pads. Then, if we measure the inductance across the (open) Primary winding pins, we will measure the effective leakage inductance L_{LK} (not just L_{LKP}).

We will find that the contribution from the Secondary-side traces can in fact make L_{LK} several times larger than L_{LKP}. L_{LKP} can of course be measured, if desired, by placing a thick conductor across the Secondary *pins* of the transformer.

The PCB used in the above-described procedure can be just a bare board with no components mounted on it other than the transformer. Or it can even be a fully assembled board (though sometimes, we may need to cut the trace connecting the Drain of the MOSFET to the transformer).

If we want to *mathematically* estimate the inductance of the Secondary-side traces, the rule of thumb we can use is **20 nH per inch**. But here, we need to include the *full* electrical path of the high-frequency output current — starting from one end of the Secondary winding, returning to its other end, *through the diode and output capacitor(s)*. We will be surprised to calculate or measure that even an inch or two of trace length can dramatically decrease the efficiency by 5–10% in low-output voltage applications.

Worked Example (7) — Designing the Flyback Transformer

A 74-W Universal Input (90–270 VAC) flyback is to be designed for an output of 5 V at 10 A and 12 V at 2 A. Design a suitable transformer for it, assuming a switching frequency of 150 kHz. Also, try to use a cost-effective 600-V-rated MOSFET.

Fixing the V_{OR} and V_z

At maximum input voltage, the rectified DC to the converter is

$$V_{\text{INMAX}} = \sqrt{2} \times \text{VAC}_{\text{MAX}} = \sqrt{2} \times 270 = 382 \text{ V}$$

With a 600-V MOSFET, we must leave at least 30-V safety margin when at V_{INMAX}. So, in our case, we do not want to exceed 570 V on the Drain. But from Figure 3.1, the voltage on the Drain is $V_{\text{IN}} + V_Z$. Therefore,

$$V_{\text{IN}} + V_Z = 382 + V_Z \leq 570$$
$$V_Z \leq 570 - 382 = 188 \text{ V}$$

We pick a standard 180-V zener.

Note that if we plot the zener dissipation equation presented earlier, as a function of V_Z/V_{OR}, we will discover that in all cases, **we get a "knee" in the dissipation curve at around $V_Z/V_{\text{OR}} = 1.4$.** So, here too, we pick this value as an optimum ratio that we would like to target. Therefore,

$$V_{\text{OR}} = \frac{V_Z}{1.4} = 0.7 \times V_Z = 0.71 \times 180 = 128 \text{ V}$$

Turns Ratio

Assuming the 5-V output diode has a forward drop of 0.6 V, the turns ratio is

$$n = \frac{V_{\text{OR}}}{V_O + V_D} = \frac{128}{5.6} = 22.86$$

Note that the 12-V output may sometimes be regulated by a linear post-regulator. In that case, we may have to make the transformer provide an output 3–5 V higher (than the final expected 12 V) — to provide the necessary "headroom" for the linear regulator to operate properly. This additional headroom not only caters to the dropout limits of the linear regulator, but in general also helps achieve a regulated 12 V *under all load conditions*. However, there are also some clever *cross-regulation* techniques available that allow us to omit the 12-V linear regulator, particularly if the regulation requirements of the 12-V rail are not too "tight," and also if there is some minimum load assured on the outputs. In our example, we are assuming there is no 12-V post-regulator present. Therefore, the required turns ratio for the 12-V output is $128/(12 + 1) = 9.85$, where we have assumed the diode drop is 1 V in this case.

Maximum Duty Cycle (theoretical)

Having verified the selection of V_Z and V_{OR} at highest input, now we need to *get back to the lowest input voltage* because we know from the previous discussions about the Buck-Boost (see the "general inductor design procedure" in *Chapter 2*) that V_{INMIN} is the worst-case point we need to consider for a Buck-Boost inductor/transformer design.

The minimum-rectified DC voltage to the converter is

$$V_{INMIN} = \sqrt{2} \times \text{VAC}_{MIN} = \sqrt{2} \times 90 = 127 \text{ V}$$

We are ignoring the voltage ripple on the input terminals of the converter, and therefore we will take this as the DC input to the converter stage. So, the duty cycle at minimum input voltage is

$$D = \frac{V_{OR}}{V_{OR} + V_{INMIN}} = \frac{128}{128 + 127} = 0.5 \quad (flyback)$$

This is clearly a "theoretical" estimate — implying 100% efficiency. We will in fact ignore this value ultimately, as we **will be estimating D more accurately by another trick**.

Note, however, that this is the *operating* D_{MAX}. When we "*power down*" our converter for example, *the duty cycle will actually increase further in an effort to maintain regulation* (unless current limit and/or duty cycle limit is encountered along the way). Then depending upon the number of *missing AC cycles* for which we may need to ensure regulation (the "holdup time" specification), we will need to select a suitable input capacitance and also the maximum duty cycle limit, D_{LIM} of our controller. Typically, D_{LIM} is set around 70%, and the capacitance is selected on the basis of the *3 μF/W rule of thumb*. For example, for our 74-W supply with an estimated 70% efficiency at low line, we will draw an input power of $74/0.7 = 106$ W. Therefore, we should use a $106 \times 3 = 318$ μF (standard value 330 μF) input capacitor. However, note that the ripple current rating of this capacitor (and its life expectancy) must be verified as described in *Chapter 6*.

Effective Load Current on Primary and Secondary Sides

Let us *lump* all the 74-W output power into an equivalent *single* output of 5 V. So, the load current for a 5-V output is

$$I_O = \frac{74}{5} \approx 15 \text{ A}$$

On the Primary side, the switch "thinks" its output is V_{OR} and the load current is I_{OR}, where

$$I_{OR} = \frac{I_O}{n} = \frac{15}{22.86} = 0.656 \text{ A}$$

Duty Cycle

The actual duty cycle is important because a slight increase in it (from the theoretical ideal efficiency value) may lead to a significant increase in the operating peak current and the corresponding magnetic fields.

The input power is

$$P_{IN} = \frac{P_O}{\text{Efficiency}} = \frac{74}{0.7} = 105.7 \text{ W}$$

The average input current is therefore

$$I_{IN} = \frac{P_{IN}}{V_{IN}} = \frac{105.7}{127} = 0.832 \text{ A}$$

The average input current tells us what the actual duty cycle "D" is, because I_{IN}/D is also the center of the Primary-side current ramp and must equal I_{LR}, that is,

$$\frac{I_{IN}}{D} = \frac{I_{OR}}{1 - D}$$

solving,

$$D = \frac{I_{IN}}{I_{IN} + I_{OR}} = \frac{0.832}{0.832 + 0.656} = 0.559$$

We thus have a more accurate estimate of duty cycle.

Actual Center of Primary and Secondary Current Ramps

The center of the Secondary-side current ramp (lumped) is

$$I_L = \frac{I_O}{1 - D} = \frac{15}{1 - 0.559} = 34.01 \text{ A}$$

The center of the Primary-side current ramp is

$$I_{LR} = \frac{I_L}{n} = \frac{34.01}{22.86} = 1.488 \text{ A}$$

Peak Switch Current

Knowing I_{LR}, we know the peak current for our selected current ripple ratio:

$$I_{PK} = \left(1 + \frac{r}{2}\right) \times I_{LR} = 1.25 \times 1.488 = 1.86 \, A$$

We may need to set the current limit of the controller, for example, based on this estimate.

Voltseconds

We have at V_{INMIN}

$$V_{ON} = V_{IN} = 127 \text{ V}$$

The on-time is

$$t_{ON} = \frac{D}{f} = \frac{0.559}{150 \times 10^3} \Rightarrow 3.727 \, \mu s$$

So, the voltsμseconds is

$$Et = V_{ON} \times t_{ON} = 127 \times 3.727 = 473 \, V\mu s$$

Primary-Side Inductance

Note that when we come to designing off-line transformers, for various reasons like reducing high-frequency copper loss, reducing size of transformer, and so on, it is more common to set *r* *at around 0.5*. So, the Primary-side inductance must then be (from the "$L \times I$" rule)

$$L_P = \frac{1}{I_{LR}} \times \frac{Et}{r} = \frac{473}{1.488 \times 0.5} = 636 \, \mu H$$

Selecting the Core

Unlike made-to-order or off-the-shelf inductors, when designing our own magnetic components, we should not forget that adding an air gap dramatically improves the energy-storage capability of a core. Without the air gap, the core could saturate even with very little stored energy. See *Chapter 5* for a deeper understanding of air gaps.

Of course, we still need to maintain the desired *L*, corresponding to the desired *r*! So, if we add too much of a gap, we will also need to add many more turns — thus increasing the copper loss in the windings. *At one point, we will also run out of window space to accommodate these windings.* So, a practical compromise must be made here, one that the following equation actually takes into account (applicable to *ferrites* in general, for any topology):

$$V_e = 0.7 \times \frac{(2+r)^2}{r} \times \frac{P_{IN}}{f} \text{ cm}^3$$

where f is in kHz.

In our case, we get

$$V_e = 0.7 \times \frac{(2.5)^2}{0.5} \times \frac{105.7}{150} = 6.17 \text{ cm}^3$$

We start looking for a core of this volume (or higher). We find a candidate in the EI-30. Its effective length and area are given in its datasheet as

$$A_e = 1.11 \text{ cm}^2$$
$$l_e = 5.8 \text{ cm}$$

So, its volume is

$$V_e = A_e \times l_e = 5.8 \times 1.11 = 6.438 \text{ cm}^3$$

which is a little larger than we need, but close enough.

Number of Turns

The voltage-dependent equation

$$B = \frac{LI}{NA} \text{ Tesla}$$

connects B to L. However we also know that a statement about r is equivalent to a statement about L — for a given frequency (the "$L \times I$ equation"). So, combining these equations, and also connecting the swing in the B-field to its peak (through r), we get a very useful form of the voltage-dependent equation, in terms of r (expressed in MKS units):

$$N = \left(1 + \frac{2}{r}\right) \times \frac{V_{ON} \times D}{2 \times B_{PK} \times A_e \times f} \quad (voltage - dependent \ equation, \ any \ topology)$$

So, even with *no* information about the permeability of the material, air gap, and so on, *we already know the number of turns required on a core with area A_e that will produce a certain B-field.* We also know that with or without an air gap, the B-field should not exceed 0.3 T for most ferrites. So, solving the equation for N (N is N_P here, the number of Primary turns),

$$N_P = \left(1 + \frac{2}{0.5}\right) \times \frac{127 \times 0.559}{2 \times 0.3 \times 1.11 \times 10^{-4} \times 150 \times 10^3} = 35.5 \text{ Turns}$$

We will have to verify that this can be accommodated in the window of the core — along with the bobbin, tape insulation, margin tape, Secondary windings, sleeving, and so on. Usually, that is no problem for a flyback.

Note that if we want to reduce N, the possible ways are — choose a larger r, or decrease the duty cycle (i.e., pick a lower V_{OR}), or allow a higher B (select new material!), or increase the area of the core — the latter, hopefully, without increasing the volume.

The number of Secondary turns (5-V output) is

$$N_S = \frac{N_P}{n} = \frac{35.5}{22.86} = 1.55 \text{ Turns}$$

But we want an *integral* number of turns. Further, approximating this to just one turn is not a good idea since there will be more leakage. We therefore prefer to set

$$N_S = 2 \text{ Turns}$$

So, with the same turns ratio (i.e., V_{OR} unchanged)

$$N_P = N_S \times n = 2 \times 22.86 \approx 46 \text{ Turns}$$

The number of turns for the 12-V output is obtained by the scaling rule

$$N_{S\_AUX} = \frac{12 + 1}{5 + 0.6} \times 2 = 4.64 \approx 5 \text{ Turns}$$

where we have assumed the 5-V diode has a drop of 0.6 V and the 12-V diode has a drop of 1 V.

Actual B-Field

So, now we can use the voltage-dependent equation to solve for B:

$$B_{PK} = \left(1 + \frac{2}{r}\right) \times \frac{V_{ON} \times D}{2 \times N_P \times A_e \times f} \text{ Tesla}$$

But in fact we don't have to use this equation anymore! We realize that B_{PK} is inversely proportional to the number of turns. So, if, with a calculated 35.5 turns, we had a peak field of 0.3 T, then with 46 turns we will have (*keeping L and r unchanged!*)

$$B_{PK} = \frac{35.5}{46} \times 0.3 = 0.2315 \text{ T}$$

The swing is related to the peak by (see section titled "Field Ripple Ratio" in *Chapter 2*)

$$\Delta B \equiv 2 \times B_{AC} = \frac{2r}{r+2} \times B_{PK} = \frac{1}{2.5} \times 0.2315 = 0.0926 \text{ T}$$

Note that in CGS units, the peak is now 2,315 G, and the AC component is half the swing, that is, 463 G (since $r = 0.5$).

> *Note: If we start with a B-field target of 0.3 T, we are likely to reach a lesser B-field after rounding up the Secondary turns to the nearest higher integer, as we did above. That of course is not only expected, but also acceptable. However, note that on power-up or power-down, for example, the B-field will increase further, as the converter tries to continue regulating. That is why we need to set the maximum duty cycle limit and/or current limit accurately, or the switch can be destroyed due to inductor/transformer saturation. Cost-effective flyback designs with fast-acting current limit and fast switches (especially those with an integrated MOSFET) generally **allow for a peak B-field of up to 0.42 T, so long as the operating field is 0.3 T or less**. But see Chapter 5 too.*

Air Gap

Finally, we need to consider the *permeability* of the material. L is related to permeability by the equation:

$$L = \frac{1}{z} \times \left(\frac{\mu \mu_0 A_e}{l_e} \right) \times N^2 H$$

Here z is the "gap factor":

$$z = \frac{l_e + \mu l_g}{l_e}$$

Note that z can range from 1 (no gap) to virtually any value. A z of 10, for example, increases the energy-handling capability of an ungapped core set by a factor of 10 (its A_L value falls by the same factor, and so does its *effective permeability* — $\mu_e = \mu \mu_0 / z$). So, large gaps certainly help, but since we are still interested in maintaining L to a certain value based on our choice of r, we will have to increase the number of turns substantially. As mentioned, at some point, we just may not be able to accommodate these windings in the available window, and further, the copper loss will also increase greatly. So, *z in the range of 10−20 is a good compromise for gapped transformers made out of ferrite material*. Let us see what it comes out to be, based on our existing requirements and choices

$$z = \frac{1}{L} \times \left(\frac{\mu \mu_0 A_e}{l_e} \right) \times N^2 = \frac{1}{636 \times 10^{-6}} \times \left(\frac{2000 \times 4\pi \times 10^{-7} \times 1.11 \times 10^{-4}}{5.8 \times 10^{-2}} \right) \times 46^2$$

So,

$$z = 16$$

This is an acceptable value too. Finally, solving for the length of air gap,

$$z = 16 = 2 = \frac{5.8 + (2,000)l_g}{5.8} \Rightarrow l_g = 0.435 \text{ mm}$$

Note: In general, if we use a center-gapped transformer, the total gap in the center must be equal to the above-calculated value, whether each center limb has been ground or not. But if spacers are being inserted on both side limbs (say on an EE or EI type of core), the thickness of the spacer on each outer limb must be half of the above-calculated value because the total air gap is then as desired. See Figure 5.17.

Selecting the Wire Gauge and Foil Thickness

In an inductor, the current undulates relatively smoothly. However, in a flyback transformer, the current in one winding stops completely to let the other winding take over. Yes, the core doesn't care (and doesn't even know) which of its windings is passing current at a given moment, *as long as the Ampere-turns is maintained* — because only the net Ampere-turns determine the field (and energy) inside the core. But as far as the windings themselves are concerned, the current is now *pulsed* — with *sharp* edges and therefore with significant high-frequency content. Because of this, "skin depth" considerations are necessary for choosing the appropriate wire thickness of the windings of a flyback transformer.

Note: We had ignored this for DC–DC inductors, but in high-frequency DC–DC designs too (or with high r), we may need to apply skin depth considerations.

At high frequencies, the electric fields between the electrons become strong enough to cause them to repel each other rather decisively and thereby cause the current to crowd on the *exterior* (surface) of the conductor (see exponential curve in Figure 3.3). This crowding worsens with frequency as per $\sqrt{f}$. There is thus the possibility that though we may be using thick wire in an effort to reduce copper loss, a good part of the cross-section of the wire (its "innards") just may not be *available* to the current. The resistance presented to the current flow is inversely proportional to the area through which the current is flowing or is able to flow. So, this *current crowding* causes an increase in the effective resistance of the copper (as compared to its DC value). The resistance now presented to the current is called the "*AC resistance*" (see lower half of Figure 3.3). This is a function of frequency because so is the skin depth. Therefore, instead of wasting precious space inside the transformer and losing efficiency, we must try to use more *optimum diameters* of wire — in which the cross-sectional area is better utilized. Thereafter, if we need to pass more current than the chosen cross-sectional area can handle, we need to *parallel* several such strands.

So, how much current can a given wire strand handle? That depends purely on the heat buildup and the need to keep the overall transformer at an acceptable temperature rise. For

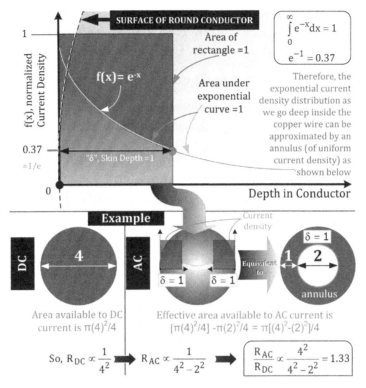

Figure 3.3: Skin depth and AC resistance explained.

this, a good guideline/rule of thumb for the current density of flyback transformers is 400 *circular mils ("cmils") per Ampere*, and that is our goal too in the analysis that follows.

> *Note: Expressing "current density" in the North American way of cmils/A needs a little getting used to. It is actually* area per unit Ampere, *not* Ampere per unit area *(as we would normally expect a "current density" to be)! So, a higher cmils/A value actually is a* lower *current density (and vice versa) — and will produce a* lower *temperature rise.*

We define the *skin depth "δ"* as the distance from the surface of a conductor at which the current density falls to 1/*e* times the value at the surface. Note that the current density at the surface is the same as the value it would have had all through the copper, were there no high-frequency effects. As a good approximation to the exponential curve, we can also imagine the current density *remaining unchanged* from the value at the surface, until the skin depth is reached, *falling abruptly to zero thereafter*. This follows from an interesting property of the exponential curve that the area under it from 0 to ∞ is equal to the area of a rectangle passing through its 1/*e* point (see Figure 3.3 and also Figures 5.21 and 5.22).

Therefore, when using round wires, if we choose the diameter as *twice* the skin depth, no point inside the conductor will be more than one skin depth away from the surface. So, no part of the conductor is unutilized. In that case, we can consider this wire as having an AC resistance equal to its DC resistance — *there is no need to continue to account for high-frequency effects so long as the wire thickness is chosen in this manner.*

If we use copper foil, its *thickness* too needs to be about twice the skin depth.

In Figure 3.4, we have a simple nomogram for selecting the wire gauge and thickness. The *upper* half of this is based on the current-carrying capability as per the usual requirement of 400 cmil/A. But the readings can obviously be linearly scaled for any other desired current density. The vertical grid on the nomogram represents wire gauges. An example based on a switching frequency of 70 kHz is presented in the figure. In a similar manner, for our previous worked example, we see that for 150-kHz operation, we should use AWG 27. But its current-carrying capacity is only 0.5 A at 400 cmils/A (and only 0.25 A at a *lower* current density of 800 cmils/A!). Therefore, since the center of the Primary current ramp was iterated and estimated to be 1.488 A, we need *three* strands of AWG 27 (twisted together) to give a combined current-carrying capability of 1.5 A (which is slightly better than what we need).

Coming to the Secondary side of the worked example, we remember we had *lumped* all the current as a 5-V equivalent load of 15 A. But in reality it is only 10 A, two-thirds of that. So, the center of its current ramp, which we had calculated was about 34 A, is actually $(2/3) \times 34 = 22.7$ A. The balance of this, that is, $34 - 22.7 = 11.3$ A, reflects as $(5.6/13) \times 11.3 = 4.87$ A into the 12-V winding. So, the center of the 12-V output's current ramp is 4.87 A. We can choose the 12-V winding arrangement using the same arguments we present below for the 5-V winding.

For the 5-V winding, we can consider using copper foil, since we have only two turns and we need a high current-carrying capability. The center of the 5-V Secondary-side current ramp is about 23 A. The appropriate thickness (2δ) at this frequency is found by projecting downward along the AWG 27 vertical line. We get about 14-mil thickness. But we still don't know if the current through it will follow our guideline of 400 cmil/A, since it is a foil. We need to check this out further.

One "cmil" is equal to 0.7854 square mils (sq.mils). Therefore, 400 cmils is $400 \times 0.7854 = 314$ sq.mils (note $\pi/4 = 0.7854$). So, for 23 A, we need $23 \times 314 = 7,222$ sq.mils. But the thickness of the foil is 14 mils. Therefore, we need the copper foil to be $7,222/14 = 515$ mils wide, that is, about half an inch. Looking at a bobbin for the EI-30 in Figure 3.5, we see it can accommodate a foil 530 mils wide. So, this is just about acceptable. Note that if the available width is insufficient, we would need to look for another core altogether — one with a "longer" (stretched-out) profile. Cores like that are available as American "EER" cores (these are EE cores with the "R" indicating a round centerpost).

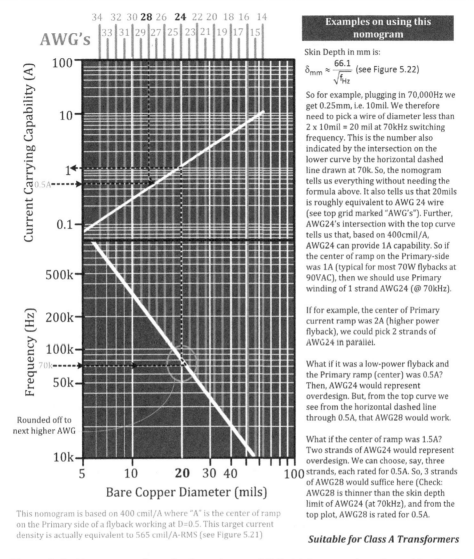

Skin Depth in mm is:

$$\delta_{mm} \approx \frac{66.1}{\sqrt{f_{Hz}}} \text{ (see Figure 5.22)}$$

So for example, plugging in 70,000Hz we get 0.25mm, i.e. 10mil. We therefore need to pick a wire of diameter less than 2 x 10mil = 20 mil at 70kHz switching frequency. This is the number also indicated by the intersection on the lower curve by the horizontal dashed line drawn at 70k. So, the nomogram tells us everything without needing the formula above. It also tells us that 20mils is roughly equivalent to AWG 24 wire (see top grid marked "AWG's"). Further, AWG24's intersection with the top curve tells us that, based on 400cmil/A, AWG24 can provide 1A capability. So if the center of ramp on the Primary-side was 1A (typical for most 70W flybacks at 90VAC), then we should use Primary winding of 1 strand AWG24 (@ 70kHz).

If for example, the center of Primary current ramp was 2A (higher power flyback), we could pick 2 strands of AWG24 in parallel.

What if it was a low-power flyback and the Primary ramp (center) was 0.5A? Then, AWG24 would represent overdesign. But, from the top curve we see from the horizontal dashed line through 0.5A, that AWG28 would work.

What if the center of ramp was 1.5A? Two strands of AWG24 would represent overdesign. We can choose, say, three strands, each rated for 0.5A. So, 3 strands of AWG28 would suffice here (Check: AWG28 is thinner than the skin depth limit of AWG24 (at 70kHz), and from the top plot, AWG28 is rated for 0.5A.

Suitable for Class A Transformers

This nomogram is based on 400 cmil/A where "A" is the center of ramp on the Primary side of a flyback working at D=0.5. This target current density is actually equivalent to 565 cmil/A-RMS (see Figure 5.21)

Figure 3.4: Nomogram for selecting wires and foil thicknesses, based on skin depth considerations.

Or we can again consider using several paralleled strands of round wire. The problem is that a bunch of 46 twisted strands (of AWG 27) is going to be bulky, difficult to wind, and will also increase the leakage inductance. So, we may like to use say 11 or 12 strands of AWG 27 twisted together into one *bunch*, and then take four of these bunches (all electrically in parallel), laid out side by side to form one layer of the transformer. For a two-turns Secondary, therefore, we would wind two layers of this.

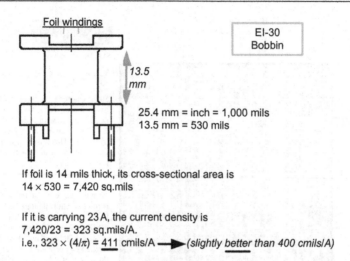

Foil windings

EI-30
Bobbin

13.5
mm

25.4 mm = inch = 1,000 mils
13.5 mm = 530 mils

If foil is 14 mils thick, its cross-sectional area is
14 × 530 = 7,420 sq.mils

If it is carrying 23 A, the current density is
7,420/23 = 323 sq.mils/A.
i.e., 323 × (4/π) = 411 cmils/A ▶ *(slightly better than 400 cmils/A)*

Figure 3.5: Checking to see if a 23-A foil can be accommodated on an EI-30 bobbin.

Forward Converter Magnetics

The procedure presented in this section applies explicitly to the *single-switch* Forward converter. However, the general procedure remains unchanged for the *two-switch* Forward converter as well.

Duty Cycle

The duty cycle of a Forward converter is

$$V_O = V_{IN} \times D \times \frac{N_S}{N_P}$$

Comparing this with the duty cycle of a Buck, we see that the only difference is the term N_S/N_P. As mentioned, this is the coarse fixed-ratio step-down function available due to transformer action. We can therefore visualize that the input voltage V_{IN} gets reflected to the Secondary side. This reflected voltage ($V_{INR} = V_{IN}/n$ where $n = N_P/N_S$) gets impressed at the Secondary-side switching node. ***From there on, we have in effect a simple DC–DC Buck stage, with an input voltage of V_{INR} and an output voltage of V_O*** (see Figure 3.6). Therefore, the design of the Forward converter's choke is *not* going to be covered here, as it is designed using the same procedure as that of any Buck inductor. However, the Forward converter's transformer is another story altogether!

> *Note: Regarding choke design, we should keep in mind that for* high-current *inductors, as would be found in a typical Forward converter, the calculated wire gauge may be too thick (and stiff) for winding easily over the core/bobbin. In that case, several thinner*

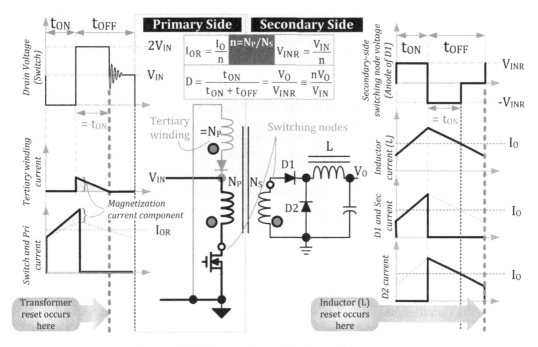

Figure 3.6: The single-ended Forward converter.

wire gauges may be twisted together to make the winding more flexible and easier to handle in production. Further, since choke and inductor design has usually little to do with high-frequency skin depth considerations, we can choose strands of almost any practical diameter, so long as we have enough net copper cross-sectional area to keep the temperature rise to within about 40–50 °C.

Unlike a flyback transformer, the Forward converter's Secondary winding conducts at the same time as the Primary winding. This leads to an *almost* complete flux cancellation inside the core. But there is one component of the Primary current waveform which remains the same, irrespective of the load. This is the *magnetization current* component — shown in gray on the left side of Figure 3.6. At zero load, this is the entire current through the Primary winding and switch (assuming duty cycle remains fixed). As soon as we try to draw some load current, the Secondary-winding current increases, and so does the Primary-winding current. Each current increases proportionally to the load current, and so their increments too are mutually proportional — the proportionality constant being the turns ratio. But more significantly, they are of *opposite sign* — that is, looking at Figure 3.6, we see that the current enters the dotted end of the transformer on the Primary side, and on the Secondary side, it leaves by the dotted end at the same time. Therefore, the net flux in the core of the transformer remains unchanged from the zero load condition (assuming *D* is fixed) because the core just never "sees" any change in the net Ampere-turns flowing through its windings. All conditions inside the core, that is, the flux, the magnetic fields, the energy stored, and even the core loss, are dependent only on the magnetization current. Of course, the windings themselves have a different story to relate — they bear the entire brunt, not only of the actual load current, but also of the *sharp edges* and consequent high-frequency content of the pulsed current waveforms.

The magnetization current component is *not* coupled by transformer action to the Secondary current. In that sense, it is like a "parallel leakage inductance." We need to subtract this component from the total switch current, and only then will we find that the Primary and Secondary currents scale according to the turns ratio. In other words, *the magnetization current does not scale* — it stays confined to the Primary side.

But in fact, the magnetization current is the only current component that is storing any energy in the transformer. So, in that sense, it is like the flyback transformer! But, if we are to achieve a steady state, even a transformer needs to be "reset" every cycle (along with the output choke). But unfortunately, the magnetization energy is effectively "uncoupled," because of the output diode direction, and so we can't transfer it over to the Secondary side. If we don't do anything about this energy, it will certainly destroy the switch by a spike similar to the leakage in a flyback. We don't want to burn it either, for efficiency reasons. Therefore, the usual solution is to use a "tertiary winding" (or "energy-recovery winding"), connected as shown in Figure 3.6. Note that this winding is in flyback configuration *with respect to the Primary winding*. It conducts only when the switch turns OFF, and thereby it freewheels the magnetization energy back into the input capacitor. There is some loss associated with this "circulating" energy term because of the diode drop and resistance of the tertiary winding. Note, however, that any bona fide leakage inductance energy also gets recycled back into the input by the tertiary winding. So, we don't need an additional clamp for it in a traditional Forward converter.

For various subtle reasons, like being able to ensure the transformer resets *predictably under all conditions*, and also for various production-related reasons, the number of turns of the tertiary winding is usually kept exactly the same as the Primary winding. Therefore by transformer action, the voltage at the Primary-side switching node (Drain of the MOSFET) must rise to $2 \times V_{IN}$ when the switch turns OFF. Therefore, in a Universal Input off-line *single-ended* (i.e., single-switch) Forward converter, we need a switch rated for at least 800 V.

As soon as the transformer is reset (i.e., the current in the tertiary winding returns to zero), the Drain voltage suddenly drops to V_{IN} — that is, *no voltage* is then present *across* the Primary winding — and therefore there is no voltage across the Secondary winding either. The catch diode of the output stage (i.e., the diode connected to the Secondary ground in Figure 3.6) then freewheels the energy contained in the choke. Note that there is actually some ringing at the Drain of the MOSFET for a while, around an average level of V_{IN}, just after transformer reset occurs. This is attributable to various undocumented parasitics. The ringing contributes significantly to the radiated electromagnetic interference (EMI).

Note that even prior to transformer reset occurring, the Secondary winding has not been conducting for a while — simply because the output diode (i.e., the one connected to the swinging end of the Secondary winding) has been reverse-biased during the time the tertiary winding was conducting.

Note also that the duty cycle of such a Forward converter can under no circumstances ever be allowed to exceed 50%. The reason for that is we have to unconditionally ensure that transformer reset will always occur, every cycle. Since we have no direct control on the transformer current waveforms, we have to just *leave enough time* for the current in the tertiary winding to ramp down to zero on its own. In other words, we have to allow *voltseconds balance* to occur naturally in the transformer. However, because the number of turns in the tertiary winding is equal to the Primary turns, the voltage across the tertiary winding is equal to V_{IN} when the switch is ON and is also equal to V_{IN} (opposite direction) when the switch is OFF. Reset will therefore occur when t_{OFF} becomes equal to t_{ON}. So, if the duty cycle exceeds 50%, t_{ON} would certainly always exceed t_{OFF}, and therefore transformer reset would never be able to occur. That would eventually destroy the switch. Therefore, just to allow t_{OFF} to be large enough, the duty cycle must always be kept to less than 50%.

We realize that the Forward converter transformer is always in discontinuous mode (DCM) (its choke, i.e. inductor L, is usually in continuous conduction mode (CCM), with an r of *0.4*). Further, since the flux in the transformer remains unchanged for all loads, we can logically deduce that *no part* of the energy flowing through it into the output must be being stored in the transformer. So, the question really is — what does the *power*-handling capability of a Forward converter transformer depend on? We intuitively realize that we can't use any size transformer for any output wattage! So, what governs the size? We will soon see that it is determined simply by *how much copper we can squeeze into the available "window area" of the core* (and more importantly, *how well we can utilize this available area*) without getting the transformer *too hot.*

Worst-Case Input Voltage End

The most basic question in design invariably is — what input voltage represents the worst-case point at which we need to start the design of the magnetics (from the viewpoint of core saturation)? For the Forward converter choke, this should be obvious — as for any Buck converter, we need to set its current ripple ratio at around 0.4 at V_{INMAX}. But coming to the transformer, we need some analysis before we can make a proper conclusion.

Note that the transformer of a Forward converter is in DCM, but the duty cycle is determined by the choke, which is in CCM. Therefore, the duty cycle of the transformer also gets "slaved" at the CCM duty cycle of $D = V_O/V_{INR}$, despite the fact that it is in DCM. This rather coincidental CCM + DCM interplay leads to an interesting observation — the *voltseconds across the Forward converter transformer is a constant irrespective of the input voltage.* The following calculation makes that clear, by the fact that V_{IN} cancels out completely:

$$Et = V_{IN} \times \frac{D}{f} = V_{IN} \times \frac{V_O}{V_{INR} \times f} = V_{IN} \times \frac{V_O \times n}{V_{IN} \times f} = \frac{V_O \times n}{f}$$

So, in fact, the swing of the transformer current (or its field) is the same at high input or at low input, or in fact at any input (as long as the choke remains in CCM). Since the transformer is in DCM, its peak is equal to its swing, and so the peak too does not depend on V_{IN}. Of course, the peak *switch* current I_{SW_PK} is the sum of the peak of the magnetization current I_{M_PK}, and the peak of the Secondary-side current waveform reflected onto the Primary side, that is,

$$I_{SW\_PK} = I_{M\_PK} + \frac{1}{n}\left[I_O\left(1 + \frac{r}{2}\right)\right]$$

So, although the **current limit of the switch must be set high enough to accommodate** I_{SW_PK} **at** V_{INMAX} (since that is where the maximum peak of the reflected output current component occurs), **as far as the transformer core is concerned, the peak current (and corresponding field) is just** I_{M_PK}, **which does not depend on** V_{IN}! This is, indeed, an interesting situation. Note also that as far as the choke is concerned, the peak inductor current is no longer equal to the (reflected) peak switch current (as in a DC–DC Buck topology), though the peak (free wheeling) diode current still is. Yes, if we subtract the magnetization current from the switch current, and then scale (reflect) it to the Secondary side according to the turns ratio, then the peak of that waveform will be equal to the peak inductor current.

So, effectively I_M has the property of *input voltage rejection*. We can understand this in the following way — as the input increases, the slope of the transformer current increases, and ΔI therefore tends to increase. However, the output choke, sensing a higher V_{INR}, decreases its duty cycle and therefore also the on-time of the transformer, and that tends to reduce its current swing. Coincidentally, these two opposing forces counterbalance each other perfectly, and so there is no net change in the current swing of the transformer.

As a corollary, the *core loss* in the transformer is independent of the input voltage too. **The copper loss, on the other hand, is always worse at low inputs** (except for the DC–DC Buck) — simply because the average input current has to increase so as to continue to satisfy the basic power requirement $P_{IN} = V_{IN} \times I_{IN} = P_O$.

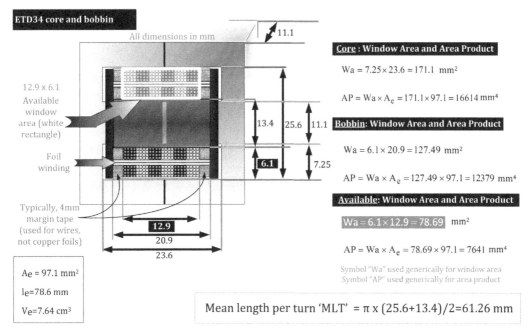

Figure 3.7: An ETD-34 bobbin analyzed.

Though we can pick any specific input voltage point for assuring ourselves that the core does not saturate *anywhere within its input range*, since the copper loss is at its worst at V_{INMIN}, we conclude that *the worst case for a Forward converter transformer is at V_{INMIN}*. For the choke, it is still V_{INMAX}.

Window Utilization

Looking at a typical winding arrangement on an "ETD-34" core and bobbin in Figure 3.7, we see that the plastic bobbin occupies a certain part of the space provided by the core — thus reducing the available window "Wa" from 171 mm² to 127.5 mm² — that is, by 74.5%. Further, if we include the 4-mm *"margin tape"* that needs to be typically provided on either side (to satisfy international safety norms regarding clearance (separation through air) and "creepage" (separation over surface of insulator) requirements between Primary and Secondary sides), we are left with an available window of only 78.7 mm² — that is a total reduction of 78.7/171 = 46%. In addition to this, looking at the left side of Figure 3.8, we see that for any given wire, only 78.5% of the square area it "physically occupies" (or will occupy in the transformer) is actually conducting (copper). So, in all, this leads to a total reduction of the available window space by $0.46 \times 0.785 = 36\%$. See also Figure 5.21.

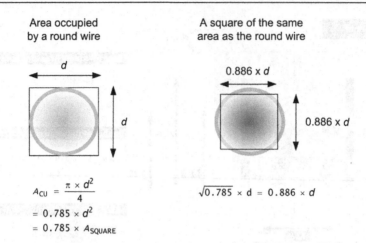

Figure 3.8: The area physically occupied by a round wire and a "square wire" of the same conducting cross-sectional area as a round wire.

We realize some more space will be lost to interlayer insulation (and any EMI screens if present), and so on. Therefore, finally, we estimate that perhaps only 30−35% of the available *core window area* will actually be occupied by copper. That is the reason why we need to introduce a "window utilization factor" K (later we will set it to an estimated value of 0.3). So,

$$K = \frac{N \times A_{\text{CU}}}{\text{Wa}}$$

and

$$N = \frac{K \times \text{Wa}}{A_{\text{CU}}}$$

Here A_{CU} is the cross-sectional area of *one* copper wire and Wa is the entire window area of the core (note that for EE, EI types of cores this is only the area of *one* of its two windows!).

Relating Core Size to Its Power Throughput

We remember that the original form of the voltage-dependent equation is

$$\Delta B = \frac{V_{\text{IN}} \times t_{\text{ON}}}{N \times A} \text{ Tesla}$$

Substituting for N, the number of Primary turns, we get

$$\Delta B = \frac{V_{\mathrm{IN}} \times t_{\mathrm{ON}} \times A_{\mathrm{CU}}}{K \times \mathrm{Wa} \times A} \text{ Tesla}$$

Performing some manipulations,

$$\Delta B = \frac{V_{\mathrm{IN}} \times I_{\mathrm{IN}} \times t_{\mathrm{ON}} \times A_{\mathrm{CU}}}{I_{\mathrm{IN}} \times K \times \mathrm{Wa} \times A} = \frac{P_{\mathrm{IN}} \times (D/f) \times A_{\mathrm{CU}}}{I_{\mathrm{IN}} \times K \times \mathrm{Wa} \times A} = \frac{P_{\mathrm{IN}} \times (D/f) \times A_{\mathrm{CU}}}{I_{\mathrm{IN}} \times D \times K \times \mathrm{Wa} \times A}$$

$$\Delta B = \frac{P_{\mathrm{IN}}}{(I_{\mathrm{SW}}/A_{\mathrm{CU}}) \times K \times f \times \mathrm{Wa} \times A} = \frac{P_{\mathrm{IN}}}{(J_{A/m^2}) \times K \times f \times \mathrm{AP}}$$

where J_{A/m^2} is the current density in A/m^2 and "AP" is called the "*area product*" (AP $= A_{\mathrm{e}} \times$ Wa). Let us now convert into CGS units for greater convenience. We get

$$\Delta B = \frac{P_{\mathrm{IN}}}{(J_{A/cm^2}) \times K \times f \times \mathrm{AP}} \times 10^8 \text{ Gauss}$$

where AP is also in cm^2 now. Finally, converting the current density into cmils/A by using

$$J_{cmils/A} = \frac{197,353}{J_{A/cm^2}}$$

we get

$$\Delta B = \frac{P_{\mathrm{IN}} \times J_{cmils/A}}{197,353 \times K \times f \times \mathrm{AP}} \times 10^8 \text{ Gauss}$$

Solving for the area us do some substitutions here. Assuming a typical current density of 600 cmil/A, utilization factor K of 0.3, and ΔB equal to 1,500 G, we get the following fundamental core-selection criterion:

$$\mathrm{AP} = 675.6 \times \frac{P_{\mathrm{IN}}}{f} \text{ cm}^4$$

Note: In a typical Forward converter, it is customary to set the swing in the B-field of the transformer at $\Delta \mathrm{B} \approx 0.15$ T. This helps reduce core loss and usually also leaves enough safety margin for avoiding hitting B_{sat} under say power-up condition at high line. Note that in a flyback, the core loss tends to be much less because ΔI is a fraction of the total current (40% typically). But since the transformer of a Forward converter is always in DCM, the swing in B is now more significant — equal to its peak value, that is, $\mathrm{B}_{PK} = \Delta \mathrm{B}$. So, if we set the peak field at 3,000 G, $\Delta \mathrm{B}$ would be 3,000 G too, roughly twice that of a flyback set to the same peak. That is why we must reduce the peak field in a Forward converter to about 1,500 G.

Worked Example (8) — Designing the Forward Transformer

We are building a 200-kHz Forward converter for an AC input range of 90−270 V. The output is 5 Vat 50 A, and the estimated efficiency is 83%. Design its transformer.

Input Power

We have

$$P_{IN} = \frac{P_O}{\text{Efficiency}} = \frac{5 \times 50}{0.83} \approx 300 \text{ W}$$

Selection of Core

We use the criterion calculated previously:

$$AP = 675.6 \times \frac{P_{IN}}{f} = 675.6 \times \frac{300}{2 \times 10^5} = 1.0134 \text{ cm}^4$$

The area product of the ETD-34 shown in Figure 3.7 is

$$AP = W \frac{[(25.6 - 11.1)/2] \times 23.6 \times 97.1}{10^4} = 1.66 \text{ cm}^4$$

This is, in theory, probably a little larger than required. But it is the closest standard size in this range. Later we will see it is in fact just about adequate.

Skin Depth

The skin depth is

$$\delta = \frac{66.1 \times [1 + 0.0042(T - 20)]}{\sqrt{f}} \text{ mm}$$

where f is in Hz and T is the temperature of the windings in °C. Therefore, assuming a final temperature of $T = 80$ °C (40 °C rise over a maximum ambient temperature of 40 °C), we get at 200 kHz

$$\delta = \frac{66.1 \times [1 + 0.0042 \times (60)]}{\sqrt{2 \times 10^5}} = 0.185 \text{ mm}$$

Thermal Resistance

An empirical formula for EE-EI-ETD-EC types of cores is

$$Rth = 53 \times V_e^{-0.54} \text{ °C}/W$$

where V_e is in cm^3. Therefore, since $V_e = 7.64$ cm^3, for the ETD-34

$$\text{Rth} = 53 \times 7.64^{-0.54} = 17.67 \,^\circ C/W$$

Maximum B-*Field*

For a 40 °C estimated rise in temperature, the maximum allowed dissipation is

$$P \equiv P_{CU} + P_{CORE} = \frac{\text{degC}}{\text{Rth}} = \frac{40}{17.67} = 2.26 \text{ W}$$

Let's divide this loss equally into copper and core losses (typical first-cut assumption). So,

$$P_{CU} = 1.13 \text{ W}$$
$$P_{CORE} = 1.13 \text{ W}$$

Therefore, the allowed core loss per unit volume is

$$\frac{\text{core loss}}{\text{volume}} = \frac{1.13}{7.64} \Rightarrow 148 \, mW/cm^3$$

Using "System B" of *Table 2.5*, we get

$$\frac{\text{core loss}}{\text{volume}} = C \times B^p \times f^d$$

where B is in Gauss and f in Hz. Therefore, solving for B,

$$B = \left[\frac{\text{core loss}}{\text{volume}} \times \frac{1}{C \times f^d} \right]^{1/p}$$

If we are using the ferrite grade "3C85" (from Ferroxcube), we see from Table 2.6 that $p = 2.2$, $d = 1.8$, and $C = 2.2 \times 10^{-14}$. Therefore,

$$B = \left[148 \times \frac{1}{2.2 \times 10^{-14} \times 2^{1.8} \times 10^{5 \times 1.8}} \right]^{1/2.2} = 720 \text{ G}$$

We note that the "B" referred to here is actually, by convention, B_{AC}. So, we get the total allowed swing as

$$\Delta B = 2 \times B = 2 \times 720 = 1,440 \text{ G}$$

Voltμseconds

Earlier, we had presented the following form of the voltage-dependent equation:

$$\Delta B = \frac{100 \times Et}{N \times A} \text{ Gauss}$$

where A is the effective area in cm^2. The **duty cycle of a typical Forward converter is set to about 0.35 at low line** so as to meet the typical 20 ms holdup time requirement without requiring an inordinately sized input capacitor. The rectified input at low line is $90 \times 2 = 127$ V. The applied voltseconds is therefore (at any line voltage)

$$Et = V_{IN} \times \frac{D}{f} = 127 \times \frac{0.35}{2 \times 10^5} = 222.25 \; V\mu s$$

Number of Turns

Since $\Delta B = 1{,}440$ G, we solve the following equation for N:

$$\Delta B = \frac{100 \times Et}{N \times A} \; \text{Gauss}$$

$$N_P = \frac{100 \times Et}{\Delta B \times A} = \frac{100 \times 222.25}{1440 \times 0.97} = 15.9 \; \text{Turns}$$

Note that this says nothing about the required *inductance*. We need this number of turns irrespective of the (Primary) inductance. Yes, changing the inductance will affect the peak magnetization and the switch current because *it changes the proportionality constant connecting B and I*. However, B still remains fixed independent of the inductance!

Assuming a 0.6-V forward drop across the diode, the required turns ratio is

$$n = \frac{N_P}{N_S} = \frac{V_{IN}}{V_{INR}} = \frac{V_{IN}}{((V_O + V_D)/D)} = \frac{127 \times 0.35}{5 + 0.6} = 7.935$$

Therefore, the number of Secondary turns is

$$N_S = \frac{15.9}{7.935} = 2.003 \; \text{Turns}$$

Note that this could have turned out to be significantly different from an integer. In that case, we would round it off to the nearest (higher) integer, and then recalculate the Primary turns, the new flux density swing, and the core loss — similar to what we did for the flyback. But at the moment, we can simply use

$$n = 8 \; \text{(turns ratio)}$$
$$N_P = 16 \; \text{Turns}$$
$$N_S = 2 \; \text{Turns}$$

Secondary Foil Thickness and Losses

The concept of skin depth presented earlier actually represents a single wire standing freely in space. For simplicity, we just ignored the fact that the field from the *nearby windings*

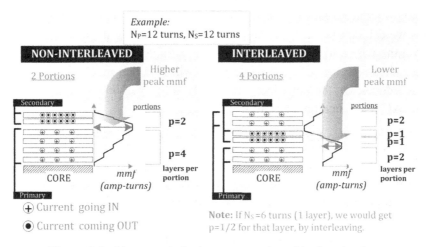

Figure 3.9: How proximity losses are reduced by interleaving.

may be affecting the current distribution significantly. In reality, ***even the annular area we were hoping would be fully available for the high-frequency current, is not***. Every winding has an associated field, and when this impinges on nearby windings, the charge distribution changes, and eddy currents are created (with their own fields). This is called the "proximity effect." It can greatly increase the *AC resistance* and thus the copper losses in the transformer.

The first thing we need to do to improve the situation is have *opposing* flux lines cancel each other. In a Forward converter, that is in fact something that tends to happen automatically because the Secondary windings pass current at the same time as the Primary, and in the opposite direction. However, even that can prove totally inadequate, especially at the higher power levels that a Forward converter is more commonly associated with. So, a further reduction in these proximity losses is achieved by *interleaving* as shown in Figure 3.9.

Basically, by splitting the sections and trying to get Primary and Secondary layers *adjacent to each other as much as possible*, we can increase cancellation of local adjoining fields. In effect, we are trying to prevent the Ampere-turns from *cumulating* as we go from one layer to the next. Note that the Ampere-turns are proportional to the local fields that are causing the proximity losses. However, it is impractical to interleave too much — because we will need several more layers of Primary-to-Secondary insulation, more terminations, and also more EMI screens at every interface (if required) — all of which will add up to higher cost and eventually lead to possibly higher, rather than lower, leakage. Therefore, ***most medium-power off-line supplies just split the Primary into two sections, one on either side of a single-section Secondary***.

The other way to reduce losses is to decrease the thickness of the conductor. But there are several ways we can do this. If, for example, we take a winding made up of single-strand wire, and split the wire into several paralleled finer strands in such a way that *the overall DC resistance does not* change in the process, we will find that the AC resistance goes up first before it reduces. On the other hand, if we take a foil winding, and decrease its thickness, the AC resistance falls before it rises again.

In Figure 3.9, we have also defined "*p*," the *layers per portion*. Note how *p* changes when we interleave.

But how do we go about actually estimating the losses? Dowell reduced a very complex multidimensional problem into a simpler, one-dimensional one. Based on his analysis, we can show that there is an *optimum thickness for each layer*. Expectedly, this turns out to be much less than $2 \times \delta$, where δ is the skin depth defined earlier.

> *Note: In the flyback, we had ignored the proximity effect for the sake of simplicity. But in any case, since the Primary and Secondary windings do* not *conduct at the same time, interleaving won't help. But interleaving is still carried out in the flyback in a manner similar to the Forward converter. However, the purpose then is to increase coupling between Primary and Secondary, and thereby reduce the leakage inductance. However, this also increases the capacitive coupling — unless grounded screens are placed at the Primary–Secondary interface. Screens are in general helpful in reducing high-frequency noise from coupling into the output and suppressing common-mode-conducted EMI. But they also increase the leakage inductance, which is of great concern particularly in the flyback. Note also that screens must be very thin, or they will develop very high eddy current losses of their own. Further, the ends of an internal screen should not be connected together, or they will constitute a shorted turn in the transformer.*

In Figure 3.10, we have plotted out Dowell's equations in a form applicable to a *square* current waveform (unidirectional) in a transformer with foil windings. Note that the original Dowell curves actually plot F_R versus X. But we have plotted F_R/X versus X, where

$$F_R = \frac{R_{\mathrm{AC}}}{R_{\mathrm{DC}}}$$

and

$$X = \frac{h}{\delta}$$

h being the thickness of the foil. The reason why we have not plotted F_R versus X is that F_R is only the *ratio* of the AC-to-DC resistance. **It is not F_R, but R_{AC} that we are really interested in minimizing**. So, the "optimum R_{AC}" point need not necessarily be the point of the lowest F_R.

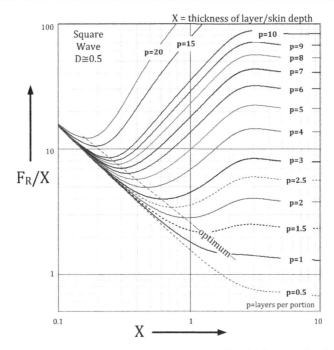

Figure 3.10: Finding the lowest AC resistance, as the thickness of a foil is varied.

Let us try to understand this for a stand-alone foil (similar to what we did in Figure 3.3). If we slowly increase the thickness of the foil, once the foil thickness exceeds 2δ, the AC resistance won't change any further, since the cross-sectional area available for the high-frequency current remains confined to δ on each side of the foil. But the DC resistance continues to decrease as per $1/h$ — and as a result F_R will increase. So, the relationship between R_{AC} and F_R is not necessarily obvious. Therefore, since $F_R = R_{AC}/R_{DC}$, with $R_{DC} \propto 1/h$, we get $R_{AC} \propto F_R/h$. And this is what we really need to minimize (for a foil). Further, since we always like to write any frequency-dependent dimension with reference to the skin depth, we have plotted F_R/X versus X in Figure 3.10.

Note that in Figure 3.10, the $p = 1$ and $p = 0.5$ curves do not really have an "optimum." For these, the F_R/X (AC resistance) can be made even smaller as we increase X (thickness). F_R will in fact become much greater than 1. However, we see that *for $p = 1$, for example, no significant reduction in AC resistance occurs if X exceeds about 2*, that is, thickness of foil equal to twice the skin depth. We can make it thicker if we want, but only for marginal improvement in the Secondary-winding losses. Further, in the process, we may also take away available area for the Primary windings (and any other Secondary windings), and that can lead to higher overall losses. *Though we are also cautioned not to fill up all "available space" with copper, especially when we come to (round) wire windings.* That can be shown, not only to increase F_R, but R_{AC} too.

Now let us apply what we have learned to our ongoing numerical example. We start by taking a copper foil wound twice on the ETD-34 bobbin — to form the 5-V Secondary winding. Since this is interleaved with respect to the Primary, only one turn "belongs" to each split section. So, the *layers per portion* for the Secondary is $p = 1$. We will calculate the losses, and if acceptable, we will stay with the resulting arrangement.

We can start with a reasonable current density (about 400 cmils/A should suffice here). We use

$$h = \frac{I_O \times J_{cmils/A} \times 10^2}{\text{width} \times 197,353} \text{ mm}$$

where h is the foil thickness in mm, I_O is the load current (50 A in our example), and "width" is the width available for the copper strip (20.9 mm for the ETD-34).

Alternatively, we can directly consult Figure 3.10 and pick an X of 2.5 for an estimated F_R/X of 1.4. Thus,

$$h = X \times \delta = 2.5 \times 0.185 = 0.4625 \text{ mm}$$

The *mean length per turn* ("MLT") of ETD-34 is 61.26 mm (see Figure 3.7), the ("hot") resistivity of copper ("ρ") is 2.3×10^{-5} Ω-mm, so we get the resistance of the Secondary winding in ohms as

$$R_{AC\_S} = \left(\frac{F_R}{X}\right) \times \frac{\rho \times \text{MLT} \times N_S}{\text{width} \times \delta} = (1.4) \times \frac{2.3 \times 10^{-5} \times 61.26}{20.9 \times 0.185} = 1.02 \times 10^{-3}$$

Note that since F_R/X is set to 1.4, the corresponding F_R is

$$F_R = 1.4 \times \frac{h}{\delta} = 1.4 \times \frac{0.4625}{0.185} = 3.5$$

This is fairly high, but as explained, it is actually helpful here, because R_{AC} goes down. Now, the current in the Secondary looks like a typical switch waveform, with its center equal to the load current (50 A), and a certain current ripple ratio set by the output choke. Its root mean square (RMS) value is

$$I_{RMS\_S} = I_O \times \sqrt{D \times \left(1 + \frac{r^2}{12}\right)} A$$

However, we do not yet know what the current ripple ratio of the choke r is *at 90 VAC*. The r has probably been set to 0.4 at V_{INMAX}, *not at* V_{INMIN}. Nevertheless, it is easy to work out the new r as follows. The duty cycle is inversely proportional to input voltage. Therefore, if D is 0.35 at 270 VAC, then at 90 VAC it is $0.35/3 = 0.117$. *Further, r varies as per (1 − D) for a Buck stage.* Therefore, the value of r at 90 VAC is

$$r = \frac{1 - 0.35}{1 - 0.117} \times 0.4 = 0.294$$

So, the RMS current in the Secondary winding is

$$I_{RMS\_S} = I_O \times \sqrt{D \times \left(1 + \frac{r^2}{12}\right)} = 50 \times \sqrt{0.35 \times \left(1 + \frac{0.294^2}{12}\right)} = 29.69 \text{ A}$$

The heat dissipated in the Secondary windings is finally

$$P_S = I_{RM\_S}^2 \times R_{AC\_S} = 29.69^2 \times 1.02 \times 10^{-3} = 0.899 \text{ W}$$

If the losses are not acceptable, we may need to look for a bobbin that will allow a *wider width* of foil. Or we can consider paralleling several thinner foils to increase p. For example, if we take four paralleled (thinner) foils in parallel (each insulated from the others), we will get four effective layers for the Secondary, and the layers per portion will then become two.

Primary Winding and Losses

For the Secondary, we have finally chosen copper foil of thickness 0.4625 mm (i.e., $0.4625 \times 39.37 = 18$ mil). Let us assume each foil is covered on both sides by a 2-mil thick Mylar® tape. Since 1 mil is 0.0254 mm, we have effectively added 4×0.0254 mm to the foil thickness. In addition, there will be three layers of 2-mil tape between each of the two Primary–Secondary boundaries (a total of 12 mil). So, in all, the *thickness occupied by the Secondary and the insulation, h_S,* is

$$h_S = (N_S \times h) + (N_S \times 4 \times 0.0254) + (12 \times 0.0254) \text{ mm}$$

or

$$h_S = N_S \times (h + 0.102) + 0.305 \text{ mm}$$

So, in our case,

$$h_S = 2 \times (0.4625 + 0.102) + 0.305 = 1.434 \text{ mm}$$

The ETD-34 has an available height inside the bobbin of 6.1 mm. That now leaves $6.1 - 1.434 = 4.67$ mm. Therefore, *each section* of the split Primary has an available winding height of 2.3 mm only. We should ultimately check that we can accommodate the Primary winding we decide on, within this space.

Note that for the Primary, the available width is only 12.9 mm (since there is 4-mm margin tape on each side — for the Secondary, since we have a foil with tape wrapped over it, we do not need margin tape). We need to find how best to accommodate eight turns into this available area with minimum losses.

> *Note: It is not mandatory to use a particular thickness of insulating tape, provided it is safety approved to withstand a specified voltage. We can, for example, use 1-mil approved tape or even ½-mil approved tape (if it suits our production, helps lower the cost, and/or improves performance in some way).*

Let us now understand the basic concept behind winding wires. For a *standalone* wire, as in Figure 3.3, as we increase the diameter of the wire, the cross-sectional area available for the high-frequency current is $(\pi \times d) \times \delta$. And since resistance is inversely proportional to cross-sectional area, we get $R_{AC} \propto 1/d$. Similarly, $R_{DC} \propto 1/d^2$. So, $F_R \propto d$. Therefore, $R_{AC} \propto 1/F_R$. This actually means that a higher F_R (bigger diameter) will decrease the AC resistance! That is not surprising because the annulus available for the high-frequency current does increase if the diameter increases. However, this is not the way to go when dealing with "non-standalone" wire. Because, by increasing the diameter, we will inevitably move to *higher number of layers*, and Dowell's equation then tells us that the losses will increase significantly on account of that alone, not decrease.

On the top left side of Figure 3.11, we have Dowell's *original* curves, which show how F_R varies with respect to X (i.e., h/δ). The parameter for each curve is *layers per portion* (i.e., p). Note that Dowell's curves talk in terms of equivalent *foils* (layers of current) only. They don't care about the actual number of turns in the Primary or Secondary (i.e., from the electrical point of view), but only about the *effective layers per portion* (from the *field point of view*). So, when we consider a layer of round wires of diameter "d," we need to convert this into an *equivalent foil*. Looking back at the right side of Figure 3.8, we see that this amounts to replacing a wire of diameter d with a foil *slightly thinner* (i.e., with the same amount of copper, but in a square shape). Alternatively, if we want to get a foil of $X = 4$ for example, we need to start with a wire of diameter $1/0.886 = 1.13$ times X. Finally, as indicated, all these copper squares then merge (from the field point of view) to give an equivalent layer of foil.

In Figure 3.11, we are also conducting a certain strategy — as an alternative way of laying out wires optimally. Suppose we have several round wires laid out side by side with a diameter $1.13 \times 4\delta$. Suppose also, that this constitutes *one layer per portion* in a certain winding arrangement. This is therefore equivalent to a single-layer foil of thickness 4δ, that is, $X = 4$. Now using Dowell's curves, the corresponding F_R is about *4* (points marked "A" in Figure 3.11). Suppose we then divide *each strand into four strands*, where each strand has a diameter half the original. Therefore, *the cross-sectional area occupied by copper remains the same* because

$$A = 4 \times \frac{\pi \times (d/2)^2}{4} = \frac{\pi \times d^2}{4}$$

The adjacent graphs share the vertical axis "F_R". The left graph is Dowell's eqns plotted out simply (versus X). The right graph plots Dowell's eqns in terms of the parameter "starting value of X (= h/δ)" --- before subdivisions. On its horizontal axis we have "sub" (the number of subdivision steps, as explained below. Point A is an example: corresponding to layers/portion "p" =1, made from round wires of diameter 4δ (i.e. X=4). This is subdivided successively to get B, C and D in order.

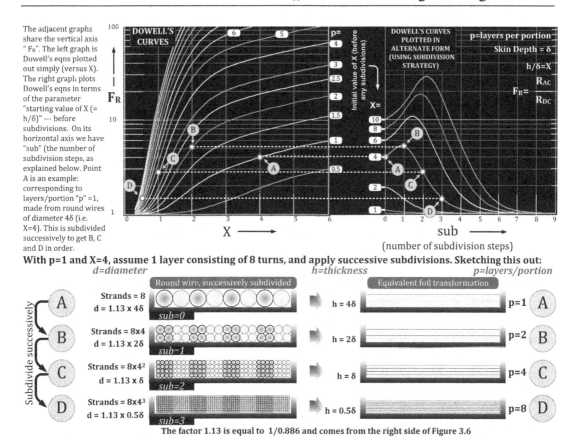

With p=1 and X=4, assume 1 layer consisting of 8 turns, and apply successive subdivisions. Sketching this out:

d=diameter	h=thickness	p=layers/portion

Figure 3.11: Understanding the process of "subdivision" — keeping the DC resistance unchanged and how the equivalent foil transformation process takes place.

However, the *equivalent foil thickness is now half of what it was* — $2δ$ (i.e., $X = 2$). And we also now have *two layers per portion* from Dowell's standpoint. Consulting Dowell's curves, we get an F_R of about **5** now (marked "B"). Since we are keeping R_{DC} fixed this subdivision strategy, $R_{AC} \propto F_R$. Therefore now, decreasing F_R is a sure way to go to decrease R_{AC}. So an F_R of *5* is decidedly worse than an F_R of *4* (not so for an actual foil winding). We now go ahead and subdivide once more in a similar manner. So, we then get four layers per portion, each with $X = 1$, and F_R has gone down to about **2.6** (points marked "C"). We subdivide once more, and we get eight layers per portion, with $X = 0.5$. This gives us an F_R of about *1.5* (marked "D"). This is an acceptable value for F_R.

Note that all these steps have been collected and plotted out in Figure 3.11 on the right side, **with the horizontal axis being the number of successive subdivision steps** (in each step we subdivided each wire into four of the same DC resistance). These steps are being called "sub" (for subdivision step), where *sub* goes from 0 (no subdivision) to 1 (one subdivision), 2 (two subdivisions), and so on. We then also realize that with each step, X and p change as per

$$X \to \frac{X}{2^{sub}}$$
$$p \to p \times 2^{sub}$$

For example, after four subdivision steps, the foil thickness will drop by a factor of 16, and the number of layers will increase by the same factor. We can then look at Dowell's curves to find out the new F_R.

However, there are a few problems with directly applying Dowell's curves to switching power regulators. For one, the original curves only talked about the ratio of the thickness to the skin depth — and we know skin depth depends on frequency. So, implicitly, Dowell's curves provide the F_R for a *sine wave*. Further, Dowell's curves do not assume the current has any DC value. So, engineers, who adapted Dowell's curves to power conversion, would usually first break up the current waveform into its *AC and DC components*, apply the F_R obtained from the curves to the AC component only, compute the DC loss separately (with $F_R = 1$), and then sum as follows:

$$P = I_{DC}^2 \times R_{DC} + I_{AC}^2 \times R_{DC} \times F_R$$

However, in our case, we have preferred to follow the more recent approach of using the actual (unidirectional) current waveform, splitting it into Fourier components, and summing to get the *effective F_R*. The losses are expressed in terms of the thickness of the foil *as compared to the δ at the fundamental frequency* (first harmonic). We also include the DC component in computing this *effective F_R*. That is the reason when calculating the Secondary-winding losses, that we were able to use the simple equation:

$$P = I_{RMS}^2 \times R_{AC} = I_{RMS}^2 \times (F_R \times R_{DC})$$

In that case, the F_R was actually the effective F_R (computed for a square wave with DC level included), though not explicitly stated. However, note that the graphs in Figure 3.11 are still based on the original sine-wave approach, and the purpose there was only to demonstrate the *subdivision technique* through the original curves.

In Figure 3.12, we have modified Dowell's original sine-wave curves. Fourier analysis has been carried out while constructing these curves, and so the designer can apply them *directly* to the typical (unidirectional) current waveforms of power conversion. We will shortly use these curves to do the calculations for the Primary winding of our ongoing numerical example.

But one question may be puzzling the reader — why are we not using the previous F_R/X curves (see Figure 3.10) that we used for the Secondary? The reason is the situation is

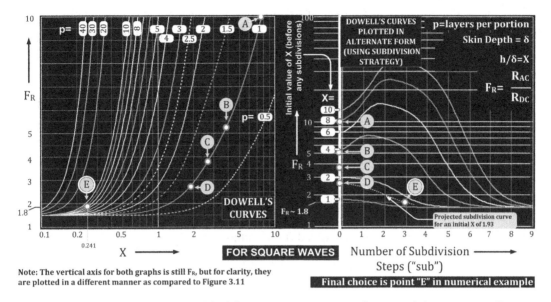

Note: The vertical axis for both graphs is still F_R, but for clarity, they are plotted in a different manner as compared to Figure 3.11

FOR SQUARE WAVES

Final choice is point "E" in numerical example

Figure 3.12: Dowell's curves modified for square current waveforms and the corresponding F_R curves for the subdivision method.

different now. The curves in Figure 3.10 were also Dowell's curves interpreted for a square wave, except that on the vertical axis we used F_R/X, not F_R. That is useful only when we are varying h and seeing when we get the lowest R_{AC}, as for a Secondary foil winding. But for the Primary (round wire) windings, we are going to fix the height of the windings in each step of the iterations that follow. We will be using the subdivision technique in each iteration, and therefore we *keep the DC resistance constant.* So, now the minimum R_{AC} (for a given iteration step) *will be achieved at the minimum F_R, not at the minimum F_R/X.*

The subdivision method was presented in Figure 3.11, but now we will use the modified curves in Figure 3.12.

First Iteration

Let us plan to try to fit eight turns on one layer. *Lesser* number of layers will usually be better. We remember that we have 12.9 mm available width on the bobbin. So, if we stack eight turns side by side (no gap between them), we will require each of these eight round wires to have a diameter of

$$d = \frac{\text{width}}{\text{turns per layer}} = \frac{12.9}{8} = 1.6125 \text{ mm}$$

We can check that the available height of 2.3 mm is big enough to accommodate this diameter of wire. The *penetration ratio* X is (using the equivalent foil transformation)

$$X = \frac{0.886 \times d}{\delta} = \frac{0.886 \times 1.6125}{0.185} = 7.723$$

The p is equal to 1. From either of the graphs in Figure 3.12, we can see that the F_R will be about 10 in this case (marked "A"). Further, from the graph on the right side, we can see that we need to subdivide the "$X = 7.7$" curve (imagine it close to the $X = 8$ curve) seven times to get the F_R below 2. That would give strands of diameter

$$d \rightarrow \frac{d}{2^{sub}} = \frac{1.6125}{2^7} = 0.0125 \text{ mm}$$

The corresponding AWG can be calculated by rounding off

$$AWG = 18.154 - 20 \ log(d)$$

So, we get

$$AWG = 18.154 - 20 \ log(0.0125) \Rightarrow 56 \text{ AWG}$$

But this is an extremely thin wire and may not even be available! Generally, from a production standpoint, *we should not use anything thinner than 45 AWG (0.046 mm)*.

Second Iteration

The problem with the first iteration is that we started with a very thick wire, with a very high F_R. So, this demanded several subdivisions to get the F_R to fall below 2. But what if we start off with a wire of diameter lesser than 1.6125 mm? We would then need to introduce some wire-to-wire spacing so that we can spread the eight turns evenly across the bobbin. However, that would be wasteful! We should remember that if a layer is already assigned and present, we might as well use it to our full advantage to lower the DC resistance — the problem only starts when we indiscriminately increase the *number of layers*. Therefore, in our case, let us try paralleling *two* thinner wires to make up the Primary. We still want to keep to one layer (without spacing). That means we will now have 16 wires placed side by side in one layer. We now define a *"bundle"* as the number of wires paralleled to make the Primary winding (we will be subdividing each of these further). So, in our case,

$$\text{bundle} = 2$$

The diameter we are starting off with is

$$d = \frac{\text{width}}{\text{turns per layer}} = \frac{12.9}{16} = 0.806 \text{ mm}$$

The penetration ratio X is

$$X = \frac{0.886 \times d}{\delta} = \frac{0.886 \times 0.806}{0.185} = 3.86$$

The p is still equal to 1. From both the graphs in Figure 3.12, we can see that the F_R will be about 5.3 in this case (marked "B"). Further, from the graph on the right side, we can see that we need to subdivide five times to get the F_R below 2. That would give strands of diameter

$$d \rightarrow \frac{d}{2^{\text{sub}}} = \frac{0.806}{2^5} = 0.025 \text{ mm}$$

This is still thinner than the practical AWG limit of 0.046 mm.

Third Iteration

So, we now parallel *three* wires to make up the Primary. That means we will have 24 wires side by side in one layer.

$$\text{bundle} = 3$$

The diameter we are starting off with is

$$d = \frac{\text{width}}{\text{turns per layer}} = \frac{12.9}{16} = 0.806 \text{ mm}$$

The penetration ratio X is

$$X = \frac{0.886 \times d}{\delta} = \frac{0.886 \times 0.538}{0.185} = 2.58$$

The p is still equal to 1. From both the graphs in Figure 3.12, we can see that the F_R will be about 3.7 in this case (marked "C"). Further, from the graph on the right side, we can see that we need to subdivide four times to get the F_R below 2. That would give strands of diameter

$$d \rightarrow \frac{d}{2^{\text{sub}}} = \frac{0.538}{2^4} = 0.034 \text{ mm}$$

But this is still too thin!

Fourth Iteration

Let us now parallel four wires to start with. We will have 32 wires in one layer.

$$\text{bundle} = 4$$

The diameter we are starting off with is

$$d = \frac{\text{width}}{\text{turns per layer}} = \frac{12.9}{32} = 0.403 \text{ mm}$$

The penetration ratio X is

$$X = \frac{0.886 \times d}{\delta} = \frac{0.886 \times 0.403}{0.185} = 1.93$$

The p is still equal to 1. From both the graphs in Figure 3.12, we can see that the F_R will be about 2.8 in this case (marked "D"). Further, from the graph on the right side, we can see that we need to subdivide three times to get the F_R below 2. That would give strands of diameter

$$d \rightarrow \frac{d}{2^{\text{sub}}} = \frac{0.403}{2^3} = 0.05 \text{ mm}$$

This corresponds to AWG 44 and would be of acceptable thickness.

Note that by the process of subdivision, the number of layers per portion goes up as

$$p \rightarrow p \times 2^{\text{sub}}$$

So, with three subdivisions, we will get

$$p \rightarrow p \times 2^{\text{sub}} = 1 \times 2^3 = 8 \quad (\textit{layers per portion})$$

that is, eight layers. The penetration ratio has similarly now become

$$X \rightarrow \frac{X}{2^{\text{sub}}} = \frac{1.93}{2^3} = 0.241$$

The F_R is now about 1.8 as can also be confirmed from the graph on the left side of Figure 3.12 (for $X = 0.241$, $p = 8$). This is point "E" in the two graphs.

The number of strands each original "bundle" has been divided into is

$$\text{strands} = 4^{\text{sub}} = 4^3 = 64$$

So, finally, *the Primary winding consists of four bundles in parallel, each bundle consisting of 64 strands, side by side in one layer, with an F_R of about 1.8.*

We can continue the process if we want to get a slightly lower F_R. But at some point, we will find the F_R *will start to go up again*. For our purpose, we will take an F_R less than 2 as acceptable to proceed with the loss estimates.

Note that further tweaking will always be required since when we bunch wires together to form a bundle, they will "stack" in a certain manner that will affect the dimensions from what we have assumed. Further, the diameter of the wire we used was for bare wire and was slightly less than the coated diameter. Note that in general, *if after winding several layers evenly, we are left with a few turns that seem to need another layer to complete, we are better off reducing the Primary number of turns and sticking to the existing completed layers, because even a few turns extra will count as a new layer from the field point of view and increase proximity losses.*

We can now calculate the losses for the two Primary sections combined, since they can be considered to be identical and with the same F_R. The AC resistance in ohms of the entire Primary winding is

$$P_{AC\_P} = (F_R) \times \frac{\rho \times \text{MLT} \times N_P}{\pi \times (d^2/4) \times \text{bundles} \times \text{strands}} = (1.8) \times \frac{2.3 \times 10^{-5} \times 61.26 \times 16}{\pi \times ((0.05)^2/4) \times 4 \times 64} = 0.08 \ \Omega$$

So, the loss is

$$P_P = I_{RMS\_P}^2 \times R_{AC\_P} = \left(\frac{I_{RMS\_S}}{n}\right)^2 \times R_{AC\_P} = \left(\frac{29.69}{8}\right)^2 \times 0.08 = 1.102 \ \text{W}$$

Had we gone further and divided the Primary into five bundles and then subdivided three times, we would get eight layers with 64 strands of 0.04-mm diameter wire per bundle and an F_R of 1.65 — which seems better than the 1.8 we got in the last step. But since the wires are so thin to start with, the DC resistance now goes up, and the dissipation will rise to 1.26 W.

Total Transformer Losses

The total dissipation in the transformer is therefore

$$P = P_{\text{CORE}} + P_{\text{CU}} = P_{\text{CORE}} + P_P + P_S = 1.13 + 1.102 + 0.899 = 3.131 \ \text{W}$$

The estimated temperature rise

$$\Delta T = \text{Rth} \times P = 17.67 \times 3.145 = 55.3 \ °\text{C}$$

What we are seeing is a typical practical situation! The temperature rise is 15 °C higher than we were expecting! However, 55 °C is perhaps still acceptable (even from the standpoint of getting safety approvals without special transformer materials). Admittedly, there is room for more optimization. However, the next time we do the process, we must note that the core loss is only a third the total loss, not half, as we had initially assumed.

Note also that calculations in related literature may predict a smaller temperature rise. But the fact is that these are usually based on the sine-wave versions of Dowell's equations, and we know that will typically underestimate the losses significantly.

CHAPTER 4

The Topology FAQ

This section serves to highlight and summarize the gamut of key topology-related design issues that should be kept in mind when actually designing converters (or when appearing for a job interview!).

Questions and Answers

Question 1: For a given input voltage, what output voltages can we get in principle, using only basic inductor-based topologies (Buck, Boost, and Buck-Boost)?

Answer: The Buck is a *step-down* topology ($V_O < V_{IN}$), the Boost only *steps-up* ($V_O > V_{IN}$), and the Buck-Boost can be used to either step-up or step-down ($V_O < V_{IN}$, $V_O > V_{IN}$). Note that here we are referring only to the *magnitudes* of the input and output voltages involved. So, we should keep in mind that the Buck-Boost inverts the polarity of the input voltage.

Question 2: What is the difference between a topology and a configuration?

Answer: We know that, for example, a "down-conversion" of 15 V input to a 5 V output is possible using a Buck *topology*. But what we are referring to here is actually a "positive-to-positive" Buck *configuration*, or simply, a "positive Buck." If we want to convert −15 V to −5 V, we need a "negative-to-negative" Buck configuration, or simply, a "negative Buck." We see that a topology is fundamental (e.g., the Buck) — but it can be *implemented* in more than one way, and these constitute its *configurations*.

Note that in the down-conversion of −15 V to −5 V, we use a Buck (step-down) topology, even though mathematically speaking, −5 V is actually a *higher* voltage than −15 V! Therefore, only *magnitudes* are taken into account in deciding what the nature of a power conversion topology is.

Similarly, a conversion of say 15 V to 30 V would require a "positive Boost," whereas −15 V to −30 V would need a "negative Boost." These are the two configurations of a Boost topology.

For a Buck-Boost, we need to always *mentally* keep track of the fact that it inverts the polarity (see next question).

Question 3: What is an "inverting" configuration?

Answer: The Buck-Boost is a little different. Although it has the great advantage of being able to up-convert or down-convert on demand, it also always ends up *inverting* the sign of the output with respect to the input. That is why it is often simply referred to as an "inverting topology." So, for example, a "positive-to-negative" Buck-Boost would be required if we want to convert 15 V to −5 V or say to −30 V. Similarly, a "negative-to-positive" Buck-Boost would be able to handle −15 V to 5 V or to 30 V. Note that a Buck-Boost *cannot* do 15 V to 5 V for example, nor can it do −15 V to −5 V. The convenience of up- or down-conversion (on demand) is thus achieved only at the expense of a polarity inversion — the conventional (inductor-based) Buck-Boost topology is useful only if we either desire, or are willing to accept, this inversion.

Question 4: Why is it that *only* the Buck-Boost gives an inverted output? Or conversely, why can't the Buck-Boost ever *not* invert?

Answer: In all topologies, there is a *voltage reversal* across the inductor when the switch turns OFF. So, the voltage at one end of the inductor "flips" *with respect to its other end.* Further, when the switch turns OFF, the voltage present at the swinging end of the inductor (i.e., the switching node) always gets "passed on" to the output, because the diode is then conducting. But in the case of the Buck-Boost, the "quiet end" of its inductor is connected to the *ground reference* (no other topology has this). Therefore, the voltage reversal that takes place at its other end (swinging end) is also a voltage reversal *with respect to ground.* And since this is the voltage that ultimately gets transmitted to the output (which is also referenced to ground), the reversal is virtually "seen" at the output. See Figure 1.15.

Of course the output rail continues to stay inverted even when the switch turns ON, because the diode then stops conducting, and there is an output capacitor present, that holds the output voltage steady at the level it acquired during the switch off-time.

Question 5: Why do we always get only up-conversion from a Boost converter?

Answer: *Inductor* voltage reversal during a switch transition occurs in all DC−DC switching topologies — it just does not necessarily lead to an *output* reversal. But in fact, inductor voltage reversal is responsible for the fact that in a Buck, the input voltage is always stepped-down, whereas in a Boost, it is stepped-up. It all depends on where the "quiet" end of the inductor connects to. In the Boost, the "quiet" end connects to the *input* rail (in the Buck, to the output rail). Therefore, since the swinging end of the Boost inductor is connected to ground during the switch on-time, it then flips *with respect to the input rail* during the switch off-time, gets connected to the output through the conducting diode, and thereby we get a boosted output voltage. See Figure 1.15.

Question 6: What is really "ground" for a DC−DC converter?

Answer: In a DC−DC converter there are two input rails and two output rails. But one of these rails is common to both the input and output. This rail is the (power) "ground." The input and output voltages are measured with respect to this (reference) rail, and this gives them their respective magnitude and polarity. See Figures 1.12 and 1.14.

Question 7: What is "ground" for the control IC?

Answer: The reference rail, around which most of the internal circuitry of the IC is built, is its local (IC) ground. This rail comes out of the package as the ground pin(s) of the IC. Usually, this is connected on the PCB directly to the power ground (the common reference rail described above). However, there are exceptions, particularly when an IC meant primarily for a certain topology (or configuration) is rather unconventionally configured to behave as another topology altogether (or just a different configuration). Then the IC ground may in fact differ from the power ground. See Figure 9.17.

Question 8: What is "system" ground?

Answer: This is the reference rail for the entire system. So in fact, all on-board DC−DC converters present in the system usually need to have their respective (power) grounds tied firmly to this system ground. The system ground in turn usually connects to the metal enclosure, and from there on to the "earth (safety) ground" (i.e., into the mains wiring).

Question 9: Why are negative-to-negative DC−DC configurations rarely used?

Answer: The voltages applied to and/or received from on-board DC−DC converters are referenced by the rest of the system to the common shared system ground. By modern convention, all voltages are usually expected to be *positive* with respect to the system ground. Therefore, all on-board DC−DC converters also need to comply with the same convention. And that makes them necessarily positive-to-positive converters.

Question 10: Why are *inverting* DC−DC converters rarely used?

Answer: We usually cannot afford to let any given on-board converter attempt to "redefine" the ground in the middle of a system. However, inverting regulators can on occasion be used, especially if the converter happens to be a "front-end" converter. In this case, since the system effectively *starts* at the output terminals of this converter, we may be able to "define" the ground at this point. In that case, the relative polarity between the input and output of the converter may become a "don't care" situation.

Question 11: Can a Buck regulator be used to convert a 15 V input to 14.5 V output?

Answer: Maybe, maybe not! Technically, this is a step-down conversion, since $V_O < V_{IN}$. Therefore, in principle, a Buck regulator should have worked. However, in practice there are some limitations regarding how *close* we can set the output of a converter in relation to the input.

Even if the switch of a Buck regulator is turned *fully* ON (say in an all-out effort to produce the required output), there will still be some remaining forward drop across the switch, V_{SW}, and this would effectively subtract from the applied input V_{IN}. Note that in this fully ON state, the switcher is basically functioning just like a "*low* dropout" regulator, or "LDO," and so the concerns expressed in *Chapter 1* regarding the minimum achievable *headroom* of an LDO apply to the switcher too in this state. As an example — if the switch drop V_{SW} is 1 V then we certainly can't get anything higher than 14 V output from an input of 15 V.

The second consideration is that even if, for simplicity, we assume *zero* forward voltage drops across both the switch and the diode, we still may not be able to deliver the required output voltage — because of *maximum duty cycle* limitations. So for example, in our case, what we need is a (theoretical) duty cycle of $V_O/V_{IN} = 14.5$ V/15 V $= 0.97$, that is, 97%. However, many Buck ICs in the market are not designed to guarantee such a high duty cycle. They usually come with an internally set maximum duty cycle limit ("D_{MAX}"), typically around 90 to 95%. And if that is so, $D = 97\%$ would be clearly out of their capability. Buck switchers with a P-channel MOSFET can usually do 100% duty cycle.

A good power supply designer also always pays heed to the tolerance or *spread* of the published characteristics of a device. This spread is usually expressed as a *range* with a specified "min" (minimum), a "max" (maximum), and a "typ" (typical, or nominal). For example, suppose a particular IC has a published maximum duty cycle range of 94–98%, we cannot *guarantee* that *all* production devices would be able to deliver a regulated 14.5 V — simply because not *all of them* are *guaranteed* to be able to provide a duty cycle of 97%. Some parts may manage 97%, but a few others won't go much beyond a duty cycle maximum of 94%. So, what we need to do is to select an IC with the published "min" of its tolerance range greater than the desired duty cycle. For example, a Buck IC with a published D_{MAX} range of 97.5–99% *may* work in our current application.

Why did we say "may" above? If we include the forward drops of the switch and diode in our calculation, we actually get a *higher* duty cycle than the 97% we got using the "ideal" equation $D = V_O/V_{IN}$. The latter equation implicitly assumes $V_{SW} = V_D = 0$ (besides ignoring other key parasitics like the inductor's DCR). So, the actual measured duty cycle in any application may well be a couple of percentage points higher than the ideal value.

In general, we should remember that whenever we get *too close to the operating limits* of a control IC, we can't afford to ignore key parasitics. We must also account for temperature variations, because temperature may affect efficiency, and thereby the required duty cycle.

Question 12: What role does temperature play in determining the duty cycle?

Answer: As mentioned in *Chapter 1*, it is generally hard to predict the overall effect of temperature on a power-supply's efficiency, and thereby on its duty cycle variation with respect to temperature. Some loss terms increase with temperature and some decrease. However, to be conservative, we should at a minimum account for the increase in the forward drop of the metal oxide semiconductor field effect transistor (MOSFET) switch. For low-voltage MOSFETs (rated ~ 30 V), the increase in R_{DS} (on-resistance) in going from room temperature to "hot" is typically 30–50%. So, we typically multiply the published room temperature on-resistance by 1.4. For high-voltage MOSFETs, as used in off-line power supplies, the increase is about 80–100%. So, we typically multiply the room temperature on-resistance by 1.8.

Question 13: How can we convert an *unregulated* input of 15 V to a regulated output of 15 V?

Answer: The term "unregulated" implies that the stated value just happens to be the "typical" (usually center) of a certain *range*, which may or may not yet have been defined. So, an "unregulated input of 15 V" could well mean say 10–20 V, or 5–25 V, or 12–18 V, and so on. Anything that includes 15 V is possible.

Of course, ultimately, we do need to know what this input range really is. But it should already be apparent that for a "15 V to 15 V conversion", if the input falls at the lower end of its range, we would need to up-convert, and if the input is at its upper end, we would need to down-convert. Therefore, we must choose a topology capable of performing *both* step-up and step-down conversions *on demand*.

How about the Buck-Boost? Unfortunately, the standard inductor-based Buck-Boost also gives us an *inverted* output, which we really don't want here. What we need is a *non-inverting* step-up/step-down topology. Looking, a suitable candidate for this is the "SEPIC" (single-ended primary inductance converter) topology (see Figures 4.1, 9.14 and 9.15). It is best visualized as *composite* topology — a Boost stage followed by a Buck cell. Though this "Boost-Buck" combination needs only one switch, it requires an additional inductor, and also entails significantly more design complexity. We may therefore wish to consider a derivative (or variant) of the conventional Buck-Boost topology, but with the inductor replaced by a *transformer*. In effect, what we are doing is — we are first separating (isolating) the input from the output, and then reconnecting the windings of the transformer in an appropriate manner so as to correct for the inversion. Thus, we get a *non-inverting* or "non-isolated transformer-based Buck-Boost" — sometimes simply called a "flyback" topology.

Question 14: It is much easier to find "off-the-shelf" inductors. So why is a transformer-based Buck-Boost even worth considering?

Answer: It is true that most designers prefer the convenience of off-the-shelf components, rather than custom-designed components (like transformers). However, *high-power* off-the-shelf inductors often come with *two identical windings* wound in parallel (on the same core), (though that may not be immediately apparent just by looking at the datasheet).

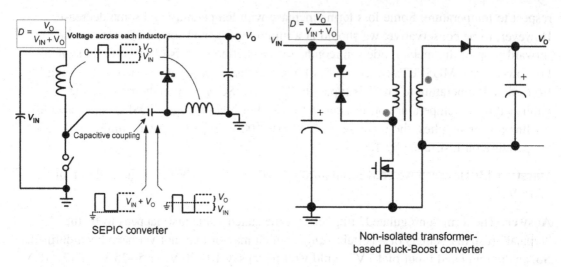

Figure 4.1: Positive-to-positive step-up/step-down converters.

Further, the ends of these two windings are sometimes completely separated from each other (no galvanic connection between the windings). The reason for this may be that from a production standpoint, it doesn't make sense to try to solder too many copper strands on to a single pin/termination. So, the intention here is that the two windings will be eventually connected to each other on the PCB itself. But sometimes, the intention of leaving separate windings on an inductor is to allow *flexibility* for the two windings to be connected to each other either in *series*, or in *parallel*, as desired. So for example, if we place the windings in series, that would reduce the current rating of the inductor (originally, each winding was expected to carry only half the rated current), but we would get a much higher inductance. If in parallel, the inductance would come down, but the current rating would increase. However, in low-voltage applications, where safety isolation is not a concern, we can also exploit this inductor structure and use it as a 1:1 transformer for correcting the polarity inversion of the Buck-Boost.

Question 15: In an inductor with split windings (1:1), how exactly does its current rating and its inductance change as we go from a parallel configuration to a series configuration?

Answer: Suppose each winding has 10 turns and a DC resistance (DCR) of 1 Ω. Now, if it is used in parallel configuration, we still have 10 turns, but the effective DCR is 1 Ω in parallel with 1 Ω, i.e., 0.5 Ω. When a series configuration is used, we get 2 Ω and 20 turns. We also know that inductance depends on the *square* of the number of turns. So, that goes up four times.

What about the current rating? This is largely determined by the amount of heat dissipation the inductor can tolerate. But its thermal resistance (in °C/W) is not determined by the winding configuration, rather by the exposed area of the inductor, and other physical characteristics. Therefore, whether in series or in parallel configuration, we have to maintain the same total I^2R loss. For example, suppose we call the current rating in parallel as "I_P", and in series "I_S," then as per the DCR in our above numerical example, we get

$$I_P^2 \times 0.5 = I_S^2 \times 2$$

So,

$$I_P = 2 \times I_S$$

Therefore, in going from a parallel to a series configuration, the inductance will *quadruple* and the current rating will *halve*.

What happens to the B-field? Don't we have to consider the possibility of saturation here? Well, B is proportional to LI/N (see Figure 2.13). So, if inductance quadruples, I halves, and N doubles, the B-field is unchanged!

Question 16: Is there any difference between the terms "Buck-Boost" and "flyback"?

Answer: The answer to that may well depend on whom you ask! These terms are often used interchangeably in the industry. However, generally, most people prefer to call the conventional inductor-based version a (true) "Buck-Boost," whereas its transformer-based version, isolated or non-isolated, is called a "flyback."

Question 17: When and why do we need isolation? And how do we go about achieving it?

Answer: We must recognize that a (transformer-based) flyback topology may or may not provide us with isolation. Isolation is certainly a natural advantage accruing from the use of a transformer. But to preserve isolation, we must ensure that *all* the circuitry connected to the switch side of the transformer ("Primary side") is kept completely independent from *all* the circuitry sitting on the output side ("Secondary side") (see Figure 1.1 in *Chapter 1*).

So for example, if in our attempt to correct the polarity inversion of a Buck-Boost we *make a connection* between the Primary and Secondary windings of the transformer, we lose isolation. But if our intention is to reset polarity, that would be acceptable.

To maintain isolation, besides making no galvanic connection between the power stages on either side of the transformer, we must not make any *signal-level* inter-connections either. However, we must carry the feedback signal (or any fault information) from the output side to the IC, via one or more "optocouplers." The optocoupler manages to preserve Primary-to-Secondary voltage isolation, but allows signal-level information to pass through. It works by first converting the Secondary-side signal into radiation (light) by means of an "LED" (light-emitting diode), beaming it over to the Primary side onto a photo-transistor, and thereby converting the signal back into electrical impulses (all this happening within the package of the device itself).

In high-voltage applications (anything over 60 V DC, e.g., off-line power supplies), it is in fact required by law, to provide electrical isolation between a *hazardous* input voltage level and any *user-accessible* ("safe") metal surfaces (e.g., output terminals of the power supply). Therefore, there is a "Primary ground" at the input side of the transformer, and a separate "Secondary ground" on the output side. The latter is tied to the ground of the system, and usually also to the earthed metal enclosure.

Question 18: In an off-line power supply, are the Primary and Secondary sides really *completely* isolated?

Answer: It is interesting to note that safety regulations specify a certain physical spacing that must be maintained between the Primary and Secondary sides — in terms of the RMS of the *voltage differential* that can be safely applied between them. The question arises — how do we define a voltage *difference* between the two sides of a transformer that are supposedly separate anyway? What is the reference level to compare their respective voltages? After all, voltage is essentially a relative term.

In fact the two sides do share a connection! As mentioned, the Secondary-side ground is usually the system ground, and it connects to the metal enclosure and/or to the ground wire of the mains supply ("earth" or "safety ground"). But further down the AC mains distribution network, the safety ground wire is connected somewhere to the "neutral" wire of the AC supply. And we know that this neutral wire comes back into the Primary side. So in effect, we have established a connection between the Primary and Secondary sides. It does not cause the user any problem, because he or she is also connected to earth. In effect, the earth potential forms the reference level to establish the voltage difference across the safety transformer, and to thereby fix the Primary-to-Secondary spacing, and also the breakdown rating of any Primary-to-Secondary insulation.

Note that in some portable equipment, only a two-wire AC cord is used to connect it to the mains supply. But the spacing requirement is still virtually unchanged, since a user can touch accessible parts on the Secondary side and complete the connection through the earth ground.

Question 19: From the standpoint of an actual power supply design procedure, what is the most fundamental difference between the three topologies that must be kept in mind?

Answer: In a Buck, the average inductor current ("I_L") is equal to the load current ("I_O"), that is, $I_L = I_O$. But in a Boost and a Buck-Boost, this average current is equal to $I_O/(1-D)$. Therefore, in the latter two topologies, the inductor current is *a function of D* (duty cycle) — and therefore indirectly a *function of the input voltage* too (for a given output).

Question 20: In the three basic topologies, how does the duty cycle change with respect to input voltage?

Answer: For *all* topologies, a high D corresponds to a low-input voltage, and a low D to a high input.

Question 21: What do we mean by the "peak current" of a DC–DC converter?

Answer: In any DC–DC converter, the terms "peak *inductor* current," "peak *switch* current," and "peak *diode* current" are all the same — referred to simply as the peak current "I_{PK}" (of the converter). The switch, diode and inductor have the same peak current value.

Question 22: What are the key parameters of an off-the-shelf inductor that we need to consider?

Answer: The inductance of an inductor (along with the switching frequency and duty cycle) determines the *peak* current, whereas the *average* inductor current is determined by the topology itself (and the specific application conditions — the duty cycle and load current). For a given application, if we decrease the inductance, the inductor current waveform becomes more "peaky," increasing the peak currents in the switch and diode too (also in the capacitors). Therefore, a typical converter design should start by first estimating the optimum *inductance* so as to *avoid saturating the inductor.* That is the most basic concern in designing/picking an inductor.

However, *inductance* by itself doesn't fully describe an inductor. In theory, by choosing a very thin wire gauge, for example, we may be able to achieve almost any inductance on a given core, just by winding the appropriate number of turns. But the current that the inductor will be able to handle without saturating is still in question, because it is not just the current, but *the product of the current and the number of turns* ("Ampere-turns") that determine the magnetic field present in the inductor core — which in turn determines whether the inductor is saturating or not. Therefore, we need to look out for an inductor with the right inductance and also the required *energy handling capability*, usually expressed in µJ (microJoules). This must be greater than or equal to the energy it needs to store in the application, $1/2 \times LI_{PK}^2$. Note that the "L" carries with it information about the number of turns too, since $L \propto N^2$, where N is the number of turns. See *Chapter 5.*

Question 23: What really determines the *current rating* of an inductor?

Answer: There are two limiting factors here. One is the heat developed (I^2R losses), which we should ensure is not excessive (typically causing a temperature rise of 50 °C or less). The second is the magnetic field it can withstand without saturating. So, most ferrites allow a maximum B-field of about 3,000 G before saturation starts.

Question 24: Does the maximum allowable B-field depend on the air gap used?

Answer: When designing (gapped) transformers, we need to remember that first, the B-field present within the core material (e.g., ferrite) is the same as the B-field in the air gap. It does not change. Second, though by changing the air gap we may end up decreasing the existing B-field, the *maximum allowable B*-field depends *only on the core material* used — it remains fixed, for example, at about 3,000 G for ferrites. Note that the H-field is defined as

$H = B/\mu$, where μ is the permeability of the material. Now, since the permeability of ferrite is much higher than that of air, and the B-field is the same in both, therefore the H-field is much lower in the ferrite than in the air gap.

Question 25: Why is it commonly stated that in a flyback transformer, the "air gap carries most of the stored magnetic energy"?

Answer: We can intuitively accept the fact that the energy stored is proportional to the volume of the magnetic material. And because of that, we also tend to think the ferrite must be carrying most of the energy, since it occupies the maximum volume — the amount of air enclosed between the ends of the ferrite being very small. However, the stored energy is also proportional to $B \times H$, and since the H-field in the gap is so much larger, it ends up storing typically two-thirds of the total energy, despite its much smaller volume.

Question 26: If air carries most of the stored energy, why do we even need the ferrite?

Answer: An air-cored coil would seem perfect as an inductor, especially since it would never saturate. However, the number of turns required to produce a given inductance would be impractically large, and so we would get unacceptable copper losses. Further, since there is nothing to "channel" (constrain) the flux lines, the air-cored inductor would spew electromagnetic interference (EMI) everywhere.

The ferrite is useful, because it is the very *means* by which we can create such high magnetic fields in the first place — without an excessive number of turns. It also provides us the "channel" for flux lines that we had been looking for. In effect, it "enables" the air gap.

Question 27: What is the basic design rule for calculating inductance for all the topologies?

Answer: To reduce stresses at various points inside a power supply, and also to generally reduce the overall size of its components, a "current ripple ratio" ("r") of about 0.4 is considered to be a good compromise for any topology, at any switching frequency.

"r" is the ratio $\Delta I/I_L$, where ΔI is the swing in the current, and I_L is the average inductor current (center of the swing ΔI). An r of 0.4 is the same as $r = 40\%$, or $r = \pm 20\%$. This means that the peak inductor current is 20% above its average value (its trough being 20% below).

To determine the corresponding inductance we use the definition $r = \Delta I/I_L$, along with the inductor equation, to get

$$V_{ON} = L \frac{\Delta I}{\Delta t} = L \frac{I_L \times r}{D/f}$$

solving

$$L = \frac{V_{ON} \times D}{I_L \times r \times f}$$

This gives us the inductance in Henrys, when f is in Hz. Note that V_{ON} is the voltage across the inductor when the switch is ON. It is therefore equal to $V_{IN}-V_O$ for a Buck, and V_{IN} for a Boost and a Buck-Boost. Also, I_L is the average inductor current, equal to I_O for a Buck, and $I_O/(1-D)$ for a Boost and a Buck-Boost.

Question 28: What is a "Forward converter"?

Answer: Just as the isolated flyback is a derivative of the Buck-Boost topology, the Forward converter is the isolated version (or derivative) of the Buck topology. It too uses a transformer (and optocoupler) for providing the required isolation in high-voltage applications. Whereas the flyback is typically suited for output powers of about 75 W or less, the Forward converter can go much higher.

The simplest version of the Forward converter uses only one transistor (switch), and is thus often called "single-ended." But there are variants of the single-ended Forward converter with either two or four switches. So, although the simple Forward converter is suited only up to about 300 W of power, we can use the "double-switch Forward converter" to get up to about 500 W. Thereafter, the half-bridge, push−pull, and full-bridge topologies can be exploited for even higher powers (see Figure 4.2 and Table 7.1). But note that all of the above topologies are essentially "Buck-derived" topologies.

Question 29: How can we tell whether a given topology is "Buck-derived" or not?

Answer: The simplest way to do that is to remember that only the Buck has a true LC filter at its output. There is nothing separating the inductor and the output capacitor.

Question 30: Which end of a given input voltage range V_{INMIN} to V_{INMAX} should we pick for starting a design of a Buck, a Boost, or a Buck-Boost converter?

Answer: Since the average inductor current for both the Boost and Buck-Boost increases as D increases ($I_L = I_O/(1-D)$) — the design of Boost and Buck-Boost inductors must be validated at the *lower* end of the given input range, that is, at V_{INMIN} — since that is where we get the highest (average and peak) inductor current. We always need to ensure that any inductor can handle the maximum peak current of the application without saturating. For a Buck, the *average* inductor current is independent of the input or output voltage. However, observing that its *peak* current increases at higher input voltages, it is preferable to design a Buck inductor at V_{INMAX}.

Question 31: Why are the equations for the *average inductor current* of a Boost and a Buck-Boost exactly the same, and why is that equation so different from that of a Buck?

Answer: In a Buck, energy continues to flow into the load (*via the inductor*) during the entire switching cycle (during the switch on-time *and* off-time). Therefore, the average inductor current must be equal to the load current, that is, $I_L = I_O$.

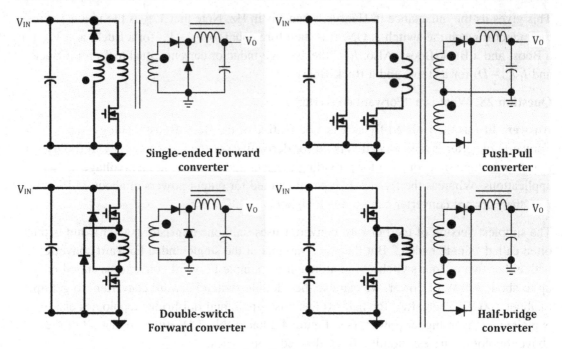

Figure 4.2: Various Buck-derived topologies.

Note that capacitors contribute nothing to *average* current flow, because, in steady state, just as the voltseconds across an inductor averages out to zero at the end of each cycle, the charge in a capacitor does likewise (charge is the integral of current over time, and has the units Amperes-seconds). If that did not happen, the capacitor would keep charging up (or discharging) on an average, until it reaches a steady state.

However, in a Boost or Buck-Boost, energy flows into the output *only* during the off-time. And it comes *via the diode*. So, the average diode current must be equal to the load current. By simple arithmetic, since the average diode current calculated over the full cycle is equal to $I_L \times (1-D)$, equating this to the load current I_O gives us $I_L = I_O/(1-D)$ for *both* the Boost and the Buck-Boost.

Question 32: What is the average *output* current (i.e., the load current) equal to for the three topologies?

Answer: This is simply the converse of the previous question. For the Buck, the average output current equals the average inductor current. For the Boost and Buck-Boost, it is equal to the average diode current.

Question 33: What is the average *input* current equal to for the three topologies?

Answer: In a *Buck*, the input current flows *only* through the switch. It stops when the switch turns OFF. Therefore, *the average input current must be equal to the average switch current*. To calculate the average of the switch current, we know it is ON for a fraction D (duty cycle) of the switching cycle, during which time it has an average value (center of ramp) equal to the average inductor current, which in turn is equal to the load current for a Buck. Therefore, the average of the switch current must be $D \times I_O$, and this must be equal to the input current I_{IN}. We can also do a check in terms of the input and output power.

$$P_{IN} = V_{IN} \times I_{IN} = V_{IN} \times D \times I_O = V_{IN} \times \frac{V_O}{V_{IN}} \times I_O = V_O \times I_O = P_O$$

We therefore get input power equal to the output power — as expected, since the simple duty cycle equation used above ignored the switch and diode drops, and thus implicitly assumed no wastage of energy, that is, an efficiency of 100%.

Similarly, the input current of a *Boost* converter flows through the inductor at all times. So, the *average input current is equal to the average inductor current* — which we know is $I_O/(1-D)$ for the Boost. Let us again do a check in terms of power

$$P_{IN} = V_{IN} \times I_{IN} = V_{IN} \times \frac{I_O}{1-D} = V_{IN} \times \frac{I_O}{1-((V_O-V_{IN})/V_O)} = V_O \times I_O = P_O$$

Coming to the **Buck-Boost**, the situation is not so clear at first sight. The input current flows into the inductor when the switch is ON, but when the switch turns OFF, though the inductor current continues to flow, its path does *not* include the input. So, the only conclusion we can make here is that the *average input current is equal to the average switch current*. Since the center of the switch current ramp is $I_O/(1-D)$, its average is $D \times I_O/(1-D)$. And this is the average input current. Let us check this out:

$$P_{IN} = V_{IN} \times I_{IN} = V_{IN} \times \frac{D \times I_O}{1-D} = V_{IN} \times \frac{(V_O/(V_{IN} + V_O)) \times I_O}{1-(V_O/(V_{IN} + V_O))} = V_O \times I_O = P_O$$

We get $P_{IN} = P_O$ as expected.

Question 34: How is the average inductor current related to the input and/or output currents for the three topologies?

Answer: For the **Buck**, we know that average inductor current is equal to the output current, that is, $I_L = I_O$. For the *Boost* we know it is equal to the input current, that is, $I_L = I_{IN}$. But for the *Buck-Boost* it is equal to the *sum* of the (average) input current and the output current. Let us check this assertion out:

$$I_{IN} + I_O = \frac{D \times I_O}{1-D} + I_O = I_O \times \left(\frac{D}{1-D} + 1 \right) = \frac{I_O}{1-D} = I_L$$

Table 4.1: Summary of Relationships of Currents for the Three Topologies.

Average Values	Buck	Boost	Buck-Boost
I_L	I_O	$I_O/(1-D)$	$I_O/(1-D)$
I_L	I_{IN}/D	I_{IN}	I_{IN}/D
I_L	I_O	I_{IN}	$I_{IN} + I_O$
I_D	$I_O - I_{IN}$	I_O	I_O
I_D	$I_O(1-D)$	I_O	I_O
I_D	$I_{IN}(1-D)/D$	$I_{IN}(1-D)$	$I_{IN}D/(1-D)$
I_{SW}	I_{IN}	$I_{IN} - I_O$	I_{IN}
I_{SW}	$I_O D$	$I_O D/(1-D)$	$I_O D/(1-D)$
I_{SW}	I_{IN}	$I_{IN}D$	I_{IN}
I_O	I_L	I_D	I_D
I_{IN}	I_{SW}	I_L	I_{SW}

It is thus proved. See Table 4.1 for a summary of similar relationships.

Question 35: Why are most Buck ICs *not* designed to have a duty cycle of 100%?

Answer: One of the reasons for limiting D_{MAX} to less than 100% is specific to *synchronous* Buck regulators (Figure 4.3) — when it utilizes a technique called "low-side current sensing."

In "low-side current sensing," to save the expense of a separate low-resistance sense resistor, the R_{DS} of the "low-side MOSFET" (the one across the "optional" diode in Figure 4.3) is often used for sensing the current. The voltage drop across this MOSFET is measured, and so if we know its R_{DS}, the current through it is also known by Ohm's law. It becomes obvious that in fact for any low-side current sense technique, we need to turn the high-side MOSFET OFF, and thereby force the inductor current into the freewheeling path, so we can measure the current therein. That means we need to set the maximum duty cycle to less than 100%.

Another reason for choosing $D_{MAX} < 100\%$ comes from the use of *N-channel MOSFETs* in any (positive-to-positive) Buck regulators. Unlike an NPN transistor, an N-channel MOSFET's Gate terminal has to be taken several volts *above* its Source terminal to turn it ON fully. So, to keep the switch ON, when the MOSFET conducts, we need to drive its Gate a few volts higher than the input rail. But such a rail is not available! The only way out is to *create* such a rail — by means of a circuit that can *pump* the input rail higher as required. This circuit is called the "bootstrap circuit," as shown in Figure 4.3.

But to work, the bootstrap circuit demands we turn the switch OFF momentarily, because that is when the switching node goes low and the "bootstrap capacitor" gets charged up to V_{IN}. Later, when the switch turns ON, the switching node (lower terminal of the bootstrap

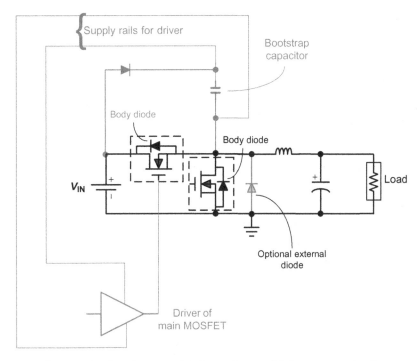

Figure 4.3: Synchronous Buck regulator with bootstrap circuit.

capacitor) rises up to V_{IN}, and in the process, literally "drags" the upper terminal of the bootstrap capacitor to a voltage higher than V_{IN} (by an amount equal to V_{IN} as per the simplified bootstrap scheme of Figure 4.3) — that happens because no capacitor loses its charge spontaneously! Therefore, the reason for setting the maximum duty cycle to less than 100% is simply to allow a bootstrap circuit (if present) to work!

We will find that a bootstrap circuit is almost always present if an *N-channel* MOSFET switch is used in a positive-to-positive (or just "positive") Buck converter, or in a positive-to-negative Buck-Boost, or in a negative-to-negative (or just "negative") Boost. Further, by circuit symmetry we can show that it will also be required (though this time to create a drive rail *below ground*) when using a *P-channel* MOSFET in a negative Buck, or in a negative-to-positive Buck-Boost, or in a positive Boost. See Figure 9.16.

Here, we should also keep in mind that the N-channel MOSFET is probably the most popular choice for switches, since it is more cost-effective as compared to P-channel MOSFETs with comparable Drain-to-Source on-resistance "R_{DS}." That is because N-channel devices require smaller die sizes (and packages). Since we also know that the ubiquitous positive Buck topology requires a bootstrap circuit when using an N-channel MOSFET switch, it becomes apparent why a good majority of Buck ICs out there have maximum duty cycles of less than 100%.

Question 36: Why are Boost and Buck-Boost ICs almost invariably designed *not* to have 100% duty cycle?

Answer: We should first be clear that the Boost and Buck-Boost topologies are so similar in nature, that any IC meant for a Boost topology can also be used for a Buck-Boost application, and vice versa. Therefore, such control ICs are generally marketed as being for *both* Boost and Buck-Boost applications.

One of the common aspects of these two topologies is that in both of these, energy is built up in the inductor during the switch on-time, during which *none* passes to the output. Energy is *delivered to the load only when the switch turns OFF*. In other words, we have to turn the switch OFF to get any energy at all delivered to the output. Contrast this with a Buck, in which the inductor, being in series with the load, delivers energy to the load even as it is being built up in the inductor itself (during the switch on-time). So, in a Buck, even if we have 100% duty cycle (i.e., switch is ON for a long time), we *will* get the output voltage to rise (smoothly). Subsequently, the feedback loop will command the duty cycle to steady out when the required output voltage is reached.

However, in the Boost and Buck-Boost topologies, if we keep the switch ON permanently, we can *never* get the output to rise, because in these topologies, energy is delivered to the output *only* when the switch turns OFF. We can thus easily get into a "Catch 22" situation, where the controller "thinks" it is not doing enough to get the output to rise — and therefore continues to command maximum duty cycle. But with a maximum 100% duty cycle, that means *zero* off-time — so how can the output *ever* rise?! We can get trapped in this illogical mode for a long time, and the switch can be destroyed. Of course, we hope that the current limit circuit is designed well enough to eventually intervene, and turn the switch OFF before the switch destructs! But generally, it is considered inadvisable to run these two topologies at 100% duty cycle. The only known $D = 100\%$ Buck-Boost IC is the LM3478 from National. Around since 2000, it still "sells" without a declaration of the problem.

Question 37: What are the "Primary" and "Secondary" sides of an off-line power supply?

Answer: Usually, the control IC drives the switch *directly*. Therefore, the IC must be located at the input side of the isolation transformer — that is called the "Primary side". The transformer windings that go to the output are therefore said to all lie on the "Secondary side." Between these Primary and Secondary sides lies "no-man's land" — the "*isolation boundary*." Safety norms regulate how strong or effective this boundary must be.

Question 38: In many off-line power supplies, we can see not one, but *two* optocouplers, usually sitting next to each other. Why?

Answer: The first optocoupler transmits *error* information from the output (Secondary side) to the control IC (Primary side). This closes the feedback loop, and tells the IC how much correction is required to regulate the output. This optocoupler is therefore often nicknamed the "regulation opto" or the "error opto." However, safety regulations for off-line power supplies also demand that no "single-point failure" anywhere in the power supply produces a hazardous voltage on the output terminals. So if, for example, a critical component (or even a solder connection) within the normal feedback path fails, there would be no control left on the output, which could then rise to dangerous levels. To prevent this from happening, an *independent* "overvoltage protection" (OVP) circuit is almost invariably required. This is usually tied to the output rail in parallel to the components of the regulation circuitry. This fault detector circuit also needs to send its sensed "fault signal" to the IC through a *separate* path altogether, so that its functioning is not compromised in the event of failure of the feedback loop. So logically, we require an *independent* optocoupler — the "fault opto." Note that by the same logic, this optocoupler must eventually connect to the IC (and cause it to shut down) using a pin *other* than the one being used for feedback. Early designs unknowingly thwarted this logic, and inadvertently got approved too by safety agencies too! Not any more though.

The reason why the two optocouplers are "sitting *next* to each other" is usually only for convenience in the PCB layout — because the isolation boundary needs to pass through these devices, and also through the transformer (see *Figure 1.1* in *Chapter 1*).

Question 39: To get safety approvals in multioutput off-line converters, do we need separate current limiting on each output?

Answer: Safety agencies regulate not only the *voltage* at user-accessible outputs, but also the maximum *energy* that can be drawn from them under a fault condition. Primary-side current sensing can certainly limit the *total* energy delivered by the supply, but cannot limit the energy (or power) from each output individually. So for example, a 300 W converter (with appropriate primary-side current limiting) may have been originally designed for 5 V at the rate of 36 A and 12 V at the rate of 10 A. But what prevents us from trying to draw 25 A from the 12 V output alone (none from the 5 V)? To avoid running into problems like this during approvals, it is wise to design separate Secondary-side current-limiting circuits for each output. We are allowed to make an exception if we are using an *integrated post-regulator* (like the LM7805) on a given output, because such regulators have built-in current limiting. Note that any overcurrent fault signal can be "OR-ed" with the OVP signal, and communicated to the IC via the fault optocoupler.

Question 40: How do safety agencies typically test for single-point failures in off-line power supplies?

Answer: Any component can be shorted or opened by the safety agency during their testing. Even the possibility of a solder connection coming undone anywhere, or a bad "via" between layers of a PCB would be taken into account. Any such single-point failure is expected to usually cause the power supply to simply shut down gracefully, or even fail catastrophically. That is fine, but in the process, no hazardous voltage is permitted to appear on the outputs, even for a moment. And no fire hazard too!

Question 41: What is a synchronous Buck topology?

Answer: In synchronous topologies, the freewheeling diode of the conventional Buck topology is either replaced, or supplemented (in parallel) with an additional MOSFET switch (see Figure 4.3). This new MOSFET is called the "low-side MOSFET" or the "synchronous MOSFET," and the upper MOSFET is now identified as being the "high-side MOSFET" or the "control MOSFET."

In steady state, the low-side MOSFET is driven such that it is "inverted" or "complementary" with respect to the high-side MOSFET. This means that whenever one of these switches is ON, the other is OFF, and vice versa — that is why this is called "synchronous" as opposed to "synchronized" which would imply both are running *in phase* (which is clearly unacceptable because that would constitute a dead short across the input). However, through all this, the effective *switch* of the *switching* topology still remains the high-side MOSFET. It is the one that effectively "leads" — dictating when to build up energy in the inductor, and when to force the inductor current to start freewheeling. The low-side MOSFET basically just follows.

The essential difference from a conventional Buck regulator is that the low-side MOSFET in a synchronous regulator is designed to present a typical forward drop of only around 0.1 V or less to the freewheeling current, as compared to a Schottky catch diode which has a typical drop of around 0.5 V. This therefore reduces the conduction loss (in the freewheeling path) and enhances efficiency.

In principle, the low-side MOSFET does not have any significant crossover loss because there is virtually no overlap between its V and I waveforms — it switches (changes state) only when the voltage across it is almost zero. Therefore, typically, the high-side MOSFET is selected primarily on the basis of its high *switching speed* (low crossover loss), whereas the low-side MOSFET is chosen primarily on the basis of its low *Drain-to-Source on-resistance*, "R_{DS}" (low conduction loss).

One of the most notable features of the synchronous Buck topology is that on decreasing the load, it typically does *not* enter discontinuous conduction mode as a diode-based (conventional) regulator would. That is because, unlike a bipolar junction transistor (BJT), the current can reverse its direction in a MOSFET (i.e., it can flow from Drain to Source or from Source to Drain). So, the inductor current at any given moment can become negative (flowing *away* from the load) — and therefore "continuous conduction mode" (CCM) is maintained — even if the load current drops to zero (nothing connected across the output terminals of the converter) (see *Chapter 1*). See also *Chapter 9* and Figure 9.1.

Question 42: In synchronous Buck regulators, why do we sometimes use a Schottky diode in parallel to the low-side MOSFET, and sometimes we don't?

Answer: We indicated above that the low-side switch is deliberately driven in such a manner that it changes its state only when the voltage across it is very small. That simply implies that during turn-off (of the high-side MOSFET), the low-side MOSFET turns ON a few nanoseconds *later*. And during turn-ON, the low-side MOSFET turns OFF just a little before the high-side MOSFET starts to conduct. By doing this, we are trying to achieve "zero-voltage (lossless) switching" (ZVS) in the low-side MOSFET. We are also trying to prevent "cross-conduction" — in which both MOSFETs may conduct simultaneously for a short interval during the transition (which can cause a loss of efficiency at best, and possible switch destruction too). However, during this brief interval when both MOSFETs are simultaneously OFF (the "dead-time"), the inductor current still needs a path to follow. However, every MOSFET contains an intrinsic "body diode" within its structure that allows reverse current to pass through it even if we haven't turned it ON (see Figure 4.3). So, this provides the necessary path for the inductor current. However, the body diode has a basic problem — it is a "bad diode." It does not switch fast, nor does it have a low forward drop. So often, for the sake of a couple of percentage points in improved efficiency, we may prefer not to depend on it, and use a "proper" diode (usually Schottky), strapped across the low-side MOSFET in particular. See *Chapter 9*.

Question 43: Why do most synchronous Buck regulators use a low-side MOSFET with an *integrated* Schottky diode?

Answer: In theory, we could just select a Schottky diode and solder it directly across the low-side MOSFET. But despite being physically present on the board, this diode may be serving no purpose at all! For example, to get the diode to take over the freewheeling current quickly from the low-side MOSFET when the latter turns OFF requires a *good low-inductance* connection between the two. Otherwise, the current may still prefer the body diode — for the critical few nanoseconds it takes before the high-side MOSFET turns ON. So, this requires we pay great attention to the PCB layout. But unfortunately, even our best efforts in that direction may not be enough — because of the significant inductive impedance that even small PCB trace lengths and internal bond wires of the devices can present when we are talking about nanoseconds. The way out of this is to use a low-side MOSFET with an *integrated* Schottky diode; that is, within the same package as the MOSFET. This greatly reduces the parasitic inductances between the low-side MOSFET and the diode, and allows the current to quickly steer away from the low-side MOSFET and into the parallel diode during the dead-time preceding the high-side turn-on.

Question 44: What limits our ability to switch a MOSFET fast?

Answer: When talking about a switching device (transistor), as opposed to a converter, the time it spends in transit *between* states is referred to as its "switching speed." The ability to switch fast has several implications, including the obvious minimization of the *V−I* crossover losses. Modern MOSFETs, though considered very "fast" in comparison to BJTs, nevertheless do *not* respond *instantly* when their drivers change state. That is because, first, the driver itself has a certain non-zero "pull-up" or "pull-down" resistance *through which* the drive current must flow and thereby charge/discharge the internal *parasitic capacitances* of the MOSFET, so as to cause it to change state. In the process, there is a certain delay involved. Second, even if our *external* resistances were zero, there still remain parasitic inductances associated with the PCB traces leading up from the Gate drivers to the Gates, that will also limit our ability to force a large Gate current to turn the device ON or OFF quickly. And further, hypothetically, even if we do achieve zero *external* impedance in the Gate section, there remain *internal impedances* within the package of the MOSFET itself — *before* we can access its parasitic capacitances (to charge or discharge them as desired). Part of this internal impedance is *inductive*, consisting of the bond wires leading from the pin to the die, and part of it is *resistive*. The latter could be of the order of several ohms in fact. All these factors come into play in determining the switching speed of the device, thereby imposing hard limits as to what transition speeds are achievable.

Question 45: What is "cross-conduction" in a synchronous stage?

Answer: Since a MOSFET has a slight delay before it responds to its driver stage, though the square-wave driving signals to the high- and low-side MOSFETs might have no intended "overlap," in reality the MOSFETs might actually be conducting simultaneously for a short duration. That is called "cross-conduction" or "shoot-through." Even if minimized, it is enough to impair overall efficiency by several percentage points since it creates a short across the input terminals (limited only by various intervening parasitics).

This situation is aggravated if the two MOSFETs have significant "mismatch" in their switching speeds. In fact, usually, the low-side MOSFET is far more "sluggish" than the high-side MOSFET. That is because the low-side MOSFET is chosen primarily for its low forward resistance, "R_{DS}." But to achieve a low R_{DS}, a larger die-size is required, and this usually leads to higher internal parasitic capacitances, which end up limiting the switching speed.

Question 46: How can we try to avoid cross-conduction in a synchronous stage?

Answer: To avoid cross-conduction, a deliberate delay needs to be introduced between one MOSFET turning ON and the other turning OFF. This is called the converter's or controller's "dead-time." Note that during this time, freewheeling current is maintained via the diode present across the low-side MOSFET (or the Schottky diode in parallel).

Question 47: What is "adaptive dead-time"?

Answer: Techniques for implementing dead-time have evolved quite rapidly as outlined below.

- **First Generation (Fixed Delay)** — The first synchronous IC controllers had a *fixed* delay between the two Gate drivers. This had the advantage of simplicity, but the set delay time had to be made long enough to cover the many possible applications of the part, and also to accommodate a wide range of possible MOSFET choices by customers. The set delay had often to be further offset (made bigger) because of the rather wide manufacturing variations in its own value. However, whenever current is made to flow through the diode rather than the low-side MOSFET, we incur higher conduction losses. These are clearly proportional to the amount of dead-time, so we don't want to set too large a fixed dead-time for all applications.

- **Second Generation (Adaptive Delay)** — Usually this is implemented as follows. The *Gate voltage* of the low-side MOSFET is monitored, to decide when to turn the high-side MOSFET ON. When this voltage goes below a certain threshold, it is assumed that the low-side MOSFET is OFF (a few nanoseconds of additional fixed delay may be included at this point), and then the high-side Gate is driven high. To decide when to turn the low-side MOSFET ON, we usually monitor the *switching node* in "real-time" and adapt to it. The reason for that is that after the high-side MOSFET turns OFF, the switching node starts falling (in an effort to allow the low-side to take over the inductor current). Unfortunately, the rate at which it falls is not very predictable, as it depends on various undefined parasitics, and also the application conditions. Further, we also want to implement something close to zero-voltage switching, to minimize crossover losses in the low-side MOSFET. Therefore, we need to wait a *varying* amount of time, until we have ascertained that the switching node has fallen below the threshold (before turning the low-side MOSFET ON). So, the adaptive technique allows "on-the-fly" delay adjustment for different MOSFETs and applications.

- **Third Generation (Predictive Gate Drive™ Technique)** — The whole purpose of adaptive switching is to intelligently switch with a delay just large enough to avoid significant cross-conduction and small enough so that the body-diode conduction time is minimized — and to be able to do that consistently, with a wide variety of MOSFETs. The "predictive" technique, introduced by Texas Instruments, is often seen by their competitors as "overkill." But for the sake of completeness it is mentioned here. Predictive Gate Drive™ technology samples and holds information from the *previous* switching cycle to "predict" the minimum delay time for the next cycle. It works on the premise that the delay time required for the next switching cycle will be close to the requirements of the previous cycle. By using a digital control feedback system to detect body-diode conduction, this technology produces the precise timing signals necessary to operate very *near the threshold* of cross-conduction.

Question 48: What is low-side current sensing?

Answer: Historically, current sensing was most often done during the *on-time* of the switch. But nowadays, especially for synchronous Buck regulators in very efficient and/or very low-output voltage applications, the current is being sensed during the *off-time*.

For example, a rather *extreme* down-conversion ratio is being required nowadays — say 28–1 V at a minimum switching frequency of 300 kHz. We can calculate that this requires a duty cycle of $1/28 = 3.6\%$. At 300 kHz, the time period is 3.3 μs, and so the required (high-side) switch on-time is about $3.6 \times 3.3/100 = 0.12$ μS (i.e., 120 ns). At 600 kHz, this on-time falls to 60 ns, and at 1.2 MHz it is 30 ns. Ultimately, that just may not give enough time to turn ON the high-side MOSFET fully, "de-glitch" the noise associated with its turn-on transition ("leading edge blanking"), and get the current limit circuit to sense the current fast enough.

Further, at very light loads we may want to be able to skip pulses *altogether*, so as to maximize efficiency (since switching losses go down whenever we skip pulses). We don't want to be forced into turning the high-side MOSFET ON every cycle — just to sense the current!

For such reasons, low-side current sensing is becoming increasingly popular. Sometimes, a current sense resistor may be placed in the freewheeling path for the purpose, or the forward drop across the low-side MOSFET is often used for the purpose. For "DCR sensing," see Figure 9.6.

Question 49: Why do some non-synchronous regulators go into an almost chaotic switching mode at very light loads?

Answer: As we decrease the load, conventional regulators operating in continuous conduction mode (CCM— see Figure 1.9) enter discontinuous conduction mode (DCM). The onset of this is indicated by the fact that the duty cycle suddenly becomes a function of load — unlike a regulator operating in CCM, in which the duty cycle depends *only* on the input and output voltages (to a first order). As the load current is decreased further, the DCM duty cycle keeps decreasing, and eventually, many regulators will automatically enter a random pulse-skipping mode. That happens simply because at some point, the regulator just *cannot* decrease its on-time further, as is being demanded. So, the energy it thereby puts out into the inductor every on-pulse starts exceeding the average energy (per pulse) requirement of the load. So, its control section literally "gets confused," but nevertheless tries valiantly to regulate by stating something like — "oops ... that pulse was too wide (sorry, just couldn't help it), but let me cut back on delivering any pulses altogether for some time — hope to compensate for my actions."

But this chaotic control can pose a practical problem, especially when dealing with current-mode control (CMC). In CMC, usually the switch current is constantly monitored, and that information is used to produce the internal ramp for the pulse-width modulator (PWM) stage to work. So, if the switch does not even turn ON for several cycles, there is no ramp either for the PWM to work off.

This chaotic mode is also a variable frequency mode of virtually unpredictable frequency spectrum and therefore unpredictable EMI and noise characteristics too. That is why *fixed-frequency* operation is usually preferred in commercial applications. And fixed frequency basically means no pulse-skipping!

The popular way to avoid this chaotic mode is to "pre-load" the converter, that is, place some resistors across its output terminals (on the PCB itself), so that the converter "thinks" there is some minimum load always present. In other words, we demand a little more energy than the minimum energy that the converter can deliver (before going chaotic).

Question 50: Why do we sometimes *want* to skip pulses at light loads?

Answer: In some applications, especially battery-powered applications, the "light-load efficiency" of a converter is of great concern. Conduction losses can always be decreased by using switches with low forward drops. Unfortunately, switching losses occur every time we actually switch. So, the only way to reduce them is by *not* switching, if that is possible. A pulse-skipping architecture, if properly implemented, will clearly improve the light-load efficiency.

Question 51: How can we implement *controlled* pulse-skipping in a synchronous Buck topology, to further improve the efficiency at light loads?

Answer: In DCM, the duty cycle is a function of the load current. So, on decreasing the load sufficiently, the duty cycle starts to "pinch off" (from its CCM value). And this eventually leads to pulse-skipping when the control runs into its minimum on-time limit. But as mentioned, this skip mode can be fairly chaotic, and also occurs only at extremely light loads. So, one of the ways this is being handled nowadays is to *not* "allow" the DCM duty cycle to pinch off below 85% of the CCM pulse width. Therefore, now more energy is pushed out into a single on-pulse than under normal DCM — and without waiting to run into the minimum on-time limits of the controller. However, now because of the much-bigger-than-required on-pulse, the control will skip even more cycles (for every on-pulse). Thereafter, at some point, the control will detect that the output voltage has fallen too much, and will command another big on-pulse. So, this forces pulse-skipping in DCM, and thereby enhances the light-load efficiency by reducing the switching losses.

Question 52: How can we quickly damage a Boost regulator?

Answer: The problem with a Boost regulator is that as soon as we apply input power, a huge inrush current flows to charge up the output capacitor. Since the switch is not in series with it, we have no control over it either. So ideally, we should *delay* turning ON our switch until the output capacitor has reached the level of the input voltage (inrush stops).

And for this, a soft-start function is highly desirable in a Boost. However, if while the inrush is still in progress, we turn the switch ON, it will start diverting this inrush into the switch. The problem with that is in most controllers, the current limit may not even be working for the first 100 ns to 200 ns after turn-on — that being deliberately done to avoid falsely triggering ON the noise generated during the switch transition ("leading edge blanking"). Now the huge inrush current gets fully diverted into the switch, with virtually no control, possibly causing failure. One way out of that is to use a diode directly connected between the input supply rail and the output capacitor (cathode of this diode being at the positive terminal of the output capacitor). So, the inrush current bypasses the inductor and Boost diode altogether. However, we have to be careful about the surge current rating of this extra diode. It need not be a fast diode, since it "goes out of the picture" as soon as we start switching (gets reverse-biased permanently).

Note also, that a proper ON/OFF function cannot be implemented on a Boost topology (as is). For that, an additional series transistor is required, to completely and effectively disconnect the output from the input. Otherwise, even if we keep the switch OFF permanently, the output rail will rise to the input level.

This FAQ was presented in an early chapter so as to not intimidate an entry-level person. For the experienced user, much more on synchronous and other topologies, plus several modern techniques is available in *Chapter 9*. Also refer to the solved examples in *Chapter 19* to seal concepts and perform design calculations.

CHAPTER 5

Advanced Magnetics: Optimal Core Selection

Part 1: Energy Transfer Principles

Overview of Topologies

We will now progress from the concepts presented in preceding chapters and develop expertise to optimize the energy storage sections of DC−DC switching power converter stages, with special attention to their magnetics.

We recapitulate briefly first. A switching topology has three key power components.

(a) An inductor in which the current undulates every cycle between two levels of current. These levels remain fixed in "steady state," that is, when power-up is complete and no changes in line or load are occurring.

(b) A switch connected to the inductor, which turns ON and OFF every cycle. Typically the ON-time interval starts under command of the clock, and the OFF-time is initiated under the command of the error-amplifier/feedback-loop.

(c) A catch ("freewheeling") diode connected to both the switch and inductor at a common node, called the "switching (or swinging) node." The diode gets reverse-biased whenever the switch is ON and forward-biased whenever the switch turns OFF.

The switch and diode have complementary actions: when one is ON, the other is OFF and vice versa. The purpose is to alternate the inductor current between the switch and diode, so that it always has a path to flow in. Otherwise the converter would get destroyed by the resulting voltage spike (see Figure 1.6 again).

In all topologies, when the switch conducts, it associates the inductor with the input voltage source. And whenever the diode conducts (i.e., when switch is OFF), it associates the inductor with the output (load). Therefore, during the ON-time, energy flows into the converter through the switch. Here we keep in mind that current always flows in a complete loop, but delivers energy only when there is a *potential difference*, since work done is, by definition, potential difference multiplied by current. The increase in energy during the

ON-time manifests itself as current ramping *up* linearly in the inductor and the switch. Similarly, during the OFF-time, inductor current flows through the diode, which conducts, and thereby establishes a path to the output. So, the converter pushes energy out into the load during the OFF-time, and the resulting decrease in inductor energy manifests itself as current ramping *down* linearly in the inductor and the diode.

The relationship between current and energy levels in the inductor is expressed by the basic relationship

$$\varepsilon = \frac{1}{2}L \times I^2 \text{ Joules}$$

One of the key questions in magnetics design is: what should the optimum value of L (i.e., the inductance) be? We already know from previous chapters that the choice of L usually depends on a very simple and almost universal criterion — that of achieving $\pm 20\%$ current ripple ($r = 0.4$) at max load. The total swing ΔI per cycle is then 40% of the average DC value (i.e., the center of ramp). But anyway, selecting L is only a secondary concern. The first step in any magnetics design process, and the most important and difficult question to answer is: what *core size* should we pick? It is energy that is the key to answering that, since we are talking about power conversion, and power is, by definition, energy per second. So, once we understand energy, we can ensure we have sized the bulky energy storage components (the inductor and the input and output capacitors) correctly to handle the energy coming their way, and at the rate at which it will come. And once we know how much energy is flowing through each stage of the converter, we can determine how much of that gets dissipated (wasted) en route, inside the switch and the diode for example. This helps us pick the power semiconductors correctly, so they do not get too hot for example. And that in turn leads us into the area of thermal management — the heatsinking, air speed, and so on. In brief, energy underlines everything that is *not* control loop design. And it is very important, because unlike control loop design, energy largely determines size and cost of the converter, and also determines the overall system reliability. We therefore need to understand energy as well as control loop theory if not better — especially in today's "green era."

Keep in mind the relationship between Watts and Joules. As mentioned, Watts is Joules per second or J/s or $J s^{-1}$. Hertz (Hz) is cycles per second, or 1/s, or s^{-1} since "cycle" is dimensionless. As an example, if we have a converter switching at 100 kHz, with an output power rating of 50 W, the energy output *per cycle*, which we are calling ε_O (expressed in Joules per cycle, or simply Joules) is

$$\varepsilon_O = \frac{P_O}{f} = \frac{50 \text{ W}}{100 \text{ k Hz}} = \frac{50 \text{ J s}^{-1}}{100 \times 10^3 \text{ s}^{-1}} = 5 \times 10^{-4} \text{ J} \Rightarrow 500 \text{ } \mu \text{ J}$$

If we pick a switching frequency of 1 MHz, the energy output per cycle would be only 50 μJ, leading to smaller magnetics for the same output power. If the efficiency was say 80%, the input power would be 50 W/0.8 = 62.5 W. The energy drawn from the input per cycle would then be 62.5 μJ for a switching frequency of 1 MHz, or 625 μJ per cycle for a frequency of 100 kHz. All this takes us back to the train terminus analogy on Page 1 of this book, where we mentioned how energy flows into a converter continuously, leaves continuously, but en route gets chopped into packets, the size of which depends on how quickly we move the trains in and out. Another way of visualizing the Joules−Watts relationship is as follows: if we are able to process 50 μJ per *event*, and we repeat that *event* a million times, we will get 50×10^6 μJ, that is, 50 J. Now, if we complete those million identical events (cycles) in exactly 1 s, that would be a frequency of 1 MHz, and we get 50 J/s, or 50 W by definition. So, we have 50 W being processed, at a switching frequency of 1 MHz. However, if we had completed those million cycles at a much slower pace, say in 10 s, we would get only 50/10 = 5 J coming out per second, which is 5 W, at a switching frequency of 10^6 cycles/10 s = 10^5 Hz, that is, 100 kHz.

In determining core sizes for use in switching power supply design, there is an underlying "topology dependency" that is often overlooked in related literature. We will now uncover this so that we can select cores more optimally.

We mentioned that during the ON-time, energy flows from the input source to the converter. Where does it go? In the case of a Boost and Buck-Boost, *all* the incoming energy (during the ON-time) gets stored *in the inductor*. But in the case of a Buck, only *part* of that gets stored in the inductor — because some of it gets delivered *directly* to the output. The reason is *in a Buck topology the inductor is in series with the output during the ON-time*. Indeed, as we mentioned, current always flows in a complete loop, but delivers energy only wherever it encounters a potential difference ($V \times I$ = Watts, $V \times I \times t$ = Joules).

Similarly, during the OFF-time, we mentioned that energy is delivered from the converter to the output. But from where exactly does it come? In the case of a Buck and Buck-Boost, all the outgoing energy (during the OFF-time) comes from the inductor, where it previously resided as stored energy $(1/2) \times LI^2$. But in the case of a Boost, only part of that is previously stored in the inductor — because some of it comes straight from the input voltage source. The reason is *in a Boost topology the inductor is in series with the input during the OFF-time*.

Note however, that for all topologies, we can always unequivocally state that the energy added to the inductor during the ON-time, whatever fraction of incoming energy it is, is exactly equal to the energy extracted from the inductor during the OFF-time (down to the last pico-Joule if you want to put it that way). The inductor current ends each cycle with *exactly* the same current *and* energy it started the cycle with. And that is by definition, a *steady state*. But we also realize that obeying that rule does not preclude delivering energy straight from the input source to the output (load) if possible — that is, *without availing of the storage capabilities of the inductor on the way*. We can visualize that in that case, we will not be placing so much demand on the inductor. Perhaps the inductor can be made *smaller* in size. In fact that situation does occur in *two* topologies — the Buck and the Boost. And of course, it is also true in all topologies *derived* from these two fundamental topologies too. For example, in the Single-ended Forward converter, Half-Bridge, Full-Bridge, 2-Switch Forward, and so on, all these being Buck derivatives. It also applies to the Boost PFC front end of high-power AC−DC power supplies. We will learn to design the magnetics of that too in *Chapter 14*.

The *Buck-Boost* is the only topology where no *direct* path of energy transfer from input to output is ever established, either during the ON-time or the OFF-time. In other words, *all* the incoming energy gets stored in the inductor during the ON-time. And during the OFF-time, all of that stored energy, *and not a pico-Joule more*, gets delivered to the output. We start to recognize clearly that that means *the size of the inductor in a Buck-Boost (or the transformer in a flyback) will always be the largest* (for a given wattage), compared to other topologies, *since a Buck-Boost has to handle **all** the energy that is pulled into the converter*. Therefore, "generalized core selection curves" or equations, that fail to connect core size *to the topology on hand*, usually miss the point altogether.

We realize it is becoming necessary to carefully understand the complete energy transfer process occurring in each of the three basic topologies, as shown in Figures 5.2−5.4. The process differs in each case. We now quickly glance through these figures, and see they include a hitherto unseen/unfamiliar form of the duty cycle equation, one that involves efficiency, η — initially an estimated or target efficiency, later we can use the actual measured one. Let us call these new equations "real-world" duty cycle equations. They are as follows:

$$D_{\text{Buck}} = \frac{V_O}{\eta V_{\text{IN}}}, \qquad D_{\text{Boost}} = \frac{V_O - \eta V_{\text{IN}}}{V_O}, \qquad D_{\text{Buck-Boost}} = \frac{V_O}{V_O + \eta V_{\text{IN}}}$$

We observe that these look remarkably similar to the very commonly used, first-estimate or "ideal" equations below:

$$D_{\text{Buck}} \approx \frac{V_O}{V_{\text{IN}}}, \qquad D_{\text{Boost}} \approx \frac{V_O - V_{\text{IN}}}{V_O}, \qquad D_{\text{Buck-Boost}} \approx \frac{V_O}{V_O + V_{\text{IN}}}$$

The apparent similarity between the above two sets of equations is actually very deceptive. It belies their enormous differences. They could not be more different from each other. They literally represent the opposite ends of the spectrum ranging from ideal to real-world. Notice that we have consciously avoided using equality signs in the "ideal" (latter) set of equations above. The reason is, we want to be very clear that those are based on an assumption of *zero losses*. They are valid only for a converter with an efficiency of 100% ($\eta = 1$), and of course we know nothing like that really exists in nature. On the other hand, the "real-world" set of equations (the former set above) are actually the most exact or accurate we can possibly encounter. Befittingly, we have used equality signs for them above (with some small reservations discussed later).

Just to refresh our memory from *Chapter 1*. The loss term can be written in terms of either input or output wattage as follows:

$$P_{\text{loss}} = P_O \times \left(\frac{1 - \eta}{\eta} \right)$$

$$P_{\text{loss}} = P_{\text{IN}} \times (1 - \eta)$$

In either case, as η decreases, the loss increases.

Now, let us see what happens to the duty cycle in two specific cases.

(a) As we lower the input voltage, the duty cycle must always increase, because that creates a longer time for instantaneous input current to flow in from the source. In terms of average input current per cycle, the product $V_{\text{IN}} \times I_{\text{IN}}$ is thus maintained. Note that in the case of the Boost and the Buck-Boost however, since energy is delivered to the output only during the OFF-time, any increase in duty cycle leaves less time for energy to flow into the output. Therefore, to keep output power fixed, the instantaneous current must simultaneously increase as D increases (and that is why the center of ramp is $I_O/(1-D)$ for these two topologies, not just I_O as for a Buck). However, so far we are assuming no change in efficiency (or output power).

(b) In the second situation, assume the input voltage is fixed but the efficiency decreases. The duty cycle must increase once again. Because that creates a longer time for additional current to be drawn in. In terms of average input current per cycle, the product $V_{\text{IN}} \times I_{\text{IN}}$ now increases, accounting for the increase in loss commensurate with the lowered efficiency.

The ideal equations are the most inaccurate, and lead to the smallest duty cycle possible for a given input/output condition. The real-world equations are the most accurate, and lead to the highest duty cycle (lowest efficiency) estimate. Between these two sets of equations, lie many other forms of duty cycle equations found in literature, all with varying degrees of accuracy. For example, we remember from *Chapter 1* that by using the fundamental principle of voltseconds balance in steady state, we too came up with the following rather similar-looking duty cycle equations ourselves:

$$D_{\text{Buck}} \approx \frac{V_O + V_D}{V_{IN} - V_{SW} + V_D}, \qquad D_{\text{Boost}} \approx \frac{V_O - V_{IN} + V_D}{V_O - V_{SW} + V_D},$$

$$D_{\text{Buck-Boost}} \approx \frac{V_O + V_D}{V_O + V_{IN} - V_{SW} + V_D}$$

Notice that this time we are qualifying them by not using equality signs. Because we realize that though these equations explicitly include the drop across the diode and switch, and therefore factor in the *conduction losses* inside those *two* components, they continue to ignore several other smaller loss terms, like the (I^2R) conduction loss in the DC resistance (DCR) of the inductor, or the various switching losses, or the AC resistance losses in the inductor, or the capacitor ESR losses, and so on — *all of which if factored in somehow, will cause duty cycle to increase further*. In the Appendix and in the solved examples in *Chapter 19*, we will see that the above equations have been extended to include the voltage drop across the DCR of the inductor. But the equality sign is still avoided, since we recognize that even those equations still leave out a whole bunch of other loss terms. Yet, admittedly, duty cycle equations with forward drops included as above, are certainly much better (more accurate) than the ideal equations presented earlier, and therefore a good starting point for most *iterative calculations*.

The final question remains: can we ever expect to provide an equation for duty cycle that is *almost*, if not perfectly, exact? For example, how do we really go about trying to model something like switching losses into a corresponding voltage drop (inside a duty cycle equation)? We really can't go that route much further. But the good news is we can get very accurate equations, if we agree to *club all the losses together* and thereby rewrite out duty cycle *in terms of overall efficiency* η, and that leads us to the *real-world* duty cycle equations presented earlier. Admittedly, those equations do not reveal *where* the losses are occurring inside the converter, but they do make a very accurate statement linking efficiency to duty cycle. We therefore need them for the subsequent number crunching in this chapter.

But first, we do ask: why are they so accurate? And, which losses *don't* they account for still? To answer these, we *derive* one of the above new equations as an example. Suppose we pick the Buck-Boost. The diode current of a Buck-Boost consists of a series of pulses of current with center of ramp I_L and duty cycle $1-D$. The average of that must equal the load current I_O. Therefore,

$$I_L \times (1 - D) = I_O$$

or

$$I_L = \frac{I_O}{1 - D}$$

On the input side, we have a series of switch current pulses with the same center of ramp I_L and duty cycle D. The average of that must equal the input current. Therefore,

$$I_L \times (D) = I_{IN}$$

or

$$I_L = \frac{I_{IN}}{D}$$

Equating the above two equations for I_L, we get

$$\frac{I_O}{1-D} = \frac{I_{IN}}{D}$$

$$\frac{I_O}{I_{IN}} = \frac{1-D}{D}$$

But we also know that efficiency is by definition

$$\eta = \frac{P_O}{P_{IN}} = \frac{V_O \times I_O}{V_{IN} \times I_{IN}}$$

So,

$$\frac{I_O}{I_{IN}} = \frac{\eta V_{IN}}{V_O}$$

Equating the above two equations for I_O/I_{IN}, we get

$$\frac{\eta V_{IN}}{V_O} = \frac{1-D}{D}$$

which simplifies to

$$D = \frac{V_O}{\eta V_{IN} + V_O}$$

It is interesting that though power supply engineers are almost conditioned to assert "duty cycle in continuous conduction mode is independent of load current," we actually *relied on current* to derive a so-called "*current-independent*" duty cycle equation above. So, is it really as current-independent as we had imagined? Quite clearly, no. But we also observe that the above derivation was somewhat surprisingly, *not* based on voltseconds balance in steady state, but on *energy balance*. Since energy is $V \times I$, the derivation naturally included current.

Such subtleties aside, having derived the real-world duty cycle equations using energy principles, we can certainly go back and apply the results to our underlying voltseconds

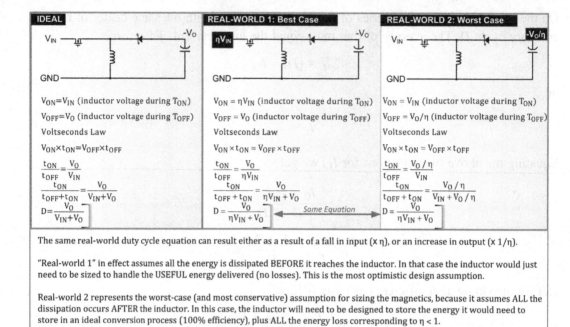

The same real-world duty cycle equation can result either as a result of a fall in input (x η), or an increase in output (x 1/η).

"Real-world 1" in effect assumes all the energy is dissipated BEFORE it reaches the inductor. In that case the inductor would just need to be sized to handle the USEFUL energy delivered (no losses). This is the most optimistic design assumption.

Real-world 2 represents the worst-case (and most conservative) assumption for sizing the magnetics, because it assumes ALL the dissipation occurs AFTER the inductor. In this case, the inductor will need to be designed to store the energy it would need to store in an ideal conversion process (100% efficiency), plus ALL the energy loss corresponding to η < 1.

The actual circuit and energy storage requirement will lie somewhere in between Real-world 1 and Real-world 2.

All Voltages above are Magnitudes

Figure 5.1: Creating equivalent ideal models for real-world converters and the effect on sizing of magnetics (example shown here is a Buck-Boost converter).

balance principle (which we recognize must always be true for any topology whichever way we look at it), and thereby create an *ideal model of our real-world converter.* That will make real-world converters much simpler to analyze, and very accurately so for most purposes. See Figure 5.1 where we have mapped our real-world converter circuit into an "equivalent" lossless (ideal) converter, by exploiting the remarkable similarity between the ideal and real-world duty cycle equations. The trick is to account for the real-world loss in two possible ways: either think of the input as having decreased from V_{IN} to $\eta \times V_{IN}$ in the corresponding ideal converter (implying that all losses occurred *prior* to the inductor), *or* imagine that the output has increased from V_O to V_O/η (implying that all losses occur after the inductor). The duty cycle is the same in both cases. And that follows from the fact that using simple arithmetic manipulation, we can write our real-world equation set above as follows:

$$D_{\text{Buck}} = \frac{V_O/\eta}{V_{IN}}, \qquad D_{\text{Boost}} = \frac{(V_O/\eta) - V_{IN}}{V_O/\eta}, \qquad D_{\text{Buck-Boost}} = \frac{V_O/\eta}{(V_O/\eta) + V_{IN}}$$

The two forms of real-world duty cycle equation *seem equivalent,* and in fact are — *but only up to a point.* Eventually, they lead to different ON-time/OFF-time *voltseconds* and therefore to *magnetics of different sizes*. The former interpretation (i.e., an effective decrease in V_{IN}) leads to an optimistic (and possibly undersized) core, whereas the latter (an effective increase in V_O) leads to a relatively larger core. In general, the *latter* model is a safer bet in design, especially if we don't know where exactly the losses corresponding to the less than unity estimated/measured efficiency are occurring inside the converter. What really happens in a practical converter lies somewhere in between the two real-world models of Figure 5.1.

We have one last question: what losses are we (still) ignoring by writing out our duty cycle equations in terms of η? In other words: how exact are our "real-world" duty cycle equations above? If we understand the sample derivation above, we will realize that in effect we have ignored any current flowing in an equivalent resistance path *parallel* to the input and the output capacitors. We assumed that *all* the average current coming in from the input source went straight into the switch, and likewise, all the average current coming from the diode went straight into the load. In doing so, we implicitly ignored ESR-related and leakage losses in the input/output capacitors. Note that we can however estimate those losses upfront, and add/subtract them from what we call "P_O" and "P_{IN}" above, then the computations with the corrected values of input and output will be as accurate as can be. For example, if at the input we measure 10 V and 1 A, and estimate that we are losing 1 W in the input cap, we need to use P_{IN} as $10 - 1 = 9$ W. That is a V_{IN} of 10 V and an I_{IN} of 9 W/10 V = 0.9 A. Similarly, if in the load we measure 5 V and 1.5 A, and we estimate we lose 0.5 W in the ESR of the output cap, we take P_O as $7.5 + 0.5 = 8$ W. That is a V_O of 5 V and I_O of 8 W/5 V = 1.6 A. So, the converter efficiency is not 7.5 W/10 W = 0.75, but actually 8 W/9 W = 0.89. Though it must be pointed out we are still ignoring leakage current paths and related losses *inside the switch and the diode*. We are also ignoring any quiescent current losses in the PWM controller IC. We can correct for all of those losses upfront too if we want, but these losses are usually considered insignificant. With these qualifications in mind, the real-world duty cycle equations provided earlier (expressed in terms of η) are as accurate as can be; the ideal equations are as inaccurate as can be. The real-world duty cycle equations factor in switching losses, conduction losses (including DCR-related losses in the inductor) and even any AC resistance losses. They are therefore the equations we have relied upon in describing the concepts related to energy transfer in Figures 5.2−5.4.

The Energy Transfer Charts

We now look at Figures 5.2−5.4. Here, we present each topology in turn, showing each step of their internal energy transfer process. To focus on energy and storage function, observe how we have split each topology into *three* reactive (energy storage) blocks — the input capacitor, the inductor (with switch and diode attached to switch its connections around), and the output capacitor.

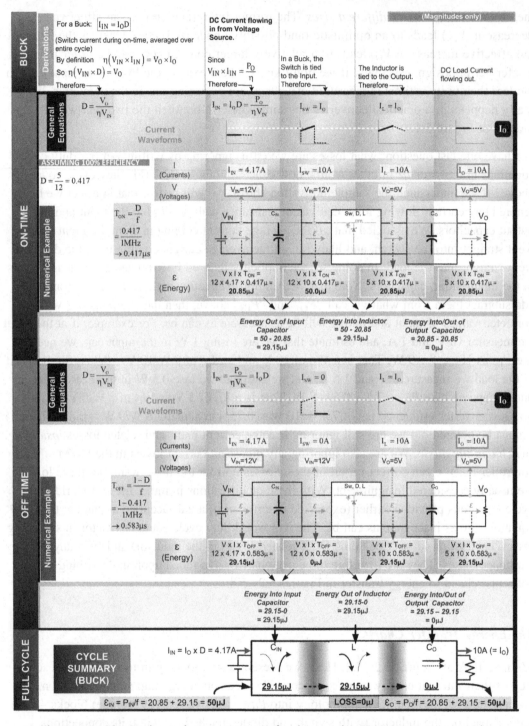

Figure 5.2: The Buck topology energy transfer chart and a 50 W converter example.

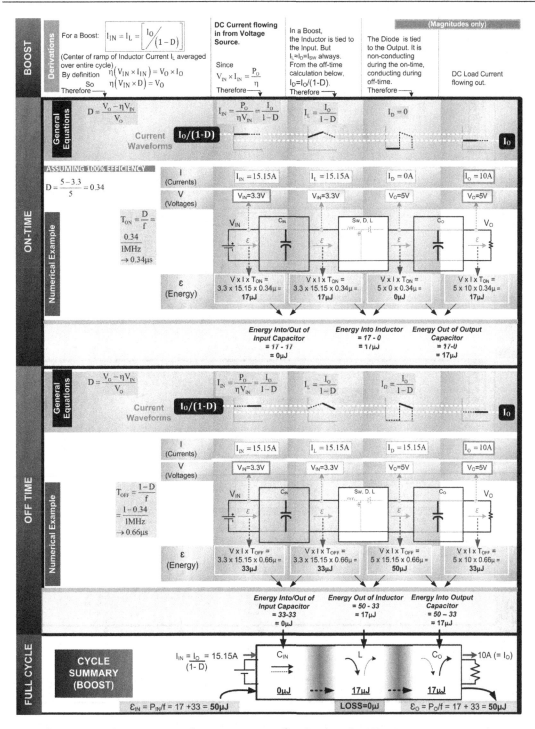

Figure 5.3: The Boost topology energy transfer chart and a 50 W converter example.

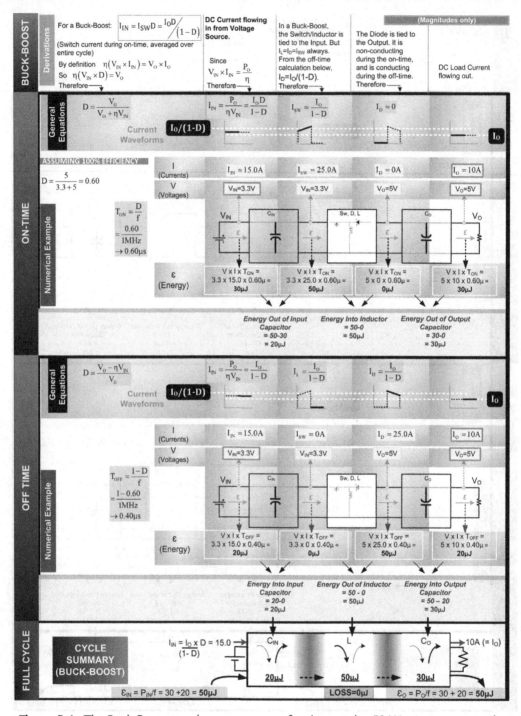

Figure 5.4: The Buck-Boost topology energy transfer chart and a 50 W converter example.

In each topology chart, we first look at what happens during the ON-time. We figure out *V* and *I* at each point of the circuit. Note that the equations for current used here are all very accurate, since they are based on the accurate (real-world) equations for *D*. We multiply *V*, *I* and the corresponding time interval ($V \times I \times$ time) to get the corresponding energy. Then from the energy *difference on either side of any given reactive block*, we can calculate how much energy is either getting *in* (i.e., stored), or coming *out* (i.e., extracted) from any particular block. We then repeat the same process for the OFF-time. Finally, we compile the "energy balance sheet": here we compare the energy in/out numbers of the OFF-time with the energy in/out numbers for the ON-time. We realize that in every case, if we store a certain amount of energy "*X* Joules" in any given block during the ON-time, then during the OFF-time, *exactly* that amount of energy, *X* Joules, must be extracted. And likewise, if we extract a certain amount of energy *X* Joules during the ON-time, exactly that very amount of energy must get stored during the OFF-time, and so on. Because, this is a *steady state* by definition — from one end of the converter to the other. We can confirm from the figures that, indeed, there is no incremental *buildup*, or decrease in energy, after one complete cycle in any of the *three* reactive elements, the inductor or the input and output capacitors. Yes, if we had *not* got this result, we should have been very worried. It would have indicated to us that either our topology was somehow flawed, or more likely, the equations we had used to calculate the energy terms (and the corresponding currents and duty cycle) were not very accurate. We do note though, that in the sample numerical examples presented in Figures 5.2–5.4, we have still assumed 100% efficiency. That has been done only for simplicity at this early conceptual stage; later, we will repeat the same process for a real-world case ($\eta < 1$), and see how the picture changes. Note also that the numerical examples within these figures all use 50 W converters of each topology, just for comparison sake.

We can learn several things from Figures 5.2–5.4. We list some of them here.

(a) A Buck-Boost inductor has to handle *all* the energy coming toward it — 50 μJ as per Figure 5.4, corresponding to 50 W at a switching frequency of 1 MHz.

> **Note:** *To be more precise for the general case of $\eta \leq 1$: the power converter has to handle P_{IN}/f if we use the conservative model in Figure 5.1, but only P_O/f if we use the optimistic model.*

In contrast, the inductors of the other 50 W converters shown in Figures 5.2 and 5.3 have to handle only a fraction of the energy flowing between input and output. But exactly what fraction is that?

(b) To answer the above question, we can derive closed-form equations relating to inductor energy storage, based on these three energy transfer charts. We have done that in Figure 5.5, from which we get the following *energy packet sizes* for each topology:

$$\Delta\varepsilon = \frac{P_{IN}}{f} \times (1-D) \quad \text{(Buck)}$$

$$\Delta\varepsilon = \frac{P_{IN}}{f} \times D \quad \text{(Boost)}$$

$$\Delta\varepsilon = \frac{P_{IN}}{f} \quad \text{(Buck-Boost)}$$

This shows us what fraction of the energy we were talking about above. Note that these derivations were based on the *conservative real-world model*, that is, where we factor in efficiency, by thinking of the *output as having increased to V_O/η rather than the input decreasing to $V_{IN} \times \eta$*.

We thus realize that the Buck and Boost inductor storage requirements are based not only on input/output power, but also on *input and output voltages (D)*. The Buck-Boost energy requirement is based on power alone — the fraction of power that it has to handle is 100% (i.e., all of it).

(c) We know that in all topologies, *low* input voltage corresponds to *high* duty cycle and *high* input to *low* duty cycle. So, we can conclude that for a Buck, $\Delta\varepsilon$ (the energy packet going in/out of the inductor every cycle) is at its maximum when $1-D$ is highest, i.e. where D is lowest, that is, when the input voltage is at its highest level. This is consistent with what we learned in *Chapter 2* regarding "worst-case input" for the Buck. For example, if we have a Buck delivering 3.3 V output, for an input range of 5–15 V, we need to design its inductor (pick the core volume) at the worst-case input of 15 V.

(d) Similarly, we recognize that for a Boost, $\Delta\varepsilon$ (the energy in/out of the inductor every cycle) is at its maximum when D is highest, that is, when input is at its lowest level. This too is consistent with what we learned in *Chapter 2* regarding "worst-case input" for the Boost. For example, if we have a Boost delivering 24 V output, for an input range of 5–12 V, we would design its inductor at the worst-case input of 5 V.

(e) Coming to the Buck-Boost, we had asserted in *Chapter 2* that we need to design its inductor at the lowest voltage of the input range. In the next section, we will see why that is true, though not very obvious so far. We are confused because we have just calculated that the energy that gets cycled through the inductor of a Buck-Boost is *fixed* — $\Delta\varepsilon = P_{IN}/f$ *irrespective of the input voltage*. And that being undeniably true, we can rightly conclude that a 50 W "universal input" flyback, for example (typical input range being 85–265VAC), does *not* need to have a bigger transformer than a 50 W flyback designed only for Europe (typical input range being 195–265VAC).

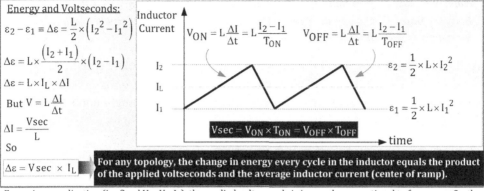

Energy and Voltseconds:

$$\varepsilon_2 - \varepsilon_1 \equiv \Delta\varepsilon = \frac{L}{2} \times \left(I_2{}^2 - I_1{}^2\right)$$

$$\Delta\varepsilon = L \times \frac{(I_2 + I_1)}{2} \times (I_2 - I_1)$$

$$\Delta\varepsilon = L \times I_L \times \Delta I$$

But $V = L\dfrac{\Delta I}{\Delta t}$

$$\Delta I = \frac{Vsec}{L}$$

So

$$\boxed{\Delta\varepsilon = V sec \times I_L}$$

Inductor Current

$$V_{ON} = L\frac{\Delta I}{\Delta t} = L\frac{I_2 - I_1}{T_{ON}} \qquad V_{OFF} = L\frac{\Delta I}{\Delta t} = L\frac{I_2 - I_1}{T_{OFF}}$$

I_2

I_L

I_1

$$\varepsilon_2 = \frac{1}{2} \times L \times I_2{}^2$$

$$\varepsilon_1 = \frac{1}{2} \times L \times I_1{}^2$$

$$Vsec = V_{ON} \times T_{ON} = V_{OFF} \times T_{OFF}$$

→ time

For any topology, the change in energy every cycle in the inductor equals the product of the applied voltseconds and the average inductor current (center of ramp).

For a given application (i.e. fixed V_{IN}, V_O, I_O), the applied voltseconds is inversely proportional to frequency. So, the energy delta (i.e. packet size "$\Delta\varepsilon$") is inversely proportional to frequency. Therefore, **keeping I_2 and I_1 unchanged as we change frequency**, the inductance required is inversely proportional to the switching frequency. We see that the required *peak* energy handling capability of the inductor ($\frac{1}{2} L \times I_2{}^2$) is also inversely proportional to frequency. [Note that for a given I_O if we keep the peak current (I_2) fixed, that is equivalent to keeping *r* (current ripple ratio) constant as we change frequency. That is therefore the implied condition here].

The center of the inductor current ramp I_L is:	$I_L = I_O$ for Buck $= \dfrac{I_O}{1-D}$ for Boost and Buck-Boost
We can take the Voltseconds ("Vsec") either during the on time or off-time, since they must always be equal. It is convenient to use these:	$Vsec = V_O/\eta$ (applied for T_{OFF}) for Buck *(see Figure 5.1)* $= V_{IN}$ (applied for T_{ON}) for Boost and Buck-Boost
The input current is:	$I_{IN} = I_O \times D$ for Buck $= \dfrac{I_O}{1-D}$ for Boost *(see Appendix)* $= \dfrac{I_O \times D}{1-D}$ for Buck-Boost *(All above equations are EXACT)*

BUCK

$$\Delta\varepsilon = Vsec \times I_L = \frac{V_O}{\eta} \times \frac{1-D}{f} \times I_O$$

But, $I_O \times D = I_{IN}$ so

$$\Delta\varepsilon = \frac{P_{IN}}{f} \times (1-D) \approx \frac{P_O}{f} \times (1-D)$$

Example: 12V to 5V@10A, η=1

$$\Delta\varepsilon = \frac{P_{IN}}{f} \times (1-D) = \frac{50}{10^6} \times (1-0.417)$$

$$\Rightarrow 29.15\mu J \quad \text{(see Figure 5.2)}$$

BOOST

$$\Delta\varepsilon = Vsec \times I_L = V_{IN} \times \frac{D}{f} \times \frac{I_O}{1-D}$$

But, $I_O \times \dfrac{1}{1-D} = I_{IN}$ so

$$\Delta\varepsilon = \frac{P_{IN}}{f} \times D \approx \frac{P_O}{f} \times D$$

Example: 3.3V to 5V@10A, η=1

$$\Delta\varepsilon = \frac{P_{IN} \times D}{f} = \frac{50 \times 0.34}{10^6}$$

$$\Rightarrow 17\mu J \quad \text{(see Figure 5.3)}$$

BUCK-BOOST

$$\Delta\varepsilon = Vsec \times I_L = V_{IN} \times \frac{D}{f} \times \frac{I_O}{1-D}$$

But, $I_O \times \dfrac{D}{1-D} = I_{IN}$ so

$$\Delta\varepsilon = \frac{P_{IN}}{f} \approx \frac{P_O}{f}$$

Example: -3.3V to 5V@10A, η=1

$$\Delta\varepsilon = \frac{P_{IN}}{f} \frac{50}{10^6}$$

$$\Rightarrow 50\mu J \quad \text{(see Figure 5.4)}$$

Compare this with the results in the Energy Transfer Charts presented previously (which used the same application conditions as above). The calculated stored inductor energy ($\Delta\varepsilon$) terms are also consistent.

Figure 5.5: The closed-form equations governing inductor energy storage requirements in the three topologies.

> *Note: We must keep in mind that the above statement applies only to a flyback/Buck-Boost. A Boost PFC stage designed only for Europe* will *have a much smaller inductor than a Universal Input Boost PFC.*

At the same time it is equally accurate to maintain that we really should design the inductor of a flyback/Buck-Boost at the lowest input voltage *of its input range.* In the next section, we will learn why *both* statements above, seemingly contradictory, are in fact simultaneously true.

Peak Energy Storage Requirements

So far, our focus has been on calculating the amount of energy going in and out of the inductor per cycle in the three topologies. In doing so, in effect, we ignored the *crest factor* (peak to average ratio) of the current waveforms. And that means we ignored the effect of the current ripple ratio *r*. Looking at Figures 5.2–5.4 once again, we realize that we had based all our calculations on the *center of the current ramp* I_L, and it was on that basis we had calculated the energy swing (packet) $\Delta\varepsilon$. However, in reality, on a real-time basis, the current ramps up and down every cycle. Therefore, there is a certain *peak energy* ε_{PEAK} associated with the peak current. We must ensure not only that the inductor can *store* a certain amount of energy every cycle, but that it can handle the *instantaneous energy* at any given part of the cycle, without saturating.

We can easily calculate the relationship between $\Delta\varepsilon$ and ε_{PEAK} as shown in Figure 5.6. The key important relationship is this

$$\varepsilon_{PEAK} = \frac{\Delta\varepsilon}{8} \times \left[r \times \left(\frac{2}{r} + 1 \right)^2 \right]$$

This equation applies to all topologies. The term $\Delta\varepsilon$ in it however depends on the specific topology. So, for a typical value of $r = 0.4$ (ripple of $\pm 20\%$), we get a peak value exactly 80% greater than the energy swing (a default factor of 1.8). In other words, in Figure 5.4, for example, the inductor energy swing for a Buck-Boost was 50 μJ, and we now realize that to store this, with an inductance selected to give us an *r* of 0.4, we actually need an inductor sized such that it can handle not 50 μJ, but the peak of $50 \times 1.8 = 90$ μJ (instantaneously). Basically, the inductor is constantly moving between the extremes of 40 μJ and 90 μJ, with a delta of 50 μJ. Similarly, in Figure 5.3, we need to pick a Boost inductor sized to store $17 \times 1.8 = 30.6$ μJ instantaneously.

> *Note: In power supplies, while calculating current stresses (as we will do in* Chapter 7*), something called the "flat-top approximation" is often used. In this approximation, we basically ignore the AC (swinging) part of any current waveform and approximate the*

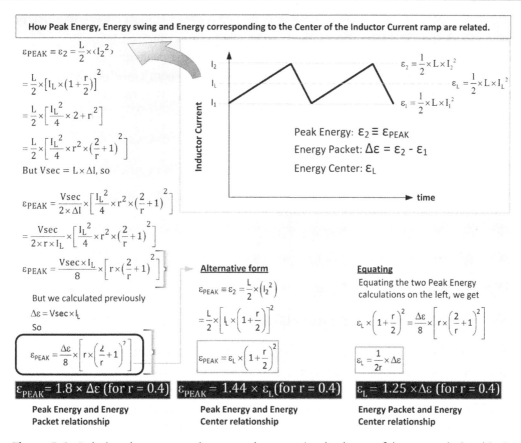

Figure 5.6: Relating the energy swing to peak energy (and other useful energy relationships).

trapezoidal current waveform with a rectangular one (extending to the center of ramp, i.e., the average inductor current). This is also called the "large inductance approximation," and corresponds to setting $r = 0$. *This approximation typically gives surprisingly accurate results for the RMS value of the switch current (within* $\pm 10\%$ *for* r *less than 0.5 and any duty cycle from 10% to 90%), but not for capacitor currents. In our present case, we ask: what does the flat-top approximation do to the selection of inductors? Is it fairly accurate? Quite the contrary as we see below.*

We now ask the seemingly obvious: why not *reduce* the current ripple ratio *r* further (i.e., increase *L*)? We intuitively feel that would lower the peak value of current and possibly the peak energy, and therefore the size of the inductor. We intuitively feel that would bring the peak value close to the value at the center of ramp. So presumably it would lead to a smaller core size (with lesser required peak energy handling capability).

But that is completely wrong. The correct answer, and one of the most counterintuitive ones in magnetics design, is that in reality, the smallest core size is obtained by *raising r* (*lowering* the inductance, not raising it). Yes that increases the peak currents, worsens the crest factor. But somehow it helps *reduce* inductor size (and energy too)! Mathematically it is simple to explain. The "culprit" is the odd term involving current ripple ratio *r* in the ε_{PEAK} equation above. We isolate that term below.

$$\varepsilon_{PEAK} = \frac{\Delta\varepsilon}{8} \times F(r)$$

where

$$F(r) = \left[r \times \left(\frac{2}{r} + 1 \right)^2 \right]$$

If *r* is very small, $2/r$ is $>> 1$. So,

$$F(r) \approx \left[r \times \left(\frac{2}{r} \right)^2 \right] = \frac{4}{r}$$

We see that the $F(r)$ function imparts an approximate inverse proportionality ($y = 1/x$) shape to all the peak energy curves, as we can confirm when we plot them out using Mathcad in Figure 5.7. All the curves have the *same shape*, and in fact are the very same normalized curve we had plotted in Figure 2.6 earlier. This basic shape shown remains unchanged for all topologies and all duty cycles (i.e., all input/output voltages), and even all switching frequencies. It is thus *fundamental*, and is perhaps the most important curve we can study for improving our overall understanding of switching power supplies. It also highlights the tremendous simplicity that results if we start to view magnetics in terms of current ripple ratio *r* rather than *L*. Because, as we can see, all the curves have a knee at $r \approx 0.4$. They do not show that property if we look only at *L*.

Realizing that the vertical axis in Figure 5.7 is effectively just the size of the core, we conclude that *size of core decreases as L decreases*. That also incidentally explains why we often say "reduce the inductance" when we really mean "reduce the inductor," *and get away with it each time.*

The question remains: how can we not mathematically, but *intuitively*, explain this seeming paradox — that is, why the energy handling requirement *decreases* if we reduce *L*, instead of increasing. The answer to this is that when we increase *L* and thereby reduce the peak value of current, we have to increase *L* by a *comparatively much bigger factor* to achieve a certain, relatively small, reduction in peak current. So, energy, which depends on *both* *L* and *I* through the equation $\varepsilon = (1/2) \times LI^2$, actually *increases* as we *increase L*. That

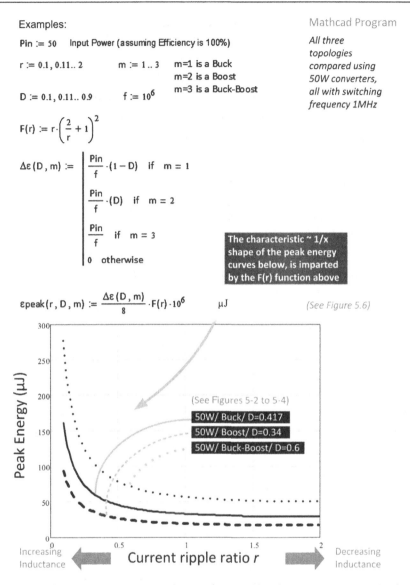

Examples:

Pin := 50   Input Power (assuming Efficiency is 100%)

r := 0.1, 0.11.. 2   m := 1 .. 3   m=1 is a Buck
                                            m=2 is a Boost
D := 0.1, 0.11.. 0.9   f := 10^6 m=3 is a Buck-Boost

$$F(r) := r \cdot \left(\frac{2}{r} + 1 \right)^2$$

$$\Delta\varepsilon(D, m) := \begin{vmatrix} \dfrac{Pin}{f} \cdot (1 - D) & \text{if} \quad m = 1 \\[2mm] \dfrac{Pin}{f} \cdot (D) & \text{if} \quad m = 2 \\[2mm] \dfrac{Pin}{f} & \text{if} \quad m = 3 \\[2mm] 0 & \text{otherwise} \end{vmatrix}$$

Mathcad Program

All three topologies compared using 50W converters, all with switching frequency 1MHz

The characteristic ~ 1/x shape of the peak energy curves below, is imparted by the F(r) function above

$$\varepsilon peak(r, D, m) := \frac{\Delta\varepsilon(D, m)}{8} \cdot F(r) \cdot 10^6 \qquad \mu J \qquad \text{(See Figure 5.6)}$$

(See Figures 5-2 to 5-4)

50W/ Buck/ D=0.417
50W/ Boost/ D=0.34
50W/ Buck-Boost/ D=0.6

Figure 5.7: Peak energy curves, core size, and optimal inductance (current ripple ratio).

is why the correct direction toward achieving core size reduction is by decreasing L (i.e., raising r), not increasing L (lowering r). The paradox is resolved. *The flat-top approximation should never be used for magnetics design or for capacitor design.*

We can now ask: if that is so, why not *decrease L even more*, to achieve the minimum possible core size? Yes indeed, we can do that. But there are penalties as was made obvious in Figure 2.6. However, despite that penalty, it is not uncommon in low-power applications to find low-cost DC–DC or AC–DC converters operating in *critical conduction mode* always (i.e., $r = 2$, exactly between CCM and DCM). We will find their inductors are really much smaller than CCM converters of the same output power. But along with high r, we will also likely see on the board some rather low-R_{DS} MOSFETs (perhaps in large packages), and low-ESR capacitors (or many capacitors in parallel) — an obvious effort to combat the adverse effects of the rather high peak and RMS currents as indicated in Figure 2.6. That is the "penalty" we may decide to pay for using smaller inductors.

Since the peak energy curves *always* have a knee at around $r \approx 0.4$, the most optimal value in almost all applications is $r \approx 0.4$ — that becomes an almost universal design target of sorts, irrespective of the topology, application conditions, and switching frequency. This is equivalent to $\pm 20\%$ ripple. Note that we could never have managed to declare a certain, fixed "optimum inductance value" for every single switching converter on earth. But certainly, talking in terms of r, we have come pretty close to doing just that. Finally, once we have fixed r, we can calculate L based on that assumption. As expected, the equations connecting r to L (see reference table in Appendix), *do* depend on topology, application conditions and switching frequency. That is where all the dependencies enter the picture.

One question remains. Previously we had stated that a flyback transformer for, say, a 50 W application designed for Europe alone, would have the same size as one designed for the entire world, including the US (disregarding any differences in efficiency here). This was because we had discovered that the amount of energy the transformer has to cycle back and forth per cycle equals P_O/f, and that has nothing to do with the input voltage. So, *why do we still insist we need to design the Buck-Boost at its lowest input voltage?* The answer is right above us actually. It is the crest factor that is responsible. As we lower the input, the peak current and energy increases. So, to be sure we avoid saturation, we need to design the transformer at the lowest input voltage (max D) of our expected input range. That is always true for a Buck-Boost/flyback. However, when we compare a European flyback transformer designed for $r = 0.4$ at 195VAC, with a North American flyback transformer designed for $r = 0.4$ at 85VAC, there would be no difference because the numerical value of the term $F(r)$ is the same in both cases. The crest factors are identical and so ε_{PEAK} is the same in both cases, and so is the required transformer size (see also Figure 14.4).

Calculating Inductance Based on Desired Current Ripple

Having understood how easy it is to start with a general design target of $r = 0.4$ (in most cases), for all topologies, applications, and frequencies, we now want to connect L and r. In particular, we want to derive the relevant equations presented in the Appendix, and also re-examine the "$L \times I$" rule which was presented earlier in Figure 2.12. Note that here, for easy reference and representation, we prefer calling Voltseconds as "Vsec" here, rather than as "*Et*" as in other chapters, but the two are the same. In Figure 5.8, we have derived the L versus r equations for each topology. We have also then shown that the "$L \times I$ rule" is universal. It applies to all topologies and can be written as

$$L = \frac{\text{Vsec}}{r \times I_L} \quad \text{(for all } topologies\text{)}$$

We now show that this is just the inductor rule $V = L \, dI/dt$ in a more exotic form.

$$\text{Since} \quad r = \frac{\Delta I_L}{I_L}$$

$$\text{we get} \quad L = \frac{\text{Vsec}}{\Delta I_L}$$

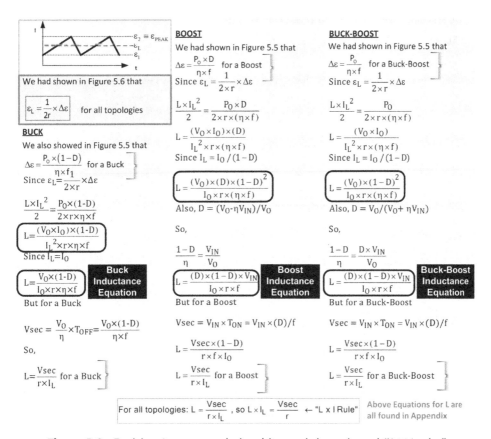

Figure 5.8: Deriving L versus r relationships and the universal "$L \times I$ rule."

In other words,

$$L = \frac{V_{ON} \times T_{ON}}{\Delta I_L} = \frac{V_{OFF} \times T_{OFF}}{\Delta I_L}$$

Solving for V (applied voltage across inductor during ON or OFF time)

$$V = L\frac{\Delta I}{\Delta t}$$

We thus get back our well-known inductor equation. They are the same equation!

In Figure 5.9, we have also calculated the DC transfer function V_O/V_{IN} for all topologies. And from that, we re-derive the equations for inductance, this time starting with the $L \times I$ rule. We have ultimately re-cast the inductor equations in terms of μH and MHz for greater ease of use. See *Chapter 19* for some relevant solved examples.

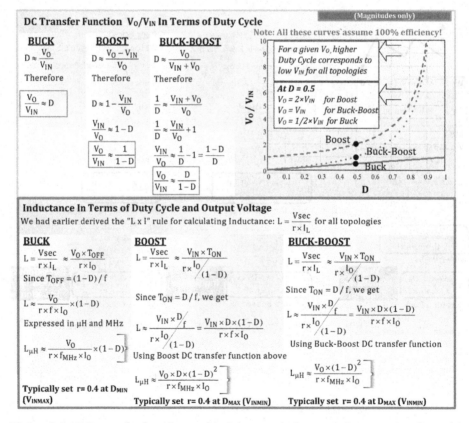

Figure 5.9: DC transfer function and inductance design equations starting from the "$L \times I$ rule."

Part 2: Energy to Core Sizes

Magnetic Circuits and the Effective Length of Gapped Cores

With energy transfer and storage concepts understood, we can start discovering how they point to the *optimum* core size in any given application. But to do so more effectively, we need to add to, and refine, some of the magnetic concepts presented in *Chapters 2* and *3*.

Let us start by looking at Figure 5.10. This is a "magnetic circuit" — the electrical equivalent of a magnetic component. We can see how reluctance in magnetics plays a part analogous to resistance in electrical circuits. Flux, defined as $\phi = B \times A$ (for uniform fields), plays the role of current. The term $N \times I$ (Ampere-turns), called magnetomotive force, is analogous to voltage source in electrical circuits.

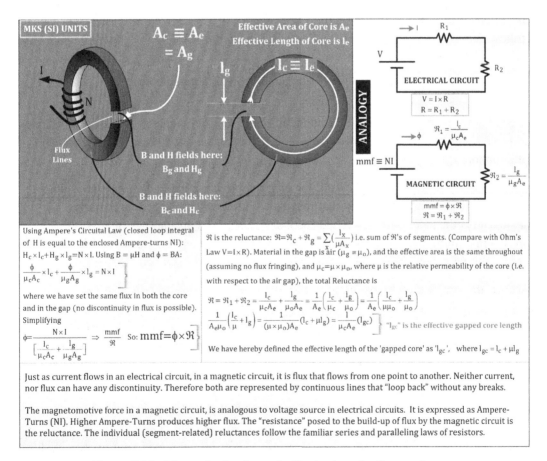

Figure 5.10: Magnetic circuits and effective length of gapped cores.

If we have an ungapped core, its effective length is basically the perimeter of an imaginary circle drawn through the center of the core. That is called "l_e" or effective length. The effective area, "A_e," is almost exactly equal to the geometrical cross-sectional area of the core. The reluctance is $l_e/\mu_c A_e$, where μ_c is the permeability of the core material (using MKS units). However, when the core is gapped, we see that the reluctance increases to $l_{gc}/\mu_c A_e$, where $l_{gc} = l_e + \mu l_g$, and l_g is the length of the gap. In effect, l_{gc} is behaving as the new effective length of the gapped core. In other words, as soon as we introduce an air gap, the effective length that appears in the magnetic circuit is not just l_e but $l_e + \mu l_g$, where μ is the *relative* permeability of the core material. Relative permeability is the permeability relative to air, that is, $\mu = \mu_c/\mu_o$. It seems that for all practical purposes, the air gap l_g has gotten multiplied by the relative permeability of the *surrounding* material μ, and then added to the effective length l_e of the core (without gap) and that has become the new effective length of the gapped core. For example, if we were using ferrite (typical relative permeability of 2,000), and the air gap was a mere 1 mm, this tiny gap actually plays the part of an effective length of *core material* equal to $2,000 \times 1 = 2,000$ mm, or 200 cm. That is the reason reluctance changed as follows:

$$\frac{l_e}{\mu_c A_e} \rightarrow \frac{l_e + \mu l_g}{\mu_c A_e} \quad \text{(ungapped core to gapped core)}$$

The good news is that the "200 cm" (magnetic length) extra magnetic length only occupies 1 mm (physically). Also, eventually, the adder μl_g becomes much larger than l_e, and so l_e is often ignored in comparison.

But what use is this extra length? Well, for one, the effective gapped *volume* goes up from $A_e \times l_e$ to $A_e \times l_{gc}$. Intuitively that means we have a much larger core volume (magnetically) than we thought we had (geometrically). The air gap has in effect become a "core volume multiplier" of sorts. And further, provided we can create strong-enough fields, we expect to be able to store much more energy too. We have a much "bigger box".

However, the simplest and most elegant way to quantify all this and discuss it mathematically-plus-intuitively is by focusing not on the adder term (μl_g), but on the ratio $(l_e + \mu l_g)/l_e$ (the new effective length, after gapping, divided by the old effective length, before gapping). If we do that, the equations of gapped cores become extremely simple to handle. That factor is called the "z-factor" (or "*gap factor*").

$$z = \frac{l_{gc}}{l_e} = \frac{l_e + \mu l_g}{l_e} \approx \mu \frac{l_g}{l_e}$$

We see that the net reluctance increases from $l_e/\mu_c A_e$ for an ungapped core to $z l_e/\mu_c A_e$ for a gapped core. So, "z" is the reluctance multiplier term.

Stored Energy in Gapped Cores and the z-Factor

Now we start to look closely at the magnetic fields, because that is where magnetic energy is ultimately stored. The effect of the air gap can be seen clearly from Figures 5.11 and 5.12. As expected, since the air gap has caused the reluctance to increase, the B-field drops from $\mu_c NI/l_e$ in an ungapped core, to $\mu_c NI/zl_e$ in a gapped core (think in terms of the electrical circuit analogy in Figure 5.10). But wait: where exactly is this reduced B-field we are referring to — are we talking about the B-field in the core or in the gap? B is related to flux through $B = \phi/A$, and since flux is continuous across boundaries, so is B. The B-field inside the core and in the air gap *is the same*. We need to remember that B (its normal

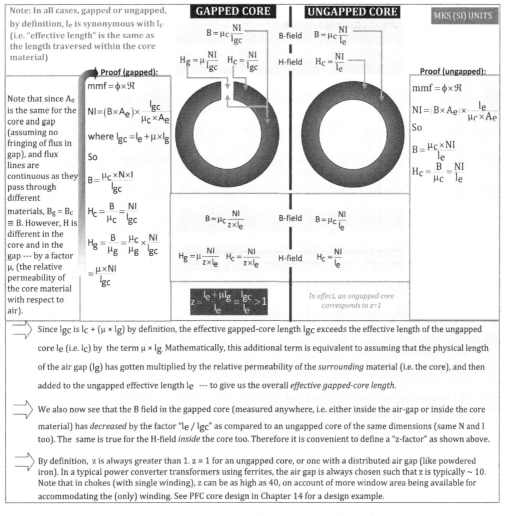

Figure 5.11: Comparing the B and H fields in gapped and ungapped cores.

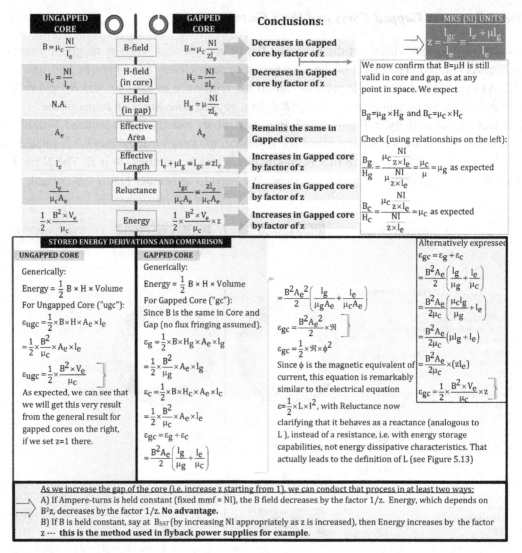

Figure 5.12: How gapped cores store more energy.

component) is continuous across material boundaries. So, the B-field is $\mu_c NI/zl_e$, both in the core and in the gap, as indicated in the figures.

However, the H-field has different values in the core and in the gap. We remember that by definition, $H_c = B_c/\mu_c$ and $H_g = B_g/\mu_g$. The H-field at any point in space is: the B-field *at that point*, divided by the permeability *at that point*. Since $B_g = B$, and the permeability of air ($\mu_g \equiv \mu_o$), is different from the permeability of the core ($\mu_c \equiv \mu\mu_o$), the H-field in the core and in the gap are different. We see that, because of the increased reluctance (due to the air gap), the *H*-field does in fact drop from its value of NI/l_e in an ungapped core, to NI/zl_e in a gapped core. *But note that this reduction occurs only inside the core.*

We ask: does the reduction in the *H*-field (within the core material) significantly reduce the overall energy storage capability? Yes, it does reduce the energy stored *in the core material*, but the energy in the air gap more than makes up for it. Fields aside, just in terms of volume, in a typical gapped core, the volume of the core material is $A_e \times l_e$. That is typically a small part of the total effective volume of the overall gapped structure, $A_e \times l_{gc}$. The bottom-line is that what happens to the *B* and *H* fields *inside the core material* is hardly important in a gapped core. The effects of the air gap predominate. Though, we must keep in mind that the gap wouldn't exist, were it not for the presence of the core around it.

In gapped ferrite *transformers*, a typical target value for *z* is 10. So, we realize that the *H*-field in the core typically drops by a factor of 10 compared to the value it would have had in an ungapped core. However, the *H*-field *in the gap* increases dramatically. For a ferrite transformer for example, with a μ (relative permeability) of around 2,000, the *H*-field in the gap is about $2{,}000/10 = 200$ times the *H*-field of an ungapped core. The density of the stored energy (i.e., energy stored per unit volume) at any given point in space is by definition $(1/2) \times B \times H$, where *B* and *H* are the magnetic fields at that point. This increase in *H* inside the air gap, combined by the fact that the effective volume of the air gap is now $\mu l_g A_e$ not just $l_g A_e$, helps in substantially increasing the energy storage capabilities of gapped cores. This is described more clearly in Figure 5.12.

> **Note:** *The effective length of the magnetic structure without any air gap is l_e. (The reluctance is $l_e/\mu\mu_O$.) With air gap added, the total effective length becomes $l_e + \mu l_g$. (The reluctance is $(l_e + \mu l_g)/\mu\mu_O$ or $zl_e/\mu\mu_O$.) In effect, the air gap has added a magnetic length equal to μl_g, instead of just l_g. Therefore, the increase in the total effective volume due to the presence of the air gap is $A_e \times \mu l_g$.*

In Figure 5.12, we have summarized and tabulated the key changes that occur when we gap a core. Of particular importance is the fact that the energy in the entire structure of a gapped core (core + gap) *is z times the energy of an ungapped core, for the same B-field.* In other words, we need to increase the Ampere-turns (the mmf) to compensate for the increased reluctance due to gapping and thereby maintain *B* at the same value as in the ungapped core. We also realize that philosophically, we always want to use whatever resources we have, to the maximum extent we can. That limit, in this case, occurs when we operate the core close to its saturation flux density (B_{SAT}). For ferrites B_{SAT} is typically 0.3 T, or 300 mT, that is, 3,000 G in CGS units. Therefore, we may want to compare the energy storage capability of, say, an ungapped EE42 core operating at 300 mT, with a gapped EE42 core operating at 300 mT. We will then find that the ratio of their energy storage capabilities is equal to *z*. That highlights the importance (and simplicity) of looking at things from the viewpoint of the *z*-factor.

How do we maintain the same *B*-field as we introduce an air gap? Well, since *B* is proportional to *NI/z*, if *z* increases as we increase the gap, we need to increase the product *NI* by the same factor. However, we now realize that, typically, in switching converters, the currents are all pre-determined — they depend on load current and duty cycle. In other words, we have no control over current. So, in such a case, we will *need to increase the number of turns* (by the same factor). And that is what ultimately bounds the maximum practical value of *z* — *because eventually we need to be able to accommodate all these additional turns* on the core. For most commercial E-type cores, the window area available is commensurate with a *z* of up to 40 if we are designing a simple *choke* (i.e., an inductor with a single winding). But for flyback transformers for example, a lower value, *z* = 10, is a more practical value for the air gap, and for picking an *optimal E-core* in most applications.

Finally, in Figure 5.13, we connect to inductance and show that

$$L = \frac{\mu\mu_o \times N^2 \times A_e}{z \times l_e}$$

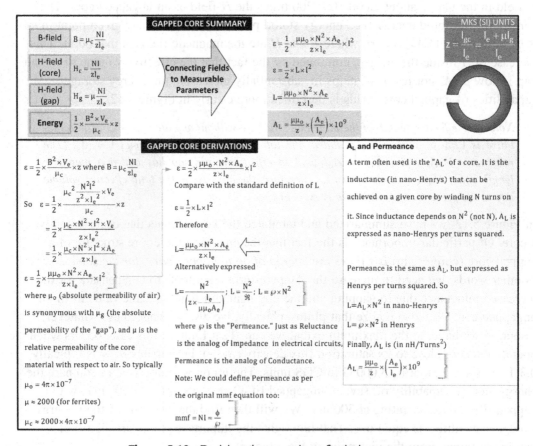

Figure 5.13: Deriving the equations for inductance.

Always keep in mind that we are using MKS units all through this discussion unless otherwise stated. So, A_e is in m^2, l_e is in m, and so on.

Note: We can now understand why in a switching power supply transformer, a z of 10 is a good target, whereas in a choke, a z of 40 is acceptable. As we start changing the number of turns with an intent to keep B fixed as we increase the gap, we need to apply another constraint — remember that we have a target inductance too, which we need to fix. Since we have just learned that L is proportional to N^2/z, for a constant L, z must be proportional to N^2. Since the maximum possible number of (Primary) turns in a transformer is half the number of turns we can put on a corresponding choke, the corresponding z is 1/4th. Therefore, if a good target for a choke is z = 40, then for a transformer it is z = 10. We will use the z = 40 condition when designing a PFC choke later in Chapter 14. *We will use z = 10 as an example later in this chapter.*

Note: In this entire core volume selection process, we are only talking in terms of energy storage capability. Which is why we are considering a flyback transformer, not a Forward converter transformer for example. In the latter case, the core really does not store any significant amount of energy, because the Primary and Secondary winding conduct simultaneously, and so useful energy is transferred directly through the transformer during the ON-time itself. The only stored energy component in the transformer core is the small excitation (magnetization) energy, which is subsequently recovered through a thin energy recovery winding during the OFF-time. In a Forward converter, the output choke is the energy storage component. But it is usually a toroid with a distributed air gap (e.g., powdered iron). That is a case of z = 1, and we should just use the stated permeability and maximum number of turns, as provided by its vendor.

Energy of a Gapped Core in Terms of the Volume of the Core

In Figures 5.14 and 5.15, we derive the final, general, closed-form energy-density equations for gapped cores, using any general magnetic material, *keeping in mind the specific demands of the topology on hand.* In the two figures, we have alternate (but equivalent) forms and derivations of the following key equations:

$$V_{e\_cm^3} = \left(\frac{P_{IN}}{f_{kHz}}\right) \times \left(\frac{(r+2)^2}{z \times r}\right) \times \left(\frac{\mu \times \pi}{B_{Tesla}^2}\right) \times \left(\frac{1-D}{10^4}\right), \text{ or}$$

(Buck)

$$V_{e\_cm^3} = \frac{31.4 \times P_{IN} \times (1-D) \times \mu}{z \times f_{MHz} \times B_{SAT\_Gauss}^2} \times \left[r \times \left(\frac{2}{r} + 1\right)\right]^2$$

In Figure 5.6, we showed that the energy packet that needs to be stored per cycle is related to the peak energy in the inductor as follows

$$\varepsilon_{PEAK} = \frac{\Delta\varepsilon}{8} \times \left[r \times \left(\frac{2}{r}+1\right)^2\right]$$

In Figure 5.12 we have shown that in terms of magnetic fields, the energy of an inductor is

$$\varepsilon = \frac{1}{2} \times \frac{B^2 \times V_e}{\mu_c} \times z$$

MKS (SI) UNITS

Cycled Energy per cycle:

$$\Delta\varepsilon = \frac{P_{IN}}{f} \times (1-D) \quad \text{(See Figure 5.5)} \quad \textbf{BUCK}$$

$$\varepsilon = \frac{1}{2} \times \frac{B^2 \times V_e \times z}{\mu_c} = \frac{\Delta\varepsilon}{8} \times \left[r \times \left(\frac{2}{r}+1\right)^2\right]$$

$$\frac{1}{2} \times \frac{B^2 \times V_e \times z}{\mu_c} = \frac{\frac{P_{IN} \times (1-D)}{f}}{8} \times \left[r \times \left(\frac{2}{r}+1\right)^2\right]$$

$$V_e = \left(\frac{P_{IN}}{f}\right) \times \left(\frac{(r+2)^2}{z \times r}\right) \times \left(\frac{\mu\mu_o}{B^2}\right) \times \left(\frac{1-D}{4}\right)$$

$$V_e = \left(\frac{P_{IN}}{f}\right) \times \left(\frac{(r+2)^2}{z \times r}\right) \times \left(\frac{\mu \times \pi}{B^2}\right) \times \left(\frac{1-D}{10^7}\right)$$

So far we have used MKS units (V_e in m^3).

Also, f was in Hz. Changing to kHz and cm^3

$$\boxed{V_{e\_cm3} = \left(\frac{P_{IN}}{f_{kHz}}\right) \times \left(\frac{(r+2)^2}{z \times r}\right) \times \left(\frac{\mu \times \pi}{B_{Tesla}^2}\right) \times \left(\frac{1-D}{10^4}\right)}$$
FOR GAPPED CORES, ANY MATERIAL

With $B_{TESLA} \le 0.3T$ (for ferrites), $\mu = 2000$,

and target z of 10 (optimum for transformers)

$$V_{e\_cm3} = 0.70 \times (1-D) \times \left(\frac{P_{IN}}{f_{kHz}}\right) \times \left(\frac{(r+2)^2}{r}\right)$$

With a typical target value r = 0.4, we get

$$\boxed{V_{e\_cm3} = [10 \times (1-D)] \times \left(\frac{P_{IN}}{f_{kHz}}\right) \text{ for a Buck}}$$
VOLUME OF A GAPPED FERRITE CORE

Cycled Energy per cycle:

$$\Delta\varepsilon = \frac{P_{IN} \times D}{f} \quad \text{(See Figure 5.5)} \quad \textbf{BOOST}$$

$$\varepsilon = \frac{1}{2} \times \frac{B^2 \times V_e \times z}{\mu_c} = \frac{\Delta\varepsilon}{8} \times \left[r \times \left(\frac{2}{r}+1\right)^2\right]$$

$$\frac{1}{2} \times \frac{B^2 \times V_e \times z}{\mu_c} = \frac{\frac{P_{IN} \times D}{f}}{8} \times \left[r \times \left(\frac{2}{r}+1\right)^2\right]$$

$$V_e = \left(\frac{P_{IN}}{f}\right) \times \left(\frac{(r+2)^2}{z \times r}\right) \times \left(\frac{\mu\mu_o}{B^2}\right) \times \left(\frac{D}{4}\right)$$

$$V_e = \left(\frac{P_{IN}}{f}\right) \times \left(\frac{(r+2)^2}{z \times r}\right) \times \left(\frac{\mu \times \pi}{B^2}\right) \times \left(\frac{D}{10^7}\right)$$

So far we have used MKS units (V_e in m^3).

Also, f was in Hz. Changing to kHz and cm^3

$$\boxed{V_{e\_cm3} = \left(\frac{P_{IN}}{f_{kHz}}\right) \times \left(\frac{(r+2)^2}{z \times r}\right) \times \left(\frac{\mu \times \pi}{B_{Tesla}^2}\right) \times \left(\frac{D}{10^4}\right)}$$
FOR GAPPED CORES, ANY MATERIAL

With $B_{TESLA} \le 0.3T$ (for ferrites), $\mu = 2000$,

and target z of 10 (optimum for transformers)

$$V_{e\_cm3} = 0.70 \times (D) \times \left(\frac{P_{IN}}{f_{kHz}}\right) \times \left(\frac{(r+2)^2}{r}\right)$$

With a typical target value r = 0.4, we get

$$\boxed{V_{e\_cm3} = (10 \times D) \times \left(\frac{P_{IN}}{f_{kHz}}\right) \text{ for a Boost}}$$
VOLUME OF A GAPPED FERRITE CORE

Cycled Energy per cycle:

$$\Delta\varepsilon = \frac{P_{IN}}{f} \quad \text{(See Figure 5.5)} \quad \textbf{BUCK-BOOST}$$

$$\varepsilon = \frac{1}{2} \times \frac{B^2 \times V_e \times z}{\mu_c} = \frac{\Delta\varepsilon}{8} \times \left[r \times \left(\frac{2}{r}+1\right)^2\right]$$

$$\frac{1}{2} \times \frac{B^2 \times V_e \times z}{\mu_c} = \frac{\frac{P_{IN}}{f}}{8} \times \left[r \times \left(\frac{2}{r}+1\right)^2\right]$$

$$V_e = \left(\frac{P_{IN}}{f}\right) \times \left(\frac{(r+2)^2}{z \times r}\right) \times \left(\frac{\mu\mu_o}{B^2}\right) \times \left(\frac{1}{4}\right)$$

$$V_e = \left(\frac{P_{IN}}{f}\right) \times \left(\frac{(r+2)^2}{z \times r}\right) \times \left(\frac{\mu \times \pi}{B^2}\right) \times \left(\frac{1}{10^7}\right)$$

So far we have used MKS units (V_e in m^3).

Also, f was in Hz. Changing to kHz and cm^3

$$\boxed{V_{e\_cm3} = \left(\frac{P_{IN}}{f_{kHz}}\right) \times \left(\frac{(r+2)^2}{z \times r}\right) \times \left(\frac{\mu \times \pi}{B_{Tesla}^2}\right) \times \left(\frac{1}{10^4}\right)}$$
FOR GAPPED CORES, ANY MATERIAL

With $B_{TESLA} \le 0.3T$ (for ferrites), $\mu = 2000$,

and target z of 10 (optimum for transformers)

$$V_{e\_cm3} = 0.70 \times \left(\frac{P_{IN}}{f_{kHz}}\right) \times \left(\frac{(r+2)^2}{r}\right)$$

With a typical target value r = 0.4, we get

$$\boxed{V_{e\_cm3} = 10 \times \left(\frac{P_{IN}}{f_{kHz}}\right) \text{ for a Buck-Boost/Flyback}}$$
VOLUME OF A GAPPED FERRITE CORE

$\Rightarrow$ **For a Buck, the maximum core volume is demanded for low D, i.e. V_{INMAX}.**
For a Boost, highest core volume is demanded when D is highest, i.e. at V_{INMIN}.
For a Buck-Boost/flyback, D does not affect core volume directly.

Figure 5.14: Useful equations for core selection for each topology.

$$V_{e\_cm^3} = \left(\frac{P_{IN}}{f_{kHz}}\right) \times \left(\frac{(r+2)^2}{z \times r}\right) \times \left(\frac{\mu \times \pi}{B_{Tesla}^2}\right) \times \left(\frac{D}{10^4}\right), \text{ or}$$

(Boost)

$$V_{e\_cm^3} = \frac{31.4 \times P_{IN} \times D \times \mu}{z \times f_{MHz} \times B_{SAT\_Gauss}^2} \times \left[r \times \left(\frac{2}{r}+1\right)^2\right]$$

$$V_{e\_cm^3} = \left(\frac{P_{IN}}{f_{kHz}}\right) \times \left(\frac{(r+2)^2}{z \times r}\right) \times \left(\frac{\mu \times \pi}{B_{Tesla}^2}\right) \times \left(\frac{1}{10^4}\right), \text{ or}$$

(Buck-Boost)

$$V_{e\_cm^3} = \frac{31.4 \times P_{IN} \times \mu}{z \times f_{MHz} \times B_{SAT\_Gauss}^2} \times \left[r \times \left(\frac{2}{r}+1\right)^2\right]$$

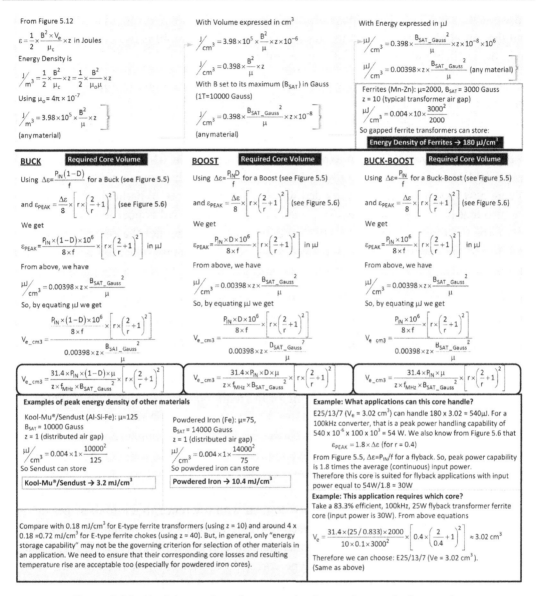

Figure 5.15: Useful equations for core selection and numerical examples.

As we expect from our understanding of energy transfer principles vis-à-vis the three topologies, the Buck and Boost equations above depend on D, and therefore on the specific input and output voltages of the application. However, the Buck-Boost demands the highest core volume, and that is independent of D as we expected.

We also see that if z is set to around 10 as suggested for ferrite transformers, then the core volume can be reduced 10 times since in all cases above, V_e is inversely proportional to z.

And we realize that was the purpose of the air gap to start with. We should also keep in mind that for chokes, as opposed to transformers, a z of around 40 is usually achievable within the available window area of E-type cores.

The alternative derivation of core volume in Figure 5.15 is based on the following approach: (a) first calculate the energy packet size for a given application, (b) connect that energy packet size to the peak energy, and then (c) look for a core that can handle that peak energy. With that approach, the question, independent of topology, is: *how much energy can a given core handle?* We can express this energy density as $\mu J/cm^3$, or J/m^3, and so on. Some useful equations for it are provided in the figure. Note that some numerical examples showing how to select cores are provided within Figure 5.15 too, in particular for a flyback. Note also that if we are using powdered metal cores, the core and copper losses are more likely to be the limiting factor and will thus determine core size, not the energy density capability of the material. So, we may need to use the core volume selection equations with caution. However, for ferrite, gapped ($z > 1$) or ungapped ($z = 1$), they are a good reference for picking the right cores.

Part 3: Toroids to E-Cores

In previous sections we derived energy equations based on toroids, and then implicitly extended them to E-type cores when talking of the required core volume. That is quite acceptable for determining volume, but things can get confusing when we try to actually implement a certain calculated air gap in particular. For example, suppose we calculate the air gap as 0.1 mm. We know that this calculation in effect assumes a toroid. So, how do we implement "0.1 mm" in a triple-limbed E-type core? To answer that we have to understand E-type cores better. For one, we keep in mind that all the terms: effective length, effective area, and effective volume, were originally defined with reference to a toroid, as per the magnetic circuit in Figure 5.10. So, we need to figure out what these quantities are with reference to the specific geometry and dimensions of cores like EE cores, ETD cores and EFD cores (generically referred to as E-type cores). In other words, we *need to map such cores into an equivalent toroid*. The way to mentally visualize this is shown in Figure 5.16.

We can visualize that in an E-core, by continuity of flux lines and symmetry, the flux ϕ going through the center limb divides up equally into the two outer limbs. Since $B = \phi/A$, if $\phi/2$ is in each outer limb, to keep the B-field the same throughout the core, the area of the outer limb needs to be half the area of the center limb. And that is how E-cores are essentially designed — with the area of each of the two outer limbs equal to half the area of the center limb — because no one typically expects, needs, or wants, the outer limbs saturating before the center limb, or the other way around (we want same B throughput). If they saturate differently, that would only represent wasted core material in most applications.

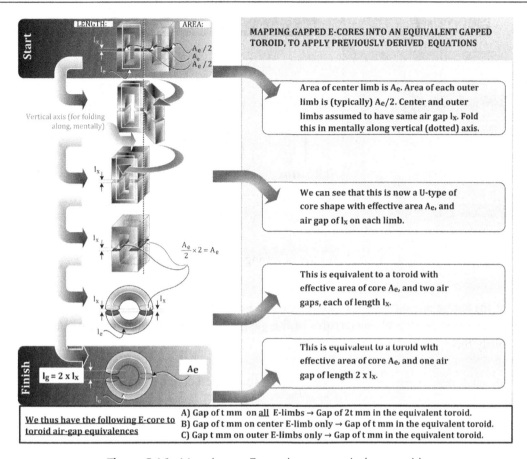

Figure 5.16: Mapping an E-core into an equivalent toroid.

Further, the cross-sectional area of the center limb is almost exactly equal to the effective area, A_e, of the core's equivalent toroid (as stated in the datasheet). The area of the outer limbs of the E-core must therefore be almost exactly half the effective area. This is shown in Figure 5.16.

We also realize that to set a calculated air gap of say 0.1 mm (with reference to an equivalent toroid), we can simply set an air gap of 0.1 mm in the center limb. We could achieve the same by grinding 0.05 mm off each center limb half, or 0.1 mm off only one half. There would be no gaps on the outer limbs; the outer halves would lie flush against each other. Alternatively, we can just take regular E-core halves (with no center or outer limb grinding), and put 0.05 mm polyester spacers on the two outer limbs. That would create a gap of 0.05 mm in all three limbs. This is illustrated in Figure 5.17, using an air gap of 0.26 mm as an example.

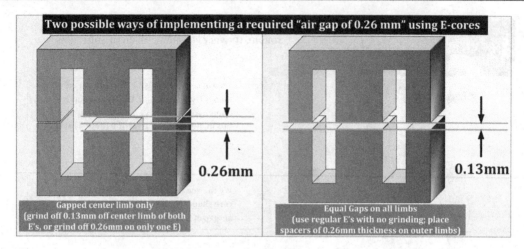

Figure 5.17: Two ways of setting an air gap of 0.26 mm.

One of the subtle advantages of using an air gap is that air starts *dominating the reluctance*, and ultimately the entire characteristics of the gapped structure. We had indicated this previously too.

$$\frac{l_e}{\mu_c A_e} \rightarrow \frac{l_e + \mu l_g}{\mu_c A_e} \approx \frac{\mu l_g}{\mu_c A_e} = \frac{l_g}{\mu_o A_e} \quad \text{(ungapped core to gapped core)}$$

This is the same as saying

$$\mathfrak{R} = \mathfrak{R}_{core} + \mathfrak{R}_{gap} \approx \mathfrak{R}_{gap} \quad (\text{since } \mathfrak{R}_{core} \ll \mathfrak{R}_{gap})$$

It is well-known that ferrite manufacture is a very complicated process, and has rather wide tolerances inherently. For example, toward the end of the process there is a stage called sintering, in which significant, and not completely predictable, volume shrinkage occurs. So the manufacturer has to guess the amount of shrinkage that will occur, and start with a higher volume. So, the final mechanical tolerances are not very good. Nor the other characteristics. However, by introducing an air gap, we can swamp out these variations to a great extent and make it all more dependent on the characteristics of the air gap. That helps make the behavior of the overall gapped structure much more predictable and repeatable. That is why even in Forward converters, a small air gap (~ 0.1 mm to 0.2 mm) is often deliberately introduced in the transformer, even though we realize that energy storage is not the purpose of the transformer of a Forward converter, unlike that of a flyback. Of course we are left with higher energy to recover and circulate through the energy recovery winding, but the increased reliability is usually considered worth it.

Finally, in Figure 5.18, we show how to connect the geometry of an E-core to its effective (published) length, area, and volume. That finally completes our deep dive into E-cores.

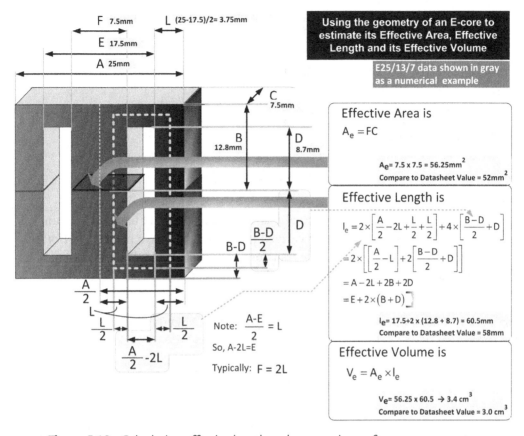

Figure 5.18: Calculating effective length, volume, and area from core geometry.

Part 4: More on AC–DC Flyback Transformer Design

Now we will apply what we have learned to an AC–DC transformer and design it down to the wire, literally. The first thing to keep in mind is that the process is iterative. For example, in the lowermost example in Figure 5.15, we coincidentally "found" a core with exactly the same volume as our requirement. But if, for example, the efficiency was somewhat lower, we would have required a larger volume. We would then "just miss out" on using the E25/13/7. That would be sad because the next available core size may be rather too large, and would represent overdesign. At this point we could compromise a little on our target current ripple ratio and select say $r = 0.55$ instead of 0.4. The required core size would then reduce and the E25/13/7 may be in the ballpark once again. Yes, there would be some impact on the input and output capacitors, in particular in terms of their higher RMS current and heating, but there would be much less impact on the switch dissipation and almost no effect on the diode ratings or its dissipation. We would need to set higher peak current limits and so on. Another possible direction is to increase the air gap just a little (without changing the inductance or r). That would demand a few more turns. But if these turns can be physically accommodated in the available window area, and if we are also willing to accept the slightly higher copper losses, that could be a good way to go too. Eventually, we can evaluate all the side-effects, and better optimize our converter design.

In Figure 5.19, we start by providing perhaps the easiest equation possible for picking a ferrite flyback transformer

$$V_{e\_cm^3} = \frac{0.01 \times P_{IN}}{f_{MHz}}$$

Its simplicity belies the fact that it is actually an *exact* equation, but based on certain defaults applied to the general equation derived previously. We have picked $z = 10$ (this corresponds to a typical air gap in ferrite transformers), $r = 0.4$ (this is the optimum target current ripple ratio and indirectly determines the inductance L), $B_{SAT} = 3,000$ G and $\mu = 2,000$ (i.e., ferrite assumed).

Going back to *Chapter 3*, we had presented the following, slightly more general, equation in the subsection titled "Selecting the Core"

$$V_{e\_cm^3} = 0.7 \times \frac{(2+r)^2}{r} \times \frac{P_{IN}}{f_{kHz}}$$

Since $0.7 \times [(2 + 0.4)^2/0.4] = 10.0$, we can see this was exactly the same equation we have presented directly above (and derived), with the difference that we have now set r to a default value of 0.4 too.

In Figure 5.19, we see that by very simple steps based on previous derivations, we now arrive at the same (~3 cm^3) volume determined in the previous example in Figure 5.15, and we again select the same core, that is, E25/13/7.

Note that the numerical examples in Figure 5.19 apply to a 30 W input. That can refer to an ideal 30 W output converter (with 100% efficiency). It also applies to a real-world converter with, say, 25 W output power and 83.3% efficiency, because $25/0.833 = 30$ W. In Example 2 of the same figure we apply the concepts to an AC−DC flyback. We have chosen an 83.3% efficient 25 W Universal Input flyback with an output of 5 V at 5 A. Here we invoke the concept of reflected output voltage, V_{OR}, as discussed in *Chapter 3*, in particular in Figure 3.2.

Note that in Figure 5.19 we discuss a "Best-Case Estimate" and a "Worst-Case Estimate." We now see more clearly what we meant in Figure 5.1. Obviously, "Best-Case" corresponds to "Real-World 1" and "Worst-Case" to "Real-World 2." We see how the

Flyback/Buck-Boost Core Size and Inductance Selection Examples

From Figure 5.15, the general equation for Buck-Boost/flyback is

$$V_{e\_cm3} = \frac{31.4 \times P_{IN} \times \mu}{z \times f_{MHz} \times B_{SAT\_Gauss}^2} \left[r \times \left(\frac{2}{r}+1\right)^2 \right]$$

With default values of r=0.4, z=10, μ=2000, B_{SAT_GAUSS}=3000

> **Simplified equation for flyback ferrite transformers**

$$V_{e\_cm3} = \frac{0.01 \times P_{IN}}{f_{MHz}}$$

Example 1: 100kHz, 30W Flyback Transformer (Ferrite)

Assuming 100% efficiency as a starting point, we have

$$V_{e\_cm3} = \frac{0.01 \times 30}{0.1} = 3.0$$

From core selection datasheets, the closest is **E25/13/7** with V_e=3.02 cm³. To account for efficiency less than 100%, we can increase z and/or increase r. For example, if efficiency is 0.8, we can raise r to 0.4/0.8 = 0.55 (by lowering L), or we can increase air gap such that z = 12.5 (check from general equation above). Or we can do both: lower L and increase air gap simultaneously.

Another way to look at this is: this is a 30W flyback. So the average energy delivered (and stored) in the inductor per cycle is from Figure 5.5: P_0/f = 30W/0.1MHz → 300 μJ (assuming η = 1). That is Δε. However, because of the ±20% current ripple (r = 0.4), the peak energy is much higher. Therefore, earlier in Figure 5-6 we had showed that for r = 0.4, that relationship is

$$\varepsilon_{PEAK} = 1.8 \times \Delta\varepsilon \quad (\text{for } r = 0.4)$$

We therefore need the inductor to handle the following peak energy, corresponding to I_{PEAK} (without saturating)

$$\varepsilon_{PEAK} = 1.8 \times \Delta\varepsilon \approx 1.8 \times \frac{P_0}{f} = 1.8 \times 300 = 540\ \mu J$$

We have shown in Figure 5-15 that a gapped ferrite transformer can handle 180μJ/cm³. So, we need a core of effective volume

$$V_{e\_cm3} = \frac{\varepsilon_{PEAK}}{180} = \frac{540}{180} = 3\ cm^3$$

This is the same result as from our simplified equation above. The closest core is E25/13/7 as before.

<u>Note</u>: It should be confirmed the required turns can be accommodated in the available window area. That will depend on input/output voltages and currents. However, z=10 is a good compromise for most applications and E25/13/7 should be OK.

Example 2: 25W Universal Flyback, 5V@5A, η=0.833

Note that P_0=25W, and P_0/η= 25/0.833=30W. So, in effect, this is almost the same example as on the left, where we took output power as 30W and assumed 100% efficiency (input was 30W in that case too). In a flyback transformer design, it is P_{IN} = P_0/η that determines Δε, and thereby determines core size (worst-case estimate).

The target reflected output voltage V_{OR} of this design is 100V as typical. So the turns ratio $N_P/N_S \equiv n$ is given 2y V_{OR}/V_O= 100/5 = 20 (ignoring diode drop here for simplicity). The duty cycle is

$$D = \frac{V_{OR}}{\eta V_{IN} + V_{OR}} = \frac{100}{(127 \times 0.833) + 100} = 0.486$$

Here we are doing the design at the lowest input voltage of the Universal Input Range i.e. at 90VAC, which is 127VDC (ignoring input voltage ripple for now). The reflected output current I_{OR} is I_0/n = 5/20 = 0.25A. This is the effective load current for the equivalent Primary-side Buck-Boost (see Figures 3.1 and 3.2). The switch "thinks" the Buck-Boost output is 100V @ 0.25A. The center of the Primary-side inductor current ramp is

$$I_L = \frac{I_{OR}}{1-D} = \frac{0.25}{1-0.486} = 0.486$$

$$I_{IN} = D \times \frac{I_{OR}}{1-D} = 0.486 \times 0.486 = 0.236$$

The total swing of current ΔI_L for a target r = 0.4 is

$$r = 0.4 = \frac{\Delta I_L}{I_L} = \frac{\Delta I_L}{0.46}$$

$$\Delta I_L = 0.486 \times 0.4 = 0.194$$

Note that as mentioned in Figure 5.1, the general duty cycle equation of a Flyback/Buck-Boost can be written in two ways:

$$D = \frac{V_{OR}}{\eta V_{IN} + V_{OR}} \quad OR \quad D = \frac{\frac{V_{OR}}{\eta}}{V_{IN} + \frac{V_{OR}}{\eta}}$$

Both are numerically identical and provide the same calculated duty cycle, average input current and center-of-ramp current. But the first form implies the worsening of efficie ncy can be modeled as if input has been reduced by a factor η for a given output. That is equivalent to assuming all the energy that is dissipated (as per the measured efficiency), is lost *preceding* the transformer i.e. in the switch for example. So the transformer had less to store only the outgoing *useful* energy (this is the best case estimate).

The second form of duty cycle above suggests that the input remained the same but the output in effect increased by a factor 1/η. This implies that all the loss occurred *after* the transformer, i.e. in the diode for example. So now the inductor has to store all the incoming energy i.e. the useful energy plus the loss term (i.e. P_{IN}). Reality is somewhere in between. But in general, the latter is the conservative way (worst case estimate) to design the transformer/inductor if we don't know upfront how the losses have actually gotten distributed inside the converter.

Best-Case Estimate: Assume that the effective *input voltage* decreases to account for the less than perfect efficiency. The on-time is D/f. During this time the voltage on the inductor is assumed to be ηV_{IN}. The inductance must be such that it causes a swing of 0.194A during the on-time (from the left-side column). So

$$V_{ON} = L \frac{\Delta I}{T_{ON}} \rightarrow L_{\mu H} = (127 \times 0.833) \times \frac{0.486}{0.1 \times 0.194} \Rightarrow \boxed{2.65\ mH}$$

$$V_{OFF} = L \frac{\Delta I}{T_{OFF}} \rightarrow L_{\mu H} = (100) \times \frac{(1-0.486)}{0.1 \times 0.194} \Rightarrow \boxed{2.65\ mH}$$

Smaller Inductance

$$I_{PEAK} = I_L \times \left(1 + \frac{r}{2}\right) = 0.486 \times 1.2 = 0.583$$

$$\varepsilon_{PEAK} = \frac{1}{2} \times L \times I_{PEAK}^2 = \frac{2.65 \times 10^{-3} \times 0.583^2}{2} \Rightarrow \boxed{450 \mu J}$$

We also know that $\varepsilon_{PEAK} = 1.8 \times \Delta\varepsilon$ (for r = 0.4). So   *Smaller Core*

$$\Delta\varepsilon = \frac{\varepsilon_{PEAK}}{1.8} = \frac{450}{1.8} = \boxed{250 \mu J}$$

At 100kHz, this is 25W. So this is only *useful* energy being stored.

Worst-Case Estimate: Here we imagine the output voltage has increased by a factor 1/η. So, as above

$$V_{OFF} = L \frac{\Delta I}{T_{OFF}} \rightarrow L_{\mu H} = \frac{100}{0.833} \times \frac{(1-0.486)}{0.1 \times 0.194} \Rightarrow \boxed{3.18\ mH}$$

$$V_{ON} = L \frac{\Delta I}{T_{ON}} \rightarrow L_{\mu H} = (127) \times \frac{0.486}{0.1 \times 0.194} \Rightarrow \boxed{3.18\ mH}$$

Larger Inductance

$$I_{PEAK} = I_L \times \left(1 + \frac{r}{2}\right) = 0.486 \times 1.2 = 0.583$$

$$\varepsilon_{PEAK} = \frac{1}{2} \times L \times I_{PEAK}^2 = \frac{3.18 \times 10^{-3} \times 0.583^2}{2} \Rightarrow \boxed{540 \mu J}$$

Larger Core

$$\Delta\varepsilon = \frac{\varepsilon_{PEAK}}{1.8} = \frac{540}{1.8} = 300 \mu J \quad (\text{for } r = 0.4)$$

(E25/13/7)

At 100kHz, this is 30W. So all the incoming energy (including losses) are now being stored in L. So this is consistent with the 100% efficient, 30W converter in Example 1 above.

Figure 5.19: AC–DC flyback design example (part 1).

answer to the question "where do the losses occur inside the converter — before or after the inductor?" affects the size of the core (peak energy), the voltseconds and the calculated inductance. Conservatively speaking, we can design the core/transformer as per the assumption that all the losses occur on the secondary side of the transformer. That is the worst-case estimate. But if we know better, we can do a more accurate estimate of peak energy requirements.

> **Note:** In *Figure 5.19, we have used a rectified voltage of 127VDC for simplicity. That will certainly be valid for a flyback with a very large input capacitor. But we must remember the input capacitor peak-charges when the input bridge conducts and discharges during the rest of the AC half-cycle. Therefore the voltage at the input of the flyback undulates at twice the AC line frequency (100 Hz or 120 Hz) between two voltage levels that are typically quite far apart. So in fact, an "average" value between the two levels should be rightly taken as the effective DC input voltage to the flyback. But that is a very involved calculation which depends on the value of the input capacitance and so on. We will take that up in* Chapter 14. *For now we stick to 127VDC as the input.*

As indicated in Figure 3.2, in the Primary-side equivalent model, the output voltage is V_{OR}. Since in Figure 5.19, we have chosen a V_{OR} of 100 V, a 25 W converter is, on its Primary side, equivalent to a Buck-Boost with an input of 127 V and an output of 100 V at 0.25 A. Note that the efficiency is not 100%. And that gives us an opportunity to "close the loop" in our unique journey, one that started with the energy transfer chart in Figure 5.4. In that chart, we had, for simplicity sake, taken a numerical example with 100% efficiency. But we had insisted we had taken *exact equations*, and had said they were valid even if the efficiency was not 100%. Now we can prove it. In Figure 5.20, we repeat the previous energy transfer calculations, but this time for the *non-ideal* case of Example 2 of Figure 5.19.

Here are the results. We see that 250 μJ comes out per cycle. At 100 kHz (0.1 MHz), that is equivalent to a power output P_O of 250 μJ $\times 10^5 = 25$ W. On the input side, we have 300 μJ coming in per cycle, corresponding to $P_{IN} = 30$ W. We can also now clearly see that exactly 50 μJ (corresponding to 5 W) is *lost somewhere in the middle block*. The energy balance sheet is already complete, vindicating all our previous equations and treatment. The final numbers add up in Figure 5.19 just as they did in Figure 5.4.

> **Note:** *The center block consists of the inductor, switch, and diode. The 5 W dissipation occurs somewhere inside this block — say, as conduction losses, or switching losses, or DCR losses, or AC resistance losses, and so on — almost anything that can take place inside this block that is directly related to the actual power conversion process (no leakage losses considered here). The energy chart is not intended to make any statement on exactly where the losses occur within this center block, and whether they can be considered before or after the inductor. The purpose of the energy transfer charts was only to clarify the concepts relating to energy storage and topology dependency.*

In brief, *provided we use the right form of duty cycle (i.e., using η), then all the equations we have presented and used for the current at different parts of the circuit are also valid, irrespective of efficiency, frequency, wattage, and so on.* The implicit assumption however, right from the beginning of this chapter, is that we are operating in continuous conduction mode (CCM).

To select wire gauge correctly, we present a primer in Figure 5.21. This has several useful relationships to figure out: skin depth, diameter, AWG and current-carrying capability. Finally, we continue the AC–DC flyback example we started in Figure 5.19, in Figures 5.22 and 5.23. By the end of that, we know the entire procedure for designing a flyback transformer.

Part 5: More on AC–DC Forward Converter Transformer Design

In *Chapter 3*, we already have a detailed numerical example for Forward converter transformer design. The only thing missing there is now provided for the advanced reader: the derivations behind the design curves in Figures 3.10 and 3.11 (see Figures 5.24 and 5.25).

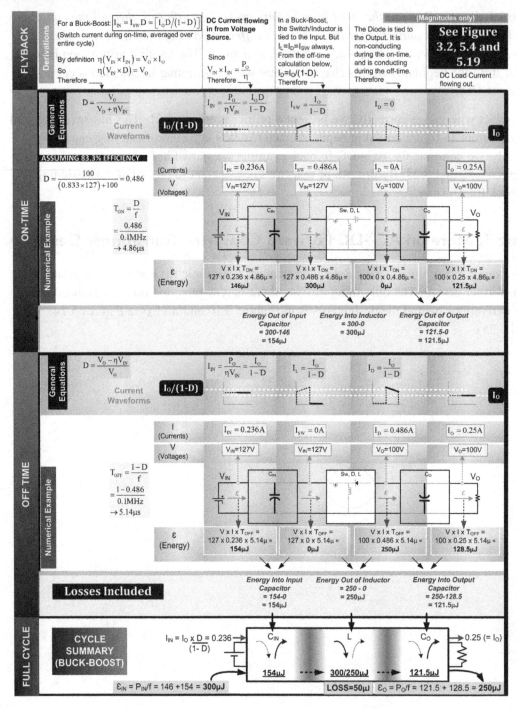

Figure 5.20: Energy transfer chart for a flyback, losses included.

Primer on Wire Gauge Selection and Examples

Circular and Square mils explained again

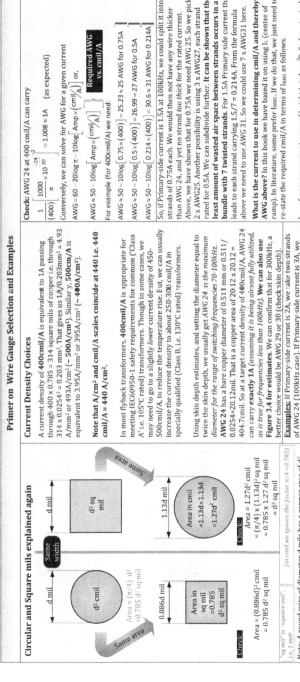

Note: A round wire of diameter d mils has a copper area of d^2 circular mils (cmils) or 0.785 d^2 sqmils. (because $\pi/4 = 0.785$).

Note: A wire of diameter d mils has the same area as a square wire of side 0.886d mil (because $\sqrt{0.785} = 0.886$).

The physical space (including the quota of air space around it) that a round wire of diameter d mils with corresponding copper area d^2 cmils occupies in a winding layer is equal to d^2 sq mils. That required space to be allocated in a winding layer (in sq mils) is thus *numerically* equal to the copper area expressed in cmils. **That highlights the significance and convenience of using circular mils in winding estimates.**

Current Density Choices

A current density of **400cmil/A** is equivalent to 1A passing through 400 × 0.785 = 314 square mils of copper i.e. through 314 × 0.0254² = 0.203 mm². That gives us 1A/0.203mm² = 4.93 A/mm² or 493A/cm² (~ **500A/cm²**). Similarly, **500cm/A** is equivalent to 3.95A/mm² or 395A/cm² (~ **400A/cm²**).

Note that A/cm² and cmil/A scales coincide at 440 i.e. 440 cmil/A = 440 A/cm².

In most flyback transformers, **400cmil/A** is appropriate for meeting IEC60950-1 safety requirements for common ['Class A' i.e. 105°C rated] transformers. Though in some cases, we may need to go to a slightly *lower* current density of 450-500cmil/A, to reduce the temperature rise. But, we can usually increase the current density to about 250-300 cmil/A in specially qualified (Class B, i.e. 130°C rated) transformers.

Using skin depth estimates, and setting the diameter equal to twice the skin depth, we usually get *AWG 24 as the maximum diameter for the range of switching frequencies 70-100kHz.* AWG 24 has a bare copper diameter of 0.511 mm or 0.511/ 0.0254=20.12mil. That is a copper area of 20.12 × 20.12 = 404.7cmil. So at a target current density of 400cmil/A, AWG 24 can carry **exactly 1A** *(assuming it is being almost fully utilized, as is true for frequencies less than 100kHz).* **We can also use Figure 3.4 for estimates.** We can confirm that for 300kHz, a better choice would be AWG 29 or 30 (check skin depth).

Examples: If Primary-side current is 2A, we take two strands of AWG 24 (100kHz case). If Primary-side current is 3A, we take three strands. But what if our Primary side current is 1.5A? Two strands of AWG 24 would be over-design. For an optimum design we prefer two identical strands *of the same AWG*, and choose a wire gauge just right to give 400cmil/A for each. We need to use the following general formula for wire gauge versus current carrying capability.

We know that $d_{\text{mils}} = \dfrac{1000}{\pi} \times 10^{-\frac{\text{AWG}}{20}}$

So $\quad \text{Amps} = (\text{cmil}/A)^{-1}\left[\dfrac{1000}{\pi}\times 10^{-\frac{\text{AWG}}{20}}\right]^2$

Current carrying capability of round wires

Check: AWG 24 at 400 cmil/A can carry

$$\frac{1}{(400)}\left[\frac{1000}{\pi}\times 10^{-\frac{24}{20}}\right]^2 = 1.008 \approx 1A \quad \text{(as expected)}$$

Conversely, we can solve for AWG for a given current

$$\text{AWG} = 60 - 20\log\pi - 10\log\left[\text{Amp}\times\left(\text{cmil}/A\right)\right] \text{ or,}$$

$$\text{AWG} \approx 50 - 10\log\left[\text{Amp}\times\left(\text{cmil}/A\right)\right]$$

For example (for 400cmil/A) we need:

$\text{AWG} \approx 50 - 10\log\left[0.75\times(400)\right] = 25.23 \approx 25$ AWG for 0.75A

$\text{AWG} \approx 50 - 10\log\left[0.5\times(400)\right] = 26.99 \approx 27$ AWG for 0.5A

$\text{AWG} \approx 50 - 10\log\left[0.214\times(400)\right] = 30.6 \approx 31$ AWG for 0.214A

So, if Primary-side current is 1.5A at 100kHz, we could split it into 2 strands of 0.75A each. We would then *not* have any wire thicker than AWG 24, and yet no strand *too thick* for the rated current. Above, we have shown that for 0.75A we need AWG 25. So we pick 2 × AWG25. Another possibility is using 3 × AWG27, each strand rated for 0.5A. We can subdivide further: **It can be shown that the least amount of wasted air space between strands occurs in a bundle with 7 twisted strands.** For 1.5A Primary side current that leads to each strand carrying 1.5/7 = 0.214A. From the formula above we need to use AWG 31. So we could use 7 × AWG31 here.

What is the current to use in determining cmil/A and thereby AWG above? In this book we have based it on using I_L (center of ramp). In literature, some prefer I_{RMS}. If we do that, we just need to re-state the required cmil/A in terms of I_{RMS} as follows.

I_{RMS} of a square current wave of height I_L is $I_L \times \sqrt{D}$. So for a typical D = 0.5, I_{RMS} is 0.707 × I_L. If I_L is 1A, I_{RMS} is 0.707A. If we pass I_L =1A through 400 cmil, we are equivalently passing 0.707A RMS through 400 cmil. That is 400/0.707 = 565 cmil/A if A is in RMS. So in terms of RMS, the flyback design target is **565 cmil/A, not 400 cmil/A.**

RULE: If using RMS, we recommend 565cmil/A. If using center of ramp (I_L), we recommend 400cmil/A. In related literature, it is usually recommended to select wire gauge for "500cmil/A" (where A is in RMS). **500cmil/A-RMS** is equivalent to 500 × 0.707 = **350 cmil/A** in terms of our terminology of using center of ramp I_L. So that is actually *more aggressive* than our recommendation.

Figure 5.21: Primer on wire gauge selection for flyback and other topologies.

AC-DC Flyback Design Example: Calculating Number of Turns, Air Gap and Wire Gauge

Example: 25W Universal Flyback, 5V@5A, η=0.833

Continuing our Universal Input 25W flyback example from Figure 5.19, we are now want to calculate the number of turns for the selected E25/13/7 core. From Chapter 2

$$\Delta B = \frac{L \times \Delta I}{N \times A_e} = \frac{V_{sec}}{N \times A_e}$$

Applying this to a flyback/Buck-Boost

$$\Delta B = \frac{V_{IN} \times D}{N \times A_e \times f} \quad \text{or} \quad (N \times A_e) = \frac{V_{IN} \times D}{\Delta B \times f}$$

But we also know from Figure 2-13 that

$$\Delta B = \frac{2r}{r+2} \times B_{PEAK}, \text{ so } (N \times A_e) = \left(1 + \frac{2}{r}\right) \times \frac{V_{IN} \times D}{2 \times B_{PEAK} \times f}$$

So far we have used MKS units. Now, expressed differently

$$N = \left(1 + \frac{2}{r}\right) \times \frac{V_{IN} \times D}{200 \times B_{PEAK} \times f_{MHz} \times A_{e\_cm2}}$$

Using our default values for converters with ferrite

$B_{PEAK} = 0.3T$ and $r = 0.4$, we get

$$N = \frac{V_{IN} \times D}{10 \times f_{MHz} \times A_{e\_cm2}}$$

For example, for our ongoing flyback example

$$N = \frac{127 \times 0.486}{10 \times 0.1 \times 0.525} = 117.6 \approx \boxed{118 \text{ Turns}}$$

[Required # of Primary Turns]

where we have used $A_e = 0.525 \text{ cm}^2$ for E25/13/7.

Note that the number of turns on the selected core, does not depend on core size (volume), output power and even air gap. At least not directly so.

Note: Iteration will usually be required because with a typical (target) turns ratio of 20, we still need to get an integral (or sometimes acceptably a half-integral) number of turns for the secondary winding. *Example*: At this point we could pick $N_P = 120$ and $N_S = 6$. So the turns ratio is 20, and 120 turns is very close to the calculated 118 turns above.

Note: We have to ensure that current limit is set just slightly higher than I_{PEAK} i.e $I_L(1+r/2)$ to avoid saturating core on power-up and/or power-down. We will consider holdup time requirements in detail in Chapter 14. But in general, we may need to pick a bigger core just to handle the higher momentary peak B-field during power-up/power-down.

In Figure 5.19, we had selected this core based on a guess value of air gap (z =10) and its corresponding energy-handling capability. Now that we know the required inductance from Figure 5.19 (i.e. 3.18 mH) and number of turns from above, we can re-affirm the value of z, and thereby calculate the actual required air gap.

$$L = \frac{\mu \mu_0 \times N^2 \times A_e}{z \times l_e} \quad (see\ Figure\ 5.13)$$

[Air Gap Factor]

$$So,\ z = \frac{\mu \mu_0 \times N^2 \times A_e}{L \times l_e}$$

$$z = \frac{2000 \times 4\pi \times 10^{-7} \times 118^2 \times (0.525 \times 10^{-4})}{(3.18 \times 10^{-3}) \times (57.5 \times 10^{-3})} = 10.048$$

where we have used $l_e = 0.0575$ m (for the chosen E25/13/7 core in MKS units). Note that final value of z has come very close to our initial ballpark/target value of 10, simply because the energy handling capability of the core, based on z=10 (180 μJ) and an effective volume of 3.02 cm³, i.e, 540 μJ) was very close to our power requirement (Δε of 300 μJ corresponding to worst-case 30W in/out of transformer at 100kHz, with a peak $\epsilon_{PEAK} \times 1.8$ Δε for r=0.4, i.e, 300 x 1.8 = 540 μJ). So, all equations provided so far are self-consistent.

Finally, we use the definition of z below, to calculate the air gap. *See Figure 5.11*

$$zl_e = l_e + \mu l g \quad [see\ Figure\ 5.11]$$

$$l_g\ _{mm} = (z-1) \times \frac{l_e\ _{mm}}{\mu}$$

Expressed in mm

$$l_g\ _{mm} = (z-1) \times \frac{\mu}{\mu} \quad \frac{l_e\ _{mm}}{}$$

$$l_g\ _{mm} = (10.048 - 1) \times \frac{57.5}{2000} = \boxed{0.26\ mm}$$

[Required Air Gap]

As shown in Figure 5.17, we can put this "0.26mm air gap" either all on the center limb, by grinding away 0.13mm off each core half center limb (no grinding on side limbs), OR we can select regular core halves with equal limbs, and put a spacer of 0.13mm thickness on the side limbs as we clamp/glue them together.

To calculate wire gauge, we first calculate the skin depth.

$$\delta = \sqrt{\frac{2\rho}{\omega \mu_0}}$$

in MKS units, where ρ is the resistivity

$Cu \approx 1.7 \times 10^{-8}$ Ωm, ω is 2πf, and μ_0 is permeability of air ($4\pi \times 10^{-7}$). Simplifying, and accounting for temperature

$$\delta_{mm} = \frac{66.1}{\sqrt{f}} \times [1 + 0.0042(T-20)]$$

[Skin Depth]

where T is the temperature of the windings in °C, f in Hz, and the term in square brackets reflects the fact that the resistance of Cu goes up 4.2% per 10°C rise. Assuming T=80°C, calculating we get $\delta = 0.262$ mm at 100kHz switching frequency. Therefore we look for a wire with radius less than or equal to the skin depth, i.e. diameter of $2 \times 0.262 = 0.524$mm. From AWG tables, the closest is AWG 24, with a bare copper diameter of 0.511mm.

Alternatively, use the approximate formula for AWG and solve

$$d_{mils} = \frac{1000}{\pi} \times 10^{-\frac{AWG}{20}}$$

[Wire gauge equations]

$$AWG = 20 \times \log\left(\frac{1000}{d_{mils} \times \pi}\right)$$

In our case we require a wire of diameter 2δ=0.524mm. In mils

$$mil = \frac{mm}{0.0254} \Rightarrow \frac{0.524}{0.0254} = 20.63$$

$$AWG = 20 \times \log\left(\frac{1000}{20.63 \times \pi}\right) = 23.8 \approx \boxed{24}$$

Note: We may pick higher AWG than this. See Figure 5.23 next.

Figure 5.22: AC−DC flyback design example (part 2).

AC-DC Flyback Design Example continued: Calculating Wire Gauge and Foil Gauge

Primary side winding

Skin Depth is

$$\delta_{mm} = \frac{66.1}{\sqrt{f}} \times [1 + 0.0042(T-20)]$$

At 100kHz switching frequency and 80°C, skin depth is

$$\delta_{mm} = \frac{66.1}{\sqrt{100000}} \times [1 + 0.0042(T-20)] = 0.262 \text{ mm}$$

We should not pick a wire of diameter exceeding d=2δ. In this case that is 2 x 0.26 = 0.52mm. Expressed in mils that is

$$\text{mils} = \frac{mm}{0.0254} \Rightarrow \frac{0.52}{0.0254} = 20.5 \text{ mils}$$

The corresponding AWG is from Figure 5-22

$$AWG = 20 \times \log\left(\frac{1000}{d_{mils} \times \pi}\right)$$

$$= 20 \times \log\left(\frac{1000}{20.5 \times \pi}\right) \approx 24 \text{ AWG}$$

Current carrying capability of AWG for a certain cmil/A is

$$Amps = \frac{1}{\left(cmil/A\right)} \left[\frac{1000}{\pi} \times 10^{\frac{-AWG}{20}}\right]^2$$

For 400 cmil/A, AWG 24 has a current capability of

$$Amps = \frac{1}{(400)} \left[\frac{1000}{\pi} \times 10^{\frac{-24}{20}}\right]^2 = 1.008 \approx 1A$$

We saw in Figure 5.19 that the center of Primary current ramp is

$$I_L = \frac{I_{OR}}{1-D} = \frac{0.25}{1-0.486} = 0.486$$

Since this is less than the maximum rated current of AWG 24, we will optimize by choosing a single strand whose AWG is

$$AWG \approx 50 - 10\log\left[Amps \times \left(cmil/A\right)\right]$$ (see Figure 5.21)

$$AWG \approx 50 - 10\log[0.486 \times (400)] = 27 \text{ AWG}$$

Secondary side winding

On the Secondary side we have an output current of 5A.

The center of ramp is

$$I_{LS} = \frac{I_O}{1-D} = \frac{5}{1-0.486} = 9.73A$$

For a target current density of 400 cmil/A, we need

$$9.73 \times 400 = 3891 \text{ cmil} \rightarrow 3891 \times \frac{\pi}{4} = 3056 \text{ sqmil}.$$

We can try to use either foils or wires. Let us see what works.

A) Foil: We have seen that at frequencies from about 70-100kHz, we should not pick a foil of thickness exceeding d=2δ=0.52mm, i.e. 0.52/25.4=0.02 inches, o~ 20 mils. This is however *not* a standard foil thickness.

Standard foil thicknesses (in mils): 1, 1.4, 3, 5, 8, 10, 16, 22. Let us therefore pick 16 mil foil to start with, i.e. 0.016 in, or 0.016 x 25.4 = 0.41 mm. This is called "26 gauge sheet", since its thickness equals the diameter of AWG 26 (round wire). The question is, is the width of the foil acceptable? The selected core is EE25/13/7. If we look at the bobbin of EE25/13/7 we realize that the maximum width of foil can be **15.2 mm**, i.e. 0.6 in, or **598 mil**. The total cross-sectional copper area is 598 mil x 16 mil = 9568 sqmil with a 16 mil sheet.

Since 1 cmil is 0.7854 sqmil, 9568 sqmil equals 9568/0.7854 = 12190 cmil. At a current density of 400 cmil/A, this allows for 12190/400 = **30.5 A**.

The current to use for current density of 400 cmil/A is the center of ramp, this time on the Secondary side. That current is I$_O$/(1-D) = 9.73A from above. We see it is well within the permissible max of 30.5A. **In fact it suggests we can decrease the foil thickness by almost a factor of 3 (30.5/ 9.73 ≈ 3).** So, instead of a 16 mil sheet, we could try a 5 mil sheet. Let us check what current density that would give us. With a width of 598 mil, the sqmils is 598 x 5 = 2990 sqmil, or 3806 cmil. The current density is 3807/9.73 = 391 cmil/ A. This is very close to our target of 400 cmil/A. We conclude: **YES, we can accept 5 mil thick Copper foil for the Secondary winding.**

B) Wire: We have seen that at frequencies from about 70-100kHz, we should not pick a wire of diameter exceeding d=2δ=0.52mm. We repeat the calculation we did for the Primary winding (on the left of this page), till almost the very last step.

In mils, 0.52 mm is

$$\text{mils} = \frac{mm}{0.0254} \Rightarrow \frac{0.52}{0.0254} = 20.5 \text{ mils}$$

The corresponding AWG is from the previous formula

$$AWG = 20 \times \log\left(\frac{1000}{d_{mils} \times \pi}\right)$$

$$= 20 \times \log\left(\frac{1000}{20.5 \times \pi}\right) \approx 24 \text{ AWG}$$

Current carrying capability of AWG for a certain cmil/A is

$$Amps = \frac{1}{\left(cmil/A\right)} \left[\frac{1000}{\pi} \times 10^{\frac{-AWG}{20}}\right]^2$$

For 400 cmil/A, AWG 24 has a current capability of

$$Amps = \frac{1}{(400)} \left[\frac{1000}{\pi} \times 10^{\frac{-24}{20}}\right]^2 = 1.008 \approx 1A$$

We saw that the center of current ramp on Secondary side is

$$I_L = \frac{I_O}{1-D} = \frac{5}{1-0.486} = 9.73$$

Since this is more than the maximum rated current of AWG 24, we can select 10 strands of AWG 24 in a single bundle, or break it up say into 2 parallel bundles with 5 strands in each and so on. A little experimentation is required by actually winding it on the bobbin to see if it fits well without excessive bulging and so on.

We have already calculated in Figure 5.22 that we would have **6 turns on the Secondary side (120 turns on the Primary side)**.

That completes the flyback transformer design.

Figure 5.23: AC–DC flyback design example (part 3).

Simplifying Dowell's Equation (for a Sine Wave) and extending it to a Rectangular Pulse

To avoid repeating errors seen in related literature, we need to look at the original August 1966 paper from Dowell himself, as a starting point. In that he provides the total loss as the sum of two terms — one related to skin effect in an infinite foil, and the other is the effect of proximity when we stack several such foils together in a transformer. The AC resistance is related to the DC resistance as follows

$$R_{AC} = R_{DC}\left[Re(M) + \frac{m^2-1}{3} Re(D)\right]$$

where

m is the number of layers per portion ("p" in this book)

Re(x) stands for Real part of (x).

$$M = \alpha h \times Coth(\alpha h) \;;\; D = 2\alpha h \times tanh(\alpha h/2)$$

where $\alpha = \sqrt{\dfrac{j\omega\mu\eta}{\rho}}$

and $j=\sqrt{-1}$, ρ is the resistivity, ω the angular frequency, and

$$\eta = N_L \frac{a}{b} \qquad \text{Note: "}\eta\text{" is not Efficiency here}$$

where

N_L is the number of turns per layer

a is the width of the square wire (same area as round wire)

b is the width of window.

Let us make some simplifications here. First, we will deal only with wires placed side to side across the entire width of window (no gaps between turns). Therefore η = 1.

Second, we realize that by definition skin depth is

$$\delta = \sqrt{\frac{2\rho}{\omega\mu}}, \text{ so } \alpha = \sqrt{\frac{j\omega\mu}{2\rho}} = \frac{\sqrt{j}}{\delta}$$

And so : $\alpha h = Y = X\sqrt{2j}$

where we have introduced "X" = $\dfrac{h}{\delta}$ and "Y" = $X\sqrt{2j}$.

(In related literature X is usually called "ξ" or "A"). So Dowell's equation in terms of our nomenclature & simple assumption is

$$R_{AC} = R_{DC}\left[Re(Y \times Coth(Y)) + \frac{p^2-1}{3} Re\left(2Y \times Tanh\frac{Y}{2}\right)\right]$$

Let us now call

$$M = Re(Y \times Coth(Y)) \;;\; D = Re\left(2Y \times Tanh\frac{Y}{2}\right)$$

Both these are functions of X. So is Y itself. We write that dependency out explicitly. It can be shown that

$$M(X) = X \frac{e^{2X}+e^{-2X}+2Sin(2X)}{e^{2X}+e^{-2X}-2Cos(2X)} \;;\; D(X) = 2X \frac{e^{X}-e^{-X}-2Sin(X)}{e^{X}+e^{-X}+2Cos(X)}$$

Without deriving it here, but using Mathcad we can confirm the above equations numerically. For example, whether we can check that M(2)=1.898, and D(2)=3.249, whether we start with Dowell's equation or the above simplified form. Note that this differs from some oft-observed erroneous equations for M & D.

So Dowell's equation simplifies to

$$F_R(X,p) = \frac{R_{AC}}{R_{DC}} = M(X) + \frac{p^2-1}{3} D(X)$$

where M(X) and D(X) are given above.
But Dowell's equation is for a pure sine wave. We now extend it to the Fourier components of a rectangular pulse of current corresponding to unity height (amplitude of 1A) and duty cycle d (using lowercase d to distinguish it from the proximity effect term D in Dowell's Equation). The magnitude of the amplitude of the nth Fourier harmonic is

$$|c_n| = \left|\frac{2}{n\pi} Sin(n\pi d)\right|$$

(see Chapter 18 for Fourier Series)

Dowell's Equation for a Sine wave

The DC component (c_0) of the Fourier series is just equal to 'd'.

The frequency of the harmonic is proportional to n. But skin depth is proportional to $1/\sqrt{f}$. So X (i.e. h divided by skin depth) for each harmonic will go up as $\sqrt{n}$. Therefore we can write M and D for each harmonic in terms of the X of the fundamental harmonic as follows

$$M(X,n) = X\sqrt{n} \frac{e^{2X\sqrt{n}}-e^{-2X\sqrt{n}}+2Sin(2X\sqrt{n})}{e^{2X\sqrt{n}}+e^{-2X\sqrt{n}}-2Cos(2X\sqrt{n})}$$

$$D(X) = 2X\sqrt{n} \frac{e^{X\sqrt{n}}-e^{-X\sqrt{n}}-2Sin(X\sqrt{n})}{e^{X\sqrt{n}}+e^{-X\sqrt{n}}+2Cos(X\sqrt{n})}$$

So dissipation due to the rectangular pulse is

$$P_{PULSE\_AC} = (d)^2 R_{DC} + \sum_{n=1} c_n^2 \times R_{AC\_n}$$

$$= d^2 R_{DC} + \sum c_n^2 \times R_{DC} \times \left[[M(X,n)] + \frac{p^2-1}{3} [D(X,n)]\right]$$

where R_{AC_n} is the AC resistance of the nth harmonic.

If we had ignored high-frequency effects, the dissipation would have been wrongly estimated to be

$$P_{PULSE\_DC} = \left(\sqrt{d}\right)^2 R_{DC} = d \times R_{DC}$$

since RMS of a square unity pulse of duty cycle d is $\sqrt{d}$.

The Ratio of the two is the effective F_R of a pulse, as opposed to Dowell's equation which only applies to a Sine wave.

$$F_R = \frac{R_{AC}}{R_{DC}} = \frac{I_{RMS}^2 \times R_{AC}}{I_{RMS}^2 \times R_{DC}} = \frac{P_{PULSE\_AC}}{P_{PULSE\_DC}}$$

$$F_R = d + \sum_n \frac{c_n^2}{d} \times \left[[M(X,n)] + \frac{p^2-1}{3} [D(X,n)]\right]$$

Dowell's Equation for a Pulsed wave

Figure 5.24: Dowell's equations simplified and applied to switching power.

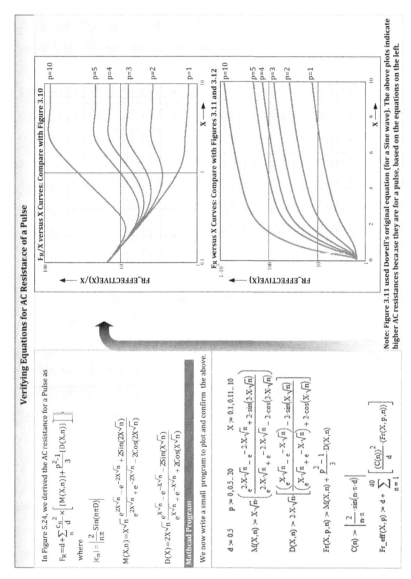

Figure 5.25: Dowell's equations plotted out in a form useful for Forward converter design.

CHAPTER 6

Component Ratings, Stresses, Reliability, and Life

Introduction

In *Chapter 5*, we looked at the selection and design of the magnetic components. Our strategy was straightforward. First, we examined different topologies and applications and understood what their requirements were in terms of *energy*. Then we studied the energy capabilities of different cores with varying materials, geometries, volumes, and air gaps. Thereafter, we proceeded to match *requirements* to *capabilities*. In this chapter and in the next, we look at the other power components involved, and try to understand how to select them using the same basic "match making" approach. However, there is a slight difference. This time, our focus is not on energy storage, but on *stresses*. Stresses form one of the major selection criteria of the non-magnetic power components, the switch, diode, and input/output capacitors. Here is what we need to know as we go down that path.

(a) Using a mechanical analogy, we have to make sure that at a minimum, the strength (rating) of the part exceeds the worst-case force (stress) applied to it in a given application, or else, the part could "break" (fail). The strength of the part may vary significantly over production lots, environmental conditions (like temperature and humidity), perhaps even over time (degradation/aging). However, the *lowest* (worst-case) strength is typically guaranteed by the vendor/manufacturer of the part. For example, a discrete MOSFET comes with Absolute Maximum ("Abs Max") ratings for its Drain-to-Source voltage (V_{DS}), and its Gate-to-Source voltage ($\pm V_{GS}$). Usually we cannot exceed the Abs Max rating *even for a moment*, without risking immediate damage. We should note that the Abs Max rating is a maximum *stress* level, not a *performance* rating. Performance is also guaranteed by the datasheet via electrical characteristics tables, up to a certain upper performance level slightly below the Abs Max level. For example, in a typical switcher or PWM controller IC, that upper performance level is referred to as its *maximum operating voltage*.

(b) Further, we need to ensure the components are able to adequately handle the stresses appearing across them *over all operating corners of the given application*. Unfortunately, especially in switching power converters, that worst-case point is not always obvious. We may need a careful pre-study to discover that. This eventually leads us to the "Stress Spider" in *Chapter 7*.

(c)   Eventually, we also want to ensure that some *safety margin* should be present. This margin may be slim for commercial products, but very generous for medical, military, or "high-rel" (high-reliability) products. Note that in electrical terminology, safety margin becomes "derating," Derating is a concept we had briefly discussed in *Chapter 1*. Now we take that discussion further.

Stresses and Derating

As power supply designers we conform to a certain "Stress Factor" defined as

$$\text{Stress factor} = \frac{\text{max applied stress}}{\text{rated stress}}, \text{ or simply} = \frac{\text{stress}}{\text{strength}}$$

The word "Design Margin" is also often used, and refers to the *reciprocal* of the Stress Factor. For example, a Design Margin of 2 means a Stress Factor of 0.5 (50%). "Margin of Safety" (or just Safety Margin) is formally

$$\text{Margin of Safety} = 1 - (\text{Design Margin})$$

where Design Margin = strength/stress.

As a rough initial guideline, a Stress Factor of about 80% is considered to be a general design target in commercial applications. That gives us $1 - 1/0.8 = 0.25$, or 25% Margin of Safety. Military or "hi-rel" (high-reliability) applications may ask for a Stress Factor below 50% (over 100% Margin of Safety), at a price of course.

Note that we have preferred to talk in terms of Stress Factor instead of using another popular term called Derating Factor. The reason is as follows. If, for example, a transistor was rated 100 V, and we applied 80 V on it, some would say that means "a Derating Factor of 80% was used." However, some engineers preferred to express the same fact by saying "20% derating was applied." Later, some engineers rather unwittingly seem to have extended that statement into "a 20% derating *factor* was applied." So, the question arose: was the Derating Factor 80% or 20% in that case? To prevent further confusion in this book, we have preferred to avoid the term Derating Factor altogether, and have used the term Stress Factor instead. However, we will continue to use the words "derating" or "stress derating" in a strictly descriptive sense. Derating to us here basically means *applying less stress than allowed*. Note that some also refer to Stress Factor as Stress Ratio. That too can be confusing since Stress Ratio is also the ratio of the minimum applied stress to the maximum applied stress in a given cycle of loading. For example, in a given operation if we applied a maximum of 80 V (rating being 100 V) and a minimum of 40 V, the Stress Ratio would be 0.5. The Stress Factor would however remain 0.8.

Derating is acknowledged to be one of the tools for enhancing overall reliability. However, not everyone agrees on exactly how it produces its obviously beneficial effects. For example, the traditional perspective is that a gradual increase in temperature induces more failures on a *statistical basis*. The oft-mentioned rule of thumb being that "every 10 °C rise leads to a doubling in failure rate." The corresponding failures, which typically occur at a slow but steady rate in the field, being considered statistical in nature, are called "random failures." But electrical engineers, being characteristically *deterministic*, are not prone to calling any event "random" and leaving it at that. Every failure has a definite cause they point out. So some have argued: is copper *more likely* to melt if it *approaches* its melting point (1,085 °C)? Alternatively put: if we heat copper up to say 800 °C, and do that a million times, will it melt *on a few occasions*? All engineers however agree that derating is certainly a very good idea for one very practical reason — it becomes a lifesaver when the "unintended" or "unanticipated" happens in the field (as usually does!). For example, there may be a temporary current/voltage overload or surge condition, perhaps some mishandling/abuse, and so on. Key culprits are lightning strikes and AC mains disturbances. In such cases, the *headroom* (safety margin) provided by derating would naturally translate into higher observed field reliability.

> *Note: Electrical energy from lightning can enter a system either by direct injection (usually the most severe), or by electrostatic/magnetic coupling, which is less severe but far more frequent. Coupling can occur even when lines are underground, because the attenuation of earth at normal cable burial levels is minimal. We note that it is considered almost impossible to survive a* direct *lightning hit, though the possibility for such an event is also extremely rare. We therefore never really design equipment to handle that severe a condition. But we do try to handle more common surge profiles as described in the European norm EN61000-4-5.*

Broadly speaking, stresses in electrical systems are considered to be *voltage, current, and temperature*. Power is sometimes considered a separate stress, but it can also be treated as a combination of stresses: voltage, current, and thermal. That is not to say any of the latter stresses are independent either, which is why their analysis is not straightforward either. For example, a certain current of I Amperes passing through a voltage difference of V Volts produces a certain dissipation $I \times V = P$ Watts that leads to a certain temperature $T = (\text{Rth} \times P) + T_{\text{AMB}}$, where Rth is the thermal resistance from the part to the ambient (i.e., its surroundings) and T_{AMB} is the ambient temperature. Note that since temperature rise is determined by rather diverse factors like PCB design, air flow, heatsinking, and so on, the topic of thermal management has been reserved as a separate chapter later in this book (*Chapter 11*). In this chapter we will focus mainly on *voltage* and *current*. An excess of either of these is called "EOS," an acronym for electrical overstress.

A common statement found in failure reports is that a certain semiconductor suffered "damage due to EOS." Eventually, all EOS failures are thermal in nature. For example under high voltage, a dielectric may break down, or a semiconductor junction may "avalanche" (like a zener diode), and allow enough current to cause a "$V \times I$" hot-spot that leads to permanent damage. Semiconductor structures can also "latchup" — in which they go into a very low-resistance or even a *negative-resistance* state (a collapsing voltage accompanied by rising current), much like a gas discharge tube (e.g., the familiar Xenon camera flash or a fluorescent tube lamp). This snapback is made to happen deliberately in the case of a discrete NPN–PNP latch, and also in an SCR (silicon-controlled rectifier, or "thyristor"). In many other cases it is unintentional, and if current is not somehow restricted and/or the snapback event lasts long enough, EOS damage can result. CMOS (complementary metal-oxide semiconductor) ICs and BiCMOS ICs (with bipolar transistors integrated) were historically very prone to latchup due to the existence of many such "parasitic thyristors." But today, an IC designer has an armory of techniques to prevent or quell latchup. Nevertheless, ICs are still routinely subjected to "latchup testing" during their development.

Note that there are no industry standards for testing products for robustness against EOS per se. Basically, we just avoid EOS by *staying within the Abs Max specifications (or ratings) of the part.*

Just as a steady excess of voltage or current constitutes overstress, *rate of change* of stress is also a possible overstress — for example, dV/dt induced stress. The most common example of this is electrostatic discharge (ESD). ESD can cause many types of failures. For example, it can also induce latchup. When we walk over a carpet, we can pick up enough electric charge to kill a semiconductor by actual physical contact ("contact discharge") or near-contact ("air discharge"). Therefore, ESD handling has become a major concern in modern manufacturing and test environments.

All modern ICs are designed with rather complex ESD protection circuitry built around their pins. The idea is to divert or dissipate electrostatic energy safely. Nowadays, all ICs also have published ESD ratings. For example, a typical datasheet will declare that an IC withstands 2 kV ESD as per "HBM" (Human Body Model), and 200 V as per "MM" (Machine Model). The Human Body Model tries to simulate ESD from humans, and actually has two versions. As per the more benevolent and more widely used (military) standard, MIL-STD-883 (now JEDEC standard JESD22-A114E), HBM is a 100 pF cap discharging into the device through a 1.5 kΩ series resistor. The rise time of the resulting current pulse is less than 10 ns and reaches a peak of 1.33 A (at 2 kV). However, the international ESD specification, IEC61000-4-2 (in Europe that is EN61000-4-2), calls for a 330 Ω resistor and a 150 pF capacitor, which gives a peak current of 7.5 A with a rise time of less than 1 ns (at 2 kV). This is actually much harsher than the MIL-STD-883 HBM profile. Note that the IEC standard was originally called IEC801-2 and was also originally intended only as an acceptance condition for end equipment (the system), but it now also does double duty as an ESD test for ICs.

To put down a popular myth, CMOS/BiCMOS chips are not the only components that are susceptible to permanent ESD damage. Bipolar and linear chips can also be damaged. PN junctions can be subjected to a hard failure mechanism called thermal secondary breakdown, in which a current spike (which can also come from ESD) causes microscopic localized spots of overheating, resulting in near-melting temperatures. Low-power TTL ICs as well as conventional op-amps can be destroyed in this manner.

The Machine Model tries to simulate ESD from production equipment, and therefore uses a 200 pF cap with a 500 nH inductor in series (instead of a resistor). Finally, data and telecom equipment also need to pass system-level (not component-level) "Cable ESD" (CESD) testing, also called CDE (Cable Discharge Event) testing. Unlike ESD, there is no industry standard for CESD/CDE testing yet. The intent of the standard is however clear: to protect the equipment under test from the following type of event: an operator pulls an unconnected cable across a carpet, and the cable develops electrostatic charge relative to earth ground. When the cable is plugged into the equipment, the stored charge gets dumped into the equipment. Modern equipment needs to typically survive up to 2 kV CESD on the output ports. Note that here there is very low limiting resistance (cable resistance), but significant line inductance/impedance to limit the peak current and its rise rime. There is also a lot of ringing due to transmission line effects as the energy goes back and forth the cable in waves. So the overall stress profile is less severe, but relatively more sustained than regular ESD.

ESD does not necessarily cause immediate failure. It is known that a latent failure in a CD4041 IC (a popular CMOS quad buffer), tucked deep inside a satellite system assembled in 1979, surfaced 5 years later in 1984 just as it was being readied for launch. Therefore, it is quite possible that we often mistake similar latent failures as "poor quality" or "bad components."

Finally, what is the difference between EOS and ESD? It is their relative profiles. ESD is of very high voltage but with a much smaller duration. In addition, the current it causes is limited by significant source impedance. However, EOS can just be a small increase over the Abs Max ratings (V and/or I), but for a comparatively longer time.

With that background, we now need to understand the ratings of components *used in power supplies* better, so as to make educated choices concerning their selection. Too much safety margin is not only expensive in terms of component cost, but can also seriously affect performance (e.g., the efficiency). Too little margin impacts reliability, and is ultimately expensive too, in the form of warranty costs as discussed later.

Part 1: Ratings and Derating in Power Converter Applications

Many corporations and engineers swear by elaborate lookup tables specifying exactly how much maximum stress can be applied on a particular component. For example, many have (perhaps rightfully so) declared that a wirewound resistor rated for "*X* Watts" must never be used at more than "*X*/2 Watts" (a 50% Stress Factor), and so on. Which begs the question: why was the resistor rated for *X* Watts anyway, why not just rate it for *X*/2 Watts? Another question is: if we faithfully follow all the derating (Stress Factor) charts, will we necessarily end up with a reliable power supply? Not at all! A badly designed flyback can get destroyed on the very first power-up — no time to even test out its vaunted derating margins. In other words, published derating factors are at best "derating *guidelines*," not "derating *rules*." In switching power supplies in particular, we must pay close attention to how the component is really being used and what truly makes sense. After all, as we mentioned in *Chapter 1*, switching power supplies use an inductor, and therefore nothing is very intuitive about them to start with. Current sources don't necessarily seem to be as tractable as we had hoped.

Operating Environments

In power supplies, there are three broad categories of events occurring during operation that we should recognize. Two of these are considered "repetitive," one is not.

(a) There is a repetitive sequence of voltages and currents based on the *switching frequency*. Let us call them "*high-frequency repetitive events*" here. They can be seen with an oscilloscope using the "Auto Trigger" or "Normal Trigger" setting. These observed stresses are what we use in complying with target Stress Factors. The ultimate idea is that under *steady-state* operation we should, under all operating conditions, be operating with a certain target Stress Factor. So, we need to look at the scope waveforms closely with an appropriate time base and also use the correct "acquisition mode" (peak capture setting if necessary) to ensure we are really capturing *the worst-case repetitive stresses*. We may need to sweep over all combinations of load current and input voltage ("min−max corners"). In multi-output power supplies we many need to sweep across *combinations* of loads too.

In estimating the voltage Stress Factor we need to include any (repetitive) *voltage spike*, even if it lasts just for a few nanoseconds. This is of particular importance in AC−DC flyback power supplies, where, for example, the narrow leakage spike at turn-off is enough to destroy the switch.

> *Note: As designers, we would usually prefer to first* reduce *the voltage spike at its root itself. Then we can try to suppress what's left of the spike, perhaps by the use of snubbers and/or clamps (which can unfortunately affect overall efficiency in the bargain). But all these steps are usually preferred to allowing the Stress Factors to be impacted. Yes, we could improve the Stress Factors by using higher rated components, but that can affect both cost and performance.*

A quick initial look at most common protection mechanisms in power supplies is found in Figure 6.1. Clamps are discussed in more detail in *Chapter 7*.

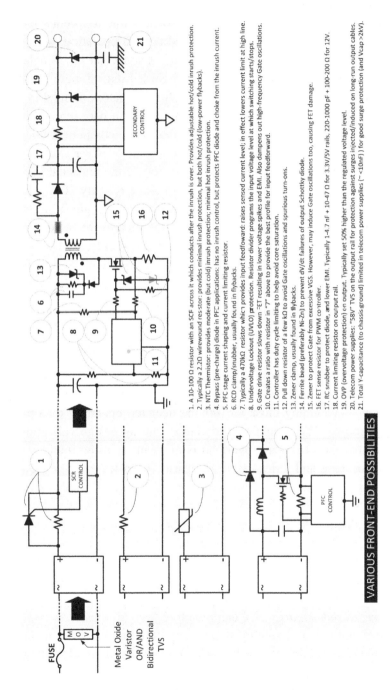

1. A 10-100 Ω resistor with an SCR across it which conducts after the inrush is over. Provides adjustable hot/cold inrush protection.
2. Typically a 2.2Ω wirewound resistor: provides minimal inrush protection, but both hot/cold (low-power flybacks).
3. NTC Thermistor: provides moderate (but cold) inrush protection; minimal hot inrush protection.
4. Bypass (pre-charge) diode in PFC applications: has no inrush control, but protects PFC diode and choke from the inrush current.
5. PFC stage current shaping and current limiting resistor.
6. RCD clamp/snubber, usually found in flybacks.
7. Typically a 470kΩ resistor which provides input feedfoward: raises sensed current level; in effect lowers current limit at high line.
8. Undervoltage Lockout (UVLO) protection. Resistor divider programs the input voltage level at which switching starts/stops.
9. Gate drive resistor slows down "ET resulting in lower voltage spikes and EMI. Also dampens out high-frequency Gate oscillations.
10. Creates a ratio with resistor in "7" above to provide the best profile for input feedforward.
11. Controller has duty cycle limiting to help avoid core saturation.
12. Pull down resistor of a few kΩ to avoid Gate oscillations and spurious turn-ons.
13. Zener clamp, usually found in flybacks.
14. Ferrite bead (preferably Ni-Zn) to prevent dV/dt failures of output Schottky diode.
15. Zener to protect Gate from excessive VGS. However, may induce Gate oscillations too, causing FET damage.
16. FET sense resistor for PWM controller.
17. RC snubber to protect diode, and lower EMI. Typically 1-4.7 nF + 10-47 Ω for 3.3V/5V rails, 220-1000 pF + 100-200 Ω for 12V.
18. Current limiting resistor on output rail.
19. OVP (overvoltage protection) on output. Typically set 50% higher than the regulated voltage level.
20. Telecom power supplies: ~58V TVS on the output rail for protection against surges injected/induced on long-run output cables.
21. Total Y-capacitance (to chassis ground) limited in telecom power supplies (~ <10nF)) for good surge protection (and Vcap >2kV).

VARIOUS FRONT-END POSSIBILITIES

FUSE

Metal Oxide
Varistor
OR/AND
Bidirectional
TVS

SCR CONTROL

PFC CONTROL

SECONDARY CONTROL

Figure 6.1: Various circuit-level protections for reducing stresses and enhancing reliability in AC–DC power supplies.

(b)   There is another category of events that happens repetitively, though at a lower and unpredictable frequency. We call them *"low-frequency repetitive events"* here. These include power-ups or power-downs, and sudden load or line transients (but conditions still assumed to be within the declared normal operating range of the power supply). These events can be captured using a scope set on "Single Acquisition" or "One-shot" mode. The stresses recorded during these events do *not* need to adhere to the recommended steady-state Stress Factors defined in (a) above since now we are dealing with momentary events. But at a bare minimum we have to ensure that accounting for overall operating conditions, component tolerances, production tolerances, ambient temperatures, and so on, we never exceed the Abs Max ratings of any component. To better account for all the expected variations upfront at the design stage itself, we should try to leave roughly 10% safety margin for such low-frequency repetitive events, and confirm that margin in initial prototype bench testing. A typical worst-case test condition for such events could involve powering up into *max load* while at high-line. That test should always be done for AC−DC flyback power supplies for example. The leakage inductance spike must be accounted for here too.

In other words, for events described in (a) we try to conform to the target Stress Factors (typically 70−80%), whereas for (b) the Stress Factors are set much higher (say 90%).

We mentioned above that we should test (a) and (b) events to maximum load at least. However, good design practice should actually involve considering *overloads as "normal" too* — including those *outside* the declared operating load range (for example output shorts). We should also remember during design validation testing ("DVT") phase that the worst-case overload is more likely to be found somewhere *between* maximum load and a dead short. Because, when we apply a dead short, a typical power supply usually goes into some sort of protective foldback mode, reducing its stresses. We should therefore sweep the load all the way from max load to dead short to identify the worst-case load (from the viewpoint of stresses). It is *usually* the load just *before* the output rail collapses rather suddenly (the "regulation knee").

(c)   There is another category of events that happens *non-repetitively* and at a much lower and *completely unpredictable rate*. The key characteristic of such events is that V_{INMAX} *is exceeded momentarily*. We can therefore just call these *"overvoltage events."* Possible culprits are lightning strikes (hopefully far away), or mains disturbances. We can capture these events with a scope set on Single Acquisition, waiting for days or months for the scope to trigger. Or we can use an IEC61000-4-5 compliant test setup with a "CWG" (combination wave generator) and recommended capacitive coupling techniques to mimic lightning surge strikes in the field.

We realize that these particular overvoltage events were not included in bullets (a) and (b) above. In (a) and (b) we eventually recommended that we should sweep the load beyond the declared operating range and call that "normal" operation too. However, we didn't do the same to the input *voltage*. Why not? There were two good reasons for that. First, the declared input voltage range usually already has some safety margin built in. For example, in AC−DC applications, the upper voltage for a universal input power supply is generally considered to be 265VAC (or 270VAC, depending on who you speak to!), even though the highest nominal voltage anywhere in the world is only 240VAC. In the European Union (EU), the harmonized range is officially 230VAC +10%, −6%, which gives us a definitive input range of 216.2VAC to 253VAC. In the UK, the range *used* to be 240VAC ±6% which in effect was 225.6VAC to 254.4VAC, but the UK is now harmonized to the European Union input range since 240VAC is well within the EU range. We thus realize that in all cases, we do have a built-in safety margin (headroom) of at least 10VAC (till we reach 265VAC). Second, we had some planned design margin (headroom) from the Stress Factors that we settled on in (a) above. That margin was meant to protect against the repetitive events mentioned in (b). Similarly, we now expect the margins from (a) to also be sufficient to protect against the overvoltage events mentioned in (c). However, the safety margins for events under (c) may eventually be even slimmer than those for (b).

With all this in mind, we may still need to do a few extra things to ensure better survival. These include increasing the input bulk capacitance to absorb more surge energy, reducing the "Y-capacitance," adding Transient Voltage Suppressor (TVS) diodes on the input and/or the output, and so on. All these go under a broader topic called "overvoltage protection" (or OVP) (see Figure 6.1 carefully).

Note that philosophically, we usually do not plan for *both* (b) and (c) type of events to get combined (i.e., to occur simultaneously). For example, we do not expect that a surge strike will occur on the AC mains at exactly the same moment as an overload gets applied at high-line to an AC−DC flyback power supply. However, in some industrial environments, huge inductive spikes in the AC mains from nearby motorized equipment may be commonplace (considered "normal"). So in that case, we may want to leave an additional 10% safety margin. Eventually, the last word must come through sound engineering judgment, not derating charts.

Recognizing the different equipments and operating environments, there is a new emerging standard for "PCDs" (power conversion devices) called **IPC-9592** (available from www.ipc.org). This classifies power supplies as

Class I: General purpose devices operating in controlled environments, with intermittent and interruptible service, with an intended life of 5 years. Examples are power supplies in consumer products.

Class II: Enhanced or dedicated service power supplies in controlled environments, with limited excursion into uncontrolled environments, having uninterrupted service, and a life expectation of 5–15 years (typical 10 years). Examples are power supplies in carrier-grade telecom equipment and network-grade computers, medical equipment, and so on.

Incidentally, IPC-9592 also carries easy-lookup derating charts both for Class I (or "5-year") Stress Factors and Class II (or "10-year") Stress Factors. Expectedly, the latter occasionally provides more headroom through lower Stress Factors. However, as indicated several times, derating charts are best treated as guidelines, not rules — especially in power supplies where matters vary widely over topologies, applications, requirements, operating conditions and so on.

Component Ratings and Stress Factors in Power Supplies

Having understood overall systems-level stresses, we now look at the key power components used in power supplies and discuss those particular ratings and characteristics of theirs *that are relevant to their proper selection in the particular application.* We will see that there are many concerns connecting cost, ratings, application, performance, and reliability, often conflicting, that we need to weigh in any good power supply design — it is not just all about derating. We will observe that eventually there are no hard-and-fast rules, just guidelines, and above all: *power supply design expertise* combining common-sense and experience.

Diodes

As an example, take the "MBR1045" diode, a popular choice for either 3.3 V or 5 V outputs of low-power universal input AC–DC flyback power supplies. Its datasheet is readily available on the web. This diode is, by popular numbering convention, a "10 A/45 V" Schottky barrier diode.

(a) *Continuous Current Rating:* The MBR1045 has a continuous average forward current rating ($I_{F(AV)}$) of 10 A. We note that in a power supply, the average (catch) diode current is equal to the load current I_O for both the Boost and the Buck-Boost/flyback topologies, and is equal to $I_O \times (1 - D)$ for the Buck/forward topologies (in CCM). Note that in the latter case, the average current is at its highest as D approaches 1, that is, as the input voltage falls. That is a rudimentary example of the "Stress Spider" we will talk about in *Chapter 7*. So finally, as an example, if we pass 8 A average current through the MBR1045, its continuous current Stress Factor is 8 A/10 A → 80%. That is acceptable from the point of view of current stress derating. However, in practice, we rarely operate a diode with that high a current. To understand why that is so, we need to understand the current rating a little better.

$I_{F(AV)}$ as specified in its datasheet is by and large a *thermal* limit. Typically, it is the current at which T_J (the junction temperature) is at its typical maximum of 150 °C. Note that we can come across diodes rated for a max T_J (T_{JMAX}) of 100 °C (very rare), 125 °C, 150 °C (most common), 175 °C, or even 200 °C (an example of the latter being the popular glass diode 1N4148/1N4448). For smaller diodes (axial or SMD), intended for direct mounting on a board, the continuous current rating is specified with the diode assumed to be virtually free-standing, exposed to natural convection, or specified as being mounted on standard FR-4 board with a specified lead length, if applicable. The current rating falls as the ambient exceeds 25 °C, to prevent T_J from exceeding T_{JMAX}. For larger packages like the TO-220 (e.g., MBR1045), the continuous rating is specified with an "infinite (reference) heatsink" attached to the metallic tab/case of the diode. So, as per the datasheet of the MBR1045, we can pass 10 A up to around 135 °C ambient. But that is only with an infinite heatsink (e.g., a water-cooled one). Much less current can be safely passed in a real-world situation with a real-world heatsink. We ask: what is the real-world maximum current rating of the diode? Unfortunately, that is for *us* to calculate as per the vendor — based on the characterization data made available to us by the vendor (the internal Rth and the forward voltage drop curves) combined with our estimate of the thermal resistance of the heatsink we are planning on using. The key limiting factor in all cases is the specified max junction temperature. It is *our responsibility* to ensure we do not exceed that value (with some safety margin too if desired).

Note that as per its datasheet, above 135 °C ambient, the continuous rating of the MBR1045 is steadily reduced with rising ambient temperature. The reason for that is when we pass 10 A continuous current through this diode, we get an estimated 15 °C internal temperature rise (from junction to case). So, with the infinite heatsink, and case held firmly at 135 °C, the junction temperature is at 150 °C. Therefore, above 135 °C ambient, we need to steadily lower the current rating of the diode to avoid exceeding the maximum specified junction temperature. We ask: what is the max continuous rating at an ambient of 150 °C? The answer is obviously zero, since we cannot afford the slightest additional heating when $T_J = T_{JMAX}$. That is how we get the almost linear sloping part of the current rating curve extending from 135 °C to 150 °C as per the datasheet of the part.

Note that this steady reduction of current rating with respect to temperature is sometimes, somewhat confusingly, referred to as a "derating curve." Resistors too, come with a similar published power derating curve (usually above 70 °C), and for much the same reason. But we should not get confused — the use of the word derating in this particular case refers strictly to a reduction in the *strength* (i.e., the rating) of the part, not to any relationship between the strength and the applied stress.

For arguments sake, we can ask: why can we *not* pass *more* than 10 A below 135 °C for the MBR1045, albeit with an infinite heatsink? In other words, why is the derating curve "capped" at 10 A? We argue: since the junction is at 150 °C with the heatsink held at 135 °C, if we lower the heatsink temperature to say 125 °C, wouldn't the junction temperature fall to 140 °C? And so, wouldn't that in turn allow for a *higher* current, so we can bring the junction back to its max allowed value of 150 °C? The reasons for the 10 A hard limit include long-term degradation/reliability concerns and also "package limitations." For example, the bond wires (connecting the die to the pins) are limited to a certain permissible current too. That rule of thumb is

$I = A \times D^{1.5}$ Amps, where D is the diameter of the wire in inches, and $A = 20{,}500$

This can also be written more conveniently in terms of mils as

$I = B \times \text{Dmils}^{1.5}$ Amps, where Dmils is the diameter of the wire in mils, and $B = 0.65$

Note that the above bond-wire equations apply to the more common case of copper or gold bond wires with length exceeding 1 mm (40 mil). For shorter bond wires, "*A*" can be increased to 30,000 ("*B*" = 0.95). For aluminum wires exceeding 1 mm, "*A*" is 15,200 ("*B*" = 0.48). For wires less than 1 mm, "*A*" is 22,000 ("*B*" = 0.7).

Returning to the typical recommended current Stress Factor for catch diodes, in a power supply design the current-related Stress Factor is set closer to 50%, that is, 5 A in the case of the MBR1045, if not lower. The reason for that is very *practical*: the forward voltage drop of the diode (called V_F or V_D) is much higher when operated close to its max rating, and therefore the resulting dissipation ($V_F \times I_{AVG}$) and the corresponding *thermal stress* can become significant. The efficiency gets adversely affected too.

(b) *Reverse Current:* Thermal issues always need to be viewed in their entirety — in particular in the context of *system requirements* (e.g., an efficiency target) and *component characteristics*, not to mention topology/application-related factors. For example, the reverse (leakage) current of Schottky diodes climbs steeply with temperature. The leakage also varies dramatically from *vendor to vendor* and we should always double-check what it is, compared to what we *assume* it is. The reverse leakage term is a major contributor to the estimated dissipation and the actual junction temperature of any Schottky diode in a switching application. After all, with 45 V reverse voltage and with just 10 mA leakage, the dissipation with $D = 0.5$, is $45 \times 10 \times 0.5 = 255$ mW. To reduce this term we want to reduce the diode temperature by better heatsinking.

In general, *thermal runaway* can occur with small heatsinks (or no heatsinks). The reason is an increase in temperature can cause more dissipation, leading to more dissipation and heating, and higher temperatures, in turn leading to higher dissipation, and so on. However, the forward voltage drop of the Schottky improves (reduces) at high temperatures (for a given current). It is a "negative temperature coefficient" device (for the range of currents it is intended for). For that reason *we actually want it to run it somewhat hot.* But its reverse leakage current can increase dramatically with temperature, leading to thermal runway for that reason. We have to strike a good compromise on temperature here.

Note the importance of application too. For example, if the Schottky diode is being used only in an OR-ing configuration (as in paralleled power supplies discussed in *Chapter 13*), its reverse current is clearly *not* an issue and we can increase the temperature-related Stress Factor closer to that of ultrafast diodes. In ultrafast diodes, reverse leakage is negligible, but its forward drop similarly decreases with temperature.

The forward voltage drop of both Schottky and standard ultrafast diodes can, in principle, increase with temperature, but that happens at very large currents, usually well beyond the continuous rating of the device. So, for all practical purposes both ultrafast diodes and regular Schottky diodes are *negative temperature coefficient* devices and we want to run them hot to improve efficiency. But as mentioned, the reverse leakage for Schottky diodes can become an issue. So we don't want to run a Schottky "*that* hot."

In Power Factor Correction (PFC) applications the silicon carbide (SiC) diode, discussed in *Chapter 14*, is becoming increasingly popular as the Boost/output diode since it has very fast recovery (~15 ns) and therefore prevents significant efficiency loss from occurring due to the reverse current spike when the PFC switch turns ON. The SiC diode is essentially a Schottky barrier diode, but a "wide bandgap device" offering very high reverse voltage ratings (up to several kV). It has about 40 times lower reverse leakage current than standard Schottky diodes (and unfortunately a higher forward voltage drop too). Note that unlike regular Schottky diodes, it is a *positive temperature coefficient device*, which means its forward voltage increases with temperature (above a certain current that is usually well within its upper operating range). So, *we do want to run this diode cool if possible*, to improve efficiency. The reverse leakage current of both SiC and ultrafast diodes is not significant.

Taking everything into account, we can target a conservative junction temperature of about 90 °C for standard Schottky diodes (on account of negative temperature coefficient but high leakage current), 105 °C as the junction temperature of SiC diodes (on account of positive temperature coefficient and low leakage), and 135 °C for ultrafast diodes (on account of negative temperature coefficient and low leakage). That gives us a temperature derating factor of $90/150 = 0.6$ for Schottky diodes, $105/150 = 0.7$ for SiC diodes, and $135/150 = 0.9$ for ultrafast diodes. Here, we are assuming all the diodes we are talking about are meant for switching applications (e.g., as catch diodes), and also that all have a $T_{J\_MAX} = 150\,°C$. If T_{J_MAX} is lower, or higher, than $150\,°C$ we can adjust the target junction temperatures based on the above-mentioned Stress Factors.

(c) *Surge/Pulsed Current Rating:* Diodes also have a *surge* current rating called I_{FSM} or I_{SURGE}. This is the maximum (safe) *momentary* current. We can visualize that a surge will not produce a steady buildup of temperature, but can cause sudden localized heat buildup inside the diode. There would be no time for any external heatsink, infinite or not, to react. It would not even depend on the bond-wire thickness. The hot-spot temperature inside the silicon junction is usually allowed to go up to around $220\,°C$, since above that threshold the typical molding compounds of the surrounding package suffer decomposition/degradation. The MBR1045 for example, has a very high *single-pulse* surge current rating of 150 A. In a *repetitive* pulse scenario, we have to combine the localized heat buildup from every pulse with a dissipation term based on the ratio of the energy dissipation pulse width to the period of repetition. The pulsed surge rating thus comes down as the repetition rate increases. Eventually, it equals the continuous current rating.

But does the surge rating of diodes really matter to us in switching power supplies? In fact it does *not*, not when we are selecting the *output/catch* diode. In that location, the inductor serves to limit the current in both the diode and the FET. After all, the diode only takes up what is an almost constant current source (the inductor current) during the off-time, and the FET takes up the same during the on-time. However, the surge rating of diodes is a major concern in the design of the *front-end* of AC–DC power supplies, because that is the place where we normally connect a voltage source almost directly across a capacitor, leading to an almost unrestricted momentary current through the diode and into the capacitor as charging current as discussed in *Chapter 1*. In particular, we need to consider the surge rating of diodes in *two front-end cases*: (a) when we select the *bridge rectifiers* for AC–DC power supplies, those *without* Power Factor Correction (PFC), and (b) when we select the pre-charge diode often found in well-designed commercial front-end Boost-PFC stages (see Figure 6.1, bullet 4). In the former case, to save the diodes, we also need to include inrush protection, either active (usually with an SCR), or passive (with an NTC varistor, but sometimes with just a $2\,\Omega$ wirewound resistor in series). In PFC stages, the pre-charge (bypass) diode can be found strapped directly across the PFC inductor and the Boost/output diode (see *Chapter 14*). Its purpose is to divert the inrush/surge current flowing into the PFC Boost output capacitor when we first connect the AC mains. That huge current then goes through this appropriately rated diode rather than possibly damaging the inductor and/or Boost diode on the way. An example of a diode explicitly intended for this bypass function is the 10ETS08S, available from several vendors. It is a 10 A/800 V standard recovery (slow) diode, with a huge 200 A non-repetitive surge rating. It also has published "$I^2 t$" and "$I^2 \sqrt{t}$" (fuse current) ratings. On the other hand, its forward drop or continuous current ratings really do not matter, since this bypass diode eventually suffers the "ignominy of getting bypassed" itself — it automatically stops conducting the moment the PFC stage starts switching. Note that the front-end of power supplies along with PFC is discussed in further detail in *Chapter 14*.

(d)   *Reverse Voltage Rating:* Coming to voltage, the MBR1045 has a maximum reverse repetitive voltage (V_{RRM}) rating of 45 V. In general, we want to keep within that rating, preferably with a Stress Factor of less than 80%. That includes any spikes and ringing of a repetitive nature. To dampen those spikes out, we may need to put in a small RC (or just a C) snubber across the diode (see Figure 6.1, bullet 17). Snubbers also greatly help in reducing EMI, though they can add significantly to the dissipation of the diode, especially if only a C-snubber is used instead of an RC-type (the C-snubber dumps most of its stored energy per cycle into the diode, instead of dumping it into the R of the RC-snubber). Note that we can often tweak the turns ratio $n = N_P/N_S$ of the transformer in flyback applications (within reason) to accommodate the diode's reverse rating. Increasing the turns ratio produces a lower reflected input voltage across the Secondary.

Though we prefer to target an 80% Stress Factor for the reverse voltage, we may have to make a strategic call in some designs and *allow* that margin to get a little worse. Because, if for example, in an effort to improve the safety margin we pick a 10 A/ 60 V diode instead of a 10 A/45 V diode, the efficiency may get degraded. Because, generally speaking, *for a given current rating, a diode with a higher voltage rating has a higher forward voltage drop at a given current*. Though one way out to increase both the safety margin and keep the forward drop low, is to consider a higher current 60 V device, like a 15 A or 20 A diode instead of the MBR1045 or a 10 A/60 V diode, cost permitting. Because, generally speaking, *for a given voltage rating, a diode with a higher current rating has a lower forward voltage drop at a given current* (though this is a *trend*; it may not be really true as we can see from Table 6.1). Unfortunately, *a diode with higher current rating also has a higher reverse leakage current than a diode with lower current rating of the same voltage rating*. Quite the contrary, *a diode with a higher voltage rating has a lower reverse leakage current than a diode with a lower voltage rating of the same current rating*.

A higher leakage current will lead to a higher dissipation in switching applications irrespective of the forward loss. See Table 6.1 for actual data extracted from one particular vendor. However, take nothing for granted, especially for reverse leakage. For example, the 3 mA reverse leakage of the MBR1045 from Fairchild compares very favorably to the 10 mA stated for the MBR1045 from On-Semi (under the same conditions), and the 100 mA for the "equivalent" SBR1045 from Diodes Inc. However,

Table 6.1: Schottky Diode Forward Voltage Drops and Reverse Leakage Currents.

Reverse Voltage Rating	Conditions	Continuous Current Rating		
		10 A	15 A	20 A
45 V	V_F at 10 A, at 25 °C	0.58 V	0.62 V	0.58 V
	I_R at 40 V, at 125 °C	3 mA	2.8 mA	3.6 mA
60 V	V_F at 10 A, at 25 °C	0.7 V	0.7 V	0.7 V
	I_R at 40 V, at 125 °C	2 mA	2.5 mA	2.8 mA

Datasheets Consulted: MBR1045, MBR1060, MBR1545CT, MBR1560CT, MBR2045CT, and MBR2060CT, all from Fairchild Semiconductor. Typical values extracted from curves.

the MBR1060 from Fairchild has an I_R of 2 mA, and that is outclassed by the 0.7 mA of the MBR1060 from On-Semi.

If we are very keen to lower the forward voltage drop, we might like considering paralleling lower current diodes. For example if we take two MBR745 diodes from Fairchild and ask them to share the 10 A current equally (5 A in each), we see from the corresponding datasheet from Fairchild that the forward drop of the MBRP745 is only 0.5 V at 5 A and at 25 °C. That is an improvement over the almost 0.6 V from a single 10 A/45 device. However, each MBRP745 has a leakage of 10 mA at 40 V and at 125 °C, so two diodes in parallel will give us a whopping 20 mA reverse leakage. Further, to get two diodes to share properly is not trivial. One way to force that is by paralleling two diodes on the same die (dual pack). Another way is shown in the context of EMI suppression in Figure 17.4.

In brief, power supply design is all about trade-offs — very careful trade-offs as exemplified in this "simple case."

There is a possible way of allowing for a voltage Stress Factor *equal to or exceeding* 1 and yet not compromising reliability. We know that Schottky diodes typically "avalanche" (behave as zeners) if the reverse voltage exceeds about 30–40% of their V_{RRM}. That property can be used to deliberately clip any spikes without using snubbers or clamps. But regular Schottky diodes can handle this zener mode of operation for a *very short time only*. For reliable operation, the Schottky diode *must have a guaranteed avalanche energy rating* (E_A expressed in µJ) stated in its datasheet, and we have to confirm from our side that the energy of the spike that it is being used to clamp in our application is well within that particular rating. An example of such a "rugged" diode is the STPS16H100CT from ST Microelectronics. This is a dual common-cathode diode package, with each diode being rated 8 A/100 V. Diodes Inc. calls such a diode a Super Barrier Rectifier (SBR®). For example, they offer a 10 A/300 V rectifier called the SBR10U300CT.

(e) *dV/dt Rating:* We all know that a diode has peak/surge/average current ratings. It also has a well-known steady reverse (blocking) voltage rating. But whereas voltage is certainly a known stress, its *rate of rise (or fall)*, dV/dt, can also produce overstress and corresponding failure modes. ESD, as discussed above, is in effect a dV/dt stress. It is well-known that MOSFETs are vulnerable to ESD, especially during handling, testing, and production. Another example is the Schottky diode. The Schottky has a maximum rated dV/dt, usually specified somewhere deep in its datasheet that engineers often miss. What could happen is that in a real application, we may have applied what we believe is a "safe" steady reverse DC voltage within the Abs Max reverse voltage rating of the diode, yet we learn that some diodes are failing "mysteriously" in large-scale production testing. One reason for that could be a *momentary* amount of excess dV/dt every switching cycle that we can capture only on an oscilloscope if we zoom in very carefully. The likelihood for this type of violation is naturally the highest when the diode is in the process of getting reverse-biased (switch turning ON). Even a little ringing in the turn-off voltage waveform for example, usually due to poor layout, can cause the instantaneous dV/dt *at some point of the waveform* to exceed its dV/dt max rating, causing damage. Nowadays, the dV/dt rating of Schottky diodes has almost universally improved to 10,000 V/μs and that has greatly helped forgive the lack of attention to this very rating. But not very long ago, diodes with 2,000 V/μs or less were being sold as "equivalents," purely based on the fact that their voltage and current ratings were the same as more expensive competitors. Perhaps we still need to be watchful for that possibility today. If necessary, we may need to slow down, dampen, or smoothen out the turn-off transition somewhat. We can do that by trying to slow down the switching transition by increasing the Gate resistor of the FET (though admittedly, that usually just creates more delay till the turn-off transition starts, rather than slowing the transition itself). An effective "trick" used in commercial flybacks is to insert a tiny *ferrite bead in series with the output diode* (see Figure 6.1, bullet 14). This can certainly adversely impact overall efficiency by a couple of percentage points, but it can raise the overall reliability significantly. Note that Ni–Zn ferrite offers higher high-frequency resistance and lower inductance than the more common Mn–Zn ferrite. So, a Ni–Zn bead affects the energy transfer (flyback) process less (as we want), but is much better at providing (resistive) damping against any high-frequency ringing that can be responsible for dV/dt violations during the diode turn-off transition.

MOSFETs

For example, let's take the "4N60" N-channel MOSFET (also called a "FET" here). This (or an equivalent) is a possible choice for low-cost universal input AC–DC flybacks up to around 50 W. By popular numbering convention the 4N60 is a 4 A/600 V MOSFET.

(a) *Continuous Current Rating:* Like diodes, FETs also have continuous/pulsed current and sometimes avalanche ratings too ("rugged" FETs). The continuous rating is once again, in effect, just a thermal limit. For the 4N60 in a TO-220 package, that rating is 4 A *when mounted on an infinite heatsink at* 25 °C. But with the case/heatsink at 100 °C, the rating is only 2.5 A (typical; depending on the vendor). Because under that condition *the junction is already at 150 °C.*

The R_{DS} is also always stated with the case held at 25 °C, passing what may often seem quite an arbitrary current through the device. The real calculations can be very tricky and iterative, because the R_{DS} of a FET depends strongly on junction temperature (and on Drain current too). But we can also take the approach of believing the data provided by a reliable vendor to reverse-estimate the R_{DS}. For example, take the 4 A/60 V TO-220 device (STP4NK60Z) from ST Microelectronics. The stated junction-to-case thermal resistance of this device is 1.78 °C/W. The vendor also states that with the case held at 100 °C, the maximum Drain current is only 2.5 A. We can assume that under this condition, the junction is already at 150 °C. So, the temperature rise from case to junction is $150 - 100 = 50$ °C. Our estimate of worst-case R_{DS} (at a junction temperature of 150 °C) is therefore

$$\Delta T_{JC} = \text{Rth}_{JC} \times P = \text{Rth}_{JC} \times \left(I_D^2 \times R_{DS}\right) \Rightarrow R_{DS} = \frac{\Delta T_{jc}}{\text{Rth}_{JC} \times I_D^2} = \frac{50}{1.78 \times 2.5^2} = 4.5 \, \Omega$$

Note that the R_{DS} stated in the datasheet is "2Ω" under rather benevolent conditions of case held at 25 °C and the device passing only 2 A current. We see that in reality, the worst-case R_{DS} is 2.25 times more. That is actually a typical "cold to hot R_{DS} factor" when dealing with high-voltage FETs for AC−DC power supplies. For logic-level FETs (say 30 V and below), the hot to cold R_{DS} factor is only about 1.4.

The 4N60 is also available in SMD packages (e.g., TO-252/DPAK, or the TO-263/D²PAK, the latter being basically a TO-220 laid down flat on the PCB). Since these do not involve an infinite heatsink, the continuous current rating is much lower. The common feature between the two 4N60s in different packages *is just their R_{DS}*. And in all cases, for all packages and any heatsinking, the max continuous rating is based on the junction temperature being at a maximum of 150 °C.

Note that whereas in diodes, it was relatively easy to estimate T_J and thereby estimate the thermal stress, in the case of a FET it is much harder to do. Even if we do know the R_{DS} accurately, what we can calculate so far is just the conduction loss term. To estimate the actual junction temperature in a switching application, we have to also carefully add switching losses as discussed in *Chapter 8*.

The 6N60 (or equivalent) is often used in universal input flyback power supplies up to 70 W. It is rated 6 A/600 V. Its R_{DS} is 1.2 Ω (typical 1 Ω). With an infinite heatsink held at 100 °C, its continuous current rating drops to 3.5 A to 3.8 A (depending on the vendor). As was the case for a diode, there are efficiency concerns that prevent us from approaching anywhere close to the continuous current rating of a FET. For example, in 70 W flybacks, the measured peak switch current may be only between 1.5 A and 2 A worst-case under steady conditions (measured with max load at 90VAC). The average switch current, assuming $D = 0.5$, is therefore 0.75 to 1 A. Nevertheless, the 6N60 is used for this application. The current-related Stress Factor is 1 A/3.8 A = 0.26, or about 25%. Therefore, we can see that significant continuous current derating has been applied *in the interests of power supply efficiency*. Flyback power supply designers therefore may declare as a rule of thumb that they use "a FET with a hot-R_{DS} of 2 Ω for 2 A peak, 4 Ω for 1 A peak, 1 Ω for 4 A peak …." and so on.

Power supply designers sometimes forget that the applied Gate-to-Source voltage (V_{GS}) can also affect the R_{DS}. The stated R_{DS} of 2.5 Ω for the 4N60 or 1.2 Ω for the 6N60 is *with $V_{GS} = 10$ V*. These FETs typically have a Gate threshold voltage of around 4 V. As power supply designers, we must in general try to ensure that the ON-pulse applied to the Gate has an amplitude greater than about 2× the Gate threshold voltage of the part. Otherwise, the R_{DS} will be higher than we had assumed.

Note: To estimate temperature and/or conduction loss, it is important to ultimately measure the R_{DS} carefully in our switching application — by calculating the ratio of the measured Drain-to-Source voltage V_{DS} when the switch is ON (its forward drop) to the corresponding measured Drain current I_D. Neither of these are trivial measurements. I_D is best measured by snapping in a current probe on a loop of wire connected to the Drain of the FET. Remember: we should never put a current probe on the Source of a FET because even the small inductance of the wire loop can cause ringing and spurious turn-on, leading to FET damage. For a V_{DS} measurement we may think it is OK to place the typical 10× voltage probe of a scope from Drain to Source while the FET is switching and zoom in to see the small voltage drop across it during the ON-time. Unfortunately, a typical scope has a vertical amplifier that will "rail" and saturate by the hundreds of volts of off-screen high voltage during the OFF-time. So the ON-time measurement will in turn be absurd, because the vertical amplifier is still trying to recover from the effects of overdrive. Therefore, the engineer may need to devise a small, innovative and non-invasive buffer circuit (i.e., a circuit with high enough impedance and minimum offset so as not to affect either the current through the FET or its observed forward voltage drop), and then place this little circuit on the Drain. The output of this buffer circuit is intended to be a true reflection of the V_{DS} during the ON-time, but clamped to a maximum of about 10−15 V during the OFF-time. The voltage probe of the scope is placed on this output node, rather than directly on the Drain of the FET to avoid saturating the scope amplifier.

Another technique that can help avoid overdrive of the scope amplifier is to use waveform averaging on the scope, then use the scope's waveform math functions to *digitally* magnify the waveform around the portions of the signal of interest. Digital magnification performs a software expansion of the captured waveform to reveal additional vertical resolution beyond the typical 8-bit resolution of the scope's ADC (when averaging is used).

(b) *Surge/Pulsed Current Rating:* As in the case of a diode, the inductor limits the current, so the surge rating of a FET is not really tested out in switching power supplies. But the rating can matter in *poorly designed* power supplies. For example, as mentioned previously, flybacks are prone to damage at initial power-up or power-down. The reason for that is for this topology, the center of the current ramp (of the inductor/switch/diode) is $I_{OR}/(1 - D)$ as explained in *Chapter 3*. So, as the input voltage is lowered, D approaches 1 and the sudden rise in current can cause core saturation, which in turn can cause a huge current spike in the FET, damaging it. A common technique to guard against power-up and power-down damage at low line in flybacks is by incorporating careful current limiting, combined with UVLO (under voltage lockout) and maximum duty cycle limiting (see Figure 6.1 once again).

At high-line, sudden overloads can cause a huge current spike at turn-off, that can lead to a voltage spike across the FET, killing it. A technique to guard against that is "Line (or Input) Feedforward." This is also discussed in *Chapter 7*. In general, without adequate protection against these "low-frequency repetitive events," no amount of steady-state stress derating will be sufficient to ensure field reliability.

(c) *Drain—Source Voltage Rating:* In a commercial flyback using cost-effective 600 V FETs, the conservative expectation of 80% voltage Stress Factor, or $0.8 \times 600 = 480$ V is *not* feasible. *In well-designed commercial power supplies* (with all the protections mentioned above and as indicated in Figure 6.1 included), the voltage Stress Factor for power MOSFETs is often set closer to 90% — because now, very few situations are considered anomalous or unanticipated. For the 4N60 or the 6N60 in flyback applications, that is a maximum of $0.9 \times 600 = 540$ V including spikes — measured at 270VAC under steady conditions *while sweeping all the way from min to max load to confirm the worst-case.* The desired minimum headroom is 60 V.

(d) *Gate—Source Voltage Rating:* MOSFETs are very easily damaged by voltages (even extremely narrow spikes) that are in excess of the specified Gate-to-Source Abs Max voltage. The Gate oxide can be easily punctured, even by electrostatic charge picked up by walking across a carpet for example. Once mounted on a board, the potential for ESD damage is obviously minimized. The 4N60 and 6N60 are both rated for ± 30 V maximum V_{GS}. For this reason we may sometimes find *protective zeners* connected from Gate to Primary Ground of the switching FET in power supplies. Note that, as indicated in Figure 6.1, because of the high impedance of the Gate, very high-frequency oscillations are possible between the trace inductance running to the Gate and the input capacitance of the FET (at the Gate pin). There is also anecdotal evidence to suggest that "protective" zeners placed at the Gate terminal can exacerbate such oscillations and lead to "mysterious" field failures. It is therefore always recommended to put in a Gate drive resistor to dampen out any potential oscillations as shown in Figure 6.1. A 4.7−22 Ω resistor is standard. A pull-down resistor of a few kΩ *placed very close to the Gate* is also recommended to avoid oscillations and spurious turn-ons, especially in AC−DC switching power supplies.

(e) *dV/dt Rating:* In the early days of MOSFET, the most common failure mode was due to very high "reapplied d*V*/d*t*." This could trigger the parasitic BJT (bipolar junction transistor) structure inside the MOSFET and lead to avalanche breakdown and snapback. It can occur even today, but it is very rare, so we virtually ignore this possibility nowadays. Modern FETs can handle 5−25 V/ns (i.e., over 5,000 V/μs).

Capacitors

In general, we have to be conscious of the voltage rating of any capacitor and stay under that limit and preferably with some typical derating. Polarized capacitors like aluminum electrolytic and solid tantalum also have *reverse* voltage ratings that we must not exceed, though aluminum caps are more tolerant than solid tantalum in this regard.

In general, in the interest of cost, we should confirm with the capacitor vendor whether their specific test/field data really support the traditional notion that *voltage stress derating* is useful in lowering the field failure rate of capacitors. That relationship seems to be in question today given the improvements in manufacturing technology. This is especially true for aluminum electrolytic caps. We can explain that specific situation as follows. The breakdown voltage of an electrolytic cap is not an abrupt threshold. It is related to the thickness of a chemically generated oxide on its electrodes. That oxide film is the dielectric that holds-off the applied voltage. If the working voltage on the cap is increased, the oxide thickness also gradually increases, raising the voltage withstand capability. On the other hand, if the working voltage is reduced, the oxide can also reduce gradually, lowering the rating (though the addition of borax greatly prevents that). That is the reason why if we take aluminum electrolytic caps out of storage after a long time, a "re-forming" phase is still recommended by many vendors, in which the applied DC voltage is slowly ramped up (with max current limited to a few mA), to let the oxide develop fully again. But this also indicates that if we operate an aluminum electrolytic cap at reduced voltages for a long period of time, its "strength" progressively lowers, and so the presumed safety margin also falls somewhat over time.

All capacitors usually have a published ripple current rating that we must not exceed. The most important parameter for aluminum electrolytic capacitors in particular, is its *life*, which is based on its "core temperature," which in turn depends on the RMS current passing through it in a given application. The max RMS (ripple) current rating is therefore a thermal rating in effect. For extending life, we may need to apply a derating to the ripple current rating. Life prediction is discussed later in this chapter.

Film capacitors are prone to dV/dt failures. The common, low-cost Mylar® (polyester/KT/MKT) capacitors are rated for only about $10-70$ V/μs. They are therefore usually not suitable for snubbers or clamps. For AC−DC flyback snubber/clamp applications, a preferred film capacitor type is polypropylene (KP/MKP), since its dV/dt rating is typically $300-1,100$ V/μs. The ceramic capacitor and the mica capacitor both have very high dV/dt ratings, and for cost reasons, the former is often used for clamps today. Note that film capacitors are generally favored in many situations over ceramic, since they are far more stable (less change in capacitance and other characteristics) with respect to applied voltage, temperature, and so on.

Keep in mind that both the dV/dt and dI/dt ratings of any selected capacitor can get fully tested at the *front end* of both AC−DC converters and DC−DC converters. We need especially high transient ratings for that particular location. This is further discussed in *Chapter 14*.

In solid tantalum (Ta-MnO$_2$) capacitors in particular, a high dV/dt can produce a large surge current due to $I = C(dV/dt)$. This can cause localized heating and immediate damage. For that reason it is often said that most tantalum capacitors rated for say "35 V," should not be used at operating voltages exceeding *half* the rated voltage (in this case 17.5 V). That is especially true at the front-end of any power supply (DC−DC converter) where even that amount of derating may not be enough. We actually need to *limit the surge current* to avoid local defects from forming in the oxide and mushrooming into failure. It is often recommended to ensure at least 1 Ω per applied Volt in the form of source impedance to limit the surge current to a maximum of 1 A. Conservative derating practices call for 3 Ω/V, which will limit the current to 333 mA. Still more conservative engineers do not even use Ta-MnO$_2$ capacitors anymore, and prefer ceramic or polymer caps.

Modern multilayer polymer (MLP) capacitors are stable, have very high dI/dt and dV/dt capabilities, and are being increasingly preferred in many applications to multilayer ceramic capacitors (MLCs), Ta-MnO$_2$ capacitors, and aluminum electrolytics. They are available up to 500 V rating and offer very low ESR for power converter output applications.

Ceramic capacitors at the front-end of a converter can themselves induce a huge voltage spike when power is first applied on the input terminals of a converter. This "input instability" phenomenon is discussed in more detail in *Chapter 17*. One solution is to place an aluminum electrolytic capacitor in parallel to the input ceramic capacitor, as a means of damping out the input oscillations.

An aluminum electrolytic capacitor, besides its well-known advantages of delivering maximum "bang for the Buck" (very high capacitance times voltage for a given volume plus cost) is extremely robust too. Its failure modes are essentially thermal in nature. So, it can withstand significant abuse for short periods of time. For example, we can typically exceed an aluminum capacitor's voltage rating by 10% for up to 30 seconds with no impact. That is called its "Surge Voltage rating." It also has an almost undefined surge current rating. It can, for example, tolerate extremely high inrush currents at the input of an AC−DC power supply with no problem at all, provided the inrush is not made to occur repetitively and too rapidly. The aluminum electrolytic has self-healing properties since its oxide layer quickly reforms. It rarely ever fails open or short unless thoroughly abused (in which case it vents). Its normal failure mode is essentially *parametric* in nature (e.g., drift of capacitance, ESR, and so on). Its Stress Factors therefore may not be of such great concern as concerns about its lifetime, as discussed later in this chapter.

PCB

A major power component frequently overlooked is the PCB itself. In power supplies this is also subject to power cycling and resulting hot-spots that we must guard against. If we open a commercial power supply, we may find power resistors mounted on *raised standoffs* rather than flush against the PCB surface. That is to protect the PCB from overheating. Standard FR-4 PCB material has a glass transition temperature T_G of around 130 °C and is therefore rated for 115 °C maximum. If we cross the glass transition threshold, the properties of the board can change in a subtle way, often permanently. For one, the TCE (Thermal Coefficient of Expansion) of the board can get affected and that can lead to subtle failures down the road, as explained a little later below.

In general, just as dV/dt can cause overstress, a high rate of change of current, dI/dt, can also cause failures. For example, we know that $V = L$ dI/dt. So if nothing else, a high dI/dt on a PCB trace can cause a voltage spike that can indirectly kill a semiconductor. It is known that very fast diodes with extremely snappy recovery characteristics can produce very large *voltage* spikes due to their abrupt current cutoff (which is in effect a large dI/dt). This voltage spike can then damage the very diode that has indirectly created it, or can even damage any weaker component in its neighborhood. Bad PCB layout can also cause similar inductive spikes, large enough to cause failures. See *Chapter 10* for more details.

Mechanical Stresses

Lastly, while focusing on electrical stresses, we must not forget the seemingly obvious: *mechanical stresses*. We could cause damage, *immediate or incipient*, due to mishandling, drops, or transportation. For that reason a typical commercial power supply PCB will have generous dabs of RTV (room temperature vulcanizing silicone) on its PCB, literally anchoring down bigger components. Every commercial power supply needs to pass a shock and vibration test during validation testing. In this context, two- or multilayer PCBs fare much better than single-sided PCBs. Because large through-hole components mounted with their pins inserted through plated-through holes (vias) and soldered through the via onto the other side of the board, get anchored much better than single-sided PCBs where the traces can rip off rather easily.

However, there are more subtle forms of failure due to mechanical stresses. In power supplies, as power and temperatures cycle up and down, significant expansions and contractions occur constantly. Generally, the "thermal coefficient of expansion" (called TCE or CTE) of SMD components differs from the TCE of the PCB on which they are mounted. So, *relative movement* occurs that can lead to severe mechanical stresses and eventual cracking. Surface-mounted multilayer ceramic capacitors, in particular, were historically extremely prone to this. Bad soldering practices can also initiate micro-cracks that develop into failures over time. Also note that large PCBs *sag*, much more than smaller PCBs, and that too produces severe stresses, especially on the relatively brittle SMD ceramic capacitors. Therefore, even today, though great efforts have been made to match the TCE of components to the TCE of standard FR-4 PCB material, many quality power supply design cum manufacturing houses still have strict internal guidelines in place restricting their power designers from using any SMD multilayer ceramic capacitor larger than size-1812 (0.18 in. × 0.12 in.), for example, even size-1210 on occasion.

"Lead forming" or "lead bending" has been used for decades in power semiconductors. One motivation for that is convenience. For example, the device may be mounted on a heatsink in such a way that to be connected to the PCB, the leads really need to be bent. However, *stress relief* is also a motivation here. With a little bend introduced in the leads, we can prevent mechanical stress due to thermal cycling from being transmitted into the package causing damage in the long run. However, we have to be cautious that in the process of performing lead bending to avoid long-term stress, we do not cause immediate stresses and incipient damage. In plastic packages, the interface between the leads and the plastic is the point of maximum vulnerability. Under no circumstances must the plastic be held or constrained while the leads are bent, because the plastic–lead interface can get damaged. And when that happens, even though the damage is not obvious, the ability of the package to resist ingress of moisture is affected. That may eventually lead to device failure through internal corrosion.

Part 2: MTBF, Failure Rate, Warranty Costs, and Life

Having provided a background on basic reliability concerns in power supply design we move on to an overview of reliability/life prediction and testing.

MTBF

The first term we should know is power-on hours or "POH." For components/devices, this may be called total device-hours ("TDH") instead, but the concept is the same. For example, one unit operating for 10^5 (100k) hours has the same power-on hours as 10 units operating for 10k hours, or 100 units operating for 1,000 h and so on. All these give us 10^5 POH. (To be statistically significant however, a larger number of units is always preferred.) Note that whenever people talk of failure rate, or MTBF (Mean Time Between Failures), they usually talk in terms of "hours," whereas in reality they mean POH or TDH. This should be kept in mind as we too undertake the specific discussion below.

Failure rate, λ, is the number of units failing in unit time. But it is expressed in so many ways and we need to know some inter-conversions. Historically, failure rate was first expressed as the % number of units/devices failing in 1,000 h of operation. Later as components with better quality emerged, people started talking in terms of the number of failures occurring in 10^6 (million) hours. That was called "ppm," for parts per million. As quality improved further, the failure rate of components was more conveniently stated as the number of failures per billion (10^9) hours. That number is called "FITs" (failures in time), and is often referred to as λ too. See Figure 6.2 for an easy look-up table for failure rate conversions.

For example, a component with a failure rate of 100 FITs is equivalent to a ppm of 0.1. That incidentally is $0.1 \times 10^{-6} = 10^{-7}$ failures/hour.

MTBF is the *inverse* of *failure rate*. Therefore, in our example here, for 100 FITs, the MTBF is 10^7 h (10 million hours). Similarly, an MTBF of 500k hours is equivalent to a failure rate of 0.2%/1,000 h, or a ppm of 2, or 2,000 FITs.

The failure rate of a system is the *sum* of the failure rates of its *components* (we are ignoring redundant systems here).

$$\lambda = \lambda_1 + \lambda_2 + \lambda_3 + \cdots + \lambda_n$$

$$\text{MTBF} = \frac{1}{\lambda} = \frac{1}{\lambda_1 + \lambda_2 + \cdots + \lambda_n}$$

An MTBF of 250k hours is a typical expectation of certain types of power supplies. Since there are 8,760 h/year, 250k hours seems immense, close to 30 years. Does that mean we can take a very large sample of these power supplies and expect only one unit to fail on an average every 30 years? Not at all. The term MTBF is vastly misunderstood, and needs to be clarified.

An MTBF of 30 years actually means that after 30 years, provided there are no wearout failures (wearout is explained later), we will be left with *a third of the power supplies that we started off with*. In other words, two-thirds *will* fail by the end of the MTBF period (30 years in our case). So out of 1,000 units, about 700 units will fail. Out of 2,000 units, 1,400 units will fail, and so on. We can thus estimate how many hundreds will fail in 5 years.

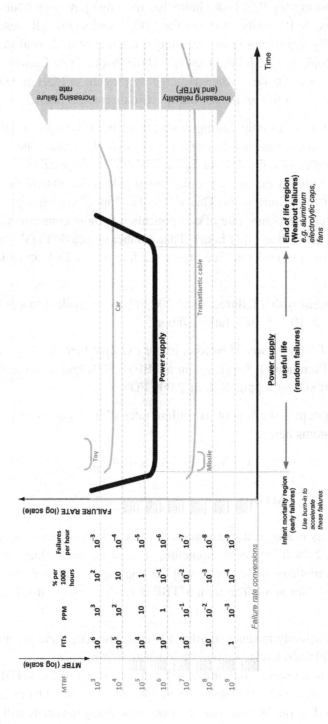

Figure 6.2: Failure rate conversions and the bathtub curve.

By definition, MTBF is in fact the *time constant* of a naturally exponentially decreasing population.

$$N(t) = N \times e^{-\lambda t} = N \times e^{-t/\mathrm{MTBF}}$$

Note that this is analogous to a capacitor discharging.

$$V(t) = V_O \times e^{-t/RC}$$

At the end of the time constant $\tau = RC$, the cap voltage is $1/e = 0.368$ of the starting voltage.

The reliability $R(t)$ is defined as

$$R(t) = \frac{N(t)}{N} = e^{-\lambda t}$$

This is in effect the probability that a given piece of equipment will perform satisfactorily for time t because $N(t)$ is the number of units left after time t and N the number we started off with. Note that reliability is a function of time. For $t = \mathrm{MTBF}$, the reliability of any system/device is only 37%. Which is another way of saying that from the start, the equipment was only 37% likely to survive until $t = \mathrm{MTBF}$. It has a lesser chance to survive longer and longer, which is why reliability falls exponentially as a function of time. Let us bullet out the different ways of stating MTBF and interpreting it:

(a) With a large sample, only 37% of the units will survive past the MTBF.
(b) For a single unit, the probability that it will survive up to $t = \mathrm{MTBF}$ is only 37%.
(c) A given unit will survive until $t = \mathrm{MTBF}$ with a 37% *confidence level*.

In our example, at the end of 5 years, we are left with

$$N(t) = 1,000 \times e^{-(5 \times 8,760)/250,000} = 839 \text{ units}$$

So $1,000 - 839 = 161$ units have failed in the first 5 years. How about after 10 years? We are then left with only

$$N(t) = 1,000 \times e^{-(10 \times 8,760)/250,000} = 704 \text{ units}$$

So, $839 - 704 = 135$ units failed between 5 and 10 years in the field.

Note: Here is the math that makes this an exponential curve: $161/1,000 = 135/839$ ($=0.161$). We should know that the ratio by which an exponential curve changes in a certain time interval is invariant. In fact it does not matter where we set $t = 0$. At any chosen starting point, N is the number of units existing at that moment. That is why the exponential is considered the most "natural" curve. On that basis we now expect that in the next 5 years (10 to 15 years) $0.161 \times 704 = 113$ additional units will fail, giving a total of 409 units failing in 15 years. There are two figures in this book that should be

looked at more closely at this juncture to get a better idea of the exponential curve —
Figure 5.3 and Figure 12.1. It will also become clear from the former figure in particu-
lar, why MTBF is perhaps an appropriate name for the time till the number of units falls
to 1/e of the initial value. Because in a sense, based on area under the curve, on an aver-
age we can consider all *the units as having failed at exactly this precise moment.*

One last misconception needs to be cleared up: if the MTBF of a power supply goes from
say 250k hours to 500k hours, does that mean "reliability has doubled"? No. First of all, the
question itself is wrong. We need to specify "*t*" while calculating $R(t)$. So suppose we pick
$t = 44$ k hours (5 years). This becomes the moment at which we are comparing reliabilities
for the two MTBF possibilities (usually set to the expected life of the equipment). So,

$$R(44\text{k}) = e^{-44k/250k} = 84\%$$
$$R(44\text{k}) = e^{-44k/500k} = 92\%$$

We see that doubling the MTBF increased reliability by roughly only 10% over 5 years
(because $92/84 = 1.095$). But warranty costs do vary in inverse proportion to the MTBF.

Warranty Costs

Why is reliability so important in a commercial environment? Cost! Engineers should know
the "$10 \times$ rule of thumb" which goes as follows: if a failure is detected at the board level
and costs $1 to fix, it will cost $10 if discovered at a system level (by a failure in
production testing), and will cost $100 to fix in the field, and so on. Several subsequent
studies actually show that the cost escalates far more rapidly than $10\times$. It therefore
becomes very clear that if the power supply design engineer understands converter stresses
and potential failure modes well in advance, and eliminates them at the design stage itself,
that is the cheapest way to go.

Example:

If there are 1,000 units with an MTBF (Mean Time Between Failures) of 250k hours, how
many units are expected to fail over 5 years? Calculate warranty costs, assuming it costs
$100 to repair one unit.

In a real-world situation, damaged equipment will be immediately replaced and put back in
the field. So, the average population in the field will not dwindle exponentially. In such a
situation, we can calculate the recurring or annual warranty cost. Over the declared 5-year
warranty period (43.8k hours), the number of failures is

$$\text{\# of failed units} = \frac{1,000 \text{ units} \times 43,800 \text{ hours/unit}}{250,000 \text{ hours/failure}} = 175.2 \text{ failures}$$

That is $175.2/5 = 35$ units per year for 1,000 units or 3.5 failures per 100 units. And that gives us the Annualized Failure Rate (AFR)

$$\text{AFR} = \frac{876,000}{\text{MTBF}} \text{ expressed as \% failures per year}$$

For our example, we have

$$\text{AFR} = \frac{876,000}{250,000} = 3.5\% \text{ failures per year}$$

Or 3.5 units failing every year for every hundred in the field. If it costs $100 to repair one unit, that gives us $350 per 100 units or $3.5 per unit every year for 5 years. The total repair cost over 5 years is an astonishing $17,520, or $17,520/5 = $3,500 per annum — for 1,000 units sold, irrespective of the proposed selling price of each unit. In other words, the *warranty repair cost per unit* is $17.52 over the stated warranty period. This cost will need to be added to the selling price of the unit by the vendor upfront, or risk going bankrupt. An alternative is to reduce the warranty period, say to the bare minimum of 90 days.

Life Expectancy and Failure Criteria

In reality, the number of units failing will often climb steeply, typically at around the 5-year mark. That is because *lifetime* issues start to come into play at this point. Engineers should not confuse life with MTBF though both eventually lead to observed failures. The concept of MTBF applies only *during the useful life of the equipment*, during which, by definition, failure rate is a *constant* (implying an exponentially decreasing curve in the absence of any repairs and/or replacements). These are also called random failures. Ultimately, the failure rate suddenly climbs as a result of *wearout* (end-of-life) failures. See Figure 6.2 for the classic bathtub curve. We have tried to indicate how some systems have high reliability but low life (missile), whereas some have relatively low reliability but long life (car), and so on.

Though, what exactly is considered a failure also needs to be defined. The equipment need not become completely non-operational as a result of the failure; it may just fall out of specified performance limits. For example, a car may continue to function even though its seat belts or audio/GPS units are not working. It depends on us whether we deem that as a failure and pull the vehicle off the road for repairs. For any electronic component, what exactly those limits are, and what therefore constitutes the set of failure criteria, are specified in the electrical tables of its datasheet.

A key culprit for wearout failures of a system/equipment is the aluminum electrolytic capacitor, if present. We will discuss life prediction of this capacitor later in this chapter. Another common contributor to life limitations is the cooling fan, if present. Note that sleeve-bearing fans are commonly said to fare worse than ball-bearing ones in terms of life.

However, ball-bearing fans also become noisier sooner, and if a certain threshold of noise is included as part of the failure criteria of fans, then sleeve-bearing fans are probably at par if not better than ball-bearing ones, and much cheaper too. Another component that can show steady degradation (wearout) is the optocoupler. If we apply a high anode current for an extended period of time, the current transfer ratio (CTR) of the opto can worsen steadily and quickly. The life on an optocoupler is said to be the point at which its CTR has fallen to 50% of its initial value. In general, overdriving a typical opto above ~10 mA causes a huge reduction in its lifetime. However, in a power supply application, we have a high-gain system, and so the slightest error-related current passing through the opto produces an immediate correction, which therefore reduces the current. So, we cannot really (continuously) *overdrive* the opto in a power supply feedback-loop application. We can therefore expect its life to be typically *over 150k hours*. Normally, we also have enough phase and gain margin built-in to prevent any instability due to CTR-related tolerances and degradation.

Component vendors often do "specmanship" around failure criteria or guaranteed performance limits. For example, Panasonic declares end-of-life for most of its leaded aluminum electrolytic capacitors on the basis of capacitance falling 20% below the initial value, but for its own SMD aluminum electrolytic, that number is 30%. Therefore, we need to *compare failure criteria* while comparing life or MTBF numbers provided by different vendors, but often even for different product families from the same vendor.

Reliability Prediction Methods

We would like to get an estimate of the prospective field reliability as early as possible. Till fairly recently, Military Handbook 217F (MIL-HDBK-217F) was widely used to predict reliability. It is no longer recommended today as it was consistently providing very pessimistic predictions based on an outdated database of component reliability figures, and was in effect obsolete. However, an effort is already underway to revise and re-invigorate it as the new VITA 51 standard (VITA stands for VMEbus International Trade Association) (see www.vita.com).

MIL-HDBK-217F went about reliability prediction in two ways — either using a simple Parts Count analysis, or a more detailed Part Stress analysis.

Parts Count was intended to be used only in initial project bidding phases when no prototype was available. It basically involved summing up the failure rates of all the components, assuming some default base failure rates and some default stress levels. This was criticized by many. First, it would clearly disadvantage companies with higher internal standards (i.e., those using higher stress derating factors) simply because it would assume the parts lists they submitted were based on the same stress level as all their competitors. Second, the logic was obviously not completely right. For example, if we add a TVS across a FET, we would clearly increase field reliability. However, as per Part Count analysis, just because now there are two parts likely to fail instead of one, the reliability has come *down*. This faulty logic would apply to any protection circuitry we may use — even the current-limiting sections of power supplies. Therefore, this approach certainly seemed mired in "trees" (components), missing the "forest" (the system).

Part Stress analysis makes more sense in terms of its underlying philosophy if not the actual numbers behind it. It is therefore still useful as a *comparative* tool between several vendors. One of the key practical inputs to Part Stress analysis is still recommended — we should actually *measure* the stresses on *all* the components. This involves taking a working prototype and putting *current, voltage and thermal* probes on each and every component. We can then check if their Stress Factors are acceptable. This hands-on exercise is very helpful in identifying the weak links in the design early on and rectifying them before they become (expensive) field failures.

The general philosophy behind MIL-HDBK-217F and other formal reliability prediction tools is as follows. The failure rate of any equipment is the sum of the failure rates of all its components. Each component has a specified *base failure rate* (at a certain reference level), which is then adjusted to the current situation (including the application itself) by multiplying it with a number of scaling factors related to the environment π_E, application π_A, quality level π_Q, secondary stresses (e.g., voltage stress π_V), and so on.

Generic/Base Failure rate of component
$$\downarrow$$
$$\lambda_i = \lambda_{base} \times \pi_Q \times \pi_E \times \pi_A \times \pi_T \times \pi_S...$$

Quality Environment Application Temperature Stress ···factors

Reducing the number of variables to a few key ones for simplicity, we can write for total equipment failure rate:

Generic/Base Failure rate
$$\downarrow$$
$$\lambda = \sum_{i=1}^{n} \left(\lambda_{ref} \times \pi_U \times \pi_I \times \pi_T \right)_i$$

Voltage   Current   Temperature

As indicated, it was common practice to get a predicted MTBF for a power supply of the order of only 100k hours by Part Stress analysis, and around 150 k hours to 200 k hours by

Table 6.2: Chi-square Lookup Table.

Number of Failures	χ^2 at 60% CL	χ^2 at 90% CL
0	1.833	4.605
1	4.045	7.779
2	6.211	10.645
3	8.351	13.362
4	10.473	15.987
5	12.584	18.549

Part Count analysis. Neither of these predicted numbers came close to demonstrated/field reliability numbers — *which were typically 3× to 6× higher.*

There are several other reliability prediction methods in use today. A popular method is the Telcordia SR-332 standard (Bellcore became Telcordia in 1997). Siemens uses SN-29500, based on IEC61709. There is also a British Telecom standard going around. Though these are in fact quite similar to the Part Stress analysis method of MIL-HDBK-217F in philosophy, they end up with very different numbers (about 3× higher MTBF). One reason for that is some of the modern methods allow for fitting test and field data to the stress models. In effect, they don't rely on an outdated database, and therefore produce more realistic predictions than MIL-HDBK-217F.

Demonstrated Reliability Testing (DRT)

Commercial power supplies are often subjected to a pre-production DRT (Demonstrated Reliability Test). Several hundred power supplies are put in a chamber and operated at maximum load and rated operating temperature (or as stated). The number of failures occurring after a certain pre-determined time tells us with a certain "confidence level" how a much larger number of units placed in the field will perform in terms of reliability. The statistical formula for MTBF, conforming to MIL-STD-781D and MIL-HDBK-338B, is

$$\text{MTBF} = \frac{2 \times \text{POH}}{\chi^2(\alpha, 2f + 2)} \text{ hours}$$

where f is the number of failures, α is the "significance level," related to the "confidence level" by

$$\text{CL} = 100 \times (1 - \alpha) \%$$

We need to refer to the Chi-square (χ^2) numbers provided in Table 6.2.

Example:

How many POH are required to demonstrate an MTBF of 250k hours with 90% confidence (temperature specified as 55 °C)?

We should have 0 failures when operated for

$$\text{POH}_0 = \frac{\chi^2 \times \text{MTBF}}{2} = \frac{4.605 \times 250,000}{2} = 575,625 \text{ Units} \times \text{hours}$$

We should have only 1 failure when operated for

$$\text{POH}_1 = \frac{\chi^2 \times \text{MTBF}}{2} = \frac{7.779 \times 250,000}{2} = 972,375 \text{ Units} \times \text{hours}$$

Example:

How many units are required for a 4-week test to demonstrate an MTBF of 250 k hours at 60% confidence level?

At 60% confidence level, for one failure maximum, we need to accumulate

$$\text{POH}_1 = \frac{\chi^2 \times \text{MTBF}}{2} = \frac{4.045 \times 250,000}{2} = 505,600 \text{ Units} \times \text{hours}$$

In 4 weeks we have 672 h. Thus, for a 4-week testing period we need to test

$$\frac{505,600 \text{ (Units} \times \text{hours)}}{672 \text{ (hours)}} = 752 \text{ Units}$$

Note that these units will all need to be operated simultaneously at maximum load, or 80% of maximum load (as specified), and at maximum ambient of 55 °C (or as specified). Typically, some of these will be run at customer's location and some at the power supply manufacturer's location. We should have at most one failure up to this point. Once that failure is analyzed and a solution put in place, it is no longer counted as a "chargeable failure" (i.e., anything of an electrical nature that can occur in the field).

Accelerated Life Testing

If failure rate rises on account of temperature, why can't we subject a batch of power supplies to high-temperature testing, and in effect "fast-forward" through the bathtub curve of Figure 6.2 at high-speed, to accumulate data quickly on what will happen in the field when the equipment is operated at normal (lower) temperatures? Hopefully, we can then estimate both the life and the MTBF we will encounter in the field — provided we know how things *scale with temperature*. In other words, we need to know the (temperature) Acceleration Factor ("AF").

That leads us to Arrhenius' equation that describes the relationship between chemical reactions and temperature. Written in a form suitable for our purpose, it says the rate of a reaction varies as

$$\text{Rate} \propto e^{-E_A/kT}$$

where E_A is the activation energy in eV (electron Volts), k is the Boltzmann constant, equal to 8.617×10^{-5} eV/K (K is Kelvin), and T is the temperature in Kelvin. Arrhenius' equation treats every reaction/failure as the act of crossing an (empirical) energy barrier with a certain height (E_A in eV). When we heat up the molecules, more and more of them get agitated enough to jump that wall and the reaction proceeds faster (i.e., more failures). Arrhenius' equation is commonly used to estimate reliability and life.

Comparing the rate at T_1 (lower) to the rate at T_2 (higher), we get the Acceleration Factor as

$$\text{AF} = e^{(E_A/k)((1/T_1) - (1/T_2))}$$

and this is the factor by which we multiply the failure rate at the lower temperature to get the failure rate at the higher temperature.

It is commonly stated that the rule of thumb is that every 10 °C rise leads to a doubling of failure rate, and also a halving of life when talking about aluminum electrolytic capacitors. That corresponds to an Acceleration Factor of 2. If we work backwards to calculate the corresponding E_A, we will see that if we set $T_1 = 273 + 50$ and $T_2 = 273 + 60$ (50 °C to 60 °C), we get AF = 2 for $E_A = 0.65$ eV. *AF is a function of temperature*. If we go from 80 °C to 90 °C, for the same E_A, we get AF = 1.8, not 2.

The doubling/halving rule every 10 °C therefore implies an assumed E_A of about 0.65 eV. But activation energies encountered, in general, can typically vary anywhere between 0.3 eV and 1.2 eV. If E_A is 0.3 eV for a particular failure mechanism, then in going from 50 °C to 60 °C, its Acceleration Factor is only 1.4, and that drops to 1.3 from 80 °C to 90 °C.

In performing actual reliability tests, there are two main categories:

(a) Accelerated Life Testing (ALT): Though the word used here is life, this test involves raising the temperature and using Acceleration Factors to predict both the life and the MTBF at lower temperatures. This has to be done cautiously so we do not introduce new failure modes that would not have been present at normal lower temperatures.
(b) Accelerated Stress Testing (AST): The purpose of this test is usually not intended to be predictive in nature. Here, we just try to precipitate failures by increasing stresses with the intention of uncovering basic weaknesses.

Under the sub-category of AST, we can perform the following tests:

(1) Highly Accelerated Life Test (HALT): This is a development tool performed on equipment (e.g., power supply). Its purpose is to identify weaknesses during the design

phase itself, so improvements can be carried out, cost permitting. It is also called STRIFE (for Stress and Life Test).

(2) Highly Accelerated Stress Screen (HASS): This is a production screen for manufactured equipment. Samples from production can be subjected to very high stresses for a short time to uncover weaknesses in the manufacturing process (or in design).

(3) Highly Accelerated Stress Test (HAST): This is a component-level test. Samples are subjected to extreme environmental stresses (temperature, pressure, and/or humidity) to uncover weaknesses. Semiconductors are commonly subjected to this during qualification.

In doing accelerated testing we have to be very careful we are only accelerating known failure modes, *not creating new ones*. In other words, our hope of doing a "fast-forward" cannot be as fast as we had perhaps hoped for.

With that, we have completed our discussion on reliability and stresses, and we now discuss life prediction of aluminum caps.

Part 3: Life Prediction of Aluminum Electrolytic Capacitors

Aluminum electrolytic capacitors, with all their advantages, are subject to slow deterioration (aging) caused by gradual evaporation of the electrolyte inside. That process is accelerated by heat generated inside the capacitor (due to self-heating) and around it too (from nearby hot components). The internal heat is the RMS current squared (calculated in *Chapter 7*) multiplied by the ESR of the capacitor. Though the capacitor is supposedly sealed to prevent escape of the electrolyte, no joint is 100% hermetic. Therefore, aging is a slow but inevitable process, and eventually determines the useful life of the capacitor. But we also have to be clear how we define "useful life." For example, we can always *choose* to declare the retirement age of an individual as 55 years, or 60 years, or even 65 years. But we have to ask what really makes sense, in terms of performance, for capacitors as for humans?

Looking at capacitor datasheets we see that a typical aluminum electrolytic capacitor has a declared life ("L_O") of 2,000 hours−5,000 hours (recently, Panasonic for example announced 105 °C caps with 10,000 hours life). Since there are $365 \times 24 = 8,760$ hours in one year, that would make any equipment containing such capacitors last less than a year. Clearly that is not enough. Typical power supply specifications call for a minimum life of 44k hours (5 years) when operated at 40 °C ambient (assuming 24-hour operation).

Now we look at the fine print of the capacitor datasheet. It says the end of useful life of an aluminum electrolytic cap is defined as a certain percentage decrease in capacitance (typically 20%) and/or a certain percentage increase in its "Dissipation Factor" (typically 200% or 300% of initial value). It clarifies that the declared useful life (2,000 hours or 5,000 hours, for example), is obtained by operating the capacitor at its maximum rated temperature (typically, the upper rated category temperature, U_R, which is 85 °C or 105 °C), and by passing a low-frequency current (typically, 120 Hz) with an RMS value set equal to its declared Ripple Current Rating I_R.

First, let us be clear what Dissipation Factor is. By definition, DF (or tan δ), is related to the ESR by

$$\text{ESR} = \frac{\tan \delta}{2\pi f \times C}$$

or

$$\tan \delta = \frac{\text{ESR}}{X_C}$$

where

$$X_C = \frac{1}{2\pi f \times C} \equiv \frac{1}{C\omega} \quad \text{(ignoring phase)}$$

So, Dissipation Factor is the ratio of the real part of the impedance (resistance, i.e., ESR) to the Reactive Impedance $(1/C\omega)$. It is therefore a measure of how *bad* a cap is, that is, its loss term versus its energy storage capability (measured at 120 Hz). It is the reciprocal of the quality factor Q. So a high DF indicates a bad cap in terms of a high ESR. Note that though electrolytic capacitors have much higher ESR than ceramic or film caps, they thankfully have much higher Capacitance/Volume.

If capacitance falls by 20% *and* DF goes up 200%, as per worst-case end-of-life limits, the ESR would have gone up 2.5 times

$$\text{ESR} \propto \frac{\tan \delta}{C} \Rightarrow \frac{200\%}{80\%} = 250\%$$

Further, though capacitance goes down from its initial value by only 20% by end-of-life, we have to keep in mind that its initial (starting) value itself may have been 10% or 20% lower than the nominal value, on account of typical capacitor tolerances. We may therefore have to keep the worst-case end-of-life capacitance value in mind upfront while selecting aluminum electrolytic capacitors, especially if capacitance is a key parameter/requirement in our selection process. But certainly, the ESR goes up significantly as the electrolyte evaporates. Further, after the end-of-life threshold, there is a rather *rapid increase* in ESR. And since higher ESR causes more heating, more heating causes more wearout (and therefore higher ESR). There can be a runaway condition thus established that could mark the end-of-life of the converter too, shortly. So, aging is a real concern when using aluminum electrolytics, and lifetime estimates are therefore important. After all, the life of the equipment is determined by the component with the shortest life. That can be the aluminum electrolytic unfortunately.

In Figure 6.3, we present a simplified thermal model of an aluminum electrolytic cap. The ripple current rating I_R is so chosen, that when I_R is passed through the cap, it produces a certain *optimum delta* between ambient and case, and also from case to the core (deep inside the capacitor). Further, these two deltas, that is, from ambient to case and from case to core, are assumed to be the same (usually). That optimum delta is set to 5 °C for 105 °C capacitors and 10 °C for 85 °C capacitors. So, if we place a 105 °C rated cap at an ambient temperature of 105 °C, and pass I_R through it, its core will be at exactly 115 °C. In other words, the stated life of the cap (L_O), 5,000 hours for example, is the life of the cap with its core held at 115 °C. Similarly, if we pick an 85 °C cap, its datasheet life figure L_O is at a core temperature of 105 °C. If we lower the ambient by 20 °C, the core temperatures will fall to $115 - 20 = 95$ °C and $105 - 20 = 85$ °C, respectively. This leads to a significant increase in life since the evaporation of the electrolyte slows down. The well-known lifetime doubling rule is that every 10 °C fall in core temperature leads to a doubling of life.

One problem is: as users, we have no way of measuring core temperature to estimate life accurately. So we need to turn to the vendor for guidance. Most vendors agree to the following simplified, *user-friendly formula* to estimate life (L). This is

$$L = L_O \times 2^{(U_R - T_{AMB})/10} \times 2^{-\Delta T_{excess}/5}$$

where

$$\Delta T_{excess} = \Delta T_{CORE-CASE} - \Delta T_{rated}$$

ΔT_{rated} is the optimum differential we talked about previously (5 °C for 105 °C caps). $\Delta T_{CORE-CASE}$ is the actual delta in the application (case to ambient as well as core to case).

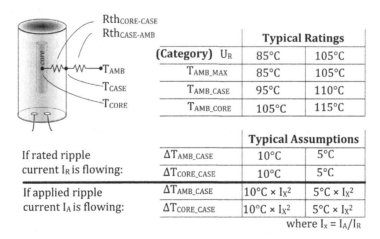

		Typical Ratings	
(Category) U_R		85°C	105°C
T_{AMB_MAX}		85°C	105°C
T_{AMB_CASE}		95°C	110°C
T_{AMB_CORE}		105°C	115°C

		Typical Assumptions	
If rated ripple current I_R is flowing:	ΔT_{AMB_CASE}	10°C	5°C
	ΔT_{CORE_CASE}	10°C	5°C
If applied ripple current I_A is flowing:	ΔT_{AMB_CASE}	10°C × I_x^2	5°C × I_x^2
	ΔT_{CORE_CASE}	10°C × I_x^2	5°C × I_x^2

where $I_x = I_A/I_R$

Figure 6.3: The capacitor model and ratings for understanding lifetime prediction.

The above formula tells us that whereas a fall in ambient temperature doubles life every 10 °C, if the delta is more than the optimum (ΔT_{rated}), then the life *halves* every 5 °C by which the delta exceeds the optimum delta. Since heating is proportional to RMS-squared, the actual delta in the application can be estimated by

$$\Delta T_{\text{CORE-CASE}} = \Delta T_{\text{rated}} \times \left(\frac{I_A}{I_R}\right)^2$$

Here, I_A is the applied RMS and I_R the rated RMS. Also, since there may be local heating from adjacent components, the T_{AMB} to be used in the above equation is taken to be T_{CASE}, and that provides roughly a 5 °C safety margin when using 105 °C capacitors. Also, for scaling purposes, it is nice to talk in terms of ratios and multipliers. So finally, a usable formula is

$$L_X = 2^{(U_R - T_{\text{CASE}})/10} \times 2^{-\Delta T_{\text{rated}}(I_X^2 - 1)/5}$$

where L_X is the lifetime multiplier, and $I_X = I_A/I_R$.

This is the formula used in the Mathcad worksheet in Figure 6.4 to plot the accompanying curve (plotted out for 105 °C caps).

To use the curves, keep in mind that life increases toward the left, and decreases towards the right. So, for example, if we have a 5,000 hours/105 °C cap at an ambient of 55 °C, and we pass 1.5 times the rated ripple current, we expect the life to *exceed* $12 \times 2,000 = 60,000$ hours. This is $60,000/8,760 = 6.85$ years. A typical requirement for most equipment is only 5 years.

Note that Chemicon is considered more conservative than most vendors. Naturally, it places one more restriction: though the cap life halves every 5 °C by which we exceed the rated ΔT, we *cannot* expect that if we reduce the delta to less than the rated ΔT, that the sign will flip in the equation above, and we can thus claim (or expect) higher life. Chemicon refuses to buy into that! That is why in Figure 6.3 we have a distinct Chemicon boundary in solid black/gray. The dashed gray lines (or something close to these) are published by other vendors, and you can use those curves at your own risk.

A numerical example is provided below.

Example:

We are using a 2,200 µF/10 V capacitor from Chemicon. Its catalog specifications are 8,000 hours at maximum rated 1.69 A, stated at 105 °C and 100 kHz. The measured case temperature in our application is 84 °C and the measured ripple current is 2.2 A. What is the expected life?

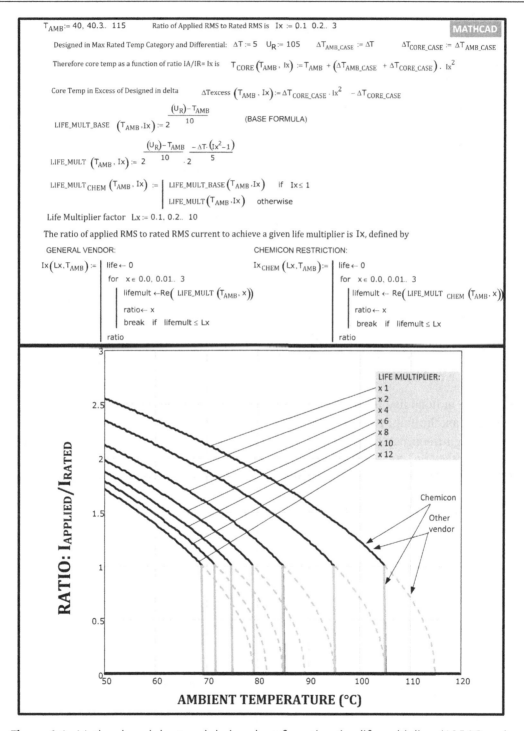

$T_{AMB} := 40, 40.3.. 115$ Ratio of Applied RMS to Rated RMS is $Ix := 0.1 \; 0.2.. \; 3$ **MATHCAD**

Designed in Max Rated Temp Category and Differential: $\Delta T := 5$ $U_R := 105$ $\Delta T_{AMB\_CASE} := \Delta T$ $\Delta T_{CORE\_CASE} := \Delta T_{AMB\_CASE}$

Therefore core temp as a function of ratio IA/IR= Ix is $T_{CORE}\left(T_{AMB}, Ix\right) := T_{AMB} + \left(\Delta T_{AMB\_CASE} + \Delta T_{CORE\_CASE}\right) \cdot Ix^2$

Core Temp in Excess of Designed in delta $\Delta Texcess \left(T_{AMB} \cdot Ix\right) := \Delta T_{CORE\_CASE} \cdot Ix^2 - \Delta T_{CORE\_CASE}$

$\text{LIFE\_MULT\_BASE} \; \left(T_{AMB}, Ix\right) := 2^{\frac{\left(U_R\right) - T_{AMB}}{10}}$ (BASE FORMULA)

$\text{LIFE\_MULT} \left(T_{AMB}, Ix\right) := 2^{\frac{\left(U_R\right) - T_{AMB}}{10}} \cdot 2^{\frac{-\Delta T \cdot \left(Ix^2 - 1\right)}{5}}$

$\text{LIFE\_MULT}_{CHEM} \left(T_{AMB} \cdot Ix\right) := \begin{vmatrix} \text{LIFE\_MULT\_BASE} \left(T_{AMB}, Ix\right) & \text{if } Ix \leq 1 \\ \text{LIFE\_MULT}\left(T_{AMB}, Ix\right) & \text{otherwise} \end{vmatrix}$

Life Multiplier factor $Lx := 0.1, 0.2.. \; 10$

The ratio of applied RMS to rated RMS current to achieve a given life multiplier is Ix, defined by

GENERAL VENDOR:

$Ix\left(Lx, T_{AMB}\right) := \begin{vmatrix} \text{life} \leftarrow 0 \\ \text{for } x \in 0.0, 0.01.. \; 3 \\ \quad \begin{vmatrix} \text{lifemult} \leftarrow Re\left(\text{LIFE\_MULT}\left(T_{AMB}, x\right)\right) \\ \text{ratio} \leftarrow x \\ \text{break if lifemult} \leq Lx \end{vmatrix} \\ \text{ratio} \end{vmatrix}$

CHEMICON RESTRICTION:

$Ix_{CHEM}\left(Lx, T_{AMB}\right) := \begin{vmatrix} \text{life} \leftarrow 0 \\ \text{for } x \in 0.0, 0.01.. \; 3 \\ \quad \begin{vmatrix} \text{lifemult} \leftarrow Re\left(\text{LIFE\_MULT}_{CHEM}\left(T_{AMB}, x\right)\right) \\ \text{ratio} \leftarrow x \\ \text{break if lifemult} \leq Lx \end{vmatrix} \\ \text{ratio} \end{vmatrix}$

Figure 6.4: Mathcad worksheet and design chart for estimating life multipliers (105 °C cap).

$$L = L_0 \times 2^{(105-84)/10} \times \overbrace{2^{(5-\Delta T)/5}} \quad \text{hours}$$

where

$$\Delta T = 5 \times \left(\frac{2.2}{1.69}\right)^2 = 8.473 \,^\circ\text{C}$$

Therefore, since $(8.473 - 5)/5 = 0.695$, we get

$$L = 8,000 \times 2^{(105-84)/10} \times \overbrace{2^{-0.695}} = 21,190 \quad \text{hours}$$

Since the internal delta is 8.473 °C, that is, greater than 5 °C, the above estimate is valid for capacitors from most other vendors too.

One question often asked is can we use forced air cooling to enhance the life of a capacitor? Some vendors provide guidelines how to estimate life under these conditions, but some staunchly resist publishing any such life estimates, or at least standing firmly behind them. In effect, most vendors say you can use life predictions under forced air flow at your own risk. The reason is it is very hard to predict the core temperature under such variable conditions. The actual air flow itself is hard to know accurately in the vicinity of the capacitor on a real board.

Lastly, keep in mind that the declared ripple current rating is usually stated at 120 Hz. Since ESR (and heating) reduces with frequency, vendors provide frequency multipliers. A typical high-frequency multiplier is 1.43 at 100 kHz. So, for example, if the capacitor is rated 1.5 A RMS at 120 Hz, its ripple current rating (in modern switching power supplies) is $1.5 \times 1.43 = 2.145$ A. That is the number we should use for I_R in the calculations above. There is no other change.

CHAPTER 7

Optimal Power Components Selection

Overview

In *Chapter 6*, we reviewed basic reliability and stress derating principles with emphasis on understanding datasheets and power component ratings. Our focus was on the *strength* aspect of the "matchmaking" process as we called it. In this chapter we turn our attention to the *stresses* aspect of that process. We thus hope to create viable and reliable component choices.

When designing and evaluating wide-input converters, we need to clearly identify the specific input voltage point at which a given stress reaches its maximum. Thereby we can deduce the worst-case operating stresses in the converter over its *entire* range of operation. That analysis will lead us to what we call the "Stress Spider" later in this chapter.

We can ask: is component selection only about matching strength (of a part) to (the applied) stress? No. That may in fact turn out to be only a *pre-requisite*. In *Chapter 13*, we will see how, for example, the *input and output ripple requirements* can eventually dictate the choice of the power components. In *Chapter 9*, we will discuss hysteretic Buck regulators for example, which depend on sufficient output voltage ripple to operate satisfactorily. In other words, various factors can affect final component choices. This chapter will help us at least shortlist eligible candidates.

The Key Stresses in Power Converters

Voltage stress is not only the most important, but also relatively the simplest to calculate and tackle. We know that if the voltage on a power semiconductor even momentarily exceeds its published Absolute Maximum (Abs Max) rating, it will very likely be destroyed almost immediately. In comparison, the current rating is usually not of such immediate concern, since it is typically (not always) *thermal* in nature, and is therefore comparatively slower-acting. In general, we can often inadvertently, or even sometimes with judicious deliberation, exceed the current ratings of a semiconductor somewhat, then "back off" quickly, without much impact. But note that this "forgiveness factor" should not be taken for granted. For example, if *core saturation* starts to occur, *destruction can be almost immediate* as shown in Figure 2.7.

Switching Power Supplies A–Z. DOI: 10.1016/B978-0-12-386533-5.00007-3

We will now summarize our key concerns.

(A) **Voltage stresses** inside a converter are invariably at their highest when the input voltage is raised to its maximum value (V_{INMAX}). We may need to carefully decipher what specific *load conditions* correspond to the highest voltage stresses, and what those stresses are. For example, in a flyback, the leakage inductance voltage spike will be worse at high loads — the spike will be higher despite the zener clamp (zener voltage depends on current), and also of longer duration (width of spike, i.e., its residual energy). In a Forward converter, the reverse voltage on the output diode may be the highest at light loads. And so on.

Summarizing: the primary voltage stress we need to worry about is
- V_{PK}, the *peak voltage (on any component)*, since that represents the maximum *instantaneous* voltage. We know from *Chapter 6* that too high a voltage can cause semiconductor junctions to (instantaneously) avalanche and/or snapback.

(B) **Current stresses** inside a converter are invariably at their highest when the load is raised to its maximum rated value (I_O). We need to carefully decipher what specific *input voltage conditions* correspond to the highest current stresses, and exactly what those stresses are. These calculations can become very complicated.

Summarizing: the primary current stresses we need to worry about are
- I_{RMS}, the root mean square (RMS) current, since that determines the conduction losses in MOSFETs (explained further below).
- I_{AVG}, the average current, since that determines the conduction losses in diodes (explained further below).
- I_{PK}, the peak current, since that can instantaneously cause core saturation and consequent destruction of the MOSFET. Remember that in any basic (inductor-based) DC–DC topology, the peak currents in the switch, diode, and inductor are all the same.

Waveforms and Peak Voltage Stresses for Different Topologies

We first look at voltage stresses. In Figures 7.1 and 7.2, we have started by plotting the waveforms *at* the switching nodes of the main conventional topologies (marked "A" or "B"). The switching node is crucial since it is a node that is a point common to both the switch and diode. Thereafter, we take the difference of the voltages on either side of the component under consideration (switch or diode), and thereby calculate the voltage *across* it. That difference voltage can be generally expressed by an equation of the form "$V_x - V_y$." Note that looking at any specific plot in the two figures, as we move along the x-axis (time increasing), in effect, a new phase of operation starts if either of the two voltages on either side of the component (V_x or V_y, or both) change. So, for each distinct phase of operation we then need to re-evaluate the difference voltage "$V_x - V_y$'." In this manner, we generate the full voltage waveform across the component over the entire switching cycle.

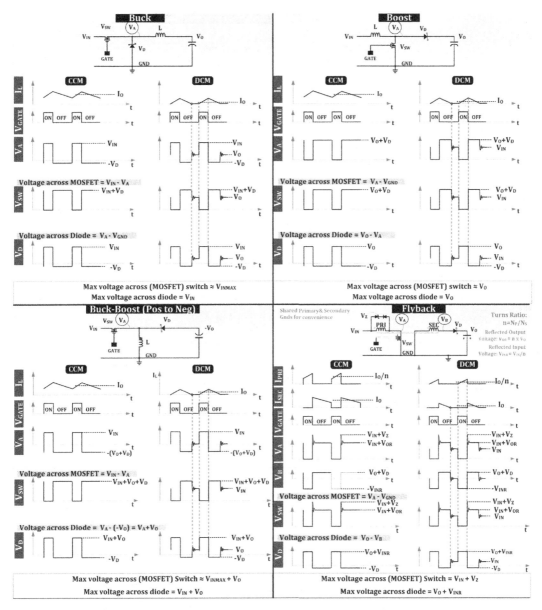

Figure 7.1: Voltage waveforms for the Buck, Boost, Buck-Boost, and Flyback.

We can thus clearly identify the worst-case voltage stress across the component (diode or switch). Note that we have repeated the above-mentioned process for *both* "continuous conduction mode" (CCM) and "discontinuous conduction mode' (DCM) (heavy and light loads), to ensure we are really discovering the worst-case voltage stress over the entire application conditions.

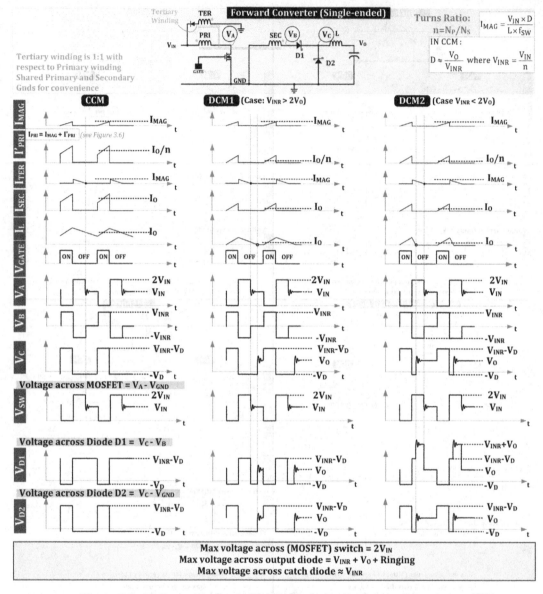

Figure 7.2: Voltage waveforms for the Single-Ended Forward Converter.

Note that we have not bothered to report any capacitor voltage stresses in the figures, because that is quite self-evident in all topologies. The simple rule is that the input capacitor must be rated for at least V_{INMAX}, whereas the output capacitor should be rated for at least V_O. There is nothing complicated there, though we may need to apply appropriate stress derating as discussed in *Chapter 6*.

We have consolidated our findings for the diode and switch in an easy lookup chart: Table 7.1. Note that in this chart, there are several "new" topologies tabulated just for completeness sake. Many of these, but not all, are discussed in more detail in *Chapter 9*.

- We notice that on the extreme right-hand side of Figure 7.2 (the Forward converter figure), we have a special case of DCM where the inductor (output choke) reaches zero current and stays there (i.e., gets de-energized) *before* the transformer de-energizes. Admittedly, that situation can happen only under some rather unusual circumstances, like excessive transformer step-down ratios combined with light loads. But, if and when that happens, the voltage stress across the output diode D1 is not just V_{INR}, as is usually stated in literature, but $V_{INR} + V_O$. The reason is the switching node on the secondary side (cathode of the diode) jumps up to V_O since the output choke "dries up" (gets de-energized), while the anode of the diode is still being dragged low by the transformer winding — down to $-V_{IN}/n$ (i.e., $-V_{INR}$), where n is the turns ratio N_P/N_S (n being much larger than 1 typically). That is the reason why we have written out the peak stresses as mentioned in Table 7.1 for the Forward converter.
- For an *active clamp* Forward converter, the high-stress situation for the output diode described above can happen much more readily, *since the transformer is always in CCM*, but the choke can go into DCM at light loads.

Note that for finding the peak voltage stresses, we are generally very interested in *transient* cases too. For example, we can hit D_{MAX} under a sudden load/line change, even though we are operating at V_{INMAX}, with a much lower steady-state duty cycle just prior to that sudden event. That explains the peak stresses reported in Table 7.1 for the active clamp Forward converter.

In an active clamp Forward converter, there is an additional MOSFET (clamp) as shown in Table 7.1. Functionally, it takes the place of the energy recovery diode in the conventional (single-ended) Forward converter. This topology and its voltage stress ratings, as presented in Table 7.1, are further discussed and derived in *Chapter 9*.

> *Note: The energy-recovery diode is in series with the tertiary (energy recovery) winding and its purpose is to ensure the transformer "resets" every cycle. In general, "reset" in any given magnetic component does not mean a return to zero current (or zero "net Ampere-turns" when talking about transformers) every cycle. As in the active clamp Forward converter transformer, reset just means the net Ampere-turns at the end of the switching cycle returns to exactly the same value it started the switching cycle off with. Intuitively, reset is simply a prerequisite for us to be able to label the magnetic component and thereby the converter as operating in a "steady (repetitive) state."*

> *Note: The tertiary winding diode can theoretically be placed either between the tertiary winding and the upper rail (as shown in Figure 7.2), or between the winding and ground as shown in Table 7.1. However, there is a preferred position as described in Chapter 9.*

Table 7.1:   Peak Voltage Stresses for Several Key Topologies.

$n = N_P/N_S$ $V_{INR} = V_{IN}/n$ $V_{OR} = nV_O$	Switch	Catch Diode	Output Diode	Coupling/ Clamp Cap	Ideal Transfer Function
Buck	V_{INMAX}	V_{INMAX}		NA	$\dfrac{V_O}{V_{IN}} = D$
Boost	V_O	V_O		NA	$\dfrac{V_O}{V_{IN}} = \dfrac{1}{1-D}$
Buck-Boost	$V_{INMAX} + V_O$	$V_{INMAX} + V_O$		NA	$\dfrac{V_O}{V_{IN}} = \dfrac{D}{1-D}$
Flyback	$V_{INMAX} + V_Z$	$V_{INRMAX} + V_O$		NA	$\dfrac{V_O}{V_{INR}} = \dfrac{D}{1-D}$
Forward	$2 \times V_{INMAX}$	V_{INRMAX}	$V_{INRMAX} + V_O$	NA	$\dfrac{V_O}{V_{INR}} = D$
2-switch Forward	V_{INMAX}	V_{INRMAX}	$V_{INRMAX} + V_O$	NA	$\dfrac{V_O}{V_{INR}} = D$
Active Clamp	$\dfrac{V_{INMAX}}{1 - D_{MAX}}$	V_{INRMAX}	$V_{INRMAX} \times \dfrac{D_{MAX}}{1 - D_{MAX}}$ $+ V_O$	$\dfrac{V_{IN} D_{MAX}}{1 - D_{MAX}}$	$\dfrac{V_O}{V_{INR}} = D$
Half Bridge	V_{INMAX}	V_{INRMAX}	V_{INRMAX}	NA	$\dfrac{V_O}{V_{INR}} = D$
Full Bridge	V_{INMAX}	$2 \times V_{INRMAX}$	$2 \times V_{INRMAX}$	NA	$\dfrac{V_O}{V_{INR}} = 2D$
Push-Pull	$2 \times V_{INMAX}$	$2 \times V_{INRMAX}$	$2 \times V_{INRMAX}$	NA	$\dfrac{V_O}{V_{INR}} = 2D$
Cuk	$V_{INMAX} + V_O$	$V_{INRMAX} + V_O$		$V_{INMAX} + V_O$	$\dfrac{V_O}{V_{IN}} = \dfrac{D}{1-D}$

(Continued)

Table 7.1:   (Continued)

$n = N_P/N_S$ $V_{INR} = V_{IN}/n$ $V_{OR} = nV_O$		Switch	Catch Diode	Output Diode	Coupling/ Clamp Cap	Ideal Transfer Function
Sepic		V_{INMAX} $+ V_O$		$V_{INMAX} + V_O$	V_{INMAX}	$\dfrac{V_O}{V_{IN}} = \dfrac{D}{1-D}$
Zeta		V_{INMAX} $+ V_O$		$V_{INMAX} + V_O$	V_O	$\dfrac{V_O}{V_{IN}} = \dfrac{D}{1-D}$

V_O is the magnitude of the output voltage

$V_{INRMAX} = \dfrac{V_{INMAX}}{n}$ (maximum reflected input voltage)

$n = \dfrac{N_P}{N_S}$ (turns ratio)

Under some conditions, additional stress may get applied on the output diode as indicated by gray lettering above.

All voltage above are magnitudes only

- The voltage ratings of any additional components of any Forward converter based topology mentioned above *are the same* as their corresponding switch voltage ratings. So, if in a universal input Forward converter, the main switch is rated > 800 V, the energy recovery diode (or the active clamp switch) must be rated > 800 V too.
- What if, instead of the single-ended Forward converter, we use a two-switch Forward converter (also called an "asymmetric half-bridge")? This has two switches on either side of the Primary winding, driven in unison (in phase). There is no tertiary winding present to add any reflected voltage on top of the input voltage rail. Therefore, in this case, both the switches need to be rated only for V_{INMAX}, not twice that as in the single-ended Forward. On the Secondary side, all the voltages are unchanged as compared to the single-ended Forward converter.
- Note that for a Buck-Boost, the diode and switch can see a maximum of $V_{INMAX} + V_O$, and must be rated accordingly (remember that we always use *magnitudes* of voltages in this book, and introduce signs only when necessary).
- The transformer-isolated version of the Buck-Boost, that is, the flyback, has an additional spike due to leakage inductance riding on top of the level $V_{INMAX} + V_{OR}$. The spike is limited by a zener of voltage V_Z referenced to V_{IN} rail. So, the maximum stress is $V_{INMAX} + V_Z$. See Figure 3.1 for more details.
- As we will learn in *Chapter 9*, the Sepic, Cuk, and Zeta are composite topologies based on the Buck-Boost. The switch/diode voltage ratings are therefore exactly the same as for a Buck-Boost. The additional power component in these composite topologies is the coupling capacitor. The voltage on that can vary from one topology to another. It is V_{INMAX} for Sepic, V_O for Zeta, and $V_{INMAX} + V_O$ for Cuk.

In the next section, we start to look at the current stresses of the power components.

The Importance of RMS and Average Currents

Work (or energy) is done when charged particles (electrons) are transported across a potential barrier (voltage). We have the basic equation for energy as $\varepsilon = V \times Q$, where Q is the amount of charge and V is the potential difference. Power (or dissipation) is, by definition, energy per second. So, we get $\varepsilon/t \equiv P = V \times Q/t = V \times I$. The last step follows because charge per second is, by definition, current. Therefore, we also get the definition of *instantaneous* dissipation as $P(t) = V(t) \times I(t)$. This is a function of time, and can change from moment to moment. However, for a repetitive waveform, we can find its average value over one cycle. In steady state, that average value remains constant. By definition it is

$$P = \frac{\int_0^T V(t) \times I(t)\, dt}{T}$$

where $T = 1/f_{SW}$ and f_{SW} is the switching frequency.

This can be further simplified based on one of the following:

(a) **V is constant:**

Assuming V is a constant (as for a diode in forward conduction)

$$P_D = V_D \times \frac{\int_0^T I(t)\, dt}{T} \equiv V_D \times I_{D\_AVG}$$

where I_{D_AVG} is the average diode current.

(b) **R is a constant:**

Assuming an equivalent resistance (as for a MOSFET in full conduction)

$$P_{SW} = \frac{\int_0^T V(t) \times I(t)\, dt}{T} = R_{DS} \times \frac{\int_0^T I(t) \times I(t)\, dt}{T}$$

so

$$P_{SW} = R_{DS} \times \frac{\int_0^T I^2(t)\, dt}{T} \equiv R_{DS} \times I^2_{SW\_RMS}$$

where I_{SW_RMS} is the RMS switch (in this case MOSFET) current.

That is why it is customary to calculate and use the average current to find the dissipation for a diode and the RMS current for the dissipation in a MOSFET.

Note that there is also a $V \times I$ (crossover loss) term occurring during every switching transition that we have neglected above in computing the dissipation. In effect, what we have calculated above is just *conduction loss*. Note that for a switch we *always* need to add a switching loss term to find the total dissipation (see *Chapter 8*). But in a diode (or synchronous FET), we usually consider its switching loss term as negligible. Note, however, that the catch diode's characteristics (slow reverse recovery) might cause significantly higher crossover losses in the switch if not in the diode itself. On the other hand, Schottky diodes, though near-ideal diodes in that sense, can have significant reverse leakage dissipation term that we need to add to the total dissipation in the diode, as discussed in *Chapter 6*.

Similarly, for a capacitor, especially an electrolytic type, we need to ensure we do not exceed its published RMS (ripple) current rating, otherwise it will have a very short life indeed as also discussed in *Chapter 6*. We need to know its I_{RMS} accurately.

At this juncture we need to gain some mastery over actually calculating RMS and average values for different topologies. We will also derive some of the key equations that appear in the Appendix of this book.

Note that several numerical examples based on this chapter, are provided in *Chapter 19*.

Calculation of RMS and Average Currents for Diode, FET, and Inductor

In Figure 7.3, we have provided the procedure for calculating the average/RMS currents of the switch, diode, and inductor, via "brute-force" integration techniques first. In general, we get

$$I^2_{RMS\_SW} = \frac{I_2^2 + I_1^2 + I_2 I_1}{3} \times D, \qquad I^2_{RMS\_D} = \frac{I_1^2 + I_2^2 + I_1 I_2}{3} \times D', \qquad I^2_{RMS\_L} = \frac{I_2^2 + I_1^2 + I_2 I_1}{3}$$

In Figure 7.4, we have provided an easy lookup formula that basically bypasses the above brute-force integration technique going forward. It provides the same results as above. The only restriction on using the simple method is that the waveform, however arbitrary, must be a combination of piecewise *linear segments*. The general rule to use is as per Figure 7.4

$$I^2_{RMS} = \frac{I_2^2 + I_1^2 + I_2 I_1}{3}(\delta_1) + \frac{I_3^2 + I_2^2 + I_3 I_2}{3}(\delta_2) + \frac{I_4^2 + I_3^2 + I_4 I_3}{3}(\delta_3) + \frac{I_5^2 + I_4^2 + I_5 I_4}{3}(\delta_4) + \cdots$$

and

$$I_{AVG} = \frac{I_2 + I_1}{2}(\delta_1) + \frac{I_3 + I_2}{2}(\delta_2) + \frac{I_4 + I_3}{2}(\delta_3) + \frac{I_5 + I_4}{2}(\delta_4) + \cdots$$

290   **Chapter 7**

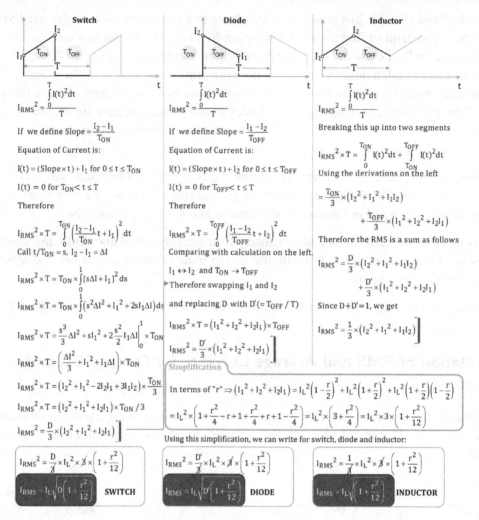

Figure 7.3: Integration method to derive RMS currents for MOSFET (Switch), diode, and inductor (any topology).

Applying the above techniques to the typical current waveforms of a power supply, we can also express the RMS in terms of the current ripple ratio r. This is shown in the embedded derivation in Figure 7.3. We thus get

$$I_{RMS\_SW}^2 = I_L^2 \times D\left(1 + \frac{r^2}{12}\right), \qquad I_{RMS\_D}^2 = I_L^2 \times D'\left(1 + \frac{r^2}{12}\right), \qquad I_{RMS\_L}^2 = I_L^2 \times \left(1 + \frac{r^2}{12}\right)$$

where I_L is the average inductor current (center of ramp). Keep in mind that as shown in *Chapter 1*, the center of ramp (I_L above) varies for different topologies.

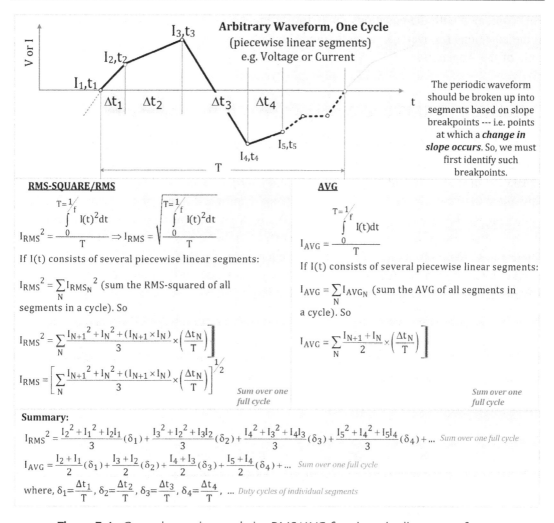

Figure 7.4: General equation to derive RMS/AVG for piecewise linear waveforms.

$$I_{L\_Buck} = I_O, \qquad I_{L\_Boost} = \frac{I_O}{1-D}, \qquad I_{L\_Buck\text{-}Boost} = \frac{I_O}{1-D}$$

We thus get the equations for RMS currents (diode, switch, and inductor) as provided in the design table of the Appendix of this book.

Similarly, for average currents, the calculation is almost self-evident, but we can consult Figure 7.4 if in doubt. We get in general (calling $D' = 1-D$)

$$I_{AVG\_SW} = I_L \times D, \qquad I_{AVG\_D} = I_L \times D', \qquad I_{AVG\_L} = I_L$$

This combined with the equations for I_L for different topologies, as provided above, leads to the equations for average currents (diode, switch, and inductor) provided in the design table of the Appendix.

Calculation of RMS and Average Currents for Capacitors

Now we discuss a key rule that helps us calculate capacitor currents. First, it should be intuitively obvious that if we take a waveform, and, *without changing its basic shape*, translate it *horizontally* (sideways), *its RMS and average values will be unchanged*. That is equivalent to changing the "0" of the *x*-axis (time), and we did just that in Figure 7.3 while evaluating the RMS of the diode waveform. We realized that for repetitive events, nature (expressed as observed heat and heat-related effects in our case) cannot possibly depend on where *we*, as mere observers, choose to start counting time, that is, where *we* decide to put a stake in the ground labeled "$t = 0$." So, *horizontal* translation ("*x*-translation") of a waveform cannot affect any results provided so far.

But what happens if we move the waveform *vertically*? Certainly, the RMS and average values will be affected. But we ask: is there any simplifying relationship, or rule, that we can identify under this "*y*-translation," and perhaps exploit in future? There is one such rule. In Figure 7.5, we numerically validate a fundamental property of vertically translated waveforms (using the general RMS calculation method of Figure 7.4). It is

$$I^2_{\text{RMS}} - I^2_{\text{AVG}} = \text{constant} \equiv I^2_{AC\_RMS}$$

This is the "AC RMS" of a waveform: it is the RMS of *only the AC part* of any given waveform. The waveform is devoid of any DC value. In other words AC RMS is the RMS of the waveform with its DC value set to zero. But why are we so interested in this AC RMS term anyway? Because a capacitor does exactly that to any current waveform. If we put a current probe in series with any cap in steady state, we will see that the DC value of that waveform is zero. But there is an AC portion to it, whose RMS value is the AC RMS mentioned above. But what happens to the DC value that *didn't* pass through the cap? It just passes it by. In effect, what the cap does to any current waveform is it *subtracts the* DC *current from the applied waveform*, retaining only the AC current part of it, and letting the rest (DC) pass. This is the *current* analog of the more familiar *voltage* expression we use when talking about caps in general: as it is commonly said that a series cap bypasses the applied AC voltage (i.e., lets that pass through), but blocks the DC voltage component across itself.

Alternatively stated, the average value of the current waveform through a capacitor *in steady state* is zero over a complete switching cycle. Otherwise it would continue to charge/ discharge a little every cycle, and that would therefore *not* be considered a steady state.

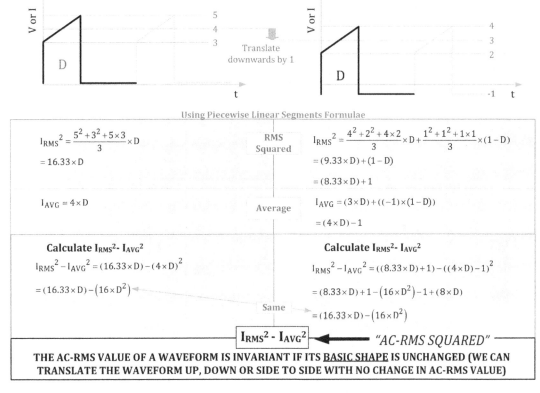

Figure 7.5: AC RMS of a waveform is invariant under translation.

This is a completely equivalent statement to the fact that the average voltage, rather the *voltseconds*, impressed across an inductor over a complete cycle is zero in steady state. Now we see that the average charge ($I \times t$ or Ampere-seconds) is similarly zero for a cap over a full cycle (in steady state).

With this background in mind, in Figure 7.6 we show how we take the "associated current waveform," remove its DC value and come up with the AC RMS (i.e., the capacitor current RMS value). Refer also to Figures 7.7−7.9 that graphically show all the current stresses in the three major topologies. We discuss that in the next section.

Some easy rules-of-thumb for selecting capacitors for a Buck converter are as follows.

We have for the input cap (of a Buck)

$$I_{CAP\_IN\_RMS} = I_O \sqrt{D\left(1 - D + \frac{r^2}{12}\right)}$$

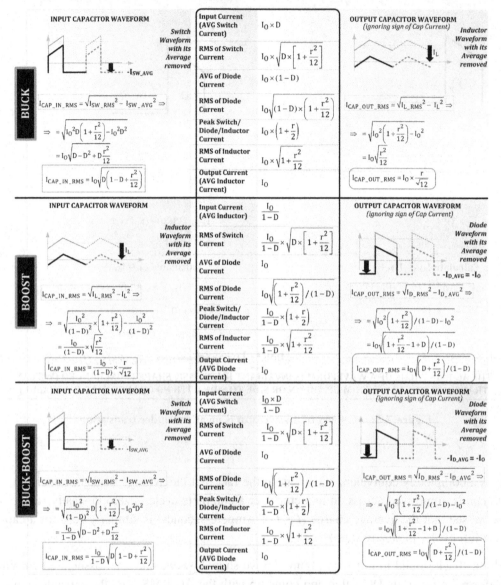

Figure 7.6: Calculating input and output capacitor current RMS values, based on tabulated current stresses.

The function $D \times (1-D)$ has a maximum at $D = 0.5$. Though the presence of the term in r affects this somewhat, we can simplify for low values of r

$$I_{CAP\_IN\_RMS} = I_O \sqrt{D\left(1 - D + \frac{r^2}{12}\right)} \approx I_O \sqrt{0.5\,(1 - 0.5)} = I_O \sqrt{0.5 \times 0.5}$$

$$I_{CAP\_IN\_RMS} \approx \frac{I_O}{2}$$

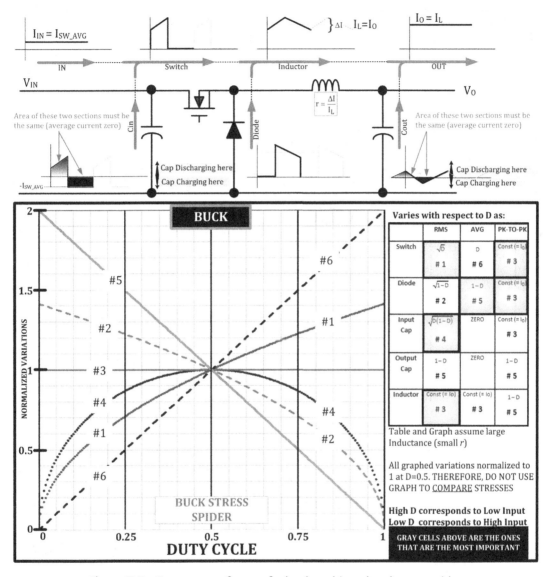

Figure 7.7: Current waveforms of a buck and its related stress spider.

For example, a 3 A Buck will need an input capacitor sized to handle at least 1.5 A, irrespective of switching frequency, output voltage and so on.

In determining the worst-case RMS, we can ask: if $D = 0.5$ is not included in the input range, what input voltage should we choose to check the suitability of the input cap of a Buck? The answer is: at the point *closest* to $D = 0.5$. For example, if D varies from 0.2 to

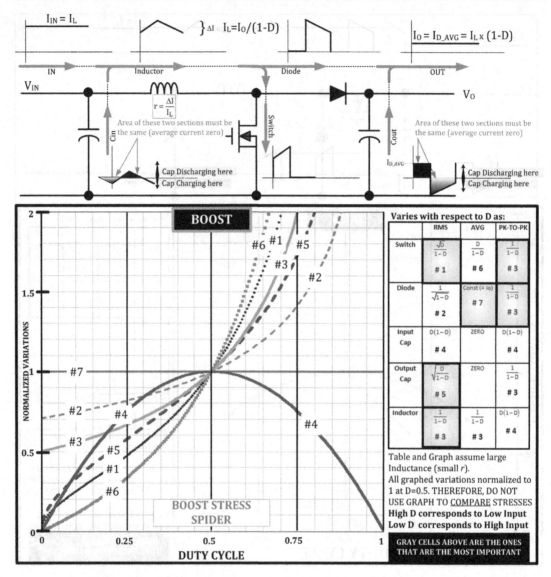

Figure 7.8: Current waveforms of a Boost and its related stress spider.

0.4 over the given input range, we will pick the input voltage end at which $D = 0.4$ (lowest input). If D varies from 0.6 to 0.8, we should pick the input voltage point at which $D = 0.6$ (highest input). Of course, if D varies from 0.3 to 0.6, we will pick the input voltage where $D = 0.5$ (i.e., $V_{IN} = 2 \times V_O$).

We also realize that since RMS stresses do not depend on switching frequency, increasing the frequency will have no effect on the size of the input cap of a Buck! We can have a

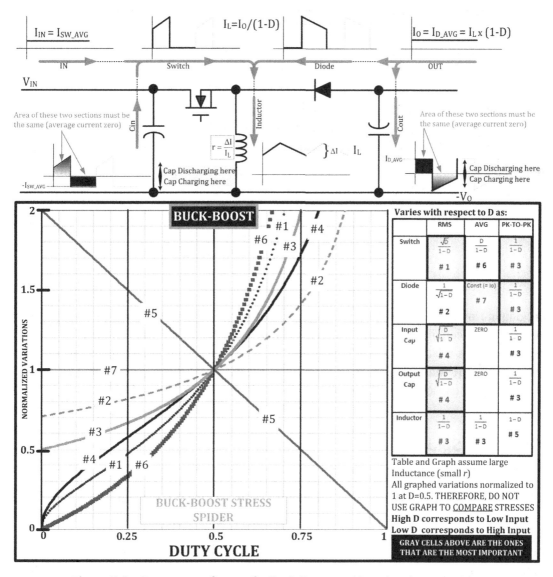

$I_{IN} = I_{SW\_AVG}$

$I_L = I_O/(1-D)$

$I_O = I_{D\_AVG} = I_L \times (1-D)$

V_{IN}

IN   Switch   Diode   OUT

Area of these two sections must be the same (average current zero)

$r = \dfrac{\Delta I}{I_L}$

$\}\Delta I \quad I_L$

Cap Discharging here
Cap Charging here

I_{D_AVG}

Area of these two sections must be the same (average current zero)

Cap Discharging here
Cap Charging here

$-I_{SW\_AVG}$

$-V_O$

BUCK-BOOST

#1 #4
#6 #3
#2
#5
#7
#2
#3
#5
#4 #1 #6

BUCK-BOOST STRESS SPIDER

NORMALIZED VARIATIONS

2
1.5
1
0.5
0

0   0.25   0.5   0.75   1

DUTY CYCLE

Varies with respect to D as:

	RMS	AVG	PK-TO-PK
Switch	$\dfrac{\sqrt{D}}{1-D}$ #1	$\dfrac{D}{1-D}$ #6	$\dfrac{1}{1-D}$ #3
Diode	$\dfrac{1}{\sqrt{1-D}}$ #2	Const (= Io) #7	$\dfrac{1}{1-D}$ #3
Input Cap	$\sqrt{\dfrac{D}{1-D}}$ #4	ZERO	$\dfrac{1}{1-D}$ #3
Output Cap	$\sqrt{\dfrac{D}{1-D}}$ #4	ZERO	$\dfrac{1}{1-D}$ #3
Inductor	$\dfrac{1}{1-D}$ #3	$\dfrac{1}{1-D}$ #3	$1-D$ #5

Table and Graph assume large Inductance (small *r*)
All graphed variations normalized to 1 at D=0.5. THEREFORE, DO NOT USE GRAPH TO <u>COMPARE</u> STRESSES
High D corresponds to Low Input
Low D corresponds to High Input
GRAY CELLS ABOVE ARE THE ONES THAT ARE THE MOST IMPORTANT

Figure 7.9: Current waveforms of a Buck-Boost and its related stress spider.

2 MHz Buck as opposed to a 100 kHz Buck, but if we were using electrolytic input capacitors, both these converters will require the same input capacitor. But things can certainly change if we use ceramic input capacitors for example — these have such high ripple ratings that the acceptable input ripple, not the RMS rating, becomes the dominant criterion for selecting the input capacitor. Ripple, current or voltage, is a result of

switching, and therefore does depend on frequency. So for example, in modern "all-ceramic solutions," the input cap of a 2 MHz Buck can be roughly half the capacitance and size of a similar 1 MHz Buck. But we have to be careful too, since the use of ceramic caps on the input can cause overshoots and instability as discussed in *Chapter 17*. We also have to be careful in drawing too much meaning from the values of input caps shown in many semiconductor vendors' typical schematics. The input cap of a Buck is rarely optimized as the output cap may be, and usually based on the "gut feel" of a typical engineer heading to the component cabinet. He tries it, it works, end of story.

Why is the RMS rating not significant in selecting the *output* cap of a Buck? Because we know that the output RMS is very small in a Buck topology. Here is the rule-of-thumb for the RMS stress on the output cap of a Buck.

$$I_{CAP\_OUT\_RMS} = I_O \frac{r}{\sqrt{12}} \approx I_O \times \frac{0.4}{\sqrt{12}}$$

$$I_{CAP\_OUT\_RMS} \approx 12\% \text{ of } I_O$$

So, output ripple, not RMS, becomes the dominant concern, whether or not we are using electrolytic or ceramic capacitors. Now, the actual capacitance, not just the capacitor's physical size as related to its RMS capability, becomes important, as also the parasitics of the cap (its ESR and ESL), which as we can see greatly affect the output ripple. Full derivations for capacitor ripple and its impact are provided in *Chapter 13*.

The Stress Spiders

One of our key learnings from previous chapters is that for a given output voltage, the duty cycle D, in effect tells us what the input voltage is. We also learned that *in all topologies, a low D corresponds to a high V_{IN}, and a high D to a low V_{IN}*. In the previous sections we have derived the RMS and average values of currents for all three fundamental topologies. We now want to know *how the stresses vary with D*, and thereby indirectly, *with respect to the input voltage*. Knowledge of that variation comes in handy in deciding at what voltage within the input range "worst-case" stress occurs and what that stress is, so we can apply the derating principles learned in *Chapter 6* and correctly select the power components.

When we look at the RMS/AVG equations derived so far, we see they include *both r* and *D*. That makes the total analysis a little complicated since r depends on D (input voltage) too. To arrive at the "Stress Spiders," in the following analysis, we have often used the "small r" (or "large L," also called the "flat-top") approximation. *But only to a certain extent* — in fact only where the term in r is insignificant — we then simply ignore it. But where r happens to be the dominant term, we do not ignore it, but use the following variations (as seen from the design table provided in the Appendix of the book).

$$r \propto (1 - D) \text{ for a Buck}$$
$$r \propto D(1 - D)^2 \text{ for a Boost}$$
$$r \propto (1 - D)^2 \text{ for a Buck-Boost}$$

Note that now we are also interested in the "peak-to-peak" currents. The peak-to-peak current in the inductor is simply ΔI. We want to know it because, for one, *core losses depend primarily on* peak-to-peak (and on the switching frequency of course, but not on I_{DC}). Note also that in all topologies, the peak switch/diode/inductor currents are the same. That is important to know since we need to confirm that the inductor is rated for the peak current through it in terms of its rated I_{SAT} or B_{SAT}, so we can be sure to avoid core saturation, as discussed in *Chapter 5*. On the other hand, the RMS of the inductor current is important because it determines if the inductor is rated for that continuous RMS current value in terms of heat and its temperature rise. Note that in Figure 7.6, we have also provided the RMS of the "diode" current. The reason for that is, we may be using a *synchronous topology*, so we may actually have a MOSFET in place of the catch diode. To know the heating in a MOSFET we need to know its its I_{RMS}, not its I_{AVG}. Therefore, both I_{AVG} and I_{RMS} are provided for the diode (freewheeling) position.

We present some examples to show how we can approximate the dependency of the stresses with respect to D as displayed via the tables and graphs of Figures 7.7–7.9.

Example: Describe the dependency with respect to D of the peak-to-peak inductor current in a Boost topology.

The equation for peak-to-peak inductor current is

$$\Delta I = r \times I_L$$

We know that the average inductor current (center of ramp) of a Boost is

$$I_L = \frac{I_O}{1 - D}$$

Therefore, since r varies as $D \times (1-D)^2$ for a Boost, we get

$$\Delta I \propto D(1 - D)^2 \times \frac{1}{1 - D} \rightarrow D(1 - D)$$

This is the value displayed in the table inside Figure 7.8 and plotted out in the adjoining graph.

Example: Describe the dependency with respect to D of the output cap RMS current in a Boost topology.

From Figure 7.6, we have

$$I_{CAP\_OUT\_RMS} = I_O\sqrt{\left(D + \frac{r^2}{12}\right)/(1 - D)}$$

Therefore, assuming large inductances, the term in r is very small compared to D. Check: $0.4^2/12 = 0.013$. If minimum D is about 10%, that is, $D = 0.1$, the term in r is 10 times smaller. We can thus approximate (for a Boost)

$$I_{CAP\_OUT\_RMS} = I_O\sqrt{\frac{D}{1 - D}}$$

$$I_{CAP\_OUT\_RMS} \propto \sqrt{\frac{D}{1 - D}}$$

This is the value displayed in the table inside Figure 7.8 and plotted out in the adjoining graph.

In this manner we get the three Stress Spiders shown in Figures 7.7–7.9. The salient points of these spiders are summarized below.

1. We had learned in *Chapter 5* that V_{INMIN} is a good point to select and design the magnetics of a Buck-Boost/flyback. Now from Figure 7.9, we see that almost all the stresses increase as D increases (low input). We realize that V_{INMIN} is a good point to select, design, and evaluate the temperatures of *all* the power components too (of a Buck-Boost). In other words, during test and evaluation, we can just set "maximum load at lowest input," and evaluate the entire Buck-Boost/flyback power supply for reliability and life requirements. One seeming exception is the peak-to-peak inductor current, which reaches a maximum at lowest D (high input). Admittedly, since core losses depend on ΔI, we may want to evaluate the choke at high input voltages too. However, core losses are usually a small part of the total choke (inductor) losses (especially in ferrites, as opposed to powdered iron for example). The dominant loss usually being copper losses, it is more common to measure the temperature of the magnetics of a Buck-Boost/flyback at its lowest input voltage, though in general, we may want to evaluate the temperature of the inductor at both the highest and lowest input voltages.

2. For a Boost, we can draw very similar conclusions as for the flyback/Buck-Boost above. V_{INMIN} is the best point to start a Boost design too, and also for *most* of the other components and stresses. There is one small surprise here — the input capacitor's RMS current has a max at $D = 0.5$, not at the high or low extremes of input voltage. In this topology, $D = 0.5$ is the point where $V_O = 2 \times V_{IN}$ (see Figure 5.9). Note that if in our application, the input range does *not* include the point $V_{IN} = V_O/2$, then for

selecting and testing the input capacitor, we need to pick that end of the input range *which lies closest to* $V_O/2$.

Note, however, that in a Boost topology, the input capacitor's RMS current value is *numerically* very small compared to its output capacitor's RMS. Because there is an inductor present between the input capacitor and the switch, which smoothens out the switch current waveform significantly, and that is what is finally presented to the input capacitor for completing the rest of the smoothening/filtering process.

That is why in all the tables in Figures 7.7–7.9, we have indicated the most important or significant stresses with a gray highlight. We can usually ignore the non-gray cells of the table. We emphasize that the Stress Spiders in the figures only express the *relative* variation of a *particular* stress *with respect to its (own normalized) value* at $D = 0.5$. So, even though all the relative variations are overlaid on a single plot, *each curve represents a completely distinct stress*. We should *not* try to use the different curves of a Stress Spider to *compare different* types of stresses.

3. In a Buck, we have several surprises in Figure 7.7. Note that we have always advocated starting the design of a Buck converter at V_{INMAX}. That indeed is a good point for designing and evaluating a Buck inductor. The inductor current has a constant center of ramp value equal to I_O, which has an AC component riding on it that increases with input voltage. So the peak current is highest at V_{INMAX}. The energy-handling capability of the Buck inductor, which depends on I_{PEAK}^2, must therefore be evaluated at V_{INMAX}. Further, from Figure 7.7, we can easily deduce that the RMS of the inductor current (its heating) will also go up as input increases. So, the RMS of the inductor current of a Buck is also at its highest at V_{INMAX}. We conclude that V_{INMAX} *is truly a good point to start the design of a Buck* (the inductor). But is that good for all the power components of a Buck?

From Figure 7.7, we see that the switch RMS/AVG is the highest at high D (V_{INMIN}) not at V_{INMAX}. Intuitively this makes sense since at low inputs, the duty cycle increases, which means the switch is ON for a longer time, so it will tend to heat up more at low input voltages (higher stresses). That means the Buck switch must be evaluated for dissipation at V_{INMIN}, not V_{INMAX}. We can also conclude that if the switch is ON for a longer time, as at V_{INMIN}, then obviously the diode must be ON for the shortest time at V_{INMIN}. And therefore, the diode would be ON for the longest time at V_{INMAX}. So, we expect the diode RMS/AVG will be worst-case at V_{INMAX}. All this is borne out in Figure 7.7 from the Buck Stress Spider. In other words, *the switch of a Buck (and its associated heatsinking) must be designed and evaluated at V_{INMIN}, whereas its diode must be selected and checked for temperature rise at V_{INMAX}*. Note also that the RMS of the input cap *is highest at $D = 0.5$* (or the input end closest to it). So, we need to be careful in selecting and testing the input cap of a Buck. We see that it has been highlighted gray in Figure 7.7, and we realize it is certainly a significant term, numerically speaking, that we need to pay close attention to.

We can ask: how is it that in Figures 7.8 and 7.9, that is, for the remaining two topologies, the worst-case point for the diode RMS current is V_{INMIN}, not V_{INMAX}? That is because though the diode is still ON for a shorter time as input falls, the instantaneous (on-time) current through it goes up steeply as the input falls.

At this stage, the reader may like to look at the solved examples in *Chapter 19*, in which most of the equations derived above have been used in numerical examples.

Stress Reduction in AC–DC Converters

The AC–DC flyback is one of the trickiest converters to design reliably in a cost-effective manner. Let us first try to understand how to apply the RMS/AVG stress equations of previous sections to it.

(a)   A flyback (transformer-based) converter can be reduced to a *Primary-side equivalent inductor-based Buck-Boost converter* for the purpose of selecting and evaluating the *Primary-side stresses* of the flyback. This means, the switch and input capacitor, in effect, "think" that they are part of a DC–DC Buck-Boost converter stage, whose input is the rectified AC input ($\sim V_{AC} \times \sqrt{2} \equiv V_{IN}$), and whose output voltage is $V_{OR} = V_O \times n$ (Volts), with load current equal to $I_{OR} = I_O/n$ (Amperes), where $n = N_P/N_S$.

(b)   Similarly, the flyback converter can be reduced to a *Secondary-side equivalent Buck-Boost* for purposes of selecting and evaluating the *Secondary-side stresses*. Which means, the diode and output capacitor, in effect, "think" that they are part of a DC–DC Buck-Boost converter stage, whose input is the *reflected* rectified AC input ($\sim V_{AC} \times \sqrt{2}/n \equiv V_{IN}/n \equiv V_{INR}$), and whose output is V_O (Volts) with load current equal to I_O (Amperes).

To understand this better, see Figure 3.2 too. With all this in mind, we realize that all the current stress equations derived so far for the Buck-Boost can be easily applied to an AC–DC flyback.

Coming to the Forward converter, we consider its output section. Here we can assume the "Buck cell" (consisting of the two common-anode diodes, choke, and output capacitor) behaves as a DC–DC Buck converter whose input is the *reflected* rectified AC input ($\sim V_{AC} \times \sqrt{2}/n \equiv V_{IN}/n \equiv V_{INR}$), and whose output is V_O (Volts), with load current equal to I_O (Amperes). So, all the equations for current stresses that we have derived for the DC–DC Buck converter, apply to the Buck cell of the Forward converter too. On the input side of the Forward converter, the switch current (Primary side) was sketched in Figure 3.6. Note that it is very similar to the switch waveform of a DC–DC Buck converter with duty cycle $D = V_O/V_{INR}$ (or equivalently V_{OR}/V_{IN}), with a load current equal to $I_{OR} = I_O/n$. The actual switch current is actually slightly more than the equivalent DC–DC Buck converter, due to the magnetization current component in the switch waveform, but since that contribution is very small compared to the overall waveform, it can be usually ignored.

Summarizing: similar to the mapping procedure described to go from an AC–DC flyback to a Buck-Boost, all the equations derived so far for RMS/AVG stresses for a Buck can also be quickly applied to the AC–DC Forward converter. So, the switch and input capacitor of the Forward converter, in effect, think that they are part of a DC–DC Buck converter stage, whose input is the rectified AC input ($\sim V_{AC} \times \sqrt{2} \equiv V_{IN}$), and whose output voltage is $V_{OR} = V_O \times n$ (Volts), with load current equal to $I_{OR} = I_O/n$ (Amperes), where $n = N_P/N_S$. On the Secondary-side, the free-wheeling diode and output capacitor of the Forward converter, in effect, think that they are part of a DC–DC Buck converter stage, whose input is the reflected rectified AC input ($\sim V_{AC} \times \sqrt{2}/n \equiv V_{IN}/n \equiv V_{INR}$), and whose output is V_O (Volts), with load current equal to I_O (Amperes).

One last component of the Forward converter that is not accounted for in the above mapping procedure is its output diode (the diode connected to the Secondary winding of the transformer). That diode conducts (only) during the on-time of the converter (with duty cycle D) and during that time it passes a current of average value I_O (since we know that I_O is the center of ramp of the Buck cell that follows). So, the average current through the output diode of the Forward converter, evaluated over the entire cycle, is $I_O \times D$. Multiplying that by the forward diode drop, gives the required diode dissipation.

If we have a two-switch ("2-switch") Forward converter, the same current passes through each of its MOSFETs. So, we have to use the same I_{RMS} value indicated above for each of the two MOSFETs, and then sum the two dissipations to calculate the total dual switch dissipation. So, in a 2-switch Forward-converter where each MOSFET has an R_{DS} of say 100 mΩ, the total switch dissipation is twice that of a single-ended Forward converter with a single MOSFET of R_{DS} 100 mΩ. The main advantage of using a 2-switch Forward is reflected in the fact that the *voltage* stress on each MOSFET is halved compared to a single-ended Forward, not the current stresses. It is also usually cheaper to find two 400 V MOSFETs with say, an R_{DS} of $x/2$ ohms each, as compared to a single 800–1000 V MOSFET with an R_{DS} of x ohms (same conduction loss).

See also Table 7.1 and Figure 7.2 to understand this better.

In *Chapter 3*, we discussed the zener clamp of a flyback as a means of reducing the leakage inductance spike of a flyback and thereby saving its switch from voltage overstress. Now, we look at another option of achieving the same basic result: *the RCD clamp*. Note that in a single-ended Forward converter, we may also have a leakage inductance spike due to the small leakage present between the Primary winding and the tertiary (energy-recovery) winding. However, these two windings are often wound bifilar (see Figure 9.22) and that helps the two windings couple well, reducing the leakage to near negligible. If not, some type of clamp or snubber may be required even for a Forward converter. For more tertiary winding aspects in the Forward converter, see *Chapter 9*.

RCD Clamps versus RCD Snubbers

Early power supplies used to invariably have an "RCD snubber" (also called a "dV/dt snubber"). RCD stands for resistor-capacitor-diode, and this little network would always be seen present across the switch. The purpose of an RCD snubber was two-fold.

(a) The RCD snubber helped reduce the transition *overlap* — between the voltage and the current *belonging to the switch* (in early days the switch was a BJT). This would therefore improve the dissipation (and temperature) *related to the switch*, though not necessarily improve the overall efficiency. For example, a lot of heat could be lost in the *R* of the RCD, instead of in the switch.

(b) The RCD snubber helped reduce the dV/dt appearing *across the switch* during the turn-off transition, thereby enhancing its reliability, especially when early MOSFETs appeared on the scene.

Today RCD snubbers are almost obsolete for several reasons: (1) BJTs are very slow and therefore rarely used, (2) modern MOSFETs are almost immune to dV/dt failures, and (3) switching transitions today are so fast, that though there is significant *V–I* overlap during the transition, the transition itself lasts for only about 50–100 ns every cycle compared to a couple of μs in early days (for overlap, see *Chapter 8*). In effect, the RCD *snubber* is history. In its place, the RCD *clamp* has surfaced.

RCD *clamps* are commonly used, especially in AC–DC flybacks. One reason is they can offer higher efficiency (and lower cost) than zener clamps (as discussed in *Chapter 3*). But that is conditional on the RCD clamp being very *carefully designed*. Note that on a schematic, an RCD snubber looks almost exactly like an RCD clamp *to the untrained eye* — the difference is a clamp uses a much *higher capacitance* (typically 10–47 nF) and a much larger *R*, whereas a snubber has a much smaller capacitance, rarely exceeding 1–2 nF, and a much smaller *R* too. Therefore, functionally speaking, the difference between an RCD snubber and an RCD clamp is as follows: whereas the capacitor of an RCD snubber fully discharges every cycle, the capacitor of an RCD clamp does *not* discharge fully between cycles, and remains always "pre-charged" — to just a little below its "clamping voltage level." The RCD clamp capacitor therefore comes into play (i.e., the RCD clamp diode conducts) only *above* a certain Drain-to-Source voltage. Below that

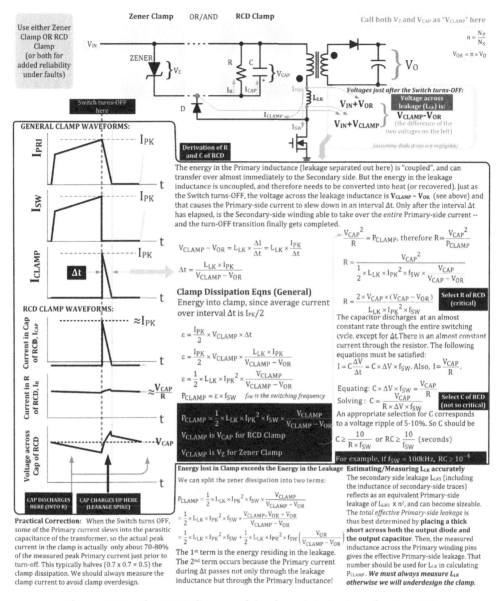

Use either Zener Clamp OR RCD Clamp (or both for added reliability under faults)

Zener Clamp OR/AND **RCD Clamp**

Call both V_Z and V_{CAP} as "V_{CLAMP}" here

$$n = \frac{N_P}{N_S}$$

$$V_{OR} = n \times V_0$$

V_{IN} ZENER $\} V_Z$ R   C $\} V_{CAP}$ V_0

I_R I_{CAP} I_{PRI} L_{LK}

Switch turns-OFF here

D I_{CLAMP}

I_{SW}

Voltages just after the Switch turns-OFF:

$V_{IN} + V_{OR}$ Voltage across leakage (L_{LK}) is:

$V_{IN} + V_{CLAMP}$ $V_{CLAMP} - V_{OR}$ (the difference of the two voltages on the left)

(assuming diode drops are negligible)

GENERAL CLAMP WAVEFORMS:

I_{PRI} ... I_{PK} ... t

I_{SW} ... I_{PK} ... t

I_{CLAMP} ... I_{PK} ... Δt ... t

RCD CLAMP WAVEFORMS:

Current in Cap of RCD, I_{CAP} ... $\approx I_{PK}$ ... t

Current in R of RCD, I_R ... $\approx \frac{V_{CAP}}{R}$ ... t

Voltage across Cap of RCD ... V_{CAP} ... t

CAP DISCHARGES HERE (INTO R)   CAP CHARGES UP HERE (LEAKAGE SPIKE)

Practical Correction: When the Switch turns OFF, some of the Primary current slews into the parasitic capacitance of the transformer, so the actual peak current in the clamp is actually only about 70-80% of the measured peak Primary current just prior to turn-off. This typically halves (0.7 x 0.7 = 0.5) the clamp dissipation. We should always measure the clamp current to avoid clamp overdesign.

Derivation of R and C of RCD

The energy in the Primary inductance (leakage separated out here) is "coupled", and can transfer over almost immediately to the Secondary side. But the energy in the leakage inductance is uncoupled, and therefore needs to be converted into heat (or recovered). Just as the Switch turns-OFF, the voltage across the leakage inductance is $V_{CLAMP} - V_{OR}$ (see above) and that causes the Primary-side current to slew down in an interval Δt. Only after the interval Δt has elapsed, is the Secondary-side winding able to take over the *entire* Primary-side current -- and the turn-OFF transition finally gets completed.

$$V_{CLAMP} - V_{OR} = L_{LK} \times \frac{\Delta I}{\Delta t} = L_{LK} \times \frac{I_{PK}}{\Delta t}$$

$$\Delta t = \frac{L_{LK} \times I_{PK}}{V_{CLAMP} - V_{OR}}$$

Clamp Dissipation Eqns (General)
Energy into clamp, since average current over interval Δt is $I_{PK}/2$

$$\varepsilon = \frac{I_{PK}}{2} \times V_{CLAMP} \times \Delta t$$

$$\varepsilon = \frac{I_{PK}}{2} \times V_{CLAMP} \times \frac{L_{LK} \times I_{PK}}{V_{CLAMP} - V_{OR}}$$

$$\varepsilon = \frac{1}{2} \times L_{LK} \times I_{PK}^2 \times \frac{V_{CLAMP}}{V_{CLAMP} - V_{OR}}$$

$$P_{CLAMP} = \varepsilon \times f_{SW} \quad f_{SW} \text{ is the switching frequency}$$

$$P_{CLAMP} = \frac{1}{2} \times L_{LK} \times I_{PK}^2 \times f_{SW} \times \frac{V_{CLAMP}}{V_{CLAMP} - V_{OR}}$$

V_{CLAMP} is V_{CAP} for RCD Clamp
V_{CLAMP} is V_Z for Zener Clamp

$$\frac{V_{CAP}^2}{R} = P_{CLAMP}, \text{ therefore } R = \frac{V_{CAP}^2}{P_{CLAMP}}$$

$$R = \frac{V_{CAP}^2}{\frac{1}{2} \times L_{LK} \times I_{PK}^2 \times f_{SW} \times \frac{V_{CAP}}{V_{CAP} - V_{OR}}}$$

$$R = \frac{2 \times V_{CAP} \times (V_{CAP} - V_{OR})}{L_{LK} \times I_{PK}^2 \times f_{SW}} \quad \boxed{\text{Select R of RCD (critical)}}$$

The capacitor discharges at an almost constant rate through the entire switching cycle, except for Δt. There is an almost constant current through the resistor. The following equations must be satisfied:

$$I = C \frac{\Delta V}{\Delta t} = C \times \Delta V \times f_{SW}. \text{ Also, } I = \frac{V_{CAP}}{R}.$$

Equating: $C \times \Delta V \times f_{SW} = \frac{V_{CAP}}{R}$

Solving: $C = \frac{V_{CAP}}{R \times \Delta V \times f_{SW}}$ $\boxed{\text{Select C of RCD (not so critical)}}$

An appropriate selection for C corresponds to a voltage ripple of 5-10%. So C should be

$$C \geq \frac{10}{R \times f_{SW}} \text{ or } RC \geq \frac{10}{f_{SW}} \text{ (seconds)}$$

For example, if $f_{SW} = 100kHz$, $RC \geq 10^{-4}$

Energy lost in Clamp exceeds the Energy in the Leakage
We can split the zener dissipation into two terms:

$$P_{CLAMP} = \frac{1}{2} \times L_{LK} \times I_{PK}^2 \times f_{SW} \times \frac{V_{CLAMP}}{V_{CLAMP} - V_{OR}}$$

$$= \frac{1}{2} \times L_{LK} \times I_{PK}^2 \times f_{SW} \times \frac{V_{CLAMP} - V_{OR} + V_{OR}}{V_{CLAMP} - V_{OR}}$$

$$= \frac{1}{2} \times L_{LK} \times I_{PK}^2 \times f_{SW} + \frac{1}{2} \times L_{LK} \times I_{PK}^2 \times f_{SW} \left(\frac{V_{OR}}{V_{CLAMP} - V_{OR}} \right)$$

The 1st term is the energy residing in the leakage. The 2nd term occurs because the Primary current during Δt passes not only through the leakage inductance but through the Primary Inductance!

Estimating/Measuring L_{LK} accurately
The secondary side leakage L_{LKS} (including the inductance of secondary-side traces) reflects as an equivalent Primary-side leakage of $L_{LKS} \times n^2$, and can become sizeable. The *total effective* Primary-side leakage is thus best determined by **placing a thick short across both the output diode and the output capacitor**. Then, the measured inductance across the Primary winding pins gives the effective Primary-side leakage. That number should be used for L_{LK} in calculating P_{CLAMP}. *We must always measure L_{LK} otherwise we will underdesign the clamp.*

Figure 7.10: The RCD clamp explained, and derivations for *R* and *C*.

voltage level, the RCD clamp is virtually non-existent. When the RCD clamp diode conducts, because the capacitance of the RCD clamp is so large, it acts to literally "clip" the leakage inductance voltage spike of the flyback to a safe value. As mentioned, in contrast, the capacitor of an RCD snubber discharges *fully* every cycle, and when the switch turns OFF, the RCD snubber capacitor is *immediately* ready to accept part of the

Efficiency Improvement at Low Line due to RCD Clamp

85-270VAC Universal Input Flyback:

Take a typical case of a 600V Fet and V_{OR} set at 100V. To protect the switch at high line we need to clamp the leakage inductance spike with a 150V clamp. If we use a zener clamp, we would have picked a 150V zener. In an RCD clamp we need to select/adjust "R" to give us a cap voltage of around 150V *at high line* (ignoring its voltage ripple). In the case of a zener clamp, the clamping voltage would have remained 150V at low-line. However, with an RCD clamp, the clamping (cap) voltage *rises to 170V* at low line because of the higher currents, as shown below, thus improving the efficiency at low line (at max load). The efficiency at high line is almost the same.

So, the voltage rating of the Switch determines the max clamping voltage at high line --- and that determines the value of "R" of the RCD Clamp. Too low a clamping voltage causes excessive clamp dissipation. The dissipation in the clamp (in R) is however at its highest at low line (max load). *Therefore the wattage rating of the chosen R is determined at what happens at low line though its value is determined at high line.*

To understand this better we need to connect the situation at high line with that at low line as shown below.

Duty Cycle Equation is: $D = \dfrac{V_{OR}}{V_{OR}+V_{IN}}$

At low line, $V_{IN} \approx \sqrt{2} \times 85VAC = 120V$

(Ignoring input bulk cap ripple here)

$D_{LO\_LINE} = \dfrac{100}{100+120} = 0.46$

At high line, $V_{IN} = \sqrt{2} \times 270VAC = 382V$

$D_{HI\_LINE} = \dfrac{100}{100+382} = 0.21$

So, D at high line is almost half of the D at low line.

Since r varies as $(1-D)^2$ (see Appendix)

$r_{HI\_LINE} = r_{LO\_LINE} \times \dfrac{(1-D_{HI\_LINE})^2}{(1-D_{LO\_LINE})^2}$

So, if r is set to 0.4 at low line, at high line:

$r_{HI\_LINE} = 0.4 \times \dfrac{(1-0.21)^2}{(1-0.46)^2} = 0.86$

Peak Current in a Flyback goes as $\dfrac{1+\frac{r}{2}}{1-D} \Rightarrow \dfrac{2+r}{1-D}$

Therefore Peak Current varies with line voltage as

$I_{PK\_HI\_LINE} = I_{PK\_LO\_LINE} \times \dfrac{2+r_{HI\_LINE}}{2+r_{LO\_LINE}} \times \dfrac{1-D_{LO\_LINE}}{1-D_{HI\_LINE}}$

$\dfrac{I_{PK\_HI\_LINE}}{I_{PK\_LO\_LINE}} = \dfrac{2+0.86}{2+0.4} \times \dfrac{1-0.46}{1-0.21} = 0.815$

So, I_{PK} at high line is ~ 20% less than at low line.

From the design equation for R in Figure 7.10

$R = \dfrac{2 \times V_{CAP} \times (V_{CAP}-V_{OR})}{L_{LK} \times I_{PK}^2 \times f_{SW}} \Rightarrow I_{PK}^2 \propto V_{CAP} \times (V_{CAP}-V_{OR})$

We can confirm that if $V_{CAP} = 150V$ at high line

and $V_{OR} = 100V$, then at low line I_{PK} falls by 20% and

$V_{CAP\_LO\_LINE}$ rises to 170V. Since $(170-150)/150 = 0.13$.

V_{CLAMP} at low line is ~ 13% higher than at high line.

From the equation of clamp dissipation, if we had used a 150V zener, the dissipation would've been more.

$\dfrac{P_{RCD\_CLAMP\_LO\_LINE}}{P_{ZENER\_CLAMP\_LO\_LINE}} = \dfrac{\frac{V_{CAP\_LO\_LINE}}{V_{CAP\_LO\_LINE}-V_{OR}}}{\frac{V_{Z\_LO\_LINE}}{V_{Z\_LO\_LINE}-V_{OR}}}$

$\dfrac{P_{RCD\_CLAMP\_LO\_LINE}}{P_{ZENER\_CLAMP\_LO\_LINE}} = \dfrac{170/(170-100)}{150/(150-100)} = 0.81$

Dissipation in an RCD clamp at low line is about 20% lower than from an equivalent zener clamp.

NUMERICAL EXAMPLE

A 50W Universal Input Flyback with 70% efficiency and switching at 100kHz uses a 600V Fet. By placing thick shorts across the output diode and the output capacitor, we measure the in-circuit effective Primary-side leakage as 20µH (about 2% of the Primary Inductance as typically expected). Find the R and C of the RCD clamp (first estimate values).

The Input Power is 50/0.72 = 70W. With a Duty Cycle of 0.46 at low line, the center of ramp on the Primary side is determined by

$P_{IN} = 70W = V_{IN} \times I_{IN} = 120V \times I_{PRI\_LO} \times 0.46$

$I_{PRI\_LO} = \dfrac{70}{120 \times 0.46} = 1.3A$

Setting r=0.4 at low line, the peak current must be

$I_{PK\_LO} = I_{PRI\_LO}\left(1+\dfrac{r}{2}\right) = 1.3 \times (1.2) = 1.56A.$

Assuming efficiency is unchanged at high line (using scaling):

$I_{PK\_HI} = 0.815 \times I_{PK\_LO} = 0.815 \times 1.56 = 1.27A.$

The current into the clamp is typically about 80% of the peak switch current since some of the current flows into the parasitic capacitance of the transformer. So we take I_{PK} as 1.27 x 0.8 below. At high line

$R = \dfrac{2 \times V_{CAP} \times (V_{CAP}-V_{OR})}{L_{LK} \times I_{PK}^2 \times f_{SW}} = \dfrac{2 \times 150 \times (150-100)}{20\mu \times (1.27 \times 0.8)^2 \times 10^5} \approx 7.3k$

$C \geq \dfrac{10}{R \times f_{SW}} = \dfrac{10}{7.3k \times 100k} \Rightarrow 14nF$

Wattage is determined at low-line

$P_{CLAMP} = \dfrac{1}{2} \times L_{LK} \times I_{PK}^2 \times f_{SW} \times \dfrac{V_{CLAMP}}{V_{CLAMP}-V_{OR}}$

$= \dfrac{1}{2} \times 20\mu \times (1.56A \times 0.8)^2 \times 100k \times \dfrac{170}{170-100} = 3.8W$

So, a possible first choice to evaluate is R=7.5k/5W, C=15nF/250V. The diode can be rated 1A/600V; some designers recommend *not* using an ultrafast diode in this position. Soft characteristics of slower diodes may help in *improving* EMI.

Figure 7.11: Connecting high Line conditions with low line, and thereby calculating *R*, *C*, and dissipation for universal input flybacks.

freewheeling current, thus lowering the dV/dt *appearing across the switch* (though also transferring the switch crossover dissipation into itself). The RCD snubber was therefore often called a "switching-aid network" — it literally aids the switching action. Whereas, an RCD clamp does just that: it *clamps*.

In Figures 7.10 and 7.11 we cover clamps in detail, in particular the RCD clamp. We derive the equations for calculating the R and C, and estimating the clamp dissipation. We also provide a numerical example. The important things to keep in mind are:

(a) The value of R is relatively independent of C, and yet is a key RCD design parameter by itself. In fact, C can be made almost as large as possible if cost permits (only its voltage ripple component will decrease; its average clamp voltage level will remain fixed since that depends on R, not on C!). The capacitance is typically chosen to be between 4 nF and 22 nF for AC–DC flyback converters switching between 70 kHz and 200 kHz. That value of C is such that the maximum voltage ripple that appears on it is below approximately $\pm 10\%$ (or the very purpose of the clamp gets defeated). However, a high value of RCD clamp capacitance comes in very handy under *sudden overloads at high line*, during which the clamp can receive a burst of excess energy, which could charge it up quickly and thereby threaten the Absolute Maximum voltage rating of the switch. If we do want to increase the C for such abnormal conditions, and/or have not properly designed the flyback protective limits (e.g., duty cycle max and current limit), we may want to play it safe by combining the RCD clamp with a paralleled zener clamp as indicated in Figure 7.10.

(b) The value of R is chosen carefully, *and usually empirically*. Its selection is purely based on not exceeding the Abs Max voltage rating of the switch under worst-case but *steady* operating conditions. R is obviously always selected at high line.

(c) However, the wattage rating of R is determined at the point at which its *dissipation* is at its maximum — and that occurs at low line.

What are the key advantages of an RCD clamp over a zener clamp? Cost is one. Efficiency is another. The reason for the efficiency improvement of an RCD is explained as follows.

For example, when using a 700 V MOSFET, a 200 V zener clamp is often used, and the V_{OR} is set to a maximum of 130 V. However, with a more cost-effective 600 V MOSFET, a 150 V zener is preferred, and the V_{OR} is set to $\sim$100 V. We saw in *Chapter 3* that, provided the voltage rating of the MOSFET is not exceeded, if V_{CLAMP} is made much bigger than V_{OR}, the dissipation in the clamp falls. This is also obvious from the equation

$$P_{CLAMP} = \frac{1}{2} \times L_{LK} \times I_{PK}^2 \times f_{SW} \times \frac{V_{CLAMP}}{V_{CLAMP} - V_{OR}}$$

($V_{CLAMP} = V_{CAP}$ for RCD Clamp, $V_{CLAMP} = V_Z$ for Zener Clamp)

For a zener clamp, the clamping voltage remains almost fixed (at 200 V and 150 V, respectively) with variations in line and load. However, using an RCD clamp, as we lower the input voltage (keeping load fixed at its max value), the clamping (capacitor) voltage *rises* on account of the higher switch currents — to $\sim$220 V and $\sim$170 V, respectively, at low line. And that lowers the clamp dissipation *by about 20%* as compared to a zener clamp under the exact same conditions. But we note that this efficiency improvement of an RCD clamp occurs only at *low line and at max load*. At lighter loads for example, the clamping voltage of an RCD clamp falls much lower than that of a fixed zener clamp, on account of the lower switch currents. Therefore, the efficiency at light loads using an RCD clamp is worse than a zener clamp. Intuitively, we can view the RCD clamp as a "bad" zener clamp whose clamping (zener) voltage depends rather steeply on how much current we push into it. That helps in reducing clamp dissipation, but it can lead to excess voltages too. Therefore the RCD clamp design is rather critical. As mentioned, some nervous engineers end up combining both the RCD and zener clamps in parallel as indicated in Figure 7.10.

In designing RCD clamps there are two key optimization details to keep in mind as also discussed in Figure 7.10.

(a) The measured current into the clamp at the instant of switch turn-off *is actually 70–80% of the peak switch current just prior to turn-off*. This is because part of the free-wheeling Primary-side current goes into the interwinding capacitance of the transformer, from where it eventually gets (largely) dissipated as heat in the windings. The actual clamp dissipation, being proportional to I_{PK}^2, is therefore *50–66% of the theoretical estimate*. This knowledge helps us to not overdesign the clamp. Very few simulation models will tell you this — bench measurements are very revealing here.

(b) The leakage inductance on *both* the Primary and Secondary sides serves to slow down the transfer of current from Primary to Secondary side, thereby extending the duration Δt (see Figure 7.10). This is the duration in which Primary-side current freewheels into the clamp waiting for all leakage inductances to either discharge or charge up as required, in going from one switching state to the other. So, the two inductances must be clubbed together to estimate Δt correctly.

The best way to determine the effective Primary-side leakage L_{LK} for use in the clamp dissipation equation is to do an *in-circuit measurement*. We need to place thick shorts across the output diode and the output capacitors, and then measure the inductance across the Primary winding pins. That reading gives us the correct L_{LK} to use. But for this method to succeed, the prototype must already be available. An alternative and surprisingly accurate estimate in the initial design stages is:

$$L_{LK} = L_{LKP} + n^2 L_{LKS} = (0.01 \times L_P) + \left(\frac{N_P}{N_S}\right)^2 \times L_{LKS}$$

where $L_{LKS} = 20$ nH/in. $\times$ (length of Secondary-side traces in inches).

Here we have used the empirical fact that a "good" AC–DC flyback transformer (say, with a split Primary, each half containing $N_P/2$ turns, sandwiching the Secondary windings containing N_S turns) has a typical leakage equal to $\sim 1\%$ of the Primary inductance (the inductance across the Primary winding with the Secondary winding open). Also, the main contribution to the Secondary-side leakage (which then reflects on to the Primary side as turns ratio squared) actually comes from the Secondary-side PCB *traces*, not the transformer (which is the reason why an in-circuit estimate of leakage is recommended). We know that PCB traces result in about 20 nH/in. We also need to numerically add up the *total* forward and return PCB trace lengths *right up to the first output capacitor* and use that to find L_{LKS} based on the 20 nH/in. rule. Note that after the first output capacitor of the flyback, we have mainly DC current and so the trace inductance is not relevant anymore.

For a given V_{OR}, the turns ratio of *low-voltage output rails* is much higher, because by definition, $V_{OR} = n \times V_O$. Since Secondary-side leakage reflects to the Primary side as n^2, it can become almost as large as the Primary-side leakage L_{LKP} — *especially where the turns ratio is high (as for low output voltages)*. Therefore, a good rule of thumb is to simply *take L_{LK} as being not just 1%, but 2% of L_P — for the purposes of determining R and P_{CLAMP}.*

Now we also understand why a 70 W flyback with only a 12 V output rail present, will "mysteriously" exhibit a much higher efficiency than a 70 W flyback with only a 5 V output rail — despite having the same V_{OR}, V_{CLAMP}, and even L_{LKP} in both cases. The difference is due to the turns ratio, which leads to much higher clamp dissipation for the 5 V output.

A way to improve overall efficiency rather dramatically, especially for low-voltage output rails, is to try to *cancel the inductance of the forward and return Secondary side traces.* That will reduce clamp dissipation significantly. For that we need to run the forward and return traces very close and parallel to each other. By doing so, the fields produced by the current flowing in opposite directions cancel, and the inductance reduces (no field — no stored energy — no inductance). That field cancelation happens automatically when using a multilayer PCB with a wide ground plane right below the traces. *And that is why double-sided PCBs automatically tend to offer superior efficiency compared to (sloppily laid-out) single-sided PCBs in low-output universal input flybacks.*

CHAPTER 8

Conduction and Switching Losses

As switching frequencies increase, it becomes of paramount importance to reduce the *switching losses* in the converter. These are the losses *associated* with the *transition* of the switch from its on-state to its off-state, and back. The higher the switching frequency, the greater the number of times the switch changes state per second. Therefore, these losses are proportional to the switching frequency. Further, of these frequency-dependent loss terms, the most significant are usually those that take place within the switch itself. Therefore, understanding the underlying sequence of events in the switch during each transition, and thereby quantifying the losses associated with each of these events, has become a key expectation of any power supply designer.

In this chapter, we are going to focus mainly on the MOSFET, since that is the most widely accepted "switch" in most high-frequency designs today. We will split its turn-on and turn-off transitions into small well-defined *subintervals*, and explain what happens in each of these. The associated design equations will also be presented. Note, however, that as in most related literature, we too will be resorting to certain *simplifications*, since modeling the MOSFET (and its interplay with the board that it is mounted on) is certainly not a trivial task, to say the least. As a result, it is possible that theoretical estimates can end up underestimating the actual switching losses by a large margin (typically 20–50%). The designer should keep that in mind, and may need to eventually incorporate some sort of a "fudge factor," to correspond with reality. In our analysis, we have included a "scaling factor" to try to minimize this error.

We will also show how to estimate MOSFET *driver* requirements and demonstrate the importance of correctly matching driver capability to the MOSFET in a given application. That should ultimately help not only applications engineers to pick better MOSFETs for their applications, but also IC designers involved in the process of designing driver stages for target applications.

A cautionary note with regard to the terminology — in most of our switching analysis, what we are calling the "load" is the load as seen by the *transistor*, it is not the load of the *DC–DC converter* stage. Similarly the "input voltage" is only the voltage *across the MOSFET* when it is OFF — it is *not* the input to the DC–DC converter stage. We will eventually make the required connections into the area of power conversion, but it should

311

be clear that initially at least, the discussion is more from *the standpoint of the MOSFET,* not the topology that it may be a part of.

Switching a Resistive Load

Before we take up inductors, it is instructive to first understand what happens when we switch a *resistive* load.

For simplicity, we are considering an *ideal* situation. So, we start with a "perfect" N-channel MOSFET in Figure 8.1. It behaves in the following manner:

- It has zero on-resistance.
- With zero Gate-to-Source voltage "Vgs" applied at its Gate, it is completely nonconducting.
- As we raise the Gate-to-Source voltage *Vgs* slightly above ground, it starts conducting, and so a Drain current "Id" flows from the Drain terminal to the Source terminal.
- The ratio of the Drain current to the Gate voltage is defined as the transconductance "*g*" of the MOSFET. It is expressed in "mhos," that is, *ohm* spelled backward. Nowadays, however, mhos is being increasingly called *Siemens*, or "S."
- We are assuming that *g* is a *constant* — equal to 1 S for this particular MOSFET. So, for example, if we apply 1 V at the Gate, the MOSFET will pass 1 A. If we apply 2 V, it will pass 2 A, and so on. Just to keep things simple here.

The application circuit shown in Figure 8.1 works as follows:

- The applied input voltage is 10 V.
- The external resistance (in series with the Drain) is 1 Ω.
- The Gate voltage is ramped up *linearly* with respect to time. So, at $t = 1$ s, it is 1 V; at $t = 2$ s, it is at 2 V; at $t = 3$ s, it is at 3 V; and so on.

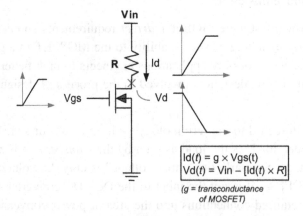

$$Id(t) = g \times Vgs(t)$$
$$Vd(t) = Vin - [Id(t) \times R]$$

(g = transconductance of MOSFET)

Figure 8.1: Switching a resistive load.

The analysis proceeds as follows ("Vds" is the Drain-to-Source voltage at any given moment, "Vgs" is the Gate-to-Source voltage, and "Id" is the Drain-to-Source current):

- At $t = 0$, Vgs equals 0 V. Therefore, from the transconductance equation, Id is 0 A. So, the drop across the 1-Ω resistor is 0 V (using Ohm's law). Therefore, the voltage at the Drain of the MOSFET "Vd" (or "Vds" in this case), equals 10 V.
- At $t = 1$ s, Vgs equals 1. Therefore, from the transconductance equation, Id is 1 A. So, the drop across the 1 Ω resistor is 1 V (using Ohm's law). Therefore, Vds equals $10 - 1 = 9$ V.
- At $t = 2$ s, Vgs equals 2. Therefore, from the transconductance equation, Id is 2 A. So, the drop across the 1 Ω resistor is 2 V (using Ohm's law). Therefore, Vds equals $10 - 2 = 8$ V.

We proceed ramping up the Gate voltage progressively in this manner. When 10 s have elapsed, Vgs is 10 V, Id is 10 A, and Vds is 0 V. *After* 10 s, no further change in Vds or Id can occur, even if Vgs is increased further.

> *Note: In general, if the Gate voltage is increased beyond what it takes to deliver a specified maximum load current, we can say that in effect, we are applying "overdrive." This is usually considered wasteful in that sense, but in practice, overdrive helps reduce the on-resistance of the MOSFET and thereby decrease its conduction losses.*

The maximum load current in our example is therefore 10 A and is "Idmax" in Figure 8.2. If we plot the Drain current and Drain voltage with respect to time, we see that the *crossover time*, "tcross," is 10 s here. Note that this time is, by definition, the time for *both* the voltage and the current to complete their transitions.

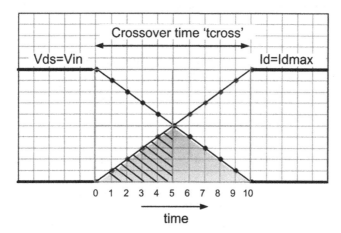

Figure 8.2: The voltage and current waveforms when switching a resistive load.

The energy lost in the MOSFET during the transition is

$$E = \int_0^{tcross} Vd(t)Id(t)dt \text{ Joules}$$

A conceptual point to keep in mind here is that in related literature, it is often stated (rather inaccurately as we will see) that the "area (jointly) enclosed by the voltage, current, and the time axis is the energy lost in the switch" (during the transition). This is the gray isosceles triangle in Figure 8.2. Half of this gray area has been *hatched*. We thus see that within the "crossover interval rectangle," there are eight triangles (in all) with the same area as the hatched triangles. Therefore, the total gray area is one-fourth the area of the crossover interval rectangle. So, *if* the statement about energy being equal to the enclosed area is true, we would have gotten

$$E = \frac{1}{4} \cdot Vin \cdot Idmax \cdot tcross \text{ Joules}$$

This is *not* correct. In fact, we would have reached the same unfortunate conclusion had we argued on the grounds that during the crossover duration, the *average* voltage is Vin/2 and the *average* current is Idmax/2, and therefore the average cross-product is equal to (Vin × Idmax)/4. This is fallacious too. In general,

$$A_{AVG} \times B_{AVG} \neq (A \times B)_{AVG}$$

So yes, this *could* in fact have turned out to be true, *if while the voltage was falling, the current had remained fixed*, and vice versa. That is what happens with an *inductive* load, as we will soon see. However, in the case of a resistive load, both the voltage and the current change *simultaneously* during the crossover interval. We clearly need another (better) way to calculate the switching loss for the resistive case.

Let us compute the instantaneous cross-product $Vds(t) \times Id(t)$ at $t = 1, 2, 3, 4, \ldots$ seconds. If we plot these points out, we get the bell-shaped curve shown in Figure 8.3. So, to get the energy lost during the crossover, we need to find the net area under *this* curve. But we can see that is not going to be easy, because this curve is rather oddly shaped. In fact, there is no other way than to carry out a formal integration/summation procedure. And for that we have to revert to the basic equations for voltage and the current (as presented in Figure 8.1). We then integrate their product over time, and we get

$$E = \frac{1}{6} \cdot Vin \cdot Idmax \cdot tcross \text{ Joules}$$

This is the correct result for the energy lost in the switch during a *resistive* turn-on transition.

If we now turn the MOSFET OFF in the same way (with the crossover time kept fixed), we will get exactly the same energy loss term again, though this time with the voltage *rising* and the current *falling*. We have two transitions per cycle.

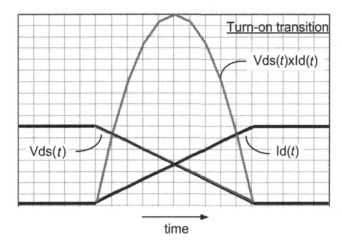

Figure 8.3: The instantaneous energy dissipation curve for resistive switching.

We can thus also conclude that if we switch *repetitively* at the rate of *fsw* Hz, the net dissipation, that is, total energy lost per unit time as heat, is equal to

$$Psw = \frac{1}{3} \cdot Vin \cdot Idmax \cdot tcross \cdot fsw \text{ Watts}$$

This is therefore the *switching loss* (in the switch) for the case of a resistive load.

> *Note: Note that to be precise, this particular term more correctly should be called the "crossover loss," as was first pointed out in* Chapter 1. *The crossover loss (i.e., specifically attributable to the V−I overlap) is not necessarily the entire switching loss taking place in the switch, as we will see.*

Now, suppose we had ramped *up* the Gate voltage at a rate of 1 V per second as before, but ramped *down faster*, say, at the rate of 2 V per second. Then the turn-on time and the turn-off transition time would be different. So, in that case we need to *split* up the crossover loss "Psw" as follows:

$$Psw = Pturnon + Pturnoff$$
$$= \frac{1}{6} \cdot Vin \cdot Idmax \cdot tcross_{ON} \cdot fsw + \frac{1}{6} \cdot Vin \cdot Idmax \cdot tcross_{OFF} \cdot fsw$$

where "tcross$_{ON}$" and "tcross$_{OFF}$" are the crossover times during turn-on and turn-off, respectively.

Now suppose the value of the external resistor was made larger, say 2 Ω instead of 1 Ω. Then the voltage at the Drain would have swung from 10 V to 0 V in only 5 s. And by that time, the Drain current would have reached only 5 A. The Gate voltage would at that moment be 5 V. However, *no further change in Id is possible* (even if we increase Vgs further). Therefore, though the crossover interval has become half of what it was, *the rise time of the current is still equal to the fall time of the voltage* (i.e., 5 s). This is a characteristic *only* of *resistive* loads (since $V = IR$ applies to them).

The rules of the game change considerably when we have an inductive load. In fact, the calculation becomes simpler — ironically because the simplicity (and predictability) of Ohm's law is lost.

Switching an Inductive Load

When we switch an inductive load (with a freewheeling path present of course!), we will get the waveforms shown in Figure 8.4 (idealized). At first sight, they may seem similar to the resistive load waveforms shown in Figure 8.2. But on closer examination, they are very different. In particular, we see that *when the current is swinging, the voltage remains fixed*, and *when the voltage is swinging, the current remains fixed.*

Let us calculate the crossover loss under these conditions. We can do a formal integration as before. But this time, we realize there is in fact an easy way out! Since *one* of the parameters (*V* or *I*) is fixed when the *other* is varying, we can now take the *average* value of the current, Idmax/2, and the *average* value of the voltage, Vin/2, to find the average cross-product. In this manner, we arrive at the energy lost (in Joules) during the turn-on transition

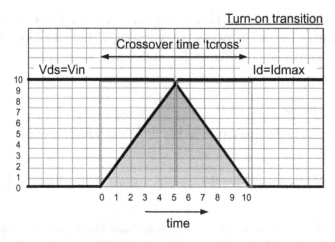

Figure 8.4: The voltage and current waveforms when switching an inductive load.

$$E = \left[\frac{Vin}{2} \cdot Idmax \cdot \frac{tcross}{2} \right] + \left[Vin \cdot \frac{idmax}{2} \cdot \frac{tcross}{2} \right]$$

$$= \frac{1}{2} \cdot Vin \cdot Idmax \cdot tcross$$

Note that for the same reason as indicated above, we can now justifiably think in terms of the *area* enclosed. By simple geometry, the gray area in Figure 8.4 is half the rectangular area, and so we get the same result as above.

We realize that our ability to avoid integration (and use simpler arguments to calculate the crossover loss) is just a piece of "good luck" here — specific to the case of an *inductive* load.

Finally, when we switch *repetitively*, the inductive switching loss is

$$Psw = Vin \cdot Idmax \cdot tcross \cdot fsw \text{ Watts}$$

Note: We may superficially conclude that switching an inductive load leads to a dissipation three times greater than a resistive load. That is indeed true, but only under the exact same conditions. In reality, the value of Idmax is fixed for the case of a resistive load (depending on the value of the resistance used). But for an inductive load, the current can be virtually anything — there is no set "Idmax" as such anymore, it is whatever current that happens to be flowing through the inductor at the instant of switching (either just before or after).

A basic question still remains — *why* are the inductive waveforms so different from the resistive case? To answer that, we have to go back to our previous analysis of the resistive load case. There we will see that we had invoked Ohm's law to find the voltage across the switch. But with an inductor, Ohm's law clearly does not apply. So, to get the waveforms shown in Figure 8.4, we have to recollect something we learned in *Chapter 1* — when we turn the switch OFF, the inductor will create *whatever voltage is necessary to maintain the continuity of current through it*. Let us now show this principle at work in an actual Buck converter, for example (see Figure 8.5).

In Figure 8.5, we first consider the *turn-on* transition (on the left). Just prior to this, the diode is obviously carrying the full inductor current (circled "1"). Then the switch starts to turn ON, trying to share some of this inductor current (circled "2"). The diode current therefore must fall correspondingly (circled "3"). However, the important point is that while the switch current is still in transit, the diode has to be able to pass *some* current (the remainder, or leftover amount of the inductor current). But, to provide even *some* of the inductor current, the diode must remain *fully* forward-biased. Therefore, nature (i.e., induced voltage in this case) *forces* the voltage at the switching node to remain slightly *below ground* — so as to keep the anode of the diode about 0.5 V higher than the cathode (circled "4"). Then, by Kirchhoff 's voltage law,

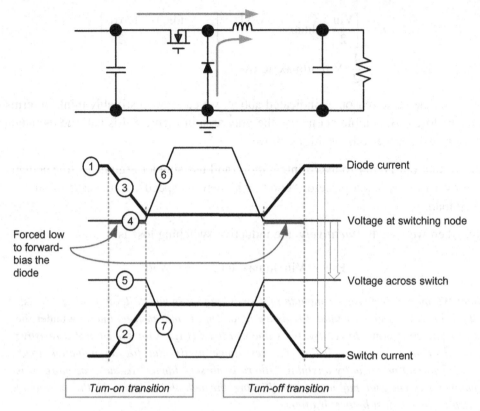

Figure 8.5: Analyzing the transitions in a Buck converter.

the voltage across the switch *stays high* (circled "5"). Only finally, when the *entire* inductor current has shifted to the switch, does the diode "let go." With that, the switching node is released, and it flies up close to the input voltage (circled "6") — and so now, the voltage across the switch is allowed to fall (circled "7").

- We therefore see that *at turn-on, the voltage across the switch does not change until the current waveform has* **completed** *its transition.* We thus get a significant *V−I* overlap.

If we do a similar analysis for the turn-off transition (right side of Figure 8.5), we will see that for the switch current to start decreasing by even a small amount, the diode must first be "positioned" to take up *any* current coming its way. So, the voltage at the switching node *must first* fall close to zero so as to forward-bias the diode. That also means the voltage across the switch must first transit fully, *before* the switch current is even allowed to decrease slightly (see Figure 8.5).

- We therefore see that *at turn-off, the current through the switch does not change until the voltage waveform has* **completed** *its transition.* We thus get a significant *V−I* overlap.

We see that the fundamental properties and behavior of an inductor, as described in *Chapter 1*, are ultimately responsible for the significant $V - I$ overlap during crossover.

The same situation is present in the case of any switching topology. Therefore, *the switching loss equation presented earlier also applies to all topologies.* What we have to remember is that in our equations, we are referring to the voltage *across* the switch (when it is OFF) and the current *through* it (when it is ON). In an actual converter, we will need to ultimately relate these V and I to the actual input/output rails and load current of the application. The procedure for that is described later.

Switching Losses and Conduction Loss

The underlying motivation for initiating *switching* in modern power conversion is often simplistically stated as follows — by switching the transistor, either the voltage across the transistor is close to zero, or the current through it is close to zero, and therefore the dissipation cross-product "$V \times I$" is also almost zero. We have seen that during the transition, that doesn't really hold true anymore (the $V-I$ *overlap*). Similarly, we should keep in mind that though the $V \times I$ losses are much closer to the ideal or "expected" value of zero when the switch is OFF, there are considerable losses when the switch is ON. That is because when the switch is OFF, it is *really* so — the *leakage current* through a modern semiconductor switch is almost negligible. However, when the switch is ON, the voltage across it is *not even close to zero* in many cases. One of the highest reported forward-drops is in the "TOPSWITCH®" (an integrated switcher IC meant for medium off-line flyback applications) — *over 15 V* across it (over rated current and temperature)! In general, there will remain a significant $V \times I$ loss term even after the inductor current has shifted entirely from the diode to the switch. This particular loss term is clearly the *conduction loss*, P_{COND} (of the switch). It can in fact be comparable to, or even greater than, the crossover loss.

However, unlike the crossover loss, the conduction loss is *not* frequency-dependent. It does depend on *duty cycle*, but not on frequency. For example, suppose the duty cycle is 0.6, then in a measurement interval of say, 1 s, the *net* time spent by the switch in the ON-state is equal to 0.6 s. But we know that conduction loss is incurred *only* when the switch is ON. So, in this case, it is equal to $a \times 0.6$, where "a" is an arbitrary proportionality constant. Now suppose the frequency is doubled. Then the net time spent in the on-state (in 1 s) *is still 0.6 s.* So, the conduction loss remains $a \times 0.6$. But now, suppose the duty cycle changes from 0.6 to 0.4 (the frequency can be even doubled in the process), the conduction loss is reduced to $a \times 0.4$. So, we realize that conduction loss can't possibly depend on frequency, only on duty cycle.

We can pose a rather philosophical question — why is it that the switching loss is frequency-dependent, but *not* the conduction loss? That is simply because the conduction loss *coincides with the interval in which power is being processed in the converter.* Therefore, as long as the *application* conditions do not change (duty cycle fixed and input and output power fixed), neither can the conduction loss.

The equation to calculate the conduction loss of a MOSFET is simply

$$P_{COND} = I_{RMS}^2 \times \text{Rds Watts}$$

where "Rds" is the on-resistance of the MOSFET. I_{RMS} is the RMS of the switch current waveform. It is equal to

$$I_{RMS} = I_O \times \sqrt{D \times \left(1 + \frac{r^2}{12}\right)} \quad \text{(Buck)}$$

$$I_{RMS} = \frac{I_O}{1 - D} \times \sqrt{D \times \left(1 + \frac{r^2}{12}\right)} \quad \text{(Boost and Buck-Boost)}$$

where I_O is now the load current of the *DC–DC converter* stage and D is its duty cycle. Note that to a first approximation (current ripple ratio assumed very small), this is equal to

$$I_{RMS} \approx I_{DC} \times \sqrt{D} \quad \text{(Buck, Boost, and Buck-Boost)}$$

where I_{DC} is the average inductor current and "I_{RMS}" is the RMS of the switch current waveform.

The *diode conduction loss* is the other major conduction loss term in a power supply. It is equal to $V_D \times I_{D\_AVG}$, where V_D is the diode forward-drop. I_{DAVG} is the average current through the diode — equal to I_O for the Boost and the Buck-Boost, and $I_O \times (1 - D)$ for the Buck. It too is independent of switching frequency.

We realize that the obvious way to reduce conduction losses is by lowering the forward-drops across the diode and switch. So, we look for diodes with a low drop — like the Schottky diode. Similarly, we look for MOSFETs with a low on-resistance "Rds." However, there are compromises involved here. The leakage current in a Schottky diode can become significant as we try to choose diodes with very low drops. We can also run into significant body capacitance, which will end up being more dissipative. Similarly, the speed at which the MOSFET switches can be adversely affected as we try to reduce its Rds.

A Simplified Model of the MOSFET for Studying Inductive Switching Losses

In Figure 8.6, on the left, we have the basic (simplified) model of the MOSFET. In particular, we observe that it has three parasitic capacitances — between its Drain, Source,

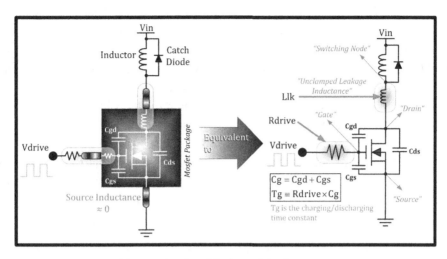

Figure 8.6: Simplified model of MOSFET.

and Gate. These "small" *interelectrode capacitances* are the key to maximizing switcher efficiency, especially at higher switching frequencies. Their role in the switching transition needs to be understood clearly.

We have seen that the basic reason why we get *any* crossover loss in the first place is that there is an unavoidable *V–I* overlap during every switching transition. That overlap occurs because the inductor keeps trying to force current, and tries to create suitable conditions for that to happen seamlessly, as we switch. But the reason why this overlap lasts *as long* as it does is mainly that these three interelectrode capacitors are *demanding to be charged or discharged* (as the case may be) at every switching event — so that they can reach their new DC levels, commensurate with the altered state of the switch and circuit. So crudely stated, if these capacitances are "big," they take a longer time to charge or discharge, thus increasing the crossover (overlap) time. And that in turn increases the crossover loss. Further, since the charging and discharging paths of these capacitors often include the *Gate resistor*, the value of the Gate resistance also considerably impacts the transition time and thereby the switching loss.

On the right side of Figure 8.6, we have further simplified our simple model. So, we have lumped the internal and external inductances present at the Drain into a single leakage inductance "Llk." Note that we are ignoring any Gate-to-Source inductance, thus implicitly assuming the printed circuit board (PCB) layout is very good in this regard. We also lump the small resistor present inside the MOSFET along with the *external* Gate resistor (if present) and the *driver resistance* (its internal pull-up or pull-down) — to give a single effective "Rdrive" or *drive resistance*.

Note that in Figure 8.6, the main inductor is "coupled" — because it has a freewheeling path available. But the leakage (or parasitic) inductance is "uncoupled" because it has no path to send forth its energy. It therefore expectedly "complains" — in the form of a voltage spike (whenever we try to change the current through it). However, in our analysis, we will be assuming this *leakage inductance is very small* (though not necessarily negligible either). We will find that this results in certain *artifacts* in the switching waveforms, which makes them appear slightly different, as compared to the idealized inductive switching waveforms shown in Figures 8.4 and 8.5. However, it turns out that these artifacts are mainly of academic interest (provided of course that Rdrive is "small"). In addition, the artifacts in question typically help *decrease* the crossover losses slightly. Therefore, the idealized waveforms are more "conservative" in that sense, and we would do fine just sticking to them.

Turning our attention to the "circuit" shown in Figure 8.6, we should be clear that this circuit doesn't really *work!* We know from our discussions in *Chapter 1* that we can never hope to achieve a *steady state* without at least an *output capacitor* present — to charge up and thereby help stabilize the voltseconds across the inductor. So, this circuit is clearly an *idealization* — it only helps us to perform a *paper-analysis* of a *particular switching transition.*

Note that ultimately, the switch cares only about the *voltage that appears across it* when it turns OFF and the *current passing through it* when it is ON. That is why this simple circuit can be safely accepted as representative of what happens in *any* topology *at the moment of transition.* For instance, we could take both the leakage and the main inductor in Figure 8.6, and place them on the *Source* side of the MOSFET instead. As long as the Gate drive is still well coupled to the Source (i.e., no inductance *between* Gate and Source), nothing really changes. That is no surprise because we know that if a certain component (or circuit block) "A" is in series with "B," we can always interchange their positions and make B in series with A, without changing a thing.

Finally, we should keep in mind that what we are calling the "Drain" in our analysis is not necessarily the *pin* of the *package* (of the same name). Nor the switching node! The inductance Llk separates these points as indicated in Figure 8.6. Therefore, for example, though the switching node is necessarily clamped close to the "Vin" rail when the diode is freewheeling, the Drain of the device may momentarily show a slightly different voltage (clearly equal to the voltage appearing across Llk).

The Parasitic Capacitances Expressed in an Alternate System

We will now progress to a detailed study of the inductive switching transitions of a MOSFET. For that, we will be splitting up the turn-on and turn-off into several subintervals of interest. We will learn that *for most of these subintervals, the Gate behaves as a simple input capacitance — that is being charged (or discharged)* through the resistor "Rdrive." The situation is identical to the simple RC circuit we discussed in *Chapter 1*. In effect, the Gate is "blind" to what all may be happening between Drain and Source (on account of the transconductance of the MOSFET).

If we look *into* the Gate, from the viewpoint of the AC drive signal, the effective input charging capacitance is the parallel combination (arithmetic sum) of Cgs and Cgd. We are going to call this simply the *Gate* or *input capacitance* "Cg" in our discussion. So,

$$Cg = Cgs + Cgd$$

The *time constant* of the charging/discharge cycles of the Gate is therefore

$$Tg = Rdrive \times Cg$$

Note: Here we seem to be indirectly suggesting that the drive resistance is the same for turn-on and turn-off. That need not be so. All the equations we will present can easily take any existing difference in the turn-on and turn-off drive resistances into account. So, in general, we will have different crossover times for the turn-on and turn-off transitions. Also note that in general, within a certain crossover interval (turn-on or turn-off), the actual time it takes for the voltage to transit need not be the same as the time the current takes (unlike the case of a resistive load).

An *alternative* system of writing the capacitances is in terms of the *effective* input, output, and reverse transfer capacitances — that is, *Ciss*, *Coss*, and *Crss*, respectively. These are related to the interelectrode capacitances as follows

$$Ciss = Cgs + Cgd \equiv Cg$$
$$Coss = Cds + Cgd$$
$$Crss = Cgd$$

So we can also write

$$Cgd = Crss$$
$$Cgs = Ciss - Crss$$
$$Cds = Coss - Crss$$

In most vendors' datasheets, we can usually find Ciss, Coss, and Crss under the section "typical performance curves." We will then see that these parasitic capacitances are a *function of voltage.* Clearly, that can significantly complicate any analysis. So, as an approximation, *we are going to assume that the interelectrode capacitances are all constants.* We will consult the typical performance curves of the MOSFET, and then pick the value of the capacitance *corresponding to the voltage that appears across the MOSFET* when it is OFF (in our given application). Later, we will show how to minimize this error, by the use of a "scaling factor."

Gate Threshold Voltage

The "perfect MOSFET" we talked about earlier (Figure 8.1) started conducting the moment we raised the Gate voltage above ground (i.e., Source). But an actual MOSFET has a certain *Gate threshold voltage "Vt."* This is typically $1-3$ V for "logic-level" MOSFETs and about $3-10$ V for high-voltage MOSFETs. So basically, we have to exceed the stated threshold voltage to get the MOSFET to conduct at all ("conduction" is defined typically as a current in excess of 1 mA).

Because Vt is not zero, the definition of transconductance also needs to be modified slightly from

$$g = \frac{Id}{Vgs} \Rightarrow g = \frac{Id}{Vgs - Vt}$$

Note that in our analysis, we are making another simplifying assumption — that the transconductance too is a constant.

Finally, with all this background information, we can start looking closely at what actually happens during the turn-on and turn-off transitions.

The Turn-On Transition

We have divided this interval into *four* subintervals as detailed individually in Figures 8.7–8.10. For quick reference and ease of understanding, the relevant explanations and comments for each subinterval are also provided within their respective figures.

Briefly, the interval *t*1 is just the time to get to the threshold Vt. During this time, we just have a simple RC charging circuit. In *t*2 also, the exponential rise continues, but this time, the Drain current starts ramping up. But for all practical purposes, the Gate doesn't "know" anything has changed because the transconductance is fully responsible for the Drain current (and further, there is no change in the Drain voltage). But in *t*3, the diode is allowed to stop conducting (since all the inductor current has by now shifted over into the switch). So, now the Drain voltage swings. But in doing so, *it injects a current through Cgd.* Note that this capacitance, despite being usually rather small, has probably the greatest effect on the crossover time — because of the fact that it *directly* injects current from a high switching voltage node (Drain) on to the Gate. Just prior to the interval *t*3, Cgd has a relatively high voltage across it. But when the switch is fully ON, the voltage across Cgd

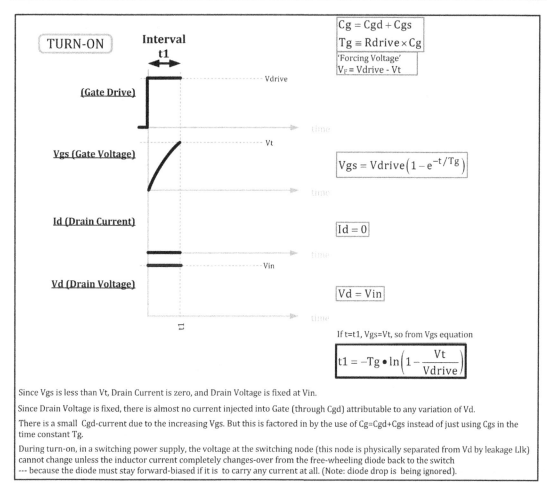

$$Cg = Cgd + Cgs$$
$$Tg \equiv Rdrive \times Cg$$
'Forcing Voltage'
$$V_F = Vdrive - Vt$$

TURN-ON   **Interval t1**

(Gate Drive) — Vdrive

Vgs (Gate Voltage) — Vt

$$Vgs = Vdrive\left(1 - e^{-t/Tg}\right)$$

Id (Drain Current)

$$Id = 0$$

Vd (Drain Voltage) — Vin

$$Vd = Vin$$

If t=t1, Vgs=Vt, so from Vgs equation

$$t1 = -Tg \bullet \ln\left(1 - \frac{Vt}{Vdrive}\right)$$

Since Vgs is less than Vt, Drain Current is zero, and Drain Voltage is fixed at Vin.

Since Drain Voltage is fixed, there is almost no current injected into Gate (through Cgd) attributable to any variation of Vd.

There is a small Cgd-current due to the increasing Vgs. But this is factored in by the use of Cg=Cgd+Cgs instead of just using Cgs in the time constant Tg.

During turn-on, in a switching power supply, the voltage at the switching node (this node is physically separated from Vd by leakage Llk) cannot change unless the inductor current completely changes-over from the free-wheeling diode back to the switch --- because the diode must stay forward-biased if it is to carry any current at all. (Note: diode drop is being ignored).

Figure 8.7: First interval of turn-on.

must decrease to its new final low value. Therefore, *during t3*, Cgd is essentially discharging. So, the question is — *what is the path the Cgd discharge current takes*! We can analyze that as follows — having reached the Gate, this discharge current has *two* choices — either to go through Cgs and/or to go through Rdrive. But the Gate is already at the constant level of $Vt + I_O/g$ — that being the Gate voltage level required by the MOSFET to support the full inductor current I_O. So, to a first approximation, the voltage across Cgs (Gate voltage) *need not*, and *does not* change. And further, since the general equation for the current through any capacitor is $I = CdV/dt$, the current through Cgs must be zero because there is no *change* in the voltage across it during this subinterval. Therefore, we conclude that *all* the current coming through Cgd into the Gate node gets

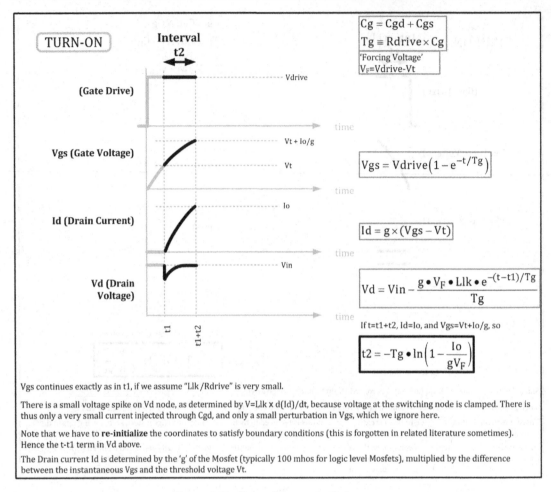

TURN-ON

Interval t2

$Cg = Cgd + Cgs$

$Tg \equiv Rdrive \times Cg$

'Forcing Voltage'
$V_F = Vdrive - Vt$

(Gate Drive)

Vdrive

time

Vgs (Gate Voltage)

Vt + Io/g

Vt

$Vgs = Vdrive\left(1 - e^{-t/Tg}\right)$

time

Id (Drain Current)

Io

$Id = g \times (Vgs - Vt)$

time

Vd (Drain Voltage)

Vin

$Vd = Vin - \dfrac{g \bullet V_F \bullet Llk \bullet e^{-(t-t1)/Tg}}{Tg}$

time

t1   t1+t2

If t=t1+t2, Id=Io, and Vgs=Vt+Io/g, so

$$t2 = -Tg \bullet \ln\left(1 - \dfrac{Io}{gV_F}\right)$$

Vgs continues exactly as in t1, if we assume "Llk /Rdrive" is very small.

There is a small voltage spike on Vd node, as determined by V=Llk x d(Id)/dt, because voltage at the switching node is clamped. There is thus only a very small current injected through Cgd, and only a small perturbation in Vgs, which we ignore here.

Note that we have to **re-initialize** the coordinates to satisfy boundary conditions (this is forgotten in related literature sometimes). Hence the t-t1 term in Vd above.

The Drain current Id is determined by the 'g' of the Mosfet (typically 100 mhos for logic level Mosfets), multiplied by the difference between the instantaneous Vgs and the threshold voltage Vt.

Figure 8.8: Second interval of turn-on.

diverted through Rdrive! But the voltage across Rdrive is fixed — one end of it is at Vdrive and the other at Vt + I_O/g. Therefore, the current through it is predetermined by Ohm's law, which means that Rdrive is actually in full control of the current through Cgd during the interval t3. However, the current through Cgd also obeys the equation $I = C \times dV/dt$. So, if I is fixed at a certain value (by Rdrive), we can calculate the corresponding dV/dt across Cgd *and thereby calculate Vd.* In effect, this means that Cgd and Rdrive are together determining the rate of fall of Drain voltage during t3 (and thus the transition time of the voltage). The *plateau* in the Gate voltage waveform during t3 is called the "Miller plateau" — referring to the effect of the reverse transfer capacitance Cgd. Finally, after the voltage too has completed its swing, the current through Cgd stops completely, and so once again, the Gate behaves as a simple RC charging circuit. Note that during t4, the

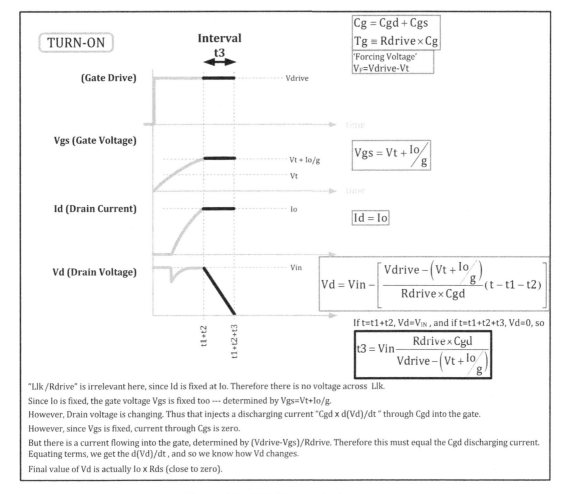

Figure 8.9: Third interval of turn-on.

Gate is in effect being *overdriven* — there is no change in the Drain current anymore (which is already at its maximum possible value). However, *driver dissipation continues during t4.*

The "crossover time," being the time during which both the current *and* the voltage are transiting, is $t2 + t3$. As indicated, to know the driver dissipation, we need to consider the *entire* duration $t1 + t2 + t3 + t4$. Note that by definition, at the end of $t4$, the Gate voltage is at 90% of its asymptotic level (Vdrive). So, we can safely assume that for all practical purposes, the driver does very little after this point. Therefore, at the end of $t4$, the transition is considered complete — from the viewpoint of the switch, and also the driver.

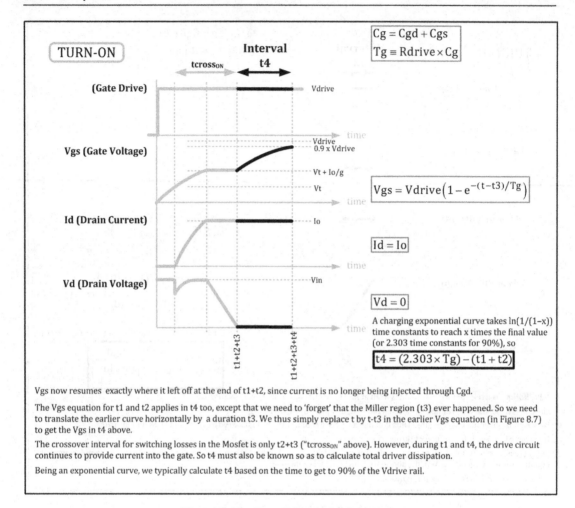

Figure 8.10: Fourth interval of turn-on.

The Turn-Off Transition

In a similar manner as for turn-on, we have divided the turn-off interval into *four* subintervals, as shown in Figures 8.11−8.14.

Briefly, the interval *T*1 is the time for the "overdrive" to cease; that is, the Gate returns to the *sustaining* level Vt + I_O/g (the minimum Gate voltage required to support the full Drain current I_O). During this time, there is no change in the Drain current, nor in the Drain voltage, and so in effect we once again have a simple RC discharging circuit. In *T*2, the Gate voltage again *plateaus*. The reason for that is that the Drain voltage *must*

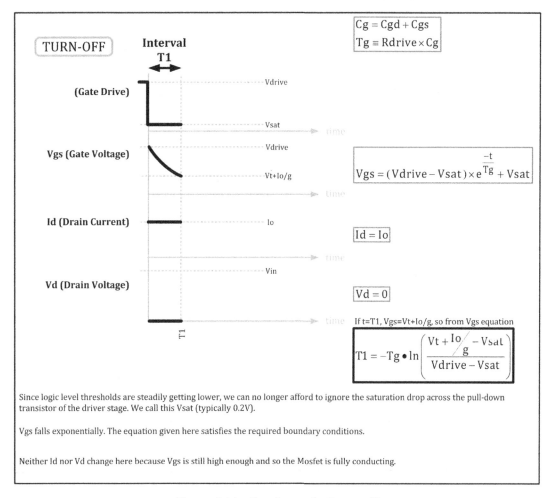

$$Cg = Cgd + Cgs$$
$$Tg \equiv Rdrive \times Cg$$

TURN-OFF

Interval
T1

(Gate Drive)

Vdrive

Vsat

time

Vgs (Gate Voltage)

Vdrive

Vt+Io/g

time

$$Vgs = (Vdrive - Vsat) \times e^{\frac{-t}{Tg}} + Vsat$$

Id (Drain Current)

Io

$$Id = Io$$

time

Vin

Vd (Drain Voltage)

$$Vd = 0$$

time

If t=T1, Vgs=Vt+Io/g, so from Vgs equation

$$T1 = -Tg \bullet \ln\left(\frac{Vt + \frac{Io}{g} - Vsat}{Vdrive - Vsat}\right)$$

T1

Since logic level thresholds are steadily getting lower, we can no longer afford to ignore the saturation drop across the pull-down transistor of the driver stage. We call this Vsat (typically 0.2V).

Vgs falls exponentially. The equation given here satisfies the required boundary conditions.

Neither Id nor Vd change here because Vgs is still high enough and so the Mosfet is fully conducting.

Figure 8.11: First interval of turn-off.

first swing close to Vin, and thereby "position" the diode to get forward-biased and be ready to start taking up the current that the switch will progressively shed (see Figure 8.5). So, *T*2 is the time for the voltage transition to complete. During *T*1 and *T*2 therefore, no change in the Drain current occurs. And with logic similar to what we presented for the turn-on subinterval *t*3, during *T*2 the rate of *rise* of the voltage Vds is once again determined (only) by Rdrive and Cgd. Finally, in *T*3, the current starts falling toward zero. The Gate voltage falls *exponentially* (as an RC circuit) — down to Vt, at which moment, the end of subinterval *T*3 is declared. The transition is now complete *as far as the switch is concerned.* But after that, during *T*4, the RC exponential discharge continues down to 10% of the initial Gate drive amplitude. As before, driver dissipation occurs over *T*1 + *T*2 + *T*3 + *T*4, whereas crossover occurs during *T*2 + *T*3.

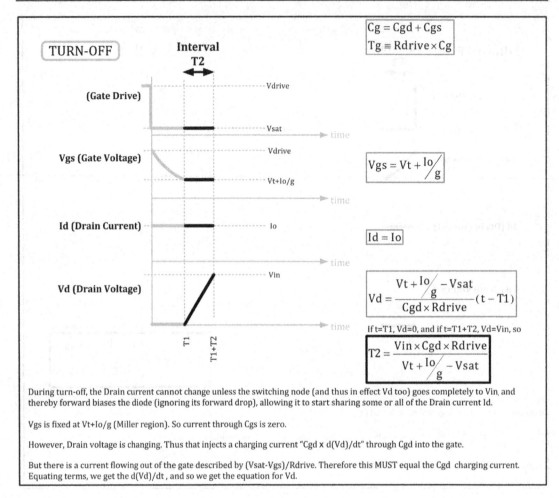

Figure 8.12: Second interval of turn-off.

Gate Charge Factors

A more recent way of describing the parasitic capacitor-based effects in a MOSFET is in terms of *Gate charge factors*. In Figure 8.15, we show how these charge factors, Qgs, Qgd, and Qg, are defined. On the right column of the table in the figure, we have given the relationships between the Gate charge factors and the capacitances, *assuming the latter are constants*. Gate charge factors represent a more accurate way of proceeding, since the interelectrode capacitances are such strong functions of the applied voltage. However, our entire analysis of the turn-on and turn-off intervals so far has been implicitly based on the assumption that the interelectrode capacitances are constants. A possible way out of this,

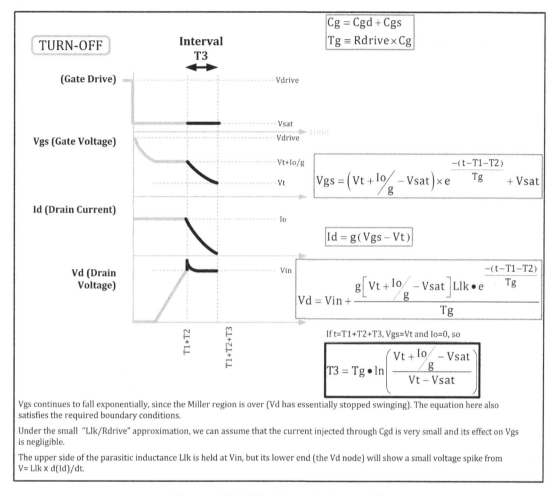

$$Cg = Cgd + Cgs$$
$$Tg \equiv Rdrive \times Cg$$

$$Vgs = \left(Vt + \frac{Io}{g} - Vsat\right) \times e^{\frac{-(t-T1-T2)}{Tg}} + Vsat$$

$$Id = g(Vgs - Vt)$$

$$Vd = Vin + \frac{g\left[Vt + \frac{Io}{g} - Vsat\right]Llk \bullet e^{\frac{-(t-T1-T2)}{Tg}}}{Tg}$$

If t=T1+T2+T3, Vgs=Vt and Io=0, so

$$T3 = Tg \bullet \ln\left(\frac{Vt + \frac{Io}{g} - Vsat}{Vt - Vsat}\right)$$

Vgs continues to fall exponentially, since the Miller region is over (Vd has essentially stopped swinging). The equation here also satisfies the required boundary conditions.

Under the small "Llk/Rdrive" approximation, we can assume that the current injected through Cgd is very small and its effect on Vgs is negligible.

The upper side of the parasitic inductance Llk is held at Vin, but its lower end (the Vd node) will show a small voltage spike from V= Llk x d(Id)/dt.

Figure 8.13: Third interval of turn-off.

one that also helps reduce the error in our switching loss estimates, is detailed in Figure 8.16, using the Si4442DY (from Vishay) as an example.

Basically, we are using the Gate charge factors to tell what the *effective* capacitances are (as the voltage swings from 0 to Vin). We see that the *effective* input capacitance (Ciss), for example, is about 50% greater than the *single-point* Ciss value that we would have read off from the typical performance curves (i.e., 6,300 pF instead of 4,200 pF). That factor accounts for the fact that as the voltage falls, the capacitance increases. Note that we could have calculated a *scaling factor* individually for each capacitance. But it is simpler to use, say Ciss, to first find a "universal" scaling factor — and then apply it across the board to *all* the capacitances. In this manner, we arrive at the effective interelectrode capacitances quoted in Figure 8.16. These are the values we should use for our switching loss

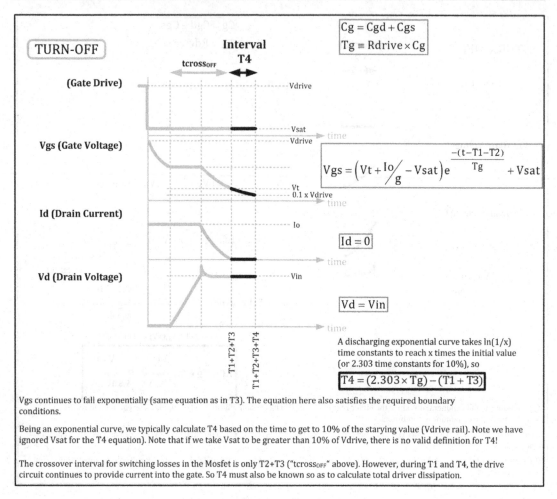

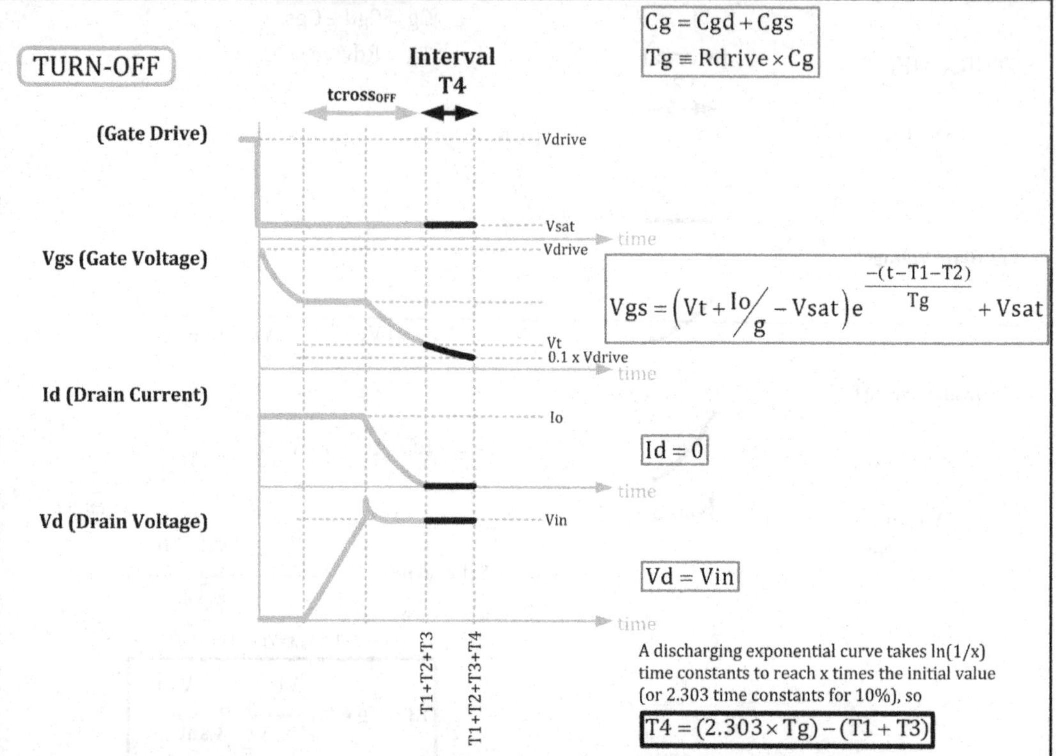
TURN-OFF
Interval
T4
tcrossOFF
(Gate Drive)
Vdrive
Vsat
Vgs (Gate Voltage)
Vdrive
time
Vt
0.1 x Vdrive
time
Id (Drain Current)
Io
time
Vd (Drain Voltage)
Vin
time
T1+T2+T3
T1+T2+T3+T4
Cg = Cgd + Cgs
Tg ≡ Rdrive × Cg
Vgs = (Vt + Io/g − Vsat)e^(−(t−T1−T2)/Tg) + Vsat
Id = 0
Vd = Vin
A discharging exponential curve takes ln(1/x) time constants to reach x times the initial value (or 2.303 time constants for 10%), so
T4 = (2.303 × Tg) − (T1 + T3)

Vgs continues to fall exponentially (same equation as in T3). The equation here also satisfies the required boundary conditions.

Being an exponential curve, we typically calculate T4 based on the time to get to 10% of the starying value (Vdrive rail). Note we have ignored Vsat for the T4 equation). Note that if we take Vsat to be greater than 10% of Vdrive, there is no valid definition for T4!

The crossover interval for switching losses in the Mosfet is only T2+T3 ("$tcross_{OFF}$" above). However, during T1 and T4, the drive circuit continues to provide current into the gate. So T4 must also be known so as to calculate total driver dissipation.

Figure 8.14: Fourth interval of turn-off.

calculations (in preference to those provided by directly reading off Ciss, Coss, and Crss from their curves). Note that for finding the scaling factor, if we had looked at Crss (Cgd) instead of Ciss, then we would find that the calculated effective capacitance is only 40% higher (than what we would read directly from the curves). So, the scaling factor can, in general, be fixed at around 1.4–1.5 typically.

Worked Example

We are switching 22 A at 15 V through a Si4442DY MOSFET, at 500 kHz. The total pull-up drive resistance, by which the Gate is driven by a pulse of amplitude 4.5 V, is

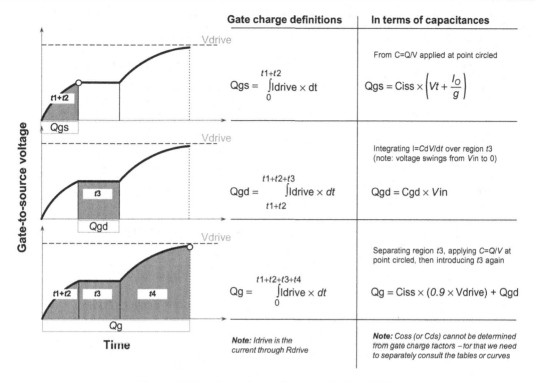

	Gate charge definitions	In terms of capacitances
	$Qgs = \int\limits_{0}^{t1+t2} Idrive \times dt$	From C=Q/V applied at point circled $Qgs = Ciss \times \left(Vt + \dfrac{I_O}{g}\right)$
	$Qgd = \int\limits_{t1+t2}^{t1+t2+t3} Idrive \times dt$	Integrating I=CdV/dt over region *t3* (note: voltage swings from *Vin* to 0) $Qgd = Cgd \times Vin$
	$Qg = \int\limits_{0}^{t1+t2+t3+t4} Idrive \times dt$	Separating region *t3*, applying *C=Q/V* at point circled, then introducing *t3* again $Qg = Ciss \times (0.9 \times Vdrive) + Qgd$
	Note: *Idrive is the current through Rdrive*	**Note:** *Coss (or Cds) cannot be determined from gate charge factors – for that we need to separately consult the tables or curves*

Figure 8.15: Gate charge factors of a MOSFET.

2 Ω. At turn-off, it is pulled-down (to source) by a total drive resistance of 1 Ω. Estimate the switching losses and the dissipation in the driver.

From Figure 8.16, we have Cg = Cgs + Cgd = 6,300 pF.

Turn-On

The time constant is

$$Tg = Rdrive \times Cg = 2 \times 6,300 \text{ pF} = 12.6 \text{ ns}$$

The time for the current to transit is

$$t2 = -Tg \times ln\left(1 - \frac{I_O}{g \times (Vdrive - Vt)}\right) = -12.6 \times ln\left(1 - \frac{22}{100 \times (4.5 - 1.05)}\right)$$

$$t2 = 0.83 \text{ ns}$$

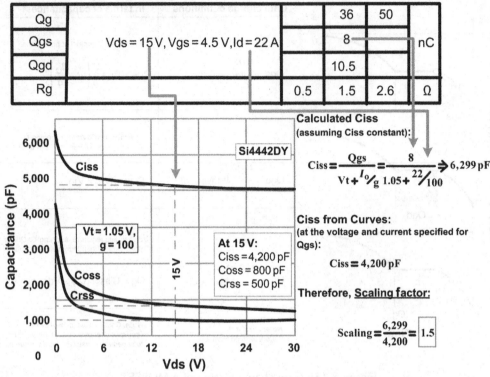

Qg				36	50	
Qgs	Vds = 15 V, Vgs = 4.5 V, Id = 22 A			8		nC
Qgd				10.5		
Rg			0.5	1.5	2.6	Ω

Calculated Ciss
(assuming Ciss constant):

$$Ciss = \frac{Qgs}{Vt + \frac{I_o}{g}} = \frac{8}{1.05 + \frac{22}{100}} \rightarrow 6,299\,pF$$

Ciss from Curves:
(at the voltage and current specified for Qgs):

$$Ciss \approx 4,200\,pF$$

Therefore, Scaling factor:

$$Scaling = \frac{6,299}{4,200} \approx \boxed{1.5}$$

Si4442DY

Vt = 1.05 V, g = 100

At 15 V:
Ciss = 4,200 pF
Coss = 800 pF
Crss = 500 pF

Final values of capacitance to use (at the voltage and current specified for Qgs):

Ciss ≈ 4,200 pF × Scaling ≈ 6,300 pF
Coss ≈ 800 pF × Scaling ≈ 1,200 pF
Crss ≈ 500 pF × Scaling ≈ 750 pF

Cgd = Crss = 750 pF
Cgs = Ciss-Cgd = 6,300-750 = 5,550 pF
Cds = Coss-Cgd =1,200-750 = 450 pF

Figure 8.16: Estimating the *effective* interelectrode capacitances from the Gate charge factors (Si4442DY as an example).

The time for the voltage to transit is

$$t3 = Vin \times \frac{Rdrive \times Cgd}{Vdrive - (Vt + (I_o/g))} = 15 \times \frac{2 \times 0.75}{4.5 - (1.05 + (22/100))}$$

$$t3 = 6.966 \text{ ns}$$

So, the crossover time during turn-on is

$$tcross\_turnon = t2 + t3 = 0.83 + 6.966 = 7.8 \text{ ns}$$

The turn-on crossover loss therefore is

$$Pcross\_turnon = \frac{1}{2} \times Vin \times I_O \times tcross\_turnon \times fsw$$

$$= \frac{1}{2} \times 15 \times 22 \times 7.8 \times 10^{-9} \times 5 \times 10^5$$

$$Pcross\_turnon = 0.64 \text{ W}$$

Turn-Off

The time constant is now

$$Tg = Rdrive \times Cg = 1 \times 6,300 \text{ pF} = 6.3 \text{ ns}$$

The time for the voltage to transit is

$$T2 = \frac{Vin \times Cgd \times Rdrive}{Vt + (I_O/g)} = \frac{15 \times 0.75 \times 1}{1.05 + (22/100)}$$

$$T2 = 8.858 \text{ ns}$$

The time for the current to transit is

$$T3 = Tg \times \ln\left(\frac{(I_O/g) + Vt}{Vt}\right) = 6.3 \times \ln\left(\frac{(22/100) + 1.05}{1.05}\right)$$

$$T3 = 1.198 \text{ ns}$$

So, the crossover time during turn-off is

$$tcross\_turnoff = T2 + T3 = 8.858 + 1.198 = 10 \text{ ns}$$

The turn-off crossover loss therefore is

$$Pcross\_turnoff = \frac{1}{2} \times Vin \times I_O \times tcross\_turnoff \times fsw$$

$$= \frac{1}{2} \times 15 \times 22 \times 10 \times 10^{-9} \times 5 \times 10^5$$

$$Pcross\_turnon = 0.83 \text{ W}$$

So finally, the **total crossover loss** is

$$Pcross = Pcross\_turnon + Pcross\_turnoff = 0.64 + 0.83 = 1.47 \text{ W}$$

Notice that *we have not even used Cds so far!* This particular capacitance does not affect the *V−I* overlap (since it is not connected to the Gate). But it still needs to be considered! Every cycle, it charges up during turn-off, and then during turn-on it dumps its stored energy inside the MOSFET. This is, in fact, *the additional loss term that needs to be added to the crossover loss term, so as to get the total switching loss in a MOSFET.* Note that in low-voltage applications, this additional term may seem insignificant, but in high-voltage/off-line applications, it does affect the efficiency noticeably. Let us calculate what it is in our case:

$$P\_Cds = \frac{1}{2} \times Cds \times Vin^2 \times fsw = \frac{1}{2} \times 450 \times 10^{12} \times 15^2 \times 5 \times 10^5 = 0.025 \text{ W}$$

So, the *total switching loss* (in the switch) is

$$Psw = Pcross + P\_Cds = 1.47 + 0.025 = 1.5 \text{ W}$$

The *driver dissipation* is

$$Pdrive = Vdrive \times Qg \times fsw = 4.5 \times 36 \times 10^{-9} \times 5 \times 10^5 = 0.081 \text{ W}$$

Note that typically, *the above driver dissipation equation underestimates the actual driver dissipation by almost 20%* — as can be confirmed by integrating the product of the drive current and the voltage across it, over each subinterval. The reason for the error is simply the Miller plateau — because during this interval, some additional current (other than from the stored charge Qg) gets injected into the drive resistor. So, our *corrected* driver dissipation estimate is $1.2 \times 0.081 = 0.097$ W. The driver supply rail current is $0.081/4.5 = 18$ mA.

Applying the Switching Loss Analysis to Switching Topologies

Now we try to understand how our preceding analysis pertains to an actual switching regulator application — in particular, what "Vin" and "I_O" are, with respect to topology.

For a *Buck*, we know that at *turn-on*, the instantaneous switch (and inductor) current is $I_O \times (1 - r/2)$, where r is the current ripple ratio and "I_O" is the load current *of the DC–DC converter.* At turn-off, the current is $I_O \times (1 + r/2)$. Usually, we can ignore the current ripple ratio and take the current as I_O for both the turn-on and the turn-off analyses. So, the load current of the DC–DC converter, I_O, becomes the same as the "I_O" used so far in the switching loss analysis. In a *Boost and Buck-Boost*, the current "I_O" in our switching loss analysis is actually the average inductor current $I_O/(1 - D)$.

Coming to the voltage across the MOSFET when it turns OFF (i.e., "Vin" in the switching loss analysis) — for the *Buck*, this is almost equal to the input rail of the DC–DC converter V_{IN} (a diode drop more in reality). Similarly, for a *Buck-Boost,* the voltage "Vin" is almost exactly equal to $V_{IN} + V_O$, where V_O is the output rail of the DC–DC converter. For a *Boost,* the voltage "Vin" is equal to V_O, that is, the output rail of the converter. Note that if we are dealing with an isolated *flyback*, the voltage at turn-off really is $V_{IN} + V_Z$, where V_Z is the voltage of the zener clamp (placed across the Primary winding). However,

Table 8.1: Connecting The Switching Loss Analysis With Actual Topologies.

	"V_{IN}"		"I_O"	
	Turn-on	**Turn-off**	**Turn-on**	**Turn-off**
Buck	V_{IN}		I_O	
Boost	V_O		$I_O/(1-D)$	
Buck-Boost	$V_{IN} + V_O$		$I_O/(1-D)$	
Flyback	$V_{IN} + V_{OR}$	$V_{IN} + V_Z$	$I_{OR}/1-D)$	
Forward	V_{IN}	$2 \times V_{IN}$	I_{OR}	
$V_{OR} = V_O \times n$ and $I_{OR} = I_O/n$, where $n = N_P/N_S$				

at turn-on, the voltage across the MOSFET is only $V_{IN} + V_{OR}$ (V_{OR} being the *reflected output voltage*, i.e., $V_O \times N_P/N_S$). In a single-ended **Forward converter**, we have $2 \times V_{IN}$ at turn-off, and only V_{IN} at turn-on. Note that in all cases discussed above, we are assuming continuous conduction mode.

We have tabulated these results in Table 8.1 for convenience.

Note that if we were in discontinuous conduction mode, there is in principle *no* switching loss at turn-on — because there is no current flowing in the inductor by that time. At turn-off, the current at transition is $I_{PK} = \Delta I$, which can be found using $V = L \times \Delta I/\Delta t$.

Worst-Case Input Voltage for Switching Losses

We must return now to the all-important question — *when we have a wide-input voltage range, what specific input voltage point represents the worst case for calculating switching losses?*

The switching loss equation is generically

$$Psw = Vin \cdot I_O \cdot tcross \cdot fsw \text{ Watts}$$

We note that in all cases, this loss depends on the *product of Vin and I_O*. But by now, we know what Vin and I_O are — from Table 8.1. So, we can analyze the situation for each topology as follows:

- For a **Buck**, "Vin $\times I_O$" $= V_{IN} \times I_O$. So, the maximum loss will obviously occur at V_{INMAX}.
- For a **Boost**, "Vin $\times I_O$" $= V_O \times I_O/(1 - D)$. So, the maximum loss will occur at D_{MAX}, that is, at V_{INMIN}.
- For a **Buck-Boost**, "Vin $\times I_O$" $= (V_{IN} + V_O) \times I_O/(1 - D)$. We also know that $D = V_O/(V_{IN} + V_O)$. So, plotting "Vin $\times I_O$," we get Figure 8.17 (a typical case).

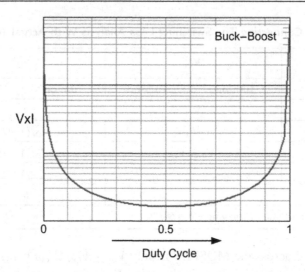

Figure 8.17: Switching loss variation with respect to duty cycle, for the Buck-Boost.

Note that the curve is *symmetrical* around $D = 0.5$ — and that is the point of *minimum* switching losses. Below that point, the *voltage* increases significantly, and above that, the *current* increases significantly. Either way, the switching losses *increase* as we move away from $D = 0.5$. Therefore, in general, we must first examine the input range of our application, and see which of its ends is *furthest* from $D = 0.5$. For example, if in our application, the input range corresponds to a duty cycle range of 0.6–0.8, we need to do the switching loss calculation at $D = 0.8$, that is, at V_{INMIN}. However, if the duty cycle range is say, 0.2–0.7, we need to do the calculation at $D = 0.2$, that is, at V_{INMAX}.

How Switching Losses Vary with the Parasitic Capacitances

In Figure 8.18, we have taken the Si4442DY, and "varied" its Ciss — just to see what can happen as a result of that. On the right vertical axis, we have the corresponding (estimated) switching loss. Note that in computing the loss curve, a "scaling factor" of 1.5 has been applied to the Ciss values given on the left vertical axis (though this is not obvious).

The gray vertical dashed line (annotated "35 nC") represents the Si4442DY *as it is*. So, under the stated conditions, we have an estimated switching loss of 2.6 W. If we increase Ciss by 50%, that is, from 4,200 pF to 6,300 pF, we see that Qg will go up to about 47 nC and the loss to 2.8 W only.

Note: In the actual calculations, using the scaling factor of 1.5, "4,200 pF" is actually 6,300 pF and "6,300 pF" is actually 9,450 pF.

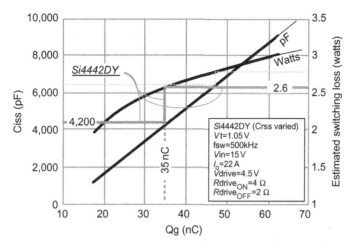

Figure 8.18: Varying the ciss of the Si4442DY.

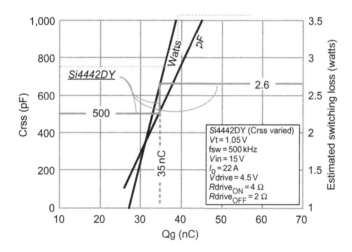

Figure 8.19: Varying the crss of the Si4442DY.

In Figure 8.19, we take the Si4442DY, and "vary" its Crss — just to see what can happen as a result of that. The gray vertical dashed line (annotated "35 nC") represents the Si4442DY *as it is.* So, under the stated conditions, we have an estimated switching loss of 2.6 W. If we increase Crss by 50%, that is, from 500 pF to 750 pF, we see that Qg will go up to about 39 nC only, but the loss goes up to 3.1 W.

In other words, Qg will certainly affect driver dissipation, but it is not necessarily a good indicator of the *switching losses* — it is more *helpful to try to minimize Qgd (or Crss) when selecting MOSFETs,* rather than just looking for a "low-Qg" MOSFET.

Note: In the worked example, we had estimated the losses to be 1.5 W. There we had a pull-up of 2 Ω and a pull-down of 1 Ω. Whereas in Figure 8.18, we have basically doubled the pull-up and pull-down resistors. However, the switching loss has not doubled — it is only 73% more.

Optimizing Driver Capability vis-à-vis MOSFET Characteristics

In Figure 8.20, we have two separate graphs. The one on the left has a fixed pull-up of 4 Ω. On the x-axis, we are therefore, in effect, varying only the pull-down. So if, for example, the x-axis is at 2, the pull-down resistor is 4 Ω/2 = 2 Ω. If the x-axis is at 4, the pull-down resistor is at 4 Ω/4 = 1 Ω. We see that as expected, the losses decrease as the pull-down is improved. We also see the effect of "varying" the threshold voltage (mentally). So, lower-threshold voltages also help lower the switching losses — *provided the pull-down is not too "weak."* On the right graph similarly, we have the results for a fixed pull-up of 10 Ω. We can thereby estimate the effect of varying the pull-up on the overall losses.

Finally, in Figure 8.21, we are keeping the *pull-up + pull-down constant*, as we vary the *ratio* of the pull-up and pull-down resistors. This is being done from the IC designer's viewpoint — suppose he or she has roughly allocated a certain die area for the driver stage, say simplistically fixed the pull-up + pull-down. Then the question is — how should the available drive capability be *distributed* between the pull-up and the pull-down sections. For example, if pull-up + pull-down = 6 Ω, is it better to split this as — pull-up = 4 Ω and pull-down = 2 Ω, or say, pull-up = 3 Ω and pull-down = 3 Ω, or pull-up = 2 Ω and pull-down = 4 Ω, and so on? We see that the answer to that *depends on the threshold voltage.* So, we need to have an idea of the MOSFETs we are planning to use, *before* we decide on the optimum ratio. From Figure 8.21, we see that *if the threshold is greater*

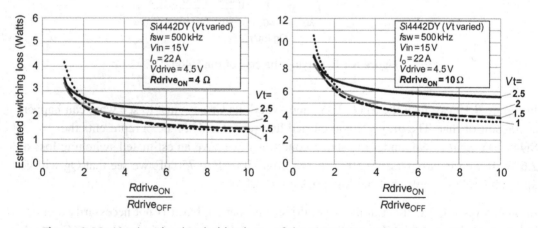

Figure 8.20: Varying the threshold voltage of the Si4442DY and the drive resistances (keeping pull-up resistance fixed).

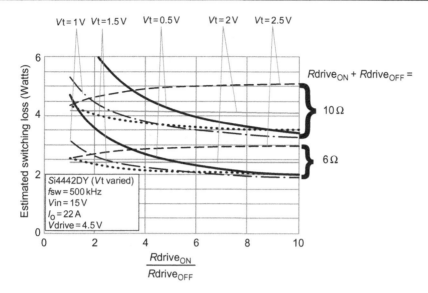

Figure 8.21: Varying the threshold voltage of the Si4442DY and the drive resistances (keeping total drive resistance, i.e., pull-up + pull-down, fixed).

than 2 V, improving the pull-up (at the expense of the pull-down) will help, and so for example — pull-up = 4 Ω and pull-down = 2 Ω will be preferable to pull-up = 5 Ω and pull-down = 1 Ω. However, *if the threshold voltage is below 2 V, we see that the reverse is true* — so now, *improving the pull-down (at the expense of the pull-up) will help.*

Note: Some vendors provide a rather wide range ("MIN" to "MAX") for threshold voltage. Often, they do not even provide a "TYP" value. But surprisingly, some do not even provide the threshold voltage at all! They simply state that their MOSFET is "capable of 4.5-V drive" (as, for example, most of the MOSFETs from www.renesas.com*).*

Please also see Chapter 19 *for a complete solved example.*

CHAPTER 9

Discovering New Topologies

It is always more interesting to learn a topic with the curiosity of a first-time explorer. That is the approach we will take too, as we journey deeper into switching power topologies and related techniques. We have garnered enough confidence and mastery over the fundamental topologies to move to the next stage of their (and our) evolution. However, the field being as large as it is, it is impossible to cover everything without lapsing into superficiality. Our focus here is therefore on developing the underlying intuition, hoping that will help us quickly understand much more as it comes our way in the future. Math, sometimes considered an antithesis to intuition, does help gel things together, especially at a later stage. So we have used it, albeit sparingly. But the emphasis is on intuition.

Part 1: Fixed-Frequency Synchronous Buck Topology

Using a FET (Safely) Instead of Diode

We realize that the catch diode has a significant amount of forward voltage drop, even when we are using a "low-drop" Schottky diode. Further, this forward-drop is relatively constant with respect to current. Looking at diode datasheets we see that, typically, reducing diode current by a factor of 10, only halves the voltage drop of a Schottky. Whereas we know that in a FET, the forward-drop is almost *proportional* to current, so typically, reducing the current by $10 \times$ reduces the voltage drop by about $10\times$. We can thus visualize that the presence of a diode, even a supposedly low-drop Schottky, will impair the converter's efficiency, *especially at light loads*. This will also be more obvious, for almost any load, when the input is high (i.e., low D), since the diode would be conducting for most of the switching cycle, instead of the FET (switch).

We keep hearing of MOSFETs with lower and lower forward-drops (low R_{DS}) every day, but diode technology seems to have remained relatively stagnant in this regard (perhaps limited by physics). Therefore, it is natural to ask: since both diodes and FETs are essentially semiconductor switches, why can't we just interchange them? One obvious reason for that is diodes do not have a third (control) terminal that we can use to turn them ON or OFF at will, as required. We conclude that diodes certainly can't replace FETs, but we should be able to replace a diode with a "synchronous FET," provided we drive it *correctly* by means of its control terminal (the Gate).

Switching Power Supplies A−Z. DOI: 10.1016/B978-0-12-386533-5.00009-7 **343**

What do we mean by driving it "correctly"? There are actually several flavors of "correct," each with its pros and cons as we will see shortly. However, there is certainly an *obvious* way to drive this synchronous FET. It is commonly called "diode emulation mode." Here we very simply attempt to make the FET *copy* basic diode behavior, but also in the process, become one with a much lower forward-drop. That means we need to drive the Gate of the synchronous FET in such a way that the FET conducts exactly when the catch diode it seeks to replace would have conducted, and stop conducting when that diode would have stopped conducting (we will likely need some fairly complex circuitry to do that, but we are not getting into detailed implementation aspects here). Conceptually at least, we presume we can't go wrong here.

At this point, we have fast-fowarded a bit, and presented "diode emulation mode" waveforms for a synchronous Buck converter in Figure 9.1, showing the Gate drives of the two FETs and the corresponding inductor current. Note that in synchronous topologies in general, commonly used FET designators such as "upper" and "lower," or "top"and "bottom" are not necessarily indicative of the actual function, and can always change. Therefore, in this chapter we have generally preferred to call the switch (i.e., the control FET) as "Q," and the synchronous FET as "Qs." That cannot change!

We note that at "high loads" (i.e., with the entire inductor current waveform above "sea-level"), the waveforms are virtually indistinguishable from the classic "non-synchronous" Buck operating in CCM (though, of course, we expect better efficiency, something not obvious looking at the waveforms). Similarly, at light loads, by applying the Gate voltage waveform for Qs as shown in the figure, we ensure the synchronous FET conducts in such a manner that the waveforms are virtually indistinguishable from a non-synchronous Buck operating in DCM. So, we ask: if these waveforms are true, does that mean we can now go ahead and remove the catch diode altogether?

Not so fast! Since we are dealing with an inductor, with almost counter-intuitive behavior on occasion, we need to be extra careful. We intuitively recognize that a catch diode is "natural" in the sense that it is *available when needed* — it presents a path to the freewheeling inductor current automatically, *without user intervention*. Basically, we can't do anything wrong here because we are not doing anything. As mentioned in *Chapter 1*, the only thing we do initially is place that diode at the right place, pointing in the right direction. Then we just sit back and rely on the inductor current to set up whatever voltages are necessary to create a path to freewheel through. With all the possible permutations, we thus get our different topologies and configurations. In all cases, if the diode path is available, the inductor will not "complain" in the form of the "killer voltage spike" discussed in Figure 1.6. However, when we use a FET instead of a diode, that is, a synchronous FET, we have an additional control terminal. And with that extra authority also comes additional

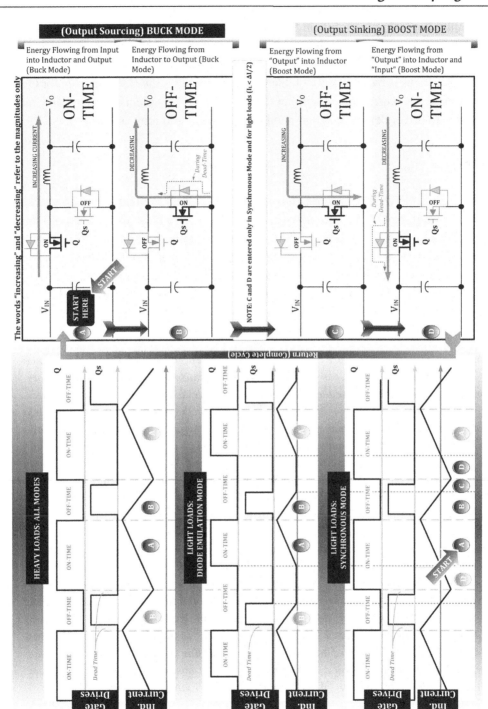

Figure 9.1: Fixed-frequency, synchronous Buck waveforms.

responsibility. For example, if this FET is somehow OFF at the wrong moment, we could conceivably resurrect the killer voltage spike. Therefore, with these sobering thoughts in mind, we now start re-examining some real-world synchronous FET scenarios with extra diligence. Our focus is based mainly on the fact that in reality, the Gate controls for Q and Qs will not *be perfectly matched*, and will not turn ON and OFF as precisely as we may have planned (due to driver delays, inherent latencies, process variations, and so on).

Birth of Dead Time

What if *Qs turns ON a little before Q has turned OFF?* That is potentially disastrous since both the FETs will be ON momentarily (overlap). Just as we were worried about a voltage spike associated with the inductor, now we should be getting equally worried about a current spike coming from the capacitor (in this case the input cap). This potentially dangerous "cross-conduction" spike of current (also called "shoot-through") will flow from the input cap through both the FETs and return via ground. Note that in a synchronous Boost, discussed later, the shoot-through current flows from the output cap, and since that is usually a higher voltage, the spike is an even bigger problem.

> *Note: Can we see or measure this current spike? Well, in severe cases, accentuated by bad PCB layout (which can mimic the effects of mismatched Gate drives), cross-conduction could easily cause the FETs to blow up. In less severe (more common) cases, it may be neither obvious nor measurable. If we place the current probe at the "right place," even the small inductance of the wire loop is usually enough to quell the cross-conduction spike, and so we may not see anything unusual. We have a current spike that we perhaps can't ever see or measure, but its existence is undeniable in the form of an increased* average *current drawn from the input voltage source (measured with a DC multimeter, not a scope) and a corresponding decrease in the measured efficiency (more significant at light loads). The quiescent current "I_Q" will be much higher than we had expected. Note that we are defining "quiescent current" here as the current drawn from the input supply when the converter is "idling" — that is, when its FETs are continuing to switch (at constant frequency), but the load connected to the converter happens to be zero. This is the point where the marketing guy often steps in and redefines quiescent current as an idling condition in which the load is zero, and the FETs are* no longer switching *(perhaps with the help of some dedicated logic pin). With that "slight change" in wording, suddenly, the above-mentioned cross-conduction loss term, along with any normal switching loss terms, is forced to zero, and therefore, the "no-load efficiency" becomes very high — on paper. To get a meaningful or realistic answer from this marketing guy, perhaps we should try asking: "What is the efficiency at 1 mA of load current?"*

The solution to the cross-conduction problem is to introduce a little "dead-time" as shown in Figure 9.1. By design, this is a small intended delay, always inserted between one FET going OFF and the other coming ON. In practice, unless excessively large, dead-time just represents a safety buffer against any accidental or unintentional overlap over manufacturing process, temperature variations, diverse PCB layouts (in the case of switchers using external FETs), and so on.

CdV/dt-Induced Turn-On

Note that there is a phenomenon called *CdV/dt*-induced turn-on that is known to cause cross-conduction, especially in low-voltage VRM-type applications — despite sufficient dead time apparently being present. For example, in a synchronous Buck, if the top N-channel FET turns ON very suddenly, it will produce a high dV/dt on the Drain of the bottom FET. That can cause enough current to flow through the Drain-to-Gate capacitance (Cgd) of the bottom FET, which can produce a noticeable voltage bump on its Gate, perhaps enough to cause it to turn ON momentarily (but may be only a partial turn-on). This will therefore produce an unexpected FET overlap, one apparent perhaps only through an inexplicably low-efficiency reading at light loads. To avoid this scenario, we may need to do one or more of the following: (a) slow down the top FET, (b) have good PCB layout (in the case of controllers driving external FETs) to ensure that the Gate drive of the lower is held firmly down, (c) design the Gate driver of the lower FET to be "stiff," (d) choose a bottom FET with slightly higher Gate threshold, if possible, (e) choose a bottom FET that has a low Cgd, (f) choose a bottom FET with a very small internal series Gate resistance, (g) choose a bottom FET with high Gate-to-Source capacitance (Cgs), and (h) perhaps even try to position the decoupling capacitor positioned on the input rail slightly far away from the FETs (even a few millimeters of trace inductance can help) so that despite slight overlap, at least the cross-conduction *current* flowing during the (voltage) overlap time gets limited by the intervening PCB trace inductances.

Counting on the Body-Diode

Dead-time not only improves efficiency by reducing chances of transistor overlap, but also impairs efficiency itself for two other reasons. But before we discuss that, a basic question arises: are we "tempting nature" by introducing this dead-time in the first place?

What if *Qs turns ON a little after Q turns OFF?* This is also what happens, in effect, when we introduce dead-time. We learned from *Chapter 1* that any delay in providing a freewheeling path for the inductor current (i.e., after the main FET turns OFF) can prove fatal — in the form of a killer voltage spike. Admittedly, there may be some parasitic capacitance somewhere in the circuit, hopefully fortuitously present at the right place, which is able to provide a temporary path for a few nanoseconds. But that capacitor being typically very small would charge (or discharge) very quickly, and then the freewheeling inductor current would have no place to go. So, even a few nanoseconds of dead-time could prove disastrous. Luckily that scenario is not a major concern here. The reason is *most*

FETs have an internal "body-diode" (shown in gray in Figure 9.1). So, for example, if the synchronous FET was not conducting exactly when required, as during the dead-times, *the inductor current will pass through its body-diode* for that tiny interval, as indicated with dashed arrows in Figure 9.1. As a result, no killer voltage spike would be seen, and we would have a switching topology that "abides by the rules of existence" presented in *Chapter 1* (aka "don't mess with the inductor"). However, to retain this advantage, and to indirectly make dead-time feasible, we need to ensure we really are using a synchronous FET with an internal body-diode. That is actually the easy part, since all discrete FETs available in the market have a body-diode inside them. However, a related question is: since a FET is a bidirectional conductor of current (when it is ON), what is the right way to "point" the FET (same question that we had posed for the catch diode)? We quickly realize that to avoid the killer voltage spike and present the body-diode to the inductor current *whenever required*, we need to go back somewhat to the way our old non-synchronous topology was working, and position the synchronous FET such that *its body-diode points in the same direction as the catch diode it seeks to replace.* This is the basic rule of all synchronous topologies that we need to keep in mind always.

Later we should observe that the body-diode of the *control FET* will also be called into play when we deal with "synchronous (complementary) mode drive" instead of the "diode emulation mode drive" currently under discussion. The correct way to connect the control FET is such that its body-diode is always pointing *away* from the inductor and toward the input source — otherwise the obvious problem would be that current will start flowing uncontrollably from the input source into the inductor, and that would clearly not qualify as a *"switching* topology" anymore.

Our conclusion is: we need a diode *somewhere* — we have been unable to replace it entirely. The last remaining question is: if the diode is present in the FET in the form of a body-diode, do we *also* need an *external one* in parallel to the FET?

> **Note:** *We mentioned above that "parasitic" capacitances can provide a temporary path for a short time. What if we actually put a physical cap at the "right place"? That actually leads to "resonant topologies." We will not be discussing those in any detail in this book, primarily because they are usually of variable frequency and therefore rarely used commercially, and also because they require a completely new set of "rules of intuition," which take much more time to develop, especially if you have just learned to think of squares, triangles, and trapezoids.*

External (Paralleled) Schottky Diode

First of all, if we place an external diode in parallel to the body-diode, for it to do anything, the current needs to "prefer" that external diode in preference to the body-diode. So, we need the forward-drop of the external diode to be less than that of the body-diode. That is actually the easy part since the body-diode has a rather high forward-drop (more than a Schottky), typically varying from about 0.5 V for small currents to 2.5 V for much higher currents (the drop being even higher for a P-channel FET as compared to an N-channel one). An external Schottky would, in principle, seem to meet our requirement.

> *Note: However, to ensure that the Schottky is "preferred" in a dynamic (switching) scenario, we need to ensure the external Schottky diode is connected with very thick short traces to the Drain and Source of the FET, otherwise the inductance of the traces will be high enough to prevent current from getting diverted from the body-diode and into the Schottky as desired. The best solution is to have the Schottky integrated into the package of the FET itself, preferably on the same die, for minimizing inductances as much as possible.*

Why does the Schottky diode help so much? In an actual test conducted in early 2007 at a major semiconductor manufacturer, adding a Schottky across the internal synchronous FET of a 2.7–16 V synchronous LED Boost IC on the author's suggestion, led to an almost 10% increase in efficiency at max load. The IC was brought back to the drawing board and redesigned with an integrated Schottky across the synchronous FET (on the same die — the vendor possessed that process technology), and was finally released a few years later as the FAN5340.

The reason for the advantage posed by the Schottky is that the body-diode of a FET has another unfortunate quality besides its high forward-drop, one that we also wish to avoid. It is a "bad" diode in the sense that its PN junction absorbs a lot of minority carriers as it starts to conduct in the forward direction. Thereafter, to get it to turn OFF, all those minority carriers need to be extracted. Till that process is over, the body-diode continues to conduct and does not reverse-bias and block voltage as a good diode is expected to. In other words the "reverse recovery characteristics" of the body-diode of a FET are very poor.

So, this is what can eventually happen in synchronous switching converters without an external Schottky. During the dead-time interval between Q turning OFF and Qs turning ON (i.e., the Q→Qs crossover), we want the body-diode of Qs to conduct immediately. We therefore inject plenty of minority carriers into it (via the diode forward current). However, the full deleterious effects of that stored charge actually show up during the *next* dead-time interval — when Qs needs to turn OFF just before Q starts to conduct (i.e., the Qs→Q crossover). Now we discover that Qs doesn't turn OFF quickly enough, and so a *shoot-through current spike* flows through Q and Qs. This unwanted reverse current ultimately does extract all the minority carriers, and the diode finally does reverse-bias. But during the rather large duration of this shoot-through, we have a significant $V \times I$ product occurring inside the body-diode, and therefore high instantaneous dissipation. The average dissipation (over the entire cycle) is almost proportional to the duration of the dead-time since the diode is just not recovering fast enough.

The only way to avoid the above reverse recovery behavior and the resulting shoot-through is to avoid even forward-biasing the body-diode, which basically means we need to bypass it completely. The way to do that is by connecting a Schottky diode across the synchronous FET, which then provides an alternative and preferred path for the current.

Whichever direction we take, we realize that we do have a diode present *somewhere* on our path — a body-diode or an external Schottky. Since the diode's forward-drop is certainly higher than that of the FET it is across, the efficiency is adversely affected on that count alone (higher conduction loss). Furthermore, if it is a body-diode, then there is an additional impact on efficiency due to the reverse recovery spike (switching loss).

Eventually, reducing the dead-time should help improve efficiency on both the above-described counts. But doing so will also increase the chances of overlap, which can then adversely affect efficiency from an entirely different angle (crossover loss). In severe cases of overlap, even overall reliability will become a major concern. We are therefore essentially walking a tight-rope here. How much dead-time is necessary and optimum? Some companies have come up with proprietary "adaptive dead-time" schemes to intelligently reduce the dead-time as much as possible, usually in real-time after several cycles of sampling, to achieve just the right amount of dead-time, thus averting overlap and also maximizing the efficiency.

Synchronous (Complementary) Drive

When we look at the top section of Figure 9.1, that is, for "heavy loads," we see that ignoring what happens during the very small dead-time, we can simply declare that *Qs is ON whenever Q is OFF*, and vice versa. Their drives are considered "*complementary.*" Therefore, design engineers just took the Gate drive as applied to Q, put an inverter to it, and applied that to the Gate of Qs (ignore dead-time circuit enhancements here). We call this a "synchronous (or complementary) mode" in contrast to the diode emulation mode discussed previously. This synchronous mode is also sometimes called the "PWM mode." It worked as expected, just like a conventional non-synchronous topology, *at least when the load was high*. However, using this simple drive, as the load was lowered, the familiar-looking DCM voltage waveforms were no longer seen. In fact the well-behaved *voltage* waveforms normally associated with CCM remained unchanged even as the inductor current dipped below "sea-level" (zero). On closer examination, stark differences had emerged in the *current* waveforms. As shown in the lowermost section of Figure 9.1, *two new phases* had appeared, both with inductor current pointing in the "negative" direction (i.e., flowing away from the positive output, toward the input). These are marked "C" and "D." During C, current flows in the reverse direction through Qs, whereas in D it flows in the reverse direction through Q.

This is how the system gets to C and D after B is over.

a. In B, the current flows in the "right" (conventional) direction, freewheeling through Qs and L. We know that the voltage across the inductor is reversed as compared to its polarity during the ON-time of the converter. So, during B, the inductor current ramps down. This is normal behavior so far. However, the current then reaches zero level. At that point, instead of "idling" at zero current as in DCM, since Qs is still conducting, and we know that FETs can conduct in either direction, the current does not stop there, *but continues to ramp linearly down, past zero, becoming negative in the process.* That is phase C. Note that in effect, we are continuing to operate in CCM since no discontinuity in current has occurred at the zero crossing level.

b. During C, the current continues to ramp linearly downward, till Qs finally turns OFF and Q turns ON. This causes the voltage across the inductor to flip once again, and with that, its polarity is now such that it causes the inductor current to ramp upward once again. This upward inflexion point, though still within the negative current region, marks the start of phase D. Note that though the current is still negative, it is now rising, commensurate with the newly applied voltage polarity. The current continues to rise linearly past zero straight into phase A (again). In A, the current is positive and continues rising, exactly as in the conventional ON-time of a non-synchronous converter. After that B starts.

The key to *not* getting confused above is to remember that a positive voltage across an inductor (ON-time) produces a positive dI/dt (upward ramp), whereas a negative voltage (OFF-time) causes a negative dI/dt (downward ramp). So, the inductor's voltage polarity determines the polarity of the *rate of change* of current (slope dI/dt) — it has nothing to do with the actual polarity of the current itself (I), which can be positive or negative as we have seen above.

What are the main advantages of "synchronous mode"? We get constant frequency, no ringing during the idling time (and therefore more predictable EMI), easier Gate drive circuitry, constant duty cycle (even at light loads), simpler stress (RMS) equations to compute (see *Chapter 19*). The key advantages of "diode emulation mode" are reduced switching losses (there are no turn-on crossover losses since the instantaneous current is zero at the moment of crossover) and generally more stable though admittedly sluggish response (single-pole open-loop gain; no low-frequency right-half-plane zero, no subharmonic instability).

Part 2: Fixed-Frequency Synchronous Boost Topology

Energy in an inductor is $\frac{1}{2}LI^2$. Its minimum value is zero, and that occurs at $I = 0$. On either side of $I = 0$, inductor energy *increases*. Stored energy doesn't depend on the direction (polarity or sign) of the inductor current. In fact, generally speaking, assigning a positive sign or a negative sign to current is an arbitrary and relative act — we only do it to *distinguish* one direction from the other, the sign by itself has no physical significance. With this in mind, we look a little more closely at the lowermost section of Figure 9.1. In terms of energy, during A, we are first *pulling in energy from the input* and building up energy in the inductor. Then, during B, the stored energy/current freewheels into the output. This is conventional operation so far. But after that, things change dramatically. Judging by the new direction of current shown in C, we now start pulling in energy from the output, but nevertheless *still building up energy in the inductor*. Finally, in D, that energy/current freewheels into the *input* cap. We ask: isn't that a Boost converter by definition? We are momentarily taking energy from what is the *lower*-voltage rail, and pumping energy into the *higher*-voltage rail. We realize that now *there is a Boost mode occurring in our "Buck converter."* Finally, after D, we start the cycle once again at A and deliver energy from the higher to the lower rail again just as a conventional Buck converter is expected to do.

We realize that the synchronous Buck operating at light loads always functions as a Boost for a brief part of its switching cycle. So we can well ask: why is the schematic in Figure 9.1 still called a "Buck," not a Boost? In fact, the schematic is the same for both the Buck and the Boost topologies as we will shortly see, the difference being purely *functional*. In Figure 9.1, the topology is a "Buck," not a "Boost," only because the Buck side is "winning" overall, *as per the current waveforms sketched*. This means that the center (average) of the inductor current waveform is shown above "sea-level" (positive), implying that net energy/current is flowing in a direction from left (higher-voltage rail) to right (lower-voltage rail). However, if the center of the inductor current waveform was below sea-level, that would imply there is a net flow of current/energy from right (lower rail) to left (higher rail). We would then have a synchronous *Boost* converter on our hands, rather than a Buck converter (though admittedly, to *sustain* the flow in that direction, we need a proper voltage source on the right side, instead of just a resistor and capacitor). This subtle topology transformation occurs as the average inductor current moves across the zero current boundary. That is shown more clearly in Figure 9.2.

To summarize: with reference to Figure 9.1, as the inductor current dips below sea-level, we get a Boost converter, but initially one which is operating as a Buck for a brief part of its switching cycle — just as above sea-level, we previously had a Buck converter, but one which was operating as a Boost for a brief part of its switching cycle. However, it remains a "Boost converter" because the Boost modes of its operation are "winning" as compared to its Buck modes — reflected in the fact that the average inductor current is negative, that is, from right to left. But as the average current dips further below, till finally no part of the inductor current waveform remains above sea-level, we get a full conventional Boost converter operating in conventional CCM, one which is not even momentarily operating as a Buck.

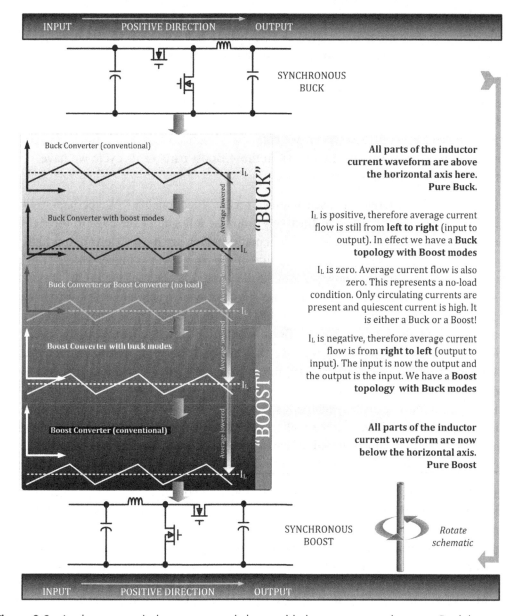

Figure 9.2: As the average inductor current is lowered below zero, a synchronous Buck becomes a synchronous Boost.

We thus realize that when we use synchronous (complementary) drives, the conventional DCM mode for both the synchronous Buck and the synchronous Boost topologies gets replaced with a new CCM-type mode in which a good amount of energy is simply being *cycled back and forth every cycle*, even though very little energy (maybe zero) is actually flowing from the input into the load. This recirculating energy does *not* promise high efficiency ("much ado about nothing"), and that is one reason the synchronous (complementary) drive mode is not used in cases where battery power, for example, needs to be conserved — in that case diode emulation mode with enhancements like pulse-skipping is more commonly used.

What happens to crossover losses? It is often stated that in a switching topology, the control FET has crossover losses, but not the synchronous FET (or catch diode). That is a fairly accurate statement generally. However, we can now visualize that when a Buck converter is operating in Boost mode (negative current), in those Boost parts of its cycle we have switching losses in Qs, not in Q.

A little bit about the difference in "schematics" versus "functionality." We perhaps viewed the schematic in Figure 9.1 rather instinctively from the left side of the page to the right. We are intuitively most comfortable with schematics which are drawn "left-in, top-high," that is, where the input source is on the left side of the page and the higher-voltage rail is placed on the top. But suppose now we look at the same schematic of Figure 9.1, from the right side to the left. Isn't that exactly how we would draw a synchronous Boost topology (based on what we have learned)? Alternatively, if we *mirror* the Buck schematic of Figure 9.1 horizontally, it will become a Boost schematic. However, functionally speaking, to make any difference in behavior, we have to actually change the polarity of the average inductor current (above or below sea-level in our example). Only then does a given schematic work as a Buck or a Boost converter, irrespective of what it may *look* like to us. In other words, in dealing with synchronous topologies, because they can source or sink current equally readily, we have to be really cautious in jumping to any conclusion about the underlying topology simply based on the way the schematic *looks*. The key is looking at the *average inductor current level* (center of ramp) — its actual functioning.

In Figure 9.1, the "negative" current of a Buck would be considered a "positive" current for a Boost. In other words, regions C and D in Figure 9.1 would be conventional (positive current) operation if we had a Boost converter on hand. And then, A and B would be regions of negative Boost inductor current. To make things clearer, we have carried out a full mirror reflection in Figure 9.3 to see how a synchronous Buck schematic gets transformed effortlessly into a synchronous Boost schematic. We observe that (for the case of complementary drives) the voltage waveforms are completely unchanged from what they were in Figure 9.1. The current waveform, however, *is not just a mirror reflection*, but is also *vertically shifted* so as to change its average value (and polarity) as explained above. Only then, functionally speaking, do we get a Boost converter, not a Buck converter.

With this insight, we can perhaps examine more formally how exactly a Buck maps into a Boost. We realize that if the average inductor current in Figure 9.1 dipped below sea-level and the synchronous Buck became a synchronous Boost, there would be no difference in the ensuing voltage waveforms. However, when we get a Boost, energy would be building

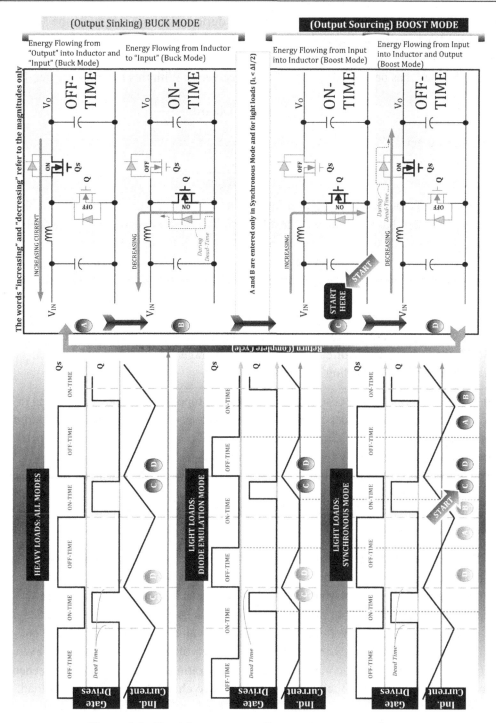

Figure 9.3: Fixed frequency, synchronous Boost waveforms.

up in the inductor (albeit with reverse current flow) during what was originally the "OFF-time" for the Buck. That interval therefore is now officially the "ON-time" for the Boost. So, in effect, T_{ON} has become T_{OFF}, and T_{OFF} is now T_{ON}, which means D has become $1 - D$. And also, "output" has become the "input," and vice versa. With this mapping we get

$$\text{Dbuck} = \frac{V_O}{V_{IN}} \Leftrightarrow 1 - \text{Dboost} = \frac{V_{IN}}{V_O}$$

Simplifying, we get

$$\frac{V_O}{V_{IN}} = \frac{1}{1 - \text{Dboost}}$$

No surprise: this is the familiar equation for the DC transfer function of a Boost topology. We now realize, that based on our new level of intuitive understanding, the Buck and Boost are just mapped versions of one to the other — with the key change being INPUT ↔ OUTPUT. No wonder, in a Buck, we say that the average *input* current equals the average current through the upper transistor (or the "switch" in non-synchronous versions), whereas in a Boost we say that the average *output* current equals the average current through the upper transistor (or the "catch diode" in non-synchronous versions). Also, in a Buck, the average inductor current equals the average *output* current, whereas in a Boost, the average inductor current equals the average *input* current. And so on. The resemblance is quite startling actually.

But to be clear: the Buck and Boost are still *independent* topologies — they just happen to be mirror images of each other under certain conditions. We speculate: is this like the electron and the anti-electron? We recall these are considered different particles, but are also "mapped" versions of each other. Similarly for a male and a female — similar, yet different, if not opposite. It therefore seems natural to ask: what happens if we put the Buck and Boost together back to back? What do we get? In fact, we get the "four-switch Buck-Boost," as described further. Later we will also generate several Boost-Buck *composites*. But before we discuss all that, some essential housekeeping is required. Here are some practical hints to start with:

a. A Buck can under severe transient conditions become a Boost for several cycles. If the output voltage is lower than its reference, it tries to get up to the reference level by turning its control FET ON. But if the output voltage is higher than its reference (as during a sudden unloading of the output), it tries to actively discharge the output cap, that is, it becomes a Boost converter. Therefore, while evaluating a Buck, we should put a voltage probe on its *input cap* too, especially in low-voltage applications where even a slight bump in input voltage may cause the voltage ratings of the FET to be exceeded.

b. Sometimes we start up a Buck converter into a "pre-charged" or "pre-biased" load. For example, we may have simply turned OFF the Buck momentarily, leaving its output cap still charged (connected to a zero/light load), and then tried to turn the converter back ON rather too quickly — only to find the output was still high. In many designs,

Buck controllers use a "soft-start" sequence while turning ON. In many of those design implementations, the reference voltage (as applied to the error amplifier) is slowly raised from a very low level up to the normal level. In such cases, a pre-biased load will appear as an overvoltage initially. The error amplifier will therefore reduce its duty cycle sharply in response. However, if we have a complementary drive, then the OFF-time of the Buck is correspondingly very large now. So now the lower transistor may be ON for a very long time. Since the OFF-time of a Buck maps into an ON-time for a Boost, in effect, we once again have a fully functioning Boost converter operating for quite some time. The pre-biased load is now serving as the input voltage source to this reverse converter. The energy being pulled in from the pre-biased load eventually goes into the input cap of the Buck. If that cap is small, and if the input voltage source (on the left) is not "stiff" enough, the cap can easily get overcharged, causing damage to the Buck controller.

c. We may therefore need to significantly oversize the input bulk capacitance, well beyond its rating based on RMS current, or an acceptable input voltage ripple, just to keep this "Boost bump" down.

d. Another concern is that when the circuit is functioning as a Boost, we need to limit the current in the *lower* transistor too, otherwise when the lower transistor turns ON (sometimes fully for a while), we will have absolutely no control over the current being pulled out of the pre-biased load, and that may damage the lower FET especially since the inductor also likely saturates in the bargain. Therefore, we may need to incorporate current limits on *both* the high- and low-side FETs of the synchronous Buck/Boost.

e. One way of handling pre-biased loads better is to *not* operate with complementary drives during soft-start, but power-up in diode emulation mode. Or, the duty cycles of both the upper and the lower FETs should be increased gradually with appropriate relative phase, moving smoothly from soft-start mode to full complementary mode without any *output glitch or discontinuity*. Quite a few proprietary techniques abound for this purpose.

f. One way of indirect current limiting, located in neither of the FETs, is called "DCR (DC resistance) sensing." This can be applied to any topology actually. It is discussed in the next section.

Part 3: Current-Sensing Categories and General Techniques

In general, current sensing/limiting has several purposes and implementations. We need to know what the basic aim is before we decide on an implementation technique. There are several possibilities to choose from here.

a. *Cycle-by-cycle current limiting*: This is used for switch protective purposes. There are situations, especially in "high-voltage" applications (defined here as anything applied above 40−50 V on an inductor/transformer), where even one cycle of excess current can cause the associated magnetic component (inductor/transformer) to saturate sharply, leading to a very steep current spike that can damage the FETs. So, we are often interested in providing a basic, but very *fast-acting current limit, threshold* which if breached, will cause the FET to turn off almost immediately — well within the ongoing ON-time of the cycle. Note that any delay in limiting the overcurrent, even for 100−200 ns, can be disastrous, as shown in Figure 2.7.

b. *Average current limiting*: This is also for protective purposes, but is more relaxed than (a), since it *spans several cycles*. It is often used in low-voltage applications because in such cases, the magnetic components usually don't see high-enough applied voltseconds to saturate too "sharply," that is, their rate of increase of *B*-field with respect to the applied Ampere-turns is quite small. And even if the magnetics do start to saturate somewhat, the typical circuit parasitics present on the board help significantly in curtailing any severe resulting current spikes. So, generally speaking, we can tolerate a few cycles of overload current without damage. Average current limiting is then considered adequate, but it is still recommended to have at least some duty cycle limiting to assist it. For example, 100% duty cycle (i.e., the switch turned ON permanently for several cycles, as in some Buck converter architectures) can be disastrous with only average current limiting present.

In AC−DC converters, there is cycle-by-cycle current limiting already present on the main switching FET. That makes the transformer relatively safe from saturation. In such cases, the purpose of providing additional average current limiting on the secondary outputs is primarily to comply with the international safety norm IEC-60950. This norm requires that most safe, nonhazardous outputs be limited to less than 240 W within 60 s. Therefore, we actually *want* a slower current limiting technique in such cases, to avoid responding too aggressively or restrictively to overloads, since those temporary overloads may just represent normal operation. In brief, average current limiting is generally geared toward providing protection against *sustained overloads*, certainly not against the rapid effects of core saturation. Therefore, RC filtering with a rather large time constant is used to filter the current waveform. We then apply that time-averaged level to the input of a current-limit comparator, and compare that to a set reference threshold applied on the other input pin of the comparator.

c. *Full current sensing*: This can be used for both control and protection. The difference as compared to (a) and (b) is that in those cases we only wanted to know when the current crossed a *certain set threshold* (and whether slowly or quickly). But we were certainly not interested in knowing the actual *shape* of the current waveform while that happened. However, in many other cases, we may want to know just that. In fact, we need to know the entire current waveshape (AC and DC levels) when using current-mode control, for example — in which method the sensed current waveform provides the ramp applied to the PWM comparator.

How can we implement the above current-sensing strategies? The most obvious way to implement any form of current sensing (monitoring), or limiting, is to insert a current-sense resistor and monitor the voltage across it. But that is obviously a lossy technique since we dissipate I^2R Watts in the resistor. Note that any attempt to lower the dissipation by reducing R is not very successful beyond a certain point because switching noise starts to ride on top of the relatively small sensed signal, drowning it out eventually. We can try to filter out the noise by introducing a small RC filter on the sensed current waveform, but we usually end up distorting the sensed signal and introducing delays. Some engineers try to avoid the I^2R loss altogether by sensing the forward voltage drop across the FET — since that is essentially a resistance called "R_{DS}." We do, however, need to compensate that sensed voltage for the well-known temperature variation of R_{DS} (the factor depends on the voltage rating of the FET). We thus realize that this FET-sensing technique can work only when the FET itself is completely identifiable in terms of its characteristics, and also we are able to monitor its temperature. This indirectly demands the FET be integrated on the controller die itself — that is, with FET-sensing technique, we should use a switcher IC, not a controller with external FETs. But there is another possible problem in using the FET-sensing technique. Nowadays, there are situations where we might not even turn-on the lower or the upper FET for several cycles (e.g. pulse skipping mode). We realize that to sense current through a FET, we have to at least turn it ON. Therefore, another attractive set of techniques has evolved for implementing "lossless" sensing. One such popular method is called DCR Sensing.

DCR Sensing

Perhaps it all started with this simple thought: instead of putting a sense resistor in series with the FET, why not put the sense resistor in series *with the inductor*? We would then obtain full information about the inductor current, rather than just the ON-time or OFF-time slice of it. This could conceivably help us in implementing various new control and protection techniques too. However, that thought process then went a step further by asking: since every inductor has a resistance "built-in" already, called its DCR, could we somehow use that resistance in place of a separate sense resistor? We would then have "lossless current sensing," since we are at least not introducing any *additional* loss term into the circuit. The obvious problem here is that the DCR is itself *inaccessible*: we just can't put an error amplifier or multimeter "directly across DCR" because DCR lies buried deep within the object we call an "inductor." However, we want to persist. We ask: is there some way to *extract* the voltage drop across this DCR from out of the total inductor voltage waveform? The answer is, yes, we can — and that technique is called DCR sensing.

To start with, in Figure 9.4 we present an unusual, but an original and innovative, method to provide basic average current limiting. Though it is not very accurate, it has been used very successfully in very large production volumes on the output of a commercial

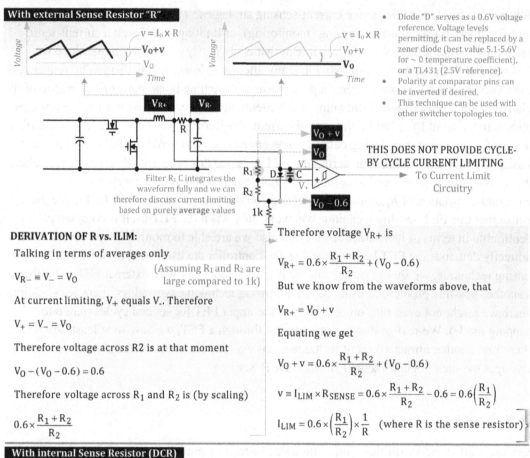

With external Sense Resistor "R"

$v = I_o \times R$

$V_O + v$

V_O

$v = I_o \times R$

$V_O + v$

V_O

V_{R+} V_{R-}

R

Filter R_1 C integrates the waveform fully and we can therefore discuss current limiting based on purely **average** values

$V_O + v$

V_O

$V_O - 0.6$

R_1 D   C

R_2

1k

THIS DOES NOT PROVIDE CYCLE-BY CYCLE CURRENT LIMITING

To Current Limit Circuitry

- Diode "D" serves as a 0.6V voltage reference. Voltage levels permitting, it can be replaced by a zener diode (best value 5.1-5.6V for ~ 0 temperature coefficient), or a TL431 (2.5V reference).
- Polarity at comparator pins can be inverted if desired.
- This technique can be used with other switcher topologies too.

DERIVATION OF R vs. ILIM:

Talking in terms of averages only

$V_{R-} \equiv V_- = V_O$

(Assuming R_1 and R_2 are large compared to 1k)

At current limiting, V_+ equals V_-. Therefore

$V_+ = V_- = V_O$

Therefore voltage across R2 is at that moment

$V_O - (V_O - 0.6) = 0.6$

Therefore voltage across R_1 and R_2 is (by scaling)

$0.6 \times \dfrac{R_1 + R_2}{R_2}$

Therefore voltage V_{R+} is

$V_{R+} = 0.6 \times \dfrac{R_1 + R_2}{R_2} + (V_O - 0.6)$

But we know from the waveforms above, that

$V_{R+} = V_O + v$

Equating we get

$V_O + v = 0.6 \times \dfrac{R_1 + R_2}{R_2} + (V_O - 0.6)$

$v \equiv I_{LIM} \times R_{SENSE} = 0.6 \times \dfrac{R_1 + R_2}{R_2} - 0.6 = 0.6 \left(\dfrac{R_1}{R_2} \right)$

$I_{LIM} = 0.6 \times \left(\dfrac{R_1}{R_2} \right) \times \dfrac{1}{R}$ (where R is the sense resistor)

With internal Sense Resistor (DCR)

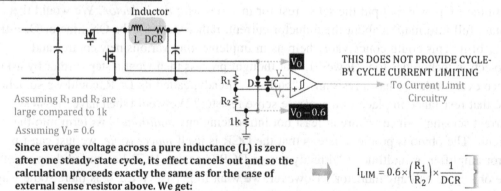

Inductor

L DCR

Assuming R_1 and R_2 are large compared to 1k

Assuming $V_D = 0.6$

V_O

$V_O - 0.6$

R_1 D   C

R_2

1k

THIS DOES NOT PROVIDE CYCLE-BY CYCLE CURRENT LIMITING

To Current Limit Circuitry

Since average voltage across a pure inductance (L) is zero after one steady-state cycle, its effect cancels out and so the calculation proceeds exactly the same as for the case of external sense resistor above. We get:

$I_{LIM} = 0.6 \times \left(\dfrac{R_1}{R_2} \right) \times \dfrac{1}{DCR}$

Figure 9.4: A DCR-based average current-limiting technique.

70-W AC/DC flyback (for meeting safety approvals). Note that it uses a filter with a very large RC time constant.

The top part of the figure shows how we can implement this current-sense technique with a discrete (external) sense resistor in series with the inductor. In the lower part of the figure, we indicate that this technique will work just fine using the DCR of a Buck inductor too. Why is that? Because, if we have a *pure* inductor (no DCR) and we average its voltage over a cycle, we will get exactly zero — because in steady state, an inductor has equal and opposite voltseconds during the on-time and off-time — by the very definition of steady state! When we use a real-world inductor, with a certain non-zero DCR, the *additional* drop during both the ON-time and the OFF-time is $I_O \times DCR$, and that is clearly the term which *remains* after we average the inductor voltage over one complete cycle or several cycles. That is the DC voltage being applied on the comparator pin in Figure 9.4.

Just for interest, we note that in the 70-W flyback we mentioned above, the "sense resistor" used was the small resistance of the small L (rod inductor) of the LC post filter of the flyback.

Note that this DCR sensing technique uses a very large RC time constant, and so we lose all information about the actual shape of the inductor waveform. That doesn't matter however, because in this particular case, we just want to implement average current limiting. But we now ask: can we extend the same technique to full current sensing?

In Figure 9.5, we present how this is done rather commonly nowadays. In this case, the RC time constant is not large, but it is in fact *matched exactly to the time constant of the L–DCR combination*, which means mathematically: $RC = L/DCR$. We see from the Mathcad graphs embedded in the figure that with this matching, and stated initial conditions, the voltage on the capacitor becomes an exact replica of the voltage across the DCR. Further, for any duty cycle other than the steady-state duty cycle, the manner in which the DCR voltage (and therefore inductor current) *changes* is also replicated exactly by the capacitor voltage, which means that the two voltages indicated will *track each other* accurately under all conditions, steady state or otherwise. For example, if the inductor current is momentarily unsteady (as during a line or load transient), the capacitor voltage will not be steady either. But it is interesting that in fact, both the DCR voltage and the cap voltage will increase or decrease *by exactly the same amount* from this point on. This indicates that both the voltages will change identically (i.e., track each other), and finally settle down identically too, at their common shared steady-state initial voltage condition. Note that the settling/initial value, as indicated in the figure, is $I_O \times DCR$.

The advantage of the above fancy version of DCR sensing is that the entire AC and DC information of the inductor current waveform is now available via the sensed capacitor voltage (assuming, however, that the inductor does not saturate somewhere in the process).

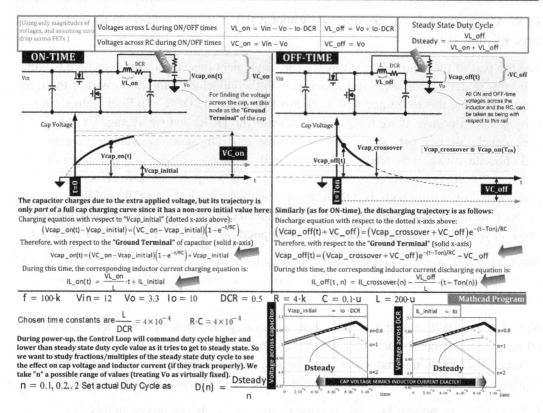

Figure 9.5: DCR current sensing explained (with exaggerated DCR).

Is there any simple intuitive math to explain this time-constant matching and the resulting tracking? Yes, but it is not a rigorous proof, as we will see. Let us start by applying the duality principle. We know that the voltage on a capacitor when charged by a current source is similar to the current through an inductor when "charged" by a voltage source. Assuming steady-state conditions are in effect, and also that the maximum voltage on the capacitor is low compared to the applied on-time and off-time voltages (V_{ON} and V_{OFF}, roughly $V_{IN}-V_O$ and V_O respectively), we get near *constant-current sources* (V_{ON}/R and V_{OFF}/R) charging and discharging the cap during the ON-time and OFF-time, respectively. The question is: what is the cap voltage swing during the ON-time and during the OFF-time. Further, are the swings identical in magnitude, or is there a "leftover delta" remaining at the end of the cycle, indicative of a nonsteady state?

Using

$$I = C\left(\frac{\Delta V}{\Delta t}\right),$$

we get

$$\Delta V_{ON} = \frac{V_{ON}}{RC} \times \frac{D}{f}, \Delta V_{OFF} = \frac{V_{OFF}}{RC} \times \frac{(1-D)}{f}$$

We know from the voltseconds law that

$$V_{ON} \times D = V_{OFF} \times (1-D)$$

Therefore,

$$\Delta V_{ON} = \Delta V_{OFF}$$

Similarly, for the inductor current, using

$$V = L\frac{\Delta I}{\Delta t},$$

we get

$$\Delta I_{ON} = \frac{V_{ON}}{L} \times \frac{D}{f}, \qquad \Delta I_{OFF} = \frac{V_{OFF}}{L} \times \frac{(1-D)}{f}$$

We know from the voltseconds law that

$$V_{ON} \times D = V_{OFF} \times (1-D)$$

Therefore,

$$\Delta I_{ON} = \Delta I_{OFF}$$

We conclude that there is *no* leftover delta remaining at the end of the cycle for either the inductor current swing or the cap voltage swing. Therefore, both inductor current and cap voltage are in steady state. That is just duality at work (as introduced in *Chapter 1*).

Let us now compare the capacitor voltage swing and the DCR voltage swing and try to *set them equal*, so we can get them to be exact copies, rather than merely scaled copies, of each other.

Set

$$\Delta V_{ON} = \frac{V_{ON}}{RC} \times \frac{D}{f}$$

equal to

$$(\Delta I_{ON} \times DCR) = \frac{V_{ON}}{L} \times \frac{D}{f} \times DCR$$

Therefore,

$$\frac{1}{RC} = \frac{DCR}{L}$$

Or

$$RC = \frac{L}{DCR}$$

That is the basic matched time-constant condition for DCR sensing, and we see it follows rather naturally from the simplified intuitive discussion above.

This simple analysis does indicate that the cap voltage is mimicing the inductor current. It tells us quite clearly that the AC portion of the inductor current is certainly being copied by the cap voltage. But what can we say about the DC level? Is the DC cap voltage copying the DC inductor current too? It actually does, but that does *not* follow from the simple intuitive analysis presented above (or from any complicated *s*-plane AC analysis you might see in literature). For that we should look at Figure 9.5.

Note that in Figure 9.5, we have used the accurate form of voltages appearing on the capacitor during the ON-time and OFF-time. Those include the small (but decisive) term $I_O \times DCR$, which eventually determines the DC value of the capacitor voltage. The DC value is easy to understand based on the analysis we presented earlier when discussing Figure 9.4. We recall that after we average out the voltseconds, *the $I_O \times DCR$ term is all we are left with*. So clearly, that must be the DC value of the capacitor voltage waveform in the present case too. Philosophically, we recognize that time constants don't enter the picture when performing DC analysis. The effects of time constants are time bound by definition. So we conclude that both Figures 9.4 and 9.5 must provide the same DC/settling values ultimately.

We realize that with good time-constant matching, the DCR voltage in Figure 9.5 becomes a carbon copy of the capacitor voltage, and vice versa — with both AC and DC values replicated. We ask: is that replication true only in steady conditions or is it also true under transient conditions? We see from Figure 9.5, for the cases of duty cycle *other than steady state* (i.e., *n* not equal to 1), that the *change* in cap voltage is also an exact replica of the *change* in DCR voltage (referring to the leftover delta at the end of the cycle). That means both the cap voltage and the DCR voltage (corresponding to inductor current) will climb or fall identically under transients, heading eventually toward their (identical) steady-state values. This implies that their relative tracking is always very good, steady state, or otherwise.

However, in reality, DCR sensing is not considered very accurate and should not be relied upon for critical applications. For example, it is well known that the nominal DCR value has a wide spread in production, and the DCR itself varies significantly with temperature.

So, the last remaining question is: what are the effects of *mismatched* time constants (L/DCR and RC)? We have seen that the inductor current reaches steady state based on an initial (DC) condition $I = I_O$, and corresponding to that, the capacitor voltage reaches steady state with an initial (DC) condition of $v = I_O \times$ DCR. We expect that to remain true for any time constant, since after several cycles, the effect of any time constant literally "fades away." In other words, the DC value of the cap voltage waveform cannot vary on account of mismatched time constants per se (but will obviously change if DCR itself varies). When subjected to a sudden line or load transients, the effect of mismatched time constants will be severe (though only *temporarily* so). For example, if the capacitance of the RC is very large, the cap voltage will change very little and rather slowly, as compared to the inductor current. But final settling value will be unchanged (for same DCR).

Note that many commercial ICs allow the user to either choose the DCR for cost and efficiency reasons, or use an external sense resistor (in series with the inductor) for higher accuracy.

The Inductorless Buck Cell

The power of "what-if" can be uniquely powerful in power conversion. So, as a variation of classic DCR Sensing, we ask: what if we *reposition* the lower terminal of the capacitor in Figure 9.5 and *connect it to ground*, as shown in Figure 9.6?

For negligible DCR, we will discover that (for fairly large capacitances) the capacitor voltage will be the same as the voltage on the output capacitor of the Buck converter. That is interesting. But we also notice that by repositioning the cap, the series RC no longer rides on the output rail, but is separate from the LC. The two branches, RC and LC, are independent and in parallel now. This implies we can

a. disconnect the LC branch completely. In other words, we can just toggle the FETs at a certain predetermined duty cycle D and the result (i.e., the cap voltage of the RC) will be a (low-power) output rail of voltage $V_O = D \times V_{IN}$ — same as for the classic Buck converter. Or we can introduce an active feedback loop too. We now have an "inductorless Buck cell" (named by the author and first published in EDN magazine on October 17, 2002). Note that this is a variation of the bucket regulators we presented in Figure 1.2. Here, by changing the duty cycle, we can use this circuit to generate any output rail, perhaps as a reference voltage. We can't really use it to give out much power, and neither is it efficient. But it can be quite useful as we will see.

b. also decide to retain the LC (inductor and the output capacitor of the Buck), but *shift the regulation point* from the output capacitor of the Buck to the capacitor of the series RC. Especially under large-signal conditions, we expect the RC-based feedback loop to have better behavior than a corresponding double-pole LC-based feedback loop.

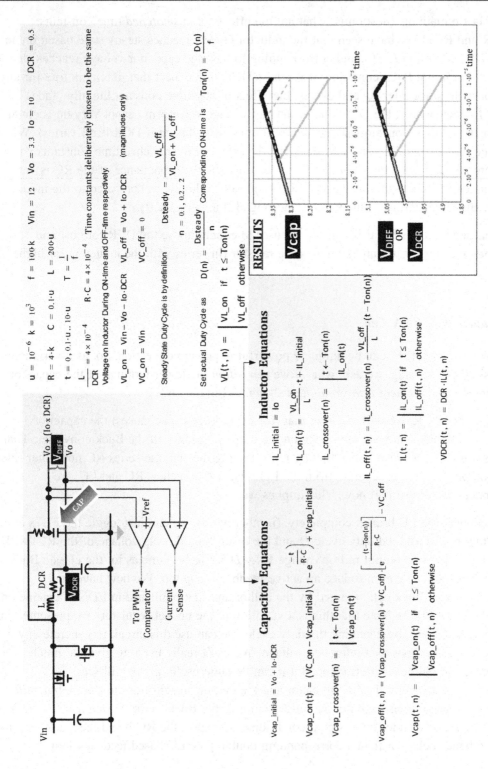

Figure 9.6: Inductorless Buck and an alternative regulation and current-sensing scheme (shown with exaggerated DCR).

For non-negligible DCR, we find that

c. with the same time-constant matching as we used in DCR sensing, the cap voltage minus the output rail is equal to the DCR voltage. So, we get back the DCR sensing we had in Figure 9.5, but now in addition, we can also implement a new well-behaved feedback loop. Both are shown in Figure 9.6. These signals can be jointly used for current-mode control.

d. we can also achieve "lossless droop regulation" as discussed next.

Note that this inductorless Buck technique can also be used in non-synchronous converters. But in that case we always need to have the LC present in parallel to the RC. Because in the absence of a synchronous FET actively forcing the switching node (close) to ground during the OFF-time, the current flowing in L becomes responsible for accomplishing that. We have to get voltage applied to the RC to actively toggle repetitively. Otherwise the cap of the RC would just charge up to the input voltage.

Lossless Droop Regulation and Dynamic Voltage Positioning

In Figure 9.7, we start with a conventional Buck. Then we move down to the next schematic and introduce a droop resistor "Rdroop." This is one of the conventional ways of implementing droop regulation. Note that we are now regulating the point marked $V_O + (I_O \times Rdroop)$ instead of regulating V_O, which means that V_O is no longer fixed. Because, for example, if the current I_O rises, $V_O + I_O \times Rdroop$ will tend to increase. But since we are holding $V_O + I_O \times Rdroop$ constant by means of the regulation loop, the only solution is for V_O to decrease as I_O increases. This is called droop regulation or "dynamic voltage positioning."

Now, we would like to do this with DCR instead of Rdroop, since that would be considered lossless. But DCR is not accessible. However, from Figure 9.6 we have learned that the inductorless Buck cell provides a rail that is exactly equal to $V_O + (I_O \times DCR)$. Further, the good news is that this point is *accessible*. So, if we connect the error amplifier to it, we will achieve the same basic effect as conventional droop regulation (proposed by the author here).

What are the benefits of droop regulation in general? In Figure 9.7, we show the transient response waveforms with and without droop. The advantage of droop is very simply explained as follows:

a. **Unload:** At high loads, the output voltage with droop is lower than without droop. This gives us additional headroom literally. So, if we suddenly unload the output, the output will jump up momentarily (its natural transient response). But since the output rail of the converter with droop implemented starts at a *lower level*, the highest voltage it goes up to under transient conditions is also lower. So, with good all-round design, we can ensure the highest point does not exceed the permissible AC + DC window, and also that the voltage finally settles down close to the *upper end* of the allowed DC window.

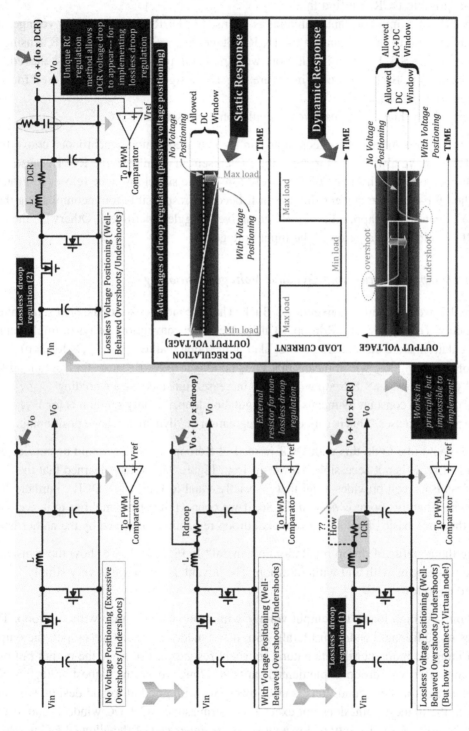

Figure 9.7: Conventional voltage positioning (droop regulation) and a novel lossless inductorless Buck implementation.

b. **Load:** Now, if we suddenly apply max load, the output will tend to momentarily go much lower (its natural transient response). However, with droop implemented, the output starts off at a *higher level* than the output of a converter without droop. So, the lowest point will now be higher, and we have a better chance of avoiding undershoot.

We see that droop regulation naturally "positions" the output at a "better level" to start the excursion with, so it is more suited to avoiding overshoots and undershoots. This is very useful especially in VRM-type applications, where the allowable regulation window is, indeed, very tight.

Part 4: The Four-Switch Buck-Boost

We had mentioned previously that the Boost and Buck are horizontally reflected and mapped versions of each other and that it would be interesting to see what happens if we place them *back to back*, like an electron with an anti-electron.

In the top schematic of Figure 9.8, we started with a Buck converter positioned at the input and follow it up with a Boost converter. However, we then realize that neither the output capacitor of a Buck nor the input capacitor of a Boost is fundamental to their respective topologies, since in both cases there is an inductor in series, and the only purpose of these capacitors is to filter out the relatively small ripple of the inductor. So, in the middle schematic of Figure 9.8, we meld the two converters together to produce our first *composite topology*: the "four-switch Buck-Boost." We can continue to view it mentally as a cascade of a Buck cell followed by a Boost cell. The cell consists of a totem-pole of two FETs driven in complementary fashion. We can visualize that the Buck stage has an output (a virtual one in this case) called "V_x," which becomes the input rail to the Boost cell. So the DC transfer function of the composite can be broken up into a product of its cascaded consitutents as follows:

$$\frac{V_O}{V_{IN}} = \left(\frac{V_x}{V_{IN}}\right) \times \left(\frac{V_O}{V_x}\right) = (\text{Dbuck}) \times \left(\frac{1}{1 - \text{Dboost}}\right) \equiv \frac{\text{Dbuck}}{1 - \text{Dboost}}$$

In the lowermost schematic of Figure 9.8, we tie the Gates of the control FETs of both the Buck and the Boost sections together. So, Dbuck = Dboost = D. In that case,

$$\frac{V_O}{V_{IN}} = \frac{D}{1 - D}$$

which is the same DC transfer function as for the classic single-transistor (or its two-transistor synchronous version) Buck-Boost topology.

The biggest underlying advantage of this composite topology is that unlike the fundamental single-switch, single-inductor Buck-Boost, the polarity of the output is not inverted with respect to the input.

Two Converters placed End-to-End:

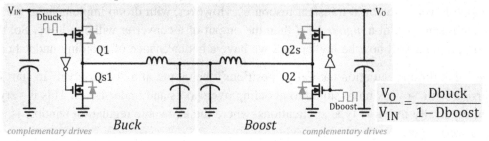

$$\frac{V_O}{V_{IN}} = \frac{Dbuck}{1 - Dboost}$$

(Composite) 4-Switch Buck-Boost (synchronous) Converter:

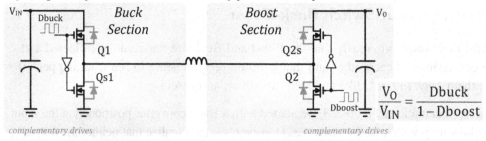

$$\frac{V_O}{V_{IN}} = \frac{Dbuck}{1 - Dboost}$$

Non-Synchronous "4-Switch" Buck-Boost Converter (basic version):

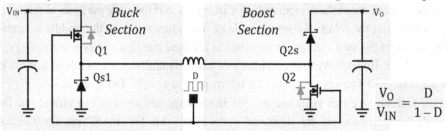

$$\frac{V_O}{V_{IN}} = \frac{D}{1 - D}$$

Figure 9.8: Evolution of the general four-switch Buck-Boost and its simpler lower-efficiency non-synchronous version.

We discuss two distinct categories of control here.

a. **Single duty cycle:** The non-synchronous version of this was presented in the lowermost schematic of Figure 9.8. We show its synchronous version in Figure 9.9. The common feature of both is that there is only a single duty cycle D applied, and the DC transfer function is $D/(1 - D)$. There are advantages and disadvantages to this simple control approach. Briefly:

Advantages: Simplest drive possible; seamless step-up/step-down operation (centered around $D = 50\%$ for the case $V_O = V_{IN}$ as for classic Buck-Boost topology); and no polarity inversion from input to output.

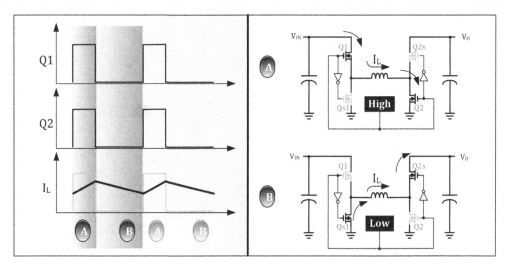

Figure 9.9: The simplest, lowest efficiency synchronous version of the general four-switch Buck-Boost.

Disadvantages: Higher conduction losses (two semiconductor switch drops during both ON-time and OFF-time); higher switching losses (both totem-poles constantly switching, e.g., no LDO-type pass-through mode when $V_O \approx V_{IN}$); both input and output caps always have high RMS currents; and overall poor efficiency (60–70%).

b. **Dual duty cycle:** As presented in the middle schematic of Figure 9.8 and further elucidated in Figure 9.10, the two totem-poles, though synchronized to the same clock, have two distinct duty cycles: "Dbuck," applied to the Buck totem-pole cell and "Dboost," applied to the Boost totem-pole. Though in all implementations, the DC transfer function is Dbuck/(1 − Dboost), there are many flavors of this approach as we will shortly see. Broadly speaking, the advantages and disadvantages of this approach are as follows:

Advantages: No polarity inversion from input to output and lower switching losses (both totem-poles are not constantly switching).

Disadvantages: Complex Gate drives; many patents in force; higher conduction losses (two semiconductor switch drops during both ON-time and OFF-time); input or output (or both) caps can have high RMS currents; nonseamless step-up/step-down operation (mode transitions occur around duty cycles limits of 0% and 100%); and potential lack of proper "retraceability" (i.e., going from $V_{IN} > V_O$ to $V_{IN} < V_O$ and back to $V_{IN} > V_O$ may involve different duty cycle combinations in either direction, and therefore varying efficiency paths).

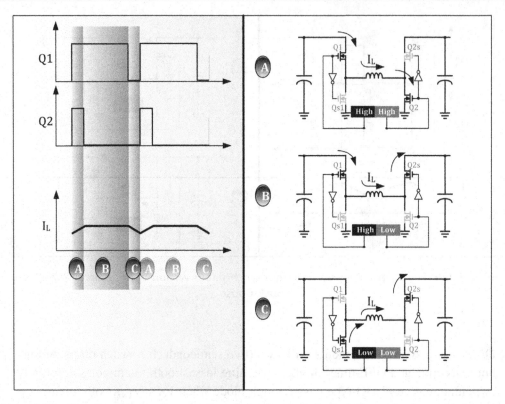

Figure 9.10: One possible implementation of the four-switch Buck-Boost in the region $V_{IN} \approx V_O$.

Note: In all discussions involving the four-switch Buck-Boost, we are focusing only on the case of fully complementary drives (no diode emulation mode). Therefore, in either the Buck or the Boost totem-poles, when one transistor of a given totem-pole is ON, the other is OFF, and vice versa. We have consequently just drawn the Gate waveforms for the control FETs Qx in the related figures and consciously omitted the Gate waveforms of the synchronous FETs Qsx. Further, for simplicity, we are assuming that the average inductor current never intersects the zero current axis (i.e., there are no recirculating energy modes as discussed in Figures 9.1–9.3).

One of the key advantages of independent Buck- and Boost- section duty cycles is that if for example, overall, step-down performance is required, say from 5 V to 3.3 V, then there is no need to ever toggle the Gates of the Boost section. We can just toggle the Buck section with duty cycle Dbuck = V_O/V_{IN}, and keep the Boost section in *pass-through mode*, that is, with Dboost = 0. So, current from the inductor will simply pass straight through the upper FET of the Boost section into the output. Similarly, if step-up performance is required, say from 3.3 V to 5 V, there is no need to toggle the Gates of the Buck section.

We can just toggle the Boost section with duty cycle Dboost = $1 - (V_{IN}/V_O)$, and keep the Buck section in pass-through mode, that is, with Dbuck = 100%. So, current from the input will pass straight through the upper FET of the Buck section into the inductor. By doing this, we save on switching losses across most of the regions of operation. All known implementations of the four-switch Buck-Boost (with dual duty cycle) agree on this aspect — there is no difference in most of the known implementations in the regions *where V_O is different from V_{IN}*. Any differences in approach are located within the region $V_{IN} \approx V_O$, for example, 5 V (nominal) to 5 V (exact) conversion. Note that on either side of this region we actually also have two FET forward-drops to account for. So, this particular region, designated "$V_{IN} \approx V_O$," can in reality be a fairly wide region. In addition, max or min duty cycle limits can affect its boundaries significantly. Further, the system may spend significant operating time in this region, so high efficiency is very desirable here. Unfortunately, this is the region where both the Buck section and the Boost section are asked to toggle (switch), and therefore, switching losses are high. In addition, as we suddenly start toggling the Gates of a previously inactive totem-pole (i.e., in pass-through mode), or stop toggling the Gates of a previously toggling (switching) totem-pole, we are liable to introduce significant output glitches, besides a potential lack of retraceability.

We now look closely in Figure 9.10 at the "special region" in which *both* the Buck and the Boost totem-poles are being switched. We can see that on either side of this region we are in pass-through mode, either in the Buck cell or in the Boost cell. In the figure, we also recognize that at a practical level, we can usually neither achieve, nor do we actually desire, duty cycles too close to 0% or 100%. So we have set a Dmin of 0.1 and a Dmax of 0.9 for *both* converter stages.

> *Note: For example, if we are using N-channel high-side FETs in a Buck, we can't go up to 100% duty cycle because we need to turn OFF the control FET momentarily, to allow the bootstrap circuitry to deliver charge into the bootstrap capacitor, which then provides power to the Gate driver. For a Boost (or a Buck-Boost) converter, it is never a good idea to use 100% duty cycle, since that leaves no time for the inductor to freewheel its stored energy into the output, besides possibly creating a sustained short across the input. A 0% duty cycle is almost impossible to achieve in general, since it takes a short duration to first turn ON the FET, then some time to turn it OFF. There are always some delays at each step, as the turn-on or turn-off commands are propagated and executed. Besides, there may be leading edge blanking in current-mode control which virtually guarantees a minimum ON-time. A minimum ON-time is also required in general, if we want to sense the current passing through any FET. And so on.*

> *Note: If we are using N-channel FETs (that is, high-side FETs), we have a problem even in pass-through mode in keeping the bootstrap rail alive. We may need to include a separate standby low-power charge-pump circuit constantly switching away. Or with low-threshold-voltage FETs, we could route power to the high-side driver of the pass-through*

stage from the bootstrap rail of the other (switching) stage, since in all cases, the pass-through stage is the one connected to the lower of the two rails V_{IN} and V_O.

The fundamental question is: how do we control *two* duty cycles (in the special region)? We cannot have two control loops, one for the Buck section and one for the Boost section, since they would likely end up fighting each other. For example, if the input is close to 5 V, and the output is 5 V, we could have a case where the intermediate voltage $V_x = 2.5$ V. The Buck would then operate at around 50% (5−2.5 V), and the Boost would also switch at around 50% (2.5−5 V). Note that this particular duty cycle combination bears strong resemblance to the simple schematic of Figure 9.9 (but only in this region). But in addition, we actually have infinite possibilities here. For example, we could also have the Buck switching at only about 80% duty cycle (intermediate voltage $V_x = 0.8 \times 5 = 4$ V), followed by the Boost switching at about 25% duty cycle (stepping up the intermediate voltage of 4−5 V). In other words, there are many ways to go even from ~5 V to 5 V. Two independent control loops will never be able to decide among themselves what the best way is. Summing up: we need to have only *one control loop*. But there are two possibilities for that too: we can either fix one of the two duty cycles (Dbuck = constant or Dboost = constant) and then allow the control loop to vary the other. Or, we need to define a certain fixed *relationship* between Dbuck and Dboost so that when the control loop determines one of them, the other gets automatically defined. It can be shown that perhaps the most optimum method of driving the Buck and Boost sections in this region, one that guarantees highest efficiency and also full retraceability, is the "constant difference duty cycle method" (US Patent number 7,804,283, inventor Maniktala and Krellner). Its operation is illustrated in Figure 9.11. This is what happens as input voltage is lowered:

a. Full Buck mode initially. Dboost = 0. Dbuck increases gradually as input is lowered, till Dbuck hits 90% duty cycle limit (Dmax). The control loop is in effect controlling and determining Dbuck.

b. At this point, we transit into the special region from the right side, and Dboost, previously in pass-through mode, is now forcibly increased from 0% to 10%.

c. The control loop responds by asking Dbuck to suddenly decrease by almost 10% to keep the output unchanged. The output glitch can be made almost negligible if we *consciously position and release* Dbuck at around the 80% mark, the moment we change Dboost from 0% to 10% in (b). There is then minimal "hunting" by the control loop and thus no significant output overshoot or undershoot.

d. As the input decreases further, the difference "Dbuck − Dboost" is enforced to be fixed at 70%. The control loop is in effect deciding both Dbuck and Dboost, but they are now tied together in this functional relationship.

e. Eventually, Dbuck hits the 90% limit again. By this time Dboost has increased to 20%. So now, the control FET Q1 of the Buck section is turned ON fully and

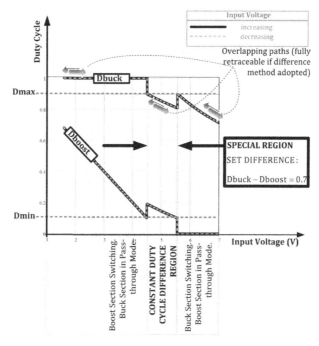

Figure 9.11: The constant difference duty cycle method for controlling the four-switch Buck-Boost (~5 V–5 V case).

Dbuck = 100%. Simultaneously, the control loop is now asked to determine the duty cycle of the Boost stage alone.

f. The control loop responds by asking Dboost to suddenly decrease by almost 10% to keep the output unchanged. The output glitch can be made almost negligible if we consciously position and release Dboost at around the 10% mark, the moment we change Dbuck from 90% to 100% in (e). There is then minimal "hunting" by the control loop.

The above pattern is completely reversible/retraceable as indicated in Figure 9.11. The duty plots shown are actually overlapping for the cases of V_{IN} falling and V_{IN} rising; the path can be fully retraced from *any input voltage point*, even within the "special region."

Part 5: Auxiliary Rails and Composite Topologies

We have just seen how the Buck and Boost topologies can be literally mated to produce the four-switch Buck-Boost. One obvious disadvantage of that approach is that we have positioned a Buck at the input and a Boost at the output. But unfortunately, we know that the Buck topology exhibits a significant RMS current in its *input* cap, due to the chopped current waveform at that location. And for the same reason, the Boost has significant RMS current passing through its *output* cap. The four-switch Buck-Boost therefore combines the worst of the two topologies in this regard. We need to ask: can we *reverse* the order? Would a Boost-Buck composite be better than a Buck-Boost composite? We know that the Boost and the Buck have low RMS currents in their input and output caps, respectively. So our hopes are high as we generate *three Boost-Buck composites* on these pages: the Cuk, the Sepic, and the Zeta topologies. We will see that in fact we do get low RMS input and output cap currents for the Cuk topology. Unfortunately, the Cuk can't shake off the "polarity inversion" weakness of its constituent Buck-Boost cell and is therefore usually not favored commercially for that reason. The two other Boost-Buck composites, the Sepic and the Zeta, are both noninverting step-up/step-down topologies. The Sepic is basically the Cuk with a *brute-force re-referencing of output rails* (with respect to the input). Unfortunately, in the course of correcting the polarity inversion of the Cuk, the Sepic ends up with high RMS current in its output cap, whereas the Zeta has a high RMS current in its input cap. We realize that the perfect switching power topology remains elusive.

The good news is we intuitively expect, and get, the same DC transfer function for all three Boost-Buck composites, as for the (four-switch) Buck-Boost composite, and for the basic (fundamental) Buck-Boost topology. In the present case, we have

$$\frac{V_O}{V_{IN}} = \left(\frac{V_x}{V_{IN}}\right) \times \left(\frac{V_O}{V_x}\right) = \frac{1}{1 - \text{Dboost}} \times \text{Dbuck} = \frac{\text{Dbuck}}{1 - \text{Dboost}}$$

One obvious problem of this rearrangement of constituent topologies to form a Boost-Buck is that the inductors of the Buck and Boost stages are no longer back to back, and therefore we are not able to meld them into one inductor as we did for the four-switch Buck-Boost. So, now we get two inductors. Luckily, it turns out that we can combine the switches of the Buck and Boost cells into one (with single duty cycle). One last question remains: how do we pass energy *between* the cells? We tap into a swinging node of the Boost cell, and inject that waveform into the Buck cell, via a "coupling capacitor." So in all, we now expect two inductors, one switch and a coupling cap, in all three Boost-Buck composites.

At this point, we should be getting worried about the stresses in the coupling cap, since all the power coming out of the converter needs to pass through this component. But we will discuss the stresses in that later. For now, we realize that the presence of a coupling cap also gives us a great opportunity. We recall that a fundamental single-inductor Buck-Boost topology has a limitation in that the output has an inverted polarity compared to the input. But we learned that if we use a transformer, we create the flyback topology which physically separates the output section from the input section. Thereafter, we could isolate those sections for safety reasons, as in AC−DC converters, but we could also reconnect them back together, so as to correct the polarity inversion. That is how we get a noninverting, nonisolated flyback.

Similarly, a capacitor also gives us the possibility of separating the input and output sections, then re-referencing the output with respect to the input. We thus derive the noninverting Sepic topology from the Cuk. Note that to keep things simple, we will not be getting into synchronous versions of any of these Boost-Buck composites here.

Is It a Boost or Is It a Buck-Boost?

Before we discuss the Cuk and Sepic, we need to understand the basic Boost and Buck-Boost topologies better. Looking at Figure 9.12, we will realize that they belong to the

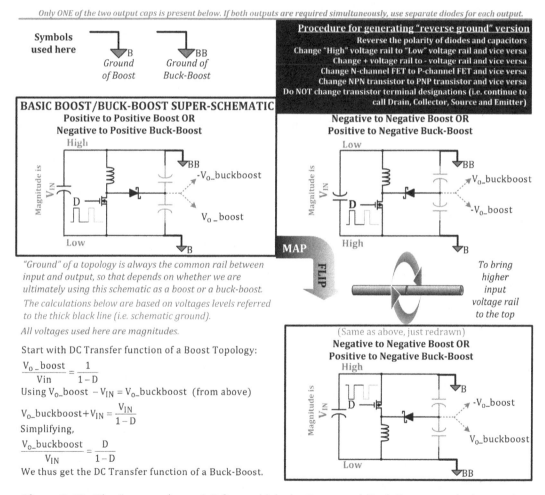

Figure 9.12: The "super-schematic" from which the Boost and Buck-Boost topologies can be derived and the method of mapping common polarity topologies to their reverse polarity counterparts.

same "super-schematic." *We* create the difference *by the way we choose to extract energy from the circuit.* In fact, by using separate output diodes *we can generate two outputs simultaneously, from the same circuit* (even though we would be able to regulate only one of them).

A numerical example will make this clearer. Suppose we take a 12-V input Boost converter and apply a duty cycle of 50% to it. We expect to get a Boosted output rail equal to twice the input voltage, that is, 24 V in our example here (use $V_O/V_{IN} = 1/(1 - D)$). Now, in a Buck-Boost, we would be drawing output energy not from the 24-V Boosted rail and ground, but from between the 24-V Boosted output rail and the 12-V input rail. Eventually, we would call that a negative-to-positive Buck-Boost because, by convention, the rail common to the input and output is the topological ground — and so the 12-V input rail would be renamed the "ground" for the Buck-Boost topology, whereas the input ground of the Boost topology would now become the -12-V input rail of the Buck-Boost. Note that in our example, the Buck-Boost output voltage would then be $24 - 12 = 12$ V. So, if Figure 9.12 is to be believed, we expect a -12-V to 12-V Buck-Boost to result when switching with a duty cycle of 50%. But that is completely consistent with what we know of a Buck-Boost topology: if we switch it with $D = 50\%$, we expect the output voltage to equal the input voltage in magnitude (use $V_O/V_{IN} = D/(1 - D)$). So, Figure 9.12 must be right. But if not fully convinced, we can repeat the above numerical example for any D, and we will realize that the Buck-Boost and Boost are truly part of the same super-schematic shown in Figure 9.12. No wonder that both the Boost and the Buck-Boost share another valuable property: *the center of inductor current in both cases is $I_O/(1 - D)$.*

In Figure 9.12, we go a step further. Starting with the super-schematic on the left, that would give us a positive-to-positive Boost and a negative-to-positive Buck-Boost, we generate a super-schematic that gives us a negative-to-negative Boost and a positive-to-negative Buck-Boost.

> *Note: Historically, voltages used to be referred to with the higher potential being labeled ground, and the rest of the circuit "hanging" from it, much like a clothesline. These were "positive-ground" systems. Nowadays, the world has shifted largely to "negative-ground" systems, where the ground is the lower rail. So a modern-day circuit generally resembles a skyline, not a clothesline. However, in creating and recognizing composite topologies, we need to understand how a negative-ground circuit, or cell, can be mapped into its positive-ground equivalent. A simple mapping procedure, outlined in Figure 9.12, is required. Using that procedure, we generate the corresponding positive-ground super-schematic. Finally, since the mapping procedure changes high voltages to low voltages, and vice versa, to get back to our familiar convention of placing higher voltages on top of the page, we flip the positive-ground super-schematic vertically, as shown in Figure 9.12, to arrive at the more conventional way to draw a negative-to-negative Boost (and positive-to-negative Buck-Boost).*

Understanding the Cuk, Sepic, and Zeta Topologies

In Figure 9.13, we first start by blindly cascading a Boost with a Buck. In the middle of the circuit, we have a DC rail bridging the two, over which power flows down, from one to the other converter. It uses two switches, and the challenge is to combine those into one.

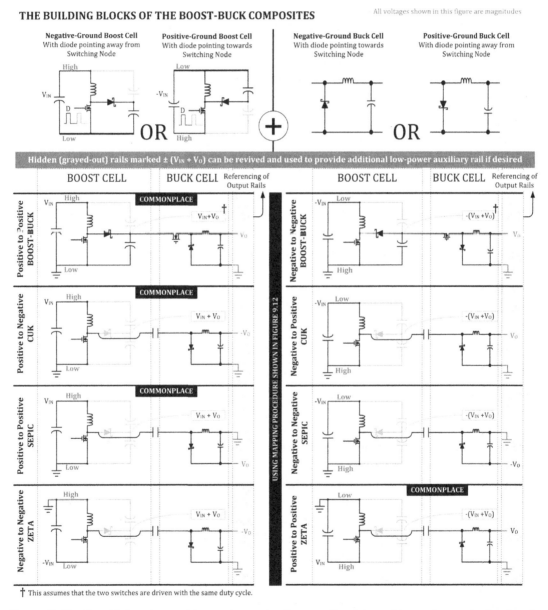

Figure 9.13: Combining the positive and negative Boost cells with positive and negative Buck cells to generate various Boost-Buck composite topologies.

In the following three schematics, we do just that, by connecting to the switching node on either side by a coupling cap. Note that on the right side, we map each composite topology to its reverse voltage counterpart as discussed above. So we can generate a negative-to-positive Cuk or a negative-to-negative Sepic if we so desire.

> *Note: Though in the standard mapping process, we change an N-channel FET to a P-channel FET, all topologies, mapped or otherwise, can be implemented with either type of FET — we just have to keep track of the Gate voltages required, and if necessary, provision for a bootstrap circuit. Also, we need to pay attention to the direction the internal body-diode is pointing so as to avoid the "killer voltage spike" we often talk about. In an N-channel FET, the body-diode (considered as an arrow from anode to cathode) points from Source to Drain. In a P-channel FET, it points from Drain to Source.*

Each circuit actually has three sections: a Boost cell, followed by a Buck cell, followed by a block labeled "Referencing of Output Rails." This latter block is where we can set virtually any output polarity as discussed previously. However, unlike transformer-based isolation as in a flyback, this re-referencing dramatically alters the path of currents as shown in Figure 9.14. The reason for that is in a transformer, there is no net flow of current between the two separated sections — and the output (Secondary) section can function on its own since both the forward and the return paths get disconnected by the transformer winding. But in capacitor coupling, the current path still exists, and so by re-referencing, that pattern can get completely changed. We therefore get three distinct topologies, with some remarkably different behaviors, but also with great similarities based on their common heritage: the Cuk, Sepic, and Zeta. As mentioned previously, we can look upon the Sepic as a brute-force attempt to correct the polarity inversion of the Cuk. And it worked. Though again, as mentioned previously, this introduced high RMS currents into the output cap.

In Figure 9.14, we calculate the DC voltage level appearing across the coupling capacitor in each composite topology, so we can pick its voltage rating better. The general "rules of construction" by which we have unequivocally generated the paths of currents shown during the ON-times and OFF-times, respectively, are bulleted out in the same figure. We thereby show that, similar to the fundamental Buck-Boost topology, the DC transfer of all these composite topologies is $D/(1-D)$, just as we had inituitively expected. And similarly, the minimum voltage rating of the FET in all cases is the same: $V_{IN} + V_O$. Note that in any switching topology, in general, if the switching FET blocks a certain voltage across it when it is OFF, then when it turns ON, that same voltage blocking responsibility gets literally transferred over to the diode, which must therefore be rated identically. In other words, the diode too must be rated $V_{IN} + V_O$ for all the three Boost-Buck composite topologies under discussion.

In Figure 9.15, we have first drawn the Sepic and Zeta topologies in a more conventional manner. We see that the similarities between the waveforms of these composite topologies

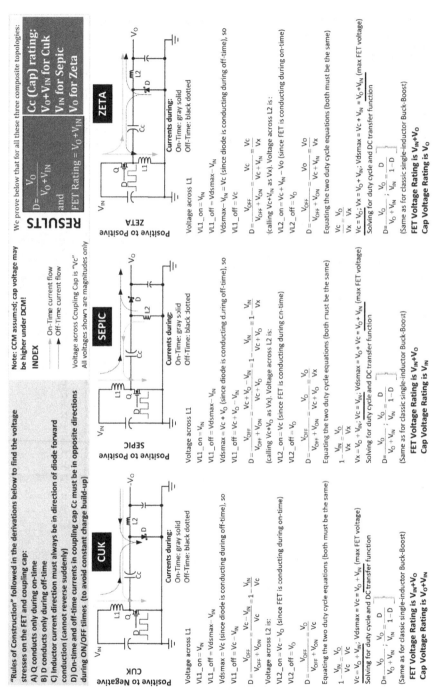

Figure 9.14: The duty cycle and voltage ratings of components of the Boost-Buck composites.

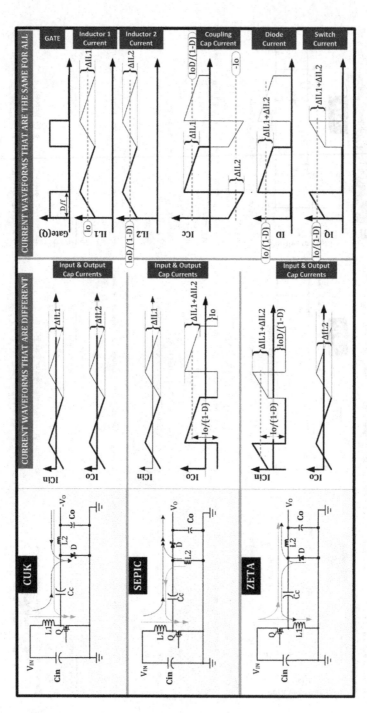

Figure 9.15:   Current waveforms in the Boost-Buck composites.

are tremendous. It is a result of their common heritage as brought out more clearly in Figure 9.14.

The current waveforms in the figure can be easily guessed based on some simple reasoning described in the next section. From Figure 9.15, we see that the only difference in the current waveforms is in the input and output capacitors. In all the other components, except for the coupling capacitor, both the required voltage and the current ratings are the same for all three topologies. This is also discussed further below.

Generating the Current Waveforms of the Cuk, Sepic, and Zeta Converters

Cuk: We refer to Figure 9.15. The center of ramp of $IL2$ must be I_O (output current) since it is the constant current flowing in from the output. So,

$$IL2 = I_O$$

During the ON-time, this current passes through Cc. In the OFF-time, $IL1$ flows through the cap in the opposite direction. Since the cap is in steady state, its change in charge during the ON-time must be equal and opposite to the change in charge during the OFF-time. Therefore, the following equality must be true:

$$IL1 \times (1 - D) = I_O \times D$$

Therefore,

$$IL1 = \frac{D}{1 - D} I_O$$

The current in Q during the ON-time is the sum of the two currents, $IL1$ and $IL2$. So,

$$IQ = IL1 + IL2 = \frac{D}{1 - D} I_O + I_O = \frac{I_O}{1 - D}$$

The current in the diode during the off-time is also the sum of $IL1$ and $IL2$. So, its center of ramp is also $I_O/(1 - D)$, as for the switch.

We see that the switch current is the same as for a fundamental Buck-Boost in terms of its height and duration. Note that in a Buck-Boost, there is a single inductor of rating $I_O/(1 - D)$. Here we have two inductors, one rated I_O, the other $I_O D/(1 - D)$. For example, if we set $I_O = 1$ A and $D = 0.5$, in the fundamental Buck-Boost, the inductor must be rated at least 2 A (ignoring the ramp portion here). In the Cuk, we need two inductors, each rated 1 A. For the same inductance, the total core volume for the Cuk is actually half that of the Buck-Boost, since core volume is proportional to LI^2. However, if we keep the same current ripple ratio r, then if the current goes from 2 A to 1 A, we need to double the inductance (the usual scaling law for inductors). So for the case of constant r, there is no net change in the total core volume in going from the Buck-Boost to the Cuk. In fact, using two inductors instead of one will provide greater surface area exposed to natural convection, which will help in high-power applications by improving the thermal dissipation. These arguments are valid for the Sepic and Zeta too.

Sepic: We refer to Figure 9.15. The output current flows in spurts through the diode during the OFF-time only (further averaged by the output cap). We have two contributions to the diode current. Together, they must average out to I_O. So,

$$(IL1 + IL2)(1 - D) = I_O$$

Coming to the coupling cap, $IL1$ flows through it in one direction during the OFF-time, and $IL2$ flows through it in the opposite direction during the ON-time. So, by equality of charge in steady state we get

$$IL1(1 - D) = IL2(D)$$

Solving the two equations above, we get as for the Cuk

$$IL2 = I_O \text{ and } IL1 = \frac{D}{1 - D} I_O$$

The current in Q during the ON-time is the sum of the two currents, $IL1$ and $IL2$. So,

$$IQ = IL1 + IL2 = \frac{D}{1 - D} I_O + I_O = \frac{I_O}{1 - D}$$

The current in the diode during the off-time is also the sum of $IL1$ and $IL2$. So, its center of ramp is also $I_O/(1 - D)$ as for the switch.

Zeta: We refer to Figure 9.15. Looking at the coupling cap first, we have two currents going through it in opposite directions during the ON-time and OFF-time, respectively. The current during the OFF-time passes through $L1$ and is the only current component in that. So, it must equal $IL1$. The current through the cap during the ON-time does not go through $L1$. But it is the only current component through $L2$ during the same time. So, it is by definition equal to $IL2$. We therefore have by charge balance in the coupling cap

$$IL1(1 - D) = IL2 \times D$$

Note that the output is in series with $L2$, and the current in an inductor cannot change suddenly. So, from the current through $L2$ during the OFF-time, we get

$$IL2 = I_O$$

Plugging this into the preceding equation we get

$$IL1 = \frac{D}{1 - D} I_O$$

The current in Q during the ON-time is thus the sum of the two currents, $IL1$ and $IL2$. So,

$$IQ = IL1 + IL2 = \frac{D}{1-D}I_O + I_O = \frac{I_O}{1-D}$$

The current in the diode during the off-time is also the sum of $IL1$ and $IL2$. So, its center of ramp is also $I_O/(1-D)$ as for the switch.

Stresses in the Cuk, Sepic, and Zeta Topologies and Component Selection Criteria

We see a remarkable pattern emerge. The current waveforms of the switch and diode in all three Boost-Buck topologies are identical and are also the same as in the fundamental single-switch single-inductor Buck-Boost. We also know their duty cycle equations and voltage ratings of switch and diode are the same. That lays credence to our initial claim that these are just composites of a Boost (or equivalently of a Buck-Boost) cell and a Buck cell.

Ignoring the AC ripple component here, we see that the switch current is a rectangle of height $I_O/(1 \quad D)$ and width D/f. So its RMS must be

$$I_{Q\_RMS} = \sqrt{\left(\frac{I_O}{1-D}\right)^2 D} \Rightarrow \frac{I_O\sqrt{D}}{1-D}, \quad \text{where } D = \frac{V_O}{V_O + V_{IN}}$$

Using $D = V_O/(V_O + V_{IN})$, we can also write this as

$$I_{Q\_RMS} = I_O \sqrt{\frac{V_O \times (V_O + V_{IN})}{V_{IN}^2}}$$

The average current through the diode is

$$I_{D\_AVG} = \frac{I_O}{1-D} \times (1-D) \Rightarrow I_O$$

Note that we have to follow the general design guidelines for a fundamental Buck-Boost topology. We realize that in this case too, the RMS current through the switch is a maximum at Dmax (lowest V_{IN}). We already know the voltage ratings of the FET and diode in all Boost-Buck composites need to be better than $V_O + V_{IN}$ (see Figure 9.14). The latter needs to be checked at Dmin (highest V_{IN}).

The RMS current through the coupling cap needs to be calculated also. In all Boost-Buck composites, we have $IL2$ flowing through the coupling cap during the ON-time and $IL1$ flowing in the OFF-time (opposite direction). However, for calculating RMS, the sign of the current does not matter since we use I^2. So, using our side-to-side segments summation formula of Figure 7.4, we get

$$I_{Cc\_RMS} = \sqrt{IL2^2 \times D + IL1^2 \times (1-D)} = \sqrt{I_O^2 \times D + \left[\left(\frac{I_O D}{1-D}\right)^2 \times (1-D)\right]} \Rightarrow I_O \sqrt{\frac{D}{1-D}}$$

Using $D = V_O/(V_O + V_{IN})$, we can also write this as

$$I_{Cc\_RMS} = I_O \sqrt{\frac{V_O}{V_{IN}}}$$

Clearly, this is the highest at lowest input (V_{INMIN}), and the cap RMS rating should be picked accordingly. We already know the voltage ratings of this cap to be $V_O + V_{IN}$ or V_{IN} or V_O respectively, in the Cuk, Sepic, and Zeta topologies (see Figure 9.14). These need to be verified at maximum input (V_{INMAX}).

For the RMS current ratings of the input and output caps of these Boost-Buck composites, we note that for a Cuk, we can ignore both these RMS currents since they are very small, whereas for the Sepic we can ignore the input cap RMS, and for a Zeta we can ignore the output cap RMS. For the output RMS of the Sepic, we can use the formula given in the Design Table of the Appendix for a Boost (or Buck-Boost) output cap. For the input RMS of the Zeta, we can use the formula for a Buck-Boost input cap. That completes the calculation of stresses for all three composite topologies, and we can pick them appropriately.

As for L1 and L2, as is our usual practice, we target a current ripple ratio r of 0.4 for both these inductors at V_{INMIM} (Dmax). That gives us the required inductance. We already know their current ratings. Note that the voltage waveform across all inductors L1 and L2 of all three Boost-Buck composites are the same (in CCM). We can see this from Figure 9.14 that in all cases

$$VL1\_on = VL2\_on = V_{IN} \quad \text{(Cuk, Sepic, and Zeta)}$$
$$VL1\_off = VL2\_off = V_O \quad \text{(Cuk, Sepic, and Zeta)}$$

This was the historical reason why Mr. Cuk decided to try to "save an inductor" by winding L1 and L2 on the same core (in the converter subsequently named after him). To his surprise, he discovered that, depending on how the windings were placed on the core (the coupling coefficients), the ripple current in either the input cap or the output cap, or both, could be reduced very close to zero. This is called "ripple steering." It is the subject of great academic attention, but is still not really used commercially on a wide scale, perhaps because (a) the input/output cap RMS currents in the Cuk converter are very low to start with, (b) it is hard to guarantee coupling coefficients in mass production (and any non-guaranteed advantage is really no advantage in the commercial arena), and (c) with the advent of low-ESR caps, the heating in the input/output caps of a Cuk converter is virtually negligible to merit any further attention. Note that if this is done, it is commonplace to use the same number of turns for both $L1$ and $L2$, and so, their ripples (not their ripple *ratios*) will become identical. This can pose a problem at light loads because one inductor will go into DCM before the other. It can also cause different voltage stresses on the components than expected. Note that we have assumed CCM all through our previous discussions. However, since ripple steering is not commonly used, we will avoid further discussion on this shared-core approach.

Part 6: Configurations and "Topology Morphology"

In this section, we return to the basics armed with our recently acquired knowledge, and see what more we can do with our basic topologies. Here we first distinguish between a *topology* and a *configuration*. For example, a regulator converting 12 V to 5 V is a positive-to-positive *configuration* of a Buck topology. But we could have a regulator converting -12 V to -5 V, and that would still be a Buck regulator, but a negative-to-negative *configuration*.

In Figure 9.12, we learned how to map a negative-ground configuration to a positive-ground one. However, we know that in power supplies we may end up redesignating the ground rail anyway, as it needs to be the common rail between the output and the input. Further, we could also end up replacing a P-channel FET with an N-channel FET, and vice versa, so long as we can drive it appropriately. For these reasons, we should not get fixated on the fact that a given circuit is either "positive ground" or "negative ground." That itself may not matter at all. But the *mapping procedure* to get from one to the other is critical in generating the different configurations of topologies, because we have learned that the mapping procedure changes positive voltages to negative voltages and vice versa. That fact is the one to remember and use.

In Figure 9.16, let us take the positive-to-positive Buck for example. We can do this either by an N-channel FET or by a P-channel FET. In the former case, we would need a bootstrap rail to drive the FET when the FET turns ON. Let us now look at the negative-to-negative Buck. We see that we can do this, once again, either by an N-channel FET, or by a P-channel FET. But this time, in the former case we do *not* need a bootstrap rail to drive the FET. However, we need a bootstrap for the P-channel FET now. We thus realize that all "high-side" configurations (where the FET is placed toward the high-side rail) are more easily driven by P-channel FETs, whereas "low-side" configurations are more easily driven using N-channel FETs. Otherwise, a bootstrap will be required in both cases.

Let us now do a mapping procedure on the positive-to-positive Buck using an N-channel FET. We realize that after mapping as per Figure 9.12, it becomes a negative-to-negative Buck using a P-channel FET. Furthermore, the drive waveform also gets flipped vertically, and positive becomes negative, and vice versa. Also, if there was a bootstrap to start with, it

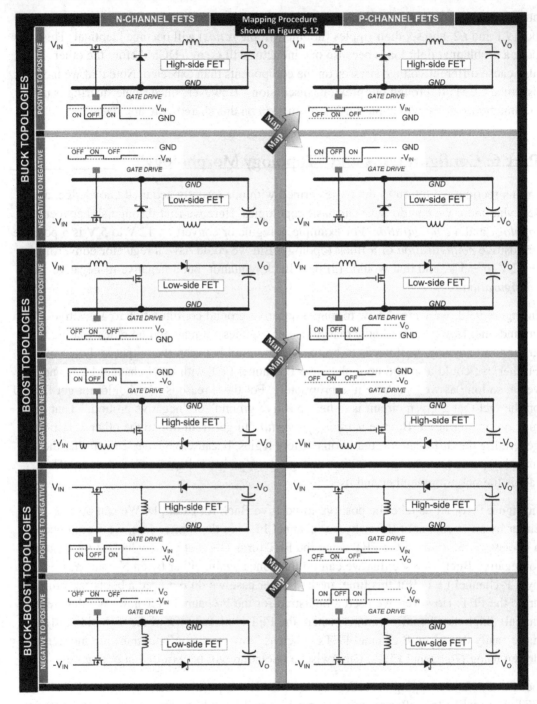

Figure 9.16: Configurations of the three basic topologies by the mapping procedure.

remains one after mapping. If no bootstrap was required initially, after mapping too, no bootstrap will be required. This way we can easily generate all the possible configurations shown in Figure 9.16.

Having understood this, we move on to "topology morphology." This relies on the basic fact that a *switcher is just a switcher*. In all cases, we are basically toggling (switching) a FET with a certain duty cycle, and then applying a feedback signal to control the output level. So, there seems no reason why we can't, at least theoretically speaking, use a switcher IC meant for one configuration, for another configuration. We may, however, need to do a voltage translation or differential sense of the output and apply it to the IC, since in all likelihood, the IC ground, which is the reference rail of its internal error amplifier and other circuitry, may be different from the ground of the topology. Is that all? No, there is one more condition. Just the way the topologies are, there is a difference in high-side and low-side topologies as far as the diode direction is concerned. If we look closely at Figure 9.16, we will see that in all *low-side* configurations, be it with an N-channel or a P-channel FET, the switching node (i.e., the common node shared by the switch, diode, and inductor) is always connected to the *anode* of the diode, whereas in *high-side* configurations, the switching node always connects to the *cathode*. This implies that we may be able to use an IC meant for a high-side configuration, in any high-side configuration, *irrespective of topology*. Similarly, an IC meant for low-side configuration may be used in any low-side configuration. We may need to bring the feedback signal to it appropriately, and also remain conscious of the new voltage and current stresses as we change topologies (thus ensuring those are still within the capability of the IC). If we ensure all that, the IC will *not know the difference* as we morph the circuitry around it into another topology. We are ignoring loop stability issues here.

We now look around and realize that, in essence, we deal with only two basic types of IC constructions commonly. One is meant for a positive-to-positive Buck converter. This is a high-side configuration. The other is a positive-to-positive Boost converter, that being a low-side configuration.

> *Note: We know that the Boost and Buck-Boost are part of the same super-schematic and also that the Cuk, Sepic, and Zeta are all Boost-Buck composites. Which is why we generally expect that, with only minor modifications, we can always use a "Boost IC" for Buck-Boost, Cuk, Sepic, and Zeta applications (with the same intended high-side or low-side configuration).*

In Figure 9.17, we take each of these two ICs and generate all the possibilities. We note that a positive Buck IC with a P-channel FET can easily do a positive-to-negative Buck-Boost or a negative Boost because all three are high-side configurations of P-channel FETs as per Figure 9.16. Similarly, a positive Boost IC with N-channel FET is a low-side configuration and can therefore do a negative Buck or a negative-to-positive Buck-Boost, simply because all these are low-side configurations. In other words, we have several

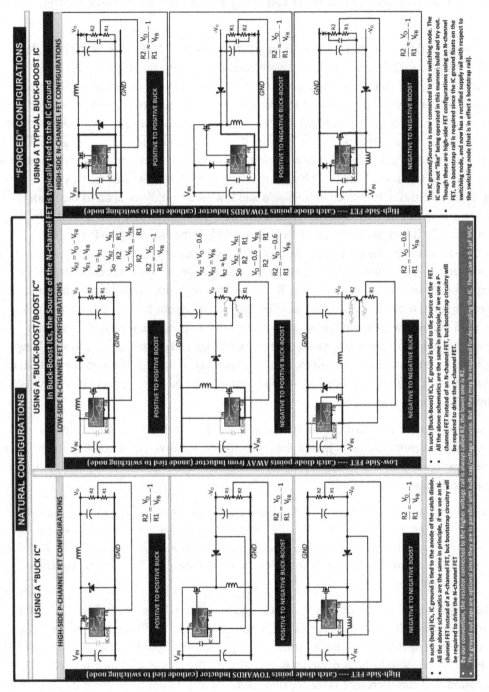

Figure 9.17: Topology morphology using the two common types of ICs in use today.

configurations that are *mutually natural* cousins. However, there is an interesting trick that we can use to force a low-side IC (e.g., a positive Boost IC) into doing high-side configurations such as the positive Buck. These are also shown in Figure 9.17 as "forced configurations". But there are limitations to this approach as indicated in the figure (poor regulation/noise).

Note that in all cases, as we morph topologies, not only do the current and voltage stresses/ratings need to be re-evaluated to confirm the new application of the IC, we should also pay attention to possible changes in stability (loop behavior) and ensure we can make the transition in a stable manner. If we have an external compensation pin, that can greatly help in our endeavor.

Part 7: Other Topologies and Techniques

Hidden Auxiliary Rails and Symmetry

In the topmost schematic of Figure 9.13, we mentioned that a brute-force composite of a Boost converter followed by a Buck converter has an intermediate rail of value $V_O + V_{IN}$. We added a footnote to say that results when the duty cycle of the Boost stage equals the duty cycle of the Buck stage. In other words, the Gates of the two FETs shown are, virtually speaking, tied together (they go ON and OFF in unison). The math behind that is shown more clearly in Figure 9.18. We see that we have a low-power regulated rail of value V_O appearing across the Boost inductor. Though it is not ground referenced, we can use it in some situations. For example, we can place an LED to indicate that the Boost stage is really switching. We can also try to regulate that auxiliary voltage rail in preference to the main output rail. This can be of use in an opto-less Primary-side sensing scheme in AC−DC power supplies, with the Buck stage replaced by a Forward converter. In the Cuk, Sepic, and Zeta converters too, we have this hidden $V_{IN} + V_O$ rail, and it can be revived for low power if we want. In the lower half of Figure 9.18, we briefly present a novel AC−DC converter idea from the author, which does not require an input bridge rectifier. It goes to show that when we understand the three basic topologies better, we may not invent a radically new topology, but there is still a lot we can do by just rearranging the building blocks.

Multiple Outputs and the Floating Buck Regulator

Another challenge is to generate multiple outputs from a single converter. In any single-switch power supply, there is always one control loop and one corresponding duty cycle. So, we get one output that is well regulated. Many attempts abound into generating multiple outputs that closely "follow" the main regulated output closely. In Figure 9.19, we collect some of the techniques used in AC−DC power supplies. They are based on the volts/turn

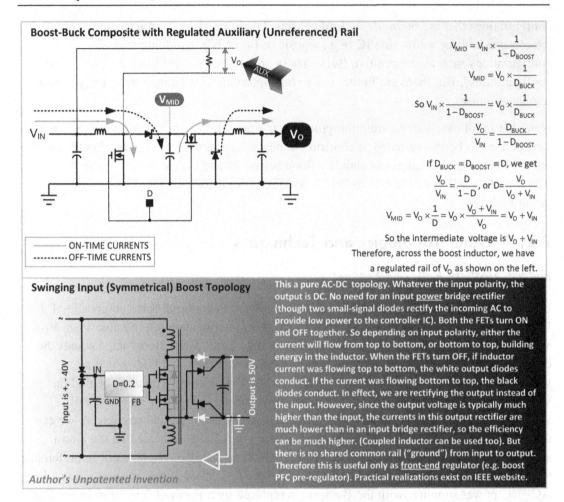

Boost-Buck Composite with Regulated Auxiliary (Unreferenced) Rail

$$V_{MID} = V_{IN} \times \frac{1}{1 - D_{BOOST}}$$

$$V_{MID} = V_O \times \frac{1}{D_{BUCK}}$$

So $V_{IN} \times \dfrac{1}{1 - D_{BOOST}} = V_O \times \dfrac{1}{D_{BUCK}}$

$$\frac{V_O}{V_{IN}} = \frac{D_{BUCK}}{1 - D_{BOOST}}$$

If $D_{BUCK} = D_{BOOST} \equiv D$, we get

$$\frac{V_O}{V_{IN}} = \frac{D}{1 - D}, \text{ or } D = \frac{V_O}{V_O + V_{IN}}$$

$$V_{MID} = V_O \times \frac{1}{D} = V_O \times \frac{V_O + V_{IN}}{V_O} = V_O + V_{IN}$$

So the intermediate voltage is $V_O + V_{IN}$

Therefore, across the boost inductor, we have

a regulated rail of V_O as shown on the left.

------- ON-TIME CURRENTS
- - - - OFF-TIME CURRENTS

Swinging Input (Symmetrical) Boost Topology

This a pure AC-DC topology. Whatever the input polarity, the output is DC. No need for an input <u>power</u> bridge rectifier (though two small-signal diodes rectify the incoming AC to provide low power to the controller IC). Both the FETs turn ON and OFF together. So depending on input polarity, either the current will flow from top to bottom, or bottom to top, building energy in the inductor. When the FETs turn OFF, if inductor current was flowing top to bottom, the white output diodes conduct. If the current was flowing bottom to top, the black diodes conduct. In effect, we are rectifying the output instead of the input. However, since the output voltage is typically much higher than the input, the currents in this output rectifier are much lower than in an input bridge rectifier, so the efficiency can be much higher. (Coupled inductor can be used too). But there is no shared common rail ("ground") from input to output. Therefore this is useful only as <u>front-end</u> regulator (e.g. boost PFC pre-regulator). Practical realizations exist on IEEE website.

Author's Unpatented Invention

Figure 9.18: The brute-force Boost-Buck composite with an auxiliary rail, and the symmetric Boost topology.

law of a transformer, which says that at any moment, a perfect transformer has identical volts per turn on any of its windings. Of course, there is no perfect transformer. We have to counter the effects of DC resistance as we pull current from the windings, and also leakage inductance effects between the windings. The former is tackled by using fairly thick wire, and the latter will include techniques to tightly couple the Secondary windings together. Also, when we draw current from the main winding, it goes through a diode, whose forward-drop depends on current and temperature. For example, if we draw a lot of current, the drop is more. So, the control loop compensates for that drop by pushing the average volts/turn in the transformer higher (by increasing the duty cycle). That causes the other outputs to increase too, even though they do not need that correction. Therefore, one way of

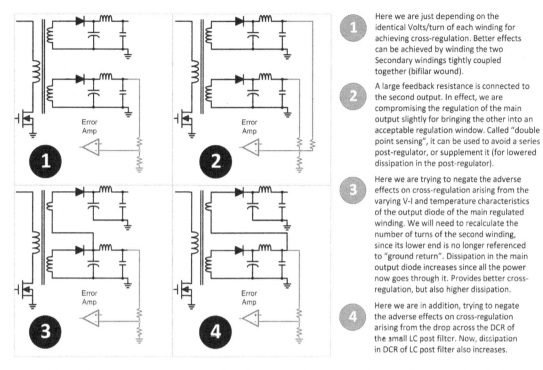

Figure 9.19: Deriving cross-regulated outputs from the main transformer in AC–DC power supplies.

compensating for this effect is to have the bottom end of the Secondary winding connect to the cathode of the main output diode, as shown in the figure. Now, as the drop across the diode increases with load, it pulls the other output down somewhat, compensating partially for the higher average volts/turn in the transformer. Other ways include trying to do a sort-of weighted control loop that is lightly affected by outputs other than the main one.

In Figure 9.20, we show a common technique to derive power for Secondary-side housekeeping circuitry in AC–DC power supplies. A winding is thrown across the output choke. That is straightforward. We can do the same with a Buck converter as shown. However, we can then try to have the Buck converter *ride on top of the very rail it generates*. That changes the Buck topology into a "floating Buck regulator." We have reduced voltage stresses on the switch and controller IC, and we also have an auxiliary rail to use if desired. In the earlier cases, if we use a turns ratio of 1:1, the voltage on the auxiliary rail will be V_O (same as the main rail). In the floating Buck regulator, the auxiliary voltage is exactly half the voltage of the main rail (i.e., $V_O/2$).

In AC–DC power supplies, we can put many windings on the transformer, and by the fact that each winding will (or should) have the same volts/turn, we can predict the voltage of

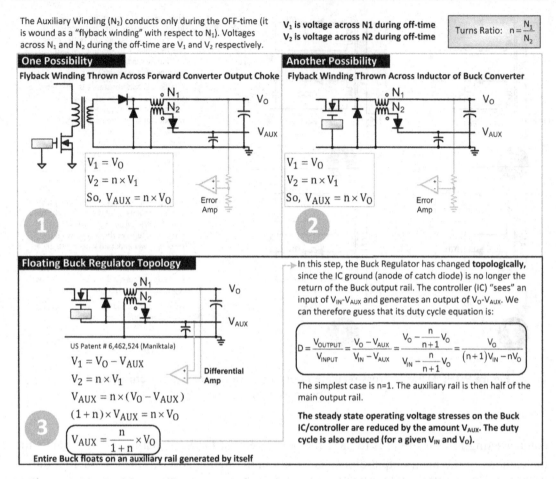

The Auxiliary Winding (N_2) conducts only during the OFF-time (it is wound as a "flyback winding" with respect to N_1). Voltages across N_1 and N_2 during the off-time are V_1 and V_2 respectively.

V_1 is voltage across N1 during off-time
V_2 is voltage across N2 during off-time

Turns Ratio: $n = \dfrac{N_1}{N_2}$

One Possibility

Flyback Winding Thrown Across Forward Converter Output Choke

$V_1 = V_0$
$V_2 = n \times V_1$
So, $V_{AUX} = n \times V_0$

Error Amp

1

Another Possibility

Flyback Winding Thrown Across Inductor of Buck Converter

$V_1 = V_0$
$V_2 = n \times V_1$
So, $V_{AUX} = n \times V_0$

Error Amp

2

Floating Buck Regulator Topology

US Patent # 6,462,524 (Maniktala)

$V_1 = V_0 - V_{AUX}$
$V_2 = n \times V_1$
$V_{AUX} = n \times (V_0 - V_{AUX})$
$(1 + n) \times V_{AUX} = n \times V_0$

$$V_{AUX} = \frac{n}{1+n} \times V_0$$

Differential Amp

3

Entire Buck floats on an auxiliary rail generated by itself

In this step, the Buck Regulator has changed **topologically**, since the IC ground (anode of catch diode) is no longer the return of the Buck output rail. The controller (IC) "sees" an input of V_{IN}-V_{AUX} and generates an output of V_0-V_{AUX}. We can therefore guess that its duty cycle equation is:

$$D = \frac{V_{OUTPUT}}{V_{INPUT}} = \frac{V_0 - V_{AUX}}{V_{IN} - V_{AUX}} = \frac{V_0 - \dfrac{n}{n+1} V_0}{V_{IN} - \dfrac{n}{n+1} V_0} = \frac{V_0}{(n+1)V_{IN} - nV_0}$$

The simplest case is n=1. The auxiliary rail is then half of the main output rail.

The steady state operating voltage stresses on the Buck IC/controller are reduced by the amount V_{AUX}. The duty cycle is also reduced (for a given V_{IN} and V_0).

Figure 9.20: Deriving auxiliary outputs from the output choke/inductor of the Forward/Buck converter and understanding the floating Buck regulator topology.

other windings placed on the same core as the main controlled-output winding. In principle, we get many regulated outputs. In practice, because of parasitics and leakage inductances, the relative regulation of the output is not very good. There are many ways to handle this as indicated in Figure 9.19.

Hysteretic Controllers

Looking back at Figure 1.2, the bucket regulator and its SCR-based version are early examples of "bang-bang regulators." A vague control loop exists that turns ON the semiconductor if the voltage falls below a certain threshold and turns it OFF if the voltage exceeds a certain threshold. There is no predictable waveform in the process. But there is also no reason why we can't try out the same technique (or lack thereof), using a conventional Buck regulator (or even other topologies). At least the new candidate uses an *inductor*, so it promises much higher efficiency. Whenever the voltage falls below a certain threshold, we would command the controller to switch with its maximum duty cycle. After a while, the output would rise and cross the upper threshold, at which point all switching would stop. This type of controller would have excellent transient response since it turns on full-blast (with full duty cycle) whenever needed, and turns OFF equally dramatically when not needed. We wouldn't have to worry about complicated poles and zeros. We would likely save a lot of die area in a controller of this type since there is no compensation circuitry or PWM comparator. But the main problem here is possible inductor saturation and "pulse-bunching." We could get a string of full-width pulses, and then none, creating almost any pattern. We would therefore likely generate highly unpredictable EMI, and audible noise too (from magnetic components and ceramic/film capacitors).

For better results, we would like to have a *steady stream* of pulses instead of pulse-bunching. Further, to minimize die area and reduce quiescent current, it would also be nice to get rid of the clock entirely. How do we start by doing the latter? We realize that we have an inherent clock present in any converter, based on its switching *natural time constants*. For example, we know that the inductor current undulates at a regular rate, determined by the applied voltages and the inductance. That is in effect a clock. So, can we use that instead of a formal clock circuit? If so, where can we extract the signal from? If we assume that the output cap of the Buck has no significant ESL, but just resistance (ESR), it has a voltage ripple riding on top of the DC output voltage. The ripple has the same periodicity as the inductor current and also *the same duty cycle* that we want to drive the switch with. It thus becomes a candidate for trying to *invert cause and effect*. This cap voltage ripple becomes the (scaled) voltage applied to the feedback pin of the controller IC. So now, if we set min and max thresholds on the feedback pin, we could generate the very inductor waveform that we are sensing — *a self-stabilizing chicken and egg situation at an electrical level,* one that could last forever. That is the only naturally stable situation that can result with the bare constraints applied. This process is shown in Figure 9.21. We also see how the frequency can be changed by varying the hysteresis between the upper and the lower thresholds. Note that any asymmetry in the thresholds will translate into a DC offset error on the output.

In reality, we know that ceramic caps in use today have very low ESR. So, the ripple on the output cap of a Buck can be very small and can also vary a lot. So, various techniques have emerged in trying to generate a proper "ESR-ramp" as outlined in Figure 9.21.

One of the problems of these hysteretic Buck regulators is that their frequency can vary a lot. Various techniques exist, some proprietary, to minimize the DC offset mentioned above and also to maintain constant frequency — all hopefully without excessively complex

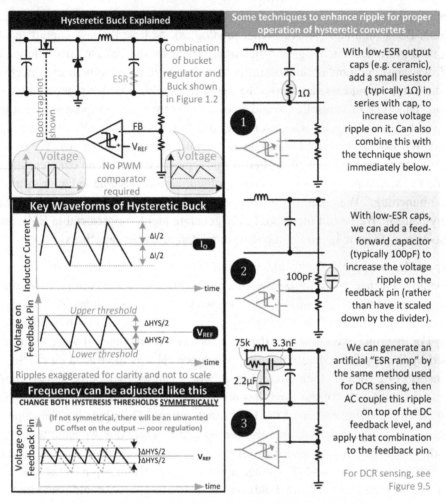

Figure 9.21: Hysteretic control and changing switching frequency.

circuitry which would end up negating the very reason for considering hysteretic regulators — low quiescent current, good transient response, and optimized silicon area and cost.

Besides varying the hysteresis band as indicated in Figure 9.21, there is another method for trying to achieve fairly constant frequency using hysteretic controllers. This is called "constant on-time" (COT) control.

For a Buck topology, we can do the following interesting analysis:

$$D = \frac{T_{ON}}{T} = \frac{V_O}{V_{IN}} \quad \text{(Buck)}$$

So,

$$T = \frac{T_{ON} \times V_{IN}}{V_O}$$

Using

$$T = \frac{1}{f}$$

$$\frac{1}{f} = \frac{T_{ON} \times V_{IN}}{V_O}$$

$$V_O = [T_{ON} \times V_{IN}] \times f$$

In other words, if we force a constant on-time, but one that is *inversely proportional to input voltage*, then for a given output voltage we will get *constant frequency*. So, in this type of control, whenever the feedback voltage falls below a set threshold, the FET is turned ON. However, the FET is not turned OFF based on any ESR-based ripple crossing an upper threshold. Instead the FET literally "times out" because the ON-pulse is generated by a simple, one-shot flip-flop (a monostable multivibrator). Remember that this hysteretic implementation also has no clock. After the FET turns OFF, then after a small, arbitrary, guaranteed OFF-time, if the feedback voltage is still below the set threshold, another one-shot pulse will follow, otherwise the pulse will be skipped. And so on. Eventually, the converter will settle down somewhat close to a steady stream of pulses. With a simple input-feedforward circuit, the width of the one-shot can be varied inversely with respect to input voltage. So finally, the frequency will be roughly constant with respect to both line and load variations.

Can we invoke constant frequency in a hysteretic Boost? Here is the math.

$$D = \frac{T_{ON}}{T} = \frac{V_O - V_{IN}}{V_O} \quad \text{(Boost)}$$

$$\text{If } \frac{A}{B} = \frac{C}{D}, \quad \text{then } \frac{B-A}{B} = \frac{D-C}{D}$$

We can thus eliminate T_{ON}

$$D = \frac{T_{ON}}{T} = \frac{V_O - V_{IN}}{V_O} \quad \text{(Boost)}$$

$$\frac{T - T_{ON}}{T} = \frac{V_O - V_O + V_{IN}}{V_O}$$

$$\frac{T_{OFF}}{T} = \frac{V_{IN}}{V_O}$$

Or

$$\frac{T}{T_{OFF}} = \frac{V_O}{V_{IN}}$$

Using

$$T = \frac{1}{f}$$

$$\frac{1}{f} = \frac{T_{OFF} \times V_O}{V_{IN}}$$

$$V_{IN} = T_{OFF} \times V_O \times f$$

Or

$$V_O = \left[\frac{V_{IN}}{T_{OFF}}\right] \times \frac{1}{f}$$

In other words, in a Boost, if we fix T_{OFF} for a given input voltage, and then vary T_{OFF} such that it is proportional with respect to the input voltage, then for a given V_O, f is a constant. This would be a constant frequency, constant off-time (also confusingly called "COT") Boost regulator.

Note that in *Chapter 12*, we discuss the causes of the RHP (right half plane) zero in the Boost and Buck-Boost topologies (operating in CCM). The intuitive reason for that is under a sudden load demand, the output dips momentarily, and therefore the duty cycle increases. But in the process, the off-time decreases. Since in both these topologies, energy is delivered to the output only during the off-time, a smaller off-time leaves less time for the new energy requirement to be met, which temporarily causes the output to dip even further before things get back to normal. So clearly, fixing a certain minimum off-time will help. The RHP zero is not present when operating in constant off-time mode. Further, in a Boost, we can get constant-frequency operation too as described above.

Don't even bother to do the math for creating a constant-frequency hysteretic Buck-Boost. The requirement is neither constant on-time nor constant off-time, but a complicated function of both V_{IN} and V_O. Therefore, its implementation, if any, will just sacrifice the expected simplicity of hysteretic control.

Pulse-Skipping Mode

All the above discussions assume CCM, in which the duty cycle is almost constant with respect to load and inversely proportional to input voltage (in a Buck). So, what happens to a COT Buck converter if the load is decreased? The one-shot generator will continue to produce CCM-based one-shot pulses, whereas in conventional DCM, the pulse width decreases rapidly with load, if not forcibly disallowed. So in effect, with COT at light loads, we would end up pumping far more energy per pulse than demanded by natural DCM. The control loop will therefore see the output rising suddenly, and try to arrest that rise by skipping several pulses in succession. We thus get "pulse-skipping mode." There are many ways to implement this feature, but the basic advantage is reduced switching losses, which are a major factor in the low-efficiency readings at light loads of synchronous converters, with complementary drives in particular. IC designers also try to take further advantage of the relatively long OFF-times, and "de-bias" some of their circuits, to reduce the quiescent current of the IC too, and maximize efficiency at light loads. For example, the FET Gate drives may no longer be held down "hard," but rather softly so. And so on. The challenge, however, is to wake up all hibernating circuits very rapidly when the load demand increases, so as not to cause output glitches as the converter moves from pulse-skipping mode to full CCM operation. One cause of output glitches is due to the system, considered purely as a power processing stage, transitioning between an artificially enforced mode and natural CCM operation. At the transition boundary, the energy levels in the capacitors and inductor need to restabilize to their new steady-state values. Till that happens, the output will either overshoot or undershoot. However, one of the best ways of avoiding this particular type of output glitch is the "duty cycle brickwall" method of implementing pulse-skipping. This is the least intrusive method and simply sets a minimum duty cycle for the converter to obey. As the load is decreased, DCM is initially entered and the duty cycle starts to progressively decrease as load is further reduced. But at some point, the duty cycle hits a brickwall, and is not allowed to shrink further. The system then, very naturally, starts skipping pulses to maintain energy balance. If however the load is increased, equally naturally, the system exits this pulse-skipping mode. The way to tradeoff output ripple and efficiency is to carefully position this brickwall, by defining the appropriate "*m*-factor" below.

$$D_{\text{DCM}} > m \times D_{\text{CCM}}$$

A typical value of m could be between 50% and 85%. This forces a natural minimum duty cycle, and the converter goes in and out of pulse-skipping mode smoothly.

Achieving Transformer Reset in Forward Converters

Finally, to close with the somewhat mundane, and also tie up some remaining loose ends, we return to the single-ended Forward converter with an energy recovery (tertiary) winding. In Figure 9.22, we indicate that the position of the energy recovery winding diode can affect converter efficiency too. This was previously touched upon in *Chapter 7*. In a conventional single-ended Forward converter, by using a 1:1 ratio between the Primary ("PRI") winding and the tertiary ("TER") winding, we ensure that the rising slope of the

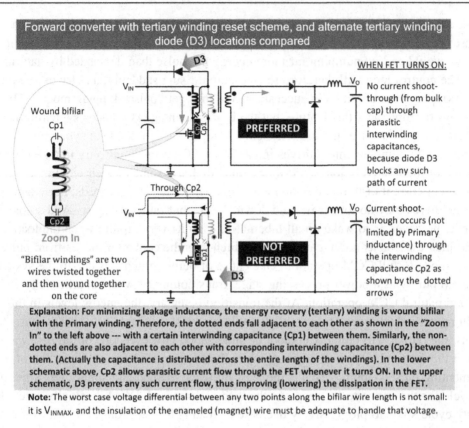

Figure 9.22: Position of the energy recovery diode can affect Forward converter efficiency.

magnetization current V_{IN}/L_{PRI} is equal in magnitude to the falling slope V_{IN}/L_{TER}, since $L_{PRI} = L_{TER}$. So, we need to ensure a maximum duty cycle of 50%, which leaves just enough time, even in the worst case, for the magnetization current to slope down *to the value it started the cycle off with* (zero in this case). That is termed transformer "reset." By achieving reset, we ensure the transformer can operate in steady state with no flux-staircasing. The impact of this is that the voltage stress on the FET is exactly $2 \times V_{INMAX}$ under all load conditions. However, a 50% max duty cycle is easily achieved by popular controllers like the UC3844, by using a frequency doubler circuit. Basically, the internal clock of the IC runs at $2 \times f_{SW}$, and every alternate cycle is blanked out. During that blanked-out interval, the FET is forced OFF. So, in effect, the max ON-time (the nonblanked-out interval) can never exceed the OFF-time, and reset is assured.

In Figure 9.23, we also show how the "active clamp Forward converter" works, and its relative pros and cons. Similar to the asymmetric half-bridge (i.e., the two-switch Forward converter, see Table 7.1), it doesn't have an additional energy recovery winding. But nor is

Explanation: The transformer is modeled as "MAG" (magnetization winding) in parallel with "PRI", the Primary winding (turns N_P), which couples with "SEC", the Secondary winding (turns N_S) to create "transformer action": typically, a *step-down* ratio of $n = N_P/N_S$. Disregarding the Primary and Secondary windings for a moment (i.e. ignoring transformer section), the Primary side circuitry (i.e. the active clamp) is **in effect a synchronous Buck-Boost converter** --- with input V_{IN}, output voltage "V_{CLAMP}" (its "output cap" is Cc) , inductance L_{MAG} and operating at (near-) **zero net load current.** Its load current I_{MAG} "cycles" back and forth with a center value (average) equal to zero --- as in any synchronous converter operating at zero load. Why is it a Buck-Boost? Because as in any Buck-Boost, inductive energy is delivered into L (L_{MAG} here) only during the ON-time (ands also, *no energy flows into* its output, which is Cc in this case). Then, during the OFF-time, *all* that stored inductive energy (*and no more*) goes directly into the output (Cc here). Hence the analogy. The difference in this case is that the associated magnetization (inductive) energy of this "synchronous Buck-Boost converter", doesn't typically get returned to the input source as in any standard synchronous Buck-Boost at zero load, but get dumped to the output of the Forward converter. This happens because, assuming a minimum pre-load on the output of the Forward converter stage, during Phase "D" below, I_{MAG} passes through the PRI winding and thus gets reflected into the SEC winding, and from there into the load connected to V_O. Wiith this analogy, the duty cycle and the max voltage on the switch are both determined as in any Buck-Boost. See equations below.

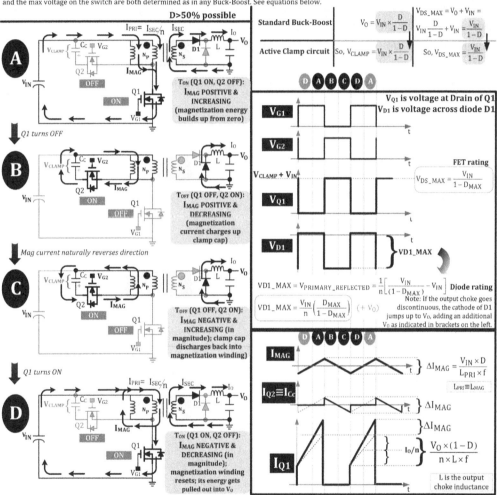

Figure 9.23: The active clamp Forward converter.

it restricted to a max of 50% duty cycle. In this case, it is of paramount importance to fix the maximum duty cycle by using a well-designed control circuit, since D_{MAX} determines the max voltage stress on the FET: $V_{INMAX}/(1 - D_{MAX})$. If D_{MAX} approaches 1, the voltage stress will approach infinity. Observe that the clamp circuit basically runs like a synchronous Buck-Boost stage, the "output" of which is the voltage of the clamp capacitor. So, as in any Buck-Boost, we can predict that the output rail (the voltage on the clamp capacitor in this case) is

$$V_{\text{CLAMP}} = V_{\text{INMAX}} \times \frac{D_{\text{MAX}}}{1 - D_{\text{MAX}}}$$

With that we close our sojourn into the exciting world of "new topologies," as we wait for more to be discovered. However one thing remains clear. Figure 1.15 is still irrefutable — there are only *three* fundamental topologies. The rest are, on deeper thought, variants or combinations of the basic three. This chapter should have made that quite clear by now. The key to understanding "new" or exotic topologies also lies therein.

CHAPTER 10

Printed Circuit Board Layout

Introduction

A great many customer "complaints" regarding switcher ICs are ultimately traced to poor printed circuit board (PCB) layout practices. When designing a PCB for a switching regulator, we need to be aware that the final product is going to be only as good as its layout. Certainly, some ICs are more noise-sensitive than others. Sometimes, the "same" part from several vendors can also have starkly varying noise sensitivities. Further, some ICs are architecturally more noise-sensitive than others (e.g., current-mode controllers are far more "layout-sensitive" than voltage-mode controllers). We also have to face the fact that virtually no semiconductor manufacturers characterize the noise sensitivity of their products (often letting the customers discover it for themselves!). However, as designers, we can certainly, with poor attention to layout, pull off the near-impossible — turn a comparatively stable IC into a jittery and nervous part — one that can malfunction and even cause catastrophic consequences (switch failure). Further, since very few of these problems can be easily corrected, or "band-aided," at a later stage, it is very important to get the layout right at the very beginning.

Most of the layout recommendations in this chapter revolve around simply ensuring basic functionality and performance. Though luckily, the beleaguered switcher designer will be happy to know, in general, the electrical aspects are all related — pointing in the same general direction. So, for example, a good layout, that is, one that helps the IC function properly, also leads to reduced electromagnetic emissions, and vice versa. There are some exceptions to this trend however, particularly when it comes to the practice of indiscriminate "copper-filling" (or copper "flooding") on PCBs, which we will touch upon later. Subsequently, the reader can try to gain more insight into the practical aspects of making switching regulators, by reading the chapters dedicated to the topic of EMI, later in this book.

Trace Section Analysis

A switch transition (crossover) occurs when the switch changes from an on-state (switch closed) to an off-state (switch open), or back. A typical transition lasts typically less than

Switching Power Supplies A–Z. DOI: 10.1016/B978-0-12-386533-5.00010-3

100 ns. But most of the trouble starts right here! In fact, the noise has comparatively little to do with the basic *switching frequency* of the converter itself — it is the *transition* that is responsible for most of the noise and all its attendant problems. The smaller the switch transition time, the more the possible consequences, as we will see.

The first requirement for the designer is to understand the flow of power-related currents in the converter. This leads to an identification of the troublesome or "critical" traces of the PCB; we must pay the closest attention to these traces. We will also see that this identification process is very "topology-dependent." So we can't, for example, design the PCB for a Buck-Boost, the same way we would do it for a Buck. The rules change significantly! We may thus also realize that very few PCB layout persons out there would understand this too well! Therefore, it really is a good idea for the power supply designer to do the layout personally, or at the very least, closely supervise the PCB person in the act.

Some Points to Keep in Mind During Layout

Let's summarize these for quick reference purposes:

- During a crossover transition, the current flow in certain trace sections has to suddenly come to a *stop*, and in certain others it has to *start* equally suddenly (within 100 ns or less typically, which is the switch transition time). These trace sections are identified as the "critical traces" in any switcher PCB layout. A very *high* dI/dt is created in them, during every switch transition (see Figure 10.1). Expectedly, these traces end up "complaining" vociferously in the form of small, but potent, voltage spikes across them. If *Chapter 1* has been fully understood by now, we realize that this is just the equation $V = L \times dI/dt$ playing out its part — with the "L" being the parasitic inductance of the PCB trace. The rule of thumb for the inductance presented by a trace is *20 nH per inch of trace length.*
- Once generated, these noise spikes can not only appear at the input/output (causing performance issues), but also infiltrate the IC control section, causing it to behave anomalously, and unpredictably. We, for example, could even end up briefly losing current-limiting function, leading to disastrous consequences.
- MOSFETs switch faster than "BJTs" (bipolar junction transistors). The transition times of a MOSFET can be of the order of 10–50 ns, as compared to a BJT's 100–300 ns. But that also makes the "spikes" far more severe in the case of the converters that use MOSFET switches — because of the much higher dI/dt's they can generate in the critical trace sections of the PCB.

  Note: One inch of trace switching, say 1 A of instantaneous current in a transition time of 30 ns, gives a spike of 0.7 V. For 3 A, and 2 in. of trace, the induced voltage tries to be 4 V!

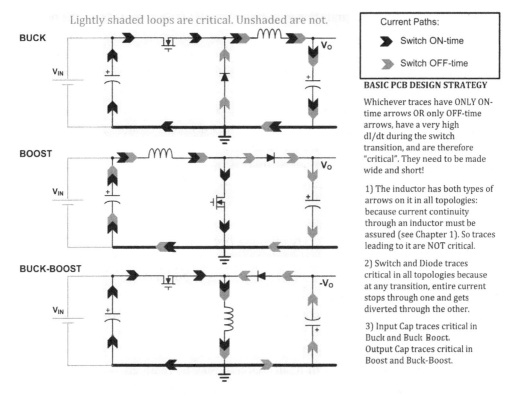

Figure 10.1: Identifying the critical trace sections for the three topologies.

Note: It is almost impossible to "see" the noise spikes. First of all, various parasitics help limit/absorb them somewhat (though they can still retain the capability to cause "controller upset"). Further, the moment we put in an oscillo-scope probe, the 10–20 pF of probe capacitance can also absorb the spikes, and we would probably see nothing significant. In addition, probes pick up so much normal switching noise through the air anyway that we are never completely sure what we are seeing!

- Integrated switchers ICs (or simply "switchers") have the switch in the same package as the control. Though that makes for convenience and low parts count, such ICs are usually more sensitive to the noise spikes generated by the parasitic trace inductances. That is because the "switching node" of the power stage (its "swinging node," i.e., the one connecting the diode, switch, and inductor) is a pin on the IC itself so that the pin conducts any unusual high-frequency noise at the switching node straight into the control sections, causing "controller upset."

- Note that while prototyping, it is a bad idea to insert a current probe (through a loop of wire) anywhere in a critical path (learn to recognize these in Figure 10.1). The current loop becomes an additional inductance that can increase the amplitude of the noise spikes dramatically. Therefore, practically speaking, it often becomes virtually impossible to measure the switch current or the diode current individually (especially in the case of switcher ICs). In such cases, only the inductor current waveform can really be measured properly.

- Note that in the Buck and the Buck-Boost, the input capacitor is also included in a critical path. That implies we need very good *input decoupling* in these topologies (for the power section). So, besides the necessary bulk capacitor for the power stage (typically a tantalum or aluminum electrolytic of large capacitance), we should also place a small ceramic capacitor (about $0.1-1$ µF) directly between the *quiet* end of the switch (i.e., at the supply side) and the ground — and also *as close as possible to the switch.*
- In Figure 10.1, the control section (IC) has not been shown. However, we should remember that the control circuitry usually needs good *local* decoupling of its own. And for that we need to provide a small ceramic capacitor *very close to the IC*. Clearly, especially when dealing with switchers, the decoupling ceramic for the power stage can often do "double duty" as the decoupling capacitor of the control too (note that this applies to the Buck-Boost and the Buck only, since the input power-decoupling capacitor is required only for them).
- Sometimes, more effective control IC decoupling may be required — in which case we can use a small resistor (typically $10-22$ Ω) from the input (supply) rail, going to a (separate) ceramic capacitor placed directly across the input and ground pins of the IC. This constitutes a small "RC filter" for the IC supply.
- Note that in all topologies, the inductor is not in the critical path. So we need not worry much about its layout, at least not from the point of view of noise. However, we have to be wary of the electromagnetic field the inductor creates, because that can impinge on nearby circuitry and sensitive traces, and cause similar (though usually not so acute) problems. So generally, it is a good idea to try to use "shielded inductors" for that reason, if cost permits. If not, it should be positioned *a little further* from the IC, in particular *keeping clear of the feedback trace.*
- In the Boost and the Buck-Boost, we see that the output capacitor is in the critical path. So this capacitor should be close to the control IC, along with the diode. A paralleled ceramic capacitor can also help, provided it does not cause loop instability issues (especially in voltage-mode control — see *Chapter 12*).

In the Buck, however, note that though the output diode needs to be positioned close to the IC/switch, the output capacitor is not critical (its current is smoothened by the inductor). If we place a ceramic capacitor in parallel to the output capacitor, it is only for the purpose of decreasing high-frequency noise and ripple at the output even further. But it is really not mandatory and can cause loop instability, particularly with voltage-mode control, especially if the effective series resistance (ESR) of the output capacitor section becomes too low (less than 100 mΩ typically).

- The position of the diode is critical in all topologies. It leads to the switching node and from there on, straight into the IC when using switcher ICs. However, in Buck converter layouts in which the diode has unfortunately been placed a little too far away from the IC, the situation can sometimes be rectified, even at a later stage, by means of a small series RC snubber connected between the switching node and ground (across the catch diode, close to the IC). This RC typically consists of a resistor (low-inductive type preferred), of value 10–100 Ω, and a capacitor (preferably ceramic), of value about 470 pF to 2.2 nF. Note that the dissipation in the resistor is $C \times V_{IN}^2 \times f$. So not only should the wattage of the resistor be appropriate for the job, but also the capacitance should not be increased indiscriminately, to avoid compromising the efficiency significantly.
- A first approximation for the inductance of a conductor (wire) having length *l* and diameter "*d*" is

$$L = 2l \times \left(ln\frac{4l}{d} - 0.75 \right) \text{ nH}$$

where *l* and *d* are in centimeters. Note that the equation for a PCB trace is not much different from that of a wire.

$$L = 2l \times \left(ln\frac{2l}{w} + 0.5 + 0.2235\frac{w}{l} \right) \text{ nH}$$

where "*w*" is the width of the trace in cm. Note that for PCB traces, the inductance hardly depends on the thickness of the copper on the board.
- The logarithmic relationship above indicates that if we halve the length of a PCB trace, we can make its inductance halve too. But we have to increase its width almost 10 times, to get its inductance to halve. In other words, simply making traces "wide" may not do much — we need to keep trace lengths *short*.
- The inductance of a "via" (through-hole) is given by

$$L = \frac{h}{5} \left(1 + ln\frac{4h}{d} \right) \text{ nH}$$

Here "*h*" is the height of the via in millimeters (equal to the thickness of the board, commonly 1.4–1.6 mm) and "*d*" is the diameter of the via in millimeters. Therefore, a via of diameter 0.4 mm on a 1.6 mm thick board gives an inductance of 1.2 nH. That may not sound like much, but it has been known to cause problems in switcher ICs, especially those using MOSFETs. Because of their fast transition times, input ceramic decoupling capacitors for such ICs become almost mandatory. Therefore, it is strongly advised that this input ceramic capacitor be placed extremely close to where the pins of the IC actually contact the board. There should be no intervening vias between this capacitor and the solder pads of the pins as this worsens decoupling significantly.

- Increasing the *width* of certain traces can in fact become counterproductive. For example, for the (positive) Buck regulator, the trace from the switching node to the diode is "hot" (swinging). Any conductor with a varying voltage on it, irrespective of the current it may be carrying, becomes an *E*-field antenna if its dimensions are large enough. Therefore, the area of the copper around the switching node needs to be reduced, not increased. That is why we need to avoid the tendency of indiscriminate "copper-filling" — the only voltage node that really qualifies for copper-filling is the ground node (or plane). All others, including the input supply rail, can start radiating significantly because of the high-frequency noise riding on them. By making large planes, we also increase the probability of that plane picking up noise from nearby traces and components, by means of inductive and capacitive coupling.

- The so-called "1-oz" board in the United States is actually equivalent to 1.4-mils copper thickness (or 35 μm) on the board. Similarly "2 oz" is twice that. For a moderate temperature rise (less than 30 °C) and currents less than 5 A, we can use a minimum 12-mils width of copper per Ampere for 1-oz board, and at least 7-mils width of copper per amp for a 2-oz board. This rule of thumb is based on the DC resistance of the trace only. So to decrease its inductive impedance and AC resistance, higher trace widths may be required.

- We have seen that *the preferred method to reduce trace inductance is to reduce length, not increase width.* Beyond a certain point, widening of traces does not reduce inductance significantly. Nor does it depend much on whether we use 1-oz or 2-oz boards. Nor if the trace is "unmasked" (to allow solder/copper to deposit and thereby increase effective conductor thickness). So, if for any reason, the trace length cannot be reduced further, another way to reduce inductance is by *paralleling the forward and return current traces.* Inductances exist because they represent stored magnetic energy. The energy resides in the magnetic field. Therefore, conversely, if the magnetic field could be canceled, the inductance vanishes. By paralleling two current traces, each carrying currents of the same magnitude but in opposite direction, the magnetic field is greatly reduced. These two traces should be parallel and very close to each other on the same side of the PCB. If a double-sided PCB is being used, the best solution is to run the traces parallel (over each other) on *opposite* sides (or adjacent layers) of the PCB. These traces can, and should, be fairly wide to improve mutual coupling and thereby the field cancellation. Note that if a ground plane is used on one side, the return path automatically "images" the forward current trace (for high frequencies), and produces the sought-after field cancellation.

- In high-power off-line flybacks, the trace inductances on the secondary side reflect on to the primary side, and can greatly increase the effective primary-side leakage inductance and impair the efficiency (see *Chapter 3*). The situation gets worse when we have to stack several output capacitors in parallel, just to handle the higher RMS currents — long traces seem inevitable here. However, one way to decrease the inductance is by the field cancellation principle discussed above. This is shown implemented in Figure 10.2. Two copper planes (or big copper islands) are allocated, starting from the output diode. One of these planes is the ground plane, the other being the output voltage rail. By using two large parallel planes carrying forward and return currents, the inductance almost completely cancels out, and leads to a very good high-frequency freewheeling path as desired. Note that in the bargain, we also get excellent current sharing between the output capacitors.

- In single-sided boards, a popular way to ensure current sharing between several paralleled output capacitors is shown in Figure 10.3. It doesn't minimize inductance, but it does ensure that the life of the first downstream capacitor does not come to a premature end (simply because of "current hogging"). Note that in the "improved" layout on the right side of the figure, the total distance from the diode through each capacitor is roughly equal in all three cases shown — thus leading to more precise sharing.

- With multilayer boards, it is a common practice to almost completely fill one layer with ground (if so, it should preferably be the layer immediately below the power components/traces). There are people who, usually rightly so, consider this a panacea for most problems. As we have seen, every signal has a return, and as its harmonics get higher, the return current, rather than trying to find the path of least DC resistance (straight line), tries to reduce the inductance by imaging itself directly under the signal path even though that may be "zigzagging" away on the

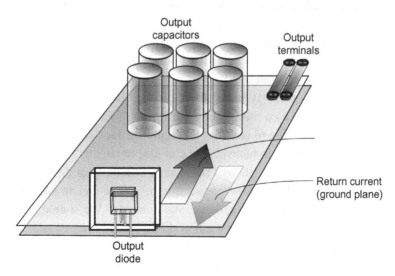

Figure 10.2: How to achieve low-inductance connections to output capacitors of a flyback.

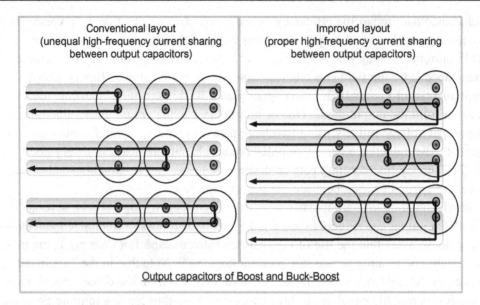

Figure 10.3: How to get output capacitors of a flyback to share current.

board. So by leaving a large ground plane, we basically "allow" nature to "do its thing" — searching and finding the path of least impedance (lowest DC resistance or lowest inductive impedance, depending upon the frequency of the harmonic). The ground plane also helps thermal management as it couples some of the heat to the other side. The ground plane can also capacitively link to noisy traces above it, causing general reduction in noise/EMI. However, it can also end up radiating if caution is not exercised. One way this can happen is to have too much capacitive coupling from noisy traces. No ground plane is perfect, and when we inject noise into it, it may get affected, especially if the copper is too thin. Also, if the ground plane is partitioned in odd ways, either to create thermal islands, or to route other traces, the current flow patterns can become irregular. No longer can return paths in the ground plane pass directly under their forward traces. The ground plane can then end up behaving as a slot antenna in terms of EMI.

• The only important *signal* trace to consider is usually the feedback trace. If this trace picks up noise (capacitively or inductively), it can lead to slightly offset output voltages — and in extreme cases (though rare), even instability or device failure. We need to keep the feedback trace short *if possible* so as to minimize pickup and keep it away from noise or field sources (the switch, diode, and inductor). We should never pass this trace *under* the inductor, or *under* the switch or diode (even if on opposite sides of the PCB). We should also not let it run close to and parallel, for more than a few millimeters at most, to a noisy (critical) trace, even on adjoining layers of the board. Though if there is an intervening ground plane that should provide enough shielding between layers.

Keeping the feedback trace short may not always be physically feasible. We should realize that keeping it short certainly is not of the highest priority. In fact, we can often deliberately make it long, just so that we can assuredly route it away from potential noise sources. We can also judiciously cut into the "quiet" ground plane to pass this particular trace through so that in effect, it is surrounded by a "sea of tranquility."

Thermal Management Concerns

Larger and larger areas of copper do not help, especially with thinner copper. A point of diminishing returns is reached for a square copper area of size 1 in. × 1 in. Some improvement continues until about 3 in. (on either side), especially for 2-oz boards and better. But beyond that, external heatsinks are required. A reasonable practical value attainable for the thermal resistance (from the case of the power device to the ambient) is about 30 °C/W. That means 30 °C rise for every Watt of dissipation.

To calculate the required copper area, we can use as a good approximation the following empirical equation for the required copper area:

$$A = 985 \times \text{Rth}^{-1.43} \times P^{-0.28} \text{ sq. in.}$$

Here P is in Watts and Rth is the desired thermal resistance in °C/W.

For example, suppose the estimated dissipation is 1.5 W. We want to ensure that, at a worst-case ambient of 55 °C, the case of the part does not rise above 100 °C (safe temperature for the PCB material — do not exceed!). Therefore, the Rth we are looking for here is

$$\text{Rth} = \frac{\Delta T}{P} = \frac{100 - 55}{1.5} = 30 \text{ °C/W}$$

Therefore, the required copper area is

$$A = 985 \times 30^{-1.43} \times 1.5^{-0.28} \text{ sq. in.}$$
$$A = 6.79 \text{ sq. in.}$$

If this area is square in shape, the length of each side needs to be $6.79^{0.5} = 2.6$ in. We can usually make this somewhat rectangular or odd-shaped too, as long as we preserve the total area. Note that if the area required exceeds 1 sq.in., a 2-oz board should be used (as in this case). A 2-oz board reduces the thermal "constriction" around the power device and allows the large copper area to be more effectively used for natural convection.

We should not think that heat is lost only from the copper side. The usual laminate (board material) used for SMT (surface mount technology) applications is epoxy-glass "FR4," which is a fairly good conductor of heat. So some of the heat from the side on which the device is mounted does get across to the other side, where it contacts the air and helps lower the thermal resistance. Therefore, just putting a copper plane on the other side also helps — but only by about 10–20%. Note that this "opposite" copper plane need not even be electrically the same point — it could, for example, just be the usual ground plane. A much greater reduction of thermal resistance (by about 50–70%) can be produced if a cluster of small vias ("thermal vias") is employed to conduct the heat from the component side to the opposite side of the PCB.

Thermal vias, if used, should be small (0.3–0.33 mm barrel diameter) so that the hole is filled up during the *plating* process. Too large a hole can cause "solder wicking" during the reflow soldering process, which leads to a lot of solder getting sucked into the holes, thereby creating bad solder joints for components in the vicinity. The "pitch" (i.e., the distance between the centers) of several such thermal vias in a given area is typically 1–1.2 mm. A grid of several such vias can be placed very close to, and alongside, a power device, and even under its tab (if present). See Chapter 11 for thermal management concerns in more detail.

Thermal Management

Thermal Resistance and Board Construction

Switching power supplies dissipate much less than linear regulators as explained in
Chapter 1. But they do. In *Chapter 6*, we saw the importance of lowering temperatures for
maximizing reliability and life. We have learned of different ways to improve efficiency of
switching power supplies, including the use of synchronous regulators as discussed in
Chapter 9. But eventually, despite all our best efforts, there will be some dissipation still
remaining. Most of this heat will be lost in the semiconductors, but some of it will be lost
in the inductors too. Especially in AC–DC power supplies, a good deal of heat will also be
lost in the EMI filter. In flyback power supplies, we typically need to parallel several output
caps, to lower the effective ESR, and thereby lower the heat generated inside C_O. The zener
clamp also gets very hot. With all these effects, not to mention some components heating
up others in their vicinity, a final qualification stage for any commercial power supply will
involve knowing the temperatures of *each and every component on the board* (usually by
connecting hundreds of thermocouples if necessary) and calculating their temperature stress
factors to ensure they are all being operated at typically better than at least 80% of their
max temperature ratings.

The relationship between the dissipation in any component and its temperature rise is
expressed as

$$\frac{\text{temperature rise}}{\text{dissipation}} \equiv \text{thermal resistance } (°C/W)$$

The thermal resistance is sometimes symbolized in literature as "θ," though we prefer to
call it Rth in this chapter. Note that the above equation implies a proportionality between
temperature rise and dissipation. For example, if the thermal resistance is 25 °C/W, and the
dissipation is 1 W, we expect a temperature rise of 25 °C, with respect to the ambient
(surrounding) temperature. So, if the ambient temperature was at 25 °C, the component will
rise to 50 °C under these conditions. However, if the dissipation in the component doubled
to 2 W, we expect a temperature rise of 50 °C, and so the temperature of the component
will be now 75 °C. Another implicit assumption made here is that the thermal resistance,
being a fixed number, does not depend on ambient temperature. So if in the 2-W case

above, the ambient temperature went up from 25 °C to say 40 °C (an increase of 15 °C), the temperature of the component will be $75 + 15 = 90$ °C.

The thermal resistance depends on several factors like the geometry of the component, and so on. But ultimately, the actual mechanism by which heat is lost is called convection. This is primarily the natural movement of air around the hot component, embodied in the phrase "hot air rises." We could literally force matters, by putting in a fan, and that would amount to "forced convection." It would significantly lower the temperatures. Note that at normal altitudes, a very small percentage of heat is lost through another mechanism, called radiation. This is just an (infrared) electromagnetic wave, and therefore it needs no air to propagate. So, it is understandable that at very high altitudes, where air is in short supply, radiation becomes the predominant mechanism for heat removal (i.e., thermal management). But we will ignore it in the initial discussion here.

One question is: at what point in the component are we actually measuring the temperature, or referring to? For example, in a semiconductor, we know from *Chapter 6* that the "junction" is of primary importance in terms of reliability. But of course, we have no access to it. What we measure on the bench is either the case or the lead/board, and then we try to correlate that to the junction temperature based on information provided by the vendor. In effect, we have several possible thermal resistances. Rth_{JA} is the thermal resistance from junction to ambient and Rth_{CA} from case to ambient. We could also define Rth_{JL} as the thermal resistance from junction to lead and Rth_{JC} from junction to case. We also have Rth_{LA} as the thermal resistance from lead to ambient and Rth_{CA} from case to ambient, and so on. We are thinking that there must be some simple math involved here, and perhaps we can add thermal resistances in series or parallel, just as we do for electrical resistance. And that is in fact true: we can create an electrical equivalent as shown in Figure 11.1. More on that in a minute.

The figure shows the proper way of soldering down a modern device with an exposed (metal) pad on its underside, on a copper island placed on the component side of a standard four-layer board. The idea is to get the heat out quickly from the device, transfer it via thermal conduction to different parts of the PCB, including the underside (which has a similar copper island right under the exposed pad). Incidentally, FR4 (standard PCB material) is not a bad conductor of heat itself. However, a large ground plane right below the component side helps push the heat out further across the board. Finally, all the exposed PCB surfaces (including all the exposed copper on the board, not necessarily even connected to the exposed pad) behave as heatsinks. Air flows past these surfaces, taking heat away by convection. Finally, the system stabilizes at a certain temperature of interest.

In Figure 11.1, we see there is a primary path for the heat to flow via thermal conduction. This is from junction to lead/exposed pad/board through which most of the dissipation P_H goes. We can usually ignore the parallel path going from junction to case and case to

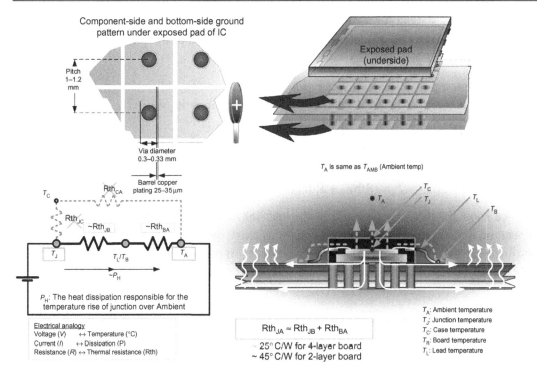

Component-side and bottom-side ground
pattern under exposed pad of IC

Pitch
1–1.2
mm

Via diameter
0.3–0.33 mm

Barrel copper
plating 25–35 μm

Exposed pad
(underside)

T_A is same as T_{AMB} (Ambient temp)

Rth_{CA}

Rth_{JC} ~Rth_{JB} ~Rth_{BA}

T_C

T_J T_L/T_B T_A

~P_H

P_H: The heat dissipation responsible for the
temperature rise of junction over Ambient

Electrical analogy
Voltage (V)   ↔ Temperature (°C)
Current (I)   ↔ Dissipation (P)
Resistance (R) ↔ Thermal resistance (Rth)

$Rth_{JA} \approx Rth_{JB} + Rth_{BA}$

~ 25° C/W for 4-layer board
~ 45° C/W for 2-layer board

T_A: Ambient temperature
T_J: Junction temperature
T_C: Case temperature
T_B: Board temperature
T_L: Lead temperature

Figure 11.1: Thermal resistance explained and the correct way to mount a
power IC on a board.

ambient (because it has a very *high* thermal resistance). In the figure, we also see the complete electrical analog. Thermal resistance is analogous to resistance and dissipation to current. The temperature (difference) is analogous to voltage (difference).

In Figure 11.1, some recommendations on the dimensions and spacing of the "thermal vias" under the exposed pad are provided. The idea is to prevent solder wicking during the soldering reflow process, which will suck solder and possibly create a bad joint under the exposed pad, thereby compromising the entire thermal performance. Note that the thermal vias are sometimes prefilled with copper. That prevents solder wicking and improves the conduction capability of the thermal vias themselves. But it is more costly.

Finally, we note that most power devices have a relatively low Rth_{JL}. So, the net junction to ambient thermal resistance Rth_{JA} is predominantly comprised of Rth_{LA}. But that is basically just the thermal resistance of the PCB (to ambient) and has almost nothing to do with the package or device itself. Which is why in Figure 11.1, we have indicated that for most modern power devices on four-layer boards, with a stackup and build as indicated in Figure 11.1, we can safely assume $Rth_{JA} \approx 25\,°C/W$ for estimating T_J. Similarly, for a two-layer board, since the inner ground plane is missing (all else remaining the same), the thermal resistance is about 45 °C/W. We can use these numbers (or better Rth data provided by the vendor) to estimate temperature rise as shown in the last solved example of *Chapter 19*.

When we mount a power semiconductor (like a TO-220 or TO-247 device in AC–DC applications) on a proper heatsink (denoted by "H") we can write similarly

$$\text{Rth}_{JA} = \text{Rth}_{JH} + \text{Rth}_{HA}$$

The final junction temperature is therefore

$$T_J = P \times (\text{Rth}_{JH} + \text{Rth}_{HA}) + T_A \approx P \times (\text{Rth}_{HA}) + T_A$$

So, if we know the thermal resistance of the heatsink, we can guess the junction temperature quite accurately. Empirical equations exist, based on the area of the heatsink, to estimate the effectiveness of heatsinks (plate types in particular). They can be applied to PCBs too with some qualifications as discussed next.

Historical Definitions

We take the simplest case of a square plate made from a very good thermally conducting material, dissipating P Watts. After some time, we will find that the plate stabilizes at a certain temperature rise of "ΔT" over the ambient.

We expect that the temperature rise will be proportional to the dissipation. The "proportionality constant" is called the thermal resistance "Rth" in °C/W. So,

$$\text{Rth} = \frac{\Delta T}{P}$$

Similarly, we expect that the thermal resistance will vary inversely with the area:

$$\text{Rth} \propto \frac{1}{A}$$

We expect to define another proportionality constant here.

Stop: What is the area we are referring to here? If we have a plate 3 in. × 3 in., we say that its area is 9 sq.in. However, the area exposed to natural convection is actually twice that — 18 sq.in. (both sides). This is one major source of confusion in using and comparing the various empirical equations provided in literature — some refer to "A" as the total exposed area, and some refer to the area of *one side*. Therefore, to avoid confusion we have adopted the following convention here.

"A" refers to the area of one side of a plate, whose both sides are exposed to cooling. The total exposed area is "$\underline{A}$" (so $\underline{A} = 2$ A).

Using our terminology, the inverse of the proportionality constant above is "*h*" in W/°C *per unit area* and is called by various names like "convection coefficient" or "heat transfer coefficient."

$$\text{Rth} = \frac{1}{h\underline{A}} = \frac{1}{2hA}$$

Finally, we have the basic equations

$$P = h \times \underline{A} \times \Delta T = 2 \times h \times A \times \Delta T = \frac{\Delta T}{\text{Rth}} \Leftarrow$$

Explicitly for *h*,

$$h = \frac{\text{dissipation}}{\text{total exposed area} \times \text{temperature rise}} \Leftarrow$$

And also,

$$h\underline{A} = \frac{1}{\text{Rth}} \quad \text{or} \quad hA = \frac{1}{2 \times \text{Rth}} \Leftarrow$$

It was originally thought that "Rth" and "*h*" were constants, and that was the intent of writing the classical equations as presented above. Later it was realized that the equations were not very accurate for various reasons. However, the equations presented above were maintained. What changed was that "*h*" or "Rth" were now "allowed" to depend on area, dissipation, and so on, all the factors they were supposedly independent of. Note that the dependency is not very severe, and so even today, we often assume that Rth and *h* are constants to a first approximation.

Empirical Equations for Natural Convection

As a first approximation, *h* is often stated in literature (at sea-level) as

$$h = 0.006 \ W/in^2\text{-}°C$$

If area is expressed in meters, this becomes

$$h = 0.006 \times (39.37)^2 = 9.3 \ W/m^2\text{-}°C$$

since there are 39.37 in. in a meter.

Nowadays we know that in reality, "*h*" can vary about 1:4 times from the commonly assumed typical values above.

So, in literature we can find the following generalized empirical equation for h, and this becomes our *standard equation no. 1*:

$$h = 0.00221 \times \left(\frac{\Delta T}{L}\right)^{0.25} \text{W/in.}^2\text{-}°C \Leftarrow \quad \text{(standard equation no. 1)}$$

where L is the length along the direction of natural convection (vertical). In the case of the simple square plate, $L = A^{0.5}$, so we can write this as

$$h = 0.00221 \times \Delta T^{0.25} \times A^{-0.125} \text{W/in.}^2\text{-}°C \quad \text{(standard equation no. 1)}$$

Observe that the above equation uses "A" which is actually half the area exposed to cooling. So, we can equivalently rewrite it in terms of the actual area involved in the cooling process:

$$h = 0.00221 \times \Delta T^{0.25} \times \left(\frac{A}{2}\right)^{-0.125} \text{W/in.}^2\text{-}°C$$

$$h = 0.00241 \times \Delta T^{0.25} \times \underline{A}^{-0.125} \text{W/in.}^2\text{-}°C$$

These are all available and published forms of the same equation for h. If the different forms are not recognized as one, it is easy to get confused and not know which equation to pick.

The above equation predicts that "h" has a specified dependency on the exposed area of the plate and also on its temperature differential with respect to ambient. This dependency (i.e., $A^{-0.125}$) implies that the cooling efficiency per unit area (i.e., "h") of large plates is worse than that of small plates. However, if this sounds surprising, we note that the overall/total cooling efficiency of a plate is $h \times A$, which depends on $A^{-0.125} \times A = A^{0.875}$. So, thermal resistance of a plate goes as $1/A^{0.875}$ and is clearly lower for a large plate than for a small plate as we would expect. Compare this to the "ideal" $1/A$ variation which was, classically speaking, expected for thermal resistance.

In literature, we often find the following "standard" formula (area in sq. in.), hereafter referred to as our *standard equation no. 2*:

$$\text{Rth} = 80 \times P^{-0.15} \times A^{-0.70} \,°C/W \quad (A \text{ in sq. in.}) \Leftarrow \quad \text{(standard equation no. 2)}$$

We notice that the first equation is written in terms of "h" and the second in terms of "Rth." How do we compare them? We can do some manipulations on these equations to bring them to a comparable format. We can rewrite our standard equation no. 1 in terms of dissipation instead of temperature rise:

$$h = 0.00221 \times \left[\frac{P}{h \times A \times 2}\right]^{0.25} \times A^{-0.125} \text{W/in.}^2\text{-}°C$$

So,

$$h = 0.00654 \times P^{0.2} \times A^{-0.3} \text{ W/in.}^2\text{-°C}$$

Or in terms of the total exposed area:

$$h = 0.008 \times P^{0.2} \times \underline{A}^{-0.3} \text{ W/in.}^2\text{-°C}$$

We can also now try to see what this will look like in MKS (SI) units. The conversion is not obvious and so we proceed as follows.

Take an imaginary plate of size 39.37 in. × 39.37 in. or 1 m × 1 m. Clearly, the thermal resistance of the plate is in °C/W and is therefore independent of the units used to measure area, and must remain unchanged by any change in the system of units used. This means that $1/(h \times \underline{A})$ is independent of units, and so is $h \times \underline{A}$. Therefore, if in MKS units we first assume a similar form for h:

$$h = C \times \Delta T^{0.25} \times A^{-0.125} \text{ Watt/m}^2\text{-°C}$$

Equating,

$$h \times A = C \times \Delta T^{0.25} \times A_{m^2}^{-0.125} \times A_{m^2} = 0.00221 \times \Delta T^{0.25} \times A_{in.^2}^{-0.125} \times A_{in.^2}$$

$$C \times A_{m^2}^{0.875} = 0.00221 \times A_{in.^2}^{0.875}$$

$$C = (39.37^2)^{0.875} \times 0.00221 = 1.37$$

So, finally in MKS units

$$h = 1.37 \times \Delta T^{0.25} \times A^{-0.125} \text{ Watt/m}^2\text{-°C} \Leftarrow$$

This is also a common form seen in literature, often thought to be a separate equation altogether.

Comparing the Two Standard Empirical Equations

We basically just have two equations to choose from. Our standard equation no. 2 is

$$h = 80 \times P^{-0.15} \times A^{-0.70} \quad \text{(Area in sq. in.)}$$

The result of manipulations on standard equation no. 1 gives us

$$\text{Rth} = \frac{1}{2hA} = 76.5 \times P^{-0.20} \times A^{-0.70} \quad \text{(Area in sq. in.)}$$

Both these use the area of one side of the plate, though it is assumed both sides are exposed to natural convection. And we thus see that the two equations, one initially expressed in

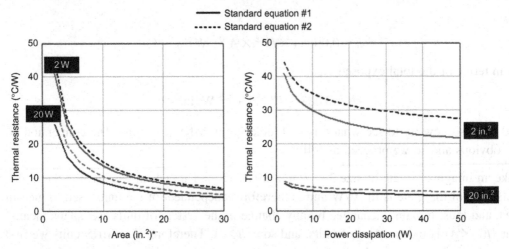

*This is the area of one side of a plate whose both sides are exposed to natural convection

Figure 11.2: Plotting out the two standard empirical natural convection equations of plate heatsinks.

terms of h and the other in terms of Rth, are not very different at all, if brought to a similar form as we have done above.

In Figure 11.2, we have compared these two commonly seen equations (with their numerous almost unrecognizable forms). We realize all the equations commonly seen in literature are actually just two equations, both of which when plotted out, as in Figure 11.2, are very close. We can pick the dotted lines (standard equation no. 2, as the more conservative).

"h" from Thermodynamic Theory

Without needing to go too deep into thermodynamic theory here is a quick check on the equations we can derive from theory. We have the dimensionless Nusselt number "Nu," which is the ratio of the convection heat transfer to the conduction heat transfer. We also have the dimensionless Grashof number "Gr," which is the ratio of buoyant flow to viscous flow. Under natural convection (laminar flow), we have the following defining equations in MKS units:

$$Nu = 3.5 + 0.5 \times Gr^{1/4}$$

where

$$Gr = \frac{g \times (1/(T_{amb} + 273)) \times \Delta T \times L^3}{\nu^2}$$

where $g = 9.8$ (acceleration due to gravity in m/s^2) and $\nu = 15.9 \times 10^{-6}$ (kinematic viscosity in m^2/s). At an ambient temperature $T_{AMB} = 40$ °C, it can be shown that this simplifies to

$$Nu = 3.5 + 52.7 \times \Delta T^{0.25} \times L^{0.75}$$

The coefficient of cooling is by definition

$$h = \frac{Nu \times K_{AIR}}{L}$$

where K_{AIR} is the thermal conductivity of air (0.026 W/m °C). So, we get our third standard equation:

$$h = 0.091 + 1.371 \times \left(\frac{\Delta T}{L}\right)^{0.25} \text{ Watt/m}^2\text{-°C}$$

or in terms of area

$$h = 0.091 + 1.371 \times \Delta T^{0.25} \times A^{-0.125} \text{ Watt/m}^2\text{-°C} \quad \Leftarrow \quad \text{(standard equation no. 3)}$$

Comparing this to the previously given empirical equations, we find that this equation too is surprisingly close, especially to a comparable form of our standard equation no. 1 derived earlier.

Unfortunately, though this third form may be more accurate because of the constant term in its equation, for that very reason it is more difficult to manipulate into all the forms the previous equations could be manipulated into. So we won't even try here. But we can use any of the equations as they are all very close when brought to the same form.

PCB Copper Area Estimate

Now, we can also provide a simple equation for estimating the copper area on a PCB. This is not a plate, but a copper island on a PCB, *and only one side is exposed to cooling*. This is not the same as using the area of one side of a plate, both sides of which are exposed to cooling. So, here we use the equation which uses the entire exposed area. For this, the standard equation no. 1 gives us

$$Rth = \frac{1}{2hA} = 76.5 \times P^{-0.20} \times A^{-0.70} \quad \text{(Area in sq. in.)}$$

In terms of area exposed to convection (calling $\underline{A}$ as Area here)

$$Rth = \frac{1}{2hA} = 76.5 \times P^{-0.20} \times \left(\frac{\text{Area}}{2}\right)^{-0.70} \quad \text{(Area in sq. in.)}$$

$$= \frac{76.5}{2^{-0.70}} \times P^{-0.20} \times (\text{Area})^{-0.70} = 124.2 \times P^{-0.20} \times (\text{Area})^{-0.70}$$

$$Rth = \frac{124.2}{P^{0.20} \times Area^{0.70}} \, °C/W \quad (\text{Area in sq. in.})$$

Solving for A, we get

$$Area = \left(\frac{124.2}{P^{0.20} \times Rth}\right)^{1/0.70}$$

$$Area = 981 \times Rth^{-1.43} \times P^{-0.29} (\text{sq. in.}) \quad \Leftarrow$$

Example:

We have a dissipation of 0.45 W from an SMT device, and we want to restrict the temperature of the PCB to a maximum of 100 °C to avoid getting too close to the glass transition of the board (which is around 120 °C for FR-4). The worst-case ambient temperature is 55 °C, let us find the amount of copper which should be made available to the device.

The required Rth of the PCB is

$$Rth = \frac{degC}{W} = \frac{100 - 55}{0.45} = 100 \, °C/W$$

So, from our equation (based on standard equation no. 1) we get

$$Area = 981 \times 100^{-1.43} \times 0.45^{-0.29} = 1.707 \, (\text{sq. in.})$$

So, we need a square copper area of side $1.707^{0.5} = 1.3$ in.

Example:

With an estimated baseline dissipation of 1 W, what should be the area of the copper on a PCB to provide about 25 °C/W?

$$Area = 981 \times 25^{-1.43} = 9.8 \, (\text{sq. in.}) \Rightarrow \sim 3.15 \, \text{in.}^2$$

Note that if the required thermal area is in excess of 1 in.^{2}, to avoid thermal constriction effects (which will make the above predictions completely erroneous) we should use 2-oz copper PCB.

Sizing Copper Traces

There are complicated curves available for copper versus temperature rise of PCB traces in the now-obsolete standard "MIL-STD-275E." These curves have also found their way into more recent standards like IPC-2221 and IPC-2222. Engineers often try to create elaborate curve fit equations to match these curves. But the truth is the earlier curves can be easily approximated by simple linear rules as follows.

The required cross-sectional area of an external trace is approximately

(a)   37 mils^2 per Ampere of current for 10 °C rise in temperature (recommended).
(b)   25 mils^2 per Ampere of current for 20 °C rise in temperature (recommended).
(c)   18 mils^2 per Ampere of current for 30 °C rise in temperature (recommended).

For the traces in inner layers, multiply the calculated width of an external trace by 2.6 to get the required width.

To calculate width of a trace from the cross-sectional area, keep in mind that 1-oz copper is 1.4 mils thick and 2-oz copper is 2.8 mils thick.

Natural Convection at an Altitude

At sea-level, over 70% of heat is transferred by natural convection and the rest by radiation. Only at very high altitudes (70,000 ft +), the ratio inverts and the heat lost by radiation could be 70–90% of the total, even though the radiated transfer is unchanged. So by about 10,000 ft the overall efficiency of cooling typically falls to 80%, at 20,000 it is only 60%, and at 30,000 it is 50%.

Knowing that the coefficient of natural convection goes as $P^{1/2}$, where P is the pressure of air, a good curve fit gives us the following useful relationship:

$$\frac{\text{Rth (feet)}}{\text{Rth (sea-level)}} = \left[(-30 \times 10^{-6} \times \text{feet}) + 1 \right]^{-0.5}$$

So, for example, we find that at 10,000 ft, all the Rth's at sea-level need to be increased by about 19.5%.

Forced Air Cooling

Fans are rated for a certain cubic feet of minute "cfm." The actual cooling, however, depends on the linear feet per minute "lfm" to which the heatsink is subjected. Two parameters are needed to find the velocity in lfm: (1) the volume of air discharged from the fan in cfm and (2) the cross-sectional area through which the cooling air passes in m^2. So lfm = cfm/Area. But, finally, we should derate the calculated lfm by 60–80% to account for backpressure.

At sea-level, the following formula gives a rough estimate of the required airflow:

$$\text{cfm} = \frac{1825}{\Delta T} \times P_{\text{kW}}$$

The ΔT is the differential between the inlet and the outlet temperatures. It is typically set to about 10–15 °C.

Note that if the inlet temperature, which is the room ambient, is 55 °C for example, then we need to add this differential ΔT as the actual local ambient inside the power supply when doing our initial calculations. However, ultimately we will be carrying out an actual temperature test by attaching thermocouples to all the components. We will thus certainly see an advantage in moving hotter components closer to the inlet during the design phase.

The linear speed is often expressed in terms of m/s. 1 m/s is equal to an lfm of 196.85. Roughly, 1 m/s is 200 lfm.

Some empirical results are as follows: at 30-W dissipation, an unblackened plate of 10 cm × 10 cm has the following Rth: 3.9 °C/W under natural cooling, 3.2 °C/W with 1 m/s, 2.4 °C/W with 2 m/s, and 1.2 °C/W with 5 m/s. Provided the air flows parallel to the fins, with speed > 0.5 m/s, the thermal resistance hardly depends on the power dissipation. That is because, on its own, even in static air, hot plates produce enough air movement around them to help in the heat transfer. Also note that blackening of plates has some effect under natural convection, but curves for forced convection depend very little on this aspect. Radiation is improved by blackening, but at sea-level that is only a small part of the overall heat transfer. In general, black anodized heatsinks still seen in some forced air designs are actually a waste and should be replaced with uncoated aluminum.

Under steady-state, roughly 2 mm thick copper is almost exactly equivalent to 3 mm thick aluminum. The only advantage of copper is its better thermal conductivity, so it may be used to avoid thermal constriction effects when using very large areas.

The curve of thermal resistance to air flow falls off roughly exponentially, and so the improvement in thermal resistance in going from still air to 200 lfm is the same as from 200 lfm to 1,000 lfm. Velocities in excess of 1,000 lfm (about 5 m/s) do not cause significant improvement.

Under forced convection, the Nusselt number at sea-level is

$$Nu_F = 0.664 \times Re^{1/2} \times Pr^{1/3} \quad \text{(laminar flow)}$$

$$Nu_F = 0.037 \times Re^{4/5} \times Pr^{1/3} \quad \text{(turbulent flow)}$$

Note that generally for natural convection, we can assume laminar flow. But under high dissipation, the hot air tends to rise so fast that it breaks up into turbulence. This is actually very useful in reducing the thermal resistance (increasing the "h"). For forced air convection, it is common to cut fingers on the sides of plate metal sinks and bend them alternately in and out. The purpose here is to actually create turbulent flow in the vicinity of the heatsink, thus lowering its thermal resistance. However, we do note from the formal analysis and equations which follow, turbulent flow provides better cooling (high "*h*") under conditions of high lfm and/or large plates only. Laminar flow will provide better cooling otherwise.

Above we have defined the Prandtl number "Pr," which is the ratio of momentum diffusion to thermal diffusion. We can take its value at sea-level to be 0.7. "Re" is the dimensionless Reynolds's number, which is the ratio of momentum flow to viscous flow. If the plate has two dimensions $L1$ and $L2$ (so that $L1 \times L2 = A$), and $L1$ is the dimension along the flow of air, then Re is

$$Re = \frac{\text{lfm}_{\text{sea-level}} \times L1_{\text{meters}}}{196.85 \times \nu}$$

where we already know $\nu = 15.9 \times 10^{-6}$ (the kinematic viscosity in m^2/s). Thus, we get the *h* under forced convection:

$$h_\text{F} = \frac{\text{Nu}_\text{F} \times K_{\text{AIR}}}{L1_{\text{meters}}} \ \text{Watts/m}^2\text{-}^\circ\text{C}$$

where K_{AIR} is the thermal conductivity of air (0.026 W/m-°C). Putting all the numbers together, we simplify to get

$$h_{\text{FORCED}} = 0.086 \times \text{lfm}^{0.8} \times L^{-0.2} \quad \Leftarrow \quad (\text{turbulent flow}, L \text{ in meters, sea-level})$$

$$h_{\text{FORCED}} = 0.273 \times \text{lfm}^{0.5} \times L^{-0.5} \ \text{Watts/m}^2\text{-}^\circ\text{C} \quad (\text{laminar flow}, L \text{ in meters, sea level})$$

At higher altitudes, we need to increase the cfm calculated at sea-level by the following factor so as to maintain the same effective cooling. This is because a fan is a constant volume mover, not a constant mass mover, and at high altitudes, the air density is much lower. Therefore, the cfm has to be increased in inverse proportion to the pressure.

$$\frac{\text{cfm(feet)}}{\text{cfm(sea-level)}} = \frac{1}{(-30 \times 10^{-6} \times \text{feet}) + 1}$$

For example, at 10,000 ft, the calculated cfm at sea-level has to be increased by 43% to maintain the same h_{FORCED}.

Radiative Heat Transfer

Radiation does not depend on air, and can take place even in vacuum, since it is electromagnetic in nature. At high altitudes, radiative heat transfer can become a significant part of the overall heat transfer. The equation for "*h*" is

$$h_{\text{RAD}} = \frac{\varepsilon \times \left(5.67 \times 10^{-8}\right) \times \left[(T_{\text{HS}} + 273)^4 - (T_{\text{AMB}} + 273)^4\right]}{T_{\text{HS}} - T_{\text{AMB}}} \ \text{Watt/m}^2\text{-}^\circ\text{C}$$

ε is the emissivity of the surface. It is 1 for a perfect blackbody, but for polished metal surfaces we should take this as 0.1. If the surface is anodized, we can take it as about 0.9.

Note that at high altitudes, under forced air cooling, the cfm falls, and so the inlet-to-outlet ΔT increases somewhat. Therefore, T_{AMB} goes up, and this affects h_{RAD}. So, it may end up looking like radiation is getting affected at higher altitudes too, but it is for a different reason altogether (rise in ambient).

Miscellaneous Issues

- A typical power supply specification will ask for meeting an altitude requirement of 10,000 ft (3,000 m). Typically, a specification will not "relax" the ambient temperature up to about 6,000 ft, after which it will allow us to reduce the upper ambient limit by about 1 °C every 1,000 ft higher.
- A typical industry thumbrule for testing power supplies at sea-level for a certain altitude requirement is to "add 1 °C every 1,000 ft to the upper limit of the maximum specified operating ambient." So, if the power supply is designed for 55 °C at sea level, we should test it at 65 °C. However, this is not always adequate. Nor do any temperature derating margins at sea-level necessarily help. A key limiting factor is not the junction temperature, but the temperature on the PCB where the device is mounted. We usually cannot exceed more than about 100–110 °C on the PCB, or it will burn.
- We can sum over all the "h's" calculated in this chapter as follows:

$$h_{\text{total}} = h_{RAD} + \left(h_{FORCED}^3 + h_{NATURAL}^3\right)^{1/3} \Leftarrow$$

- For common magnetic cores (like the E cores, ETD cores, EFD cores, etc.), thermal resistance under natural convection can be approximated by

$$\text{Rth} \cong 53 \times V_e^{-0.54}$$

where V_e is in cm^3.

- We can also use the above equation for extrusion heatsinks. Extruded heatsinks are certainly very useful under forced air cooling because then the efficiency of cooling depends on their surface area. But correlation of experimental data indicates that their cooling capabilities under natural convection conditions are a function of the volume of the space they occupy, that is, their "envelope" (ignoring the finer detail of their fin structure). That is because heat lost from one fin is largely re-acquired by the adjacent fins, and so there are very small deviations with regard to the "exoticness" of their actual shape. Typical values drawn from published curves are as follows: 0.1 in.^{3} will give about 30–50 °C/W, 0.5 in.^{3} will give about 15–20 °C/W, 1 in.^{3} will give about 10 °C/W, 5 in.^{3} will give about 5 °C/W, and 100 in.^{3} will give about 0.5–1 °C/W. The above data are for one device mounted on the heatsink. Roughly, there will be a further

20% improvement in the thermal resistance if two devices share the dissipation and are mounted *slightly apart*.

- If the fins of an extrusion heatsink are too close, they also impede the flow of air. Therefore, the recommended optimum fin spacing is about 0.25 in. for natural convection, at 200 lfm it is about 0.15 in., and at 500 lfm it is about 0.1 in. This applies for heatsinks up to 3 in. in length. We can increase the fin spacing by about 0.05 in. for heatsinks as long as 6 in.

- Finally, here is a quick run-down on fans: ball-bearing fans are more expensive. They have a longer life when the temperature (as seen by the bearing system) is higher. But they can get noisier over time. If useful life of a fan was defined as ending when the fan became noisy, the ball-bearing fan would have a smaller life than the sleeve-bearing fan. Sleeve-bearing fans are less expensive, are quieter, and easily handle any mounting attitude (angle). Their life is as good as a ball-bearing fan provided temperatures are not very high. They can sustain multiple shocks (without impacting noise or life).

CHAPTER 12

Feedback Loop Analysis and Stability

Transfer Functions, Time Constant, and the Forcing Function

In converters, we often refer to the steady-state ratio: output divided by input, V_O/V_{IN}, as the "DC transfer function" of the converter. We can define transfer functions in many ways. For example, in *Chapter 1* we discussed a simple series resistor–capacitor (RC) charging circuit (see top schematic of Figure 1.3). By closing the switch we were, in effect, applying a *step voltage* to the RC. Let us call the voltage step "v_i" (its height).

That was the "input" or "stimulus" to the system. It resulted in an "output" or "response" — which we implicitly defined as the *voltage appearing across the terminals of the capacitor*, that is, $v_O(t)$. So, the ratio of the output to the input was also a "transfer function":

$$\frac{v_o(t)}{v_i} = 1 - e^{-t/RC}$$

Note that this transfer function depends on time. In general, any output ("response") divided by input ("stimulus") is called a "transfer function."

A transfer function need not be "Volts/Volts" (i.e., dimensionless). In fact, neither the input nor the output of any such two-port network need necessarily even be a voltage. The input and output need not even be two similar quantities. For example, a two-port network can be as simple as a current sense resistor. Its input is the current flowing into it, and its output may be considered as the sensed voltage across it. So, its transfer function has the units of voltage divided by current, that is, resistance. Or we could pass a current through the resistor, but consider the response under study as its temperature. So, that would be the output now. Later, when we analyze a power supply in more detail, we will see that its pulse-width modulator (PWM) section, for example, has an input that is called the "control voltage" (output of error amplifier), but its output is a dimensionless quantity: the duty cycle (of the converter). So, the transfer function in that case has the units of Volts^{-1}. We realize the phrase "transfer function" is a very broad term.

In this chapter, we start analyzing the behavior of the converter to sudden *changes* in its DC levels, such as those that occur when we apply line and load variations. These changes cause the output to temporarily move away from its set DC regulation level V_O, and therefore give its feedback circuitry the job of correcting the output *in a manner*

Switching Power Supplies A–Z. DOI: 10.1016/B978-0-12-386533-5.00012-7

429

deemed acceptable. Note that in this "AC analysis," it is understood that what we are referring to as the output or response is actually the *change* in V_O. The input or stimulus, though certainly a change too, is defined in many different ways as we will soon see. In all cases, we are completely ignoring the DC-bias levels of the converter and focusing only on the changes *around* those levels. In effect, we are studying the converter's "AC transfer functions."

How did we actually arrive at the transfer function of the RC circuit mentioned above? For that, we first use Kirchhoff's voltage law to generate the following differential equation:

$$v_i = v_{res}(t) + v_{cap}(t) = i(t)R + \frac{q(t)}{C}$$

where $i(t)$ is the charging current, $q(t)$ is the charge on the capacitor, $v_{res}(t)$ is the voltage across the resistor, and $v_{cap}(t)$ is the voltage across the capacitor (i.e., $v_o(t)$, the output). Further, since charge is related to current by $dq(t)/dt = i(t)$, we can write the above equation as

$$v_i = R \times \frac{dq(t)}{dt} + \frac{q(t)}{C}$$

or

$$\frac{dq(t)}{dt} + \frac{1}{RC}q(t) = \frac{v_i}{R}$$

To solve this, we "cheat" a little. Knowing the properties of the *exponential function* $y(x) = e^x$, we do some educated reverse-guessing. And that is how we get the solution:

$$q(t) = C \times v_i \times \left(1 - e^{-t/RC}\right)$$

Substituting $q = C \times v_{cap}$, we arrive at the required transfer function of the RC-network given earlier.

Note that the differential equation for $q(t)$ above is in general a "*first-order*" differential equation — because it only involves the *first* derivative of time.

Later, we will see that there is a *better way* to solve such equations — it invokes a mathematical technique called the "Laplace transform." To understand and use that, we have to first learn to work in the "*frequency domain*" rather than in the "time domain" as we have been doing so far above. We will explain that soon.

Here we note that in a first-order differential equation of the above type, the term that divides $q(t)$ ("RC" in our case) is called the "time constant." Whereas, the constant term in the equation ("v_i/R" in our case) is called the "forcing function."

Understanding "e" and Plotting Curves on Log Scales

We can see that the solution to the previous differential equation brought up the exponential constant "e," where e ≈ 2.718. We can ask — why do circuits like this always seem to lead to exponential type of responses? Part of the reason for that is that the exponential function e^x does have some well-known and useful properties that contribute to its ubiquity. For example,

$$\frac{d(e^x)}{dx} = e^x \text{ and } \int (e^x)dx = e^x + c \quad \text{(where } c \text{ is a constant)}$$

But this in turn can be traced back to the observation that the exponential constant e itself happens to be one of the most *natural* parameters of our world. The following example illustrates this.

Example:

Consider 10,000 power supplies in the field with a failure rate of 10% every year. That means in 2010, if we had 10,000 working units, in 2011 we would have 10,000 × 0.9 = 9,000 units. In 2012, we would have 9,000 × 0.9 = 8,100 units left. In 2013, we would have 7,290 units left, in 2014, 6,561 units, and so on. If we plot these points — 10,000; 9,000; 8,100; 7,290; 6,561; and so on, versus time, we will get the well-known decaying exponential function. See Figure 12.1. We have plotted the same curve twice: the curve on the right has a log scale on the vertical axis. Note how it now looks like a straight line. It cannot, however, ever go to zero! The log scale is explained further.

Note that the simplest and most obvious initial assumption of a *constant* failure rate has led to an *exponential* curve. That is because the exponential curve is simply a succession of

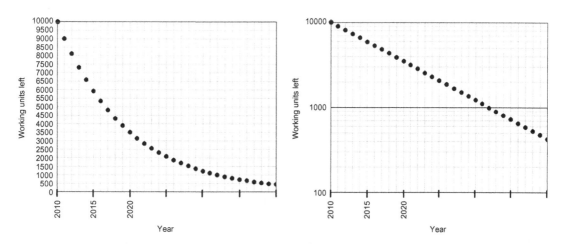

Figure 12.1: How a decaying exponential curve is naturally generated.

evenly spaced data points (very close to each other), which are in simple *geometric progression* — that is, the ratio of any point to its preceding point is a constant (equal intervals). Most natural processes behave similarly, and that is why "e" is encountered so frequently. In *Chapter 6*, we had introduced Arrhenius' equation as the basis for failures. That too was based on "e."

We recall that logarithm is defined as follows — if $A = B^C$, then $\log_B(A) = C$, where $\log_B(A)$ is the "logarithm of A to the base B." The commonly referred-to "logarithm," or "log," has an *implied* base of 10 (i.e., $B = 10$), whereas the natural logarithm "ln" is an abbreviation for a logarithm with a base "e" (i.e., where B is e = 2.718). We will be plotting a whole lot of curves in this chapter on "log scales."

Remember this: if the log of any number is multiplied by 2.303, we get its natural log. Conversely, if we divide the natural log by 2.303 we get its log. This follows from

$$\ln(10) = 2.303 \quad \text{and} \quad \frac{1}{\log(e)} = 2.303$$

Flashback: Complex Representation

Any electrical parameter is thus written as a sum of real and imaginary parts:

$$A = Re + j \times Im$$

where we have used "Re" to symbolically denote the *real* part of the number A and "Im" for its *imaginary* part. From these components, the actual magnitude and phase of A can be reconstructed as follows:

$$|A| = \sqrt{Re^2 + Im^2} \quad \text{(always positive!)} \quad (\textit{magnitude of complex number})$$

$$\varphi = \tan^{-1}\left(\frac{Im}{Re}\right) \text{ radians} \quad (\textit{argument of complex number})$$

Impedance too is broken up into a vector in this complex representation — except that though it is frequency dependent, it is (usually) not a function of time.

The "complex impedances" of reactive components are

$$Z_L = j \times L\omega$$

$$Z_C = \frac{1}{j \times C\omega}$$

To find out what happens when a complex voltage is applied to a complex impedance, we need to apply the complex versions of our basic electrical laws. So Ohm's law, for example, now becomes

$$V(\omega t) = I(\omega t) \times Z(\omega)$$

We also have the following relationships to keep in mind:

$$e^{j\theta} = \cos(\theta) + j\sin(\theta) \qquad \sin(\theta) = \frac{e^{j\theta} - e^{-j\theta}}{2j}$$

$$e^{-j\theta} = \cos(\theta) - j\sin(\theta) \qquad \cos(\theta) = \frac{e^{j\theta} + e^{-j\theta}}{2}$$

Note that in electrical analysis, we set $\theta = \omega t$. Here θ is the angle in *radians* (180° is π radians). Also, $\omega = 2\pi f$, where ω is the angular frequency in radians/s and f the (conventional) frequency in Hz.

As an example, using the above equations, we can derive the magnitude and phase of the exponential function $f(\theta) = e^{j\theta}$ as follows:

$$\text{Magnitude}(e^{j\theta}) = \sqrt{\cos(\theta)^2 + \sin(\theta)^2} = 1$$

$$\text{Argument}(e^{j\theta}) = \tan^{-1}\left(\frac{\sin(\theta)}{\cos(\theta)}\right) = \tan^{-1}\tan(\theta) = \theta$$

Repetitive and Nonrepetitive Stimuli: Time Domain and Frequency Domain Analyses

Strictly speaking, no stimulus is purely "repetitive" (periodic) in the true sense of the word. "Repetitive" implies that the waveform has been exactly that way, since "time immemorial," and remains so forever. But in the real world, there is actually a definite moment when we *apply* a given waveform (and another when we remove it). Even an applied "repetitive" sine wave, for example, is not repetitive at the *moment* it gets applied. Though, much later, the stimulus can be considered repetitive if sufficient time has elapsed from the moment of application to allow the initial transients to die out completely. This is the implicit assumption we make even when we carry out "steady-state analysis" of any circuit or converter.

But sometimes, we *do* want to know what happens at the exact *moment of application* of the stimulus. Like the case of the step voltage applied to our RC-network, we could do the same to a power supply, and we would want to ensure that its output doesn't "overshoot" (or "undershoot") too much at the instant of application of this "line transient." We could also apply sudden changes in load to the power supply, and see what happens to the output rail under a "load transient."

If we have a circuit (or network) constituted only of *resistors*, the voltage at any point in it is *uniquely and instantaneously defined* by the applied voltage. If the input varies, so does this voltage, and *proportionally* so. In other words, there is no "lag" (delay) or "lead" (advance) between the stimulus and the response. *Time* is not a variable involved in this transfer function. However, when we include reactive components (capacitors and/or inductors) in any network, it becomes necessary to start looking at how the situation *changes over* time in response to an applied stimulus. This is called *"time-domain analysis."* Proceeding along that path, as we did in the first section of this chapter with the RC circuit, can get very intimidating very quickly as the complexity of the circuit increases. We are therefore searching for simpler analytical techniques.

We know that any *repetitive* ("periodic") waveform, of almost arbitrary shape, can be decomposed into a sum of several *sine (and cosine)* waveforms of frequencies. That is what Fourier series analysis is (see *Chapter 18* for more on this topic). In Fourier series, though we do get an infinite series of terms, the series is a simple summation consisting of terms composed of discrete frequencies (the harmonics) (see *Figure 18.1* in particular). When we deal with more arbitrary waveshapes, including those that are not periodic, we need a *continuum of frequencies* to decompose that waveform, and then understandably, the summation of Fourier series now becomes an integration over frequency. Note that in the new continuum of frequencies, we also have "negative frequencies," which are clearly not amenable to intuitive visualization. But that is, how the Fourier series evolved into the "Fourier transform." In general, decomposing an applied stimulus (a waveform) into its frequency components, and understanding how the system responds to each frequency component, is called "frequency domain analysis."

> *Note: The underlying reason for decomposition into components is that the components can often be considered mutually "independent" (i.e., orthogonal), and therefore tackled separately, and then their effects superimposed. We may have learned in our physics class that we can split a vector, the applied force for example, into x and y components, F_x and F_y. Then we can apply the rule Force = mass $\times$ acceleration to each x and y component of the force separately. Finally, we can sum the resulting x and y accelerations to get the final acceleration vector.*

As mentioned, to study any nonrepetitive waveform, we can no longer decompose it into components with *discrete* frequencies as we can do with repetitive waveforms. Now we require a spread (continuum) of frequencies. That leads us to the usual simple definition of "Fourier transform" — which is simply the function $f(t)$, multiplied by $e^{-j\omega t}$ and integrated over all time (minus infinity to plus infinity).

$$\int\limits_{-\infty}^{\infty} \left(f(t) \times e^{-j\omega t} \right) \, dt$$

But one condition for using this standard definition of Fourier Transform is that the function $f(t)$ be "absolutely integrable." This means the *magnitude* of this function, when integrated over all time, remains finite. That is obviously not true even for a function as simple as $f(t) = t$ for example. In that case, we need to multiply the function $f(t)$ by an exponentially decaying factor $e^{-\sigma t}$ so that $f(t)$ is forced to become integrable for certain values of the real parameter σ. So now, the Fourier transform becomes

$$\int\limits_{-\infty}^{\infty} \left(f(t) \times e^{-\sigma t} \times e^{-j\omega t} \right) \, dt = \int\limits_{-\infty}^{\infty} \left(f(t) \times e^{-st} \right) \, dt$$

In other words, to allow for waveforms (or its frequency components) that can naturally *increase* or *decrease* over time, we need to introduce an additional (real) exponential term $e^{\sigma t}$. However, when doing steady-state analysis, we usually represent a sine wave in the form $e^{j\omega t}$, which now becomes $e^{\sigma t} \times e^{j\omega t} = e^{(\sigma + j\omega)t}$. Now we have "a sine wave with an exponentially decreasing (σ positive), or increasing (σ negative), amplitude." Note that if we are only interested in performing steady-state analysis, we can go back and set $\sigma = 0$. That takes us back to the case involving only $e^{j\omega t}$ (or sine and cosine terms), that is, repetitive waveforms.

The result of the integral involving "s" above is called the "Laplace transform," and it is a function of "s" as explained further in the next section.

The *s*-Plane

In traditional AC analysis in the complex plane, the voltages and currents were complex numbers. But the frequencies were always real, even though the frequency ω itself may have been prefixed with "j" in a manner of representation. However, now in an effort to include virtually arbitrary waveforms into our analysis, we have in effect, created a "complex-frequency plane" too, that is, $s = \sigma + j\omega$. This is called the *s-plane*. The imaginary part of this new complex-frequency number "s" is our usual (real and oscillatory) frequency ω, whereas its real part is the one responsible for the observed exponential decay of any typical transient waveform over time. Analysis in this plane is ultimately just a more *generalized* form of frequency domain analysis.

In this representation, the reactive impedances become

$$Z_L = Ls$$

$$Z_C = \frac{1}{Cs}$$

Note that resistance still remains just a pure resistance, that is, it has no dependence on frequency or on s.

To calculate the response of complex circuits and stimuli in the s-plane, we need to use the rather obvious s-plane versions of the electrical laws. For example, Ohm's law is now

$$V(s) = I(s) \times Z(s)$$

The use of s gives us the ability to solve the differential equations arising from an almost arbitrary stimulus, in an *elegant* way, as opposed to the "brute-force" method in the time domain (using t). This is the Laplace transform method.

> *Note: Any such decomposition method can be practical, only when we are dealing with "mathematical" waveforms. Real waveforms may need to be approximated by known mathematical functions for further analysis. And very arbitrary waveforms will probably prove intractable.*

Laplace Transform Method

The Laplace transform is used to map a differential equation in the "time domain" (i.e., involving "t") to the "frequency domain" (involving "s"). The procedure unfolds as explained below.

First, the applied time-dependent stimulus (one-shot or repetitive — voltage or current) is mapped into the *complex-frequency domain*, that is, the s-plane. Then, by using the *s-plane versions of the impedances*, we can transform the entire circuit into the s-plane. To this transformed *circuit*, we apply the *s-plane versions of the basic electrical laws* and thereby analyze the circuit. We will then need to solve the resultant (transformed) differential equation (now in terms of s *rather than t*). But as mentioned, we will be happy to discover that the manipulation and solution of such differential equations are much easier to do in the s-plane than in the time domain. In addition, there are also several lookup tables for the Laplace transforms of common functions available, to help along the way. We will thus get the response of the circuit in the frequency domain. Thereafter, if so desired, we can use the "inverse Laplace transform" to recover the result in the time domain. The entire procedure is shown symbolically in Figure 12.2.

A little more math is useful at this point, as it will aid our understanding of the principles of feedback loop stability later.

Suppose the input signal (in the time domain) is $u(t)$ and the output is $v(t)$, and they are connected by a general second-order differential equation of the type

$$c_2 \frac{d^2 u(t)}{dt^2} + c_1 \frac{du(t)}{dt} + c_0 u(t) = d_2 \frac{d^2 v(t)}{dt^2} + d_1 \frac{dv(t)}{dt} + d_0 v(t)$$

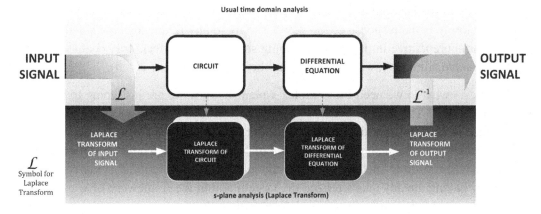

Figure 12.2: Symbolic representation of the procedure for working in the *s*-plane.

It can be shown that if $U(s)$ is the Laplace transform of $u(t)$, and $V(s)$ the transform of $v(t)$, then this equation (in the frequency domain) becomes simply

$$c_2 s^2 U(s) + c_1 s U(s) + c_0 U(s) = d_2 s^2 V(s) + d_1 s V(s) + d_0 V(s)$$

So,

$$V(s) = \frac{c_2 s^2 + c_1 s + c_0}{d_2 s^2 + d_1 s + d_0} U(s)$$

We can therefore define $G(s)$, the *transfer function* (i.e., output divided by input, now in the *s*-plane), as

$$G(s) = \frac{c_2 s^2 + c_1 s + c_0}{d_2 s^2 + d_1 s + d_0}$$

Therefore,

$$V(s) = G(s) \times U(s)$$

Note that this is analogous to the time-domain version of a general transfer function $f(t)$:

$$v(t) = f(t) \times u(t)$$

Since the solutions for the general equation $G(s)$ above are well-researched and documented, we can easily compute the response (V) to the stimulus (U).

A power supply designer is usually interested in ensuring that his or her power supply operates in a stable manner over its operating range. To that end, a sine wave is injected at a *suitable* point in the power supply, and the frequency swept, to study the response. This could be done in the lab and/or "on paper" as we will soon see. In effect, what we are looking at closely is the response of the power supply to any frequency *component* of a repetitive or nonrepetitive impulse. But in doing so, we are, in effect, only dealing with a steady sine wave stimulus (swept). So, we can then put $s = j\omega$ (i.e., $\sigma = 0$).

We can ask — why do we need the complex s-plane at all if we are just going to set $s = j\omega$ anyway at the end? The answer to that is — we *don't* always just do that. For example, we may at some later stage want to compute the *exact response of the power supply to a specific disturbance (like a step change in line or load)*. Then we would need the s-plane *and* the Laplace transform method. So, even though, we may just end up doing steady-state analysis, by having already characterized the system within the framework of s, *we retain the option to be able to conduct a more elaborate analysis of the system response to a more general stimulus if required.*

A silver lining for the beleaguered power supply designer is that he or she doesn't usually even need to know how to actually compute the Laplace transform of a function — unless, for example, the exact *step* response is required to be computed exactly — like an overshoot or undershoot resulting from a load transient. If the purpose is only to ensure sufficient stability margin is present, steady-state analysis serves the purpose. For that we simply sweep over all possible steady frequencies of input disturbance (either on paper or in the lab), and ensure there is no possibility of ever reinforcing the applied disturbance and making things worse. So, in a full-fledged mathematical analysis, it is convenient to work in the generalized s-plane. At the end, if we just want to calculate the stability margin, we can revert to $s = j\omega$. If we want to do more, we have that option too.

Disturbances and the Role of Feedback

In power supplies, we can either change the applied input voltage or increase the load. (This may or may not be done *suddenly*.) Either way, we always want the output to remain well regulated, and therefore, in effect, to "reject" the disturbance.

But in practice, that clearly does not happen as perfectly as desired. See Figure 12.3 for typical responses of converters to load transients. If instead of load, we suddenly increase the input voltage to a Buck regulator, the output *tends to follow suit* initially — since $D = V_{IN}/V_O$, and D has not *immediately* changed. This means, very briefly, V_O is proportional to V_{IN}.

To successfully correct the output and perform regulation, the control section of the IC needs to first sense the change in the output, which may take some time. After that it needs to correct the duty cycle, and that also may take some time. Then we have to wait for the inductor and output capacitor to either give up some of their stored energy or to gather some more — whatever is consistent with the conditions required for the new and

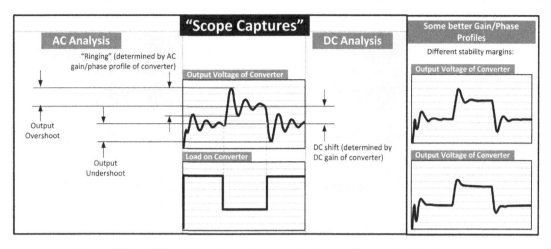

Figure 12.3: Effect of load transients, typical responses and related terms.

final *steady state*. Eventually, the output will hopefully settle down again to its new DC value. We see that there are several such *delays* in the circuit before we can get the output to stabilize. Minimizing these delays is clearly of great interest. Therefore, for example, just using smaller filter components (L and C) will often help the circuit respond faster.

Note: A philosophical question: how can the control circuit ever know beforehand, how much correction (in duty cycle) to precisely apply (when it senses that the output has shifted from its set value on account of the disturbance)? In fact, it usually doesn't! It can only be designed to know the general direction to move in, but it does not know beforehand, by how much it needs to move. Hypothetically speaking, we can do several things at our end. For example, we can command the duty cycle to change slowly and progressively, with the output being continuously monitored, and then immediately stop correcting the duty cycle at the very exact moment when the output equals its required regulation level. The duty cycle will thus never exceed the final level it is supposed to be in. However, clearly this is a slow correction process, and so though the duty cycle itself won't overshoot or undershoot, the output will certainly remain uncorrected for a rather long time. In effect, that amounts to a relative output droop or overshoot, though it is not oscillatory in nature. Another way is to command the duty cycle to change suddenly by a large arbitrary amount (though, of course, in the right direction). However, now the possibility of output overcorrection arises. The output will start getting "corrected" immediately, but because the duty cycle is far in excess of its final steady value, the output will "go the other way," before the control realizes it. After that, the control does try to correct it again, but it will likely "overreact" again. And so on. In effect, we now get "ringing" at the output. This ringing reflects a basic cause-effect uncertainty that is present in any feedback loop — the control may never fully know for sure whether the error it is seeing on the output is (a) immediate or delayed and (b) whether it is truly an external disturbance, rather than a result of its own attempted correction (coming back to haunt it, in a sense). So, if only after a lot of such avoidable ringing, the output does manage to stabilize, the converter is considered "marginally stable." In the worst case, this ringing may go on forever, even escalating, before it stabilizes at some constantly oscillating level. In effect, the control loop is now "fully confused," and the feedback loop is "unstable."

An "optimum" feedback loop is *neither too slow, nor too fast*. If it is too slow, the output will exhibit severe overshoot (or undershoot), though the output will not "ring." If it is too fast (overaggressive), the output will ring severely and even break into full instability (oscillations).

The study of how any disturbance *propagates*, either getting attenuated, or exacerbated in the process, is called "feedback loop analysis." In practice, we can test the stability margin of a feedback loop by deliberately injecting a small disturbance at an appropriate point inside it (the "*cause*"), and then seeing at what magnitude and phase it returns to the same point (the "*effect*"). If, for example, we find that the disturbance reinforces itself (at the right *phase*), cause–effect separation will be lost, and instability will result. But if the effect manages to kill or suppress the cause, we will achieve stability.

Note: The use of the word "phase" in the previous paragraph implies we are talking of sine waves *once again (there is no such thing as "phase" for a nonsinusoidal waveform). However, this turns out to be a valid assumption because, as we know, arbitrary disturbances can be decomposed into a series of sine wave components of varying frequencies. So, the disturbance/signal we "inject" (either on the bench or on a paper) can be a sine wave of arbitrary amplitude. By sweeping its frequency over a wide range, we can look for frequencies that have the* potential *to lead to instability. Because one fine day, we may receive a disturbance containing that particular frequency component, and if the margins are insufficient for that frequency, the system will break up into full-blown instability. But if we find that the system has enough margin over a wide range of (sine wave) frequencies, the system would, in effect, be stable when subjected to an arbitrarily shaped disturbance.*

A word on the *amplitude* of the applied disturbances. In this chapter, we are studying only *linear systems*. That means, if the input to a two-port network doubles, so does the output. Their *ratio* is therefore unchanged. In fact, that is why the transfer function was never thought of as say, being a *function of the amplitude* of the incoming signal. But we do know that in reality, if the disturbance is too severe, parts of the control circuit may "rail" — that means, for example, an internal op-amp's output may momentarily reach very close to its supply rails, thus affording no further correction for some time. We also do realize that there is no perfectly "linear system." But any system can be approximated by a linear system if the stimulus (and response) is "small" enough. That is why, when we conduct feedback loop analysis of power converters, we talk in terms of "small-signal analysis" and "small-signal models."

Note: For the same reason, even in bench testing, when injecting a sine wave to characterize the loop response, we must be careful not to apply too high an amplitude. The switching node voltage waveform must therefore be monitored during the test. Too large a jitter in the switching node waveform during the test can indicate possible "railing" (inside the error amplifier circuit). We must also ensure we are not operating close to the "stops" — for example, the minimum or maximum duty cycle limits of the controller and/or the set current limit. But the amplitude of the injected signal must not be too small either, otherwise switching noise is bound to overwhelm the readings (poor signal-to-noise ratio).

Note: For the same reason, most commercial power supply specifications will only ask for a certain transient response for say, from 80% load to max load, or even from 50% to max load, but not from zero to max load.

Transfer Function of the RC Filter, Gain, and the Bode Plot

We know that in general, v_o/v_i is a complex number called the transfer function. Its magnitude is defined as the "Gain." Take the simplest case of pure resistors only. For example, suppose we have two 10-k resistors in series and we apply 10 V across both of them. Suppose we define the output as the voltage at the node between the two resistors, we will get 5 V at that point. The transfer function is a real number in this case: $5/10 = 0.5$. So, we can say the gain is 0.5. That is the gain expressed as a pure ratio. We could, however, also express the gain in decibels, as $20 \times \log(|v_o/v_i|)$. In our example, that becomes $20 \times \log(0.5) = -6$ dB. In other words, gain can be expressed either as 0.5 (a ratio) or in terms of decibels (-6 dB in our case).

Note that by definition, a "decibel" or "dB" is dB $= 20 \times \log$ (ratio) — when used to express voltage or current ratios. For power ratios, dB is $10 \times \log$ (ratio).

Let us now take our simple series RC-network and transform it into the frequency domain, as shown in Figure 12.4. We can discern that the procedure for deriving its transfer function is based on a simple ratio of impedances, now extended to the s-plane.

Thereafter, since we are looking at only *steady-state* excitations (not transient impulses), we can set $s = j\omega$, and plot out (a) the *magnitude of the transfer function* (i.e., its "*gain*") and (b) the *argument of the transfer function* (i.e., its *phase*) — both in the frequency domain of course. This combined gain-phase plot is called a "Bode plot."

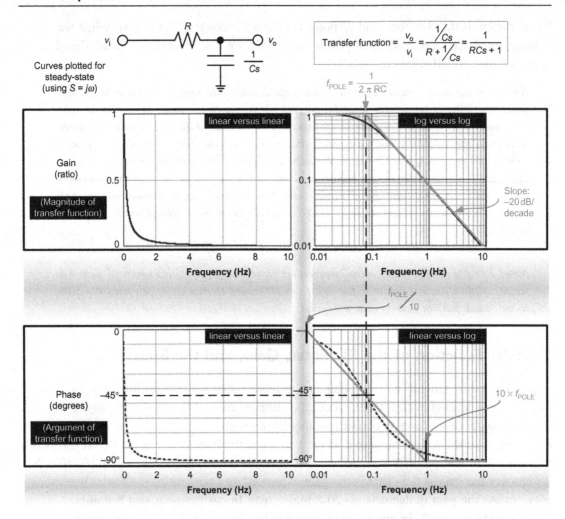

Figure 12.4: Analyzing the first-order low-pass RC filter in the frequency domain.

A word on terminology: *Note that initially, we will denote the ratio $|v_o/v_i|$ as "Gain," and we will distinguish it from $20 \times log(|v_o/v_i|)$ by calling the latter "$Gain_{dB}$." But these terms are actually often used interchangeably in literature and later in this chapter too. It can get confusing, but with a little experience it should quickly become obvious what is being meant in any particular context. Usually, however, "Gain" is used to refer to its dB version, that is, $20 \times log(|v_o/v_i|)$.*

Note that gain and phase are defined only in steady state as they implicitly refer to a sine wave ("phase" has no meaning otherwise!).

Here are a few observations based on Figure 12.4:

- We have converted the phase angle (which was originally in *radians*, $\theta = \omega t$) into degrees. That is because many engineers feel more comfortable visualizing angle in degrees instead of radians. To this end, we have used the following conversion: degrees $= (180/\pi) \times$ radians.

- Gain (on the vertical axis) is a simple ratio (not in decibels, unless stated otherwise).

- We have similarly converted from "angular frequency" (ω in radians/second) to the usual frequency (in Hz). Here we have used the equation: Hz $=$ (radians/second)/(2π).

- By varying the type of *scaling* on the gain and phase plots, we can see that the gain becomes a *straight line* if we use *log versus log* scaling. Note that in Figure 12.1, we had to use log versus *linear* scaling to get that curve to look like a straight line.

- We will get a straight-line gain plot in either of the two following cases — (*a*) if the gain is expressed as a simple ratio (i.e., Vout/Vin), and plotted on a log scale (on the y-axis) or (*b*) if the gain is expressed in decibels (i.e., $20 \times$ log Vout/Vin), and we use a linear scale to plot it. Note that in both cases, on the *x*-axis, we can either use "*f*" (frequency) and plot it using a log scale, or take $20 \times$ log(f) upfront, and plot it on a linear scale.

- In plotting logs, we must remember that the log of 0 is impossible to plot ($\log 0 \rightarrow -\infty$), and so we must not let the *origin* of a log scale ever be 0. We can set it *close* to zero, say 0.0001, or 0.001, or 0.01, and so on, but certainly not 0.

- We thus confirm by looking at the curves in Figure 12.4 that the gain at high frequencies starts decreasing *by a factor of 10 for every 10-fold increase in frequency*. Note that by the definition of decibel, a 10:1 voltage ratio is 20 dB (check 20 log(10) $=$ 20). Therefore, we can say that the gain falls at the rate of -20 dB *per decade* at higher frequencies. A circuit with a slope of this magnitude is called a "first-order filter" (in this case a low-pass one).

- Further, since this slope is *constant*, the signal must also decrease by a factor of 2 for every doubling of frequency. Or by a factor of 4 for every quadrupling of frequency, and so on. But a 2:1 ratio is 6 dB, and an "octave" is a doubling (or halving) of frequency. Therefore, we can also say that the gain of a low-pass first-order filter falls at the rate of -6 *dB per octave* (at high frequencies).

- If the *x* and *y* scales are scaled and proportioned identically, the actual angle the gain plot will make with the *x*-axis is $-45°$. The slope, that is, tangent of this angle is then $\tan(-45°) = -1$. Therefore, a slope of -20 dB/decade (or -6 dB/octave) is often simply called a "-1" slope.

- Similarly, when we have filters with *two reactive* components (i.e., an inductor *and* a capacitor), we will find the slope is -40 dB/decade (i.e., -12 dB/octave). This is usually called a "-2" slope (the actual angle being about $-63°$ when the axes are proportioned and scaled identically).

- The bold gray straight lines in the right-hand side graphs of Figure 12.4 form the "asymptotic approximation." We see that the gain asymptotes have a "break frequency" or "corner frequency" at $f = 1/(2\pi RC)$. This point can also be referred to as the "resonant frequency" of the RC filter, or a "pole" as discussed later.

- Note that the error/deviation between the actual curve and its asymptotic approximation is usually very small (only for *first-order filters*, as discussed later). For example, the worst-case error for the gain of the simple RC-network in Figure 12.4 is only -3 dB, and that occurs at the break frequency. Therefore, the asymptotic approximation is a valid "shortcut" that we will often use from now on to simplify the plots and their analysis.

- With regard to the asymptotes of the phase plot, we see that we get *two* break frequencies for it — one at one-tenth, and the other at 10 times the break frequency of the gain plot. The change in the phase angle at *each* of these break-points is 45° — giving a total phase shift of 90°. It spans *two* decades (symmetrically around the break frequency of the gain plot).

- Note that at the magnitude of the frequency where the single-pole lies, the phase shift (measured from the origin) is always 45° — that is, half the overall shift — whether we are using the asymptotic approximation or the actual curve.

- Since *both* the gain and the phase *fall* as frequency *increases*, we say we have a "pole" present. In our case, the pole is at the break frequency of $1/(2\pi RC)$. It is also called a "single-pole" or a first-order pole, since it is associated with a -1 slope.

- Later, we will see that similar to a "pole," we can also have a "zero," which is identifiable by the fact that *both* the gain and the phase start to *rise* with frequency from that location.

- In Figure 12.4, we see that the output voltage is clearly always *less* than the input voltage —that is, true for a (*passive*) RC-network (not involving op-amps yet). In other words, the gain is less than 1 (0 dB) at any frequency. Intuitively, that seems right because there seems to be no way to "amplify" a signal, without using an *active* device like an op-amp or transistor for example. However, as we will soon see, if we use passive filters involving *both* types of reactive components (*L* and *C*), we *can* in fact get the output voltage to exceed the input at certain frequencies. We then have "second-order" filters. And their response is what we more commonly refer to as "resonance."

The Integrator Op-amp ("Pole-at-Zero" Filter)

Before we go on to passive networks involving *two* reactive components, let us look at an interesting *active* RC-based (first-order) filter. The one chosen for discussion here is *the "integrator" because it happens to be the fundamental building block of any "compensation network."*

The inverting op-amp presented in Figure 12.5 has only a capacitor present in its feedback path. We know that under steady DC conditions, all capacitors essentially "go out of the picture." In our case, we are therefore left with *no* negative feedback at all *at DC* — and therefore infinite DC gain (though in practice, real op-amps will limit this to a very high, but finite value). But more surprisingly perhaps, that does not stop us from knowing the precise gain at *higher* frequencies. If we calculate the transfer function of this circuit, we will see that something "special" once again happens at the point $f = 1/(2\pi \times RC)$. However, unlike the passive RC filter, this point is not a break-point, nor a pole or zero location. It happens to be the point where the *gain is unity* (0 dB). We will denote this frequency as "fp0."

Note that so far, as indicated in Figure 12.5, *the integrator is the only stage present. So, in this particular case, "fp0" is the same as the observed crossover frequency "fcross." But in general, that will not be so. In general, in this chapter, "fp0" will refer to the crossover frequency the integrator stage* **would** *have produced* **were it present alone**.

Note that the integrator has a single-pole at "zero frequency," though 0 cannot be displayed on a log scale. We *always* strive to introduce this pole-at-zero because without it, the system would have rather poor DC (low-frequency) gain. *The integrator is the simplest way to try to get as high a DC gain as possible.* Having a high DC gain is the way to achieve good *steady-state* regulation in any power converter. This is indicated in Figure 12.3 too (labeled the "DC shift"). A high DC gain will reduce the DC shift.

On the right side of Figure 12.4, we have deliberately made the graph geometrically *square* in shape. To that end, we have assigned an equal number of grid divisions on the two axes, that is, the axes are scaled and proportioned identically. In addition, we have plotted $20 \times \log(f)$ on the y-axis (instead of just $\log(f)$). Having thus made the x- and y-axes identical in all respects, we realize why the slope is called "-1" — it really does fall at exactly 45° (now we see that visually too).

We take this opportunity to show how to do some simple math in the log-plane. This is shown in the lower part of Figure 12.5. We have derived one particular useful relationship between an arbitrary point "A" and the crossover frequency "fcross." A numerical example is also included.

$$\text{fcross} = \text{Gain}_A \times f_A$$

Note that, in general, the transfer function of any "pole-at-zero" function will always have the following form (*X* being a general real number)

$$\frac{1}{s \times (X)} \quad \text{(pole-at-zero transfer function)}$$

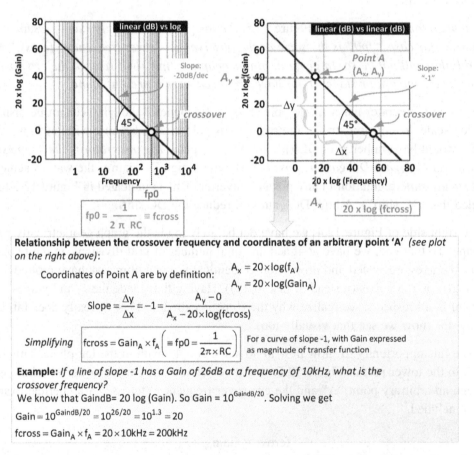

Figure 12.5: The integrator (pole-at-zero) operational amplifier and some related math.

The crossover frequency is then

$$\text{fcross} = \frac{1}{2\pi(X)} \quad \text{(crossover frequency)}$$

In our case, (X) is the time constant RC.

Mathematics in the Log-Plane

As we proceed toward our ultimate objective of control loop analysis and compensation network design, we will be *multiplying* transfer functions of *cascaded* blocks to get the *overall* transfer function. That is because the output of one block forms the input for the next block, and so on. It turns out that the mathematics of gain and phase is actually much easier to perform in the log-plane rather than in a linear plane. The most obvious reason for that is $\log(AB) = \log A + \log B$. So, we can add rather than multiply if we use logs. We have already had a taste of this in Figure 12.5. Let us summarize some simple rules that will help us later.

(a) If we take the product of two transfer functions A and B (cascaded stages), we know that the combined transfer function is the product of each:

$$\frac{v_{o2}}{v_{i1}} = \frac{v_{o2}}{v_{i2}} \times \frac{v_{i2}}{v_{i1}} \quad \Leftarrow \quad C = AB$$

But we also know that $\log(C) = \log(AB) = \log(A) + \log(B)$. In words, the gain of A in decibels plus the gain of B in decibels gives us the combined gain (C) in decibels. So, when we combine transfer functions, since *decibels add up*, that route is easier than taking the *product* of various transfer functions.

(b) The overall phase shift is the *sum of the phase shifts* produced by each of the cascaded stages. So, phase angles simply add up numerically (even in the log-plane).

(c) In Figure 12.6, we are using the term Gain_{dB} (the Gain expressed in dB), that is, $20 \log (\text{Gain})$, where Gain is the magnitude of the transfer function.

(d) From the upper half of Figure 12.6, we see that *if we know the crossover frequency (and the slope of the line), we can find the gain at any frequency.*

(e) Suppose we now shift the plotted line *vertically* (keeping the slope constant) as shown in the lower half of Figure 12.6. Then, by the equation provided therein, we can calculate *by what amount the crossover frequency shifts in the process.* Or equivalently, if we shift the crossover frequency by a known amount, we can calculate what will be the impact on the DC gain — because we will know by how much the curve has shifted either up or down in decibels.

Transfer Function of the Post-LC Filter

Moving toward power converters, we note that in a Buck, there is a *post*-LC filter present. Therefore, its filter stage can be treated as a simple "cascaded stage" immediately following the switch. The overall transfer function is very easy to compute as per the rules mentioned in the previous section (a product of cascaded transfer functions). However, when we come to the Boost and Buck-Boost, we don't have a post-LC filter — because there is a switch/ diode connected *between* the two reactive components. However, it can be shown that even the Boost and Buck-Boost can be manipulated into a "canonical model" in which an

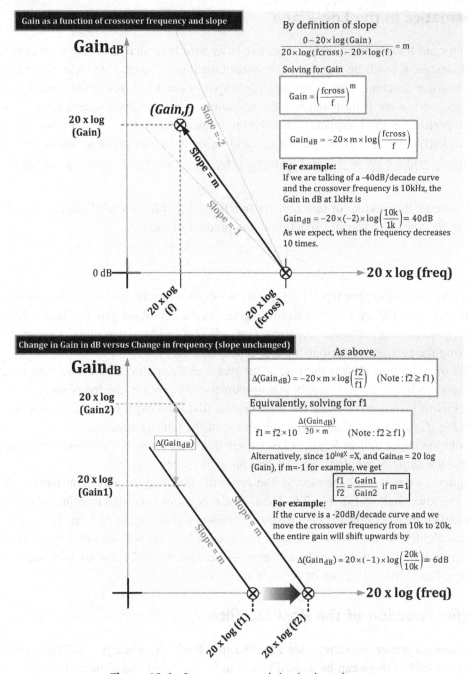

Gain as a function of crossover frequency and slope

By definition of slope

$$\frac{0-20\times\log(\text{Gain})}{20\times\log(\text{fcross})-20\times\log(\text{f})} = m$$

Solving for Gain

$$\text{Gain} = \left(\frac{\text{fcross}}{\text{f}}\right)^m$$

$$\text{Gain}_{dB} = -20\times m\times\log\left(\frac{\text{fcross}}{\text{f}}\right)$$

For example:
If we are talking of a -40dB/decade curve and the crossover frequency is 10kHz, the Gain in dB at 1kHz is

$$\text{Gain}_{dB} = -20\times(-2)\times\log\left(\frac{10\text{k}}{1\text{k}}\right) = 40\text{dB}$$

As we expect, when the frequency decreases 10 times.

Change in Gain in dB versus Change in frequency (slope unchanged)

As above,

$$\Delta(\text{Gain}_{dB}) = -20\times m\times\log\left(\frac{\text{f2}}{\text{f1}}\right) \quad (\text{Note}: \text{f2} \geq \text{f1})$$

Equivalently, solving for f1

$$\text{f1} = \text{f2}\times10^{\frac{\Delta(\text{Gain}_{dB})}{20\times m}} \quad (\text{Note}: \text{f2} \geq \text{f1})$$

Alternatively, since $10^{\log X} = X$, and $\text{Gain}_{dB} = 20\log$ (Gain), if m=-1 for example, we get

$$\frac{\text{f1}}{\text{f2}} = \frac{\text{Gain1}}{\text{Gain2}} \quad \text{if } m=1$$

For example:
If the curve is a -20dB/decade curve and we move the crossover frequency from 10k to 20k, the entire gain will shift upwards by

$$\Delta(\text{Gain}_{dB}) = 20\times(-1)\times\log\left(\frac{20\text{k}}{10\text{k}}\right) = 6\text{dB}$$

Figure 12.6: Some more math in the log-plane.

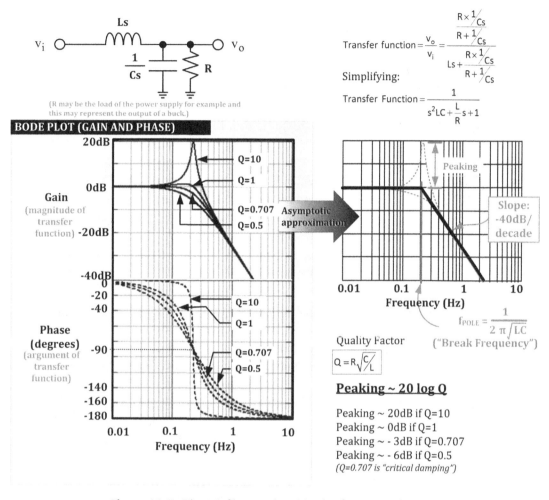

Figure 12.7: The LC filter analyzed in the frequency domain.

effective post-LC filter appears *at the output* (like a Buck) — thus making them as easy to treat as a Buck (i.e., cascaded stages). The only difference is that in this canonical model, the actual inductance L (of the Boost and Buck-Boost) gets replaced by an equivalent (or *effective*) inductance equal to $L/(1-D)^2$. The capacitor (the C of the LC) remains the same in the canonical model.

Since the simple LC post-filter now becomes representative of the output section of *any* typical switching topology, we need to understand it better as shown in Figure 12.7.

- For most practical purposes, we can assume that the break frequency (indicated in Figure 12.7) does *not depend on the load* or on any associated parasitic resistive elements of the components. In other words, the resonant frequency of the filter-plus-load combination (the break frequency, or "pole" in this case) can be taken to be simply $1/(2\pi\sqrt{(LC)})$, that is, no resistance term is included.

- The LC-filter gain *decreases* at the rate of "−2" at high frequencies. The phase also *decreases* providing a total phase shift of 180°. So, we say we have a "double-pole" (or second-order pole) at the break frequency $2\pi\sqrt{(LC)}$.

- Q is the "quality factor" (as defined in the figure). In effect, it quantifies the amount of "peaking" in the response curve at the break frequency point. Very simply put, if, for example, $Q = 20$, then the output voltage at the resonant frequency is 20 times the input voltage. On a log scale, this is written as $20 \times \log Q$, as shown in the figure. If Q is very high, the filter is considered "under-damped." If Q is very small, the filter is "over-damped." And if $Q = 0.707$, we have "critical damping." In critical damping, the gain at the resonant frequency is 3 dB below its DC value, that is, the output is 3 db below the input (similar to an RC filter). Note that −3 dB is a factor of $1/\sqrt{2} = 0.707$, that is, roughly 30% lower. Similarly, +3 dB is $\sqrt{2} = 1.414$ (i.e., roughly 40% higher).

- As indicated, the effect of resistance on the break frequency is usually minor, and therefore ignored. But the effect of *resistance on Q* (i.e., on the peaking) is significant (though eventually, that may be ignored too). However, we should keep in mind that the higher the associated *series* parasitic resistances of L and C, the *lower* is the Q. On the other hand, if we reduce the load, that is, increase the resistance across the C, Q increases. Remember that a high *parallel* resistance is in effect a small *series* resistance, and vice versa. In general, the presence of any significantly *large series* resistance ends up reducing Q, and any significantly *small parallel* resistance does just the same.

- As in Figure 12.4, we can use the "asymptotic approximation" for the LC gain plot too. However, the problem with trying to do the same with the *phase* of the LC is that there can now be a *very large error* — more so if Q becomes very large. Because if Q is very large, we can get a *very abrupt* phase shift (full 180°) in the region very close to the resonant frequency — not spread out smoothly over one-tenth to 10 times the break frequency as in Figure 12.4. This sudden phase shift can, in fact, become a real problem in a power supply, since it can induce "conditional stability" (discussed later). Therefore, a certain amount of damping helps from the standpoint of "phase-shift softening," thereby avoiding any possible conditional stability tendencies.

- Unlike an RC filter, the output voltage can in this case be *greater* than the input voltage (around the break frequency). But for that to happen, Q must be greater than 1.

- Instead of using Q, engineers often prefer to talk in terms of the "damping factor," defined as

$$\text{damping factor} = \zeta = \frac{1}{2Q}$$

So a high Q corresponds to a low ζ.

From the equations for Q and resonant frequency, we can conclude that if L *is increased, Q tends to decrease, and if C is increased, Q increases.*

> *Note: One of the possible pitfalls of putting too much output capacitance in a power supply is that we may be creating significant peaking (high Q) in its output filter's response. And we know that when that happens, the phase shift is also more abrupt, and that can induce conditional instability. So generally, if we increase C but simultaneously increase L, we can keep the Q (and the peaking) unchanged. But the break frequency changes significantly and that may not be acceptable.*

Summary of Transfer Functions of Passive Filters

The first-order (RC) low-pass filter transfer function (Figure 12.4) can be written in several different ways:

$$G(s) = \frac{(1/RC)}{s + (1/RC)} \quad \text{(RC low-pass)}$$

$$G(s) = \frac{1}{1 + (s/\omega_0)} \quad \text{(RC low-pass)}$$

$$G(s) = K\frac{1}{s + \omega_0} \quad \text{(RC low-pass)}$$

where $\omega_O = 1/(RC)$. Note that the "K" in the last equation above is a constant multiplier often used by engineers who are more actively involved in the design of filters. And in this case, $K = \omega_0$.

For the second-order filter (Figure 12.7), various equivalent forms seen in literature are

$$G(s) = \frac{(1/LC)}{s^2 + s(1/RC) + (1/LC)} \quad \text{(LC low-pass)}$$

$$G(s) = K\frac{1}{s^2 + (\omega_0/Q)s + \omega_0^2} \quad \text{(LC low-pass)}$$

$$G(s) = \frac{1}{(s/\omega_0)^2 + (1/Q)(s/\omega_0) + 1} \quad \text{(LC low-pass)}$$

$$G(s) = \frac{1}{1 + 2\zeta(s/\omega_0) + (s/\omega_0)^2} \quad \text{(LC low-pass)}$$

where $\omega_0 = 1/(LC)^{1/2}$. Note that here, $K = \omega_0^2$. Also, Q is the quality factor, and ζ is the damping factor defined earlier.

Finally, note also, that the following two relations are very useful when trying to manipulate the transfer function of the LC-filter into different forms

$$\frac{L}{R} = \frac{1}{\omega_0 Q} \quad \text{and} \quad \frac{1}{RC} = \frac{\omega_0}{Q} \quad (\text{LC } \textit{filter})$$

Poles and Zeros

Let us try to "connect the dots" now. We had mentioned in the case of both the first- and the second-order filters (Figures 12.4 and 12.7) that something called a "pole" exists. We should recognize that we got poles in both cases only because both the first- and the second-order transfer functions had terms in "s" in the *denominators* of their respective transfer functions. So, if s takes on specific values, it can force the denominator to become zero, and the transfer function (in the complex plane) then becomes "infinite." That is actually the point where we get a "pole" by definition. ***Poles occur wherever the denominator of the transfer function becomes zero.*** In general, the values of s at which the *denominator* becomes zero (i.e., the location of the poles) are sometimes called "resonant frequencies." For example, a hypothetical transfer function "$1/s$" will give us a pole at *zero frequency* (the "pole-at-zero" we talked about in the integrator shown in Figure 12.5).

Note that the gain, which is the magnitude of the transfer function (calculated by putting $s = j\omega$), won't necessarily be *really* "infinite" at the pole location as stated rather intuitively above. For example, in the case of the RC filter, we know that the gain is in fact *always* less than, or equal to unity, despite a pole being present at the break frequency.

Note that if we interchange the positions of the two primary components of each of the passive low-pass filters we discussed earlier, we will get the corresponding "high-pass" RC and LC-filters, respectively. If we calculate their transfer functions in the usual manner, we will see that besides giving us poles, we also now get single- and double-*zeros* respectively as indicated in Figure 12.8. ***Zeros occur wherever the numerator of the transfer function becomes zero.*** Note that in Figure 12.8, the zeros are not visible, only the poles are. But the presence of the zeros is indicated by the fact that we started from the left of each graph with curves rising *upward* (rather than being flat with frequency), and for the same reason, the phase started off with 90° for the first-order filter and from 180° for the second-order filter, rather than from 0°.

We had mentioned that gain-phase plots are called Bode plots. In the case of Figure 12.8, we have drawn these on the *same* graph just for convenience. Here the solid line is the gain, and to read its value, we need to look at the y-axis on the *left* side of the graph. Similarly, the dashed line is the phase, and for it, we need to look at the y-axis on the *right* side. Note that for practice, we have reverted to plotting the gain as a simple ratio (not in

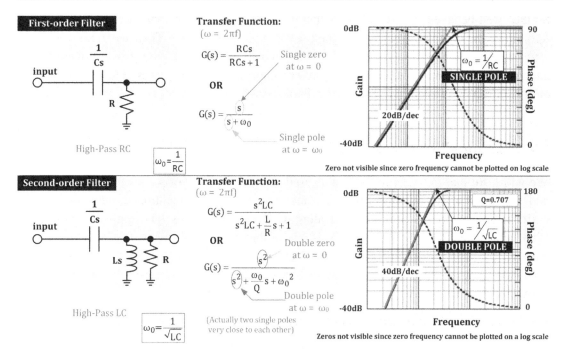

Figure 12.8: High-pass RC and LC (first-order and second-order) filters.

decibels), but we are now plotting that on a log scale. The reader should hopefully, by now, have learnt to correlate the *major* grid divisions of this type of plot with the corresponding dB. So a 10-fold increase is equivalent to +20 dB, a 100-fold increase is +40 dB, and so on.

We can now generalize our approach. A network transfer function can be described as a ratio of two polynomials:

$$G(s) = \frac{V(s)}{U(s)} = k\frac{a_0 + a_1 s + a_2 s^2 + a_3 s^3 + \cdots}{b_0 + b_1 s + b_2 s^2 + b_3 s^3 + \cdots}$$

This can be factored out as

$$G(s) = K\frac{(s - z_0)(s - z_1)(s - z_2)\cdots}{(s - p_0)(s - p_1)(s - p_2)\cdots}$$

So, the zeros (i.e., leading to the numerator being zero) occur at the complex frequencies $s = z_1, z_2, z_3, \ldots$. The poles (denominator zero) occur at $s = p_1, p_2, p_3, \ldots$

In power supplies, we usually deal with transfer functions of the form:

$$G(s) = K\frac{(s + z_0)(s + z_1)(s + z_2)\ldots}{(s + p_0)(s + p_1)(s + p_2)\ldots}$$

So the "well-behaved" poles and zeros that we have been talking about are actually in the *left-half* of the complex-frequency plane ("LHP" poles and zeros). Their locations are at $s = -z_1, -z_2, -z_3, -p_1, -p_2, -p_3, \ldots$. We can also, in theory, have right-half plane poles and zeros which have very different behavior to normal poles and zeros, and can cause almost intractable instability. This aspect is discussed later.

"Interactions" of Poles and Zeros

We will learn that in trying to find the overall transfer function of a converter, we typically add up several of its constituent transfer functions together. As mentioned, the math is easier to do on a log-plane if we are dealing with cascaded stages. The equivalent post-LC filter, which we studied in Figure 12.7, is one of those cascaded stages. However, for now we will still keep things general here, and simply show how to add up several transfer functions together in the log-plane. We just have several poles and zeros, and we must know how to add these up too.

We can break up the full analysis in two parts:

(a) *For poles and zeros lying along the same gain plot* (i.e., belonging to the same transfer function/stage) — the effect is *cumulative in going from left to right*. So, suppose we are starting from zero frequency and move right toward a higher frequency, and we first encounter a double-pole. We know that the gain will start falling with a slope of -2 beyond the corresponding break frequency. As we go further to the right, suppose we now encounter a single-zero. This will impart a *change* in slope of $+1$. So the *net* slope of the gain plot will now become $-2 + 1 = -1$, after the zero location. Note that despite a zero being present, the gain is still *falling*, though at a lesser rate. In effect, the single-zero canceled *half* the double-pole, so we are left with the response of a single-pole (to the right of the zero).

The phase angle also cumulates in a similar manner, except that in practice a phase angle plot is harder to analyze. That is because phase shift can take place slowly over *two decades around* the resonant frequency. We also know that for a double-pole (or double-zero), the change in phase may in fact be very abrupt at the resonant frequency. However, *eventually*, a *good distance away* in terms of frequency, the net effect is still *predictable*. So, for example, the phase angle plot of a double-pole, followed shortly by a single-zero, will start with a phase angle of $0°$ (at DC) which will then decrease gradually toward $-180°$ on account of the double-pole. But about a decade below the location of the single-zero, the phase angle will then gradually start increasing (though still remaining negative). It will eventually settle down to $-180° + 90° = -90°$ at high frequencies, consistent with a single net pole.

(b) *For poles and zeros lying along different gain plots* (belonging to say several cascaded stages that are being summed up) — we know that the overall gain in decibels is the sum of the gain of each (also in decibels). The effect of this math on the pole-zero interactions is therefore simple to describe. If, for example, *at a specific frequency*, we have a double-pole in one plot and a single-zero on the other plot, then the overall (combined) gain plot will have a single-pole at this break frequency. So, we see that poles and zeros tend to "destroy" (cancel) each other out. Zeros are considered to be "anti-poles" in that sense. But poles and zeros also add up *with their own type*. For example, if we have a double-pole on one plot, and a single-pole on the other plot (at the same frequency), the net gain (on the composite transfer function plot) will change slope by "−3" after this frequency. Phase angles also add up similarly. A few examples later will make this much clearer.

Closed and Open-Loop Gain

Figure 12.9 represents a general feedback controlled system. The "plant" (also sometimes called the "modulator") has a "forward transfer function" $G(s)$. A part of the output gets fed back through the feedback block, to the control input, so as to produce regulation at the

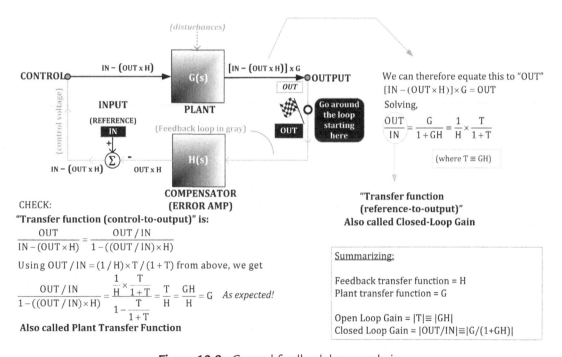

Figure 12.9: General feedback loop analysis.

output. Along the way, the feedback signal is compared with a reference level, which tells it what the desired level is for it to regulate to.

$H(s)$ is the "feedback transfer function," and we can see this goes to a summing block (or node) — represented by the circle with an enclosed summation sign.

> *Note: The summing block is sometimes shown in literature as just a simple circle (nothing enclosed), but sometimes rather confusingly as a circle with a multiplication sign (or x) inside it. Nevertheless,* it still is a summation block.

One of the inputs to this summation block is the reference level (the "input" from the viewpoint of the control system), and the other is the output of the feedback block (i.e., the part of the output being fed back). The output of the summation node is the "error" signal.

Comparing Figure 12.9 with Figure 12.10, we see that in a power supply, the plant itself can be split into several cascaded blocks. These blocks are — the PWM (not to be confused with the term "modulator" often used in general control loop theory referring to the *entire* plant), the power stage consisting of the driver-plus-switch, and the LC-filter. The feedback block, on the other hand, consists of the voltage divider (if present) and the compensating error amplifier. Note that we may prefer to visualize the error amplifier block as two cascaded stages — one that just computes the error (summation node) and another which accounts for the gain (and its associated compensation network). But in actual practice, since we apply the feedback signal to the inverting pin of the error amplifier, both functions are combined. Also note that the basic principle behind the PWM stage (which determines the duty cycle of the pulses driving the switch) is explained in the next section and in Figure 12.11.

In general, the plant can receive various "disturbances" that can affect its output. In a power supply, these are essentially the *line and load variations*. The basic purpose of feedback is to reduce the effect of these disturbances on the output voltage (see Figure 12.3, for example).

Note that the word "input" in control loop theory is not the physical input power terminal of the converter. Its location is actually marked in Figure 12.9. It happens to be the reference level we are setting the output to. The word "output" in control loop theory, however, is the same as the physical output terminal of the converter.

In Figure 12.9, we have derived the *open-loop gain* $|T| = |GH|$, which is simply the magnitude of the *product of the forward and feedback transfer functions*, that is, obtained by going *around* the loop fully once. On the other hand, the magnitude of the *reference-to-output* (i.e., input-to-output) transfer function is called the *closed-loop gain*. It is $|G/(1 + GH)|$.

Note that the word "closed" has really nothing to do with the feedback loop being *literally* "open" or "closed" as sometimes thought. Similarly, "*GH*" is called the "open-loop transfer function" — irrespective of whether the loop is literally "open," say for the purpose of

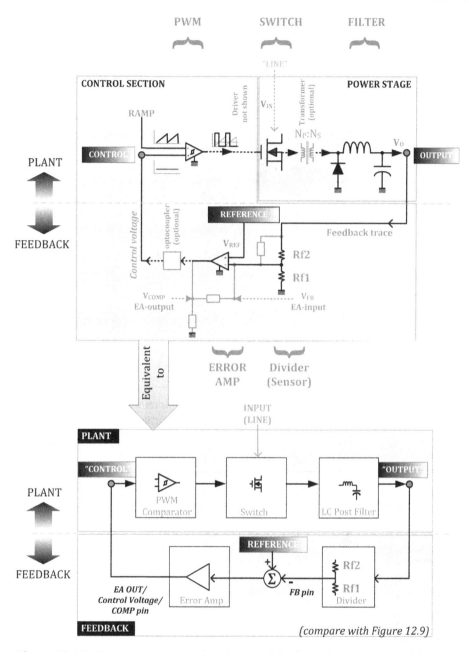

Figure 12.10: A power converter: its plant and feedback (compensator) blocks.

measurement, or "closed" as in normal operation. In fact, in a typical power supply, we can't even hope to break the feedback path for the purpose of any measurement. Because the gain is typically so high that even a minute change in the feedback voltage will cause the output to swing wildly. So, in fact, we always need to "close" the loop and thereby

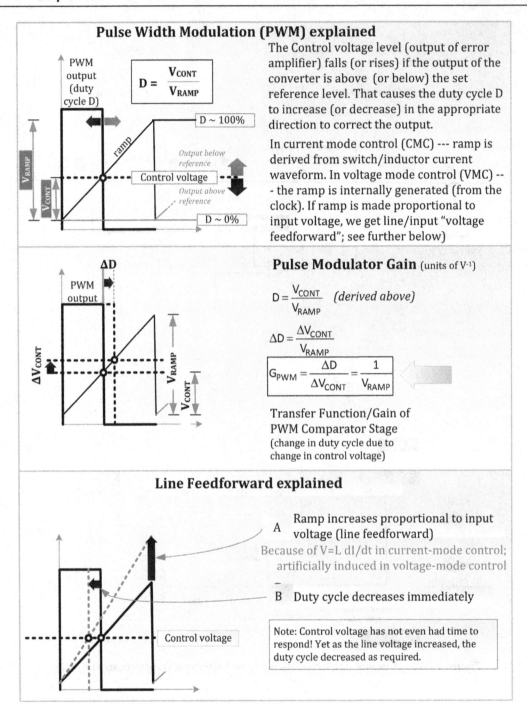

Pulse Width Modulation (PWM) explained

$$D = \frac{V_{CONT}}{V_{RAMP}}$$

The Control voltage level (output of error amplifier) falls (or rises) if the output of the converter is above (or below) the set reference level. That causes the duty cycle D to increase (or decrease) in the appropriate direction to correct the output.

In current mode control (CMC) --- ramp is derived from switch/inductor current waveform. In voltage mode control (VMC) --- the ramp is internally generated (from the clock). If ramp is made proportional to input voltage, we get line/input "voltage feedforward"; see further below)

Pulse Modulator Gain (units of V⁻¹)

$$D = \frac{V_{CONT}}{V_{RAMP}} \quad (derived\ above)$$

$$\Delta D = \frac{\Delta V_{CONT}}{V_{RAMP}}$$

$$G_{PWM} = \frac{\Delta D}{\Delta V_{CONT}} = \frac{1}{V_{RAMP}}$$

Transfer Function/Gain of PWM Comparator Stage (change in duty cycle due to change in control voltage)

Line Feedforward explained

A Ramp increases proportional to input voltage (line feedforward)

Because of V=L dI/dt in current-mode control; artificially induced in voltage-mode control

B Duty cycle decreases immediately

Note: Control voltage has not even had time to respond! Yet as the line voltage increased, the duty cycle decreased as required.

Figure 12.11: PWM action, transfer function, and line feedforward explained.

DC-bias the converter into full regulation, before we can even measure the so-called "open-loop" gain.

The Voltage Divider

Usually, the output V_O of the power supply first goes to a *voltage divider*. Here it is, in effect, just stepped-down, for subsequent comparison with the reference voltage "V_{REF}." The comparison takes place at the input of the *error amplifier*, which is usually just a conventional op-amp (voltage amplifier).

We can visualize an ideal op-amp as a device that varies its output so as to virtually equalize the voltages at its input pins. Therefore, in steady state, the voltage at the node connecting Rf2 and Rf1 (see "divider" block in Figure 12.10) can be assumed to be (almost) equal to V_{REF}. Assuming that no current flows out of (or into) the divider at this node, using Ohm's law:

$$\frac{Rf1}{Rf1 + Rf2} = \frac{V_{REF}}{V_O}$$

Simplifying,

$$\frac{Rf2}{Rf1} = \frac{V_O}{V_{REF}} - 1$$

So this tells us what *ratio* of the voltage divider resistors we must have to produce the desired output rail.

Note, however, that in applying control loop theory to power supplies, we are actually looking only at *changes* (or perturbations), **not the DC values** (though this was not made obvious in Figure 12.9). ***It can also be shown that when the error amplifier is a conventional op-amp, the lower resistor of the divider Rf1 behaves only as a DC biasing resistor and does play any (direct) part in the AC loop analysis.***

> *Note: The lower resistor of the divider Rf1 does not enter the AC analysis, provided we are considering* ideal *op-amps. In practice, it* does *affect the bandwidth of a* real *op-amp, and therefore may on occasion need to be considered.*

> *Note: If we are using a spreadsheet, we will find that changing Rf1 in a standard op-amp-based error amplifier divider does, in fact, affect the overall loop. But we should be clear that that is only because by changing Rf1, we have changed the duty cycle of the converter (via its output voltage), which thus affects the plant transfer function. Therefore, the effect of Rf1 is indirect. Rf1 does not enter into any of the equations that tell us the locations of the poles and zeros of the system.*

Note: We will see that when using a transconductance op-amp as the error amplifier, Rf1 does enter the AC analysis.

Pulse-Width Modulator Transfer Function

The *output of the error amplifier* (sometimes called "COMP," sometimes "EA-out," sometimes "*control voltage*") is applied to one of the inputs of the PWM comparator. This is the terminal marked "Control" in Figures 12.9 and 12.10. On the other input of this PWM comparator, we apply a sawtooth voltage ramp — either internally generated from the clock when using "voltage-mode control," or derived from the current ramp when using "current-mode control" (explained later). Thereafter, by standard comparator action, we get pulses of desired width with which to drive the switch.

Since the feedback signal coming from the output rail of the power supply goes to the *inverting* input of the error amplifier, if the output is below the set regulation level, the output of the error amplifier goes high. This causes the PWM to increase the pulse width (duty cycle) and thus try to make the output voltage rise. Similarly, if the output of the power supply goes above its set value, the error amplifier output goes low, causing the duty cycle to decrease (see upper third of Figure 12.11).

As mentioned previously, the output of the PWM stage is duty cycle, and its input is the "control voltage" or the "EA-out." So, as we said, the gain of this stage is not a dimensionless quantity, but has units of $1/V$. From the middle of Figure 12.11, we can see that this gain is equal to $1/V_{RAMP}$, where V_{RAMP} is the peak-to-peak amplitude of the ramp sawtooth.

Voltage (Line) Feedforward

We had also mentioned previously that when there is a disturbance, the control does not usually know beforehand how much duty cycle correction to apply. However, in the lowermost part of Figure 12.11, we have described an increasingly popular technique being used to make that a reality, at least when faced with line disturbances. This is called input-voltage/line feedforward, or simply "feedforward."

This technique requires the input voltage be sensed and the slope of the comparator sawtooth ramp increased if the input goes up. In the simplest implementation, a doubling of the input causes the slope of the ramp to double. Then, from Figure 12.11, we see that if the slope doubles, the duty cycle is immediately halved. In a Buck, the governing equation is $D = V_O/V_{IN}$. So, if a doubling of input occurs, we know that naturally, the duty cycle will eventually halve anyway. So, rather than wait for the control voltage to decrease by half to lower the duty cycle (keeping the ramp unchanged), we could also change the ramp itself — in this case, double the slope of the ramp and thereby achieve the very same result (i.e., halving of duty cycle) *almost instantaneously*.

Summarizing: the duty cycle correction afforded by this "automatic" ramp correction is exactly what is required for a Buck, since its duty cycle $D = V_O/V_{IN}$. More importantly, this correction is virtually instantaneous — we didn't have to wait for the error amplifier to detect the error on the output (through the inherent delays of its RC-based compensation network scheme), and respond by altering the control voltage. So, in effect, by input/line feedforward, we have bypassed all major delays, and therefore line correction is almost immediate — and that amounts to almost "perfect" rejection of the line disturbance.

In Figure 12.11, it is implied that the PWM ramp is created artificially from the fixed internal clock. That is called voltage-mode control. In current-mode control, the PWM ramp is basically an appropriately amplified version of the switch/inductor current. We will discuss that in more detail later. Here we just want to point out that the line feedforward technique described in Figure 12.11 is applicable only to voltage-mode control. However, the original inspiration behind the idea does come from *current-mode control* — in which the PWM ramp, generated from the inductor current, automatically increases if the line voltage increases. That partly explains why current-mode control seems to respond so much "faster" to line disturbances than traditional voltage-mode control and one of its oft-repeated advantages.

However, one question remains: how good is the "built-in" automatic line feedforward in current-mode control? In a Buck topology, the *slope* of the inductor current up-ramp is equal to $(V_{IN} - V_O)/L$. So, if we double the input voltage, we do *not* end up doubling the slope of the inductor current. Therefore, neither do we end up automatically halving the duty cycle, as we can do easily in line feedforward applied to voltage-mode control.

In other words, voltage-mode control with proportional line feedforward control, though inspired by current-mode control, provides *better* line rejection than current-mode control (for a Buck). Voltage-mode control with line feedforward is considered by many to be a far better choice than current-mode control, all things considered.

Power Stage Transfer Function

As per Figure 12.10, the "power stage" formally consists of the switch plus the (equivalent) LC-filter. Note that this is just the plant *minus* the PWM. Alternatively stated, *if we add the PWM comparator section to the power stage, we get the "plant" as per control loop theory*, and that was symbolized by the transfer function "*G*" in Figure 12.9. The rest of the circuit in Figure 12.10 is the feedback block, and this was symbolized by the transfer function *H* in Figure 12.9.

We had indicated previously that whereas in a Buck, the L and C are really connected to each other at the output (as drawn in Figure 12.10), in the remaining two topologies they are *not*. However, the small-signal (canonical) model technique can be used to transform these latter topologies into equivalent AC models — in which, for all practical purposes, a regular LC-filter does appear after the switch, just as for a Buck. With this technique, we can then justifiably separate the *power stage* into a *cascade* of two separate stages (as for a Buck):

- A stage that effectively converts the duty cycle input (coming from the output of the PWM stage) into an output voltage.
- An equivalent post-LC filter stage that takes in this output and converts it into the output rail of the converter.

With this understanding, we can finally build the final transfer functions presented in the next section.

Plant Transfer Functions of All the Topologies

Let us discuss the three topologies separately here. Note that we are assuming *voltage-mode control and continuous conduction mode* (CCM). Further, the "ESR (effective series resistance) zero" is not included here (a simple modification introduced later).

(A) **Buck Converter**

    (a) *Control-to-output transfer (plant) function*

        The transfer function of the plant is also called the "*control-to-output transfer function*" (see Figure 12.10). It is the output voltage of the converter, divided by the "control voltage" (i.e., the output of the error amplifier, or "EA-out"). We are, of course, talking only from an *AC* point of view, and are therefore interested only in the *changes* from the DC-bias levels.

        The control-to-output transfer function is a product of the transfer functions of the PWM modulator, the switch and the LC-filter (since these are *cascaded* stages). Alternatively, the control-to-output transfer function is a product of the transfer function of the PWM comparator and the transfer function of the "power stage."

        We already know from Figure 12.11 that *the transfer function of the PWM stage is equal to the reciprocal of the amplitude of the ramp.* And as discussed in the previous section, the power stage itself is a cascade of *an equivalent post-LC stage* (whose transfer function is the same as the passive low-pass second-order LC filter we discussed previously in Figure 12.7), *plus a power stage that finally converts the duty cycle into a DC output voltage V_O.*

We are now interested in finding the transfer function of the latter stage referred to above.

The overall question is — what happens to the output when we perturb the duty cycle slightly (keeping the input to the converter V_{IN} constant). Here are the steps for a Buck

$$V_O = D \times V_{IN} \quad (Buck)$$

Therefore, differentiating

$$\frac{dV_0}{dD} = V_{IN}$$

So, in very simple terms, the required transfer function of the intermediate "duty cycle-to-output stage" is equal to V_{IN} for a Buck.

Finally, the control-to-output (plant) transfer function is the product of three (cascaded) transfer functions, that is, it becomes

$$G(s) = \frac{1}{V_{RAMP}} \times V_{IN} \times \frac{1/LC}{s^2 + s(1/RC) + (1/LC)} \quad (Buck: plant\ transfer\ function)$$

Alternatively, this can be written as

$$G(s) = \frac{1}{V_{RAMP}} \times V_{IN} \times \frac{1}{(s/\omega_0)^2 + (s/(\omega_0 Q)) + 1} \quad (Buck: plant\ transfer\ function)$$

where $\omega_0 = 1/\sqrt{(LC)}$ and $\omega_0 Q = R/L$.

(b) *Line-to-output transfer function*

Of great importance in any converter design is _not_ what happens to the output when we perturb the *reference* (which is what the *closed-loop* transfer function really is), but what happens at the output when there is *a line disturbance*. This is often referred to as "audio susceptibility" (probably because early converters switching at around 20 kHz would emit audible noise under this condition).

The equation connecting the input and output voltages is simply the DC input-to-output transfer function, that is,

$$\frac{V_O}{V_{IN}} = D \quad (Buck)$$

So, D is also the factor by which the input line (V_{IN}) disturbance gets scaled, and thereafter applied at the input of the equivalent LC post-filter for further attenuation as per Figure 12.7. We already know the transfer function of the LC low-pass filter. Therefore, the line-to-output transfer function is the product of the two cascaded transfer functions, that is,

$$D \times \frac{(1/LC)}{s^2 + s(1/RC) + (1/LC)} \quad \textit{(Buck: line transfer function)}$$

where R is the load resistor (at the output of the converter).

Alternatively, this can be written as

$$D \times \frac{1}{(s/\omega_0)^2 + (s/(\omega_0 Q)) + 1} \quad \textit{(Buck: line transfer function)}$$

where $\omega_0 = 1/\sqrt{(LC)}$, and $\omega_0 Q = R/L$.

(B) **Boost converter**

(a) *Control-to-output (plant) transfer function*

Proceeding similar to the Buck, the steps for this topology are

$$V_O = \frac{V_{IN}}{1 - D}$$

$$\frac{dV_O}{dD} = \frac{V_{IN}}{(1-D)^2}$$

So the control-to-output transfer function is a product of three transfer functions:

$$G(s) = \frac{1}{V_{RAMP}} \times \frac{V_{IN}}{(1-D)^2} \times \frac{(1/\underline{L}C) \times (1 - s(\underline{L}/(R)))}{s^2 + s(1/RC) + (1/\underline{L}C)} \quad \textit{(Boost: plant transfer function)}$$

where $\underline{L} = L/(1 - D)^2$. Note that *this is the inductor in the "equivalent post-LC filter" of the canonical model. Also note that C remains unchanged.*

Alternatively, the above transfer function can be written as

$$G(s) = \frac{1}{V_{RAMP}} \times \frac{V_{IN}}{(1-D)^2} \times \frac{(1 - (s/(\omega_{RHP})))}{(s/\omega_0)^2 + (s/(\omega_0 Q)) + 1} \quad \textit{(Boost: plant transfer function)}$$

where $\omega_0 = 1/\sqrt{(\underline{L}C)}$ and $\omega_0 Q = R/\underline{L}$.

Note that we have included a *surprise term* in the numerator above. By detailed modeling, it can be shown that both the Boost and the Buck-Boost have such a term. This term represents a zero, but a different type to the "well-behaved" zero discussed so far (note the sign in front of the *s*-term is negative, so it occurs in the positive, i.e., the right-half portion of the *s*-plane). If we consider its contribution to the gain-phase plot, we will find that as we raise the frequency, the gain will *increase* (as for a normal zero), but simultaneously, the phase angle will *decrease* (opposite to a "normal" zero, more like a "well-behaved" pole).

Why is that a problem? Because, later we will see that if the overall open-loop phase angle drops sufficiently low, the converter can become unstable. That is why this zero is considered undesirable. Unfortunately, it is virtually impossible to compensate for (or "kill") by normal techniques. The usual method is to literally "push it out" — to higher frequencies where it can't affect the overall loop significantly. Equivalently, we need to reduce the bandwidth of the open-loop gain plot to a frequency low enough that it just doesn't "see" this zero. In other words, *the crossover frequency must be set much lower than the location of the RHP zero.*

The name given to this zero is the "*RHP zero,*" as indicated earlier — to distinguish it from the "well-behaved" (conventional) left-half-plane zero. For the Boost topology, its location can be found by setting the numerator of the transfer function above (see its first form) to zero, that is, $s \times (L/R) = 1$. So, the frequency location of the Boost RHP zero is

$$f_{\text{RHP}} = \frac{R \times (1 - D)^2}{2\pi L} \quad (Boost)$$

Note that the very existence of the RHP zero in the Boost and Buck-Boost can be traced back to the fact that these are the *only* topologies where an actual LC post-filter *doesn't* exist on the output. Though, by using the canonical modeling technique, we have managed to create an *effective* LC post-filter, the fact that in reality there is a switch/diode connected between the actual *L* and *C* of the topology is what is ultimately responsible for creating the RHP zero.

Note: The RHP zero is often explained intuitively as follows — if we suddenly increase the load, the output dips slightly. This causes the converter to increase its duty cycle in an effort to restore the output. Unfortunately, for both the Boost and the Buck-Boost, energy is delivered to the load only during the switch off-time. So, an increase in the duty cycle decreases the off-time, and there is now, unfortunately, a smaller interval available for the stored inductor energy to get transferred to the output. Therefore, the output voltage, instead of increasing as we were hoping, dips even further for a few cycles. This is the RHP zero in action. Eventually, the current in the inductor does manage to ramp up over several successive switching cycles to the new level consistent with the increased energy demand, and so this strange situation gets corrected — provided full instability has not already occurred!

The RHP zero can occur *at any duty cycle*. Note that its location is at a lower frequency as D approaches 1 (i.e., at lower input voltages). It also moves to a lower frequency if L is increased. That is one reason why bigger inductances are not preferred in Boost and Buck-Boost topologies.

(b) *Line-to-output transfer function*

We know that

$$\frac{V_O}{V_{IN}} = \frac{1}{1-D} \quad (Boost)$$

Therefore, we get

$$\frac{1}{1-D} \times \frac{(1/\underline{L}C)}{s^2 + s(1/RC) + (1/\underline{L}C)} \quad (Boost:\ line\ transfer\ function)$$

Alternatively, this can be written as

$$\frac{1}{1-D} \times \frac{1}{(s/\omega_0)^2 + (s/(\omega_0 Q)) + 1} \quad (Boost:\ line\ transfer\ function)$$

where $\omega_0 = 1/\sqrt{(\underline{L}C)}$ and $\omega_0 Q = R/L$.

(C) **Buck-Boost converter**

(a) *Control-to-output transfer (plant) function*

Here are the steps for this topology:

$$V_O = \frac{V_{IN} \times D}{1-D}$$

$$\frac{dV_O}{dD} = \frac{V_{IN}}{(1-D)^2}$$

(Yes, it is an interesting coincidence — the slope of $1/(1-D)$ calculated for the Boost is the same as the slope of $D/(1-D)$ calculated for the Buck-Boost!)

So, the control-to-output transfer function is

$$G(s) = \frac{1}{V_{RAMP}} \times \frac{V_{IN}}{(1-D)^2} \times \frac{(1/\underline{L}C) \times (1 - s(\underline{L}D/R))}{s^2 + s(1/RC) + (1/\underline{L}C)}$$

(*Buck-Boost: plant transfer function*)

where $\underline{L} = L/(1-D)^2$ is the inductor in the *equivalent* post-LC filter.

Alternatively, this can be written as

$$G(s) = \frac{1}{V_{\text{RAMP}}} \times \frac{V_{\text{IN}}}{(1-D)^2} \times \frac{(1-(s/\omega_{\text{RHP}}))}{(s/\omega_0)^2 + (s/(\omega_0 Q)) + 1}$$

(*Buck-Boost*: *plant transfer function*)

where $\omega_0 = 1/\sqrt{(LC)}$ and $\omega_0 Q = R/L$.

Note that, as for the Boost, we have included the RHP zero term in the numerator (in gray). Its location is similarly calculated to be

$$f_{\text{RHP}} = \frac{R \times (1-D)^2}{2\pi L \times D} \quad (\textit{Buck-Boost})$$

This also comes in at a lower frequency if D approaches 1 (lower input). Compare with what we got for the Boost:

$$f_{\text{RHP}} = \frac{R \times (1-D)^2}{2\pi L} \quad (\textit{Boost})$$

(b) **Line-to-output transfer function**
We know that

$$\frac{V_O}{V_{\text{IN}}} = \frac{D}{1-D} \quad (\textit{Buck-Boost})$$

Therefore,

$$\frac{D}{1-D} \times \frac{(1/\underline{L}C)}{s^2 + s(1/RC) + (1/\underline{L}C)} \quad (\textit{Buck-Boost}: \textit{line transfer function})$$

This is alternatively written as

$$\frac{D}{1-D} \times \frac{1}{(s/\omega_0)^2 + (s/(\omega_0 Q)) + 1} \quad (\textit{Buck-Boost}: \textit{line transfer function})$$

where $\omega_0 = 1/\sqrt{(LC)}$ and $\omega_0 Q = R/\underline{L}$.

Note that the plant and line transfer functions of all the topologies calculated above do not depend on the load current I_O. That is why gain-phase plots (Bode plots) do not change much if we vary the load current (provided we stay in CCM as assumed above).

Note also that so far we have ignored a key element of the transfer functions — the ESR of the output capacitor and its contribution to the "ESR–zero."

Whereas the DCR (DC resistance) usually just ends up decreasing the overall Q (less "peaky" at the second-order (LC) resonance), the *ESR actually contributes a zero to the open-loop transfer function*. And because it affects the gain and the phase significantly, it usually can't be ignored — certainly not if it lies *below* the crossover frequency (i.e., at a lower frequency). We will account for it later by just canceling it out with a pole.

Feedback-Stage Transfer Functions

We can now lump the entire feedback section, including the voltage divider, error amplifier, and the compensation network. However, depending on the *type of error amplifier used*, these must be evaluated rather differently. In Figure 12.12, we have shown two possible error amplifiers often used in power converters.

The analysis is as follows:

- The error amplifier can be a simple voltage-to-voltage amplification device, that is, the traditional "op-amp" (operational amplifier). This type of op-amp requires *local* feedback (between its output and inputs) to make it stable. Under steady DC conditions, both the input terminals are virtually at the same voltage level. This determines the output voltage setting. But, as discussed previously, though both resistors of the voltage divider affect the DC level of the converter's output, from the AC point of view, *only the upper resistor enters the picture*. So the lower resistor is considered just a DC *biasing resistor*, and therefore we usually ignore it in control loop (AC) analysis.
- The error amplifier can also be a voltage-to-current amplification device, that is, the "gm op-amp" (operational transconductance amplifier, or "OTA"). This is an open-loop amplifier stage with no local feedback — the loop is, in effect, completed externally. The end result still is that the voltage at its input terminals returns to the same voltage (just like a regular op-amp). If there is any difference in voltage between its input pins "ΔV," it converts that into a current ΔI flowing out of its output pin (determined by its transconductance gm = $\Delta I/\Delta V$). Thereafter, since there is an impedance Z_O connected from the output of this op-amp to ground, the *voltage* at the output pin of this error amplifier (i.e., the voltage across Z_O — also called the control voltage) changes by an amount equal to $\Delta I \times Z_O$. *For the gm op-amp, both Rf2 and Rf1 enter into the AC analysis*, because they together determine the error voltage at the pins, and therefore the current at the output of the op-amp. Note that the divider can in this case be treated as a simple (step-down) *gain block* of Rf1/(Rf1 + Rf2) (using the terminology of Figure 12.10), cascaded with the gm op-amp stage that follows.

Note: We may have noticed that we always use the inverting *terminal of the error amplifier for applying the feedback voltage. The intuitive reason for that is that an* inverting *op-amp has a DC gain of Rf/Rin, where Rf is the feedback resistor (from the output of the op-amp to its negative input terminal) and Rin is the resistor between its inverting terminal and the input-voltage source. So, the output of an inverting op-amp can be made*

Transfer Functions of Possible Error Amplifiers

This is AC (change) analysis. Therefore V_{REF} is being ignored below as it is a biasing level only
By definition, transfer function ("H(s)") is output/input $= V_{CONT}/V_O$

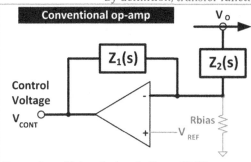

Conventional op-amp

V_O

Control Voltage V_{CONT}

Rbias

V_{REF}

Inverting Op-Amp:

$$\frac{V_{CONT}}{V_O} = -\frac{Z_1(s)}{Z_2(s)}$$

Therefore, transfer function (ignoring sign) is

$$H(s) = \frac{V_{CONT}}{V_O} = \frac{Z_1(s)}{Z_2(s)}$$

Magnitude of this is the "gain" of the error amp

Comparing with terminology in Figure 12-10:
Upper resistor Rf2 is in general Z_2 (s). The lower resistor Rf1 is "Rbias" here. However Rf1 is only a DC-biasing resistor --- it does not appear in the AC analysis, and is therefore not included in the transfer function above.

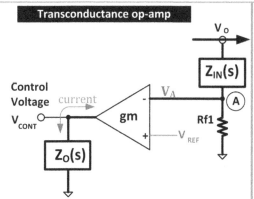

Transconductance op-amp

V_O

$Z_{IN}(s)$

Control Voltage *current* V_{CONT}

V_A

A

gm

Rf1

V_{REF}

$Z_O(s)$

Voltage at Point A is: $V_A = \dfrac{Rf1}{Rf1 + Z_{IN}(s)} \times V_O$

By definition of transconductance "g_m" (OR "gm")
(ignoring sign)

$$g_m = \frac{\text{Output Current}}{\text{Input Voltage}} = \frac{\dfrac{V_{CONT}}{Z_O(s)}}{V_A}$$

Simplifying, the transfer function (ignoring sign) is

$$H(s) = \frac{V_{CONT}}{V_O} = \frac{Rf1}{Rf1 + Z_{IN}(s)} \times g_m \times Z_O(s)$$

Usually, $Z_{IN}(s)$ is just a plain resistor (Rf2)

Comparing with terminology in Figure 12-10:
Upper resistor Rf2 is in general Z_{IN} (s). The lower resistor Rf1 is not just a DC-biasing resistor --- it <u>does</u> affect the AC signal at the input, and through gm it affects the output current, and thereby the output voltage V_{CONT}. So Rf1 is included in the transfer function above.

Conclusions:

A) If we are using a transconductance op-amp, only the *ratio* of the feedback resistors is important. We could for example have a combination of 1K/4K or 10K/40K, and so on. They would all create the same gain (attenuation), and the gain-phase plot would **not** change.

B) If we are using a conventional op-amp, the upper resistor **does** affect the gain-phase plot. If we change that, we will get entirely different gain-phase results. So just keeping the ratio unchanged does not keep the gain-phase plot unchanged in this case.

C) In adjustable regulators with a conventional op-amp, if we want to change the output voltage, it is best to change the **lower** feedback resistor, keeping the upper resistor unchanged. That way, the DC-biasing will change, but not the gain-phase (AC) characteristics of the feedback section.

Figure 12.12: Possible feedback stages and some important conclusions in their application.

smaller than its input, if so desired (i.e., gain < 1). Whereas, a non-inverting op-amp has a DC gain of 1 + (Rf/Rin), where Rin in this case is the resistor between its inverting terminal and ground. So, its output will always be greater than its input (gain > 1). The resulting restriction has been known to cause some strange and embarrassing situations in the field, especially under abnormal conditions. Therefore, we will almost never see the feedback pin of an IC as being the noninverting input of the error amp.

Lastly, note that by just using an *inverting* error amplifier, we have, in effect, also applied a −180° phase shift "right off the bat"! We will see in the following section that this increases the possibility of oscillations by itself.

Closing the Loop

We are now in a position to start tying up all the loose ends. For each of the three topologies, we now know both the forward (plant) transfer function $G(s)$ (control-to-output) and the general form of the feedback transfer function $H(s)$. Going back to the basic equation for the closed loop transfer function

$$\frac{G(s)}{1 + G(s)H(s)} \quad \text{(general closed-loop transfer function)}$$

we see that it will "explode" if

$$G(s)H(s) = -1$$

But $G(s)H(s)$ is simply the transfer function for a signal going through the $G(s)$ block, and then through the $H(s)$ block, that is, the *open-loop* transfer function. We know that the gain is the magnitude of the transfer function (using $s = j\omega$), and its phase angle is its argument. Let us calculate what these are for the transfer function −1 above.

$$\text{Gain} = |-1| = 1 \quad \text{(magnitude)}$$

$$\text{Phase} = \varphi = \tan^{-1}\left(\frac{\text{Im}}{\text{Re}}\right) = \tan^{-1}\left(\frac{0}{-1}\right) = \tan^{-1}(0) = 180° \quad \text{(argument)}$$

Note: When doing the $\tan^{-1}$ operation, we may need to visualize *where the number is actually located in the complex plane. For example, in this case, tan of 0° and tan of 180° both are zero, and we wouldn't have known which of these angles is the right answer — unless we actually* visualized *the number in the complex plane. In our case, since the number was minus 1, we correctly placed it at 180° instead of 0°.*

So, we see that **the system is unstable if a disturbance (of certain frequency) goes through the plant and feedback blocks, and returns with 180° phase shift and with exactly the same magnitude.**

There are two things surprising us here:

(a) We intuitively imagine that a signal reinforces itself only if it returns with the same phase, that is, 360°. So, why are we getting reinforcement with just 180° above? That is because the summing block that follows has one negative and one positive input (it represents a *negative feedback* system). But that also implies that another 180° shift occurs right here, that is, after the signal leaves the block designated "*H*(*s*)" in Figure 12.9 for example. So if the feedback block creates a phase shift of 180°, in effect, we get a total shift of 360°. That explains the positive reinforcement. As we mentioned earlier, the negative feedback functionality (180° shift right off the bat) is automatically included by applying the feedback voltage to the inverting pin of the error amplifier as is conventionally done.

(b) Why do we get positive reinforcement if, not only the phase shift is 180° (a total of 360°), but the returning signal is also of *exactly the same magnitude as the cause*? This is truly hard to visualize. It may become clearer if we try to draw vector diagrams in the complex plane. We will then see that only if the two above-mentioned conditions are satisfied, can a *stable* vector diagram result (i.e., we get a sustained oscillation). Otherwise we don't.

In a typical gain versus frequency plot, we will see that a gain of 1 usually occurs at only one specific frequency, and this is called the "crossover frequency" (see Figure 12.5). Beyond this point the gain becomes less than 1 (i.e., falls below the 0-dB axis).

The stability criterion above is therefore equivalent to saying that ***the phase shift of the open-loop transfer function should not be equal to 180°*** (or $-180°$) at the crossover frequency.

But we also need to ensure a certain *margin of safety*. This can be expressed in terms of the degrees of phase angle *short of 180°*, at the crossover frequency. This safety margin is called the "phase margin." But we could also talk about the safety margin in terms of the amount of gain *below 0-dB level* at the point where we get 180° phase shift. These are shown in Figure 12.13.

How much phase margin is enough? In theory, even an overall phase shift of $-179°$ (i.e., a phase margin of 1°) would not produce full instability — though there would certainly be a lot of ringing at every transient, and it would at best be very, very, marginally stable. Component tolerances, temperature variations, and even small changes in the application conditions can change the loop characteristics significantly, ushering in full-blown instability.

It is generally recommended that the phase lag introduced by the successive *G* and *H* blocks be about *45° short of $-180°$*, that is, an overall phase lag of $-135°$. That gives us a phase margin of 45°. But this target phase margin is stated *for nominal conditions*. In the

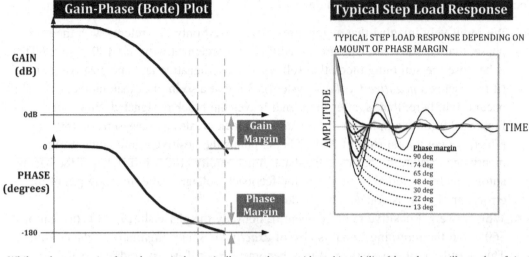

Gain-Phase (Bode) Plot

Typical Step Load Response

TYPICAL STEP LOAD RESPONSE DEPENDING ON
AMOUNT OF PHASE MARGIN

Phase margin
····· 90 deg
····· 74 deg
····· 65 deg
····· 48 deg
····· 30 deg
····· 22 deg
····· 13 deg

While a phase margin of one degree is theoretically enough to avoid total instability (though we will get a lot of ringing on transients), in practice a minimum (worst-case) margin of 20 to 30 degrees is recommended. Phase and gain margins are not independent. Typically, 30 degrees of phase margin yields about 5 dB of gain margin, and a gain margin of 10 dB corresponds to about 60 degrees phase margin.

Bench Measurement of Bode Plot

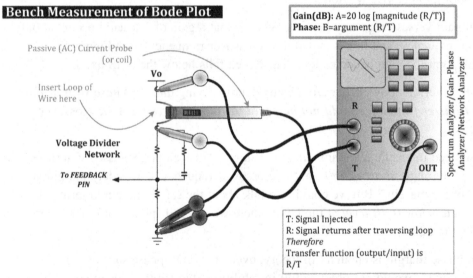

Gain(dB): A=20 log [magnitude (R/T)]
Phase: B=argument (R/T)

Passive (AC) Current Probe
(or coil)

Vo

Insert Loop of
Wire here

Voltage Divider
Network

*To FEEDBACK
PIN*

R

T   OUT

Spectrum Analyzer/Gain-Phase
Analyzer/Network Analyzer

T: Signal Injected
R: Signal returns after traversing loop
Therefore
Transfer function (output/input) is
R/T

Check SW Node on Scope while test is in Progress: The "jitter" should not be more than ~10%
of the time period of the switching cycle, and not less than ~2%. Otherwise adjust Output
Amplitude/Attenuation settings on Spectrum Analyzer.

Figure 12.13: Stability margins, measurement and responses.

worst case, we expect a minimum phase margin of around 30°. On the other hand, a phase margin of say 90° may certainly be considered "stable" because we will see no ringing as indicated in Figure 12.13, but it is usually not desirable either. Under transients, the correction may be very sluggish, and so the initial output overshoot/undershoot may be rather severe as discussed previously. A phase margin of 45° would generally be seen to cause just one or two cycles of ringing, and the overshoot/undershoot would also be minimal. However, note that besides phase margin, the crossover frequency also affects the actual step response. The "*Q*" of any second-order pole located near the crossover point can affect the phase margin significantly and thereby cause ringing. So generally, it is said that we should also ensure that *Q* of the LC is between 0.5 and 0.707.

> *Note: Under very* large *line or load steps, we will actually no longer be operating in the domain of the "small-signal" analysis, which we have been performing so far. In that case, the* initial *overshoot/undershoot at the output is almost completely determined simply by how large a bulk capacitance we have placed at the output. That capacitance is needed to "hold" the output steady, till the control loop can enter the picture and help stabilize the output. This can determine the size of the output capacitor. See the detailed solved example in Chapter 19.*

Criteria and Strategy for Ensuring Loop Stability

We should remember that phase angle can start changing gradually — starting at a frequency *even 10 times lower* than where the pole or zero may actually reside. We have also seen that a second-order double-pole (−2 slope with *two* reactive components) can cause a very sudden phase shift of about 180° at the resonant frequency if the *Q* is very high. Therefore, in practice, it is almost impossible to estimate the phase at a certain frequency, with certainty — nor therefore the phase margin — unless a certain *strategy* is followed.

One of the most popular (and simple) approaches to ensuring loop stability is as follows:

- Ensure that the *open-loop gain crosses the 0-dB axis with a −1 slope*.
- The integrator already provides this −1 slope.
- The LC post-filter poses the biggest problem, since after its LC break frequency, the second-order LC pole asks the open-loop gain to fall by an additional −2 slope (taking its slope to −3). The LC pole therefore needs to be canceled out — that is best done by introducing **two single-order zeros exactly at the location of the LC pole**.
- We also want to *maximize the bandwidth* to achieve quick response to extremely sudden load or line transients. By sampling theory, we know that we certainly need to set the crossover frequency to less than half the switching frequency.
- We also need to ensure that the crossover frequency is kept well below any *troublesome* poles or zeros — like the RHP zero for example. Keep in mind that the RHP zero occurs in CCM for the Boost and the Buck-Boost topologies, irrespective of whether we are using voltage-mode or current-mode control, and at any duty cycle. We should also try to avoid the "subharmonic instability pole" which occurs at half the switching frequency — in CCM in the Buck, Boost, and Buck-Boost topologies, when using current-mode control with $D > 50\%$. This is discussed later.

So, in practice, most designers *set the crossover frequency at about one-sixth the switching frequency* (for voltage-mode control).

In Figure 12.13, we have also presented the most common method of generating a Bode plot and measuring stability margins on the bench. Obviously, more exotic techniques are required for injecting the disturbance if the voltage divider location shown in Figure 12.13 is not available for us to insert a current loop or a small resistor.

Plotting the Open-Loop Gain for the Three Topologies

Now we want to finally start plotting the gain and phase of the *open-loop transfer function* $T(s) = G(s)H(s)$ since we know that that is the function critical to ensuring stability. As background, we have understood math in the log-plane and also the interaction of poles and zeros, whether they are on the same transfer function plot, or on more than one cascaded transfer function plots (which are being combined to provide the open-loop gain). We have also derived the plant transfer functions of all topologies and understood the basic strategy for ensuring stability. We also know that the integrator is a fundamental building block in the feedback path, without which we will have very inadequate DC regulation. Now we can put it all together.

On the left side of Figure 12.14, we have only a pure integrator in the feedback path of a Buck. We realize that the open-loop gain crosses over on the 0-dB axis with −3 slope, which is not as per our strategy. Therefore, we actually ignore the results here and introduce two zeros in the feedback loop, exactly at the location of the LC pole. This is as per our basic strategy. We see that now, indeed, the open-loop gain crosses over with a −1 slope. This is acceptable. In the same figure, we have provided the overall DC gain of the power stage and the crossover frequency (bandwidth) of the feedback loop. All this will lead to adequate phase margin. We have completed the Buck analysis. Later we will show what specific circuitry is required to place the two zeroes exactly where we have declared them to be.

In Figure 12.15, we have carried out the same calculation steps for the Boost and the Buck-Boost. Here we are assuming we have crossed over at a low-enough frequency so that the RHP zero is well outside the bandwidth of the loop, and therefore cannot affect the plots as shown.

We realize that after canceling out the LC pole, we are left with a simple −1 plot for the open-loop transfer function. This is just the transfer function of the integrator, *shifted upward by the amount "a"* as indicated in Figures 12.14 and 12.15. Note that "*a*" is the DC gain of the power stage. We already have the equations for "*a*" embedded in the two

Feedback Stage consists of just an integrator

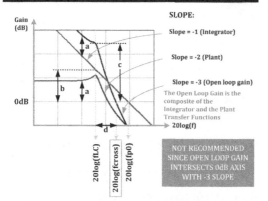

SLOPE:

Slope = -1 (Integrator)

Slope = -2 (Plant)

Slope = -3 (Open loop gain)
The Open Loop Gain is the composite of the Integrator and the Plant Transfer Functions

NOT RECOMMENDED SINCE OPEN LOOP GAIN INTERSECTS 0dB AXIS WITH -3 SLOPE

Buck Control to Output (Plant) Transfer Function as derived previously:

$$G(s) = \frac{1}{V_{RAMP}} \times V_{IN} \times \frac{1/LC}{s^2 + s\left(1/RC\right) + 1/LC} \qquad \textbf{BUCK}$$

The constant (frequency independent) part is "a" (in decibels)

$$a = 20\log\left(\frac{V_{IN}}{V_{RAMP}}\right)$$

From Figure 12.6 we know the gain at any frequency if we know the crossover frequency and the slope. We had derived.

$$Gain_{dB} = -20 \times m \times \log\left(\frac{fcross}{f}\right)$$

So we can find "b" as follows (since m=-1 for the integrator)

$$b = 20\log\frac{fp0}{fLC}$$

By geometry, we have c = a + b (ignoring LC peaking)

By geometry we also have $|Slope| = |-3| = \dfrac{c}{d} = \dfrac{a+b}{d}$

Therefore a+b=3d, where d=20log(fcross) – 20log(f_{LC})   **a+b = 3d**

Therefore, finally

$$20\log\left(\frac{V_{IN}}{V_{RAMP}}\right) + 20\log\frac{fp0}{fLC} = 3[20 \times \log(fcross) - 20 \times \log(f_{LC})]$$

$$20\log\left[\frac{V_{IN}}{V_{RAMP}} \times \frac{fp0}{fLC}\right] = 20\log\left[\left(\frac{fcross}{f_{LC}}\right)^3\right]$$

$$\frac{V_{IN}}{V_{RAMP}} \times \frac{fp0}{fLC} = \left(\frac{fcross}{f_{LC}}\right)^3$$

Solving for fp0 and equating it as per Figure 12-5

$$fp0 = \frac{V_{RAMP}}{V_{IN}} \times \left(\frac{fcross}{f_{LC}}\right)^3 \times f_{LC} \equiv \frac{1}{2\pi \times RC}$$

$$fcross = \left[\frac{V_{IN} \times f_{LC}^2}{2\pi \times RC \times V_{RAMP}}\right]^{1/3} \quad \text{(Crossover Frequency)}$$

Feedback Stage consists an integrator + 2 Zeros positioned exactly where the LC Pole is

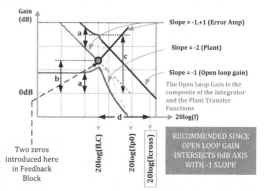

Slope = -1,+1 (Error Amp)

Slope = -2 (Plant)

Slope = -1 (Open loop gain)
The Open Loop Gain is the composite of the Integrator and the Plant Transfer Functions

RECOMMENDED SINCE OPEN LOOP GAIN INTERSECTS 0dB AXIS WITH -1 SLOPE

Two zeros introduced here in Feedback Block

Buck Control to Output (Plant) Transfer Function as derived previously:

$$G(s) = \frac{1}{V_{RAMP}} \times V_{IN} \times \frac{1/LC}{s^2 + s\left(1/RC\right) + 1/LC} \qquad \textbf{BUCK}$$

The constant (frequency independent) part is "a" (in decibels)

$$a = 20\log\left(\frac{V_{IN}}{V_{RAMP}}\right) \qquad \text{DC Gain of Power Stage}$$

From Figure 12.6 we know the gain at any frequency if we know the crossover frequency and the slope. We had derived.

$$Gain_{dB} = -20 \times m \times \log\left(\frac{fcross}{f}\right)$$

So we can find "b" as follows (since m=-1 for the integrator)

$$b = 20\log\frac{fp0}{fLC}$$

By geometry, we have c = a + b (ignoring LC peaking)

By geometry we also have $|Slope| = |-1| = \dfrac{c}{d} = \dfrac{a+b}{d}$

Therefore a+b=d, where d=20log(fcross) – 20log(f_{LC})   **a+b = d**

Therefore, finally

$$20\log\left(\frac{V_{IN}}{V_{RAMP}}\right) + 20\log\frac{fp0}{fLC} = [20 \times \log(fcross) - 20 \times \log(f_{LC})]$$

$$20\log\left[\frac{V_{IN}}{V_{RAMP}} \times \frac{fp0}{fLC}\right] = 20\log\left[\left(\frac{fcross}{f_{LC}}\right)\right]$$

$$\frac{V_{IN}}{V_{RAMP}} \times \frac{fp0}{fLC} = \left(\frac{fcross}{f_{LC}}\right)$$

Solving for fp0 and equating it as per Figure 12.5

$$fp0 = \frac{V_{RAMP}}{V_{IN}} \times \left(\frac{fcross}{f_{LC}}\right) \times f_{LC} \equiv \frac{1}{2\pi \times RC}$$

$$fcross = \left[\frac{V_{IN}}{2\pi \times RC \times V_{RAMP}}\right] \quad \text{(Crossover Frequency)}$$

MUCH SIMPLER!

Figure 12.14: Stabilizing a Buck converter and calculating its crossover frequency and DC gain of power stage (recommended method is on the right side).

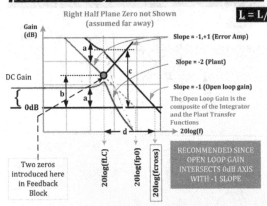

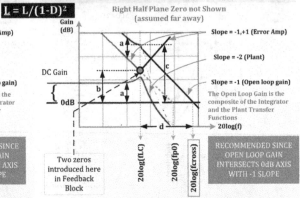

Boost Control to Output (Plant) Transfer Function as derived previously:

$$G(s) = \frac{1}{V_{RAMP}} \times \frac{V_{IN}}{(1-D)^2} \times \frac{\frac{1}{LC} \times \left(1 - s\left(\frac{L}{R}\right)\right)}{s^2 + s\left(\frac{1}{RC}\right) + \frac{1}{LC}}$$

BOOST

The constant (frequency independent) part is "a" (in decibels)

$$a = 20\log\left(\frac{V_{IN}}{V_{RAMP}(1-D)^2}\right)$$ DC Gain of Power Stage

From Figure 12.6 we know the gain at any frequency if we know the crossover frequency and the slope. We had derived:

$$Gain_{dB} = -20 \times m \times \log\left(\frac{fcross}{f}\right)$$

So we can find "b" as follows (since m=-1 for the integrator)

$$b = 20\log\frac{fp0}{fLC}$$

By geometry, we have c = a + b (ignoring LC peaking)

By geometry we also have $|Slope| = |-1| = \dfrac{c}{d} = \dfrac{a+b}{d}$

Therefore a+b=d, where d=20log(fcross) − 20log(f_{LC})   **a+b = d**

Therefore, finally

$$20\log\left(\frac{V_{IN}}{V_{RAMP}(1-D)^2}\right) + 20\log\frac{fp0}{fLC} = \left[20 \times \log(fcross) - 20 \times \log(f_{LC})\right]$$

$$20\log\left[\frac{V_{IN}}{V_{RAMP}(1-D)^2} \times \frac{fp0}{fLC}\right] = 20\log\left[\left(\frac{fcross}{f_{LC}}\right)\right]$$

So, $\dfrac{V_{IN}}{V_{RAMP}(1-D)^2} \times \dfrac{fp0}{fLC} = \left(\dfrac{fcross}{f_{LC}}\right)$

Solving for fp0 and equating it as per Figure 12.5

$$fp0 = \frac{V_{RAMP}(1-D)^2}{V_{IN}} \times \left(\frac{fcross}{f_{LC}}\right) \times f_{LC} = \frac{1}{2\pi \times RC}$$

$$fcross = \left[\frac{V_{IN}}{2\pi \times RC \times V_{RAMP}(1-D)^2}\right] \text{ (Crossover Frequency)}$$

Crossover Freq

Buck-Boost Control to Output (Plant) Transfer Function as derived previously:

$$G(s) = \frac{1}{V_{RAMP}} \times \frac{V_{IN}}{(1-D)^2} \times \frac{\frac{1}{LC} \times \left(1 - s\left(\frac{LD}{R}\right)\right)}{s^2 + s\left(\frac{1}{RC}\right) + \frac{1}{LC}}$$

BUCK-BOOST

The constant (frequency independent) part is "a" (in decibels)

$$a = 20\log\left(\frac{V_{IN}}{V_{RAMP}(1-D)^2}\right)$$ DC Gain of Power Stage

From Figure 12.6 we know the gain at any frequency if we know the crossover frequency and the slope. We had derived:

$$Gain_{dB} = -20 \times m \times \log\left(\frac{fcross}{f}\right)$$

So we can find "b" as follows (since m=-1 for the integrator)

$$b = 20\log\frac{fp0}{fLC}$$

By geometry, we have c = a + b (ignoring LC peaking)

By geometry we also have $|Slope| = |-1| = \dfrac{c}{d} = \dfrac{a+b}{d}$

Therefore a+b=d, where d=20log(fcross) − 20log(f_{LC})   **a+b = d**

Therefore, finally

$$20\log\left(\frac{V_{IN}}{V_{RAMP}(1-D)^2}\right) + 20\log\frac{fp0}{fLC} = \left[20 \times \log(fcross) - 20 \times \log(f_{LC})\right]$$

$$20\log\left[\frac{V_{IN}}{V_{RAMP}(1-D)^2} \times \frac{fp0}{fLC}\right] = 20\log\left[\left(\frac{fcross}{f_{LC}}\right)\right]$$

So, $\dfrac{V_{IN}}{V_{RAMP}(1-D)^2} \times \dfrac{fp0}{fLC} = \left(\dfrac{fcross}{f_{LC}}\right)$

Solving for fp0 and equating it as per Figure 12.5

$$fp0 = \frac{V_{RAMP}(1-D)^2}{V_{IN}} \times \left(\frac{fcross}{f_{LC}}\right) \times f_{LC} = \frac{1}{2\pi \times RC}$$

$$fcross = \left[\frac{V_{IN}}{2\pi \times RC \times V_{RAMP}(1-D)^2}\right] \text{ (Crossover Frequency)}$$

Crossover Freq

Figure 12.15: Stabilizing Boost and Buck-Boost converters and calculating their crossover frequency and DC gain of power stage.

figures. We can thus calculate the effect of this vertical shift on the crossover frequency, as per the math presented in the lower half of Figure 12.6, namely

$$f1 = f2 \times 10^{(\Delta(\text{Gain}_{\text{dB}}))/(20 \times m)} \quad (\text{Note: } f2 \geq f1)$$

Here $f1$ is fp0 and $f2$ is fcross. We therefore get

$$\text{fp0} = \text{fcross} \times 10^{-(a/20)}$$

where for a Buck, for example (Figure 12.14), we have

$$a = 20 \log \left(\frac{V_{\text{IN}}}{V_{\text{RAMP}}} \right)$$

We therefore get

$$\text{fp0} = \text{fcross} \times 10^{-((20\log(V_{\text{IN}}/V_{\text{RAMP}}))/20)} = \text{fcross} \times 10^{-\log(V_{\text{IN}}/V_{\text{RAMP}})} = \text{fcross} \times \left(\frac{V_{\text{RAMP}}}{V_{\text{IN}}} \right)$$

(since $10^{\log(x)} = x$)

So,

$$\text{fp0} = \text{fcross} \times \left(\frac{V_{\text{RAMP}}}{V_{\text{IN}}} \right) \quad (Buck)$$

Similarly, for the Boost and the Buck-Boost, we get

$$\text{fp0} = \text{fcross} \times \left(\frac{V_{\text{RAMP}}(1 - D)^2}{V_{\text{IN}}} \right) \quad (Boost\ and\ Buck\text{-}Boost)$$

This tells us where to set fp0 for the integrator section, when targeting a certain crossover frequency fcross for the open-loop gain. We will give a numerical example later.

Our overview of compensation analysis seems complete. However, there is one last complication still remaining. Besides two zeros, we may need *at least one pole* from our compensation network (besides the pole-at-zero of the integrator section). This is for canceling out the "ESR-zero" coming from the output capacitor. We have been ignoring this particular zero so far, but it is time to take a look at it now.

The ESR-Zero

We ignored the ESR of the output capacitor in Figures 12.14 and 12.15, and also in the derivation of all the transfer functions carried out earlier. For example, we had earlier provided the following control-to-output transfer function for a Buck:

$$\frac{V_{\text{IN}}}{V_{\text{RAMP}}} \times \frac{1}{(s/\omega_0)^2 + (1/Q)(s/\omega_0) + 1} \quad (Buck:\ control\text{-}to\text{-}output\ transfer\ function)$$

where $\omega_0 = 1/\sqrt{(LC)}$. The ESR-zero adds an additional term to the numerator. A full analysis shows that the control-to-output transfer function now becomes

$$\frac{V_{IN}}{V_{RAMP}} \times \frac{(s/\omega_{ESR}) + 1}{(s/\omega_0)^2 + (1/Q)(s/(\omega_0)) + 1} \qquad (Buck: complete\ control\text{-}to\text{-}output\ transfer\ function)$$

where $\omega_{ESR} = 1/((ESR) \times C)$ is the frequency (in radians per second) at which the ESR-zero is located. Judging by the sign in front of the s-term in the numerator, this is a "well-behaved" (left-half-plane) zero. But it does try to cause an increase in the slope of the open-loop transfer function by $+1$ and may thus even prevent crossover from occurring properly, besides affecting the phase significantly too. It is also based on a parasitic, which is not a guaranteed parameter. So, it is usually considered a nuisance. Though in some simpler compensation network types, *the ESR-zero may even be counted upon to provide one of the two zeros required to cancel out the LC double-pole*, as discussed previously. It may not be at the "right place," but it can still work. In general, however, the ESR-zero is considered avoidable or worth getting rid of (by a pole).

In the best case, the ESR will be very small, and so its zero will be far away (at a very high frequency). We can then simply ignore it. That situation arises when we use modern ceramic output caps for example. Otherwise, the preferred strategy is to place a pole at exactly the location of the ESR-zero, thereby canceling it out.

High-Frequency Pole

We have seen that a full-blown compensation network needs to provide

(a)   a pole-at-zero (integrator function)
(b)   two zeros at the location of the LC double-pole
(c)   one pole at the location of the ESR-zero
(d)   *a high-frequency pole*

Where did the last one come from? In general, for making the control loop less sensitive to high-frequency switching noise, *designers often put another pole roughly at about 10 times the crossover frequency (some recommend half the switching frequency)*. So, now the gain will cross the 0-dB axis with a slope of -1 as per our strategy, but at higher frequencies it will suddenly drop off more rapidly, close to a -2 slope. That will improve the Gain margin shown in Figure 12.13.

Why do some designers pick *10 times* the crossover frequency above? Because the phase introduced by this new high-frequency pole will actually start making itself felt at one-tenth the frequency of the pole, and we didn't want to adversely impact the phase angle in the vicinity of the crossover frequency (i.e., the phase margin). But we realize that the resulting open-loop gain plot is just a vertically shifted -1 plot coming from the integrator section. We also know that a single-pole provides $90°$ phase shift. So, we could be left with a phase margin of $180 - 90 = 90°$, which may be considered sloppy. Therefore, some designers try to move this high-frequency pole to a much lower frequency, *just a little higher than the crossover frequency*, to deliberately reduce the phase margin in a calculated manner — closer to the target value of $45°$.

Designing a Type 3 Op-Amp Compensation Network

Three types of error amplifier compensation schemes are used most often — called the Types $1-3$ in order of increasing complexity and flexibility. The former two are just a subset of the latter, so we will now just do a Type 3 compensation to demonstrate the full scope (though usually, Type 2 compensation should suffice).

The transfer function of a Type 3 error amplifier as shown in Figure 12.16 can be worked out easily in the manner we did before. It is given in detail in the figure, but it can also be written more generically as follows:

$$\frac{\omega p0}{s} \times \frac{(s/\omega z1) + 1(s/\omega z2) + 1}{(s/\omega p1) + 1(s/\omega p2) + 1} \quad \text{(\textit{Type 3 feedback transfer function})}$$

where $\omega p0 = 2\pi(\text{fcross})$, $\omega z1 = 2\pi(\text{fz1})$, and so on. Note that we are ignoring the minus sign in front of this transfer function, as we are separating out the $180°$ phase shift inherent in negative feedback systems.

There are two poles "$p1$" and "$p2$" (besides the pole-at-zero "$p0$") and two zeros "$z1$" and "$z2$" provided by this compensation. Note that several of the components involved play a *dual role* in determining the poles and zeros. So, the calculation can become fairly cumbersome and iterative. But a valid simplifying assumption that can be made is that **C1 is much greater than C3**. So the locations of the poles and zeros are finally

$$\text{fp0} = \frac{1}{2\pi \times R_1(C_1 + C_3)} \approx \frac{1}{2\pi \times R_1 C_1}$$

$$\text{fp1} = \frac{1}{2\pi \times R_3 C_2}$$

$$\text{fp2} = \frac{1}{2\pi \times R_2(C_1 C_3/C_1 + C_3)} = \frac{1}{2\pi \times R_2}\left(\frac{1}{C_1} + \frac{1}{C_3}\right) \approx \frac{1}{2\pi \times R_2 C_3}$$

$$\text{fz1} = \frac{1}{2\pi \times (R_1 + R_3)C_2}$$

$$\text{fz2} = \frac{1}{2\pi \times R_2 C_1}$$

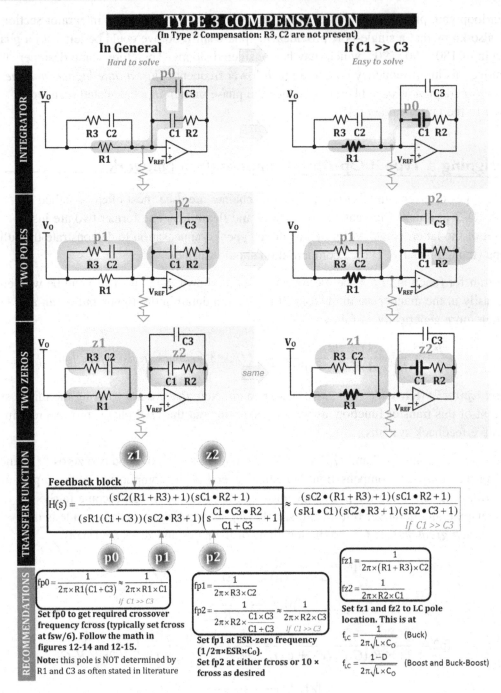

Figure 12.16: Conventional type 3 compensation using conventional voltage Op-Amp.

Note that for convenience, the reference designators of the components have changed somewhat in this section. In particular, *what we are now calling "R_1" was "Rf2" when we previously discussed the voltage divider*. Similarly, the gray unnamed resistor in Figure 12.16 was previously called "Rf1."

We can also solve for the values of the components (**with the approximation $C_1 \gg C_3$**). We get

$$C_2 = \frac{1}{2\pi \times R_1} \left(\frac{1}{\text{fz1}} - \frac{1}{\text{fp1}} \right)$$

$$R_2 = R_1 \frac{\text{fp0}}{\text{fz2}}$$

$$C_3 = \frac{1}{2\pi \times (R_2\text{fp2} - R_1\text{fp0})}$$

$$R_3 = \frac{R_1 \times \text{fz1}}{\text{fp1} - \text{fz1}}$$

Let us take up a practical example to show how to proceed in designing a feedback loop with this type of compensation.

Example:

Using a 300-kHz synchronous Buck controller we wish to step-down from 15 V to 1 V. The load resistor is 0.2 Ω (5 A). The PWM ramp is 2.14 V as per the datasheet of the part. The selected inductor is 5 µH, and the output capacitor is 330 µF, with an ESR of 48 mΩ.

We know that the plant gain at DC for a Buck is $V_{IN}/V_{RAMP} = 7.009$. Therefore, $(20 \times \log)$ of this gives us 16.9 dB. The LC double-pole is at

$$f_{LC} = \frac{1}{2\pi \times \sqrt{LC}} = \frac{1}{2\pi \times \sqrt{5 \times 10^{-6} \times 330 \times 10^{-6}}} \Rightarrow 3.918 \text{ kHz}$$

(note that for a Boost and Buck-Boost, the location of the LC pole is given in Figure 12.16, based on the canonical model). We want to set the crossover frequency of the open-loop gain at one-sixth the switching frequency, that is, at 50 kHz. Therefore, we can solve for the integrator's fp0 and thereby its "RC," by using the equation presented earlier.

$$\text{fp0} = \frac{V_{RAMP}}{V_{IN}} \times \text{fcross} \equiv \frac{1}{2\pi \times RC}$$

So, in our case,

$$R_1 C_1 = \frac{V_{IN}}{2\pi \times V_{RAMP} \times \text{fcross}} = \frac{15}{2\pi \times 2.14 \times 50 \times 10^3} = 2.231 \times 10^{-5} \text{ s}^{-1}$$

If we have selected R_1 as say 2 kΩ, C_1 is then

$$C_1 = \frac{2.231 \times 10^{-5}}{2 \times 10^3} \Rightarrow 11.16 \text{ nF}$$

The crossover frequency of the integrator section of the op-amp is

$$\text{fp0} = \frac{1}{2\pi \times R_1 C_1} = \frac{10^5}{2\pi \times 2.231} \Rightarrow 7.133 \text{ kHz}$$

The ESR-zero is at

$$\text{fesr} = \frac{1}{2\pi \times 48 \times 10^{-3} \times 330 \times 10^{-6}} \Rightarrow 10.05 \text{ kHz}$$

The required placement of zeros and poles is

$$\text{fz1} = \text{fz2} = 3.918 \text{ kHz} \quad \text{(at the LC pole location)}$$

$$\text{fp1} = \text{fesr} = 10.05 \text{ kHz} \quad \text{(pole to cancel ESR zero)}$$

$$\text{fp2} = 10 \times \text{fcross} = 500 \text{ kHz} \quad \text{(can set fp2} = \text{fcross for better results, see later)}$$

The components required to make this happen are

$$C_2 = \frac{1}{2\pi \times R_1}\left(\frac{1}{\text{fz1}} - \frac{1}{\text{fp1}}\right) = \frac{1}{2\pi \times 2 \times 10^6}\left(\frac{1}{3.918} - \frac{1}{10.05}\right) \Rightarrow 12.4 \text{ nF}$$

$$R_2 = R_1 \frac{\text{fp0}}{\text{fz2}} = 2 \times 10^3 \times \frac{7.133}{3.918} \Rightarrow 3.641 \text{ k}\Omega$$

$$C_3 = \frac{1}{2\pi \times (R_2 \text{fp2} - R_1 \text{fp0})} = \frac{10^{-6}}{2\pi \times (3.641 \times 500 - 2 \times 7.133)} \Rightarrow 88.11 \text{ pF}$$

$$R_3 = \frac{R_1 \times \text{fz1}}{\text{fp1} - \text{fz1}} = \frac{2 \times 10^3 \times 3.918}{10.05 - 3.918} \Rightarrow 1.278 \text{ k}\Omega$$

We already know C_1 is 11.16 nF and $R1$ was selected to be 2 kΩ. The results of this example are plotted in Figure 12.17.

Note that for a Boost or Buck-Boost, the only changes required in the above analysis are

$$L \Rightarrow \frac{L}{(1-D)^2} \quad \textit{(Boost and Buck-Boost)}$$

$$V_{\text{RAMP}} \Rightarrow V_{\text{RAMP}} \times (1-D)^2 \quad \textit{(Boost and Buck-Boost)}$$

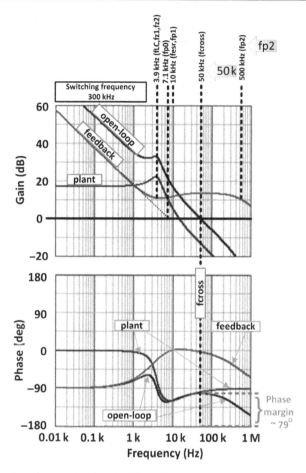

Figure 12.17: Plotting the results for the Type 3 compensation example (standard setting).

However, in the case of the Boost and Buck-Boost, we must also always ensure that the selected crossover frequency is at least an order of magnitude below the RHP zero (whose location was provided previously).

Optimizing the Feedback Loop

In Figure 12.17, we have plotted the results of the previous example, and we can see that though the crossover frequency is high enough, *the phase margin is rather too generous.* A very high phase margin may be considered "very stable" with no ringing, but the overshoot/undershoot can improve further if the phase margin is closer to 45°.

We should have realized by now that intuitively, *poles are generally responsible for making matters "worse"* — since they always introduce a phase lag, leading us closer to the danger level of $-180°$. On the other hand, *zeros boost the phase angle* (phase lead), and thereby help to cause the phase margin to increase. Therefore, to decrease the existing phase margin of $79°$ to say $45°$, we need *another pole*. The new criterion to set the high-frequency pole fp2 becomes

$$fp2 = fcross = 50 \text{ kHz}$$

We are calling this the "optimized setting" here.

We can guess that the phase shift introduced by a single-pole at its resonant frequency is $45°$, so the new phase margin should be around $79° - 45° = 34°$. We plot the gain-phase plots with this new high-frequency pole criterion (and with freshly calculated compensation component values), and we get the curve shown in Figure 12.18.

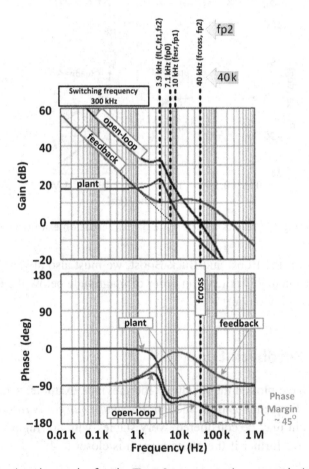

Figure 12.18: Plotting the results for the Type 3 compensation example (optimized setting).

In Figure 12.18, we see that the phase margin is now almost exactly 45°. The reason it is a little more than our initial estimate of 34° (though we desired 45°) is that the *crossover frequency has decreased slightly* to 40 kHz. It can be shown that by trying to place the high-frequency pole exactly at the crossover frequency, the crossover frequency itself shifts downward by almost exactly 20%. So the corollary to that is — *if we are starting a compensation network design in which we are going to use the high-frequency pole in this "optimized" manner, we should initially target a crossover frequency about 20% higher than we desire.*

> *Note: We can understand the lowering of the crossover frequency in the "optimized setting" case as follows. In terms of the asymptotic approximation, the open-loop gain crosses the 0-dB axis with a slope of −1, but then* immediately thereafter falls off at a *slope of −2. But since the high-frequency pole fp2 is placed very close to the crossover frequency, the gain in reality falls by 3 dB at this break-point (as compared to the asymptotic approximation). So, the actual crossover occurs a little earlier. The reason the phase is affected by almost 45° at the crossover frequency is that phase starts changing a decade below where the pole really is.*

Engineers use various other "tricks" to improve the loop response further. For example, they may "spread" the two zeros symmetrically around the LC double-pole (rather than coinciding exactly with it). One reason to put a zero (or two) slightly before the LC pole location is that the LC pole can produce a very dramatic 180° phase shift, and this can lead to "conditional stability." So, the spreading of zeros around absorbs some of the phase shift abruptness.

Conditional stability is said to occur if the phase gets rather too close to the −180° danger level at some frequency. Though oscillations do not normally occur at this point, simply because the gain is high (*crossover* is not taking place at this location), under large-signal disturbances, the gain of the converter can suddenly fall momentarily toward 0 dB, thus increasing the chance of instability. For example, if there is a very large change in line and load, the error amplifier output may "rail," that is, reach a value close to its internal supply rails. Its output transistors may then saturate, taking a comparatively long time to recover and respond. So, the gain would have effectively decreased suddenly, and it could end up crossing the 0-dB axis at the same location where the phase angle happens to be −180° — and that would meet the criterion for full-blown instability.

Input Ripple Rejection

The line-to-output transfer function of the Buck was shown to be

$$D \times \frac{(1/LC)}{s^2 + s(1/RC) + (1/LC)}$$

The plant transfer function was

$$\frac{V_{IN}}{V_{RAMP}} \times \frac{(1/LC)}{s^2 + s(1/RC) + (1/LC)}$$

We see that the *line-to-output* transfer function for the Buck is the same as its control-to-output transfer function, except that the V_{IN}/V_{RAMP} factor is replaced by D.

So, for example, if $V_{RAMP} = 2.14$ V and $D = 0.067$ (as for 1 V output from a 15 V input), then the control-to-output (plant) gain at low frequencies is

$$20 \times \log\left(\frac{V_{IN}}{V_{RAMP}}\right) = 20 \times \log\left(\frac{15}{2.14}\right) = 16.9 \text{ dB}$$

and the line-to-output transfer gain at low frequencies must be

$$20 \times \log(D) = 20 \times \log(0.067) = -23.5 \text{ dB}$$

The latter represents *attenuation*, since the response at the output is *less* than the disturbance injected into the input. But both the above-mentioned DC gains are *without feedback considered*. Alternatively, we have implicitly assumed that the error amplifier is set to a gain of 1, and there are no capacitors present anywhere in the compensation network. However, when feedback *is* present ("loop closed"), it can be shown by control loop theory that the line-to-output transfer function changes to

$$\text{Line-to-output}_{\text{with feedback}} = \left(\frac{1}{1+T}\right) \times \text{Line-to-output}_{\text{without feedback}}$$

where $T = GH$. Since T (the open-loop transfer function) at low frequencies is very large, we can write $T + 1 \approx T$. Further, since $20 \times \log(1/T) = -20 \times \log(T)$, we conclude — *at low frequencies, the additional attenuation provided, when the loop is closed, is equal to the open-loop gain*. For example, if the open-loop gain at 1 kHz is 20 dB, it attenuates a 1-kHz line disturbance by an *additional* 20 dB — over and above the attenuation already present without feedback considered. That is one reason why we are always so interested in increasing the DC gain in general (the purpose of the integrator).

For example, suppose we are interested in attenuating the 100-Hz (low-frequency) ripple component of the input voltage in an off-line power supply down to a very small value. If our crossover frequency is 500 kHz, then using the simple relationship derived in Figure 12.6, we can find the open-loop gain at 100 Hz. Here we are assuming we have carried out the recommended pole-zero cancelation compensation strategy, which leaves us with an open-loop gain plot that has a pole-at-zero type response (-1 slope). So, the gain at 100 Hz is

$$\text{Open-loop gain}_{100 \text{ Hz}} = \frac{\text{fcross}}{100 \text{ Hz}} = 500$$

Expressed in dB, this is

$$20 \times \log(\text{Open-loop gain}_{100\ \text{Hz}}) = 20 \times \log(500) = 54\ \text{dB}$$

So, the *additional* attenuation is 54 dB here. Suppose the duty cycle of the converter is 30%. Assuming it is a Forward converter with Buck-like characteristics, and its duty cycle is 30%, the line-to-output transfer function will provide a DC attenuation of $|20 \times \log(D)| = 10.5$ dB. So, by introducing feedback, the total low-frequency attenuation has increased to $54 + 10.5 = 64.5$ dB. This is equivalent to a factor of $10^{64.5/20} = 1{,}680$. So, if for example, the low-frequency ripple component at the input terminals was ± 15 V, then the output will see only $\pm 15/1680 = \pm 9$ mV of line ripple.

Load Transients

Suppose we suddenly increase the load current of a converter from 4 A to 5 A. This is a "step load" and is essentially a *nonrepetitive* stimulus. But by writing all the transfer functions in terms of s rather than just as a function of $j\omega$, we have created the framework for analyzing the response to such disturbances too. We will need to map the stimulus into the s-plane with the help of the Laplace transform, multiply it by the appropriate transfer function, and that will give us the response in the s-plane. We then apply the inverse Laplace transform and get the response with respect to time. This was the procedure symbolically indicated in Figure 12.2, and that is what we need to follow here too. However, we will not perform the detailed analysis for arbitrary load transients here, but simply provide the key equations required to do so.

The "output impedance" of a converter is the change in its output voltage due to a (small) change in the load current. With feedback not considered, it is simply the parallel combination of R, L, and C. So,

$$Z_{\text{out\_without feedback}} = R[\text{parallel}]\frac{1}{Cs}[\text{parallel}]\underline{L} \Rightarrow \frac{s\underline{L}}{1 + s(\underline{L}/R) + s^2 LC}$$

where R is the load resistance, and $\underline{L}$ is the actual inductance L for a Buck, but it is $L/(1 - D)^2$ for a Boost and a Buck-Boost.

With feedback considered, the output impedance now decreases as follows:

$$Z_{\text{out\_with feedback}} = \frac{1}{1 + T} \times Z_{\text{out\_without feedback}}$$

Even without a detailed analysis (using Laplace transform), this should tell us by how much the output voltage will *eventually* shift (settle down to), if we change the load current.

Type 1 and Type 2 Compensations

In Figure 12.19, we have also shown Type 1 and Type 2 compensation schemes (though with no particular strategy in placing the poles and zeros). These are less powerful schemes than Type 3. So, Type 3 gives us one pole-at-zero AND two poles AND two zeros, and Type 2 gives us one pole-at-zero AND one pole AND one zero. However, Type 1 gives us ONLY a pole-at-zero (simple integrator).

We know that we always need a pole-at-zero in the compensation for achieving high DC gain, good DC regulation, and low-frequency line rejection. So, the -1 slope coming from the pole-at-zero adds to the -2 slope from the double-pole of the LC-filter, and this gives us a *-3 slope* — that is, if we don't put in any more zeros and poles (as shown on the left side of Figure 12.14). But we want to intersect the 0-dB axis with a -1 overall slope. So, that means we definitely need two (single-order) zeros to force the slope to become -1.

Therefore, Type 2 compensation can also be made to work because though it provides only one zero, *we can use the zero from the ESR of the output capacitor* (despite its relative unpredictability). We remember, in Type 3, we canceled the ESR-zero out completely, citing its relative unpredictability. But now we can consider using it to our advantage, *if that is, indeed, possible*: for the Type 2 scheme to work, the ESR-zero must be located at a *lower* frequency than the intended crossover frequency.

Type 2 compensation is well suited for current-mode control, as explained later. Type 1 compensation provides only a pole-at-zero and, in fact, can only work with current-mode control, provided the ESR-zero is also below crossover.

Transconductance Op-Amp Compensation

The final stages of the analysis of voltage-mode controlled converters are reserved for the transconductance op-amp. In Figure 12.12, we had presented its transfer function generically. Now let us consider the details of implementing a compensation scheme.

We can visualize this feedback stage as a product of three cascaded transfer functions, $H1$, $H2$, and $H3$ as shown in Figure 12.19. When we plot the separate terms out as in the lower part of the figure, we see that this looks like Type 3 compensation — but in reality it is not! Because, though it provides two zeros and two poles (besides the inevitable pole-at-zero), we see a big difference — in the behavior of $H1$ (the input side). The problem is that *if we fix pole fp2 at some frequency, the location of the zero fz2 is automatically defined.* They are not independent. *There is therefore no great flexibility in using this zero and pole pair.* For example, if we try to fix *both* zeros of the overall compensation network at the LC double-pole frequency, the pole fp2 will be literally dragged along with fz2, and so the overall open-loop gain would finally fall at -2 slope again, not at -1 as desired. Therefore,

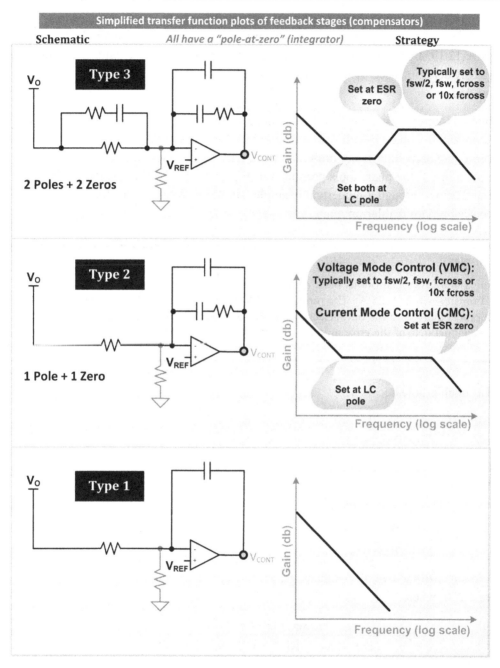

Figure 12.19: Types 1–3 compensation schemes (poles and zeros arbitrarily placed and displayed).

the zero of H1 (fz2) can only be used if the associated pole fp2 is at or beyond the crossover frequency. A possible strategy for placing the poles and zeros is indicated in Figure 12.19. But more often, Cff is just omitted, which leaves us with a simple voltage divider composed of resistors. In that case, we get $H1(s) = Rf1/(Rf1 + Rf2)$ as expected.

It actually requires a great deal of mathematical manipulation to solve the simultaneous equations and to come up with component values for a desired crossover frequency. Therefore, the derivation is not presented here, and the steps are in accordance with the basic math-in-the-log-plane tools presented in Figure 12.6. The final equations are presented below through a numerical example, similar to what we did for Type 3 compensation with regular op-amps.

Example:

Using a 300-kHz synchronous Buck controller we wish to step-down from 25 V to 5 V. The load resistor is 0.2 Ω (25 A). The ramp is 2.14 V from the datasheet of the part. The selected inductor is 5 µH, and the output capacitor is 330 µF, with an ESR of 48 mΩ. The transconductance of the error amplifier is gm = 0.3 (units for transconductance are "mhos," i.e., ohms spelled backward). The reference voltage is 1 V.

The LC double-pole occurs at

$$f_{LC} = \frac{1}{2\pi \times \sqrt{LC}} = \frac{1}{2\pi \times \sqrt{5 \times 10^{-6} \times 330 \times 10^{-6}}} \Rightarrow 3.918 \text{ kHz}$$

We choose our target crossover frequency "fcross" as 50 kHz. Suppose we pick Rf2 = 4 kΩ and Rf1 = 1 kΩ based on the voltage divider equation, the output voltage (5 V) and the reference voltage of 1 V. Then

$$Cff = \frac{(Rf1 + Rf2)}{2\pi \times (Rf1 \cdot Rf2) \times fcross} \Rightarrow 3.98 \text{ nF} \quad \text{(because pole fp2 is set at fcross)}$$

The crossover of the overall feedback gain (*H*) occurs at a frequency "fp0" as indicated in Figure 12.20, where

$$fp0 = \frac{V_{RAMP} \times (Rf1 + Rf2)}{(2\pi)^2 \times f_{LC} \times Rf2^2 \times Cff^2 \times V_{IN} \times Rf1} \Rightarrow 10.9 \text{ kHz}$$

So,

$$C_1 = \frac{1}{2\pi \times fp0} \times \frac{Rf1}{Rf1 + Rf2} \times gm \Rightarrow 0.87 \text{ µF}$$

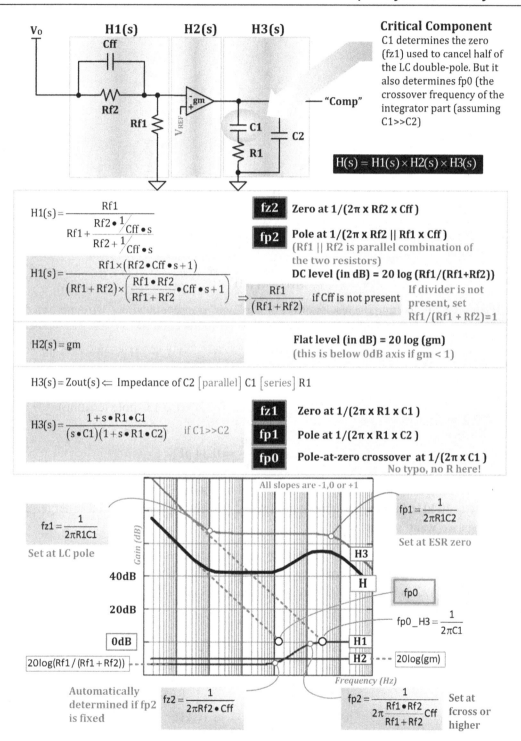

Figure 12.20: "Full-blown" transconductance operational amplifier compensation (voltage-mode control).

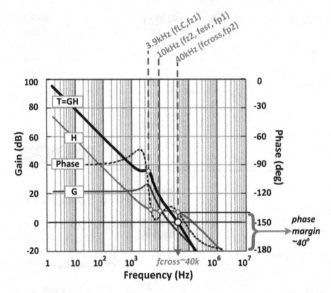

Figure 12.21: Plotting the results of the "full-blown" transconductance Op-Amp-based compensation example (voltage-mode control).

$$R_1 = \frac{1}{2\pi \times f_{LC} \times C_1} = 46.5\ \Omega \quad \text{(because fz1 is set at LC pole location)}$$

$$C_2 = C_{OUT} \times \frac{ESR}{R_1} \Rightarrow 0.34\ \mu F$$

We have presented the computed gain-phase plots in Figure 12.21. The computed crossover frequency is 40 kHz (a little less than our target of 50 kHz, so we may want to target 20% higher than we want, to start with).

Note that the location of fz2 was not fixed by us, but automatically positioned itself once we set fp2 to 50 kHz (the expected fcross, though that turned out to be ∼40 kHz). The location of fz2 in Figure 12.21 can be calculated based on the equation in Figure 12.20 as follows:

$$fz2 = \frac{1}{2\pi Rf2 \cdot Cff} \Rightarrow 10\ kHz$$

Note that the ESR-zero location is

$$fesr = \frac{1}{2\pi \cdot ESR \cdot C_{OUT}} \Rightarrow 10\ kHz$$

The above two equations are coincidentally the same in our example. But that also means, since we positioned fp1 (the other remaining pole) at the ESR-zero location, in this example, we have an overlap of fz2, fesr, and fp1 as shown in Figure 12.21. In fact, fp0 at 10.9 kHz is also very close (not shown).

From Figure 12.21, we see that we have an adequate 40° of phase margin.

> *Note: We started above example with a rather small output capacitance and large ESR than typically used for this particular application and power level. That is why C_1 is also not much larger than C_2. The intention was to shift the ESR-zero to less than the cross-over frequency, to demonstrate the principles and also to be able to plot the gain curves easily as in Figure 12.21. However, the equations and procedure presented are valid for any output capacitance and ESR.*

Simpler Transconductance Op-Amp Compensation

As mentioned, there is a practical difficulty involved in using the "full-blown" transconductance op-amp compensation scheme discussed above — because the pole and zero arising from $H1$ are *not independent*. They will even tend to coincide if say Rf2 is much smaller than Rf1 (i.e., if the desired output voltage is almost identical to the reference voltage).

So, now we try a simpler transconductance stage shown in Figure 12.22. The equations for this, based on a *new compensation strategy*, are presented in the following (new) example. Note for visual clarity in plotting that both the L and the C values used here are different from those of previous examples.

Example:

Using a 300-kHz synchronous Buck controller, we wish to step-down from 25 V to 5 V. The load resistor is 0.2 Ω (25 A). The ramp is 2.14 V from the datasheet of the part. The selected inductor is 50 μH, and the output capacitor is 150 μF, with an ESR of 48 mΩ. The transconductance of the error amplifier is gm = 0.3 (mhos), and the reference voltage is 1 V.

As before, $Rf1/(Rf1 + Rf2) = V_{REF}/V_O = 1\ V/5\ V = 0.2$.

The LC double-pole occurs at

$$f_{LC} = \frac{1}{2\pi \times \sqrt{LC}} = \frac{1}{2\pi \times \sqrt{50 \times 10^{-6} \times 150 \times 10^{-6}}} \Rightarrow 1.84\ \text{kHz}$$

We choose our target crossover frequency "fcross" as 100 kHz.

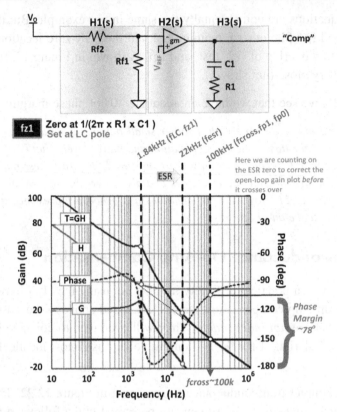

Figure 12.22: Plotting the results for the simpler transconductance Op-Amp-based compensation example (voltage-mode control).

The crossover of the feedback gain plot (H) occurs at a frequency:

$$\text{fp0} = \frac{V_{\text{RAMP}} \times \text{fcross}}{2\pi \times f_{\text{LC}} \times \text{ESR} \times C_{\text{OUT}} \times V_{\text{IN}}} \Rightarrow 103 \approx 100 \text{ kHz}$$

To achieve this fp0, we need

$$C_1 = \frac{1}{2\pi \times \text{fp0}} \times \frac{\text{Rf1}}{\text{Rf1} + \text{Rf2}} \times \text{gm} \Rightarrow 93 \text{ nF}$$

$$R_1 = \frac{1}{2\pi \times f_{\text{LC}} \times C_1} \Rightarrow 934 \ \Omega \quad \text{(since fz1 is positioned at the LC pole location)}$$

Note that the ESR-zero location is

$$\text{fesr} = \frac{1}{2\pi \cdot \text{ESR} \cdot C_{\text{OUT}}} \Rightarrow 22 \text{ kHz}$$

We have presented the computed gain-phase plots in Figure 12.22. We see we have a generous 78° of phase margin and a crossover frequency of 100 kHz.

Based on the logic presented for the Type 3 compensation scheme (non-optimized case, see later), the phase margin in this case was also expected to be around 90°. And once again, one way to reduce the phase margin closer to the optimum of 45°, is to reintroduce C_2 in Figure 12.20, which brings back fp1. We then set fp1 exactly at fcross as previously explained. We then get

$$C_2 = \frac{1}{2\pi \times R_1 \times \text{fcross}} \Rightarrow 1.7 \text{ nF}$$

By reintroducing C_2, the computed crossover again occurs slightly earlier (by about 20%) — at around 80 kHz, instead of 100 kHz, so we may want to target 20% more than we want. The phase margin is now 36° (closer to the optimal).

Also note that for this simpler compensation scheme to work, *the ESR-zero must lie between the* LC *pole frequency and the selected crossover frequency.* Otherwise this will not work with voltage-mode control.

Note that for a Boost or Buck Boost, the only changes required in the above analysis are

$$L \Rightarrow \frac{L}{(1-D)^2} \quad \text{(\textit{Boost and Buck-Boost})}$$

$$V_{\text{RAMP}} \Rightarrow V_{\text{RAMP}} \times (1-D)^2 \quad \text{(\textit{Boost and Buck-Boost})}$$

However, *we must also always ensure (for these two topologies) that the selected crossover frequency is at least an order of magnitude below the RHP zero.*

Compensating with Current-Mode Control

The plant transfer functions presented earlier were only for voltage-mode control. In current-mode control, the ramp to the PWM (for determining duty cycle) is derived from the inductor current. It can be shown that if we do that, the inductor effectively goes "out of the picture" in the sense that there is no double LC pole anymore. So, the compensation is *supposedly* simpler, and the loop can be made much faster. However, the actual mathematical modeling of current-mode control has proven extremely challenging — mainly because there are now *two* feedback loops in action — the normal voltage feedback loop (slower) and a current feedback loop (cycle-by-cycle basis; much faster). Various researchers have come up with different approaches, but they don't seem to agree completely with each other yet.

Having said that, everyone does seem to agree that *current-mode control alters the poles of the system compared to voltage-mode control, but the zeros of voltage-mode control remain unchanged.* So, the Boost and the Buck-Boost still have the same (low-frequency) RHP zero, as we discussed earlier (i.e., that applies to voltage-mode and current-mode controls when we are operating in CCM). *Therefore, care is still needed to ensure that the RHP zero is at a much higher frequency than the chosen crossover frequency.*

As mentioned, in current-mode control, the ramp to the PWM comparator is derived from the inductor current. Actually, the most common way of producing the ramp is to simply sense the forward-drop across the MOSFET (or of course by using an external sense resistor in series with it) (see Figure 12.23 (top half)). This small sensed voltage is then amplified by a "current sense amplifier" to get a voltage ramp, which is then applied to the PWM comparator. On the other pin of the PWM comparator, we have the usual output of the error amplifier (control voltage).

The inductor/switch current ramp is now obviously proportional to the voltage ramp received at the PWM comparator input. So, voltages and currents can be converted (mapped) between each other through the use of the "transfer resistance" V/I, as defined in Figure 12.23. We can look at the overall effect either in terms of ALL currents, or in terms of ALL voltages, as shown in the lower half of Figure 12.23.

Slope compensation, as discussed in more detail later, can also be expressed either as a certain applied A/s (or $A/\mu s$), or as an applied V/s. These are all *equivalent* ways of talking about the same thing, since voltages and currents are proportional to each other — *through the transfer resistance.* For example, if we know that a peak current of 1 A on the sense resistor appears as a peak voltage of 1.5 V at the PWM comparator input, the transfer resistance is simply $V/I = 1.5/1 = 1.5\ \Omega$. Yes, we do need to have inside information on the part, in particular the peak-to-peak voltage swing of its error amplifier output (which in turn determines the maximum swing of the mapped inductor current).

Note that since the ramp itself gets terminated at the exact moment when it reaches the control voltage level (because it is a comparator), in effect, we end up *regulating the peak of the inductor current ramp.* So, what we are discussing here is really just "peak current-mode control." Many experienced designers prefer "average current-mode control."

One of the subtleties of current-mode control is that (for all the topologies) we need to add a small artificial ramp to stabilize it under certain conditions. This is called *slope compensation.* Its purpose is to prevent an odd artifact of current-mode control called "subharmonic instability." Subharmonic instability usually shows up as alternate wide and narrow switching pulses (a pattern that repeats itself at the rate of half the switching frequency). In steady-state operation, we may not realize it is present. There may just be a small output ripple, one that can be further suppressed by large-enough output caps. It will,

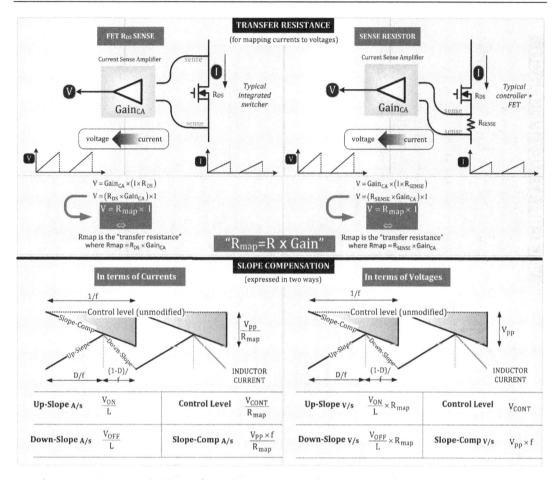

Figure 12.23: How the "transfer resistance" maps the current in the switch, into a voltage sawtooth at the comparator input, and how slope compensation can be expressed in terms of either voltages or currents.

however, manifest itself as a badly degraded transient response. Its Bode plot will also very likely not even be recognizable as one — there may be no way to even describe a phase margin in that plot. In general, we really need to look at the switching node waveform to rule out subharmonic instability conclusively.

Note: All patterns that repeat themselves at the rate of fsw/2 need not represent subharmonic instability — even noise spikes, for example (self-generated or from synchronized external sources), can cause the same effect, by producing early termination of an ongoing pulse, that forces a longer succeeding pulse, in an effort to meet the required energy demand.

What are the causes of this instability and the solution? In Figure 12.24, we have shown the control voltage (output of the error amplifier), plotted against the mapped inductor current

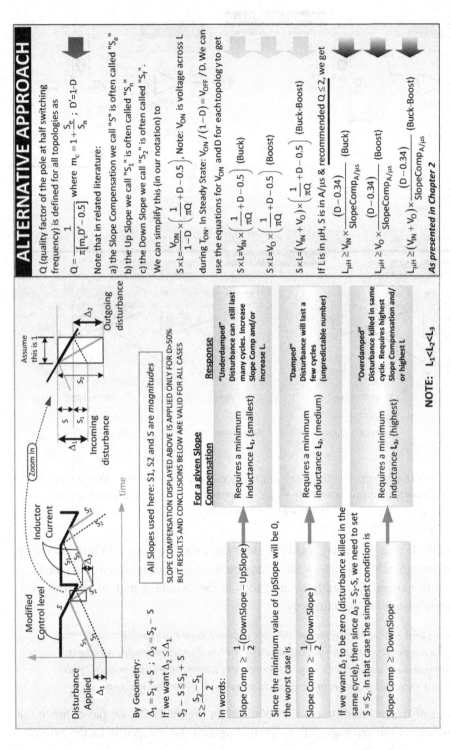

Figure 12.24: Explaining the conditions for avoiding subharmonic instability: traditional approach on the left, modern alternative method on the right (set $Q < 2$).

(these are the voltages on the two pins of the PWM comparator). Therefore, whenever the (mapped) inductor current equals the control voltage level, the pulse is terminated as shown. Note that the control voltage is no longer flat as in conventional voltage-mode control, but has a negative sawtooth superimposed on it as seen in Figures 12.23 and 12.24. This is called "slope compensation" or "ramp compensation," and it represents a possible solution to subharmonic instability. The problem itself is described in Figure 12.24. We see that a small input disturbance Δ_1, becomes Δ_2 after the given pulse ends, and that becomes the input disturbance going into the next on-time. This will become Δ_3 in the next pulse, and so on. The ratio by which the disturbance changes every cycle, using the simple geometry shown in the figure, is

$$\frac{\Delta_2}{\Delta_1} = \frac{S_2 - S}{S_1 + S}$$

If we want the disturbance to subside eventually, the condition is

$$S \geq \frac{S_2 - S_1}{2}$$

where S_2 is the down-slope of the inductor, S_1 the up-slope, and S the applied slope compensation. It should be obvious that all these three slopes must be in the same units. So, we can have them all expressed as A/μs, or all expressed as V/μs. To do that, we just need to know R_{map}.

For subharmonic instability to occur, two conditions have to be met simultaneously — the duty cycle should be close to or exceed 50%, and simultaneously, we should be in CCM. Note that the propensity to enter this subharmonic instability state increases as the duty cycle increases (i.e., as the input is lowered). So, *we should always try to rule out this instability at V_{INMIN}.* We could certainly avoid this problem altogether, by choosing DCM (discontinuous conduction mode). But otherwise, in CCM, slope compensation is the recognized sure-fix. Though it is interesting to note that by applying slope compensation, we are, in effect, *blending a little voltage-mode control with a current-mode control.* Note that to do this we could do either of the two below:

(a)   Take the sensed current ramp generated from the switch/inductor current, convert it to an equivalent sensed voltage (via R_{map}), and to that add a small fixed voltage ramp.

> *Note: In the popular off-line controller IC UC3842, designers often add a small 10–20-pF capacitor between the clock pin (Pin 4) and the current sense pin (Pin 3). This is a simple slope compensation trick, undocumented in the datasheet of the part. The purpose of that was to add a little bit of the fixed clock signal ramp to the sensed current ramp (which would go to the PWM comparator). In effect, we are thus mixing a little voltage-mode control to current-mode control, as we declared was the intuitive purpose of slope compensation. In the UC3844, the duty cycle cannot exceed 50%,*

since it is meant for Forward converters, not for flybacks. So, this trick was not necessary for that IC.

(b)   Since we are talking in terms of relative voltages at the input of a comparator, we could equivalently modify the control voltage itself (the output of error amplifier) as shown in Figures 12.23 and 12.24. The intent is, especially for duty cycles greater than 50% (where subharmonic instability can occur), we progressively decrease the control voltage steadily as the cycle progresses.

Note: Applying slope compensation as shown in Figures 12.23 and 12.24 may limit the peak current and thereby the max power of the converter when the duty cycle exceeds 50%. To avoid that, designers often design in a progressively higher current limit at large duty cycles.

Note: Though for simplicity, we have not shown it explicitly in Figure 12.23, in true current-mode control, we should not lose the "DC" (pedestal) information of the inductor/switch current.

What are the symptoms of *impending* subharmonic instability? If we take the Bode plot of any current-mode controlled converter (one that has *not* yet entered this wide-narrow-wide-narrow state), we will discover an unexplained peaking in the gain plot, at exactly *half the switching frequency* (similar to the peaking in Figure 12.7). This is the "source" of potential subharmonic instability. Note that we never consider setting the crossover frequency higher than half the switching frequency. So, in effect, this subharmonic pole will always occur at a frequency *greater* than the crossover frequency. However, we realize that the effect of this pole on the phase angle may start at a *much lower frequency*. Even strictly in terms of gain, this half-switching frequency pole remains dangerous because of the fact that if it peaks too much, *it can end up* causing the gain plot to intersect the 0-dB axis again. This represents another unintended crossover, and we know that any phase reinforcement at crossover can provoke full instability.

Subharmonic instability is nowadays modeled as a complex pole at half the switching frequency. It has a certain "Q" as described on the right side of Figure 12.24. By actual experiments, it has been shown that a Q of less than 2 typically creates stable conditions. A Q of 1 is preferred by conservative designers, and though that does quell subharmonic instability even more firmly, it does lead to a bigger inductor (producing a non-optimum current ripple ratio r of less than 0.4). Alternatively, we need to apply greater slope compensation. But too much slope compensation is akin to making the system more and more like voltage-mode control, and pretty soon, especially at light loads, the double LC pole of voltage-mode control will reappear, potentially causing instability of its own.

In Figure 12.24, we have also presented the modern method for dealing with subharmonic instability. We have thereby proven the relations presented earlier in *Chapter 2* for (minimum) inductance. These equations are

$$L_{\mu H} \geq \frac{D - 0.34}{\text{Slope comp}_{A/\mu s}} \times V_{\text{IN}} \quad \text{(Buck)}$$

$$L_{\mu H} \geq \frac{D - 0.34}{\text{Slope comp}_{A/\mu s}} \times V_O \quad \text{(Boost)}$$

$$L_{\mu H} \geq \frac{D - 0.34}{\text{Slope comp}_{A/\mu s}} \times (V_{\text{IN}} + V_O) \quad \text{(Buck-Boost)}$$

Having taken care of subharmonic instability by a suitable choice of inductance and/or slope compensation, we will no longer include it in the following analysis in which we set the poles and zeros of the compensation network.

The design equations presented for the compensation network below are based on a simpler model from Middlebrook that reduces current-mode control to something similar to voltage-mode control discussed previously. The purpose of that is to make current-mode control amenable to being handled in a familiar fashion too — as a product of several cascaded transfer functions, rather than parallel feedback loops.

The results from the Middlebrook model give a good match with far more elaborate models — provided we have taken precautionary steps. For example, we need to ensure the RHP zero (if present) is designed out (we should check its location is at least a decade away from the target crossover frequency). We should also check that the fsw/2 subharmonic pole is higher than the crossover frequency, and also that the fsw/2 pole is sufficiently damped as discussed above. If so, we can proceed as follows.

Note that in our presentation below, we are even ignoring some other poles from Middlebrook's original model, on the grounds that they usually fall well outside the crossover frequency, and are therefore of little practical interest.

In our extra-simplified model, *we are thus left with only a single-pole in the plant transfer function for all the topologies.* This pole comes from the output capacitor and the load resistor (the "output pole"). When we combine it with the inevitable pole-at-zero (from the integrator section of the op-amp), the overall (open-loop) gain will fall with a slope of −2 (after the output pole location). Therefore, we *need just one single-zero to cancel part of this slope out, and finally get a −1 slope with which to crossover as desired.* Further, this single-zero can either be deliberately introduced using Type 2 compensation (in which case we could use its available pole to cancel out the ESR-zero) — or we could simply rely on the naturally occurring *ESR-zero* of the output cap. In the latter case, we would need to ensure that the ESR-zero is at a frequency *lower* than the crossover frequency. Alternatively, that could indirectly force us to move the crossover frequency out to a higher frequency (but without getting too close to the other trouble spots mentioned above).

The design equations and steps for the transconductance op-amp are as follows (see left side of Figure 12.26).

(a)   Choose a crossover frequency "fcross." Although we would like to typically target one-third the switching frequency, we must manually confirm that this frequency is significantly below the location of the RHP zero (the equations for the RHP zero were presented earlier, and they still apply here).

(b)   We realize that, once again, while plotting the open-loop gain, the gain of the integrator will shift vertically by the amount G_O (DC gain of plant). Therefore, using the simple rule in the lower half of Figure 12.6, we can find the required *fcross* that will lead to the desired crossover frequency (of the open-loop gain). So,

$$fp0 = \frac{fcross}{A/B}$$

where the values of $G_O = A/B$ are presented in Figure 12.25.

(c)   Calculate C_1 using

$$C_1 = \frac{y \cdot gm}{2\pi \times fp0}$$

where y is the "attenuation ratio" in Figure 12.26.

(d)   Calculate R_1 using

$$R_1 = \frac{1}{2\pi \times C_1 \times f_P}$$

where f_P is the output pole of the plant, as given in Figure 12.25.

(e)   Calculate C_2 using

$$C_2 = \frac{1}{2\pi \times R_1 \times fesr}$$

where fesr is the location of the ESR-zero, that is, $1/(2\pi \times ESR \times C_O)$.

The design equations and steps for the *conventional op-amp* are as follows (see right side of Figure 12.16).

(f)   Choose a crossover frequency "fcross." Target one-third the switching frequency, if possible.

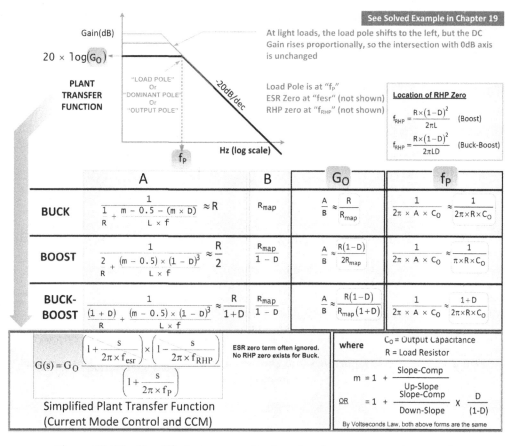

Figure 12.25: Simplified plant transfer function for current-mode control.

(g) Using the simple rule in the lower half of Figure 12.6, we can find the required fcross that will produce the desired crossover frequency (of the open-loop gain). So,

$$fp0 = \frac{fcross}{A/B}$$

where the values of $G_O = A/B$ are presented in Figure 12.25.

(h) Calculate C_1 using

$$C_1 = \frac{1}{2\pi \times R_1 \times fp0}$$

where R_1 has been chosen while setting the voltage divider.

(i) Calculate R_2 using

$$R_2 = \frac{1}{2\pi \times C_1 \times f_P}$$

where f_P is the output pole of the plant, as given in Figure 12.25.

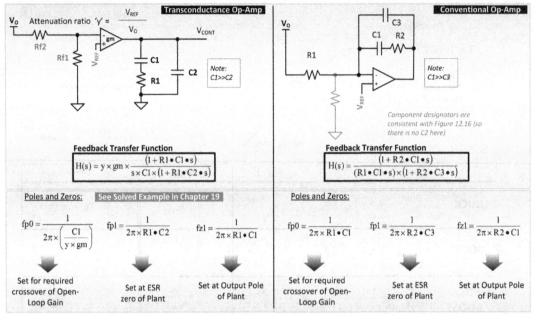

Feedback Transfer Function (Transconductance Op-Amp):

$$H(s) = y \times gm \times \frac{(1 + R1 \bullet C1 \bullet s)}{s \times C1 \times (1 + R1 \bullet C2 \bullet s)}$$

Feedback Transfer Function (Conventional Op-Amp):

$$H(s) = \frac{(1 + R2 \bullet C1 \bullet s)}{(R1 \bullet C1 \bullet s) \times (1 + R2 \bullet C3 \bullet s)}$$

Poles and Zeros (Transconductance):

$$fp0 = \frac{1}{2\pi \times \left(\dfrac{C1}{y \times gm}\right)} \qquad fp1 = \frac{1}{2\pi \times R1 \bullet C2} \qquad fz1 = \frac{1}{2\pi \times R1 \bullet C1}$$

Set for required crossover of Open-Loop Gain | Set at ESR zero of Plant | Set at Output Pole of Plant

Poles and Zeros (Conventional):

$$fp0 = \frac{1}{2\pi \times R1 \bullet C1} \qquad fp1 = \frac{1}{2\pi \times R2 \bullet C3} \qquad fz1 = \frac{1}{2\pi \times R2 \bullet C1}$$

Set for required crossover of Open-Loop Gain | Set at ESR zero of Plant | Set at Output Pole of Plant

If ESR Zero is at a much higher frequency than the crossover frequency, there is no need to "waste" the pole (fp1). We can then set the pole either to fsw, fsw/2, 10 x fcross or fcross as explained under voltage mode control. The last choice will usually give the best results.

Figure 12.26: Transconductance and conventional type 2 Op-Amp compensation for current-mode control.

(j)   Calculate C_3 using

$$C_3 = \frac{1}{2\pi \times R_2 \times \text{fesr}}$$

where fesr is the location of the ESR-zero, that is, $1/(2\pi \times \text{ESR} \times C_O)$.

The above design procedure is the same for all the topologies. We just have to use the appropriate row of the table provided in Figure 12.25. Note that for all the topologies, the "L" used is now the *actual* inductance of the converter (not the "equivalent" inductance of the canonical model).

See a full solved example in *Chapter 19*.

CHAPTER 13

Advanced Topics: Paralleling, Interleaving, and Load Sharing

Part 1: Voltage Ripple of Converters

Buck Converter Input and Output Voltage Ripple

In *Chapter 5* we learned that when we add energy to an inductor, the current through it ramps up according to $\varepsilon = (1/2) \times L \times I^2$. When we withdraw energy from it, the current ramps down. That forms the observed (AC) *current ripple*. We also defined a certain optimum ratio for that in terms of $r = \Delta I / I_L = 0.4$ ($\pm 20\%$ ripple). In an analogous manner, when we add and remove energy from the input and output capacitors, the capacitor voltage rises and falls as per $(1/2) \times C \times V^2$. That leads to an observed input or output *voltage ripple*. Just as for current in the inductor, there are general guidelines for the amount of tolerable/recommended capacitor voltage ripple relative to its DC value, the DC value being V_{IN} or V_{OUT} as the case may be. We will describe how much voltage ripple is acceptable a little later. First we need to do some math.

In the first assumption we start by ignoring equivalent series resistance (ESR) and equivalent series inductance (ESL). In Figure 13.1, we derive the basic *output* voltage ripple equation (the peak-to-peak value). In Figure 13.2, we similarly derive the basic *input* voltage ripple equation (peak-to-peak again). Note that so far, the observed ripple is purely based on energy storage, and is therefore "visible" only for small capacitances, as are typical in modern, fast-reacting converters with all-ceramic capacitors. Note also that ignoring ESR in capacitor voltage ripple is akin to ignoring the DCR of an inductor in computing or graphing its current ripple. We thus get

$$V_{O\_RIPP\_PP} = \frac{r \times I_O}{8 \times f \times C_O} = \frac{\Delta I}{8 \times f \times C_O}$$

$$V_{IN\_RIPP\_PP} = \frac{I_O \times D \times (1 - D)}{f \times C_O}$$

Since we expect that the cap voltage ripple is analogous to inductor current ripple, we can confirm that if the capacitance is increased, the voltage ripple decreases — just as when L is increased, r (the current ripple ratio) decreases. Similarly, if frequency is increased, both

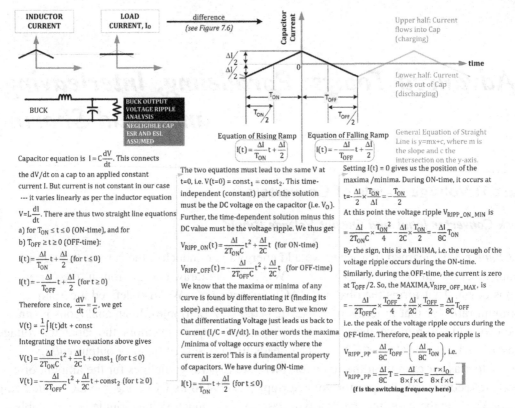

INDUCTOR CURRENT

LOAD CURRENT, I_O

difference
(see Figure 7.6)

Capacitor Current

Upper half: Current flows into Cap (charging)

time

Lower half: Current flows out of Cap (discharging)

$\frac{\Delta I}{2}$

$\frac{\Delta I}{2}$

0

T_{ON} T_{OFF}

$\frac{T_{ON}}{2}$ $\frac{T_{OFF}}{2}$

BUCK

BUCK OUTPUT VOLTAGE RIPPLE ANALYSIS
NEGLIGIBLE CAP ESR AND ESL ASSUMED

Equation of Rising Ramp
$$I(t) = \frac{\Delta I}{T_{ON}}t + \frac{\Delta I}{2}$$

Equation of Falling Ramp
$$I(t) = -\frac{\Delta I}{T_{OFF}}t + \frac{\Delta I}{2}$$

General Equation of Straight Line is y=mx+c, where m is the slope and c the intersection on the y-axis.

Capacitor equation is $I = C\frac{dV}{dt}$. This connects the dV/dt on a cap to an applied constant current I. But current is not constant in our case ··· it varies linearly as per the inductor equation $V = L\frac{dI}{dt}$. There are thus two straight line equations

a) for $T_{ON} \le t \le 0$ (ON-time), and for

b) $T_{OFF} \ge t \ge 0$ (OFF-time):

$$I(t) = \frac{\Delta I}{T_{ON}}t + \frac{\Delta I}{2} \text{ (for } t \le 0)$$

$$I(t) = -\frac{\Delta I}{T_{OFF}}t + \frac{\Delta I}{2} \text{ (for } t \ge 0)$$

Therefore since, $\frac{dV}{dt} = \frac{I}{C}$, we get

$$V(t) = \frac{1}{C}\int I(t)dt + const$$

Integrating the two equations above gives

$$V(t) = \frac{\Delta I}{2T_{ON}C}t^2 + \frac{\Delta I}{2C}t + const_1 \text{ (for } t \le 0)$$

$$V(t) = -\frac{\Delta I}{2T_{OFF}C}t^2 + \frac{\Delta I}{2C}t + const_2 \text{ (for } t \ge 0)$$

The two equations must lead to the same V at t=0, i.e. V(t=0) = $const_1$ = $const_2$. This time-independent (constant) part of the solution must be the DC voltage on the capacitor (i.e. V_O). Further, the time-dependent solution minus this DC value must be the voltage ripple. We thus get

$$V_{RIPP\_ON}(t) = \frac{\Delta I}{2T_{ON}C}t^2 + \frac{\Delta I}{2C}t \text{ (for ON-time)}$$

$$V_{RIPP\_OFF}(t) = -\frac{\Delta I}{2T_{OFF}C}t^2 + \frac{\Delta I}{2C}t \text{ (for OFF-time)}$$

We know that the maxima or minima of any curve is found by differentiating it (finding its slope) and equating that to zero. But we know that differentiating Voltage just leads us back to Current (I/C = dV/dt). In other words the maxima/minima of voltage occurs exactly where the current is zero! This is a fundamental property of capacitors. We have during ON-time

$$I(t) = \frac{\Delta I}{T_{ON}}t + \frac{\Delta I}{2} \text{ (for } t \le 0)$$

Setting I(t) = 0 gives us the position of the maxima /minima. During ON-time, it occurs at

$$t = -\frac{\Delta I}{2} \times \frac{T_{ON}}{\Delta I} = -\frac{T_{ON}}{2}$$

At this point the voltage ripple $V_{RIPP\_ON\_MIN}$ is

$$= \frac{\Delta I}{2T_{ON}C} \times \frac{T_{ON}^2}{4} - \frac{\Delta I}{2C} \times \frac{T_{ON}}{2} = -\frac{\Delta I}{8C}T_{ON}$$

By the sign, this is a MINIMA, i.e. the trough of the voltage ripple occurs during the ON-time. Similarly, during the OFF-time, the current is zero at $T_{OFF}/2$. So, the MAXIMA, $V_{RIPP\_OFF\_MAX}$, is

$$= -\frac{\Delta I}{2T_{OFF}C} \times \frac{T_{OFF}^2}{4} + \frac{\Delta I}{2C} \times \frac{T_{OFF}}{2} = \frac{\Delta I}{8C}T_{OFF}$$

i.e. the peak of the voltage ripple occurs during the OFF-time. Therefore, peak to peak ripple is

$$V_{RIPP\_PP} = \frac{\Delta I}{8C}T_{OFF} - \left(-\frac{\Delta I}{8C}T_{ON}\right), \text{ i.e.}$$

$$V_{RIPP\_PP} = \frac{\Delta I}{8C}T = \frac{\Delta I}{8 \times f \times C} = \frac{r \times I_O}{8 \times f \times C}$$

(f is the switching frequency here)

Figure 13.1: Derivation of output voltage ripple of Buck for small capacitances and no ESR/ESL.

current and voltage ripple tend to decrease as we had indicated on page 1 of this book. However, we now realize we can do some optimization too. For example, for a *given* voltage ripple target, we can decrease C_O provided we increase the frequency. Getting smaller energy-storage components seems to be a basic advantage of using higher switching frequencies. Though that is not the *only* reason for higher frequencies, as we will see when we discuss voltage regulator modules (VRMs) in more detail in Part 2 of this chapter. On the output side, as we learned in *Chapter 12*, lowering the values of L and C_O increases the bandwidth of the error−correction loop. That in turn causes the *transient response* of the converter (its response to sudden changes in line and load) to improve. That is a very important requirement in modern "POL" (point of load) converters or "VRMs" (voltage regulator modules). These are modules, typically high-frequency step-down/Buck converters, mounted on the same board as the converter's intended load. The load could be a modern microprocessor IC, requiring a very low but tightly regulated voltage rail with very high d*I*/d*t* capability.

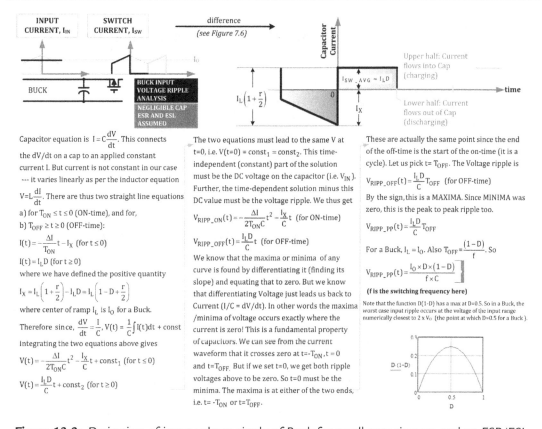

Figure 13.2: Derivation of input voltage ripple of Buck for small capacitances and no ESR/ESL.

In Figure 13.3, we plot out both the input/output voltage ripples using Mathcad, with the equations derived in Figures 13.1 and 13.2. In particular, we see how the output voltage ripple is a *composite* of two curves, one for the ON-time and one for the OFF-time — with both segments meeting "just in time," *at the exact moment* where the switch turns OFF and the diode turns ON (which is $t = 0$ as per our convention). The two segments also have the same absolute value at $t = -T_{ON}$ (just as the switch turns ON) and at $t = T_{OFF}$ (just as the diode turns OFF). That is why the cycle can repeat endlessly and conform to a "steady state". In Figure 13.3, we have also plotted out the capacitor currents (input and output) just for comparison, the individual segments and the final curve (shaded). Note that the input current waveform in Figure 13.3 is vertically flipped compared to the corresponding input cap current waveform shown earlier in Figure 7.6. By convention, cap charging current is usually considered positive, and discharging is connected to negative current. But we keep in mind that flipping any given waveform vertically does *not* change its RMS value, since RMS uses the square of current anyway. So, the sign convention, if any, doesn't really matter here.

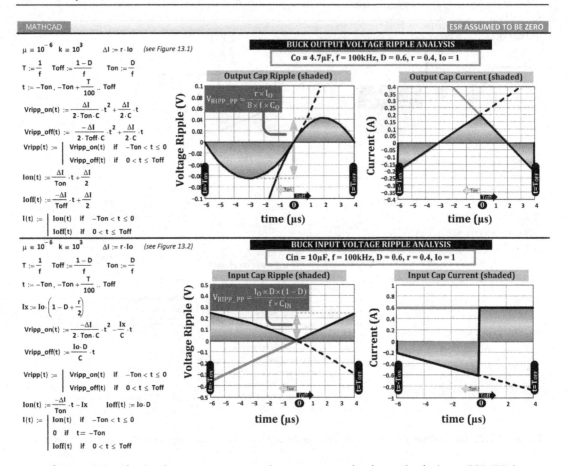

Figure 13.3: Plotting input/output capacitor current and voltage ripple (zero ESR/ESL).

In Figure 13.4, we return to the input ripple, and show how ESR affects it. The total ripple is now a simple sum of two terms, one related to ESR and one to the capacitance charging and discharging. In Figure 13.5, we take the output ripple, and add *both* the ESR and ESL terms successively to show how the waveform changes at each step. Note that in this latter case (output ripple) the total output voltage ripple is not a simple arithmetic sum of individual terms, because of the phase difference between the components.

Both input and output voltage ripples are important. For example, V_{IN} is usually used to provide power to both the converter controller IC, and the power converter stage (switch and inductor). To avoid erratic functioning of the controller IC, we need to not only add a 0.1 μF ceramic cap very close to the supply pin of the IC (decoupling), but also provide an input bulk capacitor sized such that the peak-to-peak input voltage ripple $\Delta V/V_{IN}$ is less than 1%. Sometimes, a single, large ceramic capacitor can do "double-duty," and provide both decoupling and bulk capacitance functions. The acceptable input percentage voltage

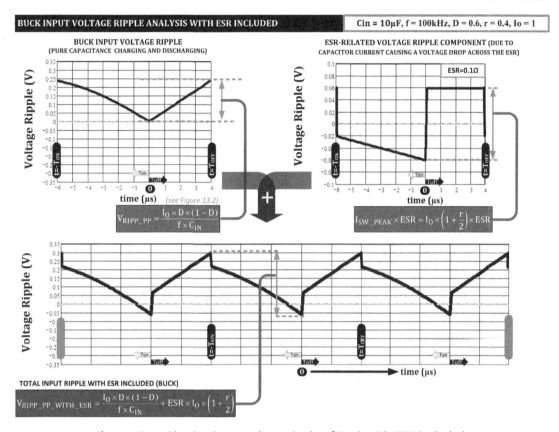

Figure 13.4: Plotting input voltage ripple of Buck with ESR included.

ripple actually depends on the particular switcher IC/controller: its design, its noise rejection capabilities, the IC package construction, the PCB layout, and so on. We may need to rely on guidance from the IC vendor here. Voltage ripple on the output cap is important too, because it can cause the circuits powered by the output rail to misbehave. For example, a typical power supply specification will call for the ripple on a 5 V output rail to be less than ± 50 mV. That is a peak to peak of 100 mV, or a percentage (peak to peak) ripple of $0.1/5 = 2\%$ — which is sometimes referred to as $\pm 1\%$, but often colloquially, simply as 1% (just as we do for "1% resistors").

With all these considerations in play, what is now the *dominant* criterion for selecting capacitors? In a Buck, the input capacitor's RMS current is much higher than the output capacitor's RMS. So, rather generally speaking, we can say that in a Buck, the input cap is determined mainly by RMS stress requirements, whereas on the output, it is simply maximum allowed voltage ripple that determines the capacitance and thereby the capacitor. We can read the sections on capacitor RMS in *Chapter 7* to refresh our memory. However,

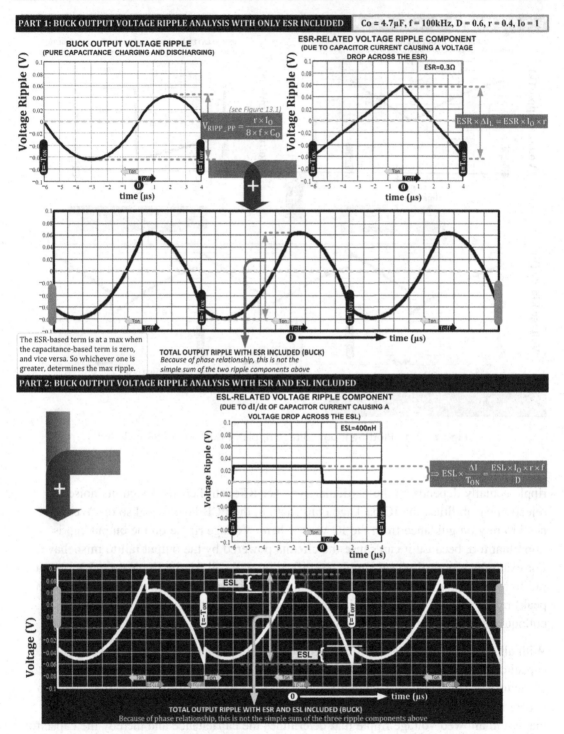

PART 1: BUCK OUTPUT VOLTAGE RIPPLE ANALYSIS WITH ONLY ESR INCLUDED   Co = 4.7µF, f = 100kHz, D = 0.6, r = 0.4, Io = 1

BUCK OUTPUT VOLTAGE RIPPLE
(PURE CAPACITANCE CHARGING AND DISCHARGING)

(see Figure 13.1)

$$V_{RIPP\_PP} = \frac{r \times I_0}{8 \times f \times C_0}$$

ESR-RELATED VOLTAGE RIPPLE COMPONENT
(DUE TO CAPACITOR CURRENT CAUSING A VOLTAGE DROP ACROSS THE ESR)

ESR=0.3Ω

$$ESR \times \Delta I_L = ESR \times I_0 \times r$$

The ESR-based term is at a max when the capacitance-based term is zero, and vice versa. So whichever one is greater, determines the max ripple.

TOTAL OUTPUT RIPPLE WITH ESR INCLUDED (BUCK)
Because of phase relationship, this is not the simple sum of the two ripple components above

PART 2: BUCK OUTPUT VOLTAGE RIPPLE ANALYSIS WITH ESR AND ESL INCLUDED

ESL-RELATED VOLTAGE RIPPLE COMPONENT
(DUE TO dI/dt OF CAPACITOR CURRENT CAUSING A VOLTAGE DROP ACROSS THE ESL)

ESL=400nH

$$\Rightarrow ESL \times \frac{\Delta I}{T_{ON}} = \frac{ESL \times I_0 \times r \times f}{D}$$

TOTAL OUTPUT RIPPLE WITH ESR AND ESL INCLUDED (BUCK)
Because of phase relationship, this is not the simple sum of the three ripple components above

Figure 13.5: Plotting output voltage ripple of Buck with ESR and ESL included.

modern ceramic caps have very high RMS ratings, in which case, on the input side of a Buck, the maximum input peak-to-peak voltage (input voltage ripple) becomes the dominant criterion for selecting the capacitor. Lifetime predictions in electrolytic caps are discussed in *Chapter 6*. In *Chapter 9*, we discussed hysteretic converters, which are based on a certain output ripple to provide the ramp. In Parts 2 and 3 of this chapter we will learn how to apply these ripple concepts to interleaved converters. The perceptive engineer will be able to go back to Figures 13.1 and 13.2 and derive the necessary ripple equations for the Boost and the Buck-Boost if so desired. Though keep in mind that in many cases, engineers simply assume a large capacitance, and base the estimated voltage ripple on the following simple generic equation

$$I_{\text{CAP\_PP\_VOLTAGE\_RIPPLE}} = I_{\text{CAP\_PP\_CURRENT\_RIPPLE}} \times \text{ESR}$$

We can use the peak-to-peak current equations found in the Appendix of this book for each topology.

See "Solved Examples" in Chapter 19, *in particular refer to Figure 19.4 for additional criteria for selection of output capacitors in modern Buck converters.*

Part 2: Distributing and Reducing Stresses in Power Converters

Overview

Intuitively, we realize that two relatively frail persons can carry a heavy suitcase if they combine forces intelligently, say by holding the suitcase *evenly* from both sides. So, now we start considering ways to distribute/share stresses by the process of *paralleling*. First we apply that concept to *components*. Thereafter, in Part 3, we try to parallel complete *sections of converters*, called "interleaving" or "multiphase operation". Finally, in Part 4, we discuss passive and active load sharing in which we parallel *entire converters*.

Many techniques have evolved over the years to distribute current and voltage stresses across several "identical" components. Discrete components are typically placed in *parallel* to split/share *current* stresses, and in *series* to split *voltage* stresses. For example, diodes are often paralleled for high current applications. It is best if they are in the same package though. The 2-switch Forward ("asymmetric half-bridge", see Table 7.1) is an example of two FETs in series, sharing voltage stress equally. The FETs just happen to be on opposite sides of the Primary winding, so the voltage sharing is perhaps not so obvious. But look at it this way: for example, if the input voltage rail is say V_{IN}, and both the switches turn OFF, the voltage across the Primary winding flips from V_{IN} to $-V_{\text{IN}}$. Though it still has seemingly (reverse) V_{IN} across it, what has really happened is that the end of the transformer winding that was initially at ground (when both switches were ON), jumps up to V_{IN}, and the end which was initially at V_{IN} falls to ground. So the total *change* in voltage is actually $V_{IN} - (-V_{IN}) = 2 \times V_{IN}$. However, *each switch sees a max of only* V_{IN} across it. Yet a voltage of $2 \times V_{IN}$ is being effectively blocked, by two FETs. Compare that with a single-ended Forward converter with a simple 1:1 energy recovery (tertiary) winding — in that case we have a single switch that blocks the entire $2 \times V_{IN}$. In other words, in a 2-switch Forward, there is subtle voltage stress sharing in effect — but more of a "what could have been" versus a "what is" situation. Similarly, capacitors are sometimes placed in series to share voltage. In a flyback, several caps may be paralleled to handle the rather severe output cap RMS current of this topology. In all cases, a key concern is to get the components to share properly. Sure: "identical" components are really not as identical as we wish. But yes, it really helps if the components were fabricated in the same production lot, preferably in the same package (for FETs or resistors), but even better: fabricated on the *same die* (or substrate). At every step, they become more and more identical, and their *relative* matching and stress-sharing capability improves. In more complicated cases, to enforce better sharing, we may need to employ techniques that are either *active* (as discussed in Part 3), or *passive* (with the help of ballasting resistors, for example). Take a look at Figure 13.6, which surveys some common component stress-sharing techniques, along with methods to ensure better sharing. Note that very often, there is a smart "correct" way to get them to share well, and a relatively "incorrect" way. Look at the bridge rectifier case for example.

Since in any converter, the actual current stresses are usually proportional to the load current (not so in synchronous converters at light loads though), the question arises: can we just split the load current (of a single power-train) into two identical paralleled converters, to achieve halving of stresses in each? If so, what would be the advantage of that? We take that up shortly (Figure 13.7).

Power Scaling Guidelines in Power Converters

First let us understand how power supplies "scale" with load (in CCM but with no negative current regions as were shown in Figure 9.1). Let us take one of the equations we derived in *Chapter 7* to illustrate something quite interesting and useful here. Let us take the RMS of the Buck converter switch current for example. We have shown in Figure 7.3 that

$$I_{SW\_RMS} = I_L \sqrt{D\left(1 + \frac{r^2}{12}\right)}, \text{ or equivalently } I_{SW\_RMS} = I_L \times \sqrt{D} \times \sqrt{\left(1 + \frac{r^2}{12}\right)}$$

We do remember that, by definition, $r = \Delta I/I_L$, where I_L is the center of ramp (DC value) of inductor current. In the case of a Buck, $I_L = I_O$. So, we can also write the above equations as

$$I_{SW\_RMS} = I_O \sqrt{D\left(1 + \frac{\Delta I^2}{12 \times I_O^2}\right)}, \text{ or equivalently } I_{SW\_RMS} = \sqrt{D\left(I_O^2 + \frac{\Delta I^2}{12}\right)}$$

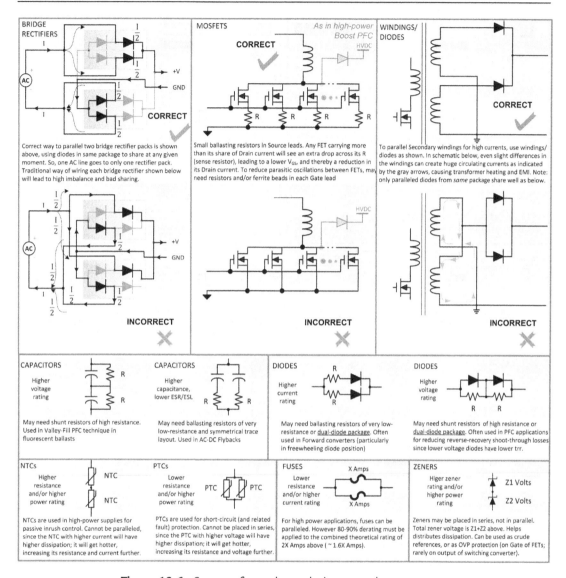

Figure 13.6: Survey of popular techniques to share stresses.

The latter equation is seen more commonly in literature. Notice that it looks "messier" than our simpler-looking equations expressed in terms of r. But cosmetics aside, the usual way of writing out the RMS currents also misses out a potentially huge simplification. Because, in contrast, by using r, we can express the switch RMS current in a more intuitive manner — as a product of three relatively *orthogonal* terms: (a) an AC term, that is, a term involving *only* r, multiplied by (b) a DC term, that is, one involving *only* load current I_O, and (c) a term related to duty cycle D, as expected. We can thus separate the terms and

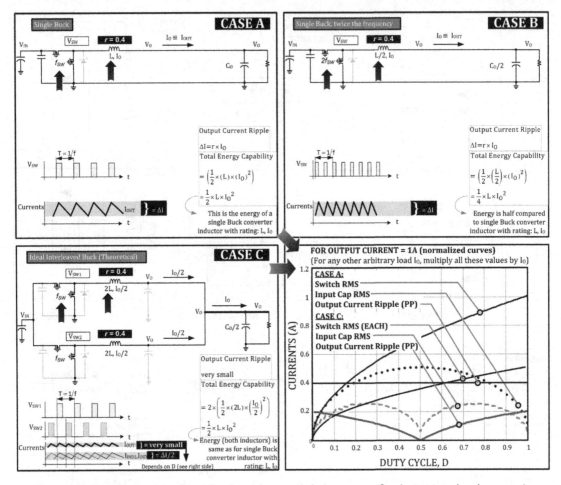

Figure 13.7: Advantages of interleaving (the graph is in terms of unit Ampere load current).

reveal the underlying concept of *power scaling in DC–DC converters*, something which is very hard to see from the usual way of writing out the RMS equations as indicated above.

Using our unique method of writing out the RMS current stress equations, we now recognize the fact that current stresses (AVG and RMS) are all *proportional to load current* (for a given *r* and fixed *D*). We thus start to realize what *scaling* implies. For example, this can mean several things:

- In terms of ability to handle stresses, a 100 W power supply will require an output capacitor roughly *twice*, in terms of capacitance and size, compared to a 50 W power supply (for the same input and output voltages). Here we are assuming that if we are using only one output capacitor, its ripple current rating is almost proportional to its capacitance. That is not strictly true though. More correctly, we can say that if a 50 W power supply has a single output capacitor of value C with a certain ripple rating I_{RIPP}, then a 100 W power supply will require two such identical capacitors — each of value C and ripple rating I_{RIPP}, *paralleled together*. That doubles the capacitance and the ripple rating (ensuring the PCB layout is conducive to good sharing too). We could then justifiably assert that *output capacitance (and its size) is roughly proportional to I_O*. Note that we are implicitly assuming switching frequency is the same for the 50 W and 100 W converters. Changing the frequency can impact capacitor selection, as we will see shortly.

- Similarly, rather generally speaking, a 100 W power supply will require an input capacitor twice that of a 50 W power supply. So, *input capacitance (and its size) will also be roughly proportional to I_O*. But more on this shortly too.

- Since heating in a FET is $I_{RMS}^2 \times R_{DS}$, and I_{RMS} is proportional to I_O, then for the same dissipation, we may initially think we would want a 100 W power supply to use a FET with one-fourth the R_{DS} of a 50 W supply. However, we are actually not interested in the *absolute* dissipation (unless thermally limited), only its percentage. In other words, if we double the output wattage of a converter, say from 50 W to 100 W, we typically expect/allow twice the dissipation too (i.e., the same efficiency). Therefore, it is good enough if the R_{DS} of the FET of the 100 W power supply is only half (not quarter) the R_{DS} of the FET used in the 50 W power supply. So in effect, **FET R_{DS} is inversely proportional to I_O**.

- We also know that for any power supply, we usually always like to set $r \approx 0.4$ for any output power. So, from the equations for L, we see that for a given r, L is inversely proportional to I_O. Which means the inductance of a 100 W power supply choke will be half that of a 50 W power supply choke. **L is therefore inversely proportional to I_O**. Note that here too, we are implicitly assuming switching frequency is unchanged. Also that the output LC product does not depend on load.

- Energy of an inductor is $(1/2) \times L \times I^2$. If L halves (for twice the wattage) and I doubles, then the required energy-handling capability of a 100 W choke must be twice that of a 50 W choke. In effect, the *size of an inductor is proportional to I_O*. Note that since L is dependent on frequency, we are again implicitly assuming that the switching frequency is unchanged here.

Concept behind Paralleling and Interleaving of Buck Converters

Hypothetically, at an abstract level, suppose we somehow implement sharing *exactly* (however we do it): for example suppose we have somehow managed to implement two paralleled, identical converters, each delivering a load current $I_O/2$. This is Case C in Figure 13.7. Both power-trains (individual converters) are connected to the same input V_{IN} and the same output V_O. They are driven at the same frequency (though the effect of synchronization, if any, between these two "phases", is only on the input/output caps as discussed later). First we show that in terms of inductor volume, paralleling *may even make matters worse.*

We know (or assume for now) from optimization principles, that for each inductor, *r* should be set to about 0.4. Since the two paralleled converters carry only $I_O/2$, we need to double the inductance of each, to achieve the same current ripple ratio for each inductor, and that is commensurate with the inductance scaling rule explained above.

Now, let us look at the energy-handling capability of each of the two inductors above. That is proportional to LI^2, and that gets halved — because *I* halves and *L* doubles. We also have two such inductors. So the total core volume (both inductors combined), is still *unchanged* from the volume of the single inductor of the original single power-train (the latter is Case A in Figure 13.7).

Note that in literature, it is often stated rather simplistically, that "interleaving helps reduce the total volume of the inductors." The logic they offer is as follows:

Single converter inductor with current "*I*":

$$\varepsilon = \frac{1}{2}LI^2$$

Two phases, each inductor carrying half the current:

$$\varepsilon = \frac{1}{2}L\left(\frac{I}{2}\right)^2 + \frac{1}{2}L\left(\frac{I}{2}\right)^2 = \frac{1}{4}LI^2$$

So, it is said the total inductor volume halves due to interleaving. Yes, it does, but only for two inductors *each with an inductance equal to that of the single converter.* However now, each phase is carrying half the current, and if we do not change *L*, the ΔI_L remains the same, but I_L (center of ramp) is half, so the current ripple ratio $\Delta I_L/I_L$ is doubled! If we are willing to accept the higher current ripple ratio in a *single*-inductor converter, we would get the same "reduction in volume." That has actually nothing to do with the concept of interleaving. It is a basic misunderstanding of energy storage concepts as detailed in *Chapter 5*. Therefore, the above "logic" is a fallacy.

What exactly do we gain in using Case C (paralleled converters) instead of Case A (single converter)? In terms of the inductor volume required to store a certain amount of energy, there seems to be no way to "cheat" physics. We learned in *Chapter 5* that for a given output power requirement and a given time interval T $(=1/f)$, the energy transferred in a certain time interval *t* is $\varepsilon = P_O \times t$. Yes, we can split this energy packet into two (or more) energy packets each handled by a separate inductor. But the total energy transferred in interval "*t*" must eventually remain unchanged, because ε/t (energy/time) has to equal P_O, and we have kept output power fixed in our current analysis. Therefore, the total inductor volume (of two inductors) in Case C is still ε (because $2 \times \varepsilon/2 = \varepsilon$). Note that just for simplicity, we assumed the logic of a flyback above, since as we had learned in *Chapter 5*, its inductor has to store *all* the energy that flows out of it (or we need to include D in the above estimates too; but the conclusions remain unchanged for any topology).

Yes, we could *double the frequency* and *spread* the energy packets more finely. We thus get Case B, and remain a single converter. Only this time, instead of two $\varepsilon/2$ packets every interval T, we need to store and transfer one packet of size $\varepsilon/2$ every $T/2$. In effect, rather intuitively speaking, we are using the *same* inductor to first deliver $\varepsilon/2$ in $T/2$; then *re-using* it to deliver the next $\varepsilon/2$ in $T/2$. "Time division multiplexing" in computer jargon. So eventually, we still get a total of ε Joules in T seconds as required (equivalently, P_O Joules in 1 s, to make up the required output power). However, (only) one inductor is present to handle $\varepsilon/2$ at any given moment. So, total inductor volume halves in Case B, as indicated in Figure 13.7.

However, going back and comparing situations belonging to the *same switching frequency* (i.e., Case A and Case C), we now point out that *in practice*, as opposed to theory, paralleling two converters may end up requiring *higher* total energy storage capability (and higher total inductor volume) — simply because there is no such thing as perfect sharing however "smart" our implementation may be. For example, to deliver 50 W output, two paralleled 25 W rated converters just won't do the job. We will have to plan for a situation where due to inherent differences, *and despite our best efforts*, one power-train (called a "phase") may end up delivering more output watts, say 30 W, and the other correspondingly, only $50 - 30 = 20$ W. But we also don't know in advance which of the two power-trains will end up carrying more current. So we will need to plan ahead for *two* 30 W rated converters, just to guarantee a 50 W combined output. In effect, we need a total inductor volume sufficient to store *60 W*, as compared to a single converter inductor that needs to store only *50 W*. Hence, we actually get an *increase* in inductor size due to paralleling (for the same current ripple ratio r).

So why don't we just stick to doubling the frequency of a *single* converter? Why even bother to consider *paralleling* converters? Nothing seems to really impress about paralleling so far. Well, we certainly want to *distribute* current stresses and the resulting dissipation across the PCB to avoid "hot-spots", especially in high-current point-of-load (POL) applications. But there is another good reason too. Looking closely at Case C in Figure 13.7 we note that the output current waveform is sketched with an AC swing that is described as "very small." No equation or number was provided here, and for good reason. We remember the output current is the *sum of the two inductor currents*. Suppose we visualize a situation where one waveform is falling at the rate of X A/µs, and the other simultaneously rising at the rate of X A/µs. If that happens, clearly *the sum of the two will remain unchanged* — we will get pure DC with no AC swing at all. How exactly can that happen? Consider the fact that in steady state, for a Buck, the only way the falling slope $(-V_O/L)$ of the inductor current can be numerically equal to the rising slope $((V_{IN} - V_O)/L)$, is if $V_{IN} - V_O = V_O$, or $V_{IN} = 2 \times V_O$, that is, $D = 50\%$. In other words, we expect that at $D = 50\%$ the output current ripple will be zero! Since for fairly large C, the voltage ripple is simply the inductor current ripple multiplied by the ESR of the output cap, we expect low output *voltage* ripple too. That is great news. See Figure 13.7 for the graphed output *current ripple* (peak to peak) on the lower right side (Mathcad generated plot). It has a minimum of zero at $D = 50\%$ as intuitively explained above.

We see that the output current ripple, and therefore the output voltage ripple, can be significantly reduced by "interleaving" — this means running the two converters with a phase shift of 180° (360° is one full clock cycle). But it also turns out that the *input RMS current is also almost zero at D = 50%* (see graph in Figure 13.7). The reason for this is that as soon as one converter stops drawing current, the other starts drawing current, so the net input current appears closer and closer to DC as duty cycle approaches 50% (except for the small AC component related to r). In other words, instead of the sharp edges of switch current waveforms that usually affect the input cap current waveforms of a Buck, we now start approaching something more similar to the smoother undulating waveforms typically found on the output cap. All these improvements are graphed in Figure 13.7 while comparing Case A to Case C (i.e., single converter of frequency f versus two paralleled ones, each with frequency f *but phase shifted as indicated by the Gate drive waveforms*). The graph in Figure 13.7 is the result of a detailed Mathcad worksheet, described in Figures 13.8 and 13.9. The waveforms from this worksheet are further shown for two cases in Figure 13.10 for a numerical example.

Note that we have made an assumption above, that the two converters switch exactly out of phase (180° apart, i.e., $T/2$ apart). As mentioned, this is called "interleaving," and in that case each power-train is more commonly referred to in literature as one "phase" of the (combined or composite) converter. So in Case C of Figure 13.7, the "converter" has two phases. We could also have more phases, and that would be generically, a *multiphase (N-phase) converter*. We have to divide T by the number of phases we desire (T/N), and start each successive converter's ON-time exactly after that sub-interval. If we run all the power-trains (i.e., phases) *in-phase* (all ON-times commencing at the same moment), the only resulting advantage in this case is we distribute the heat around. But interleaving *reduces* overall stresses and improves performance. From the output capacitor's viewpoint, the frequency effectively doubles, so the output ripple is not only much smaller, but can even be zero under the right duty cycle conditions.

At the input side, it is not the peak-to-peak ripple, but the RMS current that is very important. In general, RMS of a waveform is independent of frequency. However,

$\mu \equiv 10^{-6} \quad k \equiv 10^{3}$

$f \equiv 100 \cdot k \qquad Dtrial \equiv 0.2 \quad r \equiv 0.4 \quad Io := 1 \qquad D \equiv 0, 0.01 .. 1$

$T := \dfrac{1}{f} \quad Ton(D) := \dfrac{D}{f} \qquad Toff(D) := \dfrac{1-D}{f} \quad t := 0, 0.1 \cdot \mu .. T \quad \Delta I := r \cdot Io$

CASE A : SINGLE BUCK ANALYSIS   *Case A refers to Case A in Figure 13.7*

Inductor and Output Currents

$IonSingle(t, D) := \dfrac{\Delta I}{Ton(D)} \cdot t + \left[Io \cdot \left(1 - \dfrac{r}{2}\right) \right]$ Equation for Inductor Current during ON-time

$IoffSingle(t, D) := \dfrac{-\Delta I}{Toff(D)} \cdot (t - Ton(D)) + \left[Io \cdot \left(1 + \dfrac{r}{2}\right) \right]$ Equation for Inductor Current during OFF-time

$ISingle(t, D) := \begin{cases} IonSingle(t, D) & \text{if } 0 \le t \le Ton(D) \\ IoffSingle(t, D) & \text{if } Ton(D) < t \le T \end{cases}$ Equation for Inductor Current during ON and OFF times

Switch Currents

$IswSingle(t, D) := \begin{cases} IonSingle(t, D) & \text{if } 0 \le t \le Ton(D) \\ 0 & \text{if } Ton(D) < t \le T \end{cases}$ Equation for Switch Current during ON and OFF times

Average and RMS of Switch Current

$Isw\_avgSingle(D) := \dfrac{\displaystyle\int_0^T IswSingle(t, D) \, dt}{T}$ $Isw\_rmsSingle(D) := \left(\dfrac{\displaystyle\int_0^T IswSingle(t, D)^2 \, dt}{T} \right)^{\frac{1}{2}}$

$Isw\_avgSingle(Dtrial) = 0.193$ $Isw\_rmsSingle(Dtrial) = 0.441$

COMPARE WITH:

Closed Form Equation $Isw\_rmsSingle\_cf(D) := Io \cdot \sqrt{D \cdot \left(1 + \dfrac{r^2}{12}\right)}$ $Isw\_rmsSingle\_cf(Dtrial) = 0.45$

"cf" stands for closed-form (see equation in Appendix)

Input Capacitor Currents and Output Ripple

Input Capacitor RMS is $Icap\_rmsSingle(D) := \left(Isw\_rmsSingle(D)^2 - Isw\_avgSingle(D)^2 \right)^{\frac{1}{2}}$ $Icap\_rmsSingle(Dtrial) = 0.396$

Closed Form Equation $Icap\_rmsSingle\_cf(D) := Io \cdot \sqrt{D \cdot \left(1 - D + \dfrac{r^2}{12}\right)}$ $Icap\_rmsSingle\_cf(Dtrial) = 0.403$

Peak to Peak Output Current Ripple $Iout\_ppSingle(D) := \Delta I$ $Iout\_ppSingle(Dtrial) = 0.4$

CASE C: INTERLEAVED BUCK ANALYSIS   *Case C refers to Case C in Figure 13.7*

Now the load current of Io Amperes above is split into two power trains. So the effective load current for each power train is reset to

$Io := \dfrac{Io}{2}$ (The new AC swing (each phase))

Let us retain the same current ripple ratio for each of the two stages. So $r = 0.4$ and swing is $\Delta I := r \cdot Io$ $\Delta I = 0.2$

Inductor Current (Phase 1)

$Ion1(t, D) := \dfrac{\Delta I}{Ton(D)} \cdot t + \left[Io \cdot \left(1 - \dfrac{r}{2}\right) \right]$ Equation for Inductor Current during ON-time (Phase 1)

$Ioff1(t, D) := \dfrac{-\Delta I}{Toff(D)} \cdot (t - Ton(D)) + \left[Io \cdot \left(1 + \dfrac{r}{2}\right) \right]$ Equation for Inductor Current during OFF-time (Phase 1)

$I1(t, D) := \begin{cases} Ion1(t, D) & \text{if } 0 \le t \le Ton(D) \\ Ioff1(t, D) & \text{if } Ton(D) < t \le T \end{cases}$ Equation for Inductor Current during ON and OFF times (Phase 1)

Figure 13.8: Mathcad file (Part 1) for interleaved Buck converters as graphed in Figures 13.7 and 13.10.

Inductor Current (Phase 2)

$$\text{Ion2a}(t, D) := \frac{\Delta I}{\text{Ton}(D)} \cdot \left(t - \frac{T}{2} \right) + \left[\text{Io} \cdot \left(1 - \frac{r}{2} \right) \right] \qquad \text{Ioff2a}(t, D) := \frac{-\Delta I}{\text{Toff}(D)} \cdot \left(t - \frac{T}{2} - \text{Ton}(D) \right) + \left[\text{Io} \cdot \left(1 + \frac{r}{2} \right) \right]$$

$$\text{Ion2b}(t, D) := \frac{\Delta I}{\text{Ton}(D)} \cdot t + \text{Ion2a}(T, D) \qquad \text{Ioff2b}(t, D) := \frac{-\Delta I}{\text{Toff}(D)} \cdot t + \text{Ioff2a}(T, D)$$

$$\text{I2\_Dless50}(t, D) := \begin{vmatrix} \text{Ion2a}(t, D) & \text{if} & \frac{T}{2} < t \le \text{Ton}(D) + \frac{T}{2} \\ \text{Ioff2a}(t, D) & \text{if} & \frac{T}{2} + \text{Ton}(D) < t \le T \\ \text{Ioff2b}(t, D) & \text{otherwise} \end{vmatrix} \qquad \text{I2\_Dmore50}(t, D) := \begin{vmatrix} \text{I2\_Dless50}(t, D) & \text{if} & \text{Ton}(D) - \frac{T}{2} < t \le \frac{T}{2} \\ \text{Ion2a}(t, D) & \text{if} & \text{Ton}(D) - \frac{T}{2} + \text{Toff}(D) < t \le T \\ \text{Ion2b}(t, D) & \text{otherwise} \end{vmatrix}$$

$$\text{I2}(t, D) := \begin{vmatrix} \text{I2\_Dless50}(t, D) & \text{if} & D \le 0.5 \\ \text{I2\_Dmore50}(t, D) & \text{otherwise} \end{vmatrix}$$

Combined Current (Phase 1+2) $\qquad \text{Isum}(t, D) := \text{I1}(t, D) + \text{I2}(t, D)$

Switch and Input Capacitor Currents (each power train)

$$\text{Isw1}(t, D) := \begin{vmatrix} \text{Ion1}(t, D) & \text{if} & 0 \le t \le \text{Ton}(D) \\ 0 & \text{if} & \text{Ton}(D) < t \le T \end{vmatrix}$$

$$\text{Isw2\_Dless50}(t, D) := \begin{vmatrix} \text{Ion2a}(t, D) & \text{if} & \frac{T}{2} < t \le \text{Ton}(D) + \frac{T}{2} \\ 0 & \text{otherwise} \end{vmatrix}$$

$$\text{Isw2\_Dmore50}(t, D) := \begin{vmatrix} \text{Isw2\_Dless50}(t, D) & \text{if} & \text{Ton}(D) - \frac{T}{2} < t \le \frac{T}{2} \\ \text{Ion2a}(t, D) & \text{if} & \text{Ton}(D) - \frac{T}{2} + \text{Toff}(D) < t \le T \\ \text{Ion2b}(t, D) & \text{otherwise} \end{vmatrix}$$

$$\text{Isw2}(t, D) := \begin{vmatrix} \text{Isw2\_Dless50}(t, D) & \text{if} & D \le 0.5 \\ \text{Isw2\_Dmore50}(t, D) & \text{otherwise} \end{vmatrix}$$

Peak to Peak Current Ripple of the Combined Output

$$\text{Iout\_pp}(D) := \begin{vmatrix} \text{max} \leftarrow 0 \\ \text{for} \quad t \in 0, 0.01 \cdot \mu \,.. \, T \\ \quad \begin{vmatrix} I \leftarrow \text{Isum}(t, D) - 2 \cdot \text{Io} \\ \text{max} \leftarrow I \quad \text{if} \quad I > \text{max} \end{vmatrix} \\ 2 \cdot \text{max} \end{vmatrix}$$

Sum of Switch Current is "Input Current" $\qquad \text{Iin}(t, D) := \text{Isw1}(t, D) + \text{Isw2}(t, D)$

Average and RMS of "Sum of Switch Currents" (Iin)

$$\text{Isw\_avg}(D) := \frac{\int_0^T \text{Iin}(t, D) \, dt}{T} \qquad \text{Isw\_rms}(D) := \left(\frac{\int_0^T \text{Iin}(t, D)^2 \, dt}{T} \right)^{\frac{1}{2}}$$

Final Input Capacitor RMS

$$\text{Icap\_rms}(D) := \left(\text{Isw\_rms}(D)^2 - \text{Isw\_avg}(D)^2 \right)^{\frac{1}{2}}$$

Final Input Capacitor Waveform $\qquad \text{Icap\_IN}(t, D) := \text{Iin}(t, D) - \text{Isw\_avg}(D)$

Final Output Capacitor Waveform $\qquad \text{Icap\_OUT}(t, D) := \text{Isum}(t, D) - 2 \cdot \text{Io}$

Figure 13.9: Mathcad file (Part 2) for interleaved Buck converters as graphed in Figures 13.7 and 13.10.

interleaving does reduce the input RMS stress. That happens not due to any frequency doubling effect but because interleaving ends up changing the very shape of the input current waveform — to something closer to that of a steady stream of current (a gradual removal of the AC component depending on duty cycle).

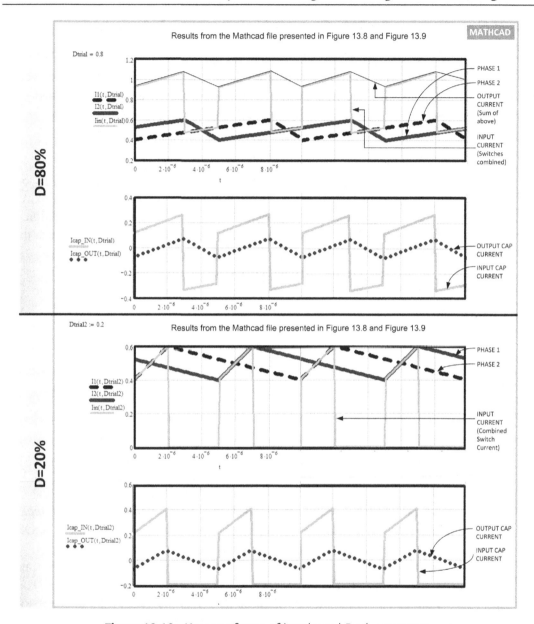

Figure 13.10: Key waveforms of interleaved Buck converters.

One drawback of interleaving as described so far, is that the inductors each still "see" a switching frequency of 1/*f*, as is obvious from their current ripple (their up-slope and down-slop durations). But later, using coupled inductors, we will discuss how we can "fool" the inductors too, into "thinking" they are at a higher switching frequency.

Closed-Form Equations for RMS Stresses of Interleaved Buck Converter

Now for some simple math to validate the key curves in Figure 13.7, and also to derive *closed-form equations*. The equations that follow can be "rigorously derived" over several rather intimidating pages if so desired. Such derivations are readily available in related literature. Here we do the same, but intuitively, and hopefully more elegantly, if not very rigorously.

(a) First let us look at the output capacitor ripple. We know that in single-phase converters, since $T_{OFF} = (1 - D)/f$, the current ripple is $\Delta I = (V_O/L) \times [(1 - D)/f]$. So the peak-to-peak ripple depends on $1 - D$. Now if we look closely at Figure 13.7, we will realize that looking at the *combined* output current, it certainly has a repetition rate of $2f$ just as we expected, but its duty cycle is not D, *but $2D$*. That is because the ON-time of each converter has remained the same, but the effective time period has been cut in half. So, the effective duty cycle for the combined output current is $T_{ON}/(0.5 \times T) = 2 \times T_{ON}/T = 2D$. Since peak to peak ripple (ΔI) is proportional to $1 - D$, for the combined output it becomes

$$I_{O\_RIPPLE\_TOTAL} = \frac{1 - 2D}{1 - D} \times I_{O\_RIPPLE\_PHASE} \quad \text{if } D \leq 50\%$$

Alternatively stated

$$\frac{I_{O\_RIPPLE\_TOTAL}}{I_{O\_RIPPLE\_PHASE}} = \frac{1 - 2D}{1 - D} \quad \text{if} \quad D \leq 50\% \quad \Leftarrow$$

This is true when the switch waveforms of both phases *do not overlap*, that is, $D < 50\%$. Which means the ON-durations of the two phases are separated in time. Of course, in that case, the OFF-durations are the ones that overlap. Recognizing this symmetry, we can actually quickly figure out what happens when the *reverse happens*, that is, when the switch waveforms overlap, or when $D > 50\%$. Because then the OFF-times do *not* overlap. We keep in mind that these are just geometrical waveforms. There is no significance to what we call the ON-time and what we call the OFF-time as far as the waveforms go. Therefore, we can easily reverse the roles of D and D', just as we did long ago in Figure 7.3. So, now we can easily guess what the peak-to-peak ripple relationship is for the case $D > 50\%$, that is, for the case $D' < 50\%$! We thus get

$$\frac{I_{O\_RIPPLE\_TOTAL}}{I_{O\_RIPPLE\_PHASE}} = \frac{1 - 2D'}{1 - D'} = \frac{2D - 1}{D} \quad \text{if } D \geq 50\% \quad \Leftarrow$$

Note that this represents a current ripple reduction for the *combined output of the interleaved Buck*, compared to the current ripple *of each of its two phases* (we are not comparing this with the single converter case anymore).

(b) Now let us look at the input cap RMS current. From the viewpoint of the input capacitor, the switching frequency is again $2f$, and the duty cycle is $2D$. The current is drawn in spikes of height I_O, which is 0.5 A for every 1 A (combined) output current. We are using the flat-top approximation here (ignoring the small term involving r). From Figure 7.6, for a Buck, we can approximate the input cap RMS as

$$I_{CAP\_IN\_RMS} = I_O \sqrt{D(1-D)} \quad \text{(single-phase converter)}$$

So for the interleaved converter, though changing effective frequency has no effect on the RMS stress of the input cap, *the effective doubling of duty cycle profoundly affects the wave shape and the computed RMS*. We get the following closed-form equation:

$$I_{CAP\_IN\_RMS} = I_O \sqrt{2D(1-2D)} \quad \text{(2-phase converter)} \quad \text{if } D \leq 50\% \quad \Leftarrow$$

where I_O is the output current of *each* phase (half the combined current output). When the waveforms overlap, using the same logic as above, we can easily guess the input cap RMS as

$$I_{CAP\_IN\_RMS} = I_O \sqrt{2D'(1-2D')} = I_O \sqrt{2(1-D)(2D-1)} \quad \text{(2-phase converter)} \quad \text{if } D \geq 50\% \quad \Leftarrow$$

To cement all this, a quick numerical example is called for.

Example:

We have an interleaved Buck converter with $D = 60\%$, rated for 5 V at 4 A. Compare the output ripple and input cap RMS to a single-phase converter delivering 5 V at 4 A.

Single-Phase Case: We typically set $r = 0.4$. So for 4 A load, the inductance is chosen for a swing of 0.4 A $\times$ 4 A = 1.6 A. That is just the peak-to-peak output current ripple. This agrees with Figure 13.7, which is for 1 A load. So, scaling that four times for a 4 A load, we get 0.4 A $\times$ 4 A = 1.6 A.

The input cap RMS with flat-top approximation is from our equation

$$I_{CAP\_IN\_RMS\_SINGLE} = I_O \sqrt{D(1-D)} = 4\sqrt{0.6(0.4)} = 1.96 \text{ A}$$

This also agrees with the Mathcad-based plot in Figure 13.7 in which we get about 0.5 A at $D = 0.6$ for 1 A load. Scaling that for 4 A load current, we get 0.5 A $\times$ 4 A = 2 A RMS; slightly higher than the 1.96 A result from our closed-form equation.

Interleaved Case: This time we split the 4 A load into 2 A per phase. The reduction in ripple equation for $D > 50\%$ gives us a "ripple advantage" of

$$\frac{I_{O\_RIPPLE\_TOTAL}}{I_{O\_RIPPLE\_PHASE}} = \frac{2D-1}{D} = \frac{2(0.6)-1}{0.6} = 0.333$$

The peak-to-peak ripple of each phase is 0.4 A × 2 A = 0.8 A. Therefore, the ripple of the combined output current must be 0.333 A × 0.8 A = 0.27 A. Comparing with the plot in Figure 13.7, we have at D = 0.6, a peak-to-peak ripple of 0.067 A. But that is for 1 A load. So, for 4 A load we get 4 A × 0.067 A = 0.268 A, which agrees closely with the 0.27 A result from the closed-form equation. Now calculating the input cap RMS from our equations, we get

$$I_{CAP\_IN\_RMS} = I_O\sqrt{2(1-D)(2D-1)} = 2\sqrt{2(1-0.6)(2(0.6)-1)} = 0.8 \text{ A}$$

Comparing with the plot in Figure 13.7, we get at D = 0.6, input cap RMS to be 0.2 A. But that is for 1 A load. So, for 4 A load we get 4 A × 0.2 A = 0.8 A, which agrees exactly with the 0.8 A result from the closed-form equation.

We have shown that the results of the Mathcad spreadsheet agree with the closed form equations above (intuitively derived). We can use either of them.

Summarizing, we see that for a single-phase converter the output current ripple (peak-to-peak) was 1.6 A, and the input cap RMS was 2 A. For the interleaved solution, output ripple fell to 0.27 A, and the input cap RMS fell to 0.8 A. This represents a significant improvement and shows the beneficial effects of interleaving (paralleling out-of-phase).

We remember that in a Buck, the dominant concern at the input cap is mainly the RMS stress it sees, whereas on the output, it is the voltage ripple that determines the capacitance. So, interleaving can greatly help in decreasing the sizes of both input and output caps. The latter reduction will *help improve loop response in the bargain* — smaller L and C components charge and discharge faster, and can therefore respond to sudden changes in load much faster too. But besides that, we can also now *revisit our entire rationale for trying to keep inductors at an "optimum" of r = 0.4*. We recall that was considered an optimum for the entire (single-phase) converter (see Figure 5.7). But now we can argue that we *don't* have a single converter anymore. And further, if we are able to reduce RMS stresses and the output ripple by interleaving, why not consciously *increase r* (judiciously though)? That could dramatically reduce the size of the inductor. In other words, *for a given output voltage ripple* (not inductor current ripple), we really can go ahead and increase r (reduce inductance) of each phase. We already know from *Chapter 5* that reducing inductance typically *reduces* the size of the inductor for a given application. Certainly, the amount of energy we need to cycle through the total inductor volume is fixed by physics as explained in *Chapter 5*. However, the *peak* energy storage requirement significantly goes down if we decrease inductance (increase r). That was also explained as part of the magnetics paradox discussed in *Chapter 5*.

One limitation of increasing r is that as we reduce load, we will approach discontinuous conduction mode (DCM) for higher and higher minimum load currents. That is why we have used synchronous Buck converters in Figure 13.7 to illustrate the principle of interleaving. As we learned in *Chapter 9*, most synchronous converters never go into DCM; they remain in CCM right down to zero load current condition (even getting down to sinking the load current if necessary). The efficiency suffers of course, but it works at constant frequency and the effects of interleaving will still apply. Another advantage of intelligent multiphase operation is that at light loads we can start to "shed" phases — for example, we can change over from say a six-phase multiphase converter to a four-phase converter at medium loads, and then go to a two-phase converter at lighter loads, and so on. That way *we reduce switching losses*, because, in effect, we are reducing overall switching frequency, and thereby greatly improving light-load efficiency.

Interleaved Boost PFC Converters

In *Chapter 9*, in particular in Figures 9.1−9.3, we had shown that a Boost is just a Buck with input and output swapped. Therefore, now that we have derived all the equations for synchronous interleaved Buck converters, there is no reason why the very same logic and equations can't be used for a synchronous interleaved Boost stage. The mapping we need to do is $D \leftrightarrow (1 - D)$, $V_{IN} \leftrightarrow V_O$, and $I_O \leftrightarrow I_O/(1 - D)$, since the ON-time of a Buck becomes the OFF-time for the corresponding Boost. We just apply this mapping to the stress equations derived above.

In *Chapter 14* we will discuss power factor correction (PFC) using a Boost converter. Now we can see that for higher power stages, not only can we parallel FETs as described in Figure 13.6, but we can get significant reduction in input and output capacitor size by interleaving (using two phases), just as we have done for the Buck converter above.

In a PFC stage the analysis is harder to do because of the sine wave input. Actual lab measurements provide the following useful rule-of-thumb: *the output cap RMS current is halved due to interleaving, as compared to a single-stage Boost PFC stage.*

Interleaved Multioutput Converters

The concept of interleaving has been arbitrarily "extended" to include cases where each power-train delivers a *different* output voltage. Check out the LM2647 for example. Clearly, now the output capacitor/s cannot be shared. But the vendors of such ICs claim dramatic improvements in the RMS of the shared *input* cap as a justification for this multioutput interleaved architecture. But how exactly do they "prove" lower input RMS current on paper? They do it by loading *both power-trains to maximum rated load*, and computing the RMS of the resulting input cap waveform. Yes, they do get a lower overall input capacitor size compared to completely separated power-trains. However, on closer examination that "RMS reduction" can be misleading — because as we pointed out, we have to be wary of stresses that do not have a maximum at extreme corners of line and load. In fact, capacitor RMS fits right into that category. In this case, if we *unload* one output, and only load the other (to its max load), we usually get much higher input cap RMS current than if both outputs were fully loaded. So what exactly did we gain? Nothing! And besides that logical problem, when we share one input cap between multiple outputs, we cannot but avoid *cross-coupling* of the two output channels. For example, if one output suddenly sees a load transient, its input cap voltage wiggles a bit, and since that is intended to be the input cap for the other power-trains too, some of that wiggle gets transmitted to the other output rail too. This is called cross-coupling, and the only solution is to actually provide completely *separate* input capacitors, placed very close to the input of each power train, and to separate those capacitors for high frequencies, either through long PCB traces, or actual input inductors, star-pointing at the input voltage source, which usually has another bulk capacitor at the point where it enters the PCB. In other words, the "advantage," if any, of this *multioutput* interleaved converter, can be almost non-existent at a practical level. Really, the product definition is itself flawed in this case. The datasheet of LM2647 was written by this author actually, and the above product definition problem was pointed out and explained in the datasheet.

Part 3: Coupled Inductors in Interleaved Buck Converters

Overview

So far we have only considered the configuration in which the two power-trains (or phases) are joined together "at their ends," thus impacting the input and output capacitors (beneficially it turned out). That is Case C in Figure 13.7. In between, where the inductors lie, the power-trains are still completely independent. That is why the inductors still had a current ripple based on f, not on $2f$ for example. We can now visualize that the effects of interleaving (coupling) can be felt on the inductors if we *somehow bring them together* instead of having them fully independent. The only way to do that practically is through a *non-physical connection*, that is, through *magnetic* coupling.

This area is of great interest in modern voltage regulator modules (VRMs). Multiphase converters, often with proprietary magnetic structures, are getting increasingly popular as a means of providing the huge dI/dt's demanded at sub-volt levels by modern microprocessor ("μP") ICs. The constraints for VRMs are slightly different. Efficiency is always important of course, since a lot of processing equipment is battery-powered. But besides that, of great importance is the very tight regulation and extremely fast transient response (minimal

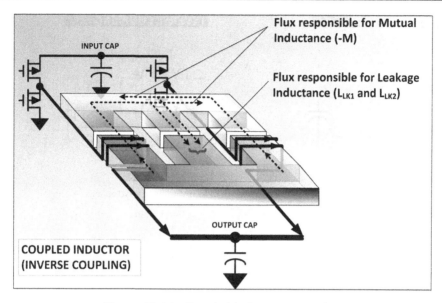

Figure 13.11: Coupled inductor magnetics.

overshoot or undershoot in response to sudden changes in load). Another driving concern is to make the VRM compact. So, putting several windings on a shared core seems to be of natural interest. Note that in fact, the output voltage ripple of the VRM (i.e., the input ripple to the μP), is not of such great concern here, since we are usually not powering any extremely noise-sensitive *analog* chips. And in any case, if we do need such a quiet rail locally for certain functions, that can always be derived by using a small post-filter (LC- or RC-based) off the main rail.

There are many ways to couple the windings of the inductors. However, to avoid a rather unnecessary and intimidating discussion, we will focus on trying to understand the currently popular technique called "inverse coupling." Also, we will focus only on two-phase converters here. But once the concepts are understood those ideas can easily be extended to *N*-phases. We will also initially restrict ourselves to more common application with *non-overlapping* switch waveforms ($D < 50\%$ for a two-phase converter). But finally, we will cover the area of overlapping switch waveforms too (i.e., $D > 50\%$).

Before diving into some fairly complex but unavoidable math, a practical implementation of inverse coupling is illustrated in Figure 13.11. Both direct and inverse coupling are explained in Figure 13.12. Here are the points to note as we go along in our development of our concepts and ideas.

- Note the direction of the windings in Figure 13.11. Part of the flux from one winding goes through the other winding and *opposes* the flux from the other winding. The

DIRECT COUPLING

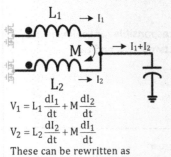

$$V_1 = L_1 \frac{dI_1}{dt} + M \frac{dI_2}{dt}$$

$$V_2 = L_2 \frac{dI_2}{dt} + M \frac{dI_1}{dt}$$

These can be rewritten as

$$V_1 = L_1 \frac{dI_1}{dt} + M \frac{dI_2}{dt} + \left(M \frac{dI_1}{dt} - M \frac{dI_1}{dt}\right)$$

$$V_2 = L_2 \frac{dI_2}{dt} + M \frac{dI_1}{dt} + \left(M \frac{dI_2}{dt} - M \frac{dI_2}{dt}\right)$$

Which can be rearranged as

$$V_1 = (L - M) \frac{dI_1}{dt} + M \frac{dI_0}{dt} = L_{LK} \frac{dI_1}{dt} + M \frac{dI_0}{dt}$$

$$V_2 = (L - M) \frac{dI_2}{dt} + M \frac{dI_0}{dt} = L_{LK} \frac{dI_2}{dt} + M \frac{dI_0}{dt}$$

So we have two uncoupled (leakage) inductors of value $L{-}M \equiv L_{LK}$ (assuming $L_1{=}L_2$), and a coupled (combined) inductance M. The equivalent diagram is shown below.

INVERSE COUPLING

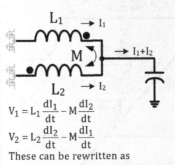

$$V_1 = L_1 \frac{dI_1}{dt} - M \frac{dI_2}{dt}$$

$$V_2 = L_2 \frac{dI_2}{dt} - M \frac{dI_1}{dt}$$

These can be rewritten as

$$V_1 = L_1 \frac{dI_1}{dt} - M \frac{dI_2}{dt} + \left(M \frac{dI_1}{dt} - M \frac{dI_1}{dt}\right)$$

$$V_2 = L_2 \frac{dI_2}{dt} - M \frac{dI_1}{dt} + \left(M \frac{dI_2}{dt} - M \frac{dI_2}{dt}\right)$$

Which can be rearranged as

$$V_1 = (L + M) \frac{dI_1}{dt} - M \frac{dI_0}{dt} = L_{LK} \frac{dI_1}{dt} - M \frac{dI_0}{dt}$$

$$V_2 = (L + M) \frac{dI_2}{dt} - M \frac{dI_0}{dt} = L_{LK} \frac{dI_2}{dt} - M \frac{dI_0}{dt}$$

So we have two uncoupled (leakage) inductors of value $L{+}M \equiv L_{LK}$ (assuming $L_1{=}L_2$), and a coupled (combined) branch with effective inductance -M. The equivalent diagram is:

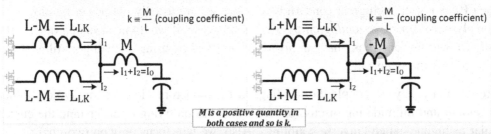

The difference between Direct Coupling and Inverse Coupling is that M appears as a positive coupled inductance in Direct Coupling, and a negative one in Inverse Coupling

Figure 13.12: Models to explain direct and inverse coupling.

opposing flux contributions (in the outer limbs), however, *do not rise and fall in unison*, because they are out-of-phase (interleaved). But their relative directions do indicate *inverse* (opposing) coupling.

* There is typically a certain air gap provided on (both) the outer limbs, but also another air gap of a different gap-length provided on the center limb. By increasing the air gap

on the center limb for example, we can make more flux go through the outer limbs, and that will increase the amount of (inverse) coupling. This is one way to adjust (fine-tune) the "coupling coefficient" *k*. Think of flux mentally as water flowing in pipes. If we block one pipe, it tries to force its way through another. The air gap amounts to a constriction in the pipe.

> *Note: In a* subsequent *figure (the Mathcad file in Figure 13.15), we have called the coupling coefficient "α" instead, to avoid confusion with the "k" used for kilo in that file.*

- Figure 13.12 represents one easy model used to represent coupled inductors. "*M*" is the mutual inductance. Keep in mind it is just one of many models out there used to help better understand coupled inductors. Other models can get even more intimidating, and besides, none of the models are considered perfect — they all have pros and cons. Some are "non-symmetrical models" and even harder to digest. For example, the question lingers why two supposedly identical windings look so different from each other. In our model too, the *node between L and M is a "fictitious node."* In a way, we have resorted to fiction to understand reality! That is why we don't really proceed much further down this path of models, except to find the starting-point equations for later analysis.

- In Figure 13.13, we write out the same equations for inverse coupling shown in Figure 13.12, but then actually analyze the actual relationships between V_1 and V_2 (the voltages *across* the inductors of the respective phases), as we switch the FETs. We examine the voltages across the inductors, and thus identify four possible states (called A, B, C, and D). Of these, one (D) is actually identical to B (look at their voltages V_1 and V_2). In other words, whatever we work out for B will apply to D. So, there are just *three* distinct states that we need to study further, not four. These are A, B, and C.

- We know the effect of the mutual inductance on the three states (A, B, and C), from the equations we wrote for inverse coupling. And we also know the actual voltages that appear across the (actual) inductors in the three states. Note that at this point we are trying to connect our real-world (actual) inductor with the inverse coupling model. And we thus find that over each of these three states, the simple equation $V = L \, dI/dt$ no longer seems to apply. *In other words, the slopes (dictated by the inverse coupling equations) do not match the assumed inductance L any more (because we realize that one winding influences the other, and vice versa).*

- We follow this through and try to *reconcile* matters. So, now we say that if we are to somehow force $V = L \, dI/dt$ to be still true (artificially), *we need to re-define or re-state inductance* (since dI/dt is known, and so is the applied voltage across the inductor). We thus generate an "equivalent inductance" for each of the three (four) states based on $L = V/(dI/dt)$. We call these L_A, L_B, L_C (and L_D which we know just equals L_B). The

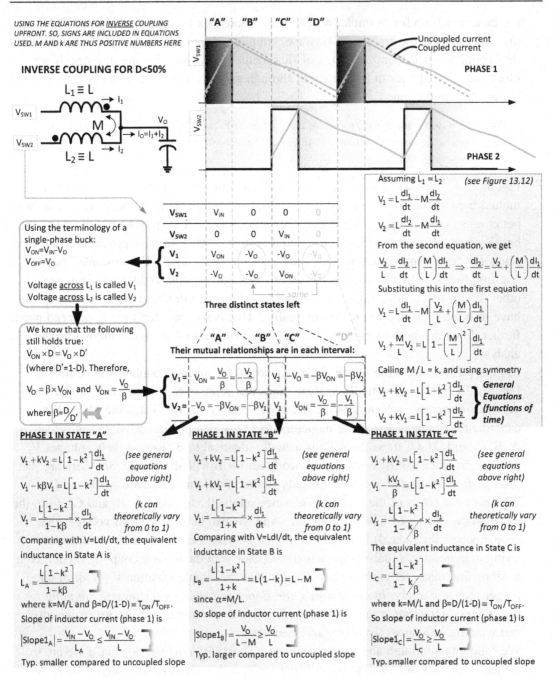

USING THE EQUATIONS FOR <u>INVERSE</u> COUPLING UPFRONT. SO, SIGNS ARE INCLUDED IN EQUATIONS USED. M AND k ARE THUS POSITIVE NUMBERS HERE

INVERSE COUPLING FOR D<50%

$L_1 \equiv L$

$L_2 \equiv L$

Uncoupled current
Coupled current

PHASE 1

PHASE 2

Using the terminology of a single-phase buck:
$V_{ON} = V_{IN} - V_O$
$V_{OFF} = V_O$

Voltage <u>across</u> L_1 is called V_1
Voltage <u>across</u> L_2 is called V_2

We know that the following still holds true:
$V_{ON} \times D = V_O \times D'$
(where $D' = 1 - D$). Therefore,

$V_O = \beta \times V_{ON}$ and $V_{ON} = \dfrac{V_O}{\beta}$

where $\beta = \dfrac{D}{D'}$

	"A"	"B"	"C"	"D"
V_{SW1}	V_{IN}	0	0	0
V_{SW2}	0	0	V_{IN}	0
V_1	V_{ON}	$-V_O$	$-V_O$	$-V_O$
V_2	$-V_O$	$-V_O$	V_{ON}	$-V_O$

same

Three distinct states left

"A" "B" "C" "D"
Their mutual relationships are in each interval:

$V_1 =$ $V_{ON} = \dfrac{V_O}{\beta} = \dfrac{V_2}{\beta}$ $\quad V_2$ $-V_O = -\beta V_{ON} = -\beta V_2$ $\quad V_1$ $V_{ON} = \dfrac{V_O}{\beta} = \dfrac{V_1}{\beta}$

$V_2 =$ $-V_O = -\beta V_{ON} = -\beta V_1$ $\quad V_1$ $V_{ON} = \dfrac{V_O}{\beta} = \dfrac{V_1}{\beta}$

Assuming $L_1 = L_2$ (see Figure 13.12)

$V_1 = L\dfrac{dI_1}{dt} - M\dfrac{dI_2}{dt}$

$V_2 = L\dfrac{dI_2}{dt} - M\dfrac{dI_1}{dt}$

From the second equation, we get

$\dfrac{V_2}{L} = \dfrac{dI_2}{dt} - \left(\dfrac{M}{L}\right)\dfrac{dI_1}{dt} \Rightarrow \dfrac{dI_2}{dt} = \dfrac{V_2}{L} + \left(\dfrac{M}{L}\right)\dfrac{dI_1}{dt}$

Substituting this into the first equation

$V_1 = L\dfrac{dI_1}{dt} - M\left[\dfrac{V_2}{L} + \left(\dfrac{M}{L}\right)\dfrac{dI_1}{dt}\right]$

$V_1 + \dfrac{M}{L}V_2 = L\left[1 - \left(\dfrac{M}{L}\right)^2\right]\dfrac{dI_1}{dt}$

Calling $M/L = k$, and using symmetry

$V_1 + kV_2 = L\left[1 - k^2\right]\dfrac{dI_1}{dt}$ } **General Equations (functions of time)**

$V_2 + kV_1 = L\left[1 - k^2\right]\dfrac{dI_2}{dt}$

PHASE 1 IN STATE "A"

$V_1 + kV_2 = L\left[1 - k^2\right]\dfrac{dI_1}{dt}$ (see general equations above right)

$V_1 - k\beta V_1 = L\left[1 - k^2\right]\dfrac{dI_1}{dt}$

$V_1 = \dfrac{L\left[1 - k^2\right]}{1 - k\beta} \times \dfrac{dI_1}{dt}$ (k can theoretically vary from 0 to 1)

Comparing with $V = L dI/dt$, the equivalent inductance in State A is

$L_A = \dfrac{L\left[1 - k^2\right]}{1 - k\beta}$

where $k = M/L$ and $\beta = D/(1-D) \equiv T_{ON}/T_{OFF}$.
Slope of inductor current (phase 1) is

$|Slope1_A| = \dfrac{V_{IN} - V_O}{L_A} \leq \dfrac{V_{IN} - V_O}{L}$

Typ. smaller compared to uncoupled slope

PHASE 1 IN STATE "B"

$V_1 + kV_2 = L\left[1 - k^2\right]\dfrac{dI_1}{dt}$ (see general equations above right)

$V_1 + kV_1 = L\left[1 - k^2\right]\dfrac{dI_1}{dt}$

$V_1 = \dfrac{L\left[1 - k^2\right]}{1 + k} \times \dfrac{dI_1}{dt}$ (k can theoretically vary from 0 to 1)

Comparing with $V = L dI/dt$, the equivalent inductance in State B is

$L_B = \dfrac{L\left[1 - k^2\right]}{1 + k} = L(1 - k) = L - M$

since $\alpha = M/L$.
So slope of inductor current (phase 1) is

$|Slope1_B| = \dfrac{V_O}{L - M} \geq \dfrac{V_O}{L}$

Typ. larger compared to uncoupled slope

PHASE 1 IN STATE "C"

$V_1 + kV_2 = L\left[1 - k^2\right]\dfrac{dI_1}{dt}$ (see general equations above right)

$V_1 - \dfrac{kV_1}{\beta} = L\left[1 - k^2\right]\dfrac{dI_1}{dt}$

$V_1 = \dfrac{L\left[1 - k^2\right]}{1 - k/\beta} \times \dfrac{dI_1}{dt}$ (k can theoretically vary from 0 to 1)

The equivalent inductance in State C is

$L_C = \dfrac{L\left[1 - k^2\right]}{1 - k/\beta}$

where $k = M/L$ and $\beta = D/(1-D) \equiv T_{ON}/T_{OFF}$.
So slope of inductor current (phase 1) is

$|Slope1_C| = \dfrac{V_O}{L_C} \geq \dfrac{V_O}{L}$

Typ. smaller compared to uncoupled slope

Figure 13.13: Derivations of equivalent inductances and slopes based on coupling coefficient k.

equations for equivalent inductance in all three states are provided in Figure 13.13. Note from the derivations of equivalent inductances, these inductances apply even if $D > 50\%$, as discussed later (Figure 13.16).

- In Figure 13.14, we take this concept of equivalent inductances and calculate the change in output (combined) ripple due to coupling ("CPL," compared to the uncoupled case denoted as "UCPL"). We get

$$\text{Output\_Ripple\_Change} = \frac{\text{Output\_Ripple\_PP\_CPL}}{\text{Output\_Ripple\_PP\_UCPL}} = \frac{1}{1 - k}$$

(This is for the combined output — into the output cap). We see that since k varies from 0 (no coupling) to 1 (max coupling), the *output ripple change will always be greater than 1*. So, (inverse) coupling will *always* cause the ripple of the *combined currents* (i.e., the *output ripple*) to worsen *as compared to the uncoupled* (interleaved) *case*. That seems understandable, if not desirable, since in a way, by inverse coupling, we cancel some of the inductance out, so we do expect higher ripple. As mentioned, that is typically acceptable in modern microprocessors so long as we stay within a tight regulation window (by good transient response and/or droop regulation methods as discussed in Part 4 of this chapter).

- Looking in from the output side into the converter, we can confirm through simulations, that it does seem that a *smaller* inductance is now driving the output current. Simulations also confirm that this lowered inductance is also accompanied by a great improvement in transient response. All that is intuitively expected, since we know smaller L and C's can charge and discharge more rapidly, and therefore don't "come in the way" of quickly dumping more energy into the load (a positive dI/dt), or removing energy (negative dI/dt). We can do some more detailed geometrical calculations (not shown here) to see how the current at the end of a switching cycle ramps up ("ΔI_X") as compared to the current at the start of the cycle, due to a small and sudden increase in duty cycle (ΔD), on account of a sudden load demand. The results of that are quoted here:

$$\left(\frac{\Delta I_X}{\Delta D}\right)_{\text{CPL}} = \frac{V_{\text{IN}}}{L_{\text{B}} \times f}$$

It is indeed surprising that the terms involving L_A and L_C have canceled out here, leaving only L_B. Researchers report they have validated this improvement of transient response through simulations. They have also tried to intuitively explain it, but it is not easy to explain or understand. Note that if the inductors were not coupled, we would have got (since L_B is equal to L if k = 0)

$$\left(\frac{\Delta I_X}{\Delta D}\right)_{\text{UCPL}} = \frac{V_{\text{IN}}}{L \times f}$$

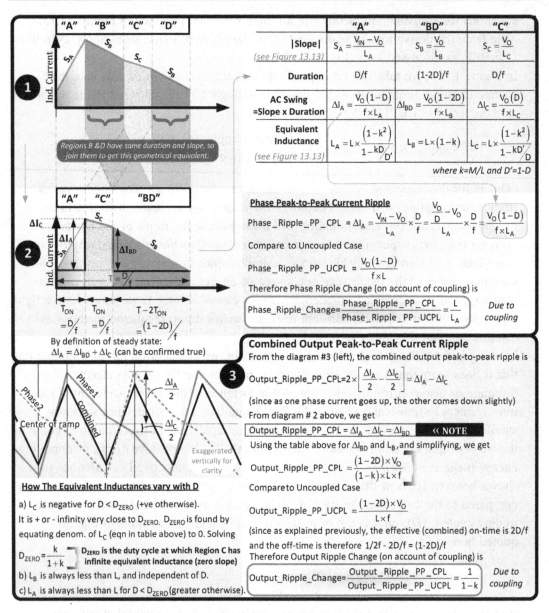

Figure 13.14: Equations for ripple in an interleaved Buck converter with coupled inductors.

Therefore, the improvement in transient response is

$$\frac{(\Delta I_X/(\Delta D))_{\text{CPL}}}{(\Delta I_X/(\Delta D))_{\text{UCPL}}} = \frac{L}{L_B} = \frac{Ł}{Ł(1-k)} = \frac{1}{1-k}$$

This is exactly the factor by which the (combined) output ripple increased. So that makes sense. Further, since k is between 0 and 1, the coupled transient response must be better than the uncoupled case. For example, if we set $k = 0.2$, we get a 25% improvement in response. For this reason, L_B is referred to in literature as the "transient equivalent inductance." We can think of it intuitively as the "output equivalent inductance." When there is a sudden change in load, it seems to be coming through a smaller inductance called L_B.

• We see that the transient response is good, but the (combined) *output ripple has worsened* on account of coupling ($1/(1 - k)$ above). *Note that this statement is true only when compared to an identical interleaved Buck, with the same inductances, but completely uncoupled (i.e., setting k = 0).* But, wouldn't we have achieved exactly the same result by simply lowering the inductance L of each phase closer to L_B and sticking to simple uncoupled inductors? That is true. *But a strange thing happens here that justifies all the trouble.* We recall that generally, the RMS on the *input* capacitor and the RMS of the switch current typically increase as we lower the inductance (because we get more "peaky" waveforms). In the case of coupled inductors, the phase ripple actually decreases on account of coupling". From Figure 13.14

$$\text{Phase\_Ripple\_Change} = \frac{\text{Phase\_Ripple\_PP\_CPL}}{\text{Phase\_Ripple\_PP\_UCPL}} = \frac{L}{L_A}$$

So, *provided L_A is greater than L*, the current ripple in each phase decreases. Surprisingly, the phase ripple does not depend on the lower equivalent *output* inductance L_B (mentioned above), but on L_A (equivalent inductance during the ON-time). This makes the phase current less peaky and leads to a reduction in the switch RMS current and the RMS of the input cap current. Now, a reduction in switch/input RMS is *not* intuitively commensurate with a smaller inductance — but a *larger* inductance. As expressed by the phase ripple expression, **we can improve efficiency, despite appearing to reduce output inductance for achieving good transient response.** For this reason, L_A is sometimes referred to in literature as the "steady-state equivalent inductance." We can think of it intuitively as the "input equivalent inductance."

By how much is L_A greater than L? From Figure 13.14, the relationship is

$$\frac{L_A}{L} = \frac{(1 - k^2)}{1 - (kD/(D'))}$$

This means we can use our usual equation for the RMS switch current in a Buck FET (see Figure 7.3)

$$I_{RMS} = I_L \sqrt{D\left(1 + \frac{r^2}{12}\right)}, \quad \text{where} \quad r = \frac{V_O}{I_O \times L \times f} \times (1 - D)$$

All we need to do here to calculate the RMS switch current for the interleaved coupled inductor Buck is to set $I_L \rightarrow I_O/2$, and $L \rightarrow L_A$.

- In Figure 13.15, we have plotted out the equivalent inductances. We see that L_B is always less than 1 (i.e., L here), indicating better transient response due to coupling. But L_A is greater than 1 (i.e., L) only above a certain threshold we have called D_{ZERO} here. More on that below.
- In Figure 13.15, we see that L_B is always less than L, as expected. Coming to L_C, that is actually negative below a certain duty cycle. That too occurs below the threshold D_{ZERO}. Note that the inductance L_C determines the slope of the wiggle in the phase ripple waveforms (Region C in Figure 13.13). The direction of the wiggle depends on the sign of L_C. Exactly at D_{ZERO}, L_C swings from positive infinite inductance to negative infinite inductance. Which means the current in Region C changes from a slope described as "minus zero" (slightly downwards) to "plus zero" (slightly upwards). At exactly D_{ZERO}, the current is flat (i.e., zero slope). Note that in Figure 13.13, we have assumed the phase current waveform in Region C continues to fall. If L_C is negative, that would not be serious in itself. However, from Figure 13.14 we see that below a certain "D_{ZERO}," L_A *is less than L*. L_A being the "steady-state equivalent inductance" as explained above, this means that now the phase ripple with coupled inductors is *worse* than the uncoupled case. We have thus lost one major advantage of coupling (other than the fast transient response which could have been achieved simply by using smaller uncoupled inductances). To achieve any improvement in efficiency (and reduction in RMS current in the switch and input cap), *we must therefore always operate only in the region $D > D_{ZERO}$*. This means that L_C must be positive, and so the phase current waveform would look like what we have sketched out in Figures 13.13 and 13.14. The wiggle in Region C should *not* be upward (positive slope). In the former figure, we have derived the equation for D_{ZERO}.

$$D_{ZERO} = \frac{k}{1 + k}$$

We can solve this for a *max coupling coefficient*

$$k_{MAX} = \frac{D}{1 - D} \equiv \frac{D}{D'}$$

In Figure 13.15 we have also provided the best value for k as a function of D.

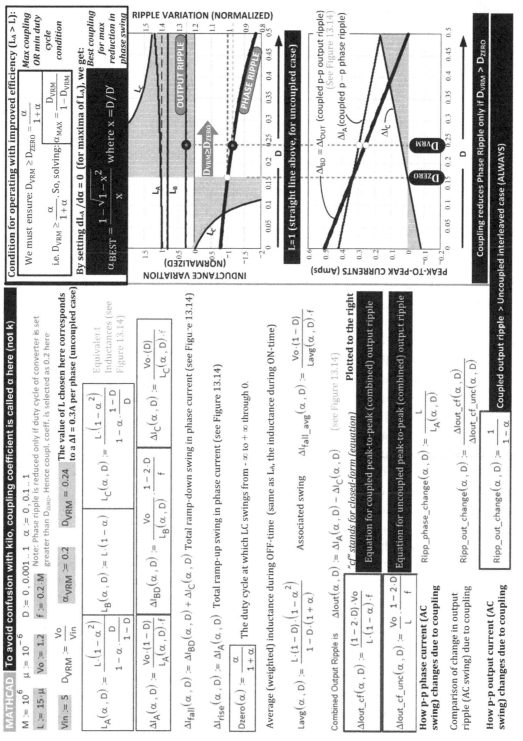

Figure 13.15: Using Mathcad to plot equivalent inductance variations a٦d ripple for coupled inductor Buck converters operating with $D < 50\%$.

- Actually, there is another practical boundary that can occur if the *denominator of L_A equals zero*. We need to avoid that singularity too. Combined with the restriction from avoiding the zero denominator of L_C, we can finally define a usable operating area when using inverse coupled inductors. The final result is that the *duty cycle of the VRM in steady state should never be set outside the following limits*

$$D_{ZERO} < D_{VRM\_SteadyState} < 1 - D_{ZERO} \quad \Leftarrow$$

  Note that just as we call $1-D$ as D' sometimes, we can call $1 - D_{ZERO}$ as D'_{ZERO} if we want.

- In Figure 13.15, we also have a numerical example. A 200 kHz VRM is stepping down 5 V to 1.2 V using coupled inductors ($k = 0.2$), each of 15 µH. So, the duty cycle is $1.2/5 = 0.24$. We can calculate from the above equation that $k_{MAX} = 0.32$. We ask: what will happen if we have coupling in excess of this value? Looking at the waveforms on a scope, we will then see that the current segment of the Phase 1 inductor current in Region "C" will rise *up* rather than *down*. This can be explained through *excessive coupling* caused by the switch of Phase 2 turning ON. Because when the switch of Phase 2 turns ON, it produces a positive rate of change of flux in its winding, but a negative change (decrease) in flux in the winding of Phase 1. Since change of flux is always to be opposed, Phase 1 suddenly decreases its rate of fall of current, or in the worst-case, even increases its current, just to try to keep the flux through it constant. But with such high coupling, L_A, the equivalent input (or steady) inductance becomes less than the uncoupled case inductance, L. So now we have a lower efficiency than the uncoupled case as explained above. That is obviously not a good way to operate.

- In Figure 13.16, we extend our treatment into the region $D > 50\%$ using the same numerical example as in Figure 13.15 to illustrate what happens over the entire input range. The key observations are as follows:
  (a) In Figure 13.13, the derivations for equivalent inductances would apply to the region $D > 50\%$, and so those equations are still valid.
  (b) The durations of each region are simply flipped according to $D \leftrightarrow D'$, since any ON-state switch waveform for $D < 50\%$, when flipped over vertically become a switch waveform for $D > 50\%$. We must note that though the current waveform for $D > 50\%$ looks very similar (vertically flipped) compared to the $D < 50\%$ waveform, *the two do **not** have the same magnitudes*. Because, with inductors we have to remember that if D increases, that always means input voltage is falling, so the applied voltseconds also decreases, which causes a smaller slope. So naturally, we expect the ripple (AC swing) to decrease *progressively* as we increase D from close to 0 toward 1. We also see that the phase ripple is not only less than the uncoupled case (looking in the region between D_{ZERO} and D'_{ZERO}), but the situation gets progressively better (as D increases).

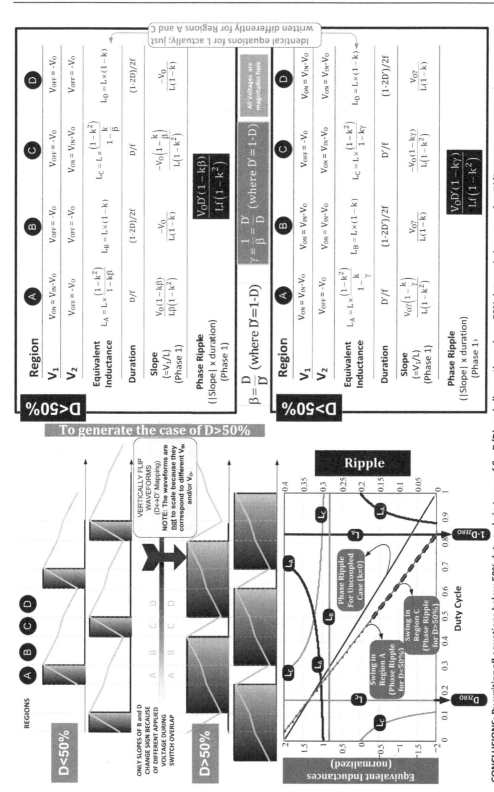

Figure 13.16: Extending our analysis of coupled inductor Buck converters to $D > 50\%$ (overlapping switch waveforms).

CONCLUSIONS: By writing all equations below 50% duty cycle in terms of $\beta = D/D'$, and all equations above 50% duty cycle in terms of $\gamma = D'/D$, we can compare the numerical results on either side of D=0.5 boundary easily, because β is *numerically the same as* γ for *symmetrical* points around D=0.5. For example, at D=0.2, β equals 0.2 and at D=0.8, γ equals 0.2. We also see that the basic equations for equivalent inductance are unchanged for D ranging from 0 to 1. But if D increases, *because of the falling input voltage (for a given output voltage)*, the Phase Ripple reduces, just as it does for the uncoupled case (but in fact falls more rapidly than the uncoupled case).

Part 4: Load Sharing in Paralleled Converters

Passive Sharing

Semiconductor vendors often release evaluation boards, such as a stand-alone "5 V at 5 A synchronous Buck converter," and so on. Many inexperienced engineers are often tempted to try to get 5 V at 10 A by just brute-force paralleling of two such "identical" boards. That never works; they find they may not even be getting 5 V at 5 A anymore, despite complaining rather indignantly to customer support.

The root cause for the strange behavior they may report is the "slight" differences between the two converters. Keep in mind that the error amplifier which is present in every feedback loop, trying to regulate each output to its set value (reference value), is always designed with very high-gain so that it can regulate the output tightly down to a few millivolts of the set value. This error amplifier therefore dramatically amplifies any existing error between the set output and the instantaneous output — much like a watch repairman who examines a "tiny problem" under a very high-magnification microscope, and thereby fixes it. Now consider this: the set output voltages of the two converters are naturally a little different because of tolerances in their resistive dividers (if present), their internal references, and so on. These set values also drift with temperature. So, when the two outputs are connected together, what may be a "natural/appropriate/correct" output voltage for one feedback loop, will likely generate a huge internal error in the feedback loop of the other converter. The two loops will end up "fighting" to determine the final output voltage. If the output "suits" one, the other will complain — by issuing forth a *huge swing in its duty cycle*, and if it suits the other, the former will complain. Depending on the actual current-sense architecture in use, things may stabilize; but usually they won't. At best, a strange and uneasy "stable condition" can occur *just after power-up*. One converter (with the highest set output) will initially try to deliver more load current than its share, but then very likely will hit its internal current (or power) limit. At that point, its output voltage will stop rising any further (if not folding back altogether, as in some current-sense architectures). The other converter then rises to the occasion, and ends up delivering the rest of the required load power — unless of course it hits current limit too on the way, in which case the output will droop — unless we have provided *another* paralleled module that can take up the slack without hitting current limit. Multiple paralleled modules can run forever like this — with several modules in current limiting (those with higher set outputs and/or lower current limits), and some not. But we don't ever get our computed/expected/ideal full power — so for example, two 25 W modules will not give 50 W; more like 30–35 W at best. Assuming even that is acceptable, with no "headroom" present to increase current in current-limited modules, the overall transient response of the combined converter is likely not very good either. Further, as indicated, if there is some sort of hysteresis or current foldback upon hitting current limit (max current reduces for a few cycles upon hitting current limit as in low-side current sensing), the two power-trains can even motorboat endlessly, each one trying relentlessly to hunt for a steady state. Current sharing under the circumstances becomes a really unlikely possibility; even some sort of bare "steady state" would be a stroke of luck.

Note that in the following sections our focus is now clearly gravitating toward simply paralleling several converters to get one high-power output rail, without any thought of synchronization or interleaving. The main intent is to just distribute the dissipation around the PCB, even across several heatsinks if necessary so as to enable high-power applications. Efficiency has taken a back seat here. But we do desire certain *features* in this case. Like *scalability* for example. Assuming we have managed to implement perfect sharing somehow, we now want to be able to parallel four 25 W modules to achieve a single 100 W output rail, hopefully six modules for 150 W, eight modules for 200 W, and so on. That is just basic scalability. But we would also like some redundancy. We could plan ahead for the possibility that one unit may fail in the field. So, we will initially use five 25 W modules for 100 W. But under normal conditions, each module is expected to deliver $100/5 = 20$ W only (automatically). However, if one module fails, we want the system to exclude that module completely (transparently), and instantly re-distribute the load so that each module now delivers its full 25 W. The transition should be effortless and automatic, to avoid down-time. Note that "excluding a failed module" means *disconnecting* it — both from the output *power rail* (so it doesn't "bring that down"), and also from any shared *signal rail* — in particular, the "load-share bus" that we will introduce shortly. The former concern typically necessitates an output Schottky OR-ing diode on each output rail before all outputs combine into one big output rail. There are ways to compensate for the additional diode drop in the regulation loop; however, the loss in efficiency is unavoidable *unless we forsake redundancy*. Now we need to get into more specifics.

Let us start where we had initially left off on the issue of sharing — back in Figure 13.7. In Case C, we had likewise *assumed* that somehow we had two power-trains that were sharing load current equally. But why would, or should, they ever share perfectly? Any specific reason why? However, before we answer that basic question, note the difference from the immediately preceding discussion above — in Case C we had implicitly assumed that there was only *one* feedback loop at work, sensing the common output rail and affecting the (identical) duty cycles of *both* converters. The drive pulses of both converters were literally tied together (ganged), though they were *delivered* in a staggered (interleaved) manner. So, there were no separate feedback loops fighting it out in that case. Nevertheless, despite the simplification of only one feedback loop in Case C, sharing still does not happen spontaneously. Because there are way too many subtle differences between even so-called "identical" power-trains.

Let us break this down by first asking: how do we intend to ensure perfectly identical *duty cycles* for the two power-trains? Because, even if the actual pulses delivered to the Gates of the respective control FETs of the two power-trains are exactly the same in width and in height, both FETs will *not* end up with the same duty cycle. Duty cycle is based on when the FET actually drives the inductor, not when *we* (try to) drive the FET. All "identical" FETs are slightly different. For example, as we saw in *Chapter 8*, there is an effective RC (internal) present at the Gate input, which produces some inherent delay. Further, the thresholds of the FETs (the Gate voltage at which they turn ON), are slightly different, and so on. The actual duty cycle that comes into play in the power conversion process will thus vary somewhat. These variations in *D* may seem numerically small to us, but are enough to cause significant differences in inductor currents.

Next we ask ourselves: for a given input and output, what is the natural duty cycle? We learned from *Chapter 5* that because of slight differences in the parasitics (R_{DS}, winding resistance and so on), the natural duty cycle linking a certain input voltage to a certain output voltage can vary. It also depends somewhat on the load current, because that affects the forward drops across the parasitics, which in turn affects the natural duty cycle. Here is the equation for a non-synchronous Buck with DCR and R_{DS} included

$$D = \frac{V_O + V_D + (I_O \times \text{DCR})}{V_{IN} + V_D - (I_O \times R_{DS})}$$

This can be also solved for I_O to find the corresponding current, commensurate with the applied duty cycle and output voltage (and a steady input).

$$I_O = \frac{DV_{IN} - V_O - (1 - D)V_D}{\text{DCR} + R_{DS}D}$$

In effect, the higher the load current, the higher the duty cycle and vice versa, *for a given output voltage*. But *for a given duty cycle*, a lower output voltage corresponds to higher current.

So, what happens in Case C of Figure 13.7 is that we have fixed the input voltage and the duty cycle. However, because of inherent differences, the natural output voltages of the two parallel converters differ. Nevertheless, we went ahead and tied their outputs together, which produced a new resulting output voltage (a sort of averaging of the two). In effect we forced a new output level that cannot obviously suit *either* of them completely. For arguments sake, assume that the applied output happens to be a little too low for the natural output level for Power-train 2, and a little too high for Power-train 1. Consider Power-train 2 first. The applied output voltage is in excess of its natural level. We know its duty cycle and input are fixed. The only thing that can possibly vary is the *current* through it. From the equation above, we see that a lower (natural) output voltage is commensurate with a higher current. So, Power-train 1 ends up passing more current than its fair share (simultaneously diminishing the current through Power-train 2 which is trying to correct its own situation at the same time). The net effect is that the natural output level of Power-train 2 gets steadily reduced and that of Power-train 1 gets steadily increased, till both coincide with the final applied voltage. Steady state ensues. But clearly, they are *no longer sharing current properly*. That was the price to pay to keep Kirchhoff happy, considering the differences in the power-trains.

A more intuitive way to look at how this situation got resolved is that Power-train 1 produces a higher natural output than the target output value. So by increasing its current, it increases the parasitic drops, thus lowering its output level — exactly to the target value. Similarly, Power-train 2 has a lower output than the target, so by reducing the parasitic drops by reducing its current, it increases its output voltage to coincide exactly with the target value.

However, rather than depend on parasitics to vary the drops and thereby adjust the output to achieve steady state, we can also add external resistances in both converters' forward paths, to help achieve quick settling. In fact, the more the resistance we add, the better the sharing becomes, though the efficiency obviously worsens. We can implement this so-called "droop method" even for cases like completely separate power converters with independent duty cycles. Droop regulation was initially discussed in the context of voltage positioning in *Chapter 9* (see Figure 9.7). However, note that we have to be careful to *allow the droop resistors to do their job*, by sensing the voltage for each converter to the *left of the droop resistor* (not the right side) (see Figure 13.17). (The droop resistors are called R_{SENSE} in the figure.) This however causes the output voltage (applied to the load) to decrease. But, since we do not know the initial efficiency of the converter upfront, we have no simple way of calculating the impact on the overall efficiency due to the introduction of these droop resistors. However, we can define the *maximum* efficiency "hit," based on assuming that the Buck converter had *100% efficiency prior to introducing these droop resistors*. We have defined "droop" in the Mathcad worksheet in Figure 13.17. We see that the equations for maximum efficiency loss and percentage droop are actually the same. So, we plot them out in the graph and we see that they do coincide. We also see that as the droop resistance increases, the sharing improves. At the same time there is a "knee" beyond which the sharing does not improve much more, but the efficiency loss continues to rise almost linearly. So, we get an optimum value of the droop resistor as follows

$$R_{SENSE\_RECOMMENDED} = \left(\frac{\Delta V}{10 \times I_{LOAD}} \right)^{1/2}$$

Note that this is an eyeballed equation — based on an initial estimate of how far apart the output voltages of the two converters can possibly be (worst case). For example, if we are holding in our hands a 5 V at 5 A module with a possible tolerance range on the output anywhere from 4.75 V to 5.25 V, the possible ΔV is as large as 0.5 V for a nominal 5 V. We will usually need a really large (and impractical) droop resistor to parallel such

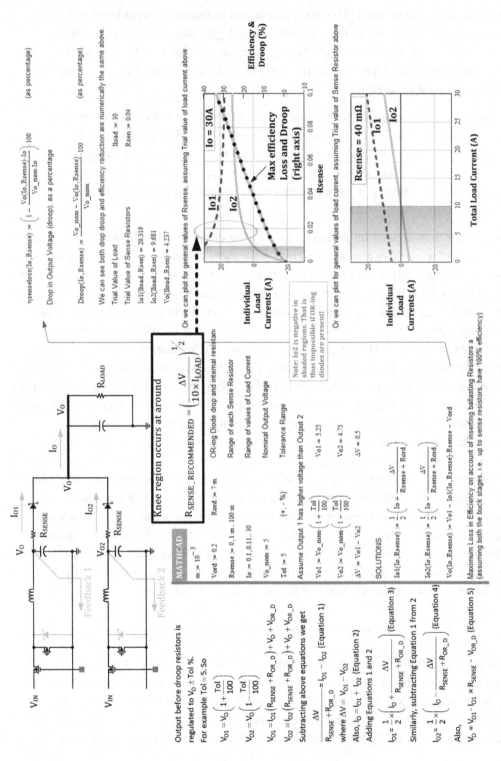

Figure 13.17: The droop method of paralleling power converters (passive load sharing).

modules. *A better option is to manually trim the outputs of such paralleled converters so that they are brought very close to each other, and then we can depend on much smaller droop resistors to equalize their currents.* Without that, passive sharing methods are inefficient and not very useful or practical.

A side note: yes, with current-mode control as discussed in *Chapter 12*, passive current sharing is much better since current is already being monitored cycle-to-cycle (though internally in the IC) as part of standard current-mode control implementation. But we do need to sense current in each converter. Also, the use of transconductance op-amps as the error amplifier of the feedback loop helps load sharing significantly, as compared to conventional voltage-based op-amps. The reason is when using transconductance op-amps, the feedback loop is completed *externally*, not locally. Locally means *directly* from the output of the error amp to its input pins, rather than through the converter. This was explained in Figure 12.12 and its related section. So, such op-amps are not so prone to "rail" just because of a small voltage difference on their input pins. In fact several transconductance op-amps belonging to multiple power-trains can be tied together at their inverting input pins without causing major issues. Transconductance op-amps are therefore recommended in droop-based (passive) load-sharing applications.

Active Load Sharing

With so many myriad causes responsible for unequal sharing, we can't afford to tackle each cause separately. But what we can do is (a) look out for the symptoms of unequal sharing, and (b) correct by brute-force. That means we need to monitor the current in each module constantly, compare it with the others, and if sharing is not so good, we can then try to enforce it by tweaking the actual duty cycle of the guilty module. And that in turn is done by tweaking the *reference voltage* of that module. We realize we need to *share* current sense information *between* modules, so each module can realize if it is running "too wide" compared to the others. We also hope to do all that with just a "single wire" running from module to module (of course all the modules share a common ground too because nothing electrical can really be over a single wire with no return path). And that takes us to the concept of a signal bus called the "load-share bus" — which is a common feature of most active load-sharing modules. There are many ways of implementing this feature. But let us start from a simplified diagram outlining the pioneering 1984 patent by Ken Small working at Boschert (subsequently called Computer Products Inc., then Artesyn Technologies, and now part of Emerson Electric). It is US Patent 4,609,828 we refer to. We have explained this in a very simplified way in Figure 13.18. The share bus represents the average of all the output currents of all the modules.

However, to avoid stepping on intellectual property, in a subsequent commercial variation of the original patent, two things were done: (A) the share resistor was replaced by a "share

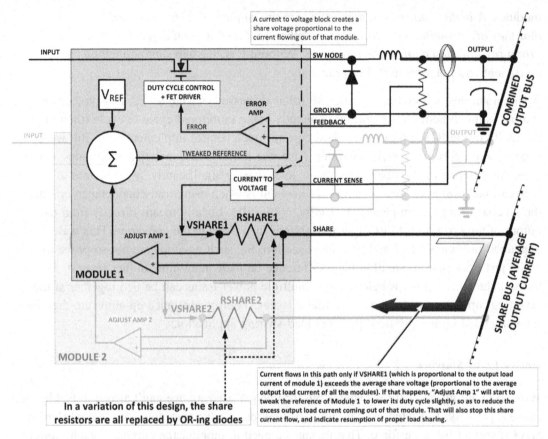

A current to voltage block creates a share voltage proportional to the current flowing out of that module.

INPUT

V REF

DUTY CYCLE CONTROL + FET DRIVER

ERROR AMP

ERROR

SW NODE

OUTPUT

GROUND

FEEDBACK

Σ

TWEAKED REFERENCE

CURRENT TO VOLTAGE

CURRENT SENSE

INPUT

COMBINED OUTPUT BUS

OUTPUT

VSHARE1   RSHARE1

SHARE

ADJUST AMP 1

MODULE 1

VSHARE2   RSHARE2

ADJUST AMP 2

MODULE 2

SHARE BUS (AVERAGE OUTPUT CURRENT)

Current flows in this path only if VSHARE1 (which is proportional to the output load current of module 1) exceeds the average share voltage (proportional to the average output load current of all the modules). If that happens, "Adjust Amp 1" will start to tweak the reference of Module 1 to lower its duty cycle slightly, so as to reduce the excess output load current coming out of that module. That will also stop this share current flow, and indicate resumption of proper load sharing.

In a variation of this design, the share resistors are all replaced by OR-ing diodes

Figure 13.18: Active load sharing explained.

diode" and (B) the reference voltage was allowed to be tweaked only in an *upward* manner (i.e., increase allowed, no decrease). Now the share bus no longer represented the average of all the output currents of all the modules, but the output current of that module *which is delivering the **highest** output current*. Because sensed current is being converted to a voltage and then OR-ed through signal diodes ("share diodes"). Therefore, the module with the highest current will be the one to forward-bias its OR-ing signal diode, thereby also simultaneously reverse-biasing the other OR-ing signal diodes. The lead module thus *automatically* becomes a "Master." Note that it becomes the Master primarily because its (unmodified) reference happens to be the highest of all the other modules. This is what happens to make it a Master. Its share diode is the first to conduct, causing a 0.6 V difference at the input pins of its adjust amplifier, forcing the output of the adjust amp low. That would leave the reference of the Master unaffected, because as mentioned, in this case the reference is allowed only to be adjusted higher, not lower, and we already know that the Master has the highest reference voltage to start with. That is exactly what we want, because the module that is forward-biasing its share diode is already providing the highest current, and we want to *retain* it as the Master to avoid constant "hunting" and any resulting unstable behavior. Note clearly that the voltage on the share bus is actually always 0.6 V (diode drop) *below* the share voltage coming out of the Master.

We recognize that the basic change in replacing share resistors with share diodes as in going from the original patent to its commercial variants, leads to a Master–Slave configuration. By means of the reverse blocking property of the share diode, this comes in handy if one module fails. In a typical fault condition in one module, the share voltage of the bad module falls, and therefore it automatically "drops out of the picture" *without pulling down the entire share bus*. The other modules can continue normal operation, and equally automatically they will "take up the slack" with a new Master being automatically appointed. This of course assumes each module has enough headroom in its basic power capability to be able to take up the slack.

Coming back to normal operation (no fault), we note that though the other modules (Slaves) follow the lead of the Master, they never quite get there. Ultimately, they all reach a mutually identical share voltage, but that voltage is still exactly one diode drop below the Master's share voltage. This is how it happens. Since their share diodes are initially reverse-biased, the outputs of their adjust amplifiers are all initially forced high. This causes their reference voltages to be tweaked higher and higher, so they start driving more and more output current on to the combined output bus. The process continues exactly to the point at which their share voltages become exactly equal to the voltage present on the share bus. At that moment their reference voltages can no longer be tweaked any higher by design. However, note that the share voltage from the Master is still 0.6 V above the voltage on the share bus, which by now exactly equals the share voltage from all the Slaves. So, there is an inherent error during steady state operation — a small mismatch between the Master's output current and of all the others. The Master will always end up pushing more current than the others. This inherent error in terms of current is basically 0.6 V divided by the actual voltage present on the share bus, because sensed voltage is proportional to the sensed current.

To minimize the above error, the Unitrode IC UC1907/2907/3907 ICs went a step further. Using some clever circuitry, they effectively canceled out the forward diode voltage drop of the share diode. In effect we have a "perfect signal diode" — it reverse-biases when expected, forward-conducts when required, and has zero forward drop. In principle, this reduces the error completely, and perfect load-sharing results. But actually, only on paper! In practice, slaves that have coincidentally similar (and slightly higher) reference voltages will constantly start fighting to become the Master. Therefore, to avoid constant hunting, a *50 mV offset* was deliberately introduced in the UC1907/2907/3907 ICs. In effect, now, instead of a diode with 0.6 V drop, or a diode with zero forward drop, we have a diode with a 50 mV forward drop. This retains all the advantages provided by replacing share resistors with share diodes, and (almost) none of the disadvantages. Which explains why the Unitrode (Texas Instruments now) load-share ICs became the work-horse of the entire load-share industry for decades. For example, the UC 3907 is still in production at the time of writing and is not obsolete. But besides its existence, a family of similar parts has also evolved from it.

CHAPTER 14

The Front End of AC−DC Power Supplies

Overview

The front end of an AC−DC power supply is often neglected or trivialized. It is however one of the trickiest parts of the design. It can define the cost, performance and reliability of the entire unit.

Of existing front end applications, the typical *low-power* application is supposed to be "the lowest" of them all. To the untrained eye this consists of a simple 4-diode (full-wave) bridge rectifier followed by a cheap aluminum electrolytic bulk capacitor. Functionally it seems self-explanatory: the bridge rectifies the AC, the bulk capacitor filters it, and that rail becomes the input of what is essentially just a high-voltage DC−DC switching converter just ahead (or a transformer-based variant thereof). We notice that in this power range, the switching converter is often a flyback. Therefore "cheap and dirty" is perhaps the first thing that springs to mind. But wait: what should the RMS rating of the bulk capacitor be? What is its expected lifetime? How do we guarantee a minimum "hold-up time"? Have we forgotten the input differential-mode (i.e. DM) filter chokes? Are they *saturating* by any chance? And why is the common-mode filter (i.e. CM) choke running so hot? (See *Chapters 15 to 18* for an understanding of input filters for tackling electromagnetic interference.) Also, does the design of the front end have any effect on the size of the transformer of the switching (PWM) stage? All these questions will be answered soon and we will realize nothing about a low-power front end design is either cheap or dirty!

Moving to medium- to high-power AC−DC power supplies, we usually have an active power factor correction (PFC) stage placed between the bridge rectifier and the bulk capacitor. That stage is also sometimes taken for granted − PFC stages are often considered to be just a plain high-voltage Boost converter stepping up a varying input (the rectified line voltage) to a steady ~ 400 V rail, which then forms the input of what is usually a Forward converter (or a variant thereof). Young engineers quickly learn to lean on some well-known PFC ICs out there, like the industry workhorse UC3854, for example. Yes, these almost turn-key PFC solutions do help a lot, especially with their abundant accompanying design notes. But do we really know everything very clearly at the end of it? For example, how do we *optimize* the PFC choke design? What is its dissipation? What about the dissipation in the PFC switch? What are the RMS equations for diode and switch

Switching Power Supplies A−Z. DOI: 10.1016/B978-0-12-386533-5.00014-0
547

stresses? What is the required RMS (ripple) current rating of the output bulk cap? And details aside, at the supposedly basic level of understanding, we can ask the following "trick question": how on earth did a DC–DC converter ever end up producing a *sine-wave* input current in the first place? Is that the natural (effortless) behavior expected of any standard DC–DC Boost converter responding to a sine-wave input voltage? If not, what changes were required to make that a reality? And can we at least say with confidence that for a given load, the *output* current of the PFC Boost converter stage is a constant as in a conventional Boost? The answers are not obvious at all as we will see in Part 2 of this chapter.

Part 1: Low-Power Applications

The Charging and Discharging Phases

Please refer to Figure 14.1 as we go along and try to understand the basic behavior here. What complicates matters is that the bridge rectifier does *not* conduct all the time. On the anode side of the diodes of the bridge rectifier is the sine-wave input, i.e. the AC (line) waveform. On the cathode side is the result of the rectified line voltage applied across a smoothing (bulk) capacitor. This relatively steady capacitor voltage is important because functionally, it forms the input rail for the PWM (switching) stage that follows, and thereby affects its design too.

> *Note: Keep in mind that even in our modern all-ceramic times, mainly because of the large "capacitance per unit volume" requirement here, the high-voltage bulk cap is still almost invariably an aluminum electrolytic. Therefore life concerns are very important here. See* Chapter 6 *for life calculations. We really need to get the RMS rating of this cap right.*

> *Note: We have a full wave bridge rectifier here. So it is convenient (and completely equivalent) to simply assume the input is a rectified sine wave as shown in Figure 14.1. We don't need to consider what "strange" things can happen when the AC sine wave goes "negative"! In effect, it doesn't. But all this should be quite obvious to the average reader.*

The appropriate diodes of the bridge get forward-biased only when the magnitude of the instantaneous AC line voltage (the anode side of the bridge) exceeds the bulk capacitor voltage (the cathode side of bridge). At that moment, the bridge conduction interval (let us call that "t_{COND}" here) commences. Fast-forwarding a bit, a few milliseconds later, the AC line voltage starts its natural sine-wave descent once again. As soon as it dips, the anode-side voltage (line input) becomes less than the cathode-side voltage (the bulk capacitor). So, at that precise moment, the diodes of the bridge get reverse-biased and the conduction time interval t_{COND} ends. After t_{COND}, we enter what we are figuratively calling the "lean season" or "winter" here, the reason for which will become obvious soon. Note that since

Equations For Lowest Bulk Capacitor Normal Operating Voltage and Bridge Conduction time in non-PFC AC-DC Power Supplies

The diode-rectified input voltage (line) is

$$V(t) = V_{AC\_PEAK} \times |Cos(\omega t)|$$

where $V_{AC\_PEAK} = \sqrt{2} \times V_{AC}$ and $\omega = 2\pi f$.

There is a normal voltage ripple, because the bridge diodes stop conducting once every half-cycle, and the bulk capacitor is left with the task of providing all the input power P_{IN} to the power supply for a while. The capacitor thus discharges from its peak value V_{AC_PEAK} to a lower value. Its instantaneous voltage at time t is called $V_{CAP}(t)$ here. The energy it delivers till elapsed time t must be related to the capacitor's stored energy by

$$\frac{1}{2} C \times \left(V_{AC\_PEAK}^2 - V_{CAP}(t)^2\right) = P_{IN} \times t \text{ (since Power is Energy/time).}$$

So, the capacitor voltage trajectory is given by

$$V_{CAP}(t) = \sqrt{2} \times \sqrt{V_{AC}^2 - \frac{P_{IN}}{C} t} \text{ where } (C/P_{IN} \text{ is Farads/W})$$

$$V_{AC} \times \sqrt{2} \times Cos(2\pi ft) = \sqrt{2} \times \sqrt{V_{AC}^2 - \frac{10^6}{(\mu FperW)} t} \quad \text{...... (1)}$$

V_{SAG} (the lowest point, where the bridge conducts again) occurs at the intersection of the AC and cap decay curves, as defined by

$$Cos(2\pi ft) = \sqrt{1 - \frac{10^6}{(\mu FperW) \times V_{AC}^2} t} \quad \text{...... (2a)}$$

We also realize that the intersection (at time "t") occurs fairly close to the peak of the AC waveform at 1/2f. We thus approximate t ≈ 1/2f.

$$Cos(2\pi ft) \approx \sqrt{1 - \frac{10^6}{2 \times f \times (\mu FperW) \times V_{AC}^2}} \equiv "\pm A" \quad \text{...... (2b)}$$

Square root always produces two solutions, one positive and one negative. In our case, the first (negative) solution gives us the estimated time t_{SAG_EST}, corresponding to the lowest point of voltage ("V_{SAG_EST}"). The second (positive) solution gives us the estimated bridge conduction time t_{COND_EST}. Both are related: they must add up to one half-cycle 1/2f.

$$t_{COND\_EST} = \frac{Cos^{-1}A}{2 \times \pi \times f} \quad \text{...... (3)}$$

$$t_{SAG\_EST} = \frac{Cos^{-1}(-A)}{2 \times \pi \times f} \text{ or equivalently}$$

$$t_{SAG\_EST} = \frac{1}{2f} - t_{COND\_EST} \quad \text{...... (4)}$$

where

$$A = \sqrt{1 - \frac{10^6}{2 \times f \times (\mu FperW) \times V_{AC}^2}} \quad \text{...... (5)}$$

The estimated lowest voltage point, corresponding to t_{SAG} is

$$V_{SAG\_EST}(t) = \sqrt{2} \times \sqrt{V_{AC}^2 - \frac{10^6}{(\mu FperW)} t_{SAG\_EST}} \quad \text{...... (6)}$$

For best results, we should now iterate, by plugging this value of V_{SAG} into the following obvious identity, to get a far more accurate closed-form equation for t_{COND} and t_{SAG}, as follows:

$$t_{COND} = \frac{1}{4f} - \frac{Sin^{-1}\left(\frac{V_{SAG\_EST}}{\sqrt{2} \times V_{AC}}\right)}{2 \times \pi \times f} \quad \text{...... (7)}$$

$$t_{SAG} = \frac{1}{2f} - t_{COND} \quad \text{...... (8)}$$

Finally, the AC voltage rises again and the two curves intersect. That is the point the diodes of the bridge get forward-biased (conduct) again, and the cap recharges. In addition, a comparatively smaller current flows straight through into the input of the converter ahead. A much larger current surge flows in to recharge the capacitor. The capacitor voltage thus rises till the peak value is once again achieved. At that point the AC voltage dips and the bridge stops conducting again. The entire process repeats every AC half-cycle. The lowest point of voltage is called V_{SAG} here. The peak to peak cap ripple is $V_{AC\_PEAK} - V_{SAG}$.

This bulk cap ripple is the steady-state input voltage ripple of the switching converter ahead, and impacts it significantly.

Example: f=50Hz, V_{AC} =90, μFperW=3

$$A = \sqrt{1 - \frac{10^6}{2 \times 50 \times (3) \times 90^2}} = 0.767$$

$$t_{COND\_EST} = \frac{Cos^{-1}0.767}{2 \times \pi \times 50} = 2.217 \times 10^{-3} \Rightarrow 2.22ms$$

$$t_{SAG\_EST} = \frac{Cos^{-1}(-0.767)}{2 \times \pi \times 50} = 7.783 \times 10^{-3} \Rightarrow 7.78ms$$

$$V_{SAG\_EST}(t) = \sqrt{2} \times \sqrt{V_{AC}^2 - \frac{10^6}{(\mu FperW)} t_{SAG\_EST}} = 104.9V$$

$$t_{COND} = \frac{1}{4f} - \frac{Sin^{-1}\left(\frac{V_{SAG\_EST}}{\sqrt{2} \times V_{AC}}\right)}{2 \times \pi \times f} = 1.92ms$$

3μF/W@90VAC

XμF/W@85VAC

3μF/W 85VAC

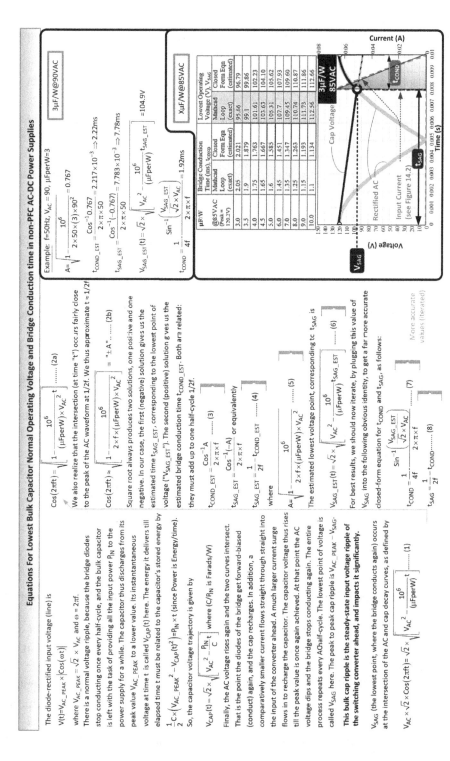

μF/W @85VAC (Peak = 120.2V)	Bridge Conduction Time (ms), t_{COND}		Lowest Operating Voltage (V), V_{SAG}	
	Mathcad Loop (exact)	Closed Form Eqn (estimated)	Mathcad Loop (exact)	Closed Form Eqn (estimated)
3.0	2.05	2.021	95.66	96.79
3.5	1.9	1.879	99.1	99.86
4.0	1.75	1.763	101.61	102.23
4.5	1.65	1.667	103.63	104.10
5.0	1.6	1.585	105.31	105.62
6.0	1.45	1.451	107.7	107.93
7.0	1.35	1.347	109.45	109.60
8.0	1.25	1.263	110.74	110.87
9.0	1.15	1.193	111.73	111.86
10.0	1.1	1.134	112.56	112.66

More accurate values (iterated)

Figure 14.1: Calculating the conduction time and lowest voltage in steady operation.

the repetition rate in our case is twice the line frequency, calling the line frequency "f" (50 Hz or 60 Hz), we see that the lean season lasts for exactly $(1/2f) - t_{COND}$, because it is basically just the rest of the time period till t_{COND} occurs again.

During this lean season, no "external help" can arrive — the diodes have literally cut off all incoming energy supplies from the AC source. The bulk capacitor therefore has to necessarily provide 100% of the power/energy requirement being constantly demanded by the switching converter section. In effect, the capacitor takes on the role of a giant reservoir of energy, and that is the main reason for its presence and size. This lean season is therefore often referred to as the "discharge interval" (of the cap), whereas t_{COND} is often called the "charging interval".

After the conclusion of this "lean season", the bridge conducts once again (when the AC voltage rises). So, t_{COND} starts all over again. But, connecting the dots, we now realize that by the required energy balance in steady state, the energy pulled in through the bridge during t_{COND} must not only provide all the instantaneous input energy being constantly demanded by the switching converter stage, but also completely *replenish* the energy of the bulk cap — i.e. to "restock" it, so to say, for the next lean season (when the bridge stops conducting again).

In terms of the widths of the two intervals involved: the diodes typically conduct for a much smaller duration than the duration for which they do not conduct. We can visualize that this ratio will get even worse for larger and larger bulk capacitances. The reason for that is that since the voltage across a very large cap will decay only slightly below its max value (relatively speaking), and since its max value is equal to the peak AC voltage, the diodes will get forward biased for only extremely short durations, very close to the peak of the AC input voltage waveform (see Figure 14.1). In such cases, we realize that since all the energy pulled out of the bulk cap during the "lean season" must be replenished *in a comparatively smaller available time* t_{COND}, the energy/current inrush spike during t_{COND} will necessarily be both very thin and very tall — since the area under the input current/ energy curve must remain almost constant.

Increasing the Capacitance, Thereby Reducing t_{COND}, Causes High RMS Currents

We present a useful analogy here that we are calling the "Arctic analogy". Consider situation where we have to provide winter food supplies for 10 scientists stationed in an Arctic camp. Suppose each scientist consumes 30 kg of food per month. So together, they need 300 kg of food per month. In the first possibility, suppose the summer is 2 months long. Also assume that summer is the only time we can transport food into the camp. So during those two summer months we need to send (a) winter supplies for 10 months (10 × 300 = 3000 kg), and (b) food for the two ongoing summer months (2 × 300 = 600 kg).

That is a total of 3600 kg (a year's supply as expected), delivered over 2 months, i.e. at the rate of 1800 kg per month on an average. In the second case, suppose summer was just 1 month long. Calculating similarly, we realize we now need to deliver a year's supply in just 1 month, i.e. at the rate of 3600 kg per month on average.

That is what happens in a low-power AC–DC power supply too. During t_{COND} (summer), energy (food) is pulled in from the input source (supermarkets). During this period, the inputted energy goes not only into sustaining the PWM section (the scientists), but also recharging the bulk cap (the Arctic warehouse). During the lean season (winter), all the energy (food) required to sustain the PWM section comes only from the bulk capacitor. And so, very similarly, if we reduce t_{COND} (the summer months) by half, the current/energy *amplitude* (the food transported per month) will double.

But why is that a problem? The problem is that though the *average* value of current does not change if we halve t_{COND}, (check: $I_{AVG} = I \times D = 2I \times D/2$), the RMS value of the current goes up by a factor $\sim$1.4. Check: $I \times \sqrt{D} \neq 2I \times \sqrt{(D/2)} = \sqrt{2} \times I \times \sqrt{D} = 1.4 \times I \times \sqrt{D}$. Generalizing, if we reduce t_{COND} by the factor "x", the RMS of the input current will increase by approximately the factor $\sqrt{x}$. Alternatively stated, the RMS of the input current must vary as $1/\sqrt{t_{COND}}$. We can check this relationship out: if t_{COND} halves, the RMS input current will increase by the factor $\sqrt{2}$ as expected. Further, we can say that the peak current is roughly proportional to $1/t_{COND}$. We can check this out too: if t_{COND} halves, the peak input current will increase by the factor 2 as expected.

Be clear that so far we are only referring to the current flowing in through the input filter chokes (the diode bridge current). We are not talking about the capacitor current, though that is very closely related to the bridge current as described in detail later, in Figure 14.2.

We know that the heating in any resistive element depends on I_{RMS}^2. So now it becomes clear why the filter chokes start to get almost "mysteriously hot" if the bulk capacitor is made injudiciously large causing t_{COND} to reduce as a result. The reason is the RMS has become much higher now.

We also know that if t_{COND} halves, the peak current doubles. But we know from *Chapter 5* that the saturation current rating (and size) of any magnetic component depends on I_{PEAK}^2. So we conclude that if t_{COND} halves, the size of the filter choke will *quadruple*. And if we didn't expect that or plan for it upfront, we can be quite sure our filter choke will start saturating, reducing its efficacy significantly.

> *Note: In Chapters 15 to 18 we will learn that common mode filter chokes have equal and opposite currents through the coupled windings, so core saturation is not actually a major concern for them, though heating certainly is. However, for differential mode*

Normal Operating Currents and Dissipation in Bulk Capacitor and Bridge in non-PFC AC-DC Power Supplies

Just past t_{SAG} (during t_{COND}), the Capacitor Charging Current is

$$I_{CAP\_CHARGE} = C\frac{dV(t)}{dt} = C\frac{d}{dt}\left(\sqrt{2}\times V_{AC}\times|Cos(\omega t)|\right)$$
$$= \sqrt{2}\times C\times V_{AC}\times\omega\times|Sin(\omega t)|$$
$$= \sqrt{2}\times C\times V_{AC}\times\omega\times|Sin[\pi-(\omega t)]|$$

where we used $Sin(\pi-\omega t)=Sin(\omega t)$. Further, since in the region of interest, $\omega t \approx \pi$, so $\pi-(\omega t)$ is very small. We also know that $Sin(x) \approx x$ for small x. So the cap current during bridge conduction becomes

$$I_{CAP\_CHARGE} \approx \sqrt{2}\times C\times V_{AC}\times\omega\times[\pi-(\omega t)]$$
$$\approx \sqrt{2}\times C\times V_{AC}\times\omega\times[\pi-(\omega t)]$$

Using $\omega = 2\pi f$, we get

$$I_{CAP\_CHARGE} \approx \sqrt{2}\times C\times V_{AC}\times(2\pi f)^2\left(t-\frac{1}{2f}\right)$$

or finally

$$I_{CAP\_CHARGE} \approx \frac{8\pi^2}{\sqrt{2}}(CV_{AC}f^2)^2\left(t-\frac{1}{2f}\right) \Leftarrow$$

[0.085A/W for 85VAC, 50Hz, 3μF/W]

Plotting this current with respect to time, we see it is if the form $y=m(x-x_0)$. So, it is a straight line which passes through the point $t=1/2f$ (max of rectified AC), with slope (dI/dt) equal to

$$Slope = \frac{8\pi^2}{\sqrt{2}}(CV_{AC}f^2)$$

This is the slope between time t_{SAG} and $1/2f$ (i.e. during t_{COND}); it is 0 elsewhere. Its peak value is therefore

$$I_{PEAK} = \frac{8\pi^2}{\sqrt{2}}(CV_{AC}f^2)\times t_{COND}$$

See adjoining plot labeled "Capacitor Current".
If we use μFperW, we get the Peak (cap) current in Amps/Watt

$$I_{PEAK\_PERWATT} = \frac{8\pi^2}{10^6\times\sqrt{2}}\times(V_{AC}f^2)\times t_{COND} \Leftarrow$$

[0.085A/W for 85VAC, 50Hz, 3μF/W]

(Note: Follow procedure in Fig14.1 to find v_{COND} and V_{SAG})

Outside t_{COND}, the capacitor provides a constant current into the input of the power supply:

$$I_{IN}(t) = 1W/V_{cap}(t) \quad Amps/W.$$

For Vcap above, use the average cap voltage value

$$V_{IN\_AVG} = \frac{(\sqrt{2}\times V_{AC})+V_{SAG}}{2}$$

[108.5V for 85VAC, 50Hz, 3μF/W]

So the additional (constant) current term flowing in from the input voltage source into the switcher stage during t_{COND}, and continuing during the rest of the half-cycle from the cap is:

$$I_{IN\_PERWATT} = \frac{1}{V_{IN\_AVG}} \quad (Amps/Watt)$$

[9.22mA/W for 85VAC, 50Hz, 3μF/W]

The full capacitor current is plotted below. It has zero DC current component through it, as any cap in steady state. The bridge current is just the cap current moved vertically to give it a non-zero DC value as shown. The DC shift is equal to I_{IN}, to make the instantaneous bridge current zero outside t_{COND} as required.

Expanded View of Plot in Figure 14.1

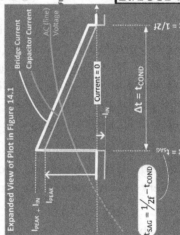

CAPACITOR RMS CURRENT
We need this value to correctly pick our input Bulk Cap. The RMS rating of the selected cap must match this RMS value. Using our geometric formula for RMS of piecewise linear curves (Fig 7.4)

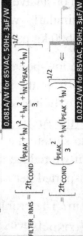

$$I_{CAP\_RMS} = \left[I_{IN}^2\times\frac{t_{SAG}}{\frac{1}{2f}}+\frac{I_{PEAK}^2}{3}\times\frac{t_{COND}}{\frac{1}{2f}}\right]^{1/2}$$
$$= \left[I_{IN}^2+2ft_{COND}\left(\frac{I_{PEAK}^2}{3}-I_{IN}\right)\right]^{1/2}$$

[0.02A/W for 85VAC, 50Hz, 3μF/W]

BRIDGE AVERAGE CURRENT
We need to know this to calculate bridge dissipation.

$$I_{BRIDGE\_AVG} = \frac{(I_{PEAK}+I_{IN})+I_{IN}}{2}\times\frac{t_{COND}}{\frac{1}{2f}}$$
$$= f\times t_{COND}\times(I_{PEAK}+2I_{IN}) \cdot$$

[0.0091 A/W for 85VAC, 50Hz, 3μF/W]

MUST BE SAME

EMI FILTER PEAK AND RMS CURRENTS
(Same as the Bridge RMS and Peak). The Peak is important because the DM and CM chokes must not saturate with this peak current. The RMS current is important because the copper AWG of the DM & CM chokes (all windings) must be able to handle the heating due to this RMS.

$$I_{FILTER\_PEAK} = I_{PEAK}+I_{IN} \Leftarrow$$

[0.081A/W for 85VAC, 50Hz, 3μF/W]

$$I_{FILTER\_RMS} = \left[2ft_{COND}\left(\frac{(I_{PEAK}+I_{IN})^2+I_{IN}^2+I_{IN}(I_{PEAK}+I_{IN})}{3}\right)\right]^{1/2}$$
$$= \left[2ft_{COND}\left(\frac{I_{PEAK}^2}{3}+I_{IN}(I_{PEAK}+I_{IN})\right)\right]^{1/2} \Leftarrow$$

[0.022A/W for 85VAC, 50Hz, 3μF/W]

Example: 25W Power Supply, n=0.833, has $P_{IN}=25/0.833=30W$. Suppose we select a 3μF/W x 30W = 90μF cap. Peak line (filter) current is 0.081A/W x 30W = 2.43A. The bridge dissipation is ~2 x 1.1V x 0.0092A/W x 30W = 0.61W. The RMS rating of the cap must exceed 0.02A/W x 30W = 0.6Arms.
For another example using a different μF/W, see Table 14.1 and the accompanying text too.

Figure 14.2: The peak/RMS current stresses in the capacitor, bridge and EMI filters.

*chokes, both saturation and heating are of great concern. So in general, the input peak
and RMS current must both be contained, by not selecting an excessively large bulk cap.*

We are seeing a familiar trend: a peaky input energy/current shape causing huge RMS/peak
stresses that can dramatically affect the cost and physical sizes of *related components*. In
general, we should always remember that *"a little ripple (both voltage and current) is
usually a good thing"* in switchers and we shouldn't be "ripple-phobic". We had received
the same lesson in *Chapters 2 and 5* when we decided to keep the current ripple ratio *r* at
an optimum of around 0.4 ($\pm 20\%$) rather than trying to increase the inductance
injudiciously to lower *r*. But, eventually, we must keep in mind it is all about *design
compromises*. For example, in the case under discussion here, *reducing* the bulk cap
injudiciously can have also led to adverse effects. For example, it can cause the PWM
switch dissipation to increase significantly, besides requiring its transformer to be bigger. In
other words, there is no clear answer − it is just optimization as usual.

The Capacitor Voltage Trajectory and the Basic Intervals

We are interested in computing the cap voltage as a function of time as it discharges
(decays) during t_{COND} (as we continue to pull energy out of it). There are many possibilities
in the way we can discharge a capacitor and the corresponding voltage trajectory it takes.
For example, if we connect a resistance to the capacitor, it will discharge as per the familiar
exponential-based capacitor discharge curve ($\sim e^{-t/RC}$) discussed in *Chapter 1*. Suppose we
were to attach a constant current source "I" across the bulk capacitor. Then its voltage would
fall in a *straight line* as per the equation $\Delta V/\Delta t = I/C$. However, in our case here, we are
connecting a *switching converter* across the cap. Its basic equation is $P_{IN} = V_O \times I_O/\eta$, where
η is the efficiency (the input power of the converter is the power being pulled out of the
capacitor). This power is clearly independent of the input voltage (assuming the efficiency η
is almost constant). In other words, P_{IN} is virtually constant irrespective of the capacitor
voltage. So, the switching converter presents itself to the cap, not as a constant resistance or
constant current load, but as a *constant power* load, which means the product "VI" is constant
as the cap decays. The corresponding trajectory can be plotted out with this mathematical
constraint, as shown in Figure 14.1.

In Figure 14.1 we try to find the corresponding diode conduction time t_{COND} and the
corresponding lowest capacitor voltage V_{SAG}. The required math and the underlying logic
are presented therein. The basis of these computations is as follows: we are trying to find
the intersection of two distinct curves, the rectified sine-wave AC input and the capacitor
discharge curve.

In the process of deriving the closed-form (estimated) equations we have needed to make
an initial assumption: that the instant at which the intersection (lowest voltage) occurs

(t = t_{SAG}) is situated just before the instant at which the AC voltage reaches its peak (t = 1/2f). However, we then iterate and get very accurate estimates from the closed form equations. To prove the point, within Figure 14.1, we have also gone and compared the closed-form estimates with the more accurate numerical results obtained from a very detailed iterative Mathcad file, one that did not make the above simplifying assumption. We see from the embedded table in Figure 14.1 that the comparison is very good indeed (after the iteration). We conclude it is OK to use the closed-form equations (judiciously). In Figure 14.1 we have also included a numerical example to calculate t_{COND}, t_{SAG} and V_{SAG}, the three key terms of interest here.

The reader will observe that we are ultimately trying to scale everything to *Capacitance per Watt* (usually called "µFperW" or "µF/W"). The advantage of doing that is the curves and numbers become *normalized* in effect. So, by scaling the capacitance proportionally to the (input) wattage, we can then apply the results *to any power level*. There will be worked examples shortly, to illustrate this procedure.

> *Note: Look hard at the plot in Figure 14.1 to understand why an error in the estimated t_{SAG} will not create a big error in V_{SAG}, provided we plug the first-estimate t_{SAG} into in the capacitor discharge curve, not in the AC curve.*

> *Note: We are consistently ignoring the two forward diode drops that come in series with the input source during t_{COND} — an assumption that can affect the calculated V_{SAG} by a couple of volts. However that is still well within component and other tolerances, so it is fair to ignore it here in the interest of simplicity.*

Tolerating High Input Voltage Ripple in AC–DC Switching Converters

The bulk capacitor voltage ripple can be expressed as $\pm(V_{AC\_PEAK} - V_{SAG})/2 \times 100\%$. For example, from Figure 14.1, we see that at 85VAC and for 3µF/W, the cap voltage will vary from a peak of 120 V (i.e. $85 \times \sqrt{2}$) to about 96 V. The average is therefore (120 + 96)/2 = 108 V. That constitutes a voltage ripple of 24 V/108 V = 0.2, i.e. 22%, or ±11%.

We realize that the input ripple to the high-voltage switching converter stage is no longer within the usually declared "rule" of < ±1% input voltage ripple, for selecting the input capacitance of a typical low-power DC–DC converter. In fact, in commercial AC–DC applications, an input voltage ripple of up to around ±15% may be considered normal, or at least acceptable/permissible, if not desirable.

As indicated previously, for several reasons, using a huge bulk capacitance, just to smooth out the voltage ripple, is really not a commercially viable option and nor does it help improve overall system performance. So it is actually preferable that we *learn to tolerate, if not welcome, this rather high input ripple* — by using *workarounds* to the problems it can cause. For example, we know that at least the controller IC certainly needs much better input filtering, or it can "misbehave". The recommended solution to that problem is to add a small low-power, low-pass RC circuit just before the supply pin of the controller IC. That brings the input ripple *as seen by the IC* down to a more acceptable level (around $\pm 1\%$). The high voltage ripple of around $\pm 10\%$ is then only "felt" only by the power stages, and that is deemed acceptable.

We realize that some of the ripple present on the input power rail gets through to the output power rail of any converter too. That aspect was discussed in *Chapter 12*. To combat this effect, an LC "post filter" (perhaps just using a cheap rod inductor followed by a medium-sized capacitor) can be added to the output of the converter.

However, eventually, it is undeniable that the input voltage ripple applied to the converter (power) section is rather large, and does affect the converter's design and overall performance (also of its input EMI filter). To re-iterate, there is no right answer to what the "correct" amount of ripple is – it is based on optimization and careful design compromises.

How the Bulk Capacitor Voltage Ripple Impacts the Switching Converter Design

We realize that the peak voltage applied to the switching converter is fixed – it is simply the (rectified) peak of the AC line voltage. Therefore the amount of input ripple we allow, indirectly determines two other important parameters: (a) the lowest instantaneous input voltage V_{SAG}, and (b) the average input voltage applied to the converter. Why are these important? We take up the latter first.

(A) As voltage at the input of the converter (bulk capacitor voltage) undulates, so does its duty cycle as the converter attempts to correct the varying input and create a steady output voltage. We know that in a flyback topology for example, as we lower the input voltage, the input current (and its center of ramp $I_{OR}/(1 - D)$) goes up on account of the increase in D at low input voltages. So the RMS current in the switch will go up significantly on account of higher input voltage ripple. Since the input voltage is undulating between two levels (V_{AC_PEAK} and V_{SAG}), the normally accepted assumption is to take the *average* value of the input voltage ripple, V_{IN_AVG}, as the input voltage applied to the switching converter. This is OK for the purpose of calculating the (average) duty cycle and thereby the (average) switch dissipation and so on. Efficiency estimates can also be done using this average input voltage.

$$V_{IN\_AVG} = \frac{V_{AC}\sqrt{2} + V_{SAG}}{2}$$

For example, in a typical universal input flyback ($3\mu F/W$ @ 85VAC), the average input voltage is typically 108 V as mentioned above (often taken to be 105 V to account for the bridge diode drops too).

(B)   The amount of ripple also determines V_{SAG}. We need to ensure that the switching converter's transformer does not saturate at V_{SAG}. In other words, using an *average voltage* for deciding transformer/inductor size is a mistake. Magnetic components can saturate over the course of just one high-frequency cycle, leave aside a *low*-frequency AC cycle. And if that happens, immediate switch destruction can follow. Therefore, the core size selection procedure for a flyback, as presented in *Chapter 5*, should actually be done *at least* as low as V_{SAG}. Even better, we should do the design typically about *10–20% lower* than V_{SAG} — or till the point of the set undervoltage lockout and/or max duty cycle limiting and/or current limiting.

> *Note: Any calculations related to required copper thickness, and any other heat-related components, can be selected using the average voltage V_{IN_AVG} given above since heat is determined on a continuous (averaged) basis. For example, the switching transformer size is determined by V_{SAG}, but its thermal design (wire gauge and so on) is determined by V_{IN_AVG}.*

General Flyback Fault Protection Schemes

This is a good time to discuss flyback protection issues briefly. We indicated above that a flyback power supply can easily blow up at power-up or power-down because of *momentary* core saturation. This was discussed briefly in *Chapter 3* too. It is just not enough to design the transformer "thinking" that the lowest input voltage is going to be V_{SAG}. Because in reality, the input of the flyback *does go down all the way to zero during every single power-down*.

The strategy to deal with the ensuing stress is as follows. We first need to know what the value of V_{SAG} is (during normal operation). Because that represents an *operating* level we do not want to affect, or inadvertently "protect" against. Thereafter, *we must place an accurate undervoltage lockout (UVLO) and/or a corresponding switch current limit just a little below V_{SAG}*. Note that if the current limit is set too high (commensurate with a much lower input voltage), we can suffer from the dangers of core saturation on power-up and power-down if the core has been designed only for V_{SAG}, or just a little lower. On the other hand, if the current limit is set too low, corresponding to a higher minimum input operating voltage, and/or the UVLO level is set too high, we start to encroach into the region of normal operation, which is obviously unacceptable.

The underlying philosophy of protection is always as follows: *any protection barrier must be built **around** normal operation and just a **little** wider*. Note that if the power supply has also to meet certain holdup requirements (to be discussed shortly), the protection will usually have to be set *at even a lower voltage*, or the bulk capacitor and/or core will need to be significantly oversized.

Note: One common misconception is that carefully controlling the maximum duty cycle *is all that is required to protect a flyback during power-up and power-down. It does help, and certainly, it should be set fairly accurately too. However, it is not enough by itself. Because, duty cycle is rather* loosely *connected to the DC level of current in CCM, as has been explained several times in previous chapters. So we also need current limiting and undervoltage lockout (UVLO) protection.*

Note: Typically, the normal operating duty cycle for an AC–DC flyback operating at its lowest rated AC input voltage (e.g. 85VAC) is set to about 50–60%, as discussed in Chapter 3 (see worked example 7 in particular). Therefore, a maximum duty cycle limit (Dmax) of about 60–65% is usually set for ensuring reliability under power-up and power-down. It is not considered wise to use a controller IC with any arbitrary Dmax. For example, there are some integrated AC–DC Flyback switchers that are built around what is to us at least, an inexplicable and arbitrary "78% max duty cycle limit" (e.g. Topswitch-FX and GX). Another mystifying example is the LM3478, a flyback switcher IC with a "100% max duty cycle" (perhaps the only one such IC in existence, because normally, 100% max duty cycle is acceptable, and even desirable, for Buck ICs, but never for Boost or Buck-Boost topologies).

We must mention what happens at *high input voltages*. At that input point, a universal input power supply (90–270VAC), will have a rectified peak DC of $270 \times \sqrt{2} = 382$ V. Even if the flyback has a good zener/RCD clamp to protect itself from voltage overstress at this input, it can still be destroyed merely by an inadvertent *current* overstress. But how can that happen at high voltages? Let us do some simple math here. To keep the equations looking simple, we simply designate V_{MIN} and V_{MAX} as the minimum and maximum input voltages going to the switching converter (i.e. the minimum and maximum cap voltages) respectively. Their corresponding duty cycles are designated D_{MAX} and D_{MIN}. Note that here D_{MAX} is not the max duty cycle limit of the controller IC but the duty cycle at the lowest operating input voltage. For a flyback we thus get

$$D_{MAX} = \frac{V_O}{V_O + V_{MIN}}, \text{ and } D_{MIN} = \frac{V_O}{V_O + V_{MAX}}$$

Eliminating V_O (or equivalently V_{OR})

$$V_O = V_{MIN}\left(\frac{D_{MAX}}{1 - D_{MAX}}\right) = V_{MAX}\left(\frac{D_{MIN}}{1 - D_{MIN}}\right)$$

So,

$$D_{MIN} = \frac{V_{MIN} D_{MAX}}{V_{MAX} - D_{MAX}(V_{MAX} - V_{IN})}$$

Note that we can divide both the numerator and denominator above by $\sqrt{2}$ and it would still be valid. So, we can just directly plug in the AC voltages in the equation above. For example, if we had designed the flyback with a D_{MAX} of 0.5 (ignoring ripple in this particular discussion), the operating duty cycle at 270VAC would be

$$D_{MIN} = \frac{90 \times 0.55}{270 - 0.55(270 - 90)} = 0.25$$

We had stated that it is always desirable any protection barrier be *just a little wider* than normal operation. So, the obvious problem we can foresee here is related to the fact that we have set the max duty cycle (protection level) at 65% for the purpose of catering to low input voltages (along with the rather stressful power-up and power-down scenario discussed above). However, now we also realize that the 65% duty cycle limit is *too wide* for protection of the flyback at 270VAC operation, because at 270VAC the normal operating duty cycle is only about 25%. It is in fact true that we can easily destroy any such poorly designed AC–DC flyback power supply by simply inducing core saturation at *high* input voltages. *All we need to do is create a sudden overload at 270VAC*, and the control loop will naturally react by pushing out the duty cycle momentarily to its maximum (65% in this case) in an effort to regulate the output. In going from 90VAC to 270VAC, the applied voltage across the transformer goes up three times. Therefore, in addition to that problem, if the time for which that high voltage gets applied to the transformer remains the same it was at low voltages, we have a major problem in the form of *three times (300%) the voltseconds* under sudden overloads at high line as compared to sudden overloads at low line. We also know that magnetic components can easily saturate purely due to excess applied voltseconds. For example, in our case here, even though the center of the inductor ramp has come down by a factor of two at high line (remember, the center of ramp varies as per $1 - D$), the increased AC current swing riding on top of it, due to the excessive applied voltseconds described above, can cause the peak instantaneous current under overloads at 270VAC to significantly exceed even the worst-case peak currents observed at 90VAC. So, this situation can cause easily transformer core saturation, and in fact, far more readily than possible at low line.

So, how do we guard against this high-line overload scenario? The easy way to close this particular vulnerability of the flyback/Buck-Boost/Boost topologies, is called 'Line Feedforward'. In many low-cost flyback power supplies using the popular UC3842 IC for example, a large additional resistor of around 470 k to 1 M is almost invariably found connected from the rectified high voltage DC line (HVDC rail) to the current sense pin (on the IC side). That way, a voltage dependent current gets summed up with the normal sensed current coming in through the typical 1 k or 2 k resistor connected to the sense resistor placed in the Source lead of the switching FET. In effect, the additional high-value resistor *raises the sensed current pedestal* (its DC value) higher and higher as the input voltage is raised. So, the current limit threshold of the IC is reached much more readily at high line, even for much smaller switch currents. We can say that, in effect, the max duty cycle has gotten limited at high line, to just a little more than required for normal operation. This input/line feedforward technique therefore provides necessary protection under the overload scenario described above, provided the high-value resistance mentioned above is rather carefully set.

The Input Current Shape and the Capacitor Current

In Figure 14.1, we had shown a *triangular-shaped* current waveform to represent the input (bridge) current. Theoretically, that is the correct shape as can be seen from the detailed derivation in Figure 14.2. In reality, because of input/line impedances along the way, the actual observed input current (bridge) waveform is perhaps closer to a triangle with rather severely rounded edges. Note that it is certainly *not* a rectangular current waveshape, as often simplistically assumed in literature (e.g. older Unitrode Application notes). The predictions of peak and RMS currents based on the theoretical triangular waveshape, as derived in Figure 14.2, are accurate, if not a little pessimistic (since they ignore line impedances), but they certainly can, and should, be used for doing a *worst-case* design.

In general, the bulk capacitor current waveform can be derived from the bridge current by subtracting the DC value I_{IN} from the bridge current as per the procedure shown in Figure 7.5 and Figure 7.6. Alternatively, the diode current is simply the capacitor discharging current with I_{IN} added to it as indicated in Figure 14.2.

In Figure 14.2, we have also provided some quick lookup numbers in gray for the most basic low-cost case of 3μF/W @ 85VAC. An embedded example in the figure shows how to use these quick lookup numbers to quickly scale and estimate the peak currents and also the capacitor ripple current, for any application. For example, we have shown that a power supply drawing 30 W at its input, and using 90μF of bulk capacitance (i.e. 3μF/W), has a peak diode current of 2.43A at 85VAC. This is based on the quick lookup number of 0.081A/W provided for the case of 3 μF/W @ 85VAC.

In Table 14.1, we have consolidated all the required equations and also provided quick look-up numbers for several μF/W cases (including 3 μF/W). This table is perhaps all we need for designing a low-power AC–DC front end. However, there are certain "holdup time" and capacitor tolerance considerations that we will discuss soon, which could affect our final choice of μF/W and also the switcher's transformer design.

Further, there is actually a much simpler and intuitive way to estimate several stress parameters, based on the "Arctic winter analogy" we had presented in a preceding section. We will look at that shortly.

Table 14.1:   Design Equations and Quick Lookup Numbers for Low-Power Front-End Design.

Description	Parameter	Equation	3 µF/W	5 µF/W	7 µF/W
(1) Required to calculate parameters below	A	$$= \sqrt{1 - \frac{10^6}{2 \times f \times (\mu FperW) \times V_{AC}^2}}$$	0.734	0.85	0.896
(2a) Bridge Conduction Duration *(first estimate)*	t_{COND} *(in seconds)*	$$\approx \frac{Cos^{-1}A}{2 \times \pi \times f}$$ *(all calculations based on this approximate value will be approximate too)*	2.38 *(ms)*	1.76 *(ms)*	1.47 *(ms)*
(3) Time coordinate of Minima of Bulk Cap Voltage *(first estimate)*	t_{SAG} *(in seconds)*	$$= \frac{1}{2f} - t_{COND}$$	7.62 *(ms)*	8.24 *(ms)*	8.53 *(ms)*
(4) Lowest Input Operating Voltage of Converter	$V_{SAG}(t)$ *(in Volts)*	$$= \sqrt{2} \times \sqrt{\left[V_{AC}^2 - \frac{10^6}{(\mu FperW)} t_{SAG} \right]}$$	96.79 *(V)*	105.62 *(V)*	109.60 *(V)*
(2b) Bridge Conduction Duration *(second estimate)*	t_{COND} *(in seconds)*	$$= \frac{1}{4f} - \frac{Sin^{-1}\left(\frac{V_{SAG}}{\sqrt{2} \times V_{AC}}\right)}{2 \times \pi \times f}$$ *(this is an iterated value, far more accurate)*	2.02 *(ms)*	1.59 *(ms)*	1.35 *(ms)*
(5) Average Bulk Cap Voltage and Average Input Voltage of Converter	V_{IN_AVG} *(in Volts)*	$$= \frac{(\sqrt{2} \times V_{AC}) + V_{SAG}}{2}$$	108.50 *(V)*	112.91 *(V)*	114.90 *(V)*
(6) Average Input Current *(in Bridge and Filter)*	I_{IN} *(in Amps)* *(for $P_{IN} = 1W$)*	$$= \frac{1}{V_{IN\_AVG}}$$ *(this is an exact equation)*	9.22 *(mA/W)*	8.86 *(mA/W)*	8.70 *(mA/W)*
(7) Peak Capacitor Charging Current	I_{PEAK} *(in Amps)(for $P_{IN} = 1W$)*	$$= \frac{8\pi^2 \times (\mu FperW)}{10^6 \times \sqrt{2}} \times (V_{AC}f^2) \times t_{COND}$$	0.072 *(A/W)*	0.094 *(A/W)*	0.112 *(A/W)*
(8) Required Ripple (RMS) Current Rating of Bulk Cap	I_{CAP_RMS} *(in Amps)*	$$= \left[I_{IN}^2 + 2ft_{COND}\left(\frac{I_{PEAK}^2}{3} - I_{IN}^2\right) \right]^{1/2}$$	0.02 *(A/W)*	0.023 *(A/W)*	0.025 *(A/W)*
(9) Average Input Current *(x $2V_D$ for Bridge Dissipation)*	I_{BRIDGE_AGE} *(in Amps)*	$$= f \times t_{COND} \times (I_{PEAK} + 2I_{IN})$$ *(would be the same as I_{IN} above if t_{COND} value used here were 100% exact)*	9.13 *(mA/W)*	8.85 *(mA/W)*	8.711 *(mA/W)*

(Continued)

Table 14.1:   (Continued)

Description	Parameter	Equation	3 µF/W	5 µF/W	7 µF/W
(10) RMS of Line Current *(in Bridge and Filter)*	I_{FILTER_RMS} *(in Amps)*	$= \left[2ft_{COND}\left(\dfrac{I_{PEAK}^{2}}{3} + I_{IN}(I_{PEAK} + I_{IN}) \right) \right]^{1/2}$	0.022 (A/W)	0.025 (A/W)	0.027 (A/W)
(11) Peak of Line Current *(in Bridge and Filter)*	I_{FILTER_PEAK} *(in Amps)*	$= I_{PEAK} + I_{IN}$	0.081 (A/W)	0.103 (A/W)	0.121 (A/W)

Notes: Calculations are meant to be done from top to bottom, in that order.

All A/W and µF/W numbers are normalized to $P_{IN} = 1W$.

All numbers are for 85VAC and Line Frequency $f = 50$ Hz.

How to Interpret µF/W correctly

Some semiconductor vendors, intending to showcase their small evaluation boards to demonstrate how "tiny" their integrated flyback switchers are, often calculate the *input* capacitance based on a supposedly "universal rule of 3 µF/W". Though we do have some reservations about this number itself as discussed later, the *way* these vendors apparently choose to interpret this number makes their recommendations really questionable.

These vendors apply the "3 µF/W rule" to the *output* wattage, not the input wattage. That is not consistent. Keep in mind that the input bulk capacitor is handling output power *plus the loss*. So the input cap really doesn't care about the output power per se, only the *input* power. In other words, a 3 µF/W rule, *if considered acceptable*, should be applied to the bulk cap based on *input watts*, not *output watts*. Otherwise we will end up with contradictory design advice.

> *Note: For example, if there is a 50 W power supply with 80% efficiency, this vendor would recommend an input capacitor of $50 \times 3 = 150$ µF (nominal). But then, along comes another 50 W power supply, this time running at say, only 50% efficiency (to exaggerate the point). The vendor would however still recommend 150 µF. However, with these recommendations in place, we will find that the two power supplies will have very different input capacitor waveforms and very different t_{COND}, t_{SAG} and V_{SAG}. So they must be different cases now, even though the vendor seemed to say the two power supplies would behave similarly with their recommendations. However, had we instead kept the capacitances per* input *watts of the two supplies the same, we would have arrived at exactly the same t_{COND}, t_{SAG} and V_{SAG} for both cases. And so, to get the same (normalized) waveform shapes, we need to actually use $62.5\,W \times 3$ µF/W $= 187.5$ µF in the first case, and $100\,W \times 3$ µF/W $= 300$ µF in the second case (assuming we are willing to accept 3 µF/W in the first place).*

Worked Example using either Quick Lookup Numbers or the "Arctic Analogy"

Example

We are designing an 85 W flyback with an estimated efficiency of 85%, for the universal input range of 85 to 270VAC. Select an input bulk cap tentatively, and estimate the key current stresses.

The Input power is 85 W/0.85 = 100 W. Let us tentatively select a 300 µF/400 V capacitor. We have thus set C_{BULK}/P_{IN} = 300 µF/100 W = 3 µF/W. We can work out the key stresses (@ worst-case 85VAC) in two ways.

Method 1: Let us use the quick lookup numbers presented in Table 14.1 for the 3 µF/W case here.

(a) t_{COND} = 2.02 ms, from Row 2b.
(b) V_{SAG} = 96.8 V, from Row 4.
(c) Average Cap voltage is 108.5 V, from Row 5.
(d) Average Input Current (and average bridge current) is 9.22 mA/W × 100 W = 922 mA, from Row 6.

> *Note: Bridge dissipation is about $2 \times 1.1\,V \times 0.922\,A = 2.0\,W$ (assuming each diode has a forward drop of 1.1 V as per datasheet of bridge). We have used the exact value of average line current here, one that is unaffected by error in t_{COND} estimate. Since average capacitor voltage is 108.5 V, we get input power as $108.5 \times 0.922 = 100\,W$ as expected.*

(e) Capacitor RMS current (predominantly a low-frequency component, to correctly pick capacitor ripple rating) is 0.02 × 100 = 2.0 A, from Row 8.
(f) RMS Input Current (to correctly pick AWG of all input filters) is 0.022 × 100 = 2.2 A, from Row 10.
(g) Peak Input Current (to rule out differential-mode input filter saturation) is 0.081 × 100 = 8.1 A, from Row 11.

Method 2: Here, we need the logic of the Arctic analogy presented earlier. We do need the closed-form equations of Figure 14.1 to estimate t_{COND} and V_{SAG}. We also refer to some of the geometric RMS equations in Table 14.1 (based on Figure 7.4).

(a) t_{COND} = 2.02 ms, from the table or equations in Figure 14.1.
(b) V_{SAG} = 96.8 V, from the table or equations in Figure 14.1.
(c) Average Cap voltage is therefore $(85 \times \sqrt{2} + 96.8)/2 = 108.5$ V.

(d) Average Input Current (and average bridge current) is $I_{IN} = 100$ W/108.5 V = 922 mA. We are using the terminology shown in the plot inside Figure 14.2.

(e) Now, using the Arctic analogy, we have to provide

$\varepsilon_{IN} = P_{IN} \times t = 100$ W $\times$ 10 ms = 1000 mJ = 1 J every half-cycle. This is analogous to one year's food supply and we have to supply all of this in $t_{COND} = 2.02$ ms (summer). Since by definition, $\varepsilon = V \times I \times t$, the average supply current during t_{COND} must therefore be $I = \varepsilon/Vt \rightarrow 1$ J/(108.5 $\times$ 2.02 ms) = 4.56 A. The average input current throughout the half-cycle is 0.922 A from above. This is the pedestal on top of which is the triangular portion of the input current (see Figure 14.2). The average of all this must equal 4.567 A. So, we get the peak input current (into cap) as

$$\frac{I_{PEAK}}{2} + I_{IN} = 4.567 \rightarrow I_{PEAK} = 2 \times (4.56 - I_{IN}) = 2 \times (4.56 - 0.922) = 7.28 \text{ A}$$

(f) The peak input current (through filter and bridge) is the peak cap current plus I_{IN} which is going to the switcher section, as shown in the plot in Figure 14.2. So the peak input current is 7.28 A + 0.922 = 8.2 A. This agrees very well indeed with the estimate of peak input current in step (g) of Method 1 above.

(g) The cap RMS current is

$$I_{CAP\_RMS} = \left[I_{IN}^2 + 2ft_{COND} \left(\frac{I_{PEAK}^2}{3} - I_{IN}^2 \right) \right]^{1/2}$$

$$= \left[0.922^2 + 2 \times 50 \times 2.02 \times 10^{-3} \times \left(\frac{7.28^2}{3} - 0.922^2 \right) \right]^{1/2} = 2.1 \text{ A}$$

This agrees very well indeed with the estimate of cap RMS current in step (e) of Method 1 above.

(h) The input filter RMS current is

$$I_{FILTER\_RMS} = \left[2ft_{COND} \left(\frac{I_{PEAK}^2}{3} + I_{IN}(I_{PEAK} + I_{IN}) \right) \right]^{1/2}$$

$$= \left[2 \times 50 \times 2.02 \times 10^{-3} \left(\frac{7.28^2}{3} + 0.922 \times 8.2 \right) \right]^{1/2} = 2.26 \text{ A}$$

This agrees very well indeed with the estimate of cap RMS current in step (f) of Method 1 above.

We now realize that armed primarily only with the Arctic analogy, we could get through all the stress estimates accurately, *without resorting to complicated equations*. Further, we are no longer tied to any specific values of μF/W. We could have chosen any capacitance at all to proceed. This illustrates the sheer power of understanding basic principles, rather than just leaning on cumbersome math and simulations. Further, we see that carefully selected intuitive analogies like the Arctic analogy do a lot to bolster basic understanding.

Accounting for Capacitor Tolerances and Life

A typical aluminum electrolytic cap may have a nominal tolerance of $\pm 20\%$. In addition, we have to account for its end-of-life capacitance, which could be 20% lower than its starting value. If we do not account for these cumulative variations upfront, we run the risk sooner or later, of encountering a much lower V_{SAG} than we had expected or planned for, and also a much lower holdup time (as discussed further below). Therefore a thorough design will calculate a certain minimum capacitance, say "X" μF, but then use an actual nominal capacitance 56% higher, i.e. $1.56 \times X$ μF (check: $1.56 \times 0.8 \times 0.8 = 1$).

For example, a 30 W universal input power supply running at 90% efficiency has an input power of $30/0.9 = 33.3$ W. Using the 3 μF/W rule, we get $33.33 \times 3 = 100$ μF. But this is just the *minimum* capacitance we need to guarantee here. The capacitor we should actually use must be at least 156 μF nominal. Therefore, we will go out and pick a standard cap value of 180 μF (nominal). The required voltage rating of the cap is obviously $265VAC \times \sqrt{2} \rightarrow 400$ V since it has to handle the highest voltage across it.

But all this introduces another complication. We now actually need to split our front-end design phase into two distinct design steps at this point — we certainly can't base any of our estimates on the starting value of 100 μF anymore.

Step 1: We realize that the initial capacitance can actually be 20% higher on account of tolerances, i.e. $C_{MAX} = 1.2 \times 180$ μF $= 216$ μF. A high capacitance leads to a *much smaller worst-case t_{COND}*, which in turn leads to *much higher peak and RMS currents* in both the input filter chokes and the RMS current through the bulk cap. So, to correctly pick the ripple current rating of the input cap, the wire gauge of the input filter chokes and also the saturation rating of the differential mode input chokes, we must now consider the *lowest possible t_{COND}* as calculated above — based on $C_{MAX} = 216$ μF. That corresponds to a maximum μF/W of $216/33.33 = 6.5$ μF/W. Looking at Figure 14.1 we see we can easily interpolate between the Mathcad values for 6 and 7 μF/W to get $t_{COND} = 1.4$ ms and $V_{SAG} = 108.6$ V. These are all we need to work out the stresses using the Arctic analogy! Or we can use the more detailed equations in Table 14.1.

Step 2: On the other hand, to find the worst-case "hold-up time" (explained further below), the worst-case dissipation in the switch and the appropriate wire gauge selection and resulting dissipation of the switching transformer, we should use the *maximum t_{COND}* (based on *lowest* capacitance). The lowest capacitance value the selected bulk capacitor can exhibit over its useful life is $C_{MIN} = 180 \, \mu F \times 0.8 \times 0.8 = 115 \, \mu F$. That is a minimum $\mu F/W$ of $115/33.33 = 3.45 \, \mu F/W$. This is the value we need to use for this step of the design. Basically, on this basis, we need to work out the maximum t_{COND} and the lowest V_{SAG} during normal operation (over the life of the product). Then we find out the lowest average input to the switching converter (V_{IN_AVG}) and use that voltage for finding the average duty cycle and corresponding RMS stresses in the switching converter stage in the usual fashion, as discussed in *Chapter 7*. Then we need to look at the holdup time charts provided below, to work out the lowest cap voltage during an input dropout event so we can correctly size the flyback transformer.

Note that in both cases above, we usually do *not* need to perform any of the stress/dissipation calculations at high line, because the worst case currents are at low line (for a flyback). Also, faced with a possibility of either 50 Hz or 60 Hz line input frequency, we note that 50 Hz gives the worst case results. So we are mostly ignoring the 60 Hz case in this chapter.

Holdup Time Considerations

Holdup time is *the duration for which the output of the power supply/converter stays within regulation on loss of input power*. The intent is simple: incoming power quality (e.g. AC mains) is not always clean and/or assured. So, to avoid frequent nuisance interruptions, we try to provide a small, but guaranteed duration, for which the input power can go away, or just sag below the declared input specifications of the power supply/ converter, and then come back up, without the load/system connected to the output of the converter from ever "knowing" that something transpired at the input of the supply/ converter. In the process, the output of the supply/converter may droop by a very small but almost unnoticeable amount (typically <5% below nominal or within the declared output regulation range of the supply). Holdup time is therefore a buffer against ever-present vagaries in input supply quality.

Note that if the input voltage is really collapsing (as in an outage), holdup time may be used to keep things alive for just a while, deliver some sort of flag or advance warning of the impending outage, and thereby provide a few milliseconds for the load/system to perform any necessary housekeeping. For example, the system may store the current state, configuration or preferences, or even a data file currently in use, so as to recover or recall them quickly when power returns.

Though the underlying concept and intent of holdup time is always the same as described above, its measurement, testing and implementation need to be treated quite differently when dealing with DC–DC converters, PFC-based AC–DC power supplies or non-PFC-based AC–DC power supplies, the latter being the topic of discussion here. Some make the mistake of trying to apply textbook equations meant for achieving a certain holdup in DC–DC converters, to AC–DC power supplies. There are major differences in all possible implementations as explained below.

Consider the case of an AC–DC power supply providing, say 12 V output, which then goes to a point-of-load (POL) DC–DC converter that then converts the 12 V to 5 V. On the 5 V rail is our load/system that we need to keep alive by guaranteeing a certain holdup time. In principle, we can try to meet the required holdup time in a variety of ways.

(a)   We can simply try to increase the *output cap of the POL converter* in a brute-force fashion, so as to reduce the output droop. Let us test this out. Suppose the POL converter is a 12 V to 5 V Buck converter delivering 3 A with 80% efficiency. With a typical 5% allowed output droop, the lowest output voltage would be $5 \times 0.95 = 4.75$ V. Using the equation for a capacitor discharging, and targeting a modest holdup time of 10 ms, we get the corresponding capacitor requirement.

$$C_{OUT} > \frac{2 \times P_O \times t_{holdup}}{(V_{initial}^2 - V_{final}^2)}$$

Solving

$$C_{OUT} > \frac{2 \times 15 \times 10 \times 10^{-3}}{(5^2 - 4.75^2)} = 0.123 \text{ F}$$

No typo here, that really is 123 000 µF! Obviously a very impractical value.

(b)   Let us try to beef up the *input capacitor of the POL converter* instead. Suppose we imagine for a moment that the input source to the POL converter just "went away" briefly, leaving the input capacitor of the POL converter the task of providing all the necessary power requirement. We also assume this Buck converter has a certain switch forward drop and a maximum duty cycle limit, due to which it needs a certain guaranteed "minimum headroom" of say 2.5 V above the output rail, to be able to regulate. In effect, we are allowing the POL converter's input capacitor voltage to droop from 12 V to 7.5 V, and we then expect that the output will stay regulated during this decay. However, remember that the input cap of this DC–DC stage has to handle the *input* power, not the *output* power. So the correct equation to use here is

$$C_{IN} > \frac{2 \times P_{IN} \times t_{holdup}}{(V_{initial}^2 - V_{final}^2)}$$

Solving

$$C_{IN} > \frac{2 \times (15/0.8) \times 10 \times 10^{-3}}{(12^2 - 7.5^2)} = 4.27 \times 10^{-3} \text{ F}$$

This is about 30 times better than trying to meet holdup time directly at the output, but still way too high.

Note: The perceptive reader will notice that there can be problems with our assumptions above. In the first case, we tried to add capacitance to the output of the switching converter, imagining it would do something to prop up the output under all cases. However, if the input rail of the POL converter is being driven by, say, a synchronous output *stage of the AC–DC supply, that stage can source or* sink *current. It can thus forcibly drive the input rail of the POL converter down to zero during an AC/line dropout. In that case, the output cap of the POL converter will also get forcibly discharged to zero through the body-diode of the switching FET of the POL converter, rendering it incapable of providing any holdup time. For the same reason, if we try to beef up the input cap of the POL converter, we could still be in trouble. The solution in such cases is to* place a diode in series with the input of the POL converter – so, *even if the output rail of the AC–DC stage gets pulled down to zero, the series diode would then get reverse-biased and prevent the input rail of the POL converter from being dragged to zero.*

(c) Since the amount of capacitance required is still too high, we realize we need to move further "up the food-chain" if we want to provide holdup time of the order of several milliseconds in a practical manner. We finally reach the *input side of the AC–DC power supply*. We discover that that leads to far more acceptable capacitance values. Why? Because to achieve a certain holdup time t_{holdup}, we basically have to provide a reservoir able to store and provide a certain fixed amount of energy equal to $P_{IN} \times t_{holdup}$. But we also know that the energy storage capability of a capacitor goes as $C \times V^2$. So as we increase V, we get a dramatic increase in energy storage capability, even with smaller C. That is why any holdup time requirement is best met *upstream* – preferably on the input side of the AC–DC power supply – i.e. at its front end.

Let us understand why the AC input voltage source sags or drops out in the first place, and what its implications are. Most input disturbances originate *locally*. For example a large load may suddenly start up nearby, like a motor or a resistive/incandescent load. That can draw a huge initial current, causing a voltage dip in its vicinity. We could also have unspecified wiring flaws, or local faults/shorts, which will eventually activate a circuit-breaker, but will produce momentary dips in the line voltage till that happens. Some relatively rare voltage sags/dropouts can originate in the utility's electric power system. The most common of those are a natural outcome of faults on distant circuits, which are eventually segregated by self-resetting circuit breakers, but only after a certain unavoidable delay during which sags/droops result. Much less common are sags/dropouts related to distant voltage regulator failures. Utilities have automated systems to adjust voltage (typically using power factor correction capacitors, or tapped switching transformers), and these also can fail on rare occasions. All these can lead to temporary sags/dropouts.

To create a level of acceptable immunity from such events, the international standard IEC 61000-4-11 (Second Edition 2004), *calls for a minimum holdup time of 10 ms.* This is actually intended to correspond to one half cycle of 50 Hz line frequency. Most commercial power supply specifications call out for a holdup time of 20 ms, and that number is intended to correspond to one complete AC cycle (two half-cycles).

However, looking at the reasons listed above for line disturbances, we realize that in almost all cases, when the input power does resume, it remains "in sync" with the previous "good" AC half-cycles. In other words, we should try and visualize line dropout in terms of *missing* AC half-cycles, not in terms of any fixed time interval of, say, 10 ms or 20 ms, and so on. We will see that that line of thinking can lead to significant cost savings.

Looking more closely at Figure 14.1, we can visualize that the situation for compliance to a certain holdup time spec depends on *when* exactly the dropout is said to have commenced. For example, if the input source drops out *just before the bridge conducts again* (i.e. close to and just before t_{SAG}), that would represent the worst-case. Because in that case we are starting off the dropout-related part of the capacitor decay from the *lowest* possible operating voltage point, and we will ultimately arrive at a much lower voltage at the end of the dropout time. In contrast, the best case occurs *just after the capacitor has been fully peak-charged*. The IEC standard does not *require* that we should test holdup time either in the worst-case condition, or in the best-case. It just recommends that any input voltage changes occur at zero crossings of the AC line voltage, though it also leaves the door open for testing at different "switching angles" *if* deemed necessary. There is also the question whether the input AC voltage should be set to 115VAC (nominal), or as a worst-case: 90VAC, or even 85VAC.

There are many OEMs that demand aggressive holdup testing of power supplies − by asking for the input to be set at 85VAC, and commencing the dropout *just before* V_{SAG}. That does increase the bulk capacitor/transformer size/cost significantly. However, especially in such cases, it is valuable to try and convince the OEM to talk of holdup in terms of the *number of half-cycles missed,* not in terms of a fixed duration expressed in ms. In doing so, we can recoup some of the higher costs.

The curves in Figure 14.3 show the trajectory for different μF/W, computed from the Mathcad program alongside. We can see that for example, one full missing AC cycle (two half-cycles) is in reality *a little less* than 20 ms (by the amount t_{COND}). Note that what we have been calling V_{SAG} so far is now designated V_{SAG0}, since it corresponds to zero missing half-cycles, or "normal operation". To account for dropouts, we now have V_{SAG1}, V_{SAG2}, V_{SAG3} and so on, corresponding to 1, 2 and 3 missing half-cycles, respectively. The numbers for V_{SAGx} are presented in the lookup tables within Figure 14.3.

Note that in Figure 14.3, we are establishing something close to 60VDC as the lowest acceptable point for flyback operation during a line dropout. So, looking at the curves and

Holdup Time versus Capacitance and Lowest Operating Voltage in non-PFC AC-DC Power Supplies with Quick Look-up Tables for 85VAC and 90VAC

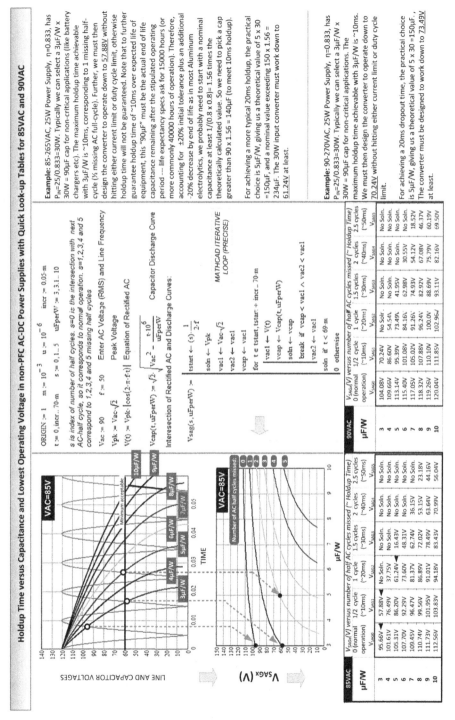

$\text{ORIGIN} := 1 \quad m := 10^{-6} \quad \text{incr} := 0.05 \cdot m$

$t := 0, \text{incr} .. 70 \cdot m \quad s := 0, 1 .. 5 \quad \text{uFperW} := 3, 3.1 .. 10$

s is index of number of half cycles. s=0 is the intersection with next AC-half cycle, so it corresponds to normal operation. s=1,2,3,4 and 5 correspond to 1,2,3,4 and 5 missing half cycles

$\text{Vac} := 90 \quad f := 50$ Enter AC Voltage (RMS) and Line Frequency

$\text{Vpk} := \text{Vac} \cdot \sqrt{2}$ Peak Voltage

$V(t) := \text{Vpk} \cdot |\cos(2 \cdot \pi \cdot f \cdot t)|$ Equation of Rectified AC

$\text{Vcap}(t, \text{uFperW}) := \sqrt{2} \cdot \sqrt{\text{Vac}^2 - \dfrac{t \cdot 10^6}{\text{uFperW}}}$ Capacitor Discharge Curve

Intersection of Rectified AC and Discharge Curves:

$$\text{Vsag}(s, \text{uFperW}) := \begin{cases} \text{tstart} \leftarrow (s) \cdot \dfrac{1}{2 \cdot f} \\ \text{vac1} \leftarrow \text{Vpk} \\ \text{vac1} \leftarrow \text{Vac} \cdot \sqrt{2} \\ \text{vac2} \leftarrow \text{vac1} \\ \text{vcap} \leftarrow \text{vac1} \\ \text{for } t \in \text{tstart, tstart} + \text{incr} .. 70 \cdot m \\ \quad \begin{vmatrix} \text{soln} \leftarrow V(t) \\ \text{vcap} \leftarrow \text{Vcap}(t, \text{uFperW}) \\ \text{soln} \leftarrow \text{vcap} \\ \text{break if vcap} < \text{vac1} \wedge \text{vac2} < \text{vac1} \\ \text{vac2} \leftarrow \text{vac1} \end{vmatrix} \\ \text{soln if } t < 69 \cdot m \\ 0 \text{ otherwise} \end{cases}$$

MATHCAD ITERATIVE LOOP (PRECISE)

Example: 85-265VAC, 25W Power Supply, η=0.833, has $P_{IN}=25/0.833=30W$. Typically we can select a 3μF/W x 30W = 90μF cap for non-critical applications (like battery chargers etc). The maximum holdup time achievable with 3μF/W is ~10ms, corresponding to 1 missing half-cycle (½ missing AC full-cycle). Further, we must then design the converter to operate down to 57.88V without hitting either current limit or duty cycle limit, otherwise holdup time will not be guaranteed. Note that to further guarantee holdup time of ~10ms over expected life of equipment, this "90μF" must be the actual end of life capacitance remaining after the stipulated operating period --- life expectancy specs ask for 15000 hours (or more commonly 40000 hours of operation). Therefore, accounting for ±20% initial tolerance plus an additional -20% decrease by end of life as in most Aluminum electrolytics, we probably need to start with a nominal capacitance at least 1/(0.8 x 0.8)= 1.56 times the theoretically calculated value. So we need to pick a cap greater than 90 x 1.56 = 140μF (to meet 10ms holdup).

For achieving a more typical 20ms holdup, the practical choice is 5μF/W, giving us a theoretical value of 5 x 30 =150μF, and a nominal value exceeding 150 x 1.56 = 234μF. The 30W input converter must work down to 61.24V at least.

Example: 90-270VAC, 25W Power Supply, η=0.833, has $P_{IN}=25/0.833=30W$. Typically we can select a 3μF/W x 30W = 90μF cap for non-critical applications. The maximum holdup time achievable with 3μF/W is ~10ms. We must then design the converter to operate down to 70.24V without hitting either current limit or duty cycle limit.

For achieving a 20ms dropout time, the practical choice is 5μF/W, giving us a theoretical value of 5 x 30 =150μF. The converter must be designed to work down to 73.49V at least.

85VAC μF/W	$V_{SAG}(V)$ versus number of half AC cycles missed (~Holdup Time)					
	0 (normal operation) V_{SAG0}	1/2 cycle (~10ms) V_{SAG1}	1 cycle (~20ms) V_{SAG2}	1.5 cycles (~30ms) V_{SAG3}	2 cycles (~40ms) V_{SAG4}	2.5 cycles (~50ms) V_{SAG5}
3	95.66V	57.88V	No Soln.	No Soln.	No Soln.	No Soln.
4	101.61V	76.49V	37.75V	No Soln.	No Soln.	No Soln.
5	105.31V	86.20V	61.24V	16.43V	No Soln.	No Soln.
6	107.70V	92.29V	73.60V	48.31V	No Soln.	No Soln.
7	109.45V	96.47V	81.37V	62.74V	36.15V	No Soln.
8	110.74V	99.56V	86.89V	72.02V	53.15V	23.18V
9	111.73V	101.95V	91.01V	78.49V	63.64V	44.16V
10	112.56V	103.83V	94.18V	83.43V	70.99V	56.04V

90VAC μF/W	$V_{SAG}(V)$ versus number of half AC cycles missed (~Holdup Time)					
	0 (normal operation) V_{SAG0}	1/2 cycle (~10ms) V_{SAG1}	1 cycle (~20ms) V_{SAG2}	1.5 cycles (~30ms) V_{SAG3}	2 cycles (~40ms) V_{SAG4}	2.5 cycles (~50ms) V_{SAG5}
3	104.08V	70.24V	No Soln.	No Soln.	No Soln.	No Soln.
4	109.66V	86.60V	54.54V	No Soln.	No Soln.	No Soln.
5	113.14V	95.39V	73.49V	41.95V	No Soln.	No Soln.
6	115.40V	101.08V	84.16V	62.98V	30.55V	No Soln.
7	117.05V	105.02V	91.26V	74.93V	54.12V	18.52V
8	118.32V	107.88V	96.24V	82.92V	67.08V	46.37V
9	119.26V	110.10V	100.00V	88.69V	75.79V	60.19V
10	120.04V	111.85V	102.96V	93.11V	82.16V	69.50V

Figure 14.3: Holdup design chart based on missing half-cycles (not time).

tables in Figure 14.3 we conclude that, based on an assumption of the most aggressive holdup testing:

(a) If the required holdup is one half-cycle (~10 ms), we can certainly use 3 μF/W, though we also should verify that the selected capacitor meets the ripple current requirement, so as to guarantee its life as per the procedure shown in *Chapter 6*.

(b) If the required holdup is two half-cycles (~20 ms), we must use at least 5 μF/W. Clearly, 3 μF/W is just not sufficient.

The above statements are based on the worst-case assumption of 85VAC input combined with a dropout commencing *just before* t_{SAG0} occurs. But, to enable the holdup time feature, we must also design the flyback such that it is able to deliver full power down to ~60VDC. For example, the protection circuitry we talked about must be moved out just past this lower operating limit.

> *Note: We do not have to size the copper of the transformer or use heatsinks for the switch/diode rated for continuous 60VDC operation, because we only need to deliver power at such a low voltage momentarily.*

> *Note: A frequently asked question on capacitor selection is: what is the dominant criterion for selecting the HVDC bulk capacitor in non-PFC universal input AC-DC flybacks? The answer is as follows. Usually, with a 10–20 ms holdup time requirement, we may discover that we need to oversize the cap somewhat (getting higher capacitance as a bonus), simply because otherwise, its ripple current rating is inadequate. So here, it the RMS rating that ultimately dominates and determines the capacitor − its size and thereby its capacitance. But the holdup time spec is not far behind. So, for a holdup time of 30–40 ms, we will usually end up selecting a cap that automatically meets the required ripple current rating. In general, we can calculate and select a capacitor that meets the ripple current rating, and then separately calculate and select a capacitor that meets the required holdup time. Finally, we simply pick the larger of the two.*

To allow a flyback to operate reliably down to 60VDC, the most important concern is to ensure that its transformer can handle the adverse stress situation gracefully (and without duty-cycle/undervoltage/current limit protections kicking in). Most universal input flybacks are designed to operate down to that low level, albeit momentarily. But it is also interesting to observe that commercial flyback transformer cores do *not* seem any bigger than what we may have intuitively expected on the basis of the worked examples and logic presented in *Chapter 5*. There are several reasons/strategies behind that. One "technique" comes from vendors of some integrated power supply ICs (e.g. Topswitch®). In their design tools they seem to "allow" designers to deliberately use a "peak B_{SAT}" of ~4200 Gauss, which we know is far in excess of the usually declared 3000 Gauss max for ferrites (see *Chapter 5*).

We do not bless that approach on these pages for that very reason. However, in the next section we present a design example that shows a legitimate way to stick to a max of 3000 Gauss, and yet keep the core size unchanged, while meeting the new holdup time requirements too. Of course, nothing comes for free, and there are some attached penalties as we will soon see.

Two Different Flyback Design Strategies for Meeting Holdup Requirements

Armed with all the detailed information on sags presented in Figure 14.3, we can do a much better job in designing a practical universal input AC–DC flyback. We go back to the design example we had initiated in *Chapter 5*, starting from Figure 5.19. In Figure 14.4 we show how to incorporate holdup time into those calculations. Our first step is to look at the right-hand side table of Figure 14.3 and pick a practical value of capacitance (e.g. 5 μF/W for meeting ∼20 ms holdup time). We realize we now need to design the flyback such that it can function down to 73.5 V instead of the 127 V number we had used earlier. Note that we are relaxing our holdup spec just a bit, by demanding we comply with the holdup requirement at a minimum of 90VAC, not at 85VAC, which is why we have gotten 73.5VDC here instead of the earlier target value of ∼60VDC.

In fact, there are *two* possible design strategies as detailed in Figure 14.4, one that leads to a bigger core and one that doesn't. Both are limited to a max B_{SAT} of 3000 Gauss. The design approaches are otherwise quite different as described below.

(a) We can design the transformer for an *r* of 0.4 at 73.5 V, in which case the transformer will be no bigger than a transformer designed for *r* = 0.4 at 127 V. We will explain the reason for that shortly.

(b) We can design the transformer for an *r* of 0.4 at 127 V, in which case, to be able to function reliably down to 73.5 V, we would need to choose a bigger core.

In Figure 14.4, using the first approach, we proceed exactly as we did in Figure 5.19, but we design the flyback at 73.5 V instead of 127 V. To our possible initial surprise, the required energy handling capability of the core (and its size) remains the same as at 127 V. Note that we haven't even had to play with the air gap. Why is that? And how do we reconcile these results with our repeated advice to design the flyback transformer at the *lowest* input voltage?

The answer to this puzzle takes us back to Figure 5.4. That tells us something very fundamental: that a flyback transformer (or a Buck-Boost inductor) is unique because it has to store *all the energy* that the power supply delivers, and no more (let us ignore losses for now). We showed the term $\Delta\varepsilon$ was directly related only to the power (via the equation $\Delta\varepsilon = P_{IN}/f$), with no direct dependence on the input voltage (or D), *unlike the other*

Flyback/Buck-Boost Core Size and Inductance Selection Examples (with Holdup Time Requirements)

Example: 25W Universal Flyback, 5V@5A, η=0.833

The AC-DC power supply is meant for 90-270VAC operation. Note that P_O=25W, and P_O/η= 25/0.833=30W. In a previous example, (Chapter 5, Figure 5.19), we had taken the rectified DC to be a steady 127V, and had designed the transformer accordingly. In effect we had assumed a very large input capacitance with negligible ripple, and also no holdup time requirements. Now we want to design the converter for ~ 20ms (two missing half cycles) holdup time with 5μF/W. We need to ensure it operates down to 73.5VDC as per the lookup table in Figure 14.3 (see under 90VAC). There are two ways to go about this. The first leads to the same transformer (E25/13/7) we had picked in Chapter 5. But that can also lead to certain performance issues: min duty cycle limiting, high peak currents, especially when operating at high line. The second method gives better performance, but leads to a larger transformer.

First Method: Suppose the target reflected output voltage V_{OR} of this design is 100V, as is typical when using 600V FETs. So the turns ratio $N_P/N_S \equiv n$ is given by V_{OR}/V_O= 100/5 = 20 (ignoring diode drop here for simplicity). The duty cycle at the lowest voltage is

$$D=\frac{V_{OR}}{\eta V_{IN}+V_{OR}}=\frac{100}{(73.5\times 0.833)+100}=0.62$$

$$I_{OR}=\frac{I_O}{n}=\frac{5}{20}=0.25 \quad \text{(reflected output current)}$$

$$I_L=\frac{I_{OR}}{1-D}=\frac{0.25}{1-0.62}=0.658 \quad \text{(center of ramp)}$$

$$r=0.4=\frac{\Delta I_L}{I_L}=\frac{\Delta I_L}{0.658} \quad \text{(current ripple ratio)}$$

$$\Delta I_L=0.658\times 0.4=0.263$$

We do a worst-case estimate by imagining the output voltage has increased by a factor $1/\eta$ (see Figure 5.1). So

$$V_{OFF}=L\frac{\Delta I}{T_{OFF}} \Rightarrow L_{\mu H}=\frac{100}{0.833}\times\frac{(1-0.62)}{0.1\times 0.263}\Rightarrow 1.735\text{ mH}$$

$$I_{PEAK}=I_L\times\left(1+\frac{r}{2}\right)=0.658\times 1.2=0.79\text{(Amps)}$$

$$\varepsilon_{PEAK}=\frac{1}{2}\times L\times I_{PEAK}{}^2=\frac{1.735\times 10^{-3}\times 0.79^2}{2}\Rightarrow 541\mu J$$

$$V_{e_cm3}=\frac{\varepsilon_{PEAK}}{180}=\frac{540}{180}=3\text{ cm}^3$$ (see Figure 5.15)

The closest core is E25/13/7

$$\text{CHECK}: \Delta\varepsilon=\frac{\varepsilon_{PEAK}}{1.8}=\frac{541}{1.8}=300\mu J$$ *(see Figure 5.6)*

(This corresponds to a 30W input wattage for a 100kHz converter). However, we see that the energy and size requirement is **exactly the same** as for the calculation at 127 VDC input done in Figure 5.19. ***If so, why do we always try to design a flyback at the lowest input voltage?*** Independence from input voltage is actually a property of a Buck-Boost (not a Boost or a Buck) because as we learned, ALL the output energy outputted from a flyback/Buck-Boost, must be stored in the inductor along the way. In other words the Δε (= 300μJ) remains the same irrespective of the voltage. So how can there be any difference in transformer size? Indeed, there is a difference in size with respect to input voltage, but it is subtle and indirect in the case of a flyback. Because the size of a magnetic component does not depend on Δε really, but on ε_{PEAK}. These two are in turn related by a function "F(r)" that involves the ripple ratio *r*.

$$\varepsilon_{PEAK}=\frac{\Delta\varepsilon}{8}\times\left[r\times\left(\frac{2}{r}+1\right)^2\right]\equiv\frac{\Delta\varepsilon}{8}\times F(r)$$ *(see Figure 5.6)*

The size of a flyback transformer (or a Buck-Boost inductor) thus depends basically on the current ripple ratio, not directly on the input voltage. So, if we designed our transformer for *r* =0.4 at 127V, we would get the same size of transformer (different inductance though) as compared to designing for the *r* = 0.4 at say 73.5V input. Or at any other voltage. So what's the catch? The problem is we have to make sure a transformer designed for operation at a certain "x" Volts does not operate at a much lower voltage, certainly not without really good protection from core saturation (using current and/or duty cycle limiting). Because if the input falls significantly lower, which does happen as during power-up and power-down, F(r) will go up steeply and the peak energy will rise steeply too. And since the transformer was perhaps not designed for that additional increase in peak energy, it would most likely saturate and cause the switch to blow. It is easy to show that F(r) rises steeply with D (falling input voltage) using two key Buck-Boost equations:

$$r=\frac{V_O}{I_O\times L\times f}(1-D)^2 \text{ and } D=\frac{V_O}{V_{IN}+V_O}$$

Suppose to meet a holdup time of 20ms, we do design the transformer for an *r* of 0.4 at 73.5V as above. This transformer will need accurate duty cycle/current limiting too, because during power-up/down, the input voltage will be momentarily lower than 73.5V. Therefore, to avoid destruction of the switch due to core saturation, in flyback/Buck-Boost design, *with or without holdup time requirement*, we need to a) calculate the peak current and duty cycle at the actual design point of input voltage ("x" Volts), and then b) then set a peak current limit and duty cycle limit ~ 10-20% higher. That margin is usually adequate and neither does it lead to a bigger transformer (gapped ferrite transformers, with a "z" of 5 to 20, saturate "softly" --- see Chapter 5 for a discussion on "z").

Second Method: The above method of meeting holdup time results in an unchanged core size. But since we designed the magnetics at a very low voltage (setting *r* of 0.4 at 73.5V), we get a much smaller duty cycle at high input voltages, and that can lead to performance issues. In this second method, we design the core for *r* = 0.4 at 127V. This gives an acceptably higher duty cycle at high inputs voltages. However, it also leads to a much lower *r* at the still-unchanged lower operating end of 73.5V. We also know that a lower *r* always results in a bigger core (see Figure 5.7). That is the disadvantage here.

$$D=\frac{100}{(127\times 0.833)+100}=0.486, \quad I_L=\frac{I_{OR}}{1-D}=\frac{0.25}{1-0.486}=0.486$$

$$r=0.4=\frac{\Delta I_L}{I_L}=\frac{\Delta I_L}{0.46} \quad \Delta I_L=0.486\times 0.4=0.194$$

$$V_{OFF}=L\frac{\Delta I}{T_{OFF}} \rightarrow L_{\mu H}=\frac{100}{0.833}\times\frac{(1-0.486)}{0.1\times 0.194}\Rightarrow 3.18\text{ mH}$$

At 73.5V, D is again 0.62. Since L is proportional to $(1-D)^2/r$, then, for a given L, *r* is proportional to $(1-D)^2$ (see Figure 5.8), so *r* at 73.5V is

$$r_{low}=r\times\left(\frac{1-D_{low}}{1-D}\right)^2=0.4\times\left(\frac{1-0.62}{1-0.486}\right)^2=0.22$$

$$I_{L_low}=\frac{I_{OR}}{1-D_{low}}=\frac{0.25}{1-0.62}=0.658$$

$$I_{PEAK_low}=I_{L_low}\times\left(1+\frac{r_{low}}{2}\right)=0.658\times\left(1+\frac{0.22}{2}\right)=0.73\text{(Amps)}$$

$$\varepsilon_{PEAK_low}=\frac{1}{2}\times L\times I_{PEAK_low}{}^2=\frac{3.18\times 10^{-3}\times 0.73^2}{2}\Rightarrow 847\mu J$$

(see Figure 5.15)

$$V_{e_cm3}=\frac{\varepsilon_{PEAK}}{180}=\frac{847}{180}=4.7\text{ cm}^3$$

The closest core is EFD30/15/9

Figure 14.4: Two practical flyback designs approaches for meeting holdup time requirements.

topologies. Then we moved on to Figure 5.6 and showed that the peak energy handling requirement of the inductor/transformer was not just the *change* in stored energy per cycle $\Delta\varepsilon$, but ε_{PEAK}. That relationship was

$$\varepsilon_{PEAK} = \frac{\Delta\varepsilon}{8} \times \left[r \times \left(\frac{2}{r} + 1 \right)^2 \right] \equiv \frac{\Delta\varepsilon}{8} \times F(r)$$

We should recognize that F(r) increases as *r* decreases.

We thus realize that for a flyback, since $\Delta\varepsilon$ does not depend on the input voltage it is *r*, and only *r* (besides of course the power rating of the converter), that determines ε_{PEAK}, and thereby the size of the core. So if we set $r = 0.4$ at 73.5 V or $r = 0.4$ at 127 V, *and go no lower* in cap voltage, the size of the core will be the same in both cases. However, if we set *r* to 0.4 at 127 V, and then a line dropout occurs, one that *we wish to ride through via a holdup time specification*, problems will arise. From Figure 2.4 we see that for a Buck-Boost, *r* decreases as the input voltage falls (D increases). In Figure 5.7 we had also explained that the term F(r) above rises steeply if *r* decreases far below 0.4. In other words, if we have chosen the inductance of the transformer such that *r* was set to 0.4 at 127 V and then a dropout occurs, the instantaneous *r* of the converter will decrease significantly, *causing ε_{PEAK} to rise steeply*. Therefore, if the core has not been sufficiently oversized to deal with this momentary stress situation, it will saturate. That is why in the two design approaches above, we will get differently sized cores.

In simple language this means, it is ***not*** necessary to increase the size of a flyback transformer, or even to change its air gap (z-factor) to meet any holdup requirements. We just need to reduce the number of turns and thereby reduce its inductance (higher r). Of course we do need to set correspondingly higher current limits, and so on.

As mentioned in Figure 14.4, the first approach does have problems. By setting *r* to 0.4 at such a low voltage of 73.5 V, as opposed to setting *r* to 0.4 at 127 V, the peaks current would increase much more at high line, and have higher RMS values too. Further, the system will go into DCM much more readily at light loads. In fact, even at the max rated load, it is most likely to be in DCM at high line. These are all the drawbacks of trying not to increase the size of the transformer while complying with the required holdup time specification. Nothing comes for free.

Finally, based on our preferences and design targets, we could decide which of the two above-mentioned transformer design approaches to pick. Better still, we may like to use a compromise solution. For example, we may want to set $r = 0.3$ instead of 0.4 at 73.5 V. That will give a slightly bigger core, but better efficiency and performance at high line.

Part 2: High-Power Applications and PFC

Overview

Let us conduct a thought experiment. Suppose we apply an arbitrary voltage waveform (with a certain RMS value V_{RMS}) across an *infinitely* large resistance. We know that the resistor stays "cold", which indicates no energy is being lost inside it. Now suppose we pass an arbitrary current waveform (with a certain RMS value, I_{RMS}) through a piece of very thick copper wire (assuming it has zero resistance). The wire still doesn't get hot, indicating no dissipation occurs inside it (ignore dissipation elsewhere). To emulate these two operations, we now substitute a mechanical switch. One moment it is an infinite resistor (switch open), the next moment a perfect conductor (switch closed). Expectedly, the switch itself remains cold whatever we do and however we switch it (any pattern). We can never dissipate any heat inside an ideal switch. The perceptive reader will recognize this was the very basis of switching power conversion as explained in *Chapter 1*.

However, if we are a little misguided and put a scope across the mechanical switch we would see an arbitrary voltage waveform with a certain RMS value. Then, if we put a current probe in series with the switch, we would see an arbitrary current waveform with a certain RMS too. Then, suppose we do some "simple math" *and multiply the RMS voltage across the switch (over a certain period of time) with the RMS current passing through it (over the same period of time)*. The product $V_{RMS} \times I_{RMS}$ would obviously be a large, *non-zero* number. But is that number equal to the power dissipated in the switch? Clearly no, since the switch is still cold. So what went wrong? In effect our "simple math" has misled us into thinking there was some dissipation in the switch. The "dissipation" we calculated above was not the real power, but the 'Apparent Power'.

$$\text{Apparent Power} = V_{RMS} \times I_{RMS}$$

When dealing with AC power distribution, in which we use sine-wave alternating current (AC) voltages, this is equivalently written as

$$\text{Apparent Power} = V_{AC} \times I_{AC}$$

Note that "AC" is another name for "RMS". For example, when we refer to the US household mains input as 120VAC, this is a sine wave with an RMS value of 120 V. Its peak value is $120 \text{ V} \times \sqrt{2} = 170$ V.

The real (or true) power is, by definition, the average power computed over a complete cycle

$$P_{REAL} = \frac{1}{T} \int_0^T V_{IN}(t) \times I_{IN}(t) \, dt$$

We actually ran into a similar potential discrepancy in the previous section where we had a worked example based on the 'Arctic analogy' though we did not point it out at that time. We recall we had shown that a 100 W (input power) converter with a 300 μF bulk cap, had an input RMS current of around 2.26 A (approximate). The apparent power was therefore, by definition, $85\text{VAC} \times 2.26\ \text{A} = 192\ \text{W}$. Whereas we know for a fact, that the real (or true) input power was only 100 W ($108.5\ \text{V} \times 0.922\ \text{A} = 100\ \text{W}$).

To document such situations, the term "power factor" was introduced. It is defined as the ratio of the real power to the apparent power. So in our example above, the power factor was $100/192 \sim 0.5$. In principle, power factor can range from zero to unity, with unity being the best possible case. We should also have connected the dots by now, and realized that higher and higher bulk capacitances will only make things worse in a low-power AC−DC front end − by causing the power factor to decrease even further, thereby causing much higher heating in related components.

Note that above, we have already indicated the underlying *problem* with low power factor − that even though only the "apparent" power is said to have increased, *the effect of this apparent power is very real* in the sense that the dissipation in nearby components increases significantly.

There is a growing demand that AC−DC power supplies, besides other mains appliances, have high power factors. A key reason is higher associated equipment and transmission costs to utility companies. But first, let us take a deep breath by recognizing that common household electricity meters don't charge us for "apparent power", only for "real power", or it is possible we would have acted with much greater personal haste in ensuring power factor correction (PFC) is implemented in all household appliances/power supplies. However, a low power factor does impact us directly, especially at higher power levels − by limiting the maximum RMS current we can pass through our household wiring, thus indirectly limiting the apparent power and also thereby the useful power we can get from it. Keep in mind, that if we have a "15A-rated" outlet, we are not allowed to exceed 15 A (RMS) *even momentarily*. Circuit breakers would likely go off to protect the building and stop us in our tracks.

Here is a specific numerical example to show the power limiting imposed by a low power factor. We are basing it on the standard 120 V/15 A outlet circuit commonly found in offices and homes in the US. In principle, this outlet should not be used to handle anything more than $120\ \text{V} \times 15\ \text{A} \times 0.8 = 1440$ Watts of power. Note that we have introduced a derating factor of 0.8 above, thus effectively maintaining a 20% safety margin (which will also prevent nuisance tripping of any circuit breakers in the bargain). Note that the computed 1440 W max in effect refers to the maximum *apparent* power, not the real power, since heating in the wiring depends only on I_{AC} (or I_{RMS}), and therefore on the apparent power, not on the useful power going through the wire (at least not directly). So, assuming that the overall efficiency of our AC–DC power supply is 75%, the power supply can be rated for a maximum *output power* of 120 V × (15 A × 0.80) × 0.75 = 1080 Watts. That would correspond to exactly 1440 W at its input (check: 1080/0.75 = 1440). *But this assumes the power factor was unity*. If the power factor was say, only 0.5 for example, as is quite typical in simple bridge-plus-cap front ends, the maximum output power rating of our power supply can only be 120 V × (15 A × 0.80) × 0.75 × 0.5 = 540 Watts. Check: 540/(0.75 × 0.5) = 1440 W. Further, in effect, we are also *wasting the current-carrying capability of the AC outlet*. Had we reduced the peak/RMS currents for a given output power, we could have raised the output power significantly. The best way to do that is to try and achieve a higher power factor.

In this chapter we are not going to go deep into concepts of reactive power versus real power and so on, other than to point out that a load consisting of a *pure resistor* (no capacitor or inductor present) dissipates *all* the energy sent its way, so its power factor is unity. The basic reason why we got a power factor of less than 1 in our low-power AC–DC front-end was very simply the input capacitor. And for the same reason, if we have a circuit consisting exclusively of pure inductors and/or pure capacitors (no resistors anywhere), the power factor will be zero. We can only store energy, never dissipate it, unless we have a resistor present somewhere. Because, though any L/C circuit would seem to initially take in energy from the AC source (judging by the observed overlap between the voltage and current), in a subsequent part of the AC cycle the relative signs between the voltage and current would flip, and at that moment, all the energy that was taken in (i.e. stored) would start being returned to the AC source. So the real (net) incoming power (computed as an average over the whole cycle) would be zero, but the apparent power would not. And the power factor would then be zero too.

We conclude that the way to introduce power factor correction in AC–DC power supplies is to make the power supply (with its load attached as usual) *appear as a pure resistor* to the AC source. That is our design target in implementing PFC.

What exactly is so special about a resistor's behavior that we want to mimic? If we apply a sine-wave voltage waveform across a resistor, we get exactly a sine-wave current waveform through it. If we apply a triangular voltage waveform, we will get a triangular current waveform. A square voltage will produce a square current and so on. And there is no time delay (phase shift) in the process either. We conclude that at any point in any arbitrary applied waveform, the instantaneous voltage is always *proportional* to the instantaneous current. The *proportionality constant is called "resistance"* (V = IR). That is what we want to mimic on our PFC circuit: it should appear as a pure resistor to the AC source.

> *Note: There are no mandatory/legal requirements that call for say the power factor to be greater than 0.9, or in fact any fixed number. Yes, we do have to keep in mind the max load limitations expressed above. But other than that, we are actually free to have any*

power factor per se. However, there are international standards that limit the amplitude of low-frequency (line-based) harmonics (50 Hz, 100 Hz, 150 Hz and so on) that we can put on the AC line. Because, if we run high-power equipment drawing huge, low-frequency surges of current from the line, in effect we are polluting *the AC environment, potentially affecting other appliances on the same line. The most common standard for power supply designers to follow for low-frequency line harmonics is IEC 61000-3-2, now accepted as the European norm EN61000-3-2. This specifies harmonic limits for most equipment, including AC−DC power supplies, drawing between 75 W to 1000 W from the AC mains. Note that the 75 W refers to the power drawn* from *the mains line, and is* not *the output power of the AC−DC power supply. For example, a "70 W fly-back" with 70% efficiency is actually a 70/0.7 = 100 W device as far as EN61000-3-2 is concerned, and it will thus be required to be compliant to the harmonic limits. EN61000-3-2 places strict limits up to the 40ᵗʰ line harmonic (2000 Hz). Indirectly, that demands nothing other than conventional* active power factor correction *(PFC). All other methods of "line harmonic reduction" (including cumbersome passive methods involving big iron chokes) have a very high risk of falling foul at the very last moment, perhaps due to just one unpredictable harmonic spike, and thereby getting stuck in qualification/pre-production forever. For similar reasons, exotic methods like "valley-fill PFC" may seem academically interesting, and a real wonder to analyze on the bench (virtual bench or a real one). But in a real production environment, we may discover at the very last stage, that they display astonishingly high electromagnetic interference (EMI) spectra. In fact the valley-fill method is usually acceptable only in lighting fixtures (e.g. electronic ballasts), because mandatory EMI limits are then generally more relaxed than the typical EN550022 Class B limits that apply to most AC−DC power supplies (lighting fixtures fall under EN55015). Standard active Boost PFC method is therefore all we will cover in this chapter, that being the best-known and most trustworthy method of complying with EN61000-3-2 without fears or tears.*

Note: We are not going to talk much about the actual nitty-gritties of fixed-frequency (CCM) Boost PFC implementation schemes here, since implementations abound, but all of them eventually lead to the same resultant behavior. It is the behavior that we are really trying to document and understand here. Because that is what really helps us understand PFC as a topic, and thereby correctly pick/design the associated power components and magnetics. We can just continue to rely on the abundant information already available concerning controller-based details, as provided by the numerous vendors of PFC ICs like the UC3854 for example.

How to get a Boost Topology to exhibit a Sine-wave Input Current?

Basics first: we start by a very simple scenario. Suppose we have a DC-DC converter in steady state, with a regulated output "V_O", and very gradually, we increase its input voltage. What happens to the input current? We are assuming that the input is changing really slowly, so for all practical purposes, at any given instant, the converter is in steady state ("quasi steady-state"). Now, what if the applied input voltage is a low-frequency rectified sine wave? Do we naturally get a low-frequency sine-wave input current too? Because if we do, we can stop right there: we are getting a sine-wave current (in phase) with a sine-wave voltage, and so by our preceding discussion, this is already behaving as a pure resistor, and therefore the power factor must be unity. We don't need to do anything more! There are no line harmonics in theory.

Unfortunately, that is just not what happens in any DC-DC switching topology. Because, we know that if the input voltage decreases, then to *keep to the same instantaneous output power*, we definitely need to increase the instantaneous input current. And, if the input voltage rises, the input current needs to decrease to keep the product $I_{IN} \times V_{IN}$ constant $(= P_{IN})$. This is shown in the upper schematic of Figure 14.5. Note that the input current goes extremely high at low input voltages, and also that the diode/output current is constant. The purpose of the output capacitor in any standard DC-DC Boost converter such as this, is to merely smoothen out the *high-frequency* content of the diode current. The low-frequency component of the diode current is a steady DC level, and needs no "smoothening".

Let us see what steps we can take to make the standard Boost converter appear as a resistor to the source. The culprit is clearly our enforced requirement of a *constant instantaneous output power*, which in turn translated into a constant input power requirement. Naturally, the current increased when the voltage was low, rather than decrease as in any resistor. In principle, we want to create some type of control loop that does the following (however it actually implements it): we want to lower the effective (instantaneous) load current requirement as seen by the Boost stage to zero when the input voltage is zero, so that $I_O/(1 - D)$, which is the input current level of a Boost (center of ramp) *remains finite*. If we can accomplish that, we ask: can we appropriately "tailor" the instantaneous load current requirement, so that besides just limiting the input current to finite values, we actually get a pure sine-wave current at the input? In principle, there is no reason why we can't do that. The only question is: *what should the load current shape be to accomplish that*? We can peek at the lower schematic of Figure 14.5 as we go along for the next part of the discussion.

Setting a time-varying instantaneous output current $I_{OE}(t)$, we get

$$I_{IN}(t) = \frac{I_{OE}(t) \times V_O}{V_{IN}(t) \times \eta}$$

$I_{OE}(t)$ is the *effective* (or instantaneous) load current as seen by the switching Boost converter. In other words, if we average out the high-frequency content of the current pulses passing through the PFC diode, we will no longer be left with a steady DC level

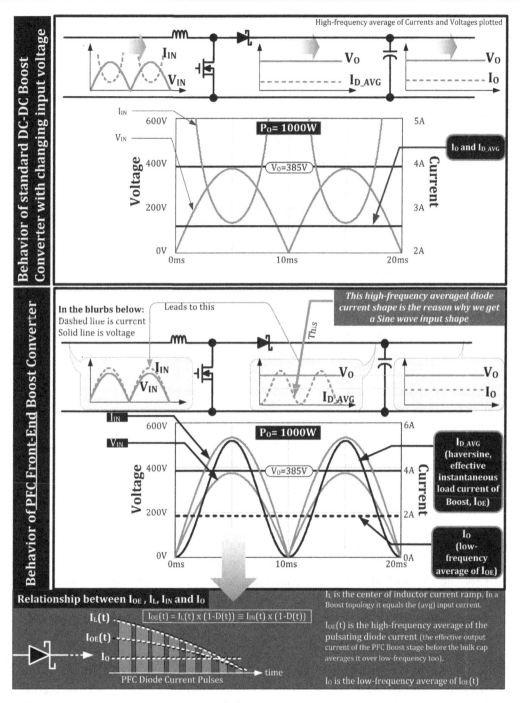

Figure 14.5: How a PFC Boost stage behaves as compared to a standard DC–DC Boost converter.

"I_O", as in any conventional Boost converter, but a *slowly varying current waveshape* — one with a significant *low-frequency component*. This current waveform is called "$I_{OE}(t)$". Of course the load, connected to the right of the Boost output cap, demands a constant current. So, the bulk cap is now responsible for smoothening out, not only the high-frequency component of the diode current pulses, but also its slowly varying low-frequency component, and then passing that smoothened current as I_O to the load.

Basically, any Boost PFC IC, irrespective of its actual implementation, ends up programming this correctly shaped load current profile, which then indirectly leads to the observed sine-wave input current. This was also indicated in the lower schematic of Figure 14.5. But keep in mind, that despite the seemingly big difference between the two schematics in Figure 14.5, at any given moment, the PFC Boost converter is (a) certainly in *quasi steady state*, and therefore, (b) if we set the appropriate load current at a given instant, all our known CCM *DC–DC Boost converter equations are still valid **at that instant***. Because, the underlying topology is *unchanged*: it is still a Boost topology, just one with a slowly varying load profile.

Here is the math that tells us the required load current waveshape to achieve PFC. We first set the requirement and then work backwards.

$$I_{IN} = K|\sin(2\pi ft)|, \text{ where f is the AC (line) frequency here}$$

K is an arbitrary constant so far. The input voltage is a rectified sine wave of peak value V_{IPK}, with the same phase. So

$$V_{IN}(t) = V_{IPK}|\sin(2\pi ft)|$$

The ratio of the input current and input voltage is thus independent of time and is called the "emulated resistance" R_E.

$$R_E = \frac{V_{IPK}|\sin(2\pi ft)|}{K|\sin(2\pi ft)|} = \frac{V_{IPK}}{K} \Rightarrow K = \frac{V_{IPK}}{R_E}$$

This is what we wanted to achieve all along: the PFC stage (with a load connected to it as usual), appears as an emulated resistance R_E to the AC source. We can also intuitively understand that if the load connected to the PFC stage starts demanding more power, R_E must decrease to allow more current to flow in from the AC Source into the PFC stage. So we expect R_E to be inversely proportional to P_O, the output power of the PFC stage.

Note that we can typically assume that the PFC stage has a very high efficiency (greater than 90%, often approximated to 100% for simplicity). So the input power of the PFC stage is almost equal to its output power. That output power then becomes the input power of the PWM stage that follows. The PWM stage has a typical efficiency of about

70–80% and its output power is thus correspondingly lower. However, here we are talking strictly in terms of the input and output power relationships of the Boost PFC stage alone. We thus write

$$\frac{V_{AC}^2}{R_E} = \frac{V_O \times I_O}{\eta} \equiv \frac{P_O}{\eta} \quad (V_{AC} \text{ is the RMS input voltage,}$$

V_O the HVDC rail and I_O the PWM input current)

Since $V_{AC} = V_{IPK}/\sqrt{2}$, we get

$$R_E = \frac{\eta V_{IPK}^2}{2P_O} \equiv \frac{\eta V_{AC}^2}{P_O} \Leftarrow \text{Remember}$$

So, the proportionality factor K is

$$K = \frac{V_{IPK}}{R_E} = \frac{V_{IPK}}{\eta V_{IPK}^2/2P_O} = \frac{2P_O}{\eta V_{IPK}}$$

Now considering each instant of the PFC stage as a Boost converter with a certain varying input, we know that the input current is

$$I_{IN} = K|\sin(2\pi ft)| = \frac{2P_O}{\eta V_{IPK}}|\sin(2\pi ft)|, \text{ where f is the AC (line) frequency here}$$

We already know that

$$I_{IN}(t) = \frac{I_{OE}(t) \times V_O}{\eta V_{IN}(t)}$$

Therefore we get

$$\frac{2P_O}{\eta V_{IPK}}\left|\sin(2\pi ft)\right| = \frac{I_{OE}(t) \times V_O}{\eta V_{IN}(t)}$$

$$\frac{2(V_O I_O)}{V_{IPK}}\left|\sin(2\pi ft)\right| = \frac{I_{OE}(t) \times V_O}{V_{IPK}\left|\sin(2\pi ft)\right|}$$

Solving, the desired equation of the instantaneous load current (required to create a sine-wave current input) is

$$I_{OE}(t) = 2I_O|\sin(2\pi ft)|^2 \Leftarrow \text{Remember}$$

In other words, we need $I_{OE}(t)$ to be of the form $\sin^2(xt)$ if we want to get a sine-wave current at the input. That is the golden requirement for any PFC Boost stage, using any controller IC.

Note that the load current I_O passes through the geometrical center of the $I_{OE}(t)$ curve (see Figure 14.5 carefully). That stands to reason, since the average of $I_{OE}(t)$ over a full AC half-cycle must equal the load current of the PFC Boost stage — a capacitor just smoothens, it cannot add any net current component.

We look more carefully at the plots inside Figure 14.5 (generated from a Mathcad file), and realize that $I_{OE}(t)$ is a $\sin^2(xt)$ function, and similar in shape to a sine (or cosine) function *translated upwards* so as to bring its lowest point coincident with the horizontal axis. This is called a "haversine". By definition

$$\text{hav}(z) = \frac{1}{2}(1 - \cos(z)) = \sin^2\left(\frac{z}{2}\right)$$

So we can say "if the load current of a Boost stage with an AC voltage input, is programmed to be a haversine, we will get a sine-wave input current". That is in a nutshell, what a CCM Boost PFC stage does, any controller IC specifics notwithstanding.

Having intuitively understood how a Boost PFC stage performs, we can look at the detailed calculations in Figure 14.6, Figure 14.7 and Figure 14.8 for all the necessary RMS/Average current equations required, including a numerical example for implementing holdup time in PFC front end stages.

Finally, in Figure 14.9 we have presented the key variations graphically plotted with respect to line voltage. On the y-axis we have the currents per Amp of load current (flowing out of the bulk capacitor). For example, at a typical setting of 385 V for the high-voltage DC rail ('HVDC'), this corresponds to 385 V/1 A = 385 W power going into the PWM stage of. On the x-axis we have the ratio of conversion V_{IPK}/V_O. A numerical example is also provided alongside.

Anti-Synchronization Technique for PFC and PWM Stages

Some power factor control ICs offer "synchronization capability" but that phrase usually means that the PFC stage is *in-phase* with the PWM stage. In other words, whenever the PFC switch turns ON, so does the PWM switch. In contrast, *out-of-phase (anti) synchronization* offers a dramatic reduction in the ripple rating (and cost) of the bulk cap.

> **Note:** *We had talked about this for the low-power AC–DC front end too. In the case of a PFC front end, because of the high bus voltage (~385 V), a typical 20–40 ms holdup time spec is met rather easily and automatically even for smaller capacitors. Therefore, the dominant factor, one that ultimately determines the final selection of the bulk cap in PFC designs, is not the holdup time, but the capacitor's ripple rating (the RMS current passing through it). So this anti-synchronization scheme does translate into significant cost savings.*

This anti-synchronization scheme was introduced as a combo IC a few years ago, and billed as the industry's first "leading edge/trailing edge modulation scheme" PFC/PWM controller

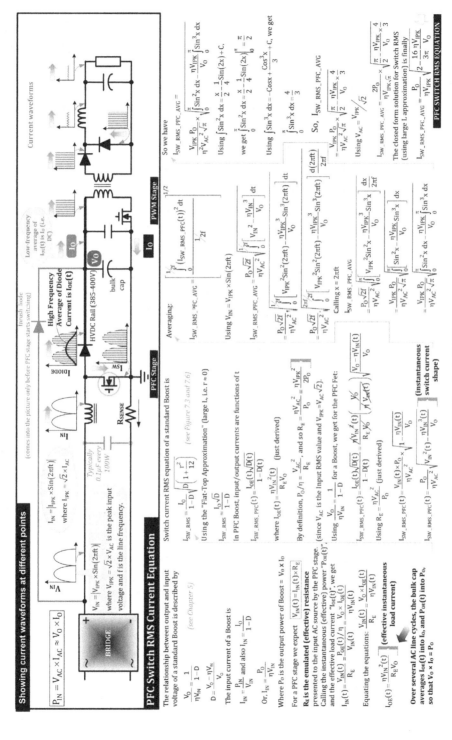

Figure 14.6: PFC switch RMS current equation.

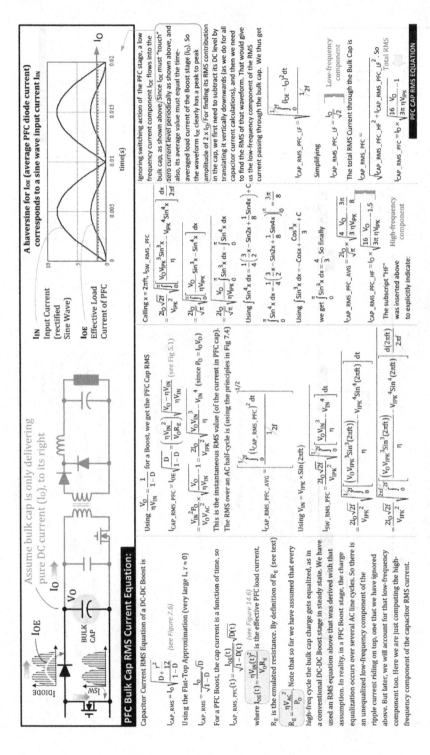

Figure 14.7: PFC input current and capacitor RMS equations.

PFC Diode AVG Current Equation

The Average PFC Boost catch diode current must be I_O (the time-averaged value of I_{OE}), where

$$I_{DIODE\_AVG\_PFC} = I_O = \frac{P_O}{V_O}$$

| PFC DIODE AVG EQUATION |

The conduction loss (~dissipation) in the PFC diode is obtained by multiplying this current with the forward drop of the diode.

Just for reference, though not usually required (unless for Synchronous Boost PFC stages), nor derived here, the RMS current through the PFC diode is

$$I_{DIODE\_RMS\_PFC} = \frac{P_O}{V_{IPK}} \sqrt{\frac{16}{3\pi} \frac{V_{IPK}}{V_O}}$$

| PFC DIODE RMS EQUATION |

PFC Input AVG Current Equation

This is required to calculate the Bridge dissipation. Each diode of that will have a typical forward drop of ~ 1.1V. At any instant, two such diodes conduct in series. So to calculate the Bridge dissipation we need to multiply ~ 2.2V with the average input current below.

$$I_{IN\_AVG\_PFC} = \frac{\int_0^{\frac{1}{2f}} I_{IPK} Sin(2\pi ft) dt}{\frac{1}{2f}} = 2\sqrt{2} \times f \int_0^{\frac{1}{2f}} I_{AC} Sin(2\pi ft) dt$$

$$= 2\sqrt{2} \int_0^{\frac{1}{2f}} \frac{P_O}{V_{AC}} Sin(2\pi ft) dt = 4f \int_0^{\frac{1}{2f}} \frac{P_O}{V_{IPK}} Sin(2\pi ft) dt$$

$$= 4f \int_0^{\frac{1}{2f}} \frac{P_O}{V_{IPK}} \frac{Sin(2\pi ft) d(2\pi ft)}{2\pi f} = \frac{2}{\pi} \times \left(\frac{P_O}{V_{IPK}}\right) \times \int_0^\pi Sin\, x\, dx$$

$$= \frac{2}{\pi} \left(\frac{P_O}{V_{IPK}}\right) \times (-Cos\, x)\Big|_0^\pi = \frac{4}{\pi} \times \left(\frac{P_O}{V_{IPK}}\right)$$

$$I_{IN\_AVG\_PFC} = \frac{4}{\pi} \times \left(\frac{P_O}{V_{IPK}}\right) = \frac{2}{\pi} \times I_{IPK}$$

| BRIDGE AVG EQUATION |

where $V_{IPK} = \sqrt{2} \times I_{AC}$, $I_{IPK} = \sqrt{2} \times I_{AC}$, $P_O = V_{AC} \times I_{AC}$

PFC Input RMS Current Equation

This is required to calculate the dissipation in the Boost choke and in any preceding EMI filter choke. For dissipation, we need to multiply the DCR (resistance) of the choke with the square of the Input RMS current calculated below.

$$I_{IN\_RMS\_PFC} = \sqrt{\frac{\int_0^{\frac{1}{2f}} I_{IPK}^2 Sin^2(2\pi ft) dt}{\frac{1}{2f}}}$$

$$= \sqrt{4f \int_0^{\frac{1}{2f}} I_{AC}^2 Sin^2(2\pi ft) dt}$$

$$= \sqrt{4f \int_0^{\frac{1}{2f}} \frac{P_O^2}{V_{AC}^2} Sin^2(2\pi ft) dt}$$

$$= \sqrt{8f \int_0^{\frac{1}{2f}} \frac{P_O^2}{V_{IPK}^2} Sin^2(2\pi ft) dt}$$

$$= \sqrt{\frac{8f}{2\pi f} \int_0^\pi \frac{P_O^2}{V_{IPK}^2} Sin^2(2\pi ft) d(2\pi ft)}$$

$$= \sqrt{\frac{4}{\pi} \int_0^\pi \frac{P_O^2}{V_{IPK}^2} Sin^2 x\, dx}$$

$$= \sqrt{\frac{4P_O^2}{\pi V_{IPK}^2} \int_0^\pi Sin^2 x\, dx}$$

$$= \sqrt{\frac{4P_O^2}{\pi V_{IPK}^2} \times \frac{\pi}{2}}$$

(using the result of the integration of $Sin^2 x$ over half a cycle as shown in the PFC Switch RMS calculation)

$$I_{IN\_RMS\_PFC} = \left(\frac{\sqrt{2} \times P_O}{V_{IPK}}\right) = I_{AC}$$

| INPUT FILTER RMS EQUATION |

(as expected)

where $V_{IPK} = \sqrt{2} \times I_{AC}$, $I_{IPK} = \sqrt{2} \times I_{AC}$, $P_O = V_{AC} \times I_{AC}$

Holdup Time Limitations

In the most typical case, the PFC stage is followed by a Forward converter which is typically designed to operate at 35% duty cycle with the bus voltage V_O set to 385V. The Forward converter is limited to 50% max duty cycle. Since D is inversely proportional to input voltage for such a (Buck-derived) converter, the lowest input voltage of operation must be

$$V_{O\_MIN} = \frac{35}{50} \times 385 = 269.5\, V$$

That is a drop of 385–269.5 = 115V. In general, since maximum allowed duty cycle is less than 50%, and there are drops across the switch etc, which increase duty cycle, a more typical number for allowed voltage drop is 100V, but the electrolytic bulk capacitors initial tolerance (up to ~20%) and aging (another ~20%) should be accounted for upfront.

$$P_O \times t_{HOLDUP} = \frac{1}{2} C\left(V_O^2 - V_{O\_MIN}^2\right)$$

$$C \ge \frac{2 \times P_O \times t_{HOLDUP}}{V_O^2 - V_{O\_MIN}^2}\ \text{Farads}$$

| HOLDUP TIME EQUATION |

PFC Bulk Cap Selection Example

QUICK REFERENCE for 20ms holdup time:

A) Holdup

$$C \ge \frac{2 \times P_O \times t_{HOLDUP}}{V_O^2 - (V_O - 100)^2}\ \text{Farads}$$

TYPICAL: 0.597μF / W (worst-case/theoretical minim.)

To meet holdup time of 20ms with a 385V bus and a 50% duty cycle limited PWM topology (e.g. Forward Converter), we need 0.597μF/W. For a 500W stage (assuming η=1) we need at least 300μF. Accounting for ~20% initial tolerance plus another ~20% due to degradation over useful life, we need to start with a nominal capacitance value 1/(0.8 × 0.8) = 1.56 times, i.e. for a 500W converter we need to pick a 470μF capacitance.

TYPICAL: 0.931μF / W (nominal C value to select)

Bulk Cap Example (contd)

B) RMS (Ripple) Current Components

$$I_{CAP\_RMS\_PFC\_HF} = \frac{P_O}{V_O} \times \sqrt{\frac{16 V_O}{3\pi V_{IPK}} - 1.5}$$

TYPICAL: 5.154 mA / W (high-freq)

For a 385V bus, at 85VAC input, this is 5.154mA/W. For a 500W stage we need 2.577A high frequency ripple rating. We also have

$$I_{CAP\_RMS\_PFC\_LF} = \frac{P_O}{V_O \sqrt{2}}$$

TYPICAL: 1.837 mA / W (low-freq)

For a 385V bus, at 85VAC input, this is 1.837mA/W. For a 500W stage we need 0.918A low frequency ripple rating.

C) Total RMS (Ripple) Current

$$I_{RMS} = \sqrt{I_{RMS\_LF}^2 + I_{RMS\_HF}^2}$$

Per Watt for a typical case of Bus Voltage 385V at 85VAC Input

$$= \sqrt{1.837^2 + 5.154^2} = 5.472mA$$

TYPICAL: 5.472 mA / W (non Al-Elko RMS rating)

For a 500W stage we need 2.736A RMS ripple rating. But this number is useful only for selecting non-Aluminum Electrolytic Caps. For Aluminum Electrolytic Caps, since ESR falls at high frequencies, heating due to high-frequency components is not that bad. So a frequency multiplier of 1.43 at 100kHz is typically provided to derate the high-frequency component as shown below.

$$I_{RIPPLE} = \sqrt{I_{RMS\_LF}^2 + \left(\frac{I_{RMS\_HF}}{1.43}\right)^2}$$

$$I_{RIPPLE\_mA\_PER\_W} = \sqrt{1.837^2 + \left(\frac{5.154}{1.43}\right)^2} = 4.07$$

Per Watt, Bus Voltage 385V, 85VAC Input we get

TYPICAL: 4.1 mA / W (Al-Elko RMS rating)

Figure 14.8: PFC diode, filter, choke and capacitor holdup equations.

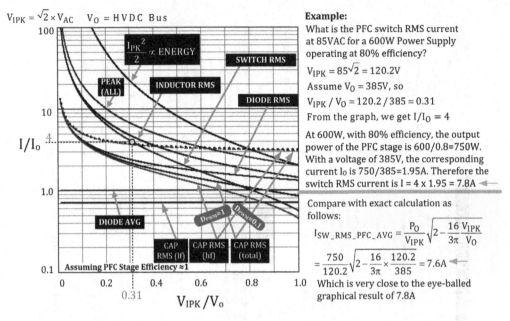

Figure 14.9: Quick lookup curves for estimating current stresses in the PFC stage.

IC. What this IC essentially tried to accomplish was based on the following intuitive thought process: we know that the PWM draws current out from the bulk capacitor, whereas the PFC dumps current into it. What if we could make these opposing currents cancel out inside the cap (i.e. no resultant current)? In other words, what we want to do is to turn ON the PWM switch at the very same moment the PFC switch turns OFF. This is an *out-of-phase* (or *anti*-synchronization). But the (varying) duty cycle of the PFC stage at any given instant is not necessarily going to be the same as the (constant) duty cycle of the PWM stage. So the cancelation cannot be perfect. However, viewed in its entirety over one AC half-cycle, we can certainly assert that *a good part* of the freewheeling (diode) current on an average, will head straight into the PWM stage *without getting recycled through the bulk capacitor, thus saving the latter from a good deal of RMS heating*. Because of the duty cycle mismatch, it is hard to provide any easy closed-form equation for calculating the net reduction in the RMS current through the capacitor. So, not surprisingly, there was in fact no such detailed application information provided for the combo IC mentioned above, and could also be the reason why it never caught on. However there are several commercial products, especially in Europe, that have since, successfully used this technique for years, possibly based on practical in-house experience and know-how.

Note: The combo IC also had a crippling design flaw actually. It had a single shutdown pin, that when activated turned OFF both the PFC and PWM stages together. In practice, we want more flexibility. For example, we may want to turn-off the switching FET of the PFC stage quickly to protect it against a momentary line fault condition, but keep the PWM section running in the interim for holdup purposes. Or we may want to turn OFF the switching FET of the PWM stage (only), to turn its output OFF when demanded. In that case we would prefer to keep the PFC stage switching along, albeit at almost zero load, so that when the output of the power supply is asked to resume again, there is no long waiting time for the PFC stage to come back up from scratch.

The combo-IC we talked about above, had an unnecessary complication too. It had a clock of fixed frequency as usual. But whereas the PWM switch would turn ON at the clock edge (normal technique), the PFC would switch OFF at the clock edge. We remember that in most power converters, regulation is achieved by varying the moment at which the switch turns OFF, whereas the turn-on is determined by a clock. That is traditional 'trailing-edge' modulation. So, in this combo IC, the PWM worked in the conventional way, but the PFC used (opposite) 'leading-edge' modulation. The moment of turn-off was pre-ordained (due to the anti-synchronization demand), so the turn-on edge had to be varied to produce regulation. In effect, we now have to regulate the PFC pulse width *knowing beforehand* that the PFC switch must turn OFF at a certain moment and then reverse-calculate the exact moment we need to turn the PFC switch ON. No doubt this is all done silently by the control loop, but it does lead to rather complicated stability characteristics. So, to avoid this, in actual lab experiments conducted by the author in Germany, it was obvious that an easier way using standard parts was also possible. A UC3854 PFC-IC was used along with a common UC3844 PWM IC. The two ICs were synchronized (with a small resistor in series with the timing capacitor of the UC3854), such that the turn-off edge of the 3844 output reset the clock of the UC3854. So now the 3854 turns ON at the exact moment the 3844 turns OFF. Both are now performing trailing-edge modulation. The only change is that, in effect, the clock (and switch turn-on moment) of the UC3854 has been somewhat handed over to the UC3844 to control. In principle we could also reverse the order: use the UC3854 as the 'master' to have the 3844 turn ON the moment the 3854 turned OFF. This also works (but remember that the designated slave must always have a slightly lower set frequency than the master, to allow the master to dominate it slightly and safely). Both schemes were tried out on the bench and verified using Mathcad. Intuitively, in the latter case, a slight inherent line frequency modulation was expected at the output of the PWM stage. However, in practice, no such modulation was observed on the bench prototype. So both techniques worked − both being trailing-edge schemes, and both with PFC-PWM anti-synchronization at work.

In Figure 14.10 we have plotted out the ripple currents for 90VAC and 270VAC respectively, vs. the PWM stage duty cycle. We see that using synchronization, we get the highest *percentage* improvement at low line (only the total RMS is shown). The corresponding numbers are provided in the embedded table for quick lookup.

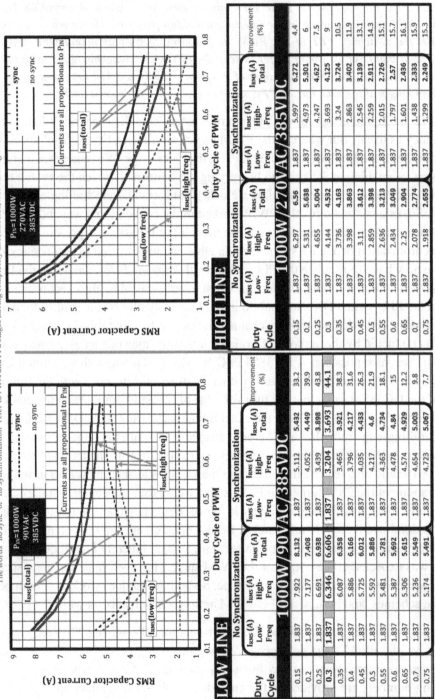

The words "sync" or "synchronization" here refer to the *anti*-synchronization technique as explained in the accompanying text.
The words "no sync" or "no synchronization" refer to PWM and PFC stages having completely unconnected, free-running clocks.

LOW LINE

1000W/90VAC/385VDC

Duty Cycle	No Synchronization			Synchronization			Improvement (%)
	I_{RMS} (A) Low-Freq	I_{RMS} (A) High-Freq	I_{RMS} (A) Total	I_{RMS} (A) Low-Freq	I_{RMS} (A) High-Freq	I_{RMS} (A) Total	
0.15	1.837	7.922	8.132	1.837	5.112	5.432	33.2
0.2	1.837	7.177	7.408	1.837	4.052	4.449	39.9
0.25	1.837	6.691	6.938	1.837	3.439	3.898	43.8
0.3	1.837	6.346	6.606	1.837	3.204	3.693	44.1
0.35	1.837	6.087	6.358	1.837	3.465	3.921	38.3
0.4	1.837	5.886	6.166	1.837	3.796	4.217	31.6
0.45	1.837	5.725	6.012	1.837	4.035	4.433	26.3
0.5	1.837	5.592	5.886	1.837	4.217	4.6	21.9
0.55	1.837	5.481	5.781	1.837	4.363	4.734	18.1
0.6	1.837	5.387	5.692	1.837	4.478	4.84	15
0.65	1.837	5.306	5.615	1.837	4.574	4.929	12.2
0.7	1.837	5.236	5.549	1.837	4.654	5.003	9.8
0.75	1.837	5.174	5.491	1.837	4.723	5.067	7.7

HIGH LINE

1000W/270VAC/385VDC

Duty Cycle	No Synchronization			Synchronization			Improvement (%)
	I_{RMS} (A) Low-Freq	I_{RMS} (A) High-Freq	I_{RMS} (A) Total	I_{RMS} (A) Low-Freq	I_{RMS} (A) High-Freq	I_{RMS} (A) Total	
0.15	1.837	6.297	6.56	1.837	5.997	6.272	4.4
0.2	1.837	5.331	5.638	1.837	4.973	5.301	6
0.25	1.837	4.655	5.004	1.837	4.247	4.627	7.5
0.3	1.837	4.144	4.532	1.837	3.693	4.125	9
0.35	1.837	3.736	4.163	1.837	3.24	3.724	10.5
0.4	1.837	3.398	3.863	1.837	2.863	3.402	11.9
0.45	1.837	3.11	3.612	1.837	2.545	3.139	13.1
0.5	1.837	2.859	3.398	1.837	2.259	2.911	14.3
0.55	1.837	2.636	3.213	1.837	2.015	2.726	15.1
0.6	1.837	2.434	3.049	1.837	1.797	2.57	15.7
0.65	1.837	2.25	2.904	1.837	1.601	2.436	16.1
0.7	1.837	2.078	2.774	1.837	1.438	2.333	15.9
0.75	1.837	1.918	2.655	1.837	1.299	2.249	15.3

Figure 14.10: Quick lookup curves and tables for PFC-PWM anti-synchronization.

We must not forget that a capacitor is chosen not on the basis of a perceived ripple current "improvement", but on the *actual ripple current through it*. We may see a big "improvement" as compared to what it would have been *at that particular input voltage* without synchronization, but the question is: what is its *absolute* value and its worst-case? We should evaluate *both extremes* of input voltage, and *pick the higher* of all the RMS currents thus reported. That would be the worst case number to check the capacitor rating against, i.e. the criterion for its selection.

The conclusions from the embedded table in Figure 14.10 are

(a) The maximum improvement (*and* the lowest absolute value) for the RMS current is at around $D_{PWM} = 0.325$. With a conventional single-ended Forward converter with a duty cycle set to about 0.3–0.35, the improvement due to synchronization is less than around 40%.

(b) In fact over the duty cycle range of 0.23–0.4, we can expect more than 32% improvement.

Capacitor RMS Current Calculations With and Without Anti-Synchronization

For the purpose of estimating the life of the Aluminum capacitor we need to know the breakup of the RMS components with respect to frequency as explained in *Chapter 6*. Note that very often, for the PFC stage, we implicitly assume $P_{IN} \approx P_O$ because of the typically (>90%) efficiency. Here is a sample calculation for capacitor selection.

Example

We have a worldwide (universal) input 70 W flyback running off the PFC stage. Its efficiency is 70%. What are the components of the capacitor RMS current at a set duty cycle of 50% at 90VAC?

In terms of the power rating of the PFC stage, we have 70 W/0.7 = 100 W. Let us first calculate the results assuming an input power of 1000 W since that is our reference baseline in Figure 14.10. With an HVDC of 385 V, the load current is (for 1000 W)

$$I_o = \frac{1000}{385} = 2.597 \text{ (Amperes)}$$

The total unsynchronized RMS current in the Bulk Cap at 90VAC is

$$I_o\sqrt{\frac{16 \times V_o}{3 \times \pi \times V_{IPK}} + \frac{1}{D_{PWM}} - 2} = 2.597 \times \sqrt{\frac{16 \times 385}{3 \times \pi \times 127} + \frac{1}{0.50} - 2} = 5.891 \text{ (Amperes)}$$

Note that we have added a term above, with the duty cycle of the PWM stage, since that also contributes to the heating in the PFC cap.

Above, we have the result for 1000 W. So for a 100 W PFC stage this current would be 0.58 A, that being the number to select the bulk capacitor *had there been no synchronization.*

If we use anti-synchronization, the stated improvement from the table in Figure 14.10 is 21.855%. So then the RMS current requirement for a 1000 W converter would be

$$I_{RMS\_SYNC} = 5.891 \times \left(1 - \frac{21.855}{100}\right) = 4.604 \text{ (Amperes)}$$

For a 100 W case we would pick an RMS rating of 0.46 A, much lower than 0.59 A (unsynchronized) above.

We also know that the low-frequency component of a 1000 W PFC stage is

$$I_{RMS\_SYNC\_LO} = \frac{I_o}{\sqrt{2}} = \frac{2.597}{\sqrt{2}} = 1.836 \text{ (Amperes)}$$

Therefore its high-frequency component is

$$I_{RMS\_SYNC\_HI} = \sqrt{I_{RMS\_SYNC}^2 - I_{RMS\_SYNC\_LO}^2} = \sqrt{4.604^2 - 1.836^2} = 4.222 \text{ (Amperes)}$$

All the above numbers assume a 1000 W converter. So for our 100 W PFC stage, the high-frequency and low-frequency components of the RMS current are therefore 0.42 A and 0.18 A respectively. Knowing the frequency multiplier for the chosen capacitor family, we can now normalize these values to be an equivalent low-frequency current as explained in *Chapter 6*, and thus select our capacitor.

> *Note: We should remember that after the above wide input comparison based calculation, we don't know (and don't need to know) whether the (worst case) synchronized RMS calculated above occurs at high line or at low line. For selection of the capacitor, the above information is sufficient.*

Interleaved Boost PFC Stages

In *Chapter 13*, we discussed interleaving as a method of reducing input and output capacitor RMS currents. So, besides the anti-synchronization scheme above, another option to save cost on the output cap of a PFC Boost stage is by interleaving. Please read *Chapter 13* for more details on interleaving, keeping in mind that in *Chapter 9* we showed that the Boost converter is a Buck converter with the input and output swapped. So it is easy to understand and quantify the benefits interleaving brings to multiple PFC stages sharing an output capacitor.

Practical Issues in Designing PFC Stages

(a) A critical component of a commercial PFC implementation is an inrush diode placed directly from the positive terminal of the bridge rectifier to the positive terminal of the PFC bulk cap (see Figure 14.6). This allows the bulk capacitor to charge up quickly when AC power is initially applied to the power supply, and helps keep the huge inrush current away from the PFC section's choke and output diode. But this inrush diode is also a key reliability issue for the entire power supply. *It is the component most likely to fail under repeated application of AC input.* It need not be a fast diode as it goes out of the picture as soon as the PFC FET starts switching. It does not get hot either, and can be an axial or SMD component. But its non-repetitive surge rating must be high. For example, a slow diode like the 1N5408 (but from a quality vendor) is usually suitable since it has a much higher surge rating than the ultrafast diodes some designers seem to prefer in this position for no obvious reason.

(b) The biggest hit in efficiency in a PFC Boost stage usually comes from a severe shoot-through current spike originating from the bulk capacitor, that passes through the still-recovering PFC output diode whenever the PFC FET turns ON. This causes very high crossover losses in the PFC FET (not much observable impact in the diode itself). Therefore the PFC diode must be an extremely fast diode. Even a diode with more than 20−30 ns recovery time is unacceptable in this position except in very low-power and non-critical applications. That is one reason why for low-power applications engineers often prefer to use Boost PFC ICs that operate in critical conduction mode as discussed later.

(c) In low- to medium-power PFC stages, the lossless (inductor-based) "turn-on snubber", which is considered virtually indispensable in high-power stages, may seem like a luxury. In that case some engineers like to reduce the reverse recovery current spike by replacing the single 600 V PFC diode with two 300 V diodes in series. Here they are relying on the fact that low-voltage diodes recover much faster than high-voltage diodes and so, despite their higher combined forward drop and consequent increased conduction losses, we actually improve efficiency by reducing the $V \times I$ crossover losses in the FET. But note, we cannot allow the full voltage to appear across either of the diodes at any moment, however brief. So they must be well-matched, especially in terms of their dynamic characteristics. That is not easy to do on a PCB, especially in mass production. Some engineers try to achieve matching by placing ballasting resistors across each diode (much as we do for capacitors placed in series for higher-voltage applications). But under dynamic conditions (as during switch transitions), this does not really help. A better option is to use two series diodes *in one package*, on the assumption that since they pass through exactly the same fabrication steps, they will automatically be well-matched. Some manufacturers even combine two series diodes on the very same chip for the best-possible dynamic matching. For example ST Microelectronics offers "Tandem diodes". This may look like any other two-terminal diode, but in reality consists of two ultrafast recovery diodes in series, exhibiting higher forward drop, but excellent reverse recovery (~12 ns). Such diodes are often called "hyperfast". An increasingly popular option nowadays is the 600 V silicon carbide (SiC) diode, as from Cree and Infineon. There is no need for any turn-on snubber either, just a SiC diode. This is becoming a popular option at medium- to high-power levels.

PFC Choke Design Guidelines

In Figure 14.11 we have presented the complete procedure for designing an air-gapped PFC ferrite choke. All the required equations are embedded in it. The key differences from previous procedures that we used (in particular in *Chapter 5*) are as follows:

(A)   Since the input voltage is constantly changing, we have to identify not only which AC voltage, but which instantaneous voltage point within its sine waveform, creates the maximum peak currents. Only then can we really ensure the PFC core does not saturate over its entire range of operation. From Figure 14.9, we see that the max peak input current occurs at low-line its very peak. We thus decide to *size the core at* $V_{AC\_MIN\_PEAK}$. That is $85 \times \sqrt{2} = 120.2\text{VDC}$ (or at 127VDC corresponding to a minimum of 90VAC) (certainly not at V_{SAG} as for low-power front end designs).

(B)   The air gap factor "z" we had defined in *Chapter 5* is

$$l_g = (z-1) \times \frac{l_e}{\mu}, \text{ where } \mu \text{ is the relative permeability } (\sim 2000 \text{ for ferrites})$$

For large air gaps (large z), we can approximate z to

$$z \approx \frac{\mu l_g}{l_e} \Rightarrow \frac{l_e}{l_g} \approx \frac{\mu}{z}$$

For E-type ferrite transformers, we had previously recommended z of 10 as a good target. That amounts to an l_e/l_g of $10/2000 = 0.5\%$. However in a choke, since there is no Secondary winding, we can virtually double the number of (Primary) turns that can be accommodated in the available window. Since z depends of turns squared (for a given inductance), what we are saying is that a *z of 40* or equivalently, *an "l_e/l_g of 2%"*, is recommended for a ferrite PFC choke. So that becomes our design target here.

(C)   The current ripple ratio *r*. This is traditionally set to 0.4, but in a PFC stage it is hard to decide its optimum value, since the input voltage is varying as a sine wave. In the

PFC Boost Inductor Design Example

Example: 500W, 85-265VAC Universal PFC, η=0.9, 100kHz

Inductance:

The maximum instantaneous inductor current occurs at low line (85VAC), at the moment the input AC voltage peaks. This is the design point at which we set r = 0.2. And that in turn determines the inductance.

The rectified AC input peaks at 85VAC × √2 = 120VDC. The duty cycle at that moment is

$$D_{INSTANT} = \frac{V_O - \eta V_{IPK}}{V_O} = \frac{385 - (0.9 \times 120)}{385} = 0.72$$

(see Figure 5.1 for worst-case estimate principle)

Note: The instantaneous load current at this moment is 2 × I_O. (see highlighted sentences on right side in Figure 14-7). We can work out inductance completely based on first principles. At the peak voltage, in terms of voltseconds

$$V_{ON} = L \frac{\Delta I}{\Delta t_{ON}}, \text{ so } L = \frac{V_{ON} \times D}{\Delta I} \times \frac{1}{f}$$

where D is 0.72 at the peak input voltage (call it "D_AT_VIPK"), f is the switching frequency in Hz. The "peak current" (center of inductor current at peak AC value) is related to the peak voltage by

$$P_{IN} = \frac{P_O}{\eta} = V_{AC} \times I_{AC} = \frac{V_{IPK} \times I_{IPK}}{2}, \text{ so } I_{IPK} = \frac{2 \times P_O}{\eta \times V_{IPK}}$$

But, since I_IPK is the center of ramp at peak input AC voltage,

$r = \frac{\Delta I}{I_{IPK}}$. We write $\Delta I = r \times I_{IPK}$, and since $V_{ON} = V_{IPK}$ at the peak

$$L = \frac{V_{ON} \times D}{r \times I_{IPK}} \bigg/ f_{SW} = \frac{V_{IPK} \times D_{AT\_VIPK}}{r \times f_{SW}} \times \frac{2 \times P_O}{\eta \times V_{IPK}}$$

$$\boxed{L_{\mu H} = \frac{\eta \times V_{IPK}^2 \times D_{AT\_VIPK} \times 10^6}{2 \times r \times P_O \times f_{SW}}} \text{ where } D_{AT\_VIPK} = \frac{V_O - \eta V_{IPK}}{V_O}$$

Using numerical values we get

$$L_{\mu H} = \frac{0.9 \times 120.2^2 \times 0.72 \times 10^6}{2 \times 0.2 \times 500 \times 100 \times 10^3} \Rightarrow L = 0.47mH$$

Size of Core:

For a DC-DC Boost converter, we worked out that the energy per cycle that goes in and out of the inductor is

$$\Delta \varepsilon = \frac{P_{IN}}{f} \times D = \frac{P_O}{\eta} \times \frac{D}{f} \approx \frac{P_O}{f} \times D$$ (see Figure 5.5)

So the inductor of a 500W@90% efficiency boost converter, running at a duty cycle of 0.72, would technically need to cycle only 500 × 0.72/0.9 = 400W. But a PFC Boost stage is different. Because its effective load current is I_OE, **which at its peak, is twice the time-averaged value "I_O"** (see Fig 14.7). In effect, the *instantaneous* power a 500W PFC Boost converter is delivering at the moment the AC voltage peaks, is **not** 500W, but 1000W. So **the PFC Boost inductor has to cycle twice the energy of a conventional DC-DC Boost stage.** The equations and numerical results are as follows

$$\Delta \varepsilon_{PEAK} = \frac{V_O \times (2 \times I_0)}{\eta f_{SW}} \times D_{AT\_VIPK} = \frac{2 P_0}{\eta f_{SW}}$$

$$\Delta \varepsilon_{PEAK} = \frac{385 \times (2 \times 1.3)}{0.9 \times 100 \times 10^3} \times 0.72 \Rightarrow 11\ mJ$$

From Figure 5-6, we get for r=0.2

$$\varepsilon_{PEAK} = \frac{\Delta \varepsilon}{8} \times \left[r \times \left(\frac{2}{r}+1\right)^2 \right] = \frac{\Delta \varepsilon}{8} \times 24.2 = 3.025 \times \Delta \varepsilon$$

So the peak energy handling capability of the PFC choke must be 3.025 × 11 → 34mJ

We can no longer use our simple rule of transformers (180 μJ/cm³) in Figure 5.15, since that was for z=10 and r=0.4. For a ferrite choke z can be 40, and for a PFC choke we set r = 0.2. Therefore, starting from our basic equation we get as follows:

$$V_{e\_cm3} = \frac{31.4 \times P_{IN} \times \mu}{z \times f_{MHz} \times B_{SAT\_Gauss}^2} \times \left[r \times \left(\frac{2}{r}+1\right)^2 \right]$$

$$V_{e\_cm3} = \frac{0.01 \times P_{IN}}{f_{MHz}} \quad \text{(for ferrite transformers: z=10, r=0.4)}$$

$$V_{e\_cm3} = \frac{0.00422 \times P_{IN}}{f_{MHz}} \quad \text{(for ferrite PFC chokes: z=40, r=0.2)}$$ (see Figure 5.19)

Using numerical values

$$V_{e\_cm3} = \frac{0.00422 \times 500}{0.1 \times 0.9} = 23.44\ cm^3$$

(As per our calculations, this needs to store a peak of 34mJ).

We can pick ETD49 (V_e = 24.0 cm3). http://www.mmgca.com/catalogue/MMG-Ferrite-ETD.pdf

Air Gap:

Using previously derived equation for air gap vs. z (see Figure 5.22)

$$l_{g\_mm} = (z-1) \times \frac{l_e\ mm}{\mu}$$ (see Figure 5.22)

So picking ETD49 (for r=0.2 case), and using z = 40, we get

$$l_{g\_mm} = (40-1) \times \frac{114}{2000} = 2.22\ mm$$ (See Figure 5.17)

So we need to grind off 1.1 mm off each half's center limb. Or put a spacer of 1.1 mm on the outer limbs.

Number of Turns:

$$(N \times A_{e\_cm2}) = \left(1 + \frac{2}{r}\right) \times \frac{V_{IN} \times D}{200 \times B_{PEAK} \times f_{MHz}}$$ (see Figure 5.22)

$$N = \left(1 + \frac{2}{r}\right) \times \frac{V_{IN} \times D}{200 \times B_{PEAK} \times f_{MHz} \times A_{e\_cm2}}$$

$$= \left(1 + \frac{2}{0.2}\right) \times \frac{120.2 \times 0.72}{200 \times 0.3 \times 0.1 \times 2.11} = 75\ Turns$$

For selecting AWG, we keep in mind that like all chokes, this is not a transformer (with sharp edges of current), but a smooth undulating current waveform. So, skin depth is not a big concern. However, since RMS input current (same as RMS inductor current for a Boost) is very high, it would demand a rather thick and *unmanageable* wire. A better option is to remember that we recommend AWG24 per Amp (see *Figure 5.22*), where the "Amp" was the center of ramp of a square pulse of about 50% duty cycle. That was equivalent to √D = 0.707 ARMS. So we were actually recommending AWG 24 per 0.70 ARMS. In our case, the RMS current in the choke is I_AC: $I_{AC} = \frac{P_O/\eta}{V_{AC}} = \frac{500}{85} = 6.54\ ARMS$

We therefore need 6.54/0.707 ~ 9 strands of bundled AWG24. That would be more flexible than a single thick strand, and easier to wind on the bobbin. It would also reduce skin effect losses in the bargain (though skin depth is not critical here).

Figure 14.11: PFC choke design equations and an example.

next section we will see that the ΔI_L of the choke is actually much higher at 180VAC than at low AC line voltages (or at high). This can lead to higher input filtering costs than expected. So, to keep costs down, when we design the choke at low-line, we leave margin for this increase in current swing at higher voltages, *by fixing r to a more optimum value of 0.2*, instead of the traditional 0.4 (i.e. set ±10% current ripple, instead of ±20%). That is our design target in the examples in Figure 14.11.

Core Losses in PFC Choke

We can refer to *Chapter 2* to see how to calculate the core loss for a steady DC input. The important thing to keep in mind is that core losses do not depend of the DC pedestal of current (I_L), only on the swing ΔI_L. And that swing is the peak to peak current swing, which in turn translates into a proportional flux density swing. But the problem with PFC is that the *input is varying even at a given line voltage V_{AC}, and further, also as that voltage varies with the applied V_{AC}. Under each condition, the current swing is so different that it seems impossible to estimate the core losses, or even their average value, for the purpose of better estimating temperature rise of the choke.*

In Figure 14.12 we have shown a mathematical simulation of the inductor current. Note that the high frequency is scaled down for visual clarity, but since the inductance used in the simulation is proportionately scaled upwards, the peaks, troughs and the peak to peak values of the currents shown are in fact the actual values we will see *under the application conditions indicated in the figure.*

Using the same Mathcad file used to plot Figure 14.12, it can be observed that at 180VAC, the peak to peak current has a shape closest to a rectangular shape (and the highest average, see the embedded table in the figure). The values in the table are for an HVDC rail of 400 V and a current of 2.5 A. So this is a 1000 W PFC stage. The way things scale is shown in the following example.

Example

We have a worldwide input 500 W PFC stage (assume 90% efficiency). It uses a choke inductance of 0.5 mH, and its switching frequency is 100 kHz. What is the average ΔI we should use in core loss estimation?

From the embedded table in Figure 14.12, we have the worst case average current swing over all AC (line) and sine-wave variations (under the stated conditions) as 3.7 A. We can scale this to our current conditions as follows

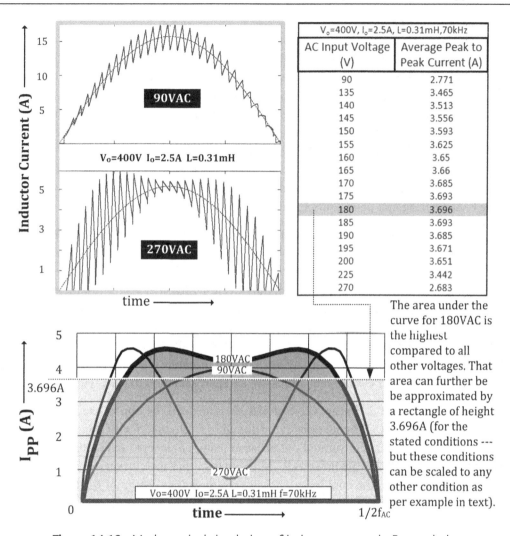

Figure 14.12: Mathematical simulation of inductor current in Boost choke.

$$\Delta I = 3.7A \times \frac{0.31 \text{ mH}}{0.47 \text{ mH}} \times \frac{70 \text{ kHz}}{100 \text{ kHz}} \times \frac{100\%}{90\%} = 1.9 \text{ A}$$

The steady DC current (or power) does not come into the picture as expected (only indirectly via the choice of inductance). The associate average ΔB can be calculated using Faraday's law with a knowledge of the core characteristics and the number of turns. The average core loss so calculated will then accurately reflect the actual averaged value, over the entire universal input range. See *Chapter 2* for the detailed core loss calculation procedure in DC–DC converters.

Borderline Active PFC using Boost

So far we have been discussing active PFC running in CCM (continuous conduction mode). For low to medium power, a popular option is "borderline PFC" (borderline conduction mode is called BCM, see *Chapter 1*). BCM-PFC is almost the same as CCM-PFC, except that the PFC FET waits till the inductor current ramps down to zero before turning ON again. In effect we are forcing the inductor to operate in critical conduction mode. The main advantage of doing that is the PFC diode current is zero whenever the PFC FET turns ON. So, borderline Boost PFC stages have no "PFC diode reverse recovery shoot-through" problem. They do not need turn-on snubbers. But they do work with variable frequency and so their EMI filtering can pose a problem.

What are their stress equations? To answer that, keep in mind that BCM (also called critical conduction mode) is a special case of CCM with $r = 2$. Admittedly BCM is on the extreme end, but still a member of CCM. Further, we recognize that RMS/Avg equations do not depend on the switching frequency, only on duty cycle D. In other words, all the stress equations we have calculated so far, should still apply — in principle. But they really don't! Because on closer examination we see that in all the previous CCM-PFC derivations, we used the "flat-top approximation". This assumes a very small r (large L). Which is why none of the stress equations we derived (or any available in literature) include "r" (or equivalently L). In other words, our PFC stress equations are an approximation to start with.

In BCM, r has a value of 2. But we have already assumed r to be almost zero! That is the reason why, though our CCM equations are still valid in principle for BCM, *they will not give accurate answers*. This is pointed out because many engineers (and textbooks) mistakenly apply the CCM equations to BCM. Note also that the current swing is much more in BCM, and so are the core losses, along with the entire design of the PFC choke. The peak currents are much higher. Borderline PFC design is just not trivial. We will therefore avoid it here, justified on the grounds that in any case, PFC is really required for fairly high power applications. Further, with the growing popularity of SiC diodes, our opinion is we should now focus exclusively on CCM-PFC, which is far more predictable in terms of its EMI spectrum and required input filtering. Certainly no passive PFC or flyback-based PFC, or valley-fill PFC as discussed previously.

CHAPTER 15

EMI Standards and Measurements

Part 1: Overview and Limits

Sooner or later, every power supply designer finds out the hard way that if anything has the potential to cause a return to the drawing board at the very last moment, it is either a *thermal* issue, a *safety*-related issue, or a stubborn *EMI* (electromagnetic interference) problem. Of these, EMI may be the least predictable and time-bound of all. It turns out to be a veritable "balloon" — if we try to "push" in the emissions spectrum at one frequency, it "bulges" out at another. If we manage to achieve compliance with *conducted* emission limits, we may find it was at the expense of *radiated* limits, and so on.

EMI in power supplies is admittedly a challenging area, partly because a lot of *uncharacterized parasitics* enter the stage, each vying for attention. So, bench tweaking is not going to be completely avoidable. But with a clear insight into the principles, any major redesign should never be required. We have to be cautious however, in applying the terminology and concepts relating to EMI from the area of digital networks (*signal integrity issues*), directly to switching power supplies. Sure there are great similarities, but the devil is in the details.

The Standards

Electromagnetic interference (EMI) is just one aspect of the area of electromagnetic compatibility (EMC, see Figure 15.1). In our electromagnetic world, all electrical devices need to coexist with each other, and be "good neighbors" — not causing too much interference to others, and not being overly sensitive to others' interference either. For example, we need to assure a prospective buyer that his or her new set-top box is not going to malfunction whenever someone in the adjacent building just turns on an electric shaver or vacuum cleaner. Both aspects of EMC, emissions and susceptibility, are, therefore, regulated by various international EMC laws. These are the two sides of the coin called EMC. As we can easily foresee, switching power supplies are best labeled as culprits, not victims.

Switching Power Supplies A–Z. DOI: 10.1016/B978-0-12-386533-5.00015-2 **597**

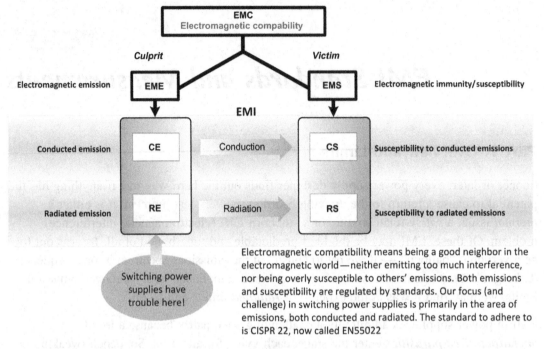

Figure 15.1: The EMI/EMC "tree" (emissions and susceptibility).

Recognizing that the local environment plays a role in defining the amount of received interference and its level of acceptability, EMI limits are split into two basic application categories.

- *Class A*, corresponding to commercial/industrial equipment/environment. Their corresponding limits are relatively relaxed.
- *Class B*, corresponding to domestic or residential equipment. Their corresponding limits are relatively stringent.

Rather broadly stated, Class B limits are roughly about 10 dB lower than Class A limits. This represents a ratio of about 1:3 in terms of the amplitudes of the emission levels $(20 \times \log(3) \approx 10\,\text{dB})$.

Note that when in doubt as to where a certain piece of equipment may be used eventually, we need to design it for Class B requirements.

There is another major compliance requirement for us to be able to sell a product in the marketplace, and that is based on *safety*. In many countries, EMC and safety compliance are clubbed together under a regional conformity mark. The CE mark (i.e., European Conformity mark) is one such example. Another is the CCC mark (China Compulsory

Certification, required by the People's Republic of China. i.e.. mainland China). Such marks indicate compliance to both the required EMC standard and the safety standard.

The generally accepted international EMI standard has been historically called CISPR-22. In the European Union, applicable products need to comply with EN55022 (this is basically just CISPR-22 after getting ratified and mandated for use in the European Union). On the other hand, the generally accepted international safety standard is IEC 60950-1 (historically called IEC 950, then IEC 60950, and so on). In Europe, applicable products need to comply with EN60950 (the 'Low Voltage Directive'), which is essentially the same as IEC 60950-1 (after getting ratified and passed into law). Though there are certain parts of Europe, especially the Scandinavian regions, where compliance to IEC 60950-1 is required with some additional requirements referred to as national/regional "deviations." In the US, the IEC 60950-1 safety standard has been adopted as UL 60950-1 (currently in 2nd Edition).

As indicated, in the US, the issues of safety and EMC are taken up separately, and there is no combined conformity mark. The "UL mark" (Underwriters Laboratory Inc.) indicates compliance to *product safety* standards, whereas "FCC certification" (Federal Communications Commission) reflects compliance with EMI standards.

There are differences between the EMI standard CISPR-22 and the US standard (FCC Part 15). We will discuss the differences later. But very generally speaking, it is said that if a power supply meets CISPR-22, it will likely meet FCC standards. Furthermore, FCC often accepts certification to CISPR-22. So, in general, CISPR-22 has become the underlying standard to comply with across the world (for IT-related equipment and power supplies).

EMI Limits

In Figure 15.2, we have plotted out the EMI limits imposed by the CISPR-22 and FCC standards. We see there are "conducted" EMI limits expressed in μV or $dB\,\mu V$ as a function of frequency. This is the actual voltage drop measured across a certain resistor within special receiving equipment as we will see. Then there are "radiated" EMI limits expressed in $\mu V/m$ or $dB\,\mu V/m$, representing the measured electric field "*E*" *at a specified distance from the emitter* (culprit), measured by an antenna in a special chamber. Why measure only electric fields, not magnetic fields? Because the two are proportional to each other at great distances as we will learn.

In Figure 15.3, we have collected the numbers from Figure 15.2 into easy lookup tables. Alongside, are sample calculations to show how to calculate μV from $dB\,\mu V$ and vice versa. We still need to explain certain key terms contained in these two figures and in the accompanying sample calculations. We will do that further below.

We observe that standard conducted EMI emission limits are typically only up to 30 MHz. We can ask — why weren't the limits set higher? The reason is that by 30 MHz, any

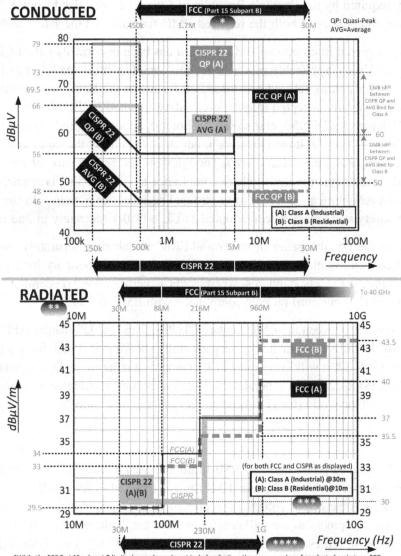

Figure 15.2: Plots of conducted and radiated EMI limits as per CISPR-22 and FCC part 15.

conducted noise is expected to automatically suffer severe attenuation in the mains wiring, and therefore won't really be able to travel far enough to cause interference "further down the road." However, since cables can radiate and send electromagnetic fields over great distances, typical EMI radiation limits cover the range from 30 MHz to 1 GHz.

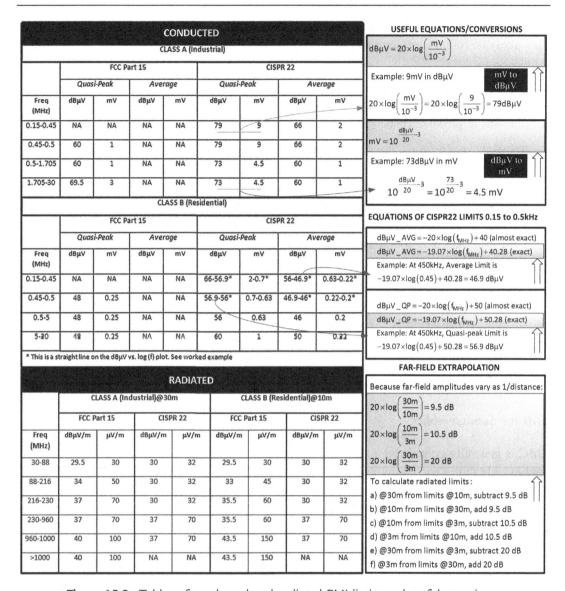

Figure 15.3: Tables of conducted and radiated EMI limits and useful equations.

Comparing the (conducted emissions) Class A quasi-peak limits of FCC and the CISPR-22 in Figures 15.2 and 15.3, and then doing the very same comparison for the Class B limits, we can justifiably ask — do the numbers imply that the FCC standards are more stringent than those of CISPR-22? Not really. The first difference is that FCC measurements are done at much lower (US) line voltage levels, whereas CISPR measurements are done at roughly twice that voltage. So, we may be comparing apples to oranges. Further, though FCC has no defined average detection limits (only quasi-peak), the language allows for a relaxation of the quasi-peak limits (by 13 dB) if a quasi-peak reading exceeds the average by more than 6 dB. Therefore, *practically speaking*, equipment compliant to CISPR will very likely be found compliant to FCC limits.

Some Cost-Related Rules-of-Thumb

A brief look at possible costs:

- The FCC spectrum for digital equipment (currently) begins at 450 kHz, while the equivalent CISPR/EN regulations start at 150 kHz. So, FCC compliance can be achieved with a relatively small and inexpensive filter.
- CISPR/EN Class A compliance often requires a filter with at least twice the volume of the FCC-level unit. This filter can therefore be up to 50% more expensive.
- CISPR/EN Class B compliance can require a filter with 3−10 times the volume of the FCC unit, and could cost up to four times more.

Note: CISPR limits apply to line voltages of 230 VAC, whereas FCC limits are tested at US line voltage (115 VAC). For a given output power, the input operating current is higher if the input voltage is less. Therefore, if any equipment is designed to operate at US line voltages, thicker copper is required in the filter chokes, and that is somewhat of a cost adder.

EMI for Subassemblies

EMC is generally considered a system-level concern, since from the legal perspective, it applies only to the end-equipment. So, a *component power supply* (also called an "OEM" power supply or a "subassembly," e.g., the one inside our desktop computer) does not usually have to meet any EMI/EMC standard per se, unlike a stand-alone power supply. The ultimate EMC responsibility rests with the system manufacturer. However, take the case of a component off-line power supply (the front-end power converter for the system). Here, a major component of the observed EMI measured at the input of the system will clearly be coming from the power supply. So, it certainly won't help if the power supply itself is producing more EMI than the limits that apply to the overall system. We should then keep in mind that when the power supply is integrated with the equipment, there are always some hard-to-predict *interactions* between the power supply and the rest of the system — through the connectors, wiring, chassis, grounding, and so on. So, the final EMI spectrum is not necessarily just the arithmetic sum (in dB) of the different subassemblies. Keeping this in mind, the system manufacturer would most likely call out for a front-end converter to maintain its EMI to less than 6−10 dB *below* the legal limits. That would usually leave enough headroom for the rest of the system, as also for unexpected interactions between the power supply and the rest of the system. In addition, certification labs themselves may require submitted prototypes to be at least 2–3 dB below certification limits — so as to leave margin for variations in subsequent production lots. Summing all this up — the practical resulting situation for (front-end) OEM power supplies is simply this — yes, they don't *need* to comply with the legal EMI limits, in fact they need to be *better* (than component power supplies).

What about DC–DC converters that happen to be positioned deep *inside* the equipment (like point of load converters and bricks)? Again, there are no legally applicable EMI/EMC standards for these per se. Further, since they are likely to be preceded by various circuits and filters, surge suppressors, fuses, capacitors, inrush limiters, and so on (e.g., inside the front-end AC–DC power supply too), there is usually an adequate (and fortuitous) EMI barrier already present, that prevents noise from the DC–DC converter from getting on to the AC mains lines via conduction. Further, if we assume there is in effect an EMI radiation shield present (e.g., the grounded metal enclosure), radiated EMI may also not be of great concern. Therefore, typically, for low-power on-board DC–DC converters, no dedicated input filter stages may ever be required. However, if such a filter does become necessary, it can usually just be a simple single-stage LC circuit — possibly even using a small ferrite bead inductor as the "*L*" of the LC filter. Sometimes, just one such LC filter stage may be good enough to service several paralleled DC–DC converters.

Despite the above natural EMI buffers that are usually present, manufacturers of DC–DC converter modules are often going through the trouble of profiling the EMI spectrum present at the (unfiltered) inputs of their products. The purpose is that, whether legally required or not, the information will come in handy for the system designer when he or she makes EMI-related decisions later. Further, very often nowadays, even the *outputs* of DC–DC converters are being EMI-profiled.

Electromagnetic Waves and Fields

Light, radio-frequency (RF) waves, infrared (IR) radiation, microwaves, and so on, are all electromagnetic waves. For all these, the basic relationship connecting their wavelength λ (in m), their frequency f (in Hz), and the speed of the wave u in the medium of propagation (in m/s), is given by $\lambda = u/f$. For a wave propagating in free space (or air), the speed u is called "c" and has the value 3×10^8 m/s. An easy form to remember is

$$\lambda_{\text{meters}} = \frac{300}{f_{\text{MHz}}}$$

The ratio c/u is always greater than 1, and is called the *index of refraction* of the material (through which the wave travels at the speed u). Note that though c is popularly called the velocity of *light*, it is the same for any electromagnetic wave. It can be shown that $c = 1/\sqrt{(\mu_o \varepsilon_o)}$, where μ_o is the *permeability* of free space (vacuum or air) and ε_o is the *permittivity* of free space. μ_o and ε_o are fundamental constants, since they represent the properties of our universe.

We may remember from our physics class that if a piece of electronic equipment has any dimension close to $\lambda/4$, it can end up radiating (or receiving) the corresponding frequency very effectively. This is the principle behind a radio antenna. (Note that a symmetrical antenna has a total physical length of $\lambda/2$, but that comes from balanced sections of length $\lambda/4$ placed on each side). So, what if an antenna is much shorter than the "optimum length" of $\lambda/4$? Antennas are actually quite effective down to less than $\lambda/10$ — which explains why we can pick up almost all the FM stations well enough from a (fixed length) whip antenna on our car. What if the antenna is much longer than $\lambda/4$? In that case, we can intuitively consider the antenna as being in effect, clamped at $\lambda/4$ — the remaining length is basically superfluous. Therefore, it seems wise to never judge an antenna by its length alone, and we need to keep this in mind especially when designing a PCB for a switching power supply. There, we must minimize large areas of copper with swinging *voltages* on them (e.g., the switching node), and also reduce the enclosed area of any *current* loop, especially those containing high-frequency harmonics. This was discussed in *Chapter 10*.

When we plug a piece of equipment into the AC power lines, its input cable (AC line cord) can combine with the wiring of the building to form a giant antenna. This can produce strong radiated interference that can affect the operation of other devices in the vicinity. In addition to the radiation process, the emissions can also just keep conducting down the mains wiring, thereby directly entering other similarly plugged-in devices. Therefore, there are distinct *radiated emission* limits and *conducted emission* limits specified within all EMI regulatory standards.

We have realized that it would be a mistake to jump to the conclusion that a certain cable length or PCB (printed circuit board) trace is either "too short" or "too long," and therefore *not* contributing to a certain stubborn EMI peak that we may be observing. Further, we should keep in mind that any antenna is as good a *receiver*, as it is a *transmitter*. So, we could have a situation where radiation is originally generated by the *output* cables, but then picked up by the *input* cables (by radiation), from which point onward it gets *conducted* into the wiring of the building (or/and radiated once again). In fact, we will find that the input and output cables are often responsible for a lot of high-frequency EMI noise, both in the radiated spectrum and the conducted spectrum.

Concerns about cable length take on a whole new meaning when they are coupled with circuits containing modern *high-speed* digital chips. Such chips are themselves powerful EMI emitters, but with the help of inadvertent antennas like the surrounding PCB traces and cables, and also with the inadvertent help of various board, component, and enclosure parasitics, they can put on quite a show, courtesy Maxwell.

Maxwell showed that whenever an electric field (the "E-field" — dimensions V/m) varies with time it produces a magnetic field (the "H-field" — dimensions A/m), and vice versa. In fact, the better-known Faraday's law of induction (without which no transformer in the world would exist) is actually the first of the set of four Maxwell's unifying equations. So, we learn that the *E* and *H* fields appear simultaneously, the moment the original magnetic or electric source has a *time variance*. At some distance away, these fields combine to form an electromagnetic wave — that propagates out into space (at the speed of light).

We can ask — what really makes modern digital chips, and modern switching converters, so much worse than their predecessors, from the standpoint of EMI? That is because of the escalating *frequencies* involved. Smaller and smaller PCB traces and lead lengths can become effective antenna at very high frequencies. So, we are nowadays getting painfully aware of the fact that as the frequency increases, the intensity of these fields (at a certain distance) also increases. Note however, that when talking about *switching power converters*, the "frequency" that we are talking about is not necessarily the basic PWM switching frequency (which is only of the order of say 100–1000 kHz). We are referring more to the exceedingly fast *transition* times — which are of the order of 10–100 ns. The Fourier analysis of such a switching waveform will reveal a large amount of very *high-frequency content*, associated with the actual *switch transitions*. These sharp voltage and current edges are what really exacerbate the problem as we will see in *Chapter 18*.

Maxwell's equations are usually written out in a way that doesn't fully reveal the following fact clearly: it really does not matter whether we are talking about waveforms of switched voltages (time-varying *E*-fields) or of switched currents (time-varying *H*-fields) — eventually, their respective equations are *complementary*, and very similar. More important, *these two fields become proportional to each other* at a large distance away, constituting an *electromagnetic wave*, one that becomes self-sustaining and can, therefore, travel great distances on its own (yes, across galaxies). It also follows that if we "kill" one component of an electromagnetic wave (either its *E*-field or *H*-field), we will manage to kill the entire electromagnetic wave. Therefore, we often use RF (radio-frequency) shielding or electromagnetic shielding. However, if we want to suppress slowly varying fields, and/or near-fields, we often use electrostatic shields and/or magnetic shields.

To go back to our physics class for a moment, in general, circuits that cause fields can be sorted into four basic classes:

1. Electrostatic
2. Magnetostatic
3. Electric, time variant
4. Magnetic, time variant

The electrostatic class is simply a fixed distribution of charges. Since the charges do not move, no current flows. The basic building block here is the "charge dipole," where two equal and opposite charges are placed a certain distance apart (one may be at infinity), or we could have a wire held at some fixed voltage. In either case, we have an electric field E that does not vary with time. There is no associated magnetic field (H is zero) based on Maxwell's equations. There is therefore no concept of wave impedance here, and the ratio of E to H is infinite.

Magnetostatic circuits consist of DC current loops. This is the dual of the electrostatic case. There is a constant magnetic field H present that is time invariant. Field information doesn't propagate in this case either.

The third class above is a time-variant electric circuit. We could start by thinking of a slowly varying electrostatic circuit. Consider these to be more or less equivalent cases:

(a)   A charge dipole where the charges vary sinusoidally.

(b)   A current element where current flows back and forth sinusoidally along a line (charges would build up and reverse at the ends, so this is equivalent to the previous example).

(c)   Any collection of *open-ended* wires driven by arbitrary voltage sources, including dipole and whip antennas, as well as low-speed leads exiting circuit boards and driven by common mode voltages ("common mode" is explained a little later).

(d)   A short, sinusoidally varying current element known as a "Hertzian dipole." "Short" just means small *in comparison with a wavelength at the drive frequency*. If it is short, we can assume the current is uniform over the wire at any instant.

By Maxwell's laws close to such an electric source, we get not only an electric field, but also an associated magnetic field. The electric field contains components that vary as $1/r^3$, $1/r^2$, and $1/r$, where r is the distance from the dipole. The magnetic field contains components that vary as $1/r^2$, and $1/r$. Far away, the $1/r^2$ and $1/r^3$ components have decayed sharply due to the increasing distance. Finally, both E and H fields are left with an approximate $1/r$ variation, so their ratio E/H becomes a constant. Note that from our physics class, we remember that the field from a point source (charged particle) varies as $1/r^2$. That is clearly not true for time-varying fields.

The dual to the Hertzian dipole in our fourth case above, is a sinusoidally excited *current loop*. The electric and magnetic fields for a sinusoidally driven infinitesimal current loop mirror those for the Hertzian dipole. Here, the near field magnetic field exhibits $1/r^3$, $1/r^2$, and $1/r$ behavior, while the electric strength has components that fall off as $1/r^2$ and $1/r$. Far away, both E and H exhibit $1/r$ behavior.

The ratio of the E and H fields is called the wave impedance. Far away, it is a constant. This is shown in Figure 15.4. In general, the proportionality constant between E and H

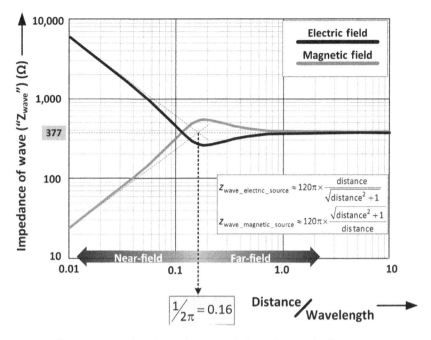

Figure 15.4: Electric and magnetic impedances in free space.

depends on the material of propagation, and equals $E/H = \sqrt{(\mu/\varepsilon)}$, where μ in this case is the (absolute) *permeability* of the material and ε its *permittivity*. Note that ε is the electrical analog of the magnetic parameter μ that as we know describes the extent to which a given material allows itself to become magnetized by an external magnetic field. We also note that the units of E are V/m, and H is A/m. Therefore, the ratio E/H has the units V/A, which is simply *resistance* (ohms). If the material of propagation is air or vacuum (free space), $E/H = \sqrt{(\mu_o/\varepsilon_o)} = 120 \times \pi = 377$ ohms. $E/H = 377\,\Omega$ is called the *wave impedance* in free space, or the *intrinsic impedance* of free space.

From Figure 15.4, it also becomes clear why it is often colloquially said that "electric fields have high impedance," whereas "magnetic fields have low impedance." A small circular current loop of trace on a PCB produces magnetic fields, but a strip of copper or metal with a swinging voltage on it (e.g., a heatsink) forms a source of electric fields. Of course, once there is time variance involved, the H-field leads to an associated E-field, and an E-field produces H-fields. At a great distance, the E and H fields become proportional to each other and form an electromagnetic wave. We define the boundary between what is considered a "far-field" and what is a "near field," as the distance $\sim \lambda/6 \sim 0.16\lambda$ away from the EMI source.

Extrapolation

We see that far fields decay as per $1/r$ (inversely proportional to distance). So, if the field at distance r_1 is E_1 and at r_2 it is E_2, they are related as

$$\frac{E_1}{E_2} = \frac{r_2}{r_1}$$

The ratio of the fields expressed in decibels, is by definition

$$dB \Rightarrow 20 \times \log\left(\frac{E_1}{E_2}\right) = 20 \times \log\left(\frac{r_2}{r_1}\right)$$

For example, if x_1 is 10 m, and x_2 is 100 m, the field strengths are different by

$$dB \Rightarrow 20 \times \log\left(\frac{100}{10}\right) = 20\,dB$$

Or E_1 is 20 dB greater than E_2. Alternatively expressed, fields decay as per $1/r$, or 20 dB per decade (of distance). So, as distance changes by a factor of 10, the field falls by 20 dB — *provided it is a far-field*. And if we can confirm that it indeed is a far-field, then the amplitude of the field becomes entirely predictable. For example, if we know the far-field at a given point in space, we can use the preceding physics to accurately predict its value at any another point in space (again, provided that point is also within the far-field region). We call this technique "inverse linear distance extrapolation."

In standard radiated EMI tests, the specified measurement range of frequencies is 30 MHz and above (usually up to 1 GHz). We ask: from what distance can these frequencies be considered far-fields? For 30 MHz, the distance in meters that marks the boundary between near-field and far-field is, as per Figure 15.4

$$\frac{\lambda}{2\pi} = \frac{c}{f} \times \frac{1}{2\pi} = \frac{3 \times 10^8}{30 \times 10^6 \times 2\pi} = \frac{10}{2\pi} = 1.6 \text{ m}$$

For 1 GHz it is proportionately smaller. In other words, if an antenna is placed at least 1.6 m away from the radiating source (say at 3 m), and we measure the field at that point, we can then use inverse linear distance extrapolation to predict the field amplitude at any other (further) distance — say 10 m, 30 m, and so on.

For confirming compliance to Class B limits, FCC generally requires the antenna distance to be 3 m, whereas CISPR requires 10 m. For confirming compliance to Class A limits, FCC specifies the distance to be 10 m, whereas CISPR requires 30 m. The good news is that 3 m is certainly far-field for all frequencies in the radiated range of interest, and we should be able to apply extrapolation to ultimately compare apples to apples. These normalized limits are shown side by side in Figure 15.2. Here, we have done two things to the original FCC limits. (A) The original FCC limits are in μV and μV/m, not dB μV and dB μV/m as used by CISPR-22. In Figure 15.3, we have included sample calculations showing how we can convert between the two. (B) We have extrapolated the Class A FCC limits to 30 m and the Class B FCC limits to 10 m, to compare them with the CISPR limits.

> *Note: From Figure 15.2, it might seem that both Class A and Class B FCC limits are identical, at 29.5 dB μV/m, over the range 30–88 MHz. But don't forget that Class A limits are shown at 30 m, whereas Class B limits are at 10 m. We know from the extrapolation factor calculations in Figure 15.3 that the field at 30 m will be exactly 9.5 dB lower than its value at 10 m. So, in fact, the FCC Class A limits are exactly 9.5 dB (~10 dB) higher, and therefore more relaxed than FCC Class B limits, when compared at the same distance from the EMI emitter.*

> *Note: The near-field/far-field definition in Figure 15.4 assumes the EMI emitter is a point source. Therefore, especially in the case of a 3-m test, some further validation may be necessary to prove that we really do have only far-fields.*

Quasi-Peak, Average, and Peak Measurements

We have not yet explained what the rationale is behind the two types of limits — average and quasi-peak (for the conducted EMI limits in Figure 15.2).

Historically, quasi-peak (or almost-peak) was meant to *simulate human responses* to noise. Humans have a slowly increasing level of aggravation or annoyance to a persistent disturbance. Therefore, to simulate this (subjective) response, there are built-in attack and release rates in quasi-peak detection. Conceptually, it works like a peak detector followed by a lossy integrator. Applied to a switching power supply, we note that in quasi-peak detection, the signal level is effectively weighted according to the repetition frequency of the spectral components constituting the signal. So, the result of a quasi-peak measurement will always be dependent on the repetition rate. The higher this repetition frequency (e.g., switching frequency), the higher the measured quasi-peak level.

Because of the finite charge and discharge time constants involved in quasi-peak detection, the spectrum analyzer must sweep considerably slower in quasi-peak setting. The entire EMI measurement process thus becomes very slow. To avoid delay, *peak detection* can be carried out, as it is much faster. However, we will then always get the *highest reading*, followed by quasi-peak and then by average (see Figure 15.5).

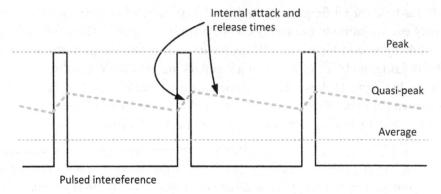

Figure 15.5: Average, quasi-peak, and peak readings of a pulsed wave.

Part 2: Measurements of Conducted EMI

Differential Mode and Common Mode Noise

Here, we clarify the concepts of "common mode" and "differential mode" noise in conducted EMI. Initially, we are going to stick to more conventional descriptions of these parameters. But in *Chapter 16*, we will start discussing certain nuances/differences that can arise in applying the concepts to the area of power conversion.

Conducted emissions fall into two basic categories

• Differential mode (DM), also called *symmetric* mode or *normal* mode.
• Common mode (CM), also called *asymmetric* mode or *ground leakage mode*.

Looking at Figure 15.6, "L" stands for Live (or "Line" or "Phase"), "N" for Neutral, and "E" is the "Safety Ground" or simply, "Earth" wire. "EUT" stands for Equipment Under Test. Note that the Earth is shown represented by the IEC symbol for Protective Earth (ground with a circle around it), occasionally labeled "PE" in literature. The DM noise generator is across the L and N pair. It tries to push/pull a current I_{dm} through these two wires. No current flows through the Earth connection on account of this noise source.

To avoid confusion, we should note that the net common mode current going through the Earth is called "I_{cm}" in our case ($I_{cm}/2$ in each line). However, in related literature, this is often called "$2I_{cm}$" (I_{cm} in each line).

> *Note: There is nothing special about the DM noise current direction as indicated in the figure. It can well be the other way around — that is, going in through either L or N, and coming out of the other. In off-line power supplies, we will see that in fact, the direction reverses every AC half cycle.*

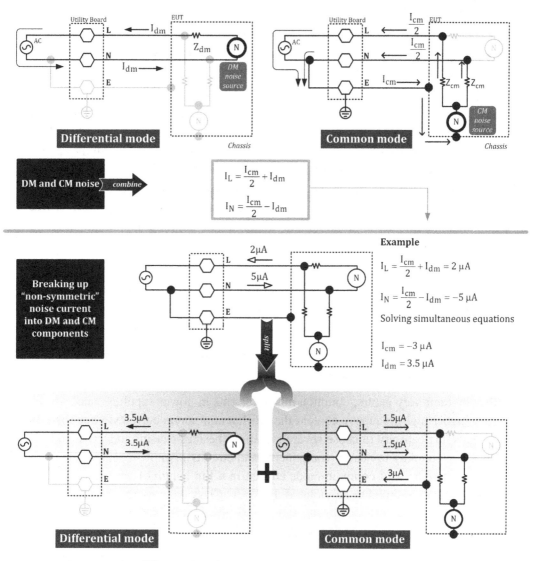

Figure 15.6: Differential and common mode noise with a worked example.

Note: The designer may realize that the basic AC input operating current of the power supply is also differential in that sense — since it flows in through one of the L or N wires and leaves by the other. However, the I_{dm} shown in Figure 15.6 does not include this component. That is because the operating current, though differential, is not considered to be "noise." Further, its main components (and also its key harmonics) are of very low-frequency, being virtually DC, and well below the range of standard conducted EMI limit curves (150 kHz to 30 MHz). The standards regulating line harmonics were *discussed in Chapter 14. However, it must not be forgotten that the operating current of the power supply can DC-bias the EMI chokes, and can thereby adversely affect the performance of the EMI filtering and also of any current probes being used to gather data. So, though we can certainly ignore AC (line) while discussing EMI, we should realize it can have a major though indirect effect on the performance of the filter.*

In Figure 15.6, the CM noise source is shown connected at one end to Earth. On its other side, it is assumed that the noise source sees equal impedances on each of the L and N lines. It will therefore drive equal noise currents into these two wires, and in the same direction. We realize that if the impedances are unbalanced, we will get "mixed mode" (MM) noise current distribution (in the L and N wires). And that is in fact a common scenario in actual power supplies. Note that this mode is equivalent to a mixture of true-CM noise mixed with some DM noise as demonstrated in the worked example in Figure 15.6.

Just as common mode noise generated by a power supply flows into the mains wiring, it can also flow into the output. Engineers often instinctively tend to disregard common mode noise present in the output of their power supplies, instead focusing at the input only. But it is important to understand what both components are, for the following two key reasons:

(a)   As mentioned previously, in telecom networks and distributed power applications, the common mode noise will use the long output cables as a giant antenna.

Also, by their very nature, common mode currents in power supplies usually have much higher high-frequency content than differential mode currents. They therefore also have the capacity to cause severe radiation (besides causing inductive and capacitive coupling to nearby components and circuits). An oft-repeated rule-of-thumb is that a mere $5\,\mu A$ of common mode current in a 1 m length of wire can cause FCC Class B radiation limits to be violated. For FCC Class A limits this number goes up to $15\,\mu A$. Note also that the shortest standard AC power cord is 1 m in length. We thus see the importance of reducing common mode noise currents both at the input and output of a power supply.

(b)   In terms of qualifying the power supply itself, engineers typically have a target spec for output noise and ripple measurement. That is actually a *differential* measurement. Engineers spend a long time trying to get the oscilloscope probe positioned correctly on the output terminals (with minimum length of probe ground wire), simply to avoid picking up common mode noise *via radiation*. But common mode noise can still affect the differential ripple measurement. Because, if the power supply is providing power to a real subsystem (not a resistive "dummy test load"), then looking into the input of this subsystem, we will rarely (if ever), see *equal* (balanced) impedances (i.e., from each of its input terminals to the Earth ground). So, what really happens is that any "common mode" noise existing previously on the output rails of the power supply, now becomes a *differential* input voltage ripple (of high frequency). It takes on the profile of mixed mode currents as shown in Figure 15.6. We will see that no amount of common mode rejection ratio (CMRR) in the subsystem will help completely, since the erstwhile pure common mode noise is now partly differential. The subsystem could start misbehaving as a result of that. Summarizing: *common mode noise gets converted into differential mode noise if the line impedances are unequal.* For the very same reason, even at the input of the power supply, we always use *balanced* filters. In general, reducing common mode noise *at the point of creation* is a high priority. But after that, *equalizing* the line impedances becomes important. The latter can often be achieved by placing *balanced* filters at the input and output of the power supply, and also at the input of the subsystem that it powers — for example, two inductors, one on each input line, instead of just one inductor.

Note: There is nothing special about showing the CM noise current in Figure 15.6 coming out of the equipment (through both the L and N wires). It could well be in the reverse direction. And like DM noise in an AC–DC power supply, it too could be sloshing back and forth, depending on what part of the incoming AC half cycle we are on, at a given moment.

Note: We will see that in an actual power supply, differential mode noise is initiated by a swinging (pulsating) current — but the DM noise generator is itself closer to a voltage source. On the other hand, common mode noise is initiated by a swinging voltage, but the CM noise generator itself behaves more like a current *source. That is actually what makes common mode noise so much more "stubborn" — like any other current source, it demands a path to flow through. And since its path can include the chassis, the enclosure can then become a large high-frequency antenna too.*

Measuring Conducted EMI with a LISN

For measuring EMI, we need to use an ISN (Impedance Stabilization Network). In off-line power supplies, this is called a LISN (Line Impedance Stabilization Network) — also called an AMN (Artificial Mains Network) (see Figure 15.7 for a simplified schematic). Note that the LISN, as recommended for CISPR-22 compliance, is detailed in another standard called CISPR-16. It provides the following functions:

- It is a source of clean AC power to the power supply, being virtually transparent to the line frequency component. This allows normal operation of the equipment under test (EUT, e.g., the power supply).
- It blocks any extraneous noise from the AC line from entering the measurement area (by two $50\,\mu H$ chokes), and thereby keeps it ready for clean measurements on the EUT.

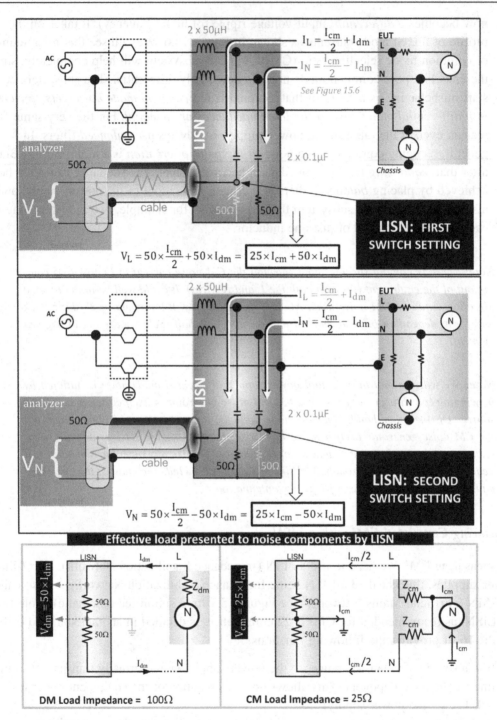

Figure 15.7: Simplified schematic of LISN and the load it presents to the CM and DM noise.

- It similarly blocks noise from the power supply from entering the AC line, and instead, diverts it into the measurement area.
- It blocks the AC line component from entering the measurement area by means of two 0.1 µF blocking capacitors.
- It provides a stable and *balanced* impedance to the emanating noise, attempting to replicate the typical impedance normally presented by the AC wiring to the noise.
- The diverted noise is then measured by the measurement receiver/spectrum analyzer.
- Most importantly, the LISN makes the measurements repeatable, anywhere in the world.

Note that we have implicitly assumed that

- The inductance (50 µH) is low enough not to impede (AC) line current (50/60 Hz) at all — but high enough to be considered "open" over the frequency range of interest (150 kHz to 30 MHz).
- The blocking capacitance (0.1 µF) is low enough *not* to pass the AC (line) voltage — but high enough to appear as a "dead short" over the frequency range of interest.

The combined noise current is diverted into 50 Ω resistors as seen from Figure 15.7. To ensure signal integrity, the receiver (measuring instrument) is set to 50 Ω impedance setting, and that is what the high-frequency noise "sees" when looking into the coaxial cable. Note that on the LISN, we have two switch settings, one for measuring the voltage across the resistor in one phase (L), and the other position for measuring the other phase (N). But since the receiver already presents 50 Ω to the noise currents on the channel being measured, the discrete 50 Ω resistor is disconnected on that channel. Then, to maintain balance, a discrete 50 Ω resistor is placed on the other channel as indicated in the figure.

In the lower part of Figure 15.7, we follow the paths of the DM and CM components separately, and realize that **the LISN presents a load impedance of 100 Ω (two 50 Ω in series) to the DM component, and a load impedance of 25 Ω (two 50 Ω in parallel) to the CM component**. Note that in both cases, one of the "two 50 Ω" resistors is discrete, while the other is the input impedance of the receiver.

As we flick the switch on the front panel of the LISN, we will measure the following noise voltages

$$V_L = 25 \times I_{cm} + 50 \times I_{dm}$$

$$V_N = 25 \times I_{cm} - 50 \times I_{dm}$$

Both the V_L scan and the V_N scan obviously need to comply individually with the limits in Figure 15.2.

How different can the V_L and V_N scans be? In fact, the above two equations may have inspired a rather misleading statement in related literature. It is often stated: "*if the noise* *emission is predominantly DM, the V_L and V_N scans will look almost the same. The scans also look identical if the noise is predominantly CM. And if the V_L and V_N scans look very different, that implies that **both** CM and DM emissions are present.*" However, in the case of an off-line (AC−DC) power supply, this statement is not true. Because, it would imply that somehow the emissions on the L and N lines are *different.* However, we know that in any typical off-line power supply (with an input bridge rectifier), the L and N lines are essentially *symmetrical* — both from the viewpoint of the operating current *and*, therefore, the noise spectrum. So, every successive AC half cycle, the operating current, *and* the noise distribution get transposed from one line to the other. True, at any *given moment*, the noise on L will be quite different from that on N, but when *averaged over several AC cycles* (as any spectrum analyzer would do), equality (symmetry) is restored. Any remnant differences between the V_L and V_N scans can be traced back to some undocumented asymmetries between the two halves of the test circuit, or some severe radiation source impinging asymmetrically on the cables, or on traces/wires very close to the inlet of the power supply (between the EMI filter and the AC inlet socket).

In *Chapter 17*, we will see that the frequency response (sensitivity) of the LISN is not "perfect" because its inductors and blocking caps are not ideally large. Understanding the actual frequency response of the LISN helps us a great deal in designing *cost-effective* EMI filters.

Simple Math for Estimating Maximum Conducted Noise Currents

From Figures 15.2 and 15.3, we see that over the range 0.15−0.5 MHz, CISPR Class A restricts the average voltage to 2 mV. We already know that the impedance presented by the LISN to the CM noise is 25 Ω. So, if we know voltage and resistance, we know current. In this case the maximum permissible CM current is 2 mV/25 Ω = 0.08 mA or 80 μA. This, however, assumes no DM noise component. So, in general, if we assume equal contributions coming from CM and DM noise, we should halve the target to 40 μA. Over the range 0.5−30 MHz, CISPR Class A restricts the voltage to 1 mV, or 40 μA. We can target 20 μA. Similarly, FCC Class A restricts the voltage to 1 mV from 0.45 MHz to 1.7 MHz. Therefore, the maximum permissible CM current is 40 μA over this range. From 1.7 MHz to 30 MHz FCC Class A allows 3 mV, or a maximum of 120 μA, and so on. For DM noise, we should use 100 Ω instead of 25 Ω to calculate the maximum permissible noise current. In *Chapter 18*, we will do a more detailed calculation.

Separating CM and DM Components for Conducted EMI Diagnostics

We observe that the standards do not require us to measure the CM and DM components individually, but rather a certain sum as described in the preceding equations in Figure 15.7. However, there are times when engineers do want to see both the CM and DM components separately — for troubleshooting and/or diagnostic purposes. So, various people have come up with clever ideas to separate the CM and DM components. Some of these are presented in Figure 15.8.

- The first is a device called the "LISN MATE," which is quite rare now. It was invented by an engineer named M.J. Nave. It provides about 50 dB attenuation for the DM component, but the CM component comes right through (slightly attenuated — by about 4 dB).
- The next one in the figure is a transformer-based device. It exploits the fact that common mode voltages cannot cause transformer action — because transformer action requires a *differential* voltage be applied, so as to produce current in the windings, and thereby cause the flux to swing within the core. Unlike the LISN MATE, in this case both CM and DM noise components are outputted.

  Note: Both methods above unfortunately require modifications to the standard LISN — because they invoke a certain math between the V_L and V_N components. However, a LISN normally provides either V_L or V_N at any given moment — not both (at the same time as required here). We can modify the traditional LISN, but that is not only tricky to do, but also hazardous because of the high voltages involved. Therefore, a completely different approach is to simply buy a LISN explicitly designed for the purpose of providing separate CM and DM noise scans (besides providing the necessary "summed up" scan for achieving compliance).

- Finally, we have shown two current probes, wired up in such a way that they are actually solving two "simultaneous equations" on the L and N wires. They separate the CM and DM components. Note that by doing these two measurements at the same time (using two probes rather than one), we have retained valuable information about the relative phase relationship between the CM and DM components.

Note: The bandwidth and current capability of the current probes used for noise measurements are important. For very high currents (up to thousands of Amperes if necessary), a possible choice is current probes based on the "Rogowski principle." The output from a Rogowski probe depends not on the instantaneous current enclosed, but on the rate of change of current. So, instead of just placing several turns around the wire to be sensed, as in a typical current transformer, the Rogowski probe effectively takes an air-cored solenoid and then bends that in a circle around the sensed wire (like a doughnut). Such probes are also considered virtually non-invasive. The usual lab active current probes (which also measure DC, and therefore include a Hall sensor), are usually just not suited for these high-bandwidth noise measurements.

Note: When viewing pulse transition times below 100 ns, or emission noise frequencies above a few Megahertz, it is advisable to keep the cable length small. Thereafter, we must terminate the cable at the oscilloscope, or measuring instrument, with a 50 Ω resistor.

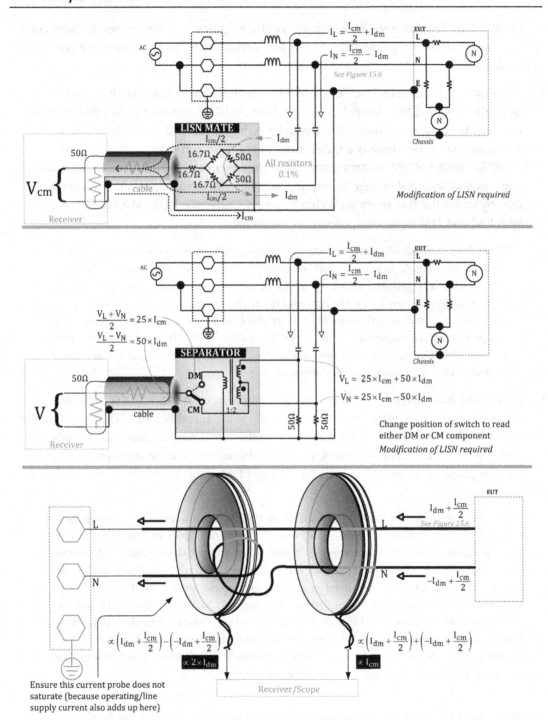

Figure 15.8: Ways to separate CM and DM components for troubleshooting.

However, most modern oscilloscopes incorporate a selectable 50 Ω input impedance. Correct termination of cables prevents standing wave effects. Note that, with this 50 Ω termination, the measured voltage is approximately half of what it really is *because we essentially have a voltage divider formed by the cable and the terminating resistor. Oscilloscopes will usually automatically correct for this, if they "know" that there is a 50 Ω termination present. Also note that fast-rising pulses can produce spurious ringing, due to high-frequency current crowding on the* surface of the cable shield. *This can be suppressed by threading the measurement cable through one or more ferrite beads (or toroids). For example, some report that they obtained good results by placing three turns through four ferrite cores of about 1 in. inside diameter, 2 in. outside diameter and 1/2 in. thickness.*

In the upper part of Figure 15.9, we show a practical technique to separate that part of the conducted emissions spectrum attributable to radiation pickup occurring within the power supply itself. We see how to identify these as *E*-fields or *H*-fields. For this experiment, we need to cut the PCB traces just before the input bridge, and then route the AC power from a canned filter outside the enclosure. The ends of the existing filter are kept either open (to receive *E*-fields) or connected together through a small loop (for seeing the *H*-fields). The other end of this EMI filter is then routed as usual to the LISN and spectrum analyzer. We can thus see the "extraneous" radiation-based noise being picked up by the internal EMI filter via radiation inside the power supply. It will give us an indication if a heatsink for example is causing severe *E*-fields, or if a certain magnetic component is causing severe *H*-fields. We can also wave a small plate of thick copper (connected to Earth) in suspected areas to see which component may be the actual source of the fields. For analyzing the source of magnetic near-fields, a slab of ferrite (from a typical EMI suppression kit) works better than a copper plate and that can be waved around similarly (no need to Earth it).

Caution: AC power is **NOT** to be applied through the LISN in the above experiment. This will cause a serious hazard to the user. Also, any plate/slab must be well-wrapped in insulating tape to prevent accidental contact with nearby components.

Near-Field Sniffers for Radiated EMI Diagnostics

Radiated EMI measurements for compliance, are always done in the far-field zone of Figure 15.4. That means the receiving antenna is placed so far away that radiation irregularities resulting from the actual geometry of the product are ignored. All the fields are in effect, aggregated and lumped together, and the overall spectrum subjected to a test. So, though a far-field test can tell whether the product passes or fails as a whole, it cannot point to, leave aside pinpoint, the actual source of a problem. For example, it cannot tell if there is an opening in the metal enclosure that is leaking too much radiation. And if so, where it may be. In order to locate the source of a problem, why not just come closer and look? That is why near-field sniffers are good diagnostic tools. We did that in the upper part of Figure 15.9 for correlating with a conducted EMI scan. We created, what was in

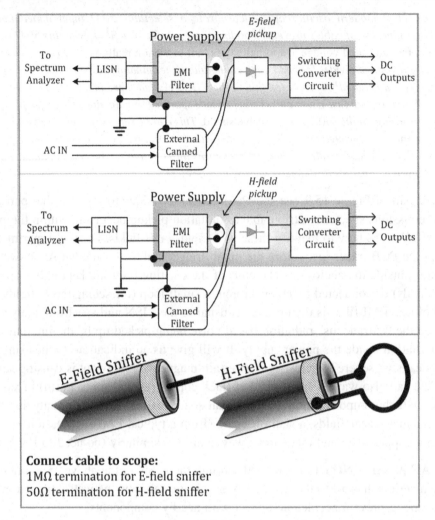

Figure 15.9: Analyzing magnetic and electric field sources inside a power supply and near-field sniffers made from a coaxial cable.

effect, a near-field sniffer so we know the radiated fields in a particular location (around the EMI filter), which may be causing us to fail the conducted EMI test. That technique points to a more general way to "sniff" out locations elsewhere in the power supply with high local electric fields and high magnetic fields. In the lower part of Figure 15.9, we show how a simple coaxial cable, connected to a receiver or scope, can be turned into either an *E*-field sniffer or an *H*-field sniffer. We move it around inside the power supply to locate strong sources of EMI. Then, on the basis of frequency, we try to correlate a certain stubborn EMI peak (in the far-field spectrum) to an abnormally radiating component or section of the power supply.

CHAPTER 16

Practical EMI Line Filters and Noise Sources in Power Supplies

Part 1: Practical Line Filters

Having understood common-mode (CM) and differential-mode (DM) noise concepts in *Chapter 15*, we can look at the overall strategy for designing conducted-mode EMI filters. We will focus mainly on filters for off-line (AC–DC) power supplies. However, several of the tips will be obviously applicable to DC–DC converters too.

Safety issues, thermal issues, and even loop stability concerns are intricately linked to the issue of EMI filter design. Particularly in designing EMI filters for off-line applications, safety becomes a major concern because the voltages are high enough to cause injury. We will briefly cover that aspect first.

Basic Safety Issues in EMI Filter Design

The concept of safety (as per IEC 60950-1), and how it impacts the filter section are explained rather simplistically in the following steps.

- Any exposed metal (conducting) part (e.g., the chassis or output cables) is capable of causing an electrical shock to the user. To prevent a shock, such parts must be earthed and/or isolated from the high-voltage parts of the power supply in some way.
- No *single point failure* anywhere in the equipment should lead the user to be exposed to an electrical shock. There should be *two* levels of protection, so if one gives way, there is still one level of protection available. That is the basic premise underlying safety: the probability of two unconnected/independent faults occurring at the same point and at the same instant is considered negligible.
- Levels of protection that are considered essentially equivalent are (a) earthing of the exposed metal surface, (b) physical separation (typically up to 4 mm) between any exposed metal and parts of the circuit containing high voltage, and (c) a layer of approved insulator (or dielectric) between any exposed metal and the high voltage. Note that in (c) this (single-level protection) insulator/dielectric must have a minimum dielectric withstand capability of 1,500 VAC or 2,121 VDC.

- To qualify the above slightly — connecting the metal enclosure of the equipment to Earth is *not* always considered an acceptable level of safety and protection, simply because earthing may still not be guaranteed. For example, the wiring of older houses may not even contain an Earth wire inside the common household power outlet. However, assuming for now that earthing is acceptable, we then need one more level of protection as per the safety concept. This could be "4 mm" of separation for example (IEC 60950-1 provides tables for knowing exactly how many millimeters are required, and where). But consider the case of a high-voltage MOSFET (switch) mounted directly on the (earthed) metal enclosure (say, to provide better heatsinking). We obviously can't provide "4 mm" of physical separation in this case (between the MOSFET and the earthed enclosure). So, now we need to place one layer of approved insulator between them. In this position, the insulator is said to serve as "basic insulation."
- If earthing is not considered acceptable to be recognized as a valid level of safety, or if we are dealing with equipment with a two-wire AC cord (no Earth wire), then, besides the layer of *basic insulation*, we need an additional level of protection between exposed metal (e.g., output leads) and high voltage (AC). This could be another insulating layer, with identical dielectric withstand capability. It is then "supplementary insulation." Together, the two layers (basic + supplementary) are said to constitute "double insulation." We could also use a single layer of insulation, with dielectric withstand properties equivalent to double insulation (i.e., 3,000 VAC or 4,242 VDC). That would then be called "reinforced insulation."
- Why do we connect the enclosure to Earth in the first place, if that is not always acceptable from a safety standpoint? The main reason for earthing a metal enclosure is that *we want to prevent radiation from inside the equipment from spilling out*. Without a metal enclosure, there is very little chance that a typical off-line switching power supply can comply with radiated-mode (and perhaps even possibly conducted-mode) emission limits. By earthing, the enclosure is held at a fixed potential and therefore forms an excellent shield around the power supply for meeting radiated emission limits. But the metal enclosure is also, rather expectedly, eyed by engineers as an excellent and fortuitous heatsink. So, in practice, power semiconductors are often going to be mounted on the enclosure directly (with appropriate insulation ratings as indicated above). However, by doing this, we also create leakage paths (resistive/capacitive) from the internal subsystems/circuitry to the metal chassis. And even if we ensure that these leakage currents (CM noise currents) are small enough not to constitute a safety hazard, they can present a major EMI headache. We realize that if these leakage currents are not "drained out" in some way, the enclosure will charge up to some unpredictable/ indeterminate voltage, and will ultimately start radiating (a dipole, or an electric field source). That would clearly be contrary to the very purpose of using a metal enclosure. So, we really need to connect the enclosure to Earth, other than for safety reasons. We note that even if we didn't have power devices mounted on the enclosure, there could

be other leakage paths present from inside circuitry to the enclosure. And besides that, an unearthed enclosure would also inductively pick up and re-radiate the strong internal electric/magnetic fields.

- To help in diverting CM currents away, capacitors are often connected between various parts of the power supply and the earthed enclosure (or earth terminal). These are called "Y-capacitors" or Y-caps. Similarly, to reduce DM currents, high-voltage capacitors are often placed *between* L and N of the incoming AC mains. These line-to-line components are called "X-capacitors" or X-caps. All Y-capacitors on the Primary side of an AC−DC power supply must be rated for a certain minimum voltage as discussed later. And, of the Y-caps on the Primary side, *those that are before the bridge rectifier* (i.e., on the line side) also carry low-frequency AC, and therefore safety agencies in effect regulate their total capacitance value. We will discuss this later.

- The bottom-line is that (a) providing a good metal enclosure and (b) properly connecting it to Earth are the most effective methods of preventing radiated EMI. However, by creating this galvanic connection (to Earth), we also now provide a "freeway" for conducted (CM) noise to flow "merrily" into the wiring of the building. So, in trying to suppress radiated EMI, we might end up increasing the conducted EMI, and vice versa. We thus realize that to be able to stay within the applicable *conducted* emission limits, we now need to provide a conducted EMI (CM noise) filter somewhere. Such are the reasons why, in *Chapter 15*, when we first introduced the topic of EMI, we had warned that EMI is a veritable balloon — push it in on one side only to find it bulging out on the other.

- Generally speaking, if the equipment is designed *not* to have any Earth connection (e.g., a two-wire AC cord), there will usually be no metal enclosure present either. Ignoring the problem of meeting radiation limits for now, the good news here is that *no significant CM noise* can be created either — simply because CM noise needs an Earth connection by definition. Therefore, a CM filter need not be present in this case. However, we must remember that conducted noise limits include not only CM noise, but DM noise too. So, irrespective of the type of enclosure and earthing scheme, DM filters are always required.

- In plastic enclosures, to comply with radiation limits, the inside of the plastic box may be shielded, say by an insulated wrap-around metal foil, or by depositing metal on its insides.

Four different coating processes are commonly used, each with its strengths and weaknesses.

(a) *Vacuum deposition:* A metal (e.g., aluminum) is melted in a vacuum chamber and its droplets sputtered onto the inner surface of the enclosure, gradually building up a continuous metallic layer. This process is ideal when very thin coatings are required on relatively detailed moldings. It is less effective if used on an enclosure with poorly fitting joints because the thin metal layer does not provide gap-filling properties, making electrical continuity difficult to maintain. Therefore, RF leakage around joints and seams is a real problem. Tooling costs are also high because only very precise masks and fixtures will ensure that metal is applied to only the inner surfaces of the enclosure.

(b) *Loaded paints:* This is the most cost-effective method. A silicone rubber shields the external surfaces, and the metallic-loaded paints are applied using a wet spray process using either an automated or a manual setup. A finished coating thickness of 50–75 μm gives good coverage without obliterating fine details of the molding. Copper- or silver-loaded paint is suitable for most applications requiring commercial levels of attenuation.

(c) *Zinc arc spray:* This results in a relatively thick layer over the molding surface. It is very effective in applications where a high magnetic field is expected, because of the direct relationship between material thickness and signal absorption. However, this process involves elevated temperatures, so it is most suitable for use with polycarbonate enclosures. The thickness of the coating can also obliterate fine details of enclosure molding.

(d) *Electroless plating:* This produces the best performance, but it is very expensive and not particularly suitable for high-volume applications. Masking is very difficult, and the screening material can easily get deposited on both the internal and the external surfaces. Further, a secondary finishing operation is required to remove the excess material and perhaps to apply a final painted-over finish. But since colors are intrinsic to the materials used in plastic enclosures, the secondary paint operation is a major disadvantage. All this gets complicated and costly. Therefore, this method is justified only in applications where high attenuation levels are the overriding concern.

Safety Restrictions on the Total Y-Capacitance

Y-caps don't just let high-frequency noise pass through. If positioned before the bridge rectifier, they also conduct some of the low-frequency line current. This AC-related leakage current flows into the protective Earth/chassis where a person may contact it, providing an involuntary path to ground, at obvious risk to personal health. Therefore, safety agencies limit the total RMS current introduced into the Earth by equipment to a maximum of 0.25 mA, 0.5 mA, 0.75 mA, or 3.5 mA, depending on the type of equipment and its "installation category," that is, its enclosure, its earthing scheme, and its internal isolation. In general, 0.5 mA seems to have become the standard value for most off-line power supplies. However, allowing higher leakage current allows for higher Y-capacitances, and thereby smaller CM chokes. So, it may be worth investigating if we can increase the earth leakage current above the so-called default value of 0.5 mA.

We can calculate how much leakage current we get per nF of Y-capacitance. The reactance of a 1-nF cap is

$$X_C = \frac{1}{2\pi \times f \times C} = \frac{1}{2\pi \times 50 \times 10^{-9}} \Rightarrow 3.183 \text{ M}\Omega$$

$$I = \frac{V}{X_C} = \frac{250}{3.183 \times 10^6} \Rightarrow 79 \text{ }\mu\text{A}$$

Every nF gives 79 μA at 250 VAC/50 Hz. This gives us a maximum allowed capacitance of 500/79 = 6.4 nF for 0.5 mA. Note that this is the sum of all the Y-capacitances present on the board (before the bridge rectifier). For example, in a typical two-stage EMI filter, we will often find four Y-caps, each one being 1 nF. This totals 4 nF. We can increase all the Y-caps to 1.2 nF, and we will still comply with the 0.5-mA limit. However, if we use four 1.5-nF caps, we may be in trouble if we are using caps with ±20% tolerance. Because we could then get a total lumped Y-capacitance of 1.2 × 4 × 1.5 = 7.2 nF, which will give us an earth leakage current of 7.2 × 79 = 569 μA, which exceeds the limit of 500 μA. Yes, we could use only two Y-capacitors, each of value 2.2 nF.

Besides a formal Y-capacitor, we may have earth leakage currents injected from high-frequency CM noise currents passing through small parasitic caps elsewhere on the board. These should also be accounted for in computing the total ground leakage current, and thereby correctly selecting the Y-caps of the line filter. In general, if a Y-cap is connected from the rectified DC rails to Earth (or from the output rails to Earth), there is no significant AC-related earth leakage current through these capacitors, so we can usually use much larger Y-caps in such locations.

Practical Line Filters

We now look at a typical power supply line filter, as shown in Figure 16.1. Its ultimate purpose is to control *conducted* emissions in general, and therefore it has *two* discernible stages — one for *DM* and one for *CM*. Let us make some relevant observations:

- Both the CM and DM stages are symmetrical (balanced). From the viewpoint of the noise emerging from the bridge rectifier and flowing towards the LISN, there are in effect two LC filters in cascade (both for DM and CM noise). This filter configuration can provide good high-frequency attenuation (roll-off). We know that any LC filter provides 40-dB/decade attenuation, so two in series can, in principle, provide up to 80 dB/decade.
- Typical practical values for the inductance of a CM choke in medium-power converters range from 10 mH to 50 mH (per leg). The large values result because we are limited by safety considerations in the amount of total Y-capacitance we can use. On the other

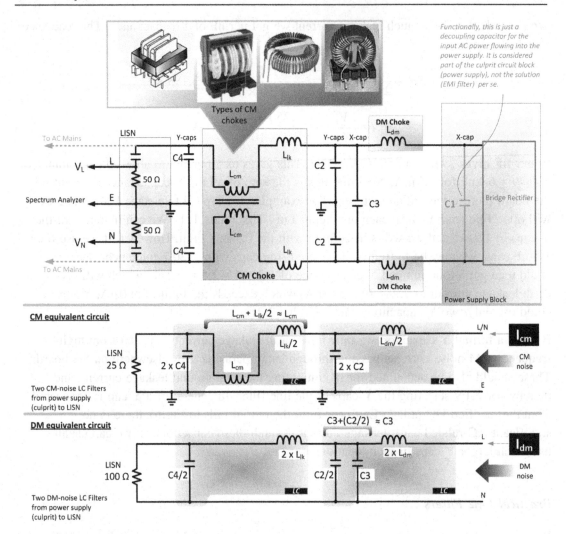

Figure 16.1: Practical line filter and the corresponding CM and DM equivalent circuits.

hand, there being no similar concern for DM noise, the DM choke is always much smaller (in inductance, but not necessarily in size as we will see). Typical values for the DM choke are $500\,\mu H$ to $1\,mH$.

- The leakage inductance of the CM choke L_{lk} acts as a DM choke, albeit one with a rather small inductance. The leakage inductance of the CM choke is roughly $1-3\%$ of L_{cm}, depending on its construction. That can serve as an unintentional, but effective DM choke.

- However, since the capacitance associated with a DM filter can be very large, and is not restricted by safety concerns, low-power flybacks (up to $70-80\,W$) rarely even use a discrete DM choke. Instead they usually have two CM chokes in cascade, relying on the leakage of each to provide DM noise filtering.

- *If the chosen CM choke is toroidal in shape rather than a U-core or E-core, it will have almost no leakage inductance. In that case, we would most likely need to provide discrete DM chokes to create a DM filter.*

- In Figure 16.1, we have shown both the CM and the DM filter stages as being symmetrical (balanced). For example, we have placed *identical* DM chokes on *each* of the L and N lines. However, the *DM choke is also a part of the CM equivalent circuit* as we can see from the CM equivalent circuit. As mentioned in *Chapter 15*, since line impedance imbalance can cause CM noise to get converted into DM noise, it is generally advised to keep *both* the CM and DM stages symmetrical (balanced).

- Occasionally, *unbalanced* DM filters provide acceptable overall EMI scans — for example, a single DM choke placed on one line only. It is worth trying out for sure. Sometimes, in very low-power applications, or in the case of DC–DC converters/ bricks, a plain decoupling capacitor (e.g., C1) may suffice too. Tuned filter stages are also sometimes seen in commercial off-line power supplies (e.g., from *Weir Lambda*, UK). But there are some anecdotal industry experiences that suggest that under severe line transients or under input surge waveforms, as those typically used for immunity testing, tuned filters can display unexpected oscillations (resonances), ultimately provoking failure of the power supply itself. Therefore, tuned filters are usually avoided in commercial designs.

- One obvious way to maintain equal CM inductances in both lines is to *wind them on the same core* (e.g., a toroid). That automatically assures a good inductance match (assuming of course that there are an equal number of windings per leg). Note that if we are winding the CM choke ourselves (as during prototyping), we must note the relative direction of the windings, as indicated in Figure 16.1 (see the embedded third and fourth sample CM choke pictures from the left). With such a winding arrangement, the magnetic field inside the core will cancel out completely (in principle) for DM noise. For the same reason, the flux due to the operating AC line current will also cancel out (that too being differential in nature). Therefore, the CM choke will be basically "visible" only to the CM noise component. This choke does not need to be "large" based on DC-bias considerations, but in reality, is in fact quite bulky — due to the high inductance required, and because of the associated large number of turns (all rated for the AC operating current).

- While winding toroidal CM chokes, we need to keep other safety requirements in mind too. We cannot simply wind the two windings carelessly overlapping each other — we need to maintain a specified physical separation between the windings of each AC phase. Nor can we just use a *bare* ferrite toroid to wind them on because ferrite can be a rather good electrical conductor. Neither should we rely on the enamel coating of a typical copper magnet wire as it is typically not a safety-approved coating. What we really need is a safety-approved coating for the toroid and/or a suitable bobbin.

> *Note: The reader is cautioned that there are several widely used but confusing symbols for the CM choke existing in schematics found in related literature. Whatever the symbol, as long as it is meant to serve as a CM choke, the direction of the windings must be as in the embedded toroid pictures in Figure 16.1. The correct winding polarity is also indicated by dots in the same figure.*

- We should also keep in mind that in theory, the AC operating current flux cancels out in the CM choke. In reality, on account of small imbalances in symmetry of the windings the choke could "topple over" (i.e., progressively flux-staircase over toward one side). This would expectedly degrade the EMI performance, but in extreme cases, the core may even saturate. Though that is not a catastrophic event, it still needs to be fixed since the filter is likely no longer EMI-compliant. Sometimes we may need to oversize the CM choke, just to give it a higher-than-zero DC-bias rating. We may also need to provide some air gap. This may be an actual air gap (between split halves), or it may be a *distributed* gap, as in powdered iron cores. See *Chapter 5* for a better understanding of air gaps and why they help smoothen out production tolerances in core materials and magnetic assemblies. But all such steps do lead to an increase in size of the choke.

- The inter-winding capacitance of a choke affects its characteristics significantly at high frequencies. This can be intuitively visualized as an AC path providing an easy detour for noise to flow past the windings rather than through them. To minimize the end-to-end capacitance of a toroidal winding, it is recommended that the winding be single layer. Coming to U-core type of CM chokes, in Figure 16.1, the sample CM choke picture second from the left is better than the one to its left, in terms of minimizing end-to-end capacitance. That is because of the *split* introduced in each winding section by the special bobbin used. The split also helps increase the leakage inductance (which helps reduce DM noise). CM choke bobbins with multiple splits are also available, at a higher price of course.

- If we *reverse* the current (or winding) direction in one of the windings of a CM choke, then it becomes a DM choke (for both lines). However, now it is also subject to the flux produced by the AC line input current (no flux cancellation occurs). Therefore, DM chokes, in general, should always be put through a "core-saturation check." We may see that DM chokes may need to be quite large, just to avoid core saturation — despite the fact that their inductance is usually much less than that of CM chokes. See *Chapter 14* for the peak currents in the input lines. That should help ensure the DM choke is really effective.

- We can consider spending some more money and avail of magnetic materials like "amorphous" cores or "Kool Mu®" if we want to achieve higher inductance (with higher saturation flux densities) in a smaller size.

- Note that the EMI filter stage is usually placed *before* the input bridge (i.e., toward the incoming AC line input) — because in that position it also suppresses any noise originating from the bridge diodes. Diodes are known to produce a significant amount of medium- to high-frequency noise, especially at the moment they are just turning OFF. Small RC snubbers (or sometimes just a "C") are therefore often placed across each diode of the input bridge. Though sometimes, we can get away simply by choosing diodes with softer recovery characteristics.

- Note that input rectifier bridges using ultrafast diodes are often touted as offering a significant reduction in EMI. Opinions about them remain mixed. Some people claim that with these ultrafast diode packs, it is possible to remove the small ceramic capacitors often placed across the four diodes of a typical bridge rectifier. However, in actual tests conducted by the author, the ultrafast diodes didn't seem to make much difference. If anything, the conducted EMI spectrum actually worsened. We know that very fast diodes can have very "snappy" characteristics too, producing rather sharp spikes of reverse current and forward voltage at turn-off and turn-on, respectively. And that could explain the rather poor results with them.

- Line-to-line capacitors (the X-capacitors or "X-caps"), placed *before* the input bridge, must be safety approved. After the bridge (i.e., on the rectified side), it's basically a "don't care" situation from the safety point of view. At that position, we can use any suitably rated cap really. Note that since an approved X-cap is essentially a front-end component, it can see rather huge voltage spikes coming in from the AC mains. For example, the voltage spikes could be from motorized equipment connected to the same wiring, turning ON and OFF repeatedly. That is why an approved X-cap, even though rated "250 VAC," is actually 100% impulse-tested up to 2.5-kV peak.

- Line-to-earth capacitors, or "Y-capacitors," as discussed previously, used anywhere on the Primary side (in an off-line application) *must always be safety approved.* Any Y-cap placed as a bridging component from anywhere on the Primary side to anywhere on the Secondary side must be safety approved. Since Y-caps are critical in terms of having the potential to cause electrocution if they fail, *approved* Y-caps are typically impulse-tested up to 5-kV peak. They should also be guaranteed to fail open, not shorted. Depending on the equipment's expected region of operation, some safety agencies (e.g., in Scandinavian regions) may require us to assume no earthing is present, and therefore we will need to create two levels of protection, by placing two Y-caps in series (basically corresponding to *double* insulation).

- Note that Y-caps placed between the Secondary side and the Earth/enclosure are often just standard (non-approved) 0.1-μF/50-VDC ceramic capacitors. Sometimes, the low-voltage ($<$60-V) outputs can even be directly connected to the enclosure. Safety agencies don't care. However, if the power supply is intended to provide the 48-V rail in PoE (Power over Ethernet) applications, by IEEE requirements, we need to provide 1,500-VRMS ($\sim$2,500-VDC) functional isolation between Earth/enclosure and the

48-V output rail plus its return. The reason is that severe voltage spikes may be picked up by the cables carrying data and power across the building. And if we connect those directly to the enclosure, we can cause the user an unpleasant (though not fatal) shock. So, in off-line power supplies meant for telecom applications in general, we cannot use low-voltage Y-caps on the outputs anymore. To support PoE, we therefore use either approved or nonapproved 2.5-kV-rated Y-capacitors on the outputs.

> *Note: What were traditionally called X- and Y-capacitors are more accurately "X2" and "Y2" capacitors respectively. Historically, X2 and Y2 caps were meant for single-phase equipment, whereas X1 and Y1 caps were meant for three-phase equipment. From the viewpoint of safety regulations (like impulse voltage rating and so on), X1 and Y1 caps are considered equivalent to two X2 and Y2 capacitors in series respectively. Therefore, for example, Y1 caps are typically 100% impulse-tested to 8 kV (Y2 caps to 5 kV). We could use Y1 caps for meeting Scandinavian safety deviations.*

- Traditionally, off-line X-caps were of special metallized film + paper construction, whereas Y-caps were a specially constructed disc ceramic type. However, we can also find X-caps that are ceramic, as we can find Y-caps which are film type. It's a choice dictated by cost, performance, and stability concerns. Film capacitors are known to always provide much better stability over temperature, voltage, time, and so on — than most ceramics. In addition, if they are of "metallized" construction, they also possess *self-healing* properties. Note that ceramic capacitors, in general, do not have any inherent self-healing property. However, ceramic caps used as approved Y-caps are specifically constructed in such a way that they will not fail *shorted* under any condition.

- If for any reason (e.g., filter bandwidth or cost), ceramic is preferred for a Y-cap position, then we need to carefully account for its basic tolerance, the variation of capacitance with respect to temperature and applied voltage, and all other long-term variations and drifts. That is because we need a certain filtering efficacy over the life of the product. But at the same time we can't increase the leakage current into the chassis or we will violate ground leakage current limits. We should therefore keep in mind that the capacitance value stated in the datasheet is not just a nominal (or typical) value, but in fact can be a rather misleading value. For example, the fine print may reveal that the test voltage at which the capacitance is stated is close to, or equal to, zero volts. So, the actual capacitance value the capacitor presents in a real working circuit may be very different from its declared (nominal) value. This is, in general, especially true for ceramic capacitors which use a high dielectric constant ("high-K") material (e.g., Z5U, Y5V, etc.). *We should also know that ceramic capacitors age slowly*, except for COG/NP0 types. But COG/NP0 capacitors are very expensive and bulkier too. A typical X7R capacitor ages 1% for every decade of time (in hours). So, its capacitance after 1,000 h will be 1% less than what it was after 100 h, and so on. Higher dielectric constant ceramics like Z5U can age 4–6% for every decade of time.

Table 16.1:   Practical Limitations in Selecting Components and Materials for EMI Filters.

X-Capacitors		Y-Capacitors	
Capacitance (pF)	**Resonant Frequency (MHz)**	**Capacitance (μF)**	**Resonant Frequency (MHz)**
1,000	53	0.01	13
1,500	42	0.022	9
2,200	35	0.047	6.5
3,300	29	0.1	4.5
4,700	21	0.22	2.7
6,800	19	0.47	1.9

Magnetic Materials for EMI Chokes			
		Initial Permeability	**Bandwidth (MHz)**
Powdered iron		60	10
		33	50
		22	100
		10	>100
Ferrite		15,000	0.17
		10,000	0.3
		5,000	1.0
		3,000	1.2
		2,500	1.5
		1,500	3.0

So, in effect our filter stage too gets less effective with time. And we need to account for this upfront, in the *initial* design.

- Theoretical filter performance is based on the assumption that we are using "ideal" components. However, real-life inductors are always accompanied by some winding resistance (DCR) and some inter-winding capacitance. Similarly, real capacitors have an equivalent series resistance (ESR) and an equivalent series inductance (ESL). At high frequencies, the inductance will start to dominate, and so a capacitor will basically no longer be functioning as one (from the signal point of view). However, capacitors with *smaller* capacitances generally remain capacitive up to much higher frequencies than do larger capacitances. See Table 16.1 for some typical self-resonant frequencies (the point above which, capacitors start becoming inductive). Therefore, quite often, a *smaller* Y-cap may help in situations where a larger Y-cap is not yielding the required results. We can also consider paralleling a larger value Y-cap with a small Y-cap.

- Surface mount ("SMD") versions of off-line safety capacitors are available — for example from Wima in Germany and Syfer in UK. But especially for SMD caps, we must keep in mind that it is not enough for a manufacturer to spontaneously declare their capacitor "complies" with a certain safety standard — the capacitor should actually be *approved*. Only after being tested by various safety agencies, is it allowed to carry their respective certification marks. From the electrical point of view, one of the great advantages of SMD caps is their very low ESL. This improves their high-frequency performance in any filter application.

- Note that in the relentless drive toward reducing parasitics, we may forget that some ESR or DC winding resistance ("DCR") is often useful in helping *damp* out oscillations. Without any resistance to burn up the energy, oscillations will last forever. That is one of the reasons why engineers sometimes pass one or both of the leads of a standard through-hole Y-cap through a small ferrite bead (preferably of a material with lossy characteristics, like Ni–Zn). This can often help suppress a particular high-frequency resonance involving the Y-cap that is perhaps showing up in the conducted EMI scan. But we must be careful that in doing so, we do not land up with a radiation problem instead.

- Designers of low-voltage, low-power DC–DC converters may find the "X2Y" patented product range available from Syfer (and from the company X2Y itself) useful if they need to miniaturize and lower the component count. This is a three-terminal integrated SMD capacitor-based EMI filter that simultaneously provides line-to-line decoupling and also line-to-ground decoupling. From X2Y's website quoted verbatim: "The patented X2Y® Technology consists of proprietary electrode arrangements that are embedded in passive components. Components with X2Y technology can be manufactured in a variety of dielectric materials including ceramic, metal oxide varistor (MOV), and ferrite. The main embodiment of X2Y Technology is in a multilayer ceramic capacitor, this allows capacitor manufacturers to license X2Y. End-users can then purchase X2Y® components just as they would any other passive component. Currently, there are six licensed manufacturers who make and sell X2Y components."

- Picor (a subsidiary of Vicor) sells active input EMI filter stages for standard 48-V bricks and off-line power supplies. These are expensive (typically ~$50 per unit), but may be a viable choice if board space is at a premium.

- We note that a Y-cap is always tested to higher safety standards than an X-cap. So, we can always use a Y-cap at an X-cap position, but not vice versa. For example, we can consider placing a ceramic Y-cap in parallel with a film X-cap to improve the DM filter bandwidth. However, nowadays, to keep things simple, manufacturers are selling safety approved caps designated for use either as an X-cap or a Y-cap.

- We often run into the problem of needing to improve filter performance in the low-frequency region of the CISPR-22 Class B limits. We do this by trying to increase the "LC" product as much as practically possible (in the process, lowering the resonant frequency). Given a choice, we would prefer to realize the large target value of the LC product, by using larger capacitances instead of impractically sized inductors. But as we know, the total Y-capacitance (before the bridge rectifier) is limited by safety considerations. Note that X-caps too seemed to be limited for many years to a maximum value of $0.22\,\mu\text{F}$, or occasionally $0.47\,\mu\text{F}$. But that was actually based on availability and component technology limitations. Nowadays, we can get approved X-caps up to $10\,\mu\text{F}$. We should, however, be conscious that large input capacitances can cause undesirably high inrush surge currents at power-up. This may also cause eventual failure of the X-cap, especially if it is the very first component after the AC input inlet. The X-cap may not fail outright since we know that film caps can self-heal from such events. However, eventually, the capacitance will degrade over time. Therefore, despite EMI concerns, we should try to place X-caps *after* any input surge protection element, for example, after the NTC (negative temperature coefficient) thermistor, or a wirewound resistor, perhaps even after a front-end choke.

Examining the Equivalent DM and CM Circuits and Filter Design Hints

The filter in the upper part of Figure 16.1 reduces to the CM and DM equivalents shown in the lower part of the same figure. Note that C2 and C4 are small (being Y-caps), but C3 is large. Some observations are:

CM filter: We see that the discrete DM choke L_{dm} acts as a CM filter element too. The CM filter stage consists of two LC stages. One stage has a large LC product (low-frequency pole) corresponding to the product $2 \times L_{\text{cm}} \times \text{C4}$. The other provides high-frequency filtering since its LC product is relatively small: $L_{\text{dm}} \times \text{C2}$.
DM filter: We see that the CM capacitor (C4) acts as a DM filter element too. The DM filter stage consists of two LC stages. One stage has a large LC product (low-frequency pole) corresponding to the product $2 \times L_{\text{dm}} \times \text{C3}$. The other provides high-frequency filtering since its LC product is small: L_{lk}. Note that if we use a toroidal CM choke, this high-frequency DM filter stage may barely exist since L_{lk} will be almost zero.

Though CM chokes usually have a high inductance (and that is certainly needed — particularly for complying with CISPR-22 limits below $500\,\text{kHz}$), a good part of CM noise is usually found in the frequency range of $10-30\,\text{MHz}$. So, we must consider the fact that not all ferrites have sufficient *bandwidth* to be able to maintain their inductance (A_{L}) at such high frequencies. In fact, materials with a *high* permeability tend to have a *lower* bandwidth, and vice versa ("Snoek's law"). Therefore, a "high-inductance" CM filter may look good on paper, it but may not be as effective as we had thought, at high frequencies. See Table 16.1 for typical values of initial permeability versus bandwidth (bandwidth being defined here as a 6-dB fall in permeability).

In the latter sections of this chapter, we will see that a DM noise generator is more like a *voltage* source. So putting in an LC filter works well for a DM source, as it presents a "wall" (mismatch) of impedance which serves to block the DM emissions from entering the mains. But a CM noise source behaves more like a *current* source. And we know that current sources *demand* to keep current flowing. They can surmount any "wall" of impedance by increasing voltage if necessary. Therefore, for dealing with CM sources, rather than try to only block CM noise from flowing into the mains, we should also close the current loop locally inside the power supply itself. *Plus we should try to dissipate that energy rather than have it circulate constantly (or slosh back and forth).* Only then would the CM filter stage as shown in Figure 16.1 really work. This is part of the pattern of nonintuitive behavior of inductors and current sources at large, as discussed in *Chapter 1*. This is also the point we start understanding why EMI issues in power supplies are so much different from those faced by signal integrity engineers.

To dissipate CM energy effectively, a "lossy" ferrite material for the CM choke often works well. The usual ferrite used for power transformers and inductors is of manganese–zinc composition. Lossy ferrites (using nickel–zinc) have higher AC resistance at high frequencies, so they can be more helpful in "killing" high-frequency CM noise components. Unfortunately, lossy ferrites also have very low initial permeabilities, so it is almost impossible to get the desired high inductance (as sorely needed to filter low-frequency noise). So, the functions are separated — one CM choke is used to "block" and another is introduced to "burn." The latter CM choke's inductance is not important anymore, so it could be just a small bead/toroid/sleeve made of lossy material, with both the L and N wires passing through it together. Many commercial power adapters, for example, have a big ferrite sleeve on their input AC cord.

Engineers are often mystified to find that making the DM choke out of (low permeability) powdered iron or lossy ferrite helps too, when all else has failed — despite all the talk about DM noise being essentially a "low-frequency emission" and so on. The reason is as follows — the CM noise in a power supply is actually mixed-mode (MM) noise at its point of creation (this will be discussed shortly). Though ultimately, by cross-coupling, it does tend to spread onto both lines almost equally. We had shown in *Chapter 15* that MM noise can be considered a mix of CM and DM components. Therefore, in practice, we do get a fair amount of high-frequency DM noise too — arising out of the mixed-mode noise. That is why high-bandwidth/low-permeability/lossy materials often help in DM noise suppression too.

The DM and CM filters are often, but not always, laid out in the order shown in Figure 16.1. The CM filter stage is shown closest to the AC inlet. The idea is that the very last stage the noise encounters (as it travels from the power supply into the mains) should be a CM filter. Because, if, for example, this last stage was a DM stage, then if it was not very well balanced in impedance from the viewpoint of the noise emerging from the preceding CM filter, the CM noise could get converted into DM noise as previously explained. However, admittedly, many successful commercial designs have reversed the order shown in Figure 16.1. There seems to be no universally accepted rule for which stage should come before which one.

A possible location for an additional X-cap is directly on the prongs of the AC inlet socket (at the entrance to the power supply). We remember that in this position any line-to-line capacitor will be exposed to a huge current surge at power-up, and could degrade, if not fail immediately. So if this X-cap position seems to be the last resort, it should at least be made as small as possible (typically $0.047-0.1\,\mu\text{F}$). Or we can try ceramic capacitors in this position (approved ceramic X-caps or Y-caps can be tried here), since they have high surge current capabilities arising from the very low heat generated in them (due to very low ESR).

Similarly, the two front-end Y-caps ("C4" in Figure 16.1), or two additional Y-caps, can also be connected directly on to the prongs of the AC inlet socket, rather than on the PCB. This can help a great deal if the wires going from the PCB to the mains inlet socket are themselves picking up stray fields (radiation).

Sealed chassis mountable line filters (sometimes with integrated standard "IEC 320" line inlets) are available from several companies like Corcom (now part of Tyco Electronics) and Schaffner in Germany. Such filters perform well, but they are less flexible to subsequent tweaking, and also far more expensive than board-mounted solutions. Note that the performance of most commercially available line filters is specified with $50\,\Omega$ at both ends of the filter. Therefore, its actual performance in a real power supply may be quite different from what its datasheet says.

To prevent the enclosure from radiating, it is important to make a good high-frequency connection from the enclosure to the Earth terminal (the middle prong of the AC inlet). A thick wire (preferably braided) can be used to connect the two. Better still, standard AC inlets with built-in metal brackets can be used for the purpose. These are easily available nowadays, as from Methode Electronics. The connection from the CM filter on the board can be made to the AC Earth terminal via the enclosure. Metal standoffs can be positioned on the enclosure to provide mechanical stability and connect to the correct PCB traces on the EMI filter. However, since we want to prevent CM noise from entering the AC mains, making a good ground connection may serve the opposite purpose. However, if we don't make a good connection to Earth ground, we may reduce the amount of CM noise directly thrown into the AC wiring, but we will have a major radiation problem instead. And that could easily convert itself into a conducted-mode EMI problem eventually, through radiated pickup. However, the ambivalence associated with the basic question of how "good" should the connection be from the midpoint of the two C2/C4 caps in Figure 16.1 to the Earth prong of the AC inlet has resulted in something called the "ground choke."

The Ground Choke

We ask — is it really a good idea to insert a small inductor (e.g., a bead or small toroid with a few turns on it) on the wire connecting the on-board EMI filter to the Earth prong? This is called a "ground choke" or "Earth choke." It is commonly found on low-power evaluation boards (from vendors promoting their integrated power IC solutions), but rarely seen on a commercial power supply. Why is that?

When we place the ground choke, we are basically trying to prevent conducted CM noise from flowing into the mains wiring. But in return, we may have a radiation problem. In addition to that, there are industry-documented cases where the ground choke has caused severe *system* problems. For example, if a power supply is suddenly connected to the mains at the peak of the input AC waveform, there is a high surge of current through its Y-caps too. If there is a ground choke present, it causes the voltage on the Earth traces *and the enclosure* to locally "bump up." Now in some cases, the return of the output rails of the power supply is also connected directly to the enclosure, and forms the ground plane for the entire system. The system would also typically connect to the chassis/enclosure at several points downstream. So, this surge-induced bump, around the power supply, causes severe imbalances across the system ground plane — leading to data upsets and even destruction of the subsystems. A similar situation will arise during ESD testing and conducted immunity testing, in which surge voltages are applied from line to line, or from line to Earth. So, however, tempting it may seem to the power supply designer (who is focused only in solving his conducted-mode EMI problem) *a ground choke should be avoided at all costs.* Some high-voltage semiconductor companies, who are only making *open-frame* (enclosure-less/standalone) evaluation boards, seem to have nothing to lose, and everything to gain, by putting in a ground choke. Perhaps they know that being open-frame anyway, no one expects them to comply with *radiation* limits. So, they may just have pushed the problem from the conducted emissions plot toward a future radiation emissions' saga for the systems designer.

Some Notable Industry Experiences in EMI Filter Design

One of the most stubborn cases of conducted EMI failure encountered by the author while working in Germany was ultimately (and rather mysteriously) solved by simply reversing the orientation of the CM choke (i.e., turning it by 180° on the PCB). It was later deduced that the leakage from the core was being picked up by a nearby trace or component, and so the phase of the coupling had somehow become an issue (interference pattern). But since most inductors/chokes are symmetrically built, and also do not carry any marking to distinguish one side from the other, implementing such a fix was not easy in production. However nowadays, with so many similar "orientation-sensitive" cases being reported (even relating to the main inductor of the converter itself), some key inductor manufacturers have taken the step of placing a "polarity mark" on their inductors/chokes.

In another well-documented EMI problem at a leading power supply manufacturing house, it was discovered that the CM choke had to be rotated by 90° (not 180°) to achieve compliance. That clearly spells "bad news" if the unit is already in production, because it means the PCB layout has to be redesigned (and perhaps the power supply needs to be requalified too).

Part 2: DM and CM Noise in Switching Power Supplies

Main Source of DM Noise

Now we turn our attention to a real power supply to see for ourselves where all the buzz is really coming from. First consider what would happen if the input bulk capacitor of the power supply had been a "perfect" capacitor, that is, with zero effective series resistance (ESR) (ignoring all other capacitor parasitics too). Then any possible differential noise source inside the power supply would be completely bypassed/decoupled by this capacitor. Clearly, the reason this does not happen is the *non-zero ESR of the bulk capacitor.*

So, the ESR of the input capacitor is the major portion of the impedance "Z_{dm}" seen by the DM noise generator (see Figure 15.6). The input capacitor, besides being refreshed by the operating current flowing in through the supply lines, also tries to provide the high-frequency pulses of current demanded by the switcher. But whenever current passes through any resistance, such as the ESR in this case, there must be a corresponding voltage drop. So, we will see a high-frequency voltage ripple across the terminals of the input capacitor as seen in Figure 16.2. The high-frequency voltage ripple shown in the figure is, in effect, the DM noise generator. It is essentially a voltage source (V_{ESR_PP}). In theory, this high-frequency ripple-related noise is supposed to appear on the left (input) side of the bridge rectifier only when the bridge is conducting. In reality, it may spill past the parasitic capacitances of the diodes of the bridge, during the OFF-time of the bridge too. In the next section, we will see that DM noise can also be generated when the bridge rectifier is OFF, but this time it appears as a current source, not a voltage source.

The Main Source of CM Noise

Various inadvertent paths may be present carrying high-frequency current into the enclosure. A key culprit is shown in Figure 16.3. As the Drain of the FET swings high and low, current flows through the parasitic capacitance between the FET and the heatsink (the heatsink is connected to, or is, the enclosure in our case). We see that when the AC line current is holding the bridge rectifier ON (diodes D2 and D4 first ON, followed by D1 and

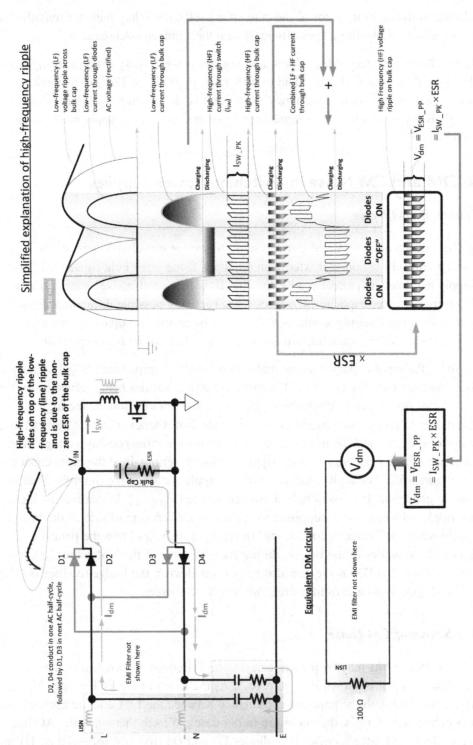

Figure 16.2: How DM noise is created.

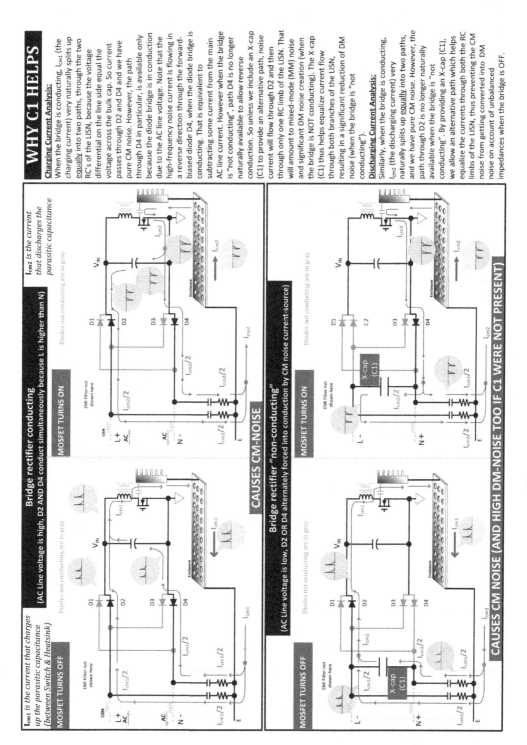

Figure 16.3: How CM noise is created.

WHY C1 HELPS

Charging Current Analysis:

When the bridge is conducting, I_{cm1} (the charging current) very naturally splits up equally into two paths, through the two RC's of the LISN, because the voltage differential on the line side equal the voltage across the bulk cap. So current passes through D2 and D4 and we have pure CM noise. However, the path through D4 in particular, is available only because the diode bridge is in conduction due to the AC line voltage. Note that the high-frequency noise current is flowing in a reverse direction through the forward-biased diode D4, when the diode bridge is conducting. That is equivalent to subtracting some current from the main AC line current. However when the bridge is "not conducting", path D4 is no longer naturally available to allow reverse conduction. So unless we include an X-cap (C1) to provide an alternative path, noise current will flow through D2 and then through only one RC limb of the LISN. That will amount to mixed-mode (MM) noise and significant DM noise creation (when the bridge is NOT conducting). The X-cap (C1) thus helps equalize current flow through both branches of the LISN, resulting in a significant reduction of DM noise (when the bridge is "not conducting").

Discharging Current Analysis:

Similarly, when the bridge is conducting, I_{cm2} (the discharging current) very naturally splits up equally into two paths, and we have pure CM noise. However, the path through D2 is no longer naturally available when the bridge is "not conducting". By providing an X-cap (C1), we allow an alternative path which helps equalize the currents through both the RC limbs of the LISN, thus preventing the CM noise from getting converted into DM noise on account of unbalanced impedances when the bridge is OFF.

Table 16.2: Typical Mounting Capacitances.

Package	Area (cm^2)	Material	K	Thickness (mm)	Capacitance (pF)
TO-3	5	Silicone rubber	5	0.2	111
		Mica	3.5	0.1	155
TO-220	1.644	Silicone rubber	5	0.2	36
		Mica	3.5	0.1	51
TO-3P	3.25	Silicone rubber	5	0.2	72
		Mica	3.5	0.1	101
TO-247F	2.8	Silicone rubber	5	0.2	62
		Mica	3.5	0.1	87

D3 in the next AC half-cycle, and so on), the noise injected into the enclosure sees almost equal impedances (and voltages) and therefore flows equally on the L and N phases. So, it constitutes pure CM. However, when the bridge turns OFF, the noise forces (only) one diode of the bridge to turn ON, since the inductor current is behind the noise current too. However, this creates unequal impedances, and so the noise current does return, but through *either* the L or the N phase, not split equally between the two. As a result, as we have shown in Figure 15.6, this MM noise is in effect a pure CM noise component plus a very large, high-frequency DM noise component. The way to equalize the CM noise flow into both phases, and thereby reduce this contribution to the DM noise source, is to provide an X-cap right before the bridge. Any required EMI filter can be placed between this cap and the AC mains, as I have indicated in Figure 16.1 too.

Typical values of parasitic capacitance that can be created in a power supply by the insulator are presented in Table 16.2. Here we are comparing a traditional insulator material, mica, with a modern choice, silicone rubber. K is the dielectric constant.

Chassis-Mounting of Semiconductors

Even seasoned engineers are often extremely nervous about chassis-mounting of power devices. Often they can be coaxed into mounting the output diodes in this manner (since the diodes have lower voltages), but *not* the high-voltage FET. However, there is a way to do this. If the Y-cap shown in Figure 16.4 (marked "Y-cap") is placed, along with a high-frequency ceramic cap across the bulk cap, together they will return the injected noise very close to the FET. Note that we need a metal standoff from the enclosure to the PCB very close to where the FET is mounted (PCB is not shown in the figure). Small current loops minimize radiated *H*-fields. So EMI is reduced. This is also a good place to insert a Ni−Zn

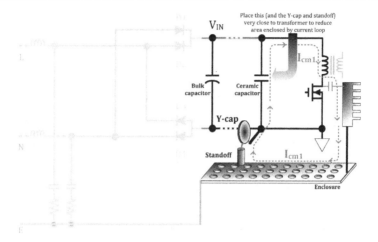

Figure 16.4: How to mount power devices on the enclosure.

ferrite bead somewhere in the path, to help burn up the circulating heatsink-injected noise energy.

The CM Noise Source

We were implicitly assuming by using the X-cap C1 in Figure 16.3, that we were dealing with pure CM noise, not MM noise. In Figure 16.5, we draw the exact injected current previously indicated vaguely in Figure 16.3. We have a current spike associated with the rising voltage edge and another spike coinciding with the falling edge. (This is just one way engineers model this; there are other treatments in literature, but yielding similar results.)

The shape of the current waveform in Figure 16.5 provides the entire *harmonic content* of the injected current. The measured noise at the LISN when $t <$ tcross is

$$V_{cm1} = 25 \times I_{cm} = \frac{25 \times A \times C_P}{\text{tcross}} \times \left(1 - e^{-t/25C_P}\right)$$

and for $t >$ tcross it is (all this based on M.J. Nave's paper in APEC 1989)

$$V_{cm2} = 25 \times I_{cm} = \frac{25 \times A \times C_P}{\text{tcross}} \times \left(e^{-(t - \text{tcross})/25C_P}\right)$$

The same situation occurs when the FET turns ON — only the directions are reversed. We can do a Fourier analysis of the measured voltage waveform, and the result is (in frequency domain)

$$V_{cm} = \frac{50 \times A \times C_P}{T} \times \left[\frac{sin\{(n \times \pi \times \text{tcross})/T\}}{(n \times \pi \times \text{tcross})/T}\right] \times \left[e^{-jn\pi(\text{tcross}/T)} - e^{-jn\pi((\text{tcross}/T)+2D)}\right]$$

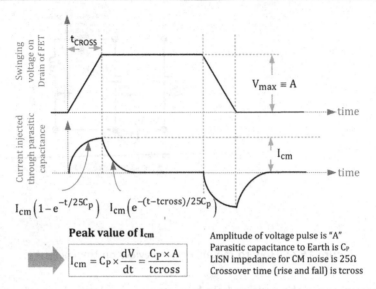

Peak value of I_{cm}

$$I_{cm} = C_P \times \frac{dV}{dt} = \frac{C_P \times A}{tcross}$$

Amplitude of voltage pulse is "A"
Parasitic capacitance to Earth is C_P
LISN impedance for CM noise is 25Ω
Crossover time (rise and fall) is tcross

Figure 16.5: CM noise current injected through the FET-to-Heatsink mounting capacitance.

In *Chapter 18*, we will discuss Fourier series in greater detail. We will see that the $\sin x/x$ term (called the "sinc function") always contributes an (additional) 20 dB/decade roll-off after the breakpoint which is located at $x = 1$. Its value is close to unity till the breakpoint (i.e., for $x < 1$). The term in the rightmost bracket above (involving the imaginary terms with $j = \sqrt{-1}$) has no roll-off. Its magnitude changes between the limits 0 and 2 as the harmonic number changes or/and duty cycle changes. So, for all practical purposes, if we are just interested in the envelope, we can take its maximum value of 2 to get the voltage picked up on the LISN.

$$V_{cm} = \frac{100 \times A \times C_P}{T} \times \left[\frac{\sin\{(n \times \pi \times tcross)/T\}}{(n \times \pi \times tcross)/T} \right]$$

This plot of V_{cm} is flat till the break frequency ($x = 1$), after which it rolls off at 20 dB/decade. The flat part (the "pedestal") can be found using the approximation $\sin x/x \cong 1$ for small x. We thus get

$$V_{cm} = \frac{100 \times A \times C_P}{T}$$

For example, if $A = 200$ V (amplitude of voltage on Drain of FET), $C_P = 200$ pF, $f_{SW} = 100$ kHz, we get

$$V_{cm} = \frac{100 \times 200 \times 200 \times 10^{-12}}{10^{-5}} = 0.4 \text{ V}$$

This is without an EMI filter. With a filter, we will need to provide a certain attenuation to bring it within the acceptable limits described in *Chapter 15*. A complete design procedure is provided in *Chapter 18*.

The Road to Cost-Effective Filter Design

In *Chapter 18*, we will see that the envelope of the switching harmonics falls off with a slope of −20 dB/decade as frequency increases. Alternatively stated, the harmonic amplitudes rise at the rate of 20 dB/decade as the frequency decreases. Note that by definition, that is, in effect, a 10-fold increase (20 dB) in amplitude of the harmonics every 10-fold decrease in frequency (decade). Looking at the CISPR-22 Class B EMI limits in Figure 15.2, from the dashed gray extrapolations, we see that from 150 kHz to 500 kHz, the CISPR limits allow for a 20-dB/decade increase in switching harmonics. The original purpose of that was to create an allowance for switching power supplies. However, by designing a typical one-stage low-pass LC filter for EMI, we find that its attenuation decreases at the rate of 40 dB/decade as the frequency decreases. In other words, it becomes less effective at lower frequencies. That is the reason it becomes imperative that we design the filter such that we are below the limits at the *lowest* frequency of interest (150 kHz, or the switching frequency of the converter, whichever is greater). Once we do that, as the frequency increases, the CISPR22 limit lines demand a 20-dB/decade reduction in

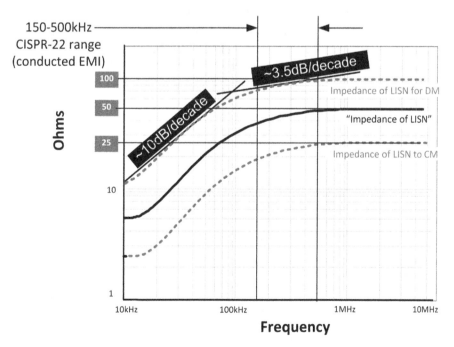

Figure 16.6: LISN impedance at low frequencies.

harmonics as frequency increases toward 500 kHz. However, we get that automatically since the envelope of our switching harmonics also falls off at the rate of 20 dB/decade. But there are several factors affecting us here as described below.

(a) As the frequency increases, the LC filter becomes more and more effective, at the rate of 40 dB/decade. So, we get an additional advantage of 40 dB/decade over what the limits allow.

(b) The LISN sensitivity actually increases with frequency (see Figure 16.6). That will give higher readings as the frequency increases. It is, in effect, a disadvantage of 3.5–10 dB/decade depending on frequency.

Together, from (a) and (b) above, we get a theoretical advantage of about 40 −3.5 = 36.5 dB/decade as frequency increases. However, we are going to need that margin. As the frequency increases, the response of the filter is again going to worsen due to parasitics and ferrite bandwidth. Also, CM noise effects will start to dominate. In addition, conducted-mode readings will get affected by radiation pickup. There will also be some additional spikes in the EMI scan due to parasitics we didn't model. Note that spikes should be dealt with individually at the *board level*, rather than try to bring the entire EMI spectrum down by a brute-force over-designed filter.

In general, it is important to be aware of the "trends" described above. In *Chapter 18*, we will do a formal EMI filter design. In effect, because we have "margin" at higher frequencies, the objective of filter design is to primarily achieve compliance at the lowest frequencies of interest. That is also the hardest to do. Which is why, compliance to FCC Part 15 Subpart B (at 450 kHz) is usually far easier than for CISPR-22 (at 150 kHz).

CHAPTER 17

Fixing EMI Across the Board and Input Filter Instability

Part 1: Practical Techniques for EMI Mitigation

Here we first look at some of the practical design aspects involved in controlling EMI. These supplement the basic PCB layout guidelines presented in *Chapter 10*. We first emphasize one aspect of that chapter: that the most potent and cost-effective method of reducing EMI is the ground plane.

The Ground Plane

The ground plane is a very effective method of bringing down the overall level of the EMI emissions. On a multilayer board, if the very next layer to the side containing the power components (and their associated traces) is this ground plane, the EMI can drop by about 10−20 dB. This is more cost-effective than opting initially for a "cheap" one- or two-sided board, and then having to use bulky filters later. However, the integrity of a ground plane should be maintained, as far as possible.

We should remember that return currents tend to travel by the shortest *straight line* path at low frequencies. But at high frequencies (or the higher harmonics of the switching waveform), the return currents tend to image themselves directly under their respective forward traces (on the opposite layer). Therefore, currents, given a chance, automatically try to reduce the area they enclose — as this lowers the self-inductance, and thereby offers the current, the lowest impedance route possible (at low frequencies, trace impedances are resistive, but at high frequencies they are inductive). So, for example, if we make ill-considered cuts in the ground plane (possibly with the intention of "conveniently" routing some other trace), the return currents of the power converter stage will get diverted along the sides of any intervening cuts, and will thereby effectively create *slot antennas* on the PCB.

The Role of the Transformer in EMI

Very often a young engineer resolves a stubborn EMI problem by just "playing" with the transformer. We can learn a lot from similar excursions. With magnetics in general, nothing is perhaps completely known or obvious.

The transformer comes into the picture in the following ways:

- With its windings carrying high-frequency current, it becomes an effective H-field antenna. These fields can impinge upon nearby traces and cables, and enlist their help in getting transported out of the enclosure, via conduction or radiation.
- Since parts of the windings have a swinging voltage across them, they can also become effective E-field antennas.
- The parasitic capacitance between the Primary and Secondary windings transfers noise across the isolation boundary. Since the Secondary-side ground is usually connected to the chassis, this noise returns via the Earth plane, in the form of CM noise. The situation is very similar to the tradeoffs required in heatsink mounting issues. In this case, we wish to couple the Primary and Secondary very close to each other in order to reduce leakage inductance (especially in flyback transformers), but this also increases their mutual capacitance, and thus the CM noise.

Here are some standard techniques that help prevent the above:

- In a safety-approved transformer, there are three layers of safety-approved polyester ("Mylar®") tape between the Primary and Secondary windings, for example, the popular #1298 from 3M. In addition to these layers, a copper "Faraday shield" may be inserted to "collect" the noise currents arriving at the isolation boundary, and divert them (usually to the Primary ground) (see Figure 17.1). Note that this shield should be a very thin strip of copper foil so as to avoid eddy current losses and also keep the leakage inductance down. So, it is typically 2–4 mils thick, consisting of one turn wound around the center limb. A wire is soldered close to its approximate geometric center and goes to the Primary ground. Note that the ends of the copper shield should not be galvanically connected together, as that would constitute a shorted turn from the viewpoint of the transformer. Some designs also use another Faraday shield, on the Secondary side (after the three layers of insulation). This is connected to the Secondary ground. However, most commercial ITE (information technology equipment) power supplies don't need either of these shields, provided adequate thought has gone into the winding and construction, as we will soon see.
- There is usually also a circumferential copper shield (or "flux band") around the *entire* transformer (see Figure 17.1). The ends of this shield can be, and are usually, shorted (soldered) together. It serves primarily as a radiation shield. It is often left floating in low-cost designs. However, it may (and should) be connected to the Secondary ground.

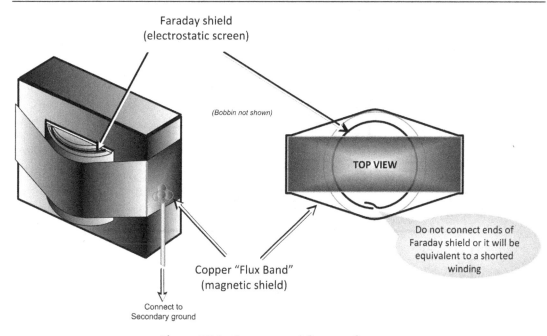

Figure 17.1: Screens used for transformers.

Safety issues will need to be considered, in regards to IEC 60950-1 requirements in terms of insulation between Primary and Secondary and separations (the required Primary to Secondary "creepage," i.e., distance along the insulating surfaces, and "clearance," i.e., shortest distance through air) as applicable. When the transformer uses an air gap on its outer limbs, the fringing flux emanating from the gap causes eddy current losses in the band. So, this band is also usually only 2−4 mils thick. Like the Faraday shield, this too can often be omitted by good winding techniques.

• To reiterate, from the point of view of EMI, a flyback transformer should be preferably *center-gapped*, that is, no gap on its *outer* limbs. The fringing fields from exposed air gaps become strong sources of radiated EMI besides causing significant eddy current losses in the surrounding copper band.

• There is usually an auxiliary winding present on the Primary side, which provides a low-voltage rail for the controller and related circuitry. One end of this is connected to Primary ground. Therefore, it can actually double over as a crude Faraday shield if we (a) wind it evenly and spread it out over the available bobbin width and (b) help it collect and divert noise by AC-coupling its opposite end (i.e., the diode end) to Primary ground, through a small 22- to 100-pF ceramic capacitor as shown in the topmost schematic of Figure 17.2.

Figure 17.2 also reveals low-noise construction techniques as applied to a typical flyback transformer. We should compare the right-hand schematics with their equivalent "winding"

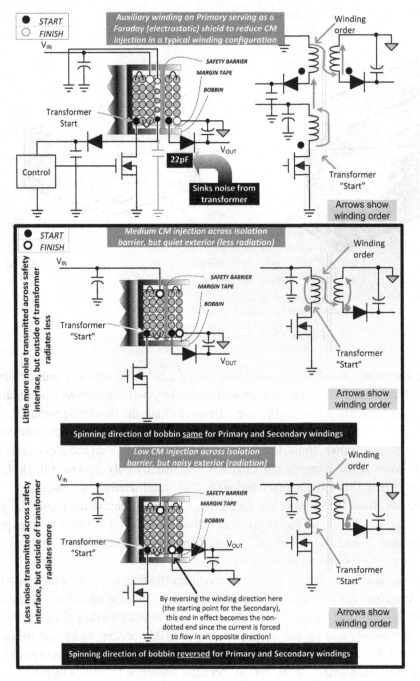

Figure 17.2: Low-noise transformer winding techniques.

versions on the left. In the discussion below, we note that though transformers with split windings are not being explicitly discussed here, the same principles can be easily extended and applied to them too. Here are some observations based on Figure 17.2:

• Since the Drain of the FET is swinging, it is a good idea to keep the corresponding end of the Primary winding buried as deep as possible, that is, it should be the *first layer* to be wound on the bobbin. The outer layers tend to shield the fields emanating from the layers below. For sure, the Drain end of this winding should *not* be adjacent to the "safety barrier" (the three layers of polyester tape) because the injected noise current is proportional to the net dV/dt across the two "plates" of the parasitic capacitor (formed by the windings on either side of the interface). Since we really cannot reduce the capacitance much (without adversely impacting the leakage inductance), we should at least try to reduce the net dV/dt across this interface capacitor.

• Comparing any diagram on the left with its corresponding schematic on the right, we see that the "start" and "finish" ends of any winding have also been indicated. In particular, all the start ends have been shown with dots in the schematic. Note that in a typical production sequence, the coil winding machine always spins the bobbin in the same direction, for every layer and winding placed successively. Therefore, since all the start ends (i.e., dotted ends) are magnetically equivalent, if one dotted end goes high, the other dots also go high at the same moment (as compared to their opposite ends). We can also see that from the point of view of the actual physical proximities involved, every dotted end of a winding automatically falls close to the nondotted end of the next winding (with the usual fixed winding direction). This means that for the flyback transformer of Figure 17.2, the diode end of the Secondary winding will then necessarily fall adjacent to the safety barrier. Yes, because of that we will have a certain amount of dV/dt still present across the barrier. But note that this dV/dt is much smaller than if the Drain end of the Primary winding was brought adjacent to the safety barrier (because of the bigger voltage swing on the Primary side, due to the large turns ratio). However, the transformer as shown in the top two schematics of Figure 17.2 now has the advantage that the "quiet end" (ground) of the Secondary winding is now the outermost layer. That is by itself a good shield. So, we can safely drop the ubiquitous circumferential shield (flux band). Consider the alternative — suppose we had wound the transformer the "wrong" way, that is, by reversing all the start and finish ends shown in Figure 17.2. That would have brought the Drain end of the Primary winding right next to the safety barrier, with the Secondary ground end (which is usually connected to the chassis) directly across the isolation boundary. With this winding arrangement, we would have a healthy dose of CM noise injected directly into the chassis/Earth — not the best way to achieve compliance for sure.

• When we go through the same reasoning for a Forward converter transformer, we will find that with the described winding sequence, we will automatically have the quiet ends of both Primary and Secondary windings "overlooking each other" across the safety barrier (isolation boundary). That is because the relative polarities between the Primary and Secondary windings in a Forward converter are opposite to those of a flyback transformer. So now, very little noise will be injected through the parasitic capacitance. That is good. But the outermost layer is not "quiet" anymore, and we could have a radiation problem. So, in this case, the circumferential shield may become necessary.

- Another way out of the Forward converter "outer surface radiation problem" is to ask our production team to reverse the direction of the Secondary winding (only). So, for example, if up to the finish of the Primary winding, the machine was spinning clockwise, for the Secondary we should specify an anticlockwise direction (with expected resistance coming, not from the transformer, but our production staff!). If we do that, the reasoning given previously for the flyback will now apply unchanged to the Forward converter transformer too. So, we would now have a "quiet" exterior (without any flux band necessary), though some more common-mode noise will be transferred across the isolation boundary due to the dV/dt. Note that in general, aiming for a "quiet" exterior (low radiation) seems to be a better option than trying solely to prevent noise injection through the interface capacitance, because the latter can be overcome by various tricks — like *having the auxiliary winding double over as a Faraday shield*, and so on. But a radiation problem can be hard to manage. We do note, however, that a Forward converter transformer has no (or very small) air gap, so it is generally considered "quieter" in terms of radiation to start with (as compared to a flyback).

- In the lowermost schematic of Figure 17.2, we have an alternative winding technique for those flyback cases where we are troubled by conducted EMI noise, in particular CM noise. The way to minimize that is to then reduce the dV/dt across the safety barrier, by bringing the quiet end of the Secondary winding next to the safety interface.

Tip: We don't need to draw any current at all from the "Faraday winding" (uppermost schematic of Figure 17.2) to make it work. So, it need not even be required by our circuitry (for an auxiliary power rail). In that case, we could just wrap a few turns of thin wire (spread out evenly), with one end of it connected to Primary ground and the other end via a small 22-pF capacitor to Primary ground. This technique certainly saves production costs associated with the making and placing of a formal Faraday shield — not to mention the improvement in efficiency due to the reduced leakage inductance (as compared to what a formal Faraday shield may lead to). In that sense, this informal Faraday shield is a very useful idea, worth trying out.

- When the transistor is mounted on the chassis for thermal reasons, there is a technique that is used to actually try to cancel the current injected through the heatsink capacitance. This is done by placing another winding, equivalent to the main winding, and opposite in phase — when one winding turns OFF, this noise recovery winding turns ON. Note that the noise recovery winding can be of much thinner wire (see Figure 17.3).

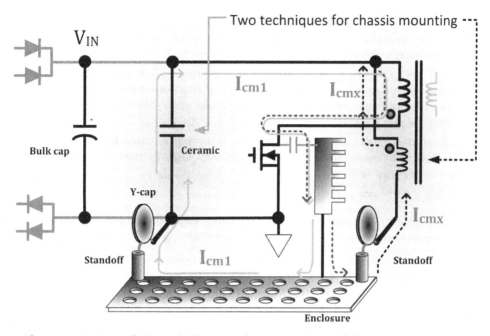

Figure 17.3: Cancellation winding to reduce CM noise and direct return method.

The idea is that if the noise current (I_{cmx}) is being pushed out from the Primary winding, the cancellation winding gets the same current pulled in. Therefore, in effect, the injected current does a quick "U-turn" back to its noise source. Note that this additional cancellation winding should be very closely coupled to the main winding. Often it is wound bifilar with the Primary winding (i.e., both wound simultaneously, rather than one on top of the other). However, we should be aware that in that case, we will have a high-voltage differential between the two windings at points along their length. So if, for example, there are pinholes in the enamel insulation, there is a danger of flashover and resulting failure of the power supply. The solution is to use wires with "double insulation." In Figure 17.3, the cancellation winding method is shown along with another technique we had presented in Figure 16.4. Both are independent, and either one, or both, can be used for good results.

Note: The above technique does nothing to cancel the noise injected through the interface capacitance (i.e., between the Primary and Secondary windings). But despite that limitation, a 5–10-dB μV reduction in conducted EMI is still possible (at various points in the EMI spectrum). So, this could certainly be worth trying out, if there is a last-minute problem and a major redesign of the board needs to be avoided. It may be therefore prudent to plan for this winding in advance, including a PCB placeholder for the additional Y-capacitor.

Note: The above idea can clearly be applied to any *off-line topology (and also all high-power DC–DC converters) — whenever the switch needs to be mounted on the chassis/enclosure (and its Drain is swinging). A similar technique may be useful on the Secondary side too, if the catch diode is to be mounted on the chassis. However, this Secondary-side heatsink noise injection is of concern only when the tab of the diode (which is almost invariably the cathode of the diode) happens to be the switching node for that particular topology/configuration. So, we can work out that the normal Boost and flyback topologies don't have this problem, since the cathode end of their diodes is "quiet." However, the (positive-to-positive) Buck and the Forward converter do have swinging cathodes (tabs), so we should be careful when chassis-mounting their diodes.*

- *Rod inductors* are often used in LC post-filtering stages on the output. Because of their open magnetic structure, they have been called "EMI cannons." But they are nevertheless still popular because of their low cost, and also the low "real estate" they need. Some tricks have therefore been developed to control their ill-effects. They should be placed vertically (as they normally are). Then, if two such rods are being used on a given output, we should wind the two rods identically, but reverse the current flow in one of them, as compared to the other (by suitable modification of the PCB) (see Figure 17.4). Looking from the top, one rod should be carrying current clockwise and the other anticlockwise. This helps redirect the flux from one, back into the other ("U-turn"). In that way, much less "EMI-spilling" occurs.

EMI from Diodes

Here we list some of the things to keep in mind and try out, as regards diodes:

- Diodes are a potent source of low- to high-frequency noise. Slow diodes (like those in a typical input bridge) can also contribute significant wideband noise.
- Input bridges which use ultrafast diodes are available, and their vendors claim significant reduction in EMI. But in practice they don't seem to provide much advantage. They also typically have much lower input surge current ratings. In fact, in a front-end position, any component always needs to be able to handle a lot of stress (if not abuse), such as the inrush stresses occurring during power-up at high line.
- To minimize EMI, ultrafast diodes should be selected on the basis of softer reverse recovery characteristics. For medium- to high-power converters, RC snubbers are also often placed across these diodes (at the expense of some efficiency). In low-voltage applications, Schottky diodes are often used. Though these diodes have no reverse recovery time in principle, their body capacitance can be relatively high, and can end up resonating with PCB trace inductances. So an RC snubber is also often helpful for Schottkys. Note that if any diode has been fully recovered (i.e., zero current) before the

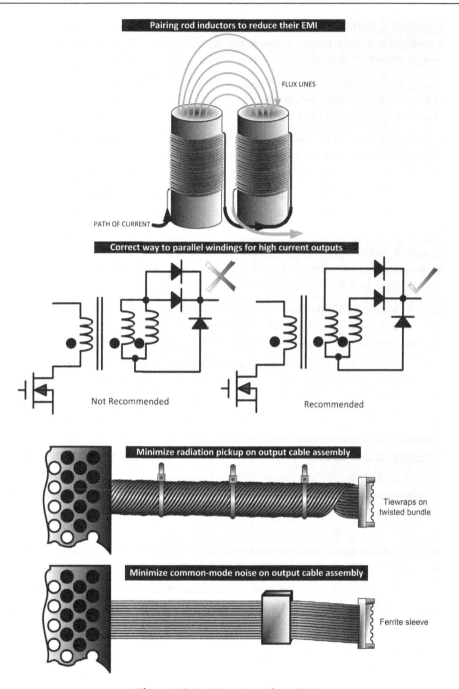

Figure 17.4: Ways to reduce EMI.

voltage across it starts to swing, there is no reverse recovery current. In that case, diodes really don't have to be "super-super-fast." In fact, many engineers have reported much lower EMI by choosing slower diodes for snubbers/clamps. A popular choice for snubber applications is the soft-recovery fast diode BYV26C (or BYM26C for medium power) from NXP (formerly Philips).

- It is advisable to have the FET switch roughly two to three times *slower* than the reverse recovery time of the catch diode — to avoid shoot-through currents — which will produce strong *H*-fields (in addition to causing dissipation). Therefore, it is not uncommon to intentionally degrade the FET switching speed by adding a resistor (typically $10-100\,\Omega$ in off-line applications) in series with the Gate — maybe with a diode across the Gate resistor so as to leave the turn-off speed unaffected (for efficiency reasons).

- Small capacitors may often be placed across the FET (Drain-to-Source). But this can create a lot of dissipation inside the FET, since every cycle the capacitor energy is dumped into the switch.

- Ultrafast diodes also exhibit high forward-voltage spikes at turn-on. So momentarily, the diode forward voltage may be $5-10\,V$ (rather than the expected 1 V or so). Usually, the snappier the reverse recovery, the worse is the forward spike too. Therefore, at FET turn-off, the diodes become strong *E*-field sources (voltage spikes), whereas at FET turn-on, the diodes will generate strong *H*-fields (current spikes). A small RC snubber across the diode will help control the forward-voltage spike.

- In integrated switchers, access to the Gate of the FET may not be available. In that case, the turn-on transition can be slowed by inserting a resistor of about $10-50\,\Omega$ in series with the bootstrap capacitor. The bootstrap capacitor is in effect the voltage source for the internal floating driver stage. At turn-on, it is asked to provide the high-current spike required to charge up the Gate capacitance of the FET. So, a resistor placed in series with this bootstrap capacitor limits the Gate charging current somewhat, and thereby slows the turn-on.

- To control EMI, ferrite beads (preferably of lossy nickel-zinc material) are sometimes placed in series with catch diodes (often slipped on to their leads), such as at the output diode of a typical off-line flyback. However, these beads must be very small, as they can have a significant effect on the efficiency of the power supply.

> *Note: In multi-output off-line flyback converters, we may find larger beads (possibly with more than one turn, and made of the more common manganese–zinc ferrite) in series with the output diodes belonging to some of the auxiliary outputs (i.e., those not being directly regulated). But the purpose of these beads is not EMI suppression, but to block some of the voltseconds and thereby improve the "centering" of the outputs.*

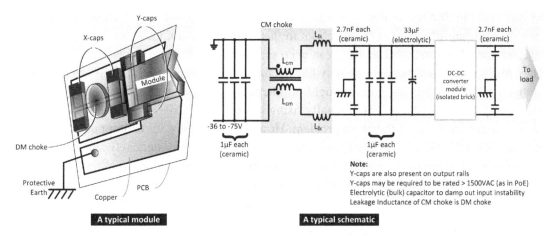

Figure 17.5: Typical EMI filter for DC–DC modules (Bricks).

- A comment about split/sandwich windings. In general, the Primary winding may be broken up into two windings, which are then positioned on either side of the Secondary winding — so as to reduce leakage inductance in flybacks, and proximity effect losses in Forward converters. This is acceptable for EMI provided the two-split winding sections are *in series*. In general, putting windings in parallel is *not* a good idea (especially from the EMI point of view). In high-current power supplies, the Secondary winding is also sometimes broken up into two windings (or foils). The intention is usually to increase the current handling capability (see Figure 17.5). But these split Secondaries are also usually placed *physically apart*, on either side of the Primary winding. However, in paralleled windings, the two supposedly "equal" sections are actually always magnetically slightly *different* — because of their different physical positions inside the transformer. Plus, their DCR is also just a little different (different lengths), creating the possibility of an *internal current loop*. The designer may be completely unaware of the current loop, except for severe tell-tale ringing present on the voltage waveform and a "mysteriously" bad EMI scan. There could also be a lot of unexpected heating. So, if paralleling is *really needed*, it is better to use the scheme shown in Figure 17.4 on the right-hand side. Here the forward drops of the two diodes help "ballast" the windings, and this also helps "iron out" any inequality between the two halves.

Are We Going to Fail the Radiation Test?

Most of the smaller companies cannot afford a precompliance setup for radiated emission tests. However, a few of them have a fairly good idea beforehand, whether they are going to be successful in that test or not — just by looking hard at the conducted EMI scan. What they do is to carefully look at the spectrum in the third region of CISPR-22. This is the flat region from 5 MHz to 30 MHz. They can even scan higher to higher frequencies, if possible. They realize that even though they may have achieved compliance with the conducted limits in this third region, it is not good enough! So, they look at the overall shape of the plot in this region. If they find that it is gradually rising toward the 30-MHz end, they are quite confident that they have a radiation problem. However, if the plot starts drooping, or remains generally flat as 30 MHz approaches, they are likely to immediately submit the prototype to a lab for the formal radiated limit compliance certification. In other words, one can actually "see" the energy level in the 5–30-MHz region. If there is an unexpected amount of conducted noise energy in this region, radiation can't be too far off either!

A quick diagnostic test for understanding a particular high-frequency-conducted EMI problem is to twist the *output* cables of the power supply tightly together (along with their respective return wires). This induces *field cancellation* (also called flux *containment*), thus reducing radiation effects related to the output cables (if present) (see Figure 17.4). If the conducted EMI scan really improves by twisting, we may have a radiation problem — either from the enclosure or the output cables themselves, or from both.

The above-described twisting procedure was actually implemented in full production on a particular high-volume commercial design. A few tie-wraps were used to hold the bunch of wires tightly together along the twisted position. This happened to be a last-ditch effort to avoid costly last-minute redesign just before full production. This "twist-and-tie-wrap" technique is admittedly not very practicable or desirable in production. But it is cheap. Note that a ferrite sleeve, slipped over the entire output cable bunch, was also found to be working well. But it was disqualified simply because it was far more expensive than three or four tie-wraps! However, it is interesting to note here that though a ferrite sleeve may look like a radiation shield and even produce almost similar results as twisting the cables, it actually works by reducing the common-mode noise currents themselves, not merely by "shielding" the EMI due to these tiny currents. Twisting, on the other hand, tries to cancel the fields of adjacent wires (with their returns). Looking back, in this particular case, the root cause was that there was obviously a significant amount of common-mode noise already present on the output, which was causing the output cables to radiate. The radiation was thereafter being picked up by the input cables, leading to a failed conducted EMI test.

Part 2: Modules and Input Instability

There are certain things we may do unintentionally at the input of the converter that can have a major impact on the performance of the EMI filter, and also the converter itself. If we don't know the rules of the game, we can end up saturating our filter chokes and even inducing basic converter instability.

Practical Line Filters in DC–DC Converter Modules

See Figure 17.5c for an example of how EMI suppression techniques are applied to DC–DC converters. We have shown an industry standard isolated "brick" along with its external EMI filtering. The input to this particular module is a coarsely regulated "−48VDC" or "−60VDC" bus, forming part of a distributed power architecture for a data/telecom network. Its output is isolated and regulated (e.g., 3.3 V/50 A, 12 V/10 A, or 48 V/2 A, and so on). The −48VDC input is usually derived from an off-line telecom power supply (called a "rectifier").

See how the traces are laid out in the module's external EMI filter. Note, in particular, the placement of the Y-caps. We should also keep in mind that one of the most effective methods of suppressing EMI, especially in board-mounted DC–DC converters, is a good ground plane. On a multilayer board, best results are usually obtained by having this plane be the internal layer just below the top (power) component side. Up to 20-dB reduction in noise is possible.

Note: As per typical safety regulations, voltages below 60 VDC are generally considered non-hazardous and therefore not subject to the isolation/earthing requirements described earlier. The Y-caps on the output rail can be just 100-V standard caps or less. However, in distributed power networks, such as Power over Ethernet, we require 1,500-VAC isolation from the power cable network to the enclosure to protect the user from spikes picked up on the long cables. So, the Y-caps shown in the figure may be required to be (standard) 2-kV-rated components. However, since there is no AC, we are not governed by AC leakage safety restrictions. Therefore, very large Y-caps can be used if desired.

Note: For protection against ESD (electrostatic discharge) upsets, 0.01-µF caps between the terminal block contacts and Earth are often also included. These are essentially Y-caps. But note that there have been cases, particularly when these caps were ordinary 50-V multilayer ceramic ("MLCs"), which got destroyed during the course of an ESD test — simply because they got charged up to excessive voltages! Therefore, these capacitors, and any other Y-caps present, must be evaluated under abnormal but likely disturbances too. Eventually, we may need to increase the capacitance and/or the voltage rating and/or size of the caps just to protect them from/against overcharging.

Since around 1971 the phenomenon of "input oscillations" or "input instability" has received quite a lot of attention. It has been shown that instability can occur if the output impedance of the filter is not within a certain "safe" window, as related to the input impedance of the converter (we are talking about the impedances presented to the power flow now — not to the CM or DM noise). So, with the modern trend of low-impedance "all-ceramic" solutions in DC–DC converters, the possibility of this particular type of instability is becoming more and more real.

One of the easiest ways to see the impact of the *negative input impedance* of a typical converter is to set it up with *only* ceramic input capacitors (about $10\,\mu F$ or less) — and then do a "hard power-up." In this type of power-up test, the dV/dt of the applied input is kept intentionally very high. On the bench, this can be done by simply slamming the banana plug from the input of the converter into the output terminals of a (low-impedance/high-current) lab DC power supply. Then, if we monitor the input (supply) pin of the converter with a digital oscilloscope (triggered correctly, and in one-shot acquisition mode), we will see an initial overshoot — that can be as high as 1.5–2.5 times the supposed DC voltage level (as set on the lab supply). Note that if the input capacitance is large enough (beyond a certain value), the dV/dt (and overshoot) gets automatically reduced, due to the higher charging current required for this input capacitor. On the other hand, if the ceramic capacitor is replaced by an aluminum electrolytic (even with a lower capacitance), the overshoot is dramatically reduced. Tantalum capacitors also produce overshoots under hard power-up, but these are less pronounced than with ceramic.

> *Note: We should remember that in any case, it's never a very good idea to use tantalums at the* input *of any converter — due to tantalum's inherent surge current limitations. However, if for some reason, tantalums must be used at the input (in any topology), or at the outputs (for a Boost or Buck-Boost), we must ensure that they are "100% surge-tested" by their vendors. And even for such surge-tested tantalum capacitors, it is recommended that the maximum voltage applied across them in our application be less than half their voltage rating, that is, a voltage derating of 50%. This was also discussed in Chapter 6.*

We see that it is possible to damage a DC–DC converter, which uses only small-value ceramic capacitors at its input — more so when we already happen to be operating rather close to its maximum input voltage rating.

The reader should note that in Figure 17.5, we have placed an *electrolytic capacitor in parallel to the ceramic input capacitors* — for the purpose of damping out "input instability." This needs further explaining. To understand the underlying causes associated with this phenomenon, we need to start with the well-known Buck converter equations and see what happens if we (hypothetically) "jiggle" the duty cycle, just a little bit, around its steady-state value. Note that in a practical situation, this could happen very easily under normal line or load transients. Therefore, expressing the input voltage and the input current as a function of duty cycle, we get (for a Buck)

$$V_{IN}(D) = \frac{V_O}{D}$$

$$I_{IN}(D) = I_O \times D$$

So,

$$dV_{IN} = -\frac{V_O}{D^2} dD \quad \text{and} \quad dI_{IN} = I_O dD$$

Dividing the above equations, we get (for a Buck converter)

$$\frac{dV_{IN}}{dI_{IN}} = -\frac{V_O}{I_O \times D^2}$$

V/I, above, is resistance. It is the *incremental resistance* at the input. Let us call it "R_{IN}." So, for a Buck converter, the incremental resistance in ohms is

$$R_{IN} = -\frac{R_L}{D^2}$$

Here R_L is the load resistance (ohms), and is assumed constant. Note that both the input voltage and the input current always have *positive* values in a (positive-to-positive) Buck converter. Therefore, the ratio V_{IN}/I_{IN} is also certainly a positive quantity. It's only the *relative change* that is in opposite directions — hence the minus sign in the equation above. Mathematics aside, all this means is that since the input of a converter is a constant power input ($P_O \approx P_{IN}$), if input voltage falls, the current increases, just the opposite of a pure resistance.

Example:

What is the input resistance of a 3.3-V/50-V brick, with an input range of 36–75 V?

Output power is $3.3 \times 50 = 165$ W. R_L is $3.3/50 = 0.066\,\Omega$. Duty cycle is $3.3/36 = 0.092$ at 36-V input. So, R_{IN} is $-0.066/(0.092)^2 = -7.8\,\Omega$. In terms of decibels, we get $-20 \times \log (7.8) \cong -18$ dB Ω (do not try to take log of a negative number!). A similar calculation at 75-V input gives -31 dB Ω. We will see that this means that *input instability is more likely to occur at low input voltages.*

What is it about the interaction of the impedances at the filter–converter interface that causes this instability? Let us see what is really happening as we jiggle the input to the filter (V_{IN}) in Figure 17.6.

V_{INC} is the voltage that appears at the terminals of the converter. The filter impedance and the converter impedance form a voltage divider.

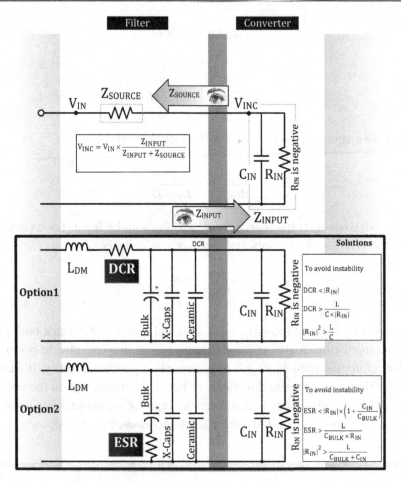

Figure 17.6: Input interaction and two possible solutions to increase damping.

$$V_{INC} = V_{IN} \times \frac{Z_{INPUT}}{Z_{INPUT} + Z_{SOURCE}}$$

Using a regular voltage divider, there would be no problem. We use such a divider to set the voltage on the feedback pin of our controller. In that case, if we raise V_{IN}, we expect V_{INC} to rise too — *provided both the resistors of the divider are "normal."* But in our case, one of them, Z_{INPUT}, is not "normal" — it is a *negative* resistance. So what really happens is if we increase V_{IN}, then V_{INC} falls! The control loop of the converter will "think" that the input has *fallen* rather than increased, so it will respond incorrectly to the change. And isn't that the usual recipe for output oscillations?

Note that above we are talking not about DC values, but changes (increments), that is, AC values. So, we can identify that the problem really starts *when V_{INC} is negative.* Because a

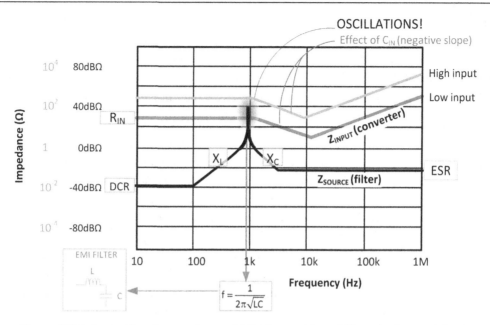

Figure 17.7: Input filter interaction and Middlebrook's stability criterion explained.

negative V_{INC} simply means that V_{INC} is moving in an *opposite* direction to V_{IN}, confusing the feedback loop. Note that in the equation for V_{INC} above, the numerator is already negative. So, the only way to get a positive sign for V_{INC} is to make the denominator negative too. This means *the basic criterion for avoiding input instability is*

$$Z_{SOURCE} < |Z_{INPUT}|$$

Now, in reality, the input impedance of the converter is *frequency dependent* (R_{IN} was just the low-frequency value of Z_{INPUT}). In the more detailed converter model, a parallel capacitance C_{IN} (see Figure 17.6) appears across the input of the converter, mainly due to the *output filter components of the converter being reflected into the input*. This causes the downward slope in Figure 17.7. It increases the chance of input instability. In Figure 17.7, we have shown a typical input impedance plot with respect to frequency. Note that only the magnitude of the converter input impedance has been displayed, primarily because the *y*-axis is in log scale, and we know that log scales cannot be negative.

Z_{SOURCE} (the output impedance of the filter) is also changing with frequency. Looking into the output terminals of the filter (from converter), we see basically a simple parallel LC filter stage. Therefore, Z_{SOURCE} has the shape indicated in Figure 17.7.

The stability criterion means that we are demanding that the output impedance of the filter must be always less than the input impedance of the converter for any frequency. But what happens if the LC filter has *insufficient damping* and therefore has a resonance peak? This is the oval highlighted problem area in Figure 17.7, and we can see that in this region we are violating the basic stability criterion. This resonant peak needs to be suppressed. Therefore, in addition to the basic stability criterion, a follow-up criterion must be added to ensure that at the LC corner frequency, the *LC filter peak is properly damped out* ($Q = 1$). This applies to the EMI filter design procedures described in *Chapter 18* too.

For ensuring damping, we could simply add some more resistance to the choke (DCR) as shown in Figure 17.6. But that is not a very good idea since the entire operating current also passes through this choke, and the overall efficiency would suffer. Instead, it is preferable to add a slight resistance (ESR) to the capacitor as also shown in Figure 17.6. We know that any capacitor in steady-state blocks any DC voltage completely. So, the input capacitor sees only the AC component of the input current flowing into the switching FET. This therefore correspondingly reduces the dissipation required to achieve a given target of damping. However, we also need to maintain good decoupling at the input of the converter (to keep its control sections from getting affected, as also to suppress EMI). Therefore, the usual commercially implemented solution for such bricks is to place an additional *high-ESR capacitor in parallel to the existing low-ESR decoupling capacitors*. The stability conditions for each option are also presented in Figure 17.6. One of them is

$$C_{BULK} \gg C_{IN}$$

where C_{IN} is *total effective capacitance at the input terminals of the converter* (including an actual input cap, any ceramic input capacitors, any X-caps, supply decoupling caps, and so on). C_{IN} is typically a few μF, but without elaborate modeling of the converter, or some type of measurement, its value may be unknown to most designers since it also depends on the output capacitors. But generally speaking, if C_{BULK} is chosen to be much larger than the discrete low-ESR input cap, it effectively "swamps" out the effect of C_{IN}, and so the system is stable. The rule of thumb is that C_{BULK} *should be four to five times the total effective low-ESR input capacitance present at the input to the converter*, that is, before C_{BULK} was added.

CHAPTER 18

The Math Behind the Electromagnetic Puzzle

Fourier Series in Power Supplies

In previous chapters we have seen that differential and common mode noise (DM and CM) emissions are basically either voltage or current sources with arbitrary waveshapes. What they have in common is that they are repetitive, and therefore involve the basic switching frequency of the converter. When we design a filter to suppress these emissions, we have to be conscious of the fact that the efficacy of a filter is best characterized in terms of the impedance it presents to a *sine wave of a given frequency* passing through it. In other words, to know the efficacy of the filter to CM and DM emissions, it is best to "decompose" the DM and CM waveforms into a sum of infinite sine wave components of varying amplitude and phase. As per the litmus test, when all the components are summed up, we should get back the original waveshape we started off with. The procedure of decomposition and reconstruction of an arbitrary waveform, either current or voltage, into sine wave components that are multiples of a basic (fundamental) frequency, is called Fourier series analysis. In general, we start with a "fundamental frequency" or "first harmonic" (invariably the switching frequency of the converter in our case), and add components of varying (and invariably diminishing) amplitudes. The frequencies are multiples of the fundamental frequency f, that is, f, $2f$, $3f$, $4f$, and so on. The nth harmonic corresponds to a frequency nf.

In Figure 18.1, we have included a short course to refresh our collective memory on Fourier series. We remember in school we were used to dealing with Fourier series in terms of a certain angle "θ" expressed in radians (radians being a dimensionless quantity). All the functions we dealt with were based on θ, and repeated every 2π radians. We have a very similar situation in power supplies, where all voltages and currents are functions of "t" (time), repeating every time period (T) in steady state. We realize we can replace θ in all our familiar Fourier series expansions with the dimensionless quantity "$2\pi t/T$." This way, we will achieve migration of all the well-known Fourier series textbook equations to the world of power supplies. This is the required substitution to remember as explained in Figure 18.1

$$\theta \to \frac{2\pi}{T}t$$

Switching Power Supplies A–Z. DOI: 10.1016/B978-0-12-386533-5.00018-8
663

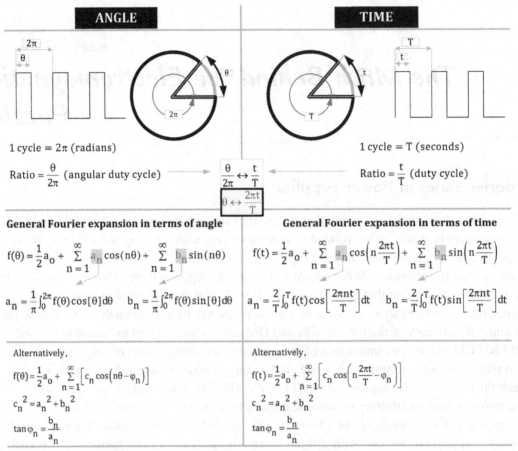

Figure 18.1: Applying Fourier series in power supplies.

Note that angle θ itself may be rather confusingly called "x" or "2α" in textbooks, but nevertheless, we recognize it is an angle (in radians).

The Rectangular Wave

We use the above-mentioned mapping procedure to analyze a rectangular waveform of amplitude "A" in terms of its Fourier series. Assume we are, for example, talking about the rectangular voltage "Vds" (same as "V_{DS}") across the switching FET. The results are shown in Figure 18.2. We have first analyzed the wave by assuming it is "ideal," that is, with zero

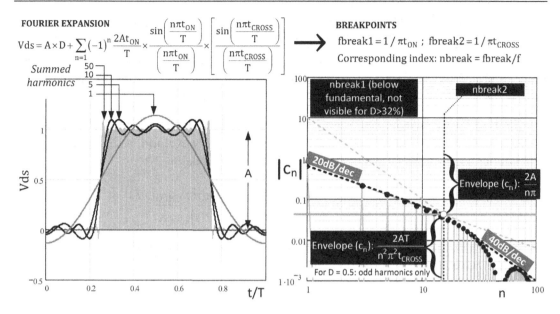

Figure 18.2: Fourier series in power supplies and plotting the Fourier coefficients.

crossover (switch transition) time. We then get the following Fourier coefficients (with $2\pi t/T$ replacing θ, and pulse width designated as t_{ON}):

$$|c_n| = 2A \times \frac{t_{ON}}{T} \times \left|\frac{\sin(n\pi t_{ON}/T)}{(n\pi t_{ON}/T)}\right| \equiv 2A \times \left|\frac{\sin(n\pi D)}{n\pi}\right| \quad \text{or} \quad |c_n| \equiv 2AD \times \left|\frac{\sin\gamma}{\gamma}\right|$$

where $\gamma = n\pi t_{ON}/T$ and $|c_n|$ represents the magnitude of the amplitudes of the nth harmonic. We are interested in magnitudes because a conducted EMI scan doesn't care about the signs. However, we do need the signs to correctly reconstruct the original waveform. In general, we can sum only a few terms of the Fourier series to get fairly close to the original waveform. Summed up to "$n$max" harmonics, we get Vds (as a function of time)

$$Vds = AD + \sum_{n=1}^{n\text{max}}\left[c_n \cos\left(n\frac{2\pi t}{T}\right)\right]$$

Note that the first term is AD (amplitude times duty cycle), is just the average (DC) value of the waveform over the full time period T. We should remember that the average value is always the first term (the "$n = 0$" term) in any Fourier expansions, and also that it is in fact irrelevant to EMI calculations since the CISPR-22 range starts at 150 kHz. However, besides the signs, the DC term is also required to correctly reconstruct the original waveform, as we have done in Figure 18.2 with a unity amplitude wave of 50% duty cycle as an example. We have summed over 1, 5, 10, 50, and finally 10,000 harmonics (not shown). We see how we move from an offset sine wave (when we sum only the $n = 0$ and $n = 1$ terms) to a perfectly rectangular waveform (with a very large n, we get the exact waveform we started with).

However, so far we have assumed zero crossover time. If we have a *real-world situation* with nonzero crossover times, then, instead of a rectangular waveform, we get a trapezoid. The Fourier coefficients get modified by an additional term below

$$|c_n| = 2A \times \frac{t_{ON}}{T} \times \left| \frac{\sin(n\pi t_{ON}/T)}{n\pi t_{ON}/T} \right| \times \left| \frac{\sin(n\pi t_{CROSS}/T)}{n\pi t_{CROSS}/T} \right|$$

We now have a product of two functions of the type $\sin(x)/x$, instead of one. We need to understand what the properties of this function are.

The Sinc Function

The function $\sin(\gamma)/\gamma$ is called the sinc function. Its key properties are shown in Figure 18.3. On a log versus log plot (lowermost plot), it appears "flat-topped" at lower frequencies, with a unity value initially. In reality it is actually sloping rather gently downward, and at $\gamma = 1$ its value is $\sin(1) = 0.84$. We call that a "breakpoint" because right after that, the slope changes dramatically. At higher frequencies it falls off with a slope of -20 dB/decade, a straight line on a log versus log plot. Note that the sinc function can be considered "clamped" to 1, for all frequencies less than the break frequency.

When we have two breakpoints, as in the equation for the real-world c_n's above (one for t_{ON}, the other for t_{CROSS}) we eventually see a downward slope of -40 dB/decade (obvious in the lowermost right-hand plot of Figure 18.2). Each breakpoint has clearly contributed -20 dB/decade to that slope. The actual breakpoints in terms of frequency are obtained by setting γ to 1. We thus get

$$f\text{break}1 = \frac{1}{\pi \times t_{ON}}; \qquad f\text{break}2 = \frac{1}{\pi \times t_{CROSS}}$$

In terms of the harmonic index "n" — imagining "n" varies smoothly, rather than just having integral values

$$n\text{break}1 = \frac{1}{\pi \times t_{ON} \times f}; \qquad n\text{break}2 = \frac{1}{\pi \times t_{CROSS} \times f}$$

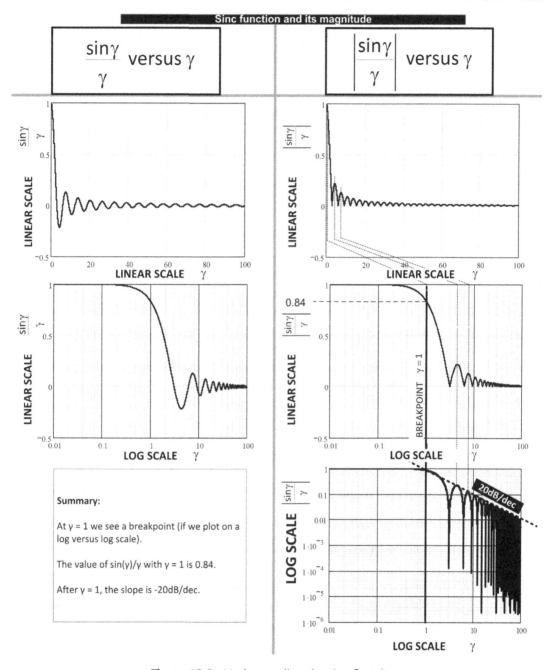

Figure 18.3: Understanding the sinc function.

Note that *n*break1 is less than 1 for duty cycles greater than 32%. Which means for all practical purposes it then becomes "invisible" since we start off with $n = 1$.

$$1 \geq \frac{1}{\pi \times t_{ON} \times f} = \frac{1}{\pi D}, \quad \text{that is} \quad D \geq \frac{1}{\pi} = 0.32$$

In other words, the *duty cycle needs to be below 0.32 for the first breakpoint to be "visible"* (in terms of affecting any amplitudes of the Fourier terms). Otherwise, it occurs at frequencies below the fundamental frequency, and therefore can't determine any of the actual harmonic amplitudes (i.e., $n \geq 1$). Of course, the *effect* of this first breakpoint is felt at almost *all* frequencies, since it imparts a downward slope of -20 dB/decade for all frequencies exceeding it (and likewise, clamps harmonic amplitudes below it, *if any*).

The Envelope of the Fourier Amplitudes

For designing the conducted EMI filters, we are not concerned with the granularity of the actual harmonic amplitudes. What matters to us is their *envelope*. That is what we really try to adjust and try to keep within the EMI limit lines, because even a single point in excess of the limits represents a failed EMI test. We see from Figure 18.2 ($D = 0.5$) and from Figure 18.4 ($D = 0.2$) that there is one breakpoint which is dependent on the *switching frequency*. After that breakpoint, the envelope of the amplitudes falls off as per $1/n$, which is equivalent to a slope of -20 dB/decade. As per the equations in Figure 18.2, we have described the envelope as

$$c\_envelope_n = \frac{2A}{n\pi} \quad \text{(for zero crossover time, or between first and second breakpoints)}$$

This equation is easy to understand since we know that the actual harmonic amplitudes (for a square waveform) are

$$|c_n| = 2A \times \frac{t_{ON}}{T} \times \frac{\sin(n\pi t_{ON}/T)}{n\pi t_{ON}/T} = 2A \times \frac{\sin(n\pi D)}{n\pi}$$

Since for all angles (and all duty cycles), the maximum value of the sine term (in the rectangle border above) is unity, the envelope must be described by "$2A/n\pi$." And that is the envelope equation we have provided in Figure 18.2.

However, we see that there is a second breakpoint, after which the envelope falls off as per $1/n^2$, which is basically equivalent to -40 dB/decade. The location of this breakpoint depends on the crossover time. We can visualize the second breakpoint as adding another -20 dB/decade to the already falling -20 dB/decade curve, leading to a cumulative slope of $(-20) + (-20) = -40$ dB/decade. The envelope after the second breakpoint is

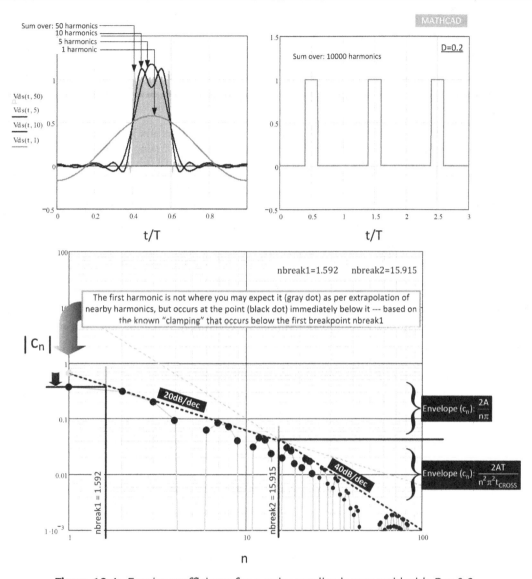

Figure 18.4: Fourier coefficients for a unity amplitude trapezoid with $D = 0.2$.

$$creal\_envelope_n = \frac{2AT}{n^2\pi^2 t_{CROSS}} \quad \text{(for non-zero crossover times, after second breakpoint)}$$

Note that the original equation for c_n included duty cycle, whereas the envelope equations do not (because we have "maxed out" the terms). So yes, the harmonic amplitudes do go up and down with respect to D, but the envelope remains fixed and independent of D, as we can see in Figures 18.2 and 18.4. *Note that D **does** enter the picture if it is below 32% as we see in the example below.*

Example 1

What are the amplitudes of the fundamental frequency (first harmonic), the second and the third harmonics of the V_{DS} waveform of a 100 kHz universal input (upper limit 270VAC) AC−DC flyback converter with 5 V output and turns ratio of 19?

From *Chapter 3* we know that V_{OR} = turns ratio $\times V_O = 19 \times 5 = 95$ V. The duty cycle at high line (270VAC, rectified 382VDC) is $D = V_{OR}/(V_{IN} + V_{OR}) = 0.2$. In Figure 18.4, we have plotted out the results of the Mathcad file originally used for Figure 18.2, but this time *performed for D = 0.2*. Note that we have assumed unit amplitude in these plots, knowing that everything will scale proportionally to the amplitude.

Note that if the converter is operating in "continuous conduction mode" (CCM) at high line (unlikely) we can assume a simple rectangular V_{DS} waveform. That assumption will give accurate results for conducted mode EMI calculations even if the converter is in DCM.

The height (amplitude) of the rectangular V_{DS} (same as Vds in this chapter) pulse is $V_{IN} + V_{OR} = 382 + 95 = 477$ V (ignoring the leakage inductance spike). From Figure 18.2 we know that the magnitude of the harmonic Fourier coefficients is

$$|c_n| = 2A \times \frac{t_{ON}}{T} \times \frac{\sin(n\pi t_{ON}/T)}{n\pi t_{ON}/T} = 2A \times \frac{\sin(n\pi D)}{n\pi}$$

We thus get three harmonic amplitudes based on the exact Fourier coefficient equations above

$$|c_1| = 178.5 \text{ V}; \quad |c_2| = 144.4 \text{ V}; \quad |c_3| = 96.3 \text{ V}$$

That is what we essentially plotted in Figure 18.4 *for unit amplitude*. We can confirm the graph matches the results of the calculations above, by scaling the numbers "eye-balled" from the graph (i.e., 0.38, 0.3, and 0.2 for the first three harmonics, i.e., the solid black dots), for the case of $A = 477$ V as follows:

$$|c_1| = 0.38 \times 477 = 181 \text{ V}; \quad |c_2| = 0.3 \times 477 = 143 \text{ V}; \quad |c_3| = 0.2 \times 477 = 95.4 \text{ V}$$

We see the graphical-based results are close enough to the accurate calculation above.

However, now, coming to the equation for the *envelope* (before the second breakpoint), we get

$$c\_envelope_n = \frac{2A}{n\pi}$$

Using this simplified equation, we get for the three harmonics

$$|c_1| = 303.7 \text{ V}; \quad |c_2| = 151.8 \text{ V}; \quad |c_3| = 101.2 \text{ V}$$

We see a close match with the terms $n = 2$ and $n = 3$, *but not for n = 1*. The actual harmonic amplitude is only about 180 V based on our exact and graphical calculations above, not 304 V. In other words, the envelope equation is now giving misleading results for the first (fundamental) harmonic. To avoid overdesign of the filter we should understand the reason for this carefully. It is explained in Figure 18.4 too. The basic reason is that since the duty cycle was lower than 32%, the first breakpoint is "visible" at an effective n of 1.592, and it therefore clamps the amplitude of the $n = 1$ term. In Figure 18.4, as per the envelope equation, we would have got the solid gray dot as the first harmonic amplitude (roughly between 0.6 and 0.7). In reality, we get the solid black dot (a little below 0.4).

However, we can actually rather cleverly, use the envelope equation itself to predict the amplitude of the first harmonic. This is shown as follows. The equivalent "breakpoint index" is

$$n\text{break1} = \frac{1}{\pi \times t_{ON} \times f} = \frac{1}{\pi \times D} = \frac{1}{\pi \times 0.2} = 1.592$$

Note that in effect we are now implicitly assuming that "n" can be a *continuum of values*, not just a range of integers. It helps us in the math that follows, but keep in mind it has no physical significance in terms of the Fourier series. At this point, the coordinate of the envelope is

$$c\_envelope_n = \frac{2A}{n\pi} = \frac{2 \times 477}{1.592 \times \pi} = 190.7 \text{ V}$$

For unit amplitude, the amplitude needs to be $190.7/477 = 0.4$. That is close enough to the actual calculation (we had gotten 0.374 for unit amplitude based on the graph in Figure 18.4 for the first harmonic). *Our conclusion is just from the two equations below, we can estimate the amplitude of the first harmonic quite accurately.* To re-emphasize: it is important to know the amplitude of the first harmonic most accurately, since our conducted EMI filter design is usually based on it. Here is a summary of what we need to do to quickly and generally (for any D) estimate, based on simple envelope calculations.

(a) Find $n\text{break1}$ using

$$n\text{break1} = \frac{1}{\pi \times D}$$

(b) If $n\text{break1}$ is less than 1 (duty cycle greater than 32%), the following equation gives the amplitude of the first harmonic

$$c\_estimated_1 = \frac{2A}{\pi}$$

(c) If $n$break1 is greater than 1 (duty cycle less than 32%), the following equation gives the amplitude of the first harmonic

$$c\_estimated_1 = \frac{2A}{n\text{break1} \times \pi}$$

Practical DM Filter Design

The trapezoid we have looked at so far is a voltage waveform. But it can equally well be a current waveform if we use the "flat-top approximation" (for large inductance). In DM filter design as applied to a Buck or a Buck-Boost (or their AC−DC equivalents, the Forward converter and the flyback converter), the input current switch current is of the shape we had previously indicated in Figure 16.2. So, the DM noise source is

$$V_{dm} = I_{SW} \times ESR$$

where I_{SW} here refers to the center of ramp. The Boost is not considered here as its DM filter design is almost trivial, since in effect, it has a natural LC filter on its input. In Table 18.1, we have provided the height of the switch current trapezoid for all the relevant topologies (with the flat-top approximation).

If there was no EMI filter present, the switching noise current received by the LISN would be

$$I_{LISN} = \frac{V_{dm}}{Z_{LISN\_dm}} = \frac{I_{SW} \times ESR}{100} \text{ Amperes}$$

(since the LISN has an impedance of 100 Ω for DM noise).

However, the analyzer itself only measures the noise across *one* of the two effective series 50 Ω resistors in the LISN. So, the measured level of noise is

$$V_{LISN\_DM\_NOFILTER} = I_{LISN} \times 50 = \frac{I_{SW} \times ESR}{2} \text{ Volts}$$

We have assumed that C_{BULK} is very large, and that it has no ESL, and also that its ESR is much less than 100 Ω. All reasonable assumptions of course.

Table 18.1: Switch Currents (Center of Ramp) for the Relevant Topologies.

Topology	I_{SW} (Switch Current)
Buck	I_O
Forward	$I_O \times (N_S/N_P)$
Buck-Boost	$I_O/(1-D)$
Flyback	$I_O \times (N_S/N_P)/(1-D)$

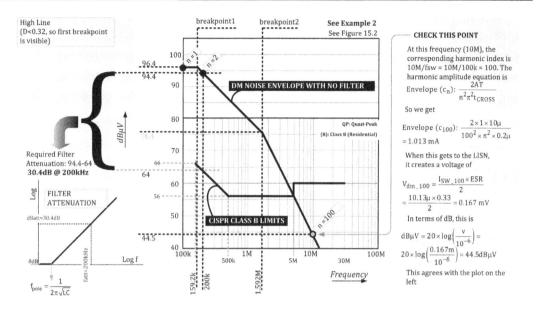

Figure 18.5: DM filter calculation for universal input flyback (see solved example).

Note that to be more accurate and avoid overdesign, we need to split I_{SW} into harmonic components as in the following example.

Example 2

What is the DM noise spectrum at 270VAC as measured at the LISN, for a 100 kHz universal input flyback with output of 5 V at 15.2 A? The transformer turns ratio is 19. Assume 200 ns rise and fall times. We are using an aluminum electrolytic bulk capacitor whose datasheet states that it has a capacitance of 270 μF, a dissipation factor (tangent of loss angle) of $\tan \delta = 0.15$ as measured at 120 Hz, and a frequency multiplier factor of 1.5 at high frequencies.

See Figure 18.5 for this entire example. Also read the section in *Chapter 16*, titled "The Road to Cost-Effective Filter Design."

ESR Estimate

The ESR is to be first computed at the 120 Hz test frequency. By definition

$$\text{ESR}_{120} = \frac{\tan \delta}{2\pi f \times C} = \frac{0.15 \times 10^6}{2 \times 3.142 \times 120 \times 270} = 0.74 \ \Omega$$

At a high frequency, the ripple current is allowed to increase by the frequency multiplier of 1.5. If the ESR had not changed, the heating, that is, I^2R would have gone up by the factor $1.5 \times 1.5 = 2.25$. Clearly, the dissipation has remained the same (which is the reason for allowing the current to increase). This implies that the ESR falls by the factor $1/2.25 = 0.44$ at high frequencies. Therefore, for our purpose, the high-frequency ESR value to use is

$$\text{ESR} = \frac{0.74}{1.5^2} = 0.33 \ \Omega$$

DM Calculations at High Line

See Figure 18.5 for this entire calculation.

(a) **Duty Cycle and Current Trapezoid**

The maximum peak rectified voltage is $270 \times 1.414 = 382$ V. The V_{OR} is by definition $5 \ V \times 19 = 95$ V. The lowest duty cycle (highest input) is

$$D = \frac{V_{OR}}{V_{IN} + V_{OR}} = \frac{95}{382 + 95} = 0.2$$

Load current of 15.2 A translates to the following switch current pedestal based on Table 18.1.

$$I_{SW} = \frac{N_S}{N_P} \times \frac{I_O}{1 - D} = \frac{1}{20} \times \frac{15.2}{1 - 0.2} = 1 \ A$$

(b) **Breakpoints**

As in Example 1, we see that the index and corresponding frequency breakpoints are

$$nbreak1 = \frac{1}{\pi \times D} = \frac{1}{\pi \times 0.2} = 1.592$$

$$nbreak2 = \frac{1}{\pi \times t_{CROSS} \times f} = \frac{1}{\pi \times 0.2\mu \times 100\,k} = 15.92$$

$$fbreak1 = \frac{1}{\pi \times D} \times f = \frac{1}{\pi \times 0.2} \times 100\,k = 159.2\,k$$

$$fbreak2 = \frac{1}{\pi \times t_{CROSS}} = \frac{1}{\pi \times 0.2\,\mu} = 1.592\,M$$

We see that $f$break1 exceeds the fundamental frequency of 100 kHz, so its amplitude will get clamped. We use the envelope equations previously provided, to find out the exact amplitude of the first harmonic. It is

$$c\_estimated_1 = \frac{2A}{nbreak1 \times \pi} = \frac{2 \times 1}{1.592 \times \pi} = 0.4$$

The second Fourier harmonic (based on envelope) is at 200 kHz and of amplitude

$$c\_estimated_2 = \frac{2A}{2 \times \pi} = \frac{2 \times 1}{2 \times \pi} = 0.32$$

(c) **EMI Spectrum with no Filter**

Above, we have calculated the harmonic current amplitudes based on an envelope estimate. When these current harmonics reach the LISN, still assuming no EMI filter, we get the following voltages:

$$V_{dm\_1} = \frac{I_{SW\_1} \times ESR}{2} = \frac{0.4 \times 0.33}{2} = 0.066 \text{ V} \Rightarrow 20 \log(0.066/10^{-6}) = 96.4 \text{ dB } \mu V$$

$$V_{dm\_2} = \frac{I_{SW\_2} \times ESR}{2} = \frac{0.32 \times 0.33}{2} = 0.053 \text{ V} \Rightarrow 20 \log(0.053/10^{-6}) = 94.4 \text{ dB } \mu V$$

(d) **Required Filter Attenuation**

From the equation in Figure 15.3, over the range 150–500 kHz, the equation for the quasi-peak limit of CISPR-22 Class B is

$$dB \text{ } \mu V\_QP = -20 \times \log(f_{MHz}) + 50 \quad \text{(almost exact)}$$

At the second harmonic of the no-filter spectrum (200 kHz), this limit is

$$dB \text{ } \mu V\_QP = -20 \times \log(0.2) + 50 = 64 \text{ dB } \mu V$$

So, looking at Figure 18.5, we get the required filter attenuation as 94.4 − 64 = 30.4 dB at 200 kHz. We are ignoring the fundamental since it lies outside the CISPR range.

(e) **Calculating Filter Components at Stated Line Condition**

We therefore need to pick a low-pass LC filter with an appropriate corner frequency ("pole" in Figure 18.5) that provides this attenuation. For example, if we are using a one-stage LC low-pass filter, we know it has an attenuation characteristic of about 40 dB/decade above its corner frequency (i.e., $1/2\pi\sqrt{(LC)}$). So, the equation of the required response is (where "att" stands for "with attenuation")

$$slope = \frac{dB_{att}}{\log f_{att} - \log f_{pole}} \Rightarrow \frac{dB_{att}}{slope} = \log f_{att} - \log f_{pole}$$

$$\log f_{pole} = \log f_{att} - \frac{dB_{att}}{slope} \Rightarrow f_{pole} = 10^{[\log f_{att} - (dB_{att}/slope)]}$$

Therefore, solving we get

$$f_{pole} = 10^{[\log f_{att} - (dB_{att}/slope)]} = 10^{[\log 200 \text{ k} - (30.4/40)]} = 34.8 \text{ kHz}$$

We therefore need a filter that has an LC of

$$LC = \left(\frac{1}{2\pi \times 34,800}\right)^2 = 2.1 \times 10^{-11} \text{ s}^2$$

For example, if we have picked the X-cap C3 in Figure 16.1 as 0.22 µF, the net DM inductance is L (twice the individual DM inductances in each line). We get

$$L \equiv 2L_{dm} = \frac{2.1 \times 10^{-11}}{0.22 \times 10^{-6}} = 95 \ \mu H \Rightarrow L_{dm} = 48 \ \mu H$$

Before we build this filter, we need to repeat all the above steps at low line too (90VAC). We get another inductance recommendation as seen below.

DM Calculations at Low Line

(a) **Duty Cycle and Current Trapezoid**

$$D = \frac{V_{OR}}{V_{IN} + V_{OR}} = \frac{95}{127 + 95} = 0.43$$

$$I_{SW} = \frac{N_S}{N_P} \times \frac{I_O}{1 - D} = \frac{1}{20} \times \frac{15.2}{1 - 0.43} = 1.33 \ A$$

(b) **Breakpoints**

$$nbreak1 = \frac{1}{\pi \times D} = \frac{1}{\pi \times 0.43} = 0.74$$

$$nbreak2 = \frac{1}{\pi \times t_{CROSS} \times f} = \frac{1}{\pi \times 0.2 \ \mu \times 100 \ k} = 15.92$$

$$f\,break1 = \frac{1}{\pi \times D} \times f = \frac{1}{\pi \times 0.43} \times 100 \ k = 74 \ k$$

$$f\,break2 = \frac{1}{\pi \times t_{CROSS}} = \frac{1}{\pi \times 0.2 \ \mu} = 1.592 \ M$$

We see that $f\,break1$ is below the fundamental frequency of 100 kHz. So, it will not affect any harmonic amplitudes. The amplitude of the first harmonic (at 100 kHz) is then just based on the simple equation below.

$$c\_estimated_1 = \frac{2A}{\pi} = \frac{2 \times 1.33}{\pi} = 0.85$$

The second Fourier harmonic (based on envelope) is at 200 kHz and of amplitude

$$c\_estimated_2 = \frac{2A}{2 \times \pi} = \frac{2 \times 1.33}{2 \times \pi} = 0.42$$

(c)   EMI Spectrum with no Filter

$$V_{dm\_1} = \frac{I_{SW\_1} \times ESR}{2} = \frac{0.85 \times 0.33}{2} = 0.14 \text{ V} \Rightarrow 20 \log(0.14/10^{-6}) = 102.9 \text{ dB } \mu V$$

$$V_{dm\_2} = \frac{I_{SW\_2} \times ESR}{2} = \frac{0.42 \times 0.33}{2} = 0.069 \text{ V} \Rightarrow 20 \log(0.069/10^{-6}) = 96.8 \text{ dB } \mu V$$

(d)   Required Filter Attenuation

At the second harmonic of the no-filter spectrum (200 kHz), the CISPR Class B limit is

$$\text{dB } \mu V\_QP = -20 \times \log(0.2) + 50 = 64 \text{ dB } \mu V$$

So, the required filter attenuation is $96.8 - 64 = 32.8$ dB at 200 kHz.

(e)   Calculating Filter Components at Stated Line Condition

$$f_{pole} = 10^{[\log f_{att} - (dB_{att}/slope)]} = 10^{[\log 200 \text{ k} - (32.8/40)]} = 30.3 \text{ kHz}$$

We therefore need a filter that has an LC of

$$LC = \left(\frac{1}{2\pi \times 30,300}\right)^2 = 2.76 \times 10^{-11} \text{ s}^2$$

For example, if we have picked the X-cap C3 in Figure 16.1 as 0.22 μF, the net DM inductance required is L (twice the individual DM inductances in each line). We get

$$L \equiv 2L_{dm} = \frac{2.76 \times 10^{-11}}{0.22 \times 10^{-6}} = 126 \text{ } \mu H \Rightarrow L_{dm} = 63 \text{ } \mu H$$

At high line we had 48 μH, at low line we get 63 μH on account of the higher currents at low line. Our final choice is the greater of the two, that is, 63 μH.

The chosen inductor must not saturate on account of the very high peak AC currents, and we should evaluate that once again, both at low line and at high line. To find the worst-case peak AC current in the filter, we need to look at the equations in *Chapter 14*. That helps us in picking the right size of DM inductor. We also need to ensure the quality factor of the inductor is lower than unity, to avoid ringing. For that we need to ensure enough DCR. If that makes it too lossy, we can consider raising the Q to 1.5 or 2, provided we have some additional headroom in the EMI scan.

Filter Safety Margin

We will need to include some safety margin that we have not included above. Typically, a margin of about 10–12 dB may be necessary. So far we have assumed no CM noise, whereas the CISPR limit lines constitute a mix of both CM and DM noise. Therefore, if we want to leave margin for that, and make the simplest assumption that the CM and DM noise contributions are equal, *we then need to lower the DM noise level to half of what we have so far allowed above.* Further, since $20 \times \log(2) = 6$ dB, this implies we need to leave a DM margin of 6 dB just for possible CM noise encroaching on our measurements. In addition, the certification lab or OEM may want us to demonstrate say, 3 dB margin, to ensure compliance over production lots. *So, in all we should plan for a margin of 10 dB.* We may therefore go back to "step d" and set the required attenuation to $30.4 + 10 = 40.4$ dB instead. Then all the steps are the same.

> **Note:** *We can ask — since the breakpoint associated with the rise and fall times didn't enter the picture here, does that mean that it doesn't matter how fast we turn-on and turn-off the FET? Yes, from the DM noise viewpoint it really doesn't matter (much). However, there are parasitics that we have ignored (chiefly the ESL and trace inductances). And since, unlike the ESR, these will produce frequency-dependent voltage spikes, it is in our interest not to keep the FET crossover (transition) times too small.*

> **Note:** *If our switching frequency was greater than 150k, then the first harmonic at low line, which is very high indeed (being unclamped by the first breakpoint), would have appeared in the CISPR range. That would have made the DM filter much bigger.*

Practical CM Filter Design

Having understood that the worst-case for DM filter design is at low line, on account of the higher currents, we can sense that the worst-case common mode noise will occur at high line since the voltages are highest; and as described in Figure 16.5 that will lead to the highest injected currents into the enclosure. Let us therefore do this calculation at high line only. Let us also ignore the second breakpoint as we have already seen it is too far away to affect the filter design per se. Note that we are assuming that by using the X-cap C1 in Figure 16.3, this truly is CM noise, not mixed mode (MM) noise.

Example 3

This is a continuation of Example 2. What is the corresponding CM noise spectrum assuming a mounting capacitance of 100 pF (to earth)?

CM Calculations at High Line

See Figure 18.6 for this entire example.

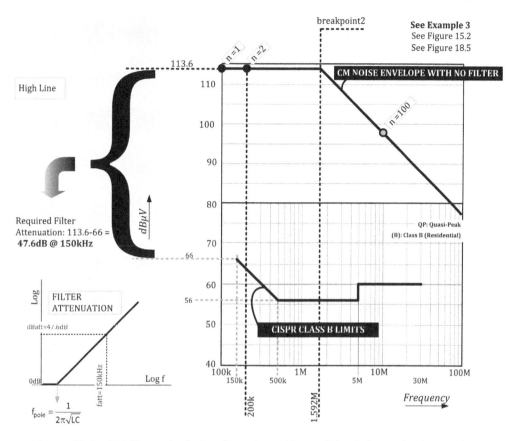

Figure 18.6: CM filter calculation for universal input flyback (see solved example).

(a) **Duty Cycle and Current Trapezoid**

The maximum peak rectified voltage is $270 \times 1.414 = 382$ V. The V_{OR} is by definition 5 V $\times 19 = 95$ V. The lowest duty cycle (highest input) is

$$D = \frac{V_{OR}}{V_{IN} + V_{OR}} = \frac{95}{382 + 95} = 0.2$$

The V_{DS} amplitude for a flyback is $V_{IN} + V_{OR} = 382 + 95 = 477$ V. If this were a single-ended Forward converter, we could use $2 \times V_{IN} = 764$ V. This is "A" (pulse amplitude) in the Fourier expansion.

(b) **Breakpoints**

As in Example 1, we see that the index and corresponding frequency breakpoints are as follows:

$$n\text{break}1 = \frac{1}{\pi \times D} = \frac{1}{\pi \times 0.2} = 1.592$$

$$f\text{break}1 = \frac{1}{\pi \times D} \times f = \frac{1}{\pi \times 0.2} \times 100\,\text{k} = 159.2\,\text{k}$$

$$c_1 = \frac{2A}{n\text{break}1 \times \pi} = \frac{2 \times 477}{1.592 \times \pi} = 190.75 \ \text{V}$$

The second Fourier harmonic (based on envelope) is at 200 kHz and of amplitude

$$c_2 = \frac{2A}{2 \times \pi} = \frac{2 \times 477}{2 \times \pi} = 151.8 \ \text{V}$$

(c)   EMI Spectrum with no Filter

Above, we have calculated the harmonic voltage amplitudes based on an envelope estimate. These generate harmonic currents in the line, based on the impedance in the path including the LISN.

$$I_{\text{cm\_1}} = \frac{c_1}{25 - (j/2\pi \times f\text{break}1 \times C_P)} = \frac{(2\pi \times f\text{break}1 \times C_P) \times c_1}{(50\pi \times f\text{break}1 \times C_P) - j}$$

Similarly for the second harmonic (frequency f_{SW}),

$$I_{\text{cm\_2}} = \frac{c_2}{25 - (j/(2 \times f_{\text{SW}} \times 2\pi C_P))} = \frac{4\pi \times f_{\text{SW}} \times C_P \times c_2}{(50\pi \times 2f_{\text{SW}} \times C_P) - j}$$

Using the fact that the magnitude of an imaginary number $c/(a - jb)$ equals $c/(a^2 + b^2)^{0.5}$, we get the following magnitudes of harmonic currents

$$\left| I_{\text{cm\_1}} \right| = \frac{2\pi \times f\text{break}1 \times C_P \times c_1}{\sqrt{(50\pi \times f\text{break}1 \times C_P)^2 + 1}} = \frac{2\pi \times 159.2 \times 10^3 \times 10^{-10} \times 190.75}{\sqrt{(50\pi \times 159.2 \times 10^3 \times 10^{-10})^2 + 1}} = 0.019 \ (\text{Amperes})$$

$$\left| I_{\text{cm\_2}} \right| = \frac{4\pi \times f_{\text{SW}} \times C_P \times c_2}{\sqrt{(100\pi \times f_{\text{SW}} \times C_P)^2 + 1}} = \frac{4\pi \times 100 \times 10^3 \times 10^{-10} \times 151.8}{\sqrt{(100\pi \times 100 \times 10^3 \times 10^{-10})^2 + 1}} = 0.019 \ (\text{Amperes})$$

Note the rather astonishing result: the amplitudes of the different harmonic CM currents are the same — there is no 20 dB/decade roll-off anymore.

These harmonic currents flow through the 25 Ω equivalent CM LISN impedance and get converted into voltage picked up by the spectrum analyzer as per $V_{\text{cm}} = 25 \times I_{\text{cm}}$. So we get

$$V_{\text{cm\_1,2}} = 25 \times I_{\text{cm\_1,2}} = 0.019 \times 25 = 0.477 \ \text{V} \Rightarrow 20 \ \log(0.477/10^{-6}) = 113.6 \ \text{dB} \ \mu\text{V}$$

Note that in *Chapter 16* we had stated that

$$V_{\text{cm}} = \frac{100 \times A \times C_P}{T}$$

Let us check if this is borne out. We get

$$V_{cm} = \frac{100 \times A \times C_P}{T} = \frac{100 \times 477 \times 10^{-10}}{10\mu} = 0.477 \text{ V (i.e., } 113.6 \text{ dB/}\mu\text{V as above)}$$

Summary: The CM noise spectrum is flat-topped (clamped) at exactly $100 \times A \times C_P/T$, right up till the second breakpoint (not discussed above). We know that all the harmonic amplitudes fall off by (an additional) 20 dB/decade after f break2 $= 1/\pi t_{CROSS}$, and therefore, so does V_{cm_n} (as per the property of the sinc function). We thus conclude our analysis with the CM envelope plot shown in Figure 18.6.

(d) Required Filter Attenuation

At 50 kHz, the CISPR Class B (QP) limit is 66 dB μV.

So, looking at Figure 18.6, we get the required filter attenuation as 47.6 dB at 150 kHz.

Note: As explained in Chapter 16 *under "The Road to Cost-effective Filter Design," even though the CM noise spectrum is flat, since the pole of the LC filter is always well below 150 kHz, the CM noise spectrum after inserting the filter will fall at −40 dB/decade. Since the CISPR-22 Class B limit is falling at −20 dB/decade, the CM noise spectrum will accrue an "advantage" of 20 dB/decade as the frequency rises. So, if we ensure that we are compliant at the lowest frequency (150 kHz), we are assured (theoretically) that we will be automatically compliant at higher frequencies.*

(e) Calculating Filter Components at Stated Line Condition

We therefore need to pick a low-pass LC filter with an appropriate corner frequency ("pole" in Figure 18.6) that provides this attenuation. For example, if we are using a one-stage LC low-pass filter, we know it has an attenuation characteristic of about 40 dB/decade above its corner frequency (i.e., $1/2\pi\sqrt{(LC)}$). So, the equation of the required response is

$$f_{pole} = 10^{[\log f_{att} - (dB_{att}/slope)]} = 10^{[\log 150 \text{ k} - (47.6/40)]} = 9.7 \text{ kHz}$$

We therefore need a filter that has an LC of

$$LC = \left(\frac{1}{2\pi \times 9,700}\right)^2 = 2.7 \times 10^{-10} \text{ s}^2$$

For example, suppose we have finally picked two Y-caps marked "C4" in Figure 16.1 as 2.2 nF each (no "C2" caps are present in view of the safety restrictions on total Y-capacitance as also discussed in *Chapter 16*). Then, as per the equivalent diagram, the effective C is therefore 4.4 nF. We can thus find L_{cm} as follows.

$$L \equiv L_{cm} = \frac{2.7 \times 10^{-10}}{4.4 \times 10^{-9}} = 0.061 \text{ H} \Rightarrow L_{cm} = 61 \text{ mH}$$

This is a very high inductance. We may therefore consider two identical CM filters (LC stages) in cascade instead. Let us also split the Y-cap into four 1.2 nF caps, so each stage gets a total Y-capacitance of 2.4 nF. That way we will not exceed the safety requirements.

Now, *two LC filters in cascade* will give us 80 dB/decade attenuation. Therefore, we get

$$f_{\text{pole}} = 10^{[\log f_{\text{att}} - (\text{dB}_{\text{att}}/\text{slope})]} = 10^{[\log 150 \text{ k} - (47.6/80)]} = 38.1 \text{ kHz}$$

$$LC = \left(\frac{1}{2\pi \times 38,100}\right)^2 = 1.745 \times 10^{-11} \text{ s}^2$$

$$L \equiv L_{\text{cm}} = \frac{1.745 \times 10^{-11}}{2.4 \times 10^{-9}} = 0.0073 \text{ H} \Rightarrow L_{\text{cm}} = 7.3 \text{ mH}$$

This is a readily available value, with an inductance about 10 times smaller than our first calculation with only one CM stage. Though we have two CM stages instead of one, the net savings in terms of total volume occupied by the CM chokes is about $10/2 = 5$ times. Our conclusion is we really need a *two-stage CM filter* to comply with CISPR-22 Class B conducted EMI requirements. Note that the "accidental" CM stage in Figure 16.1 consisting of $L_{\text{cm}}/2$ and C3 usually has a pole at a much higher frequency, so it will not be able to do much here. We should plan on a full extra CM stage.

CHAPTER 19

Solved Examples

Example 1

We have a non-synchronous Buck converter operating in continuous conduction mode (CCM). Its input is 12 V and output is 5 V. It uses a BJT switch with a forward drop $V_{CE}(\text{sat}) \equiv V_{SW} = 0.2$ V. The catch diode is a Schottky device with forward drop $V_D = 0.4$ V. The load current is 1.5 A. What is the duty cycle? What is the dissipation in the BJT and in the diode? What is the estimated efficiency?

We set: $V_O = 5$ V, $V_{IN} = 12$ V, $V_D = 0.4$ V, $V_{SW} = 0.2$ V, $I_O = 1.5$ A.

$$D = \frac{V_O + V_D}{V_{IN} + V_D - V_{SW}} = \frac{5 + 0.4}{12 + 0.4 - 0.2} = 0.4426$$

Dissipation in the BJT and diode, and the total losses are

$$P_{BJT} = I_O \times D \times V_{SW} = 0.1328 \text{ W}$$
$$P_D = I_O \times (1 - D) \times V_D = 0.3344 \text{ W}$$
$$P_{LOSS} = P_{BJT} + P_D = 0.4672 \text{ W}$$

Note that we averaged the switch dissipation over a complete cycle by multiplying it by D, and similarly, averaged the diode dissipation by multiplying it by $1 - D$.

The output power, input power and efficiency are

$$P_O = V_O \times I_O = 7.5 \text{ W}$$
$$P_{IN} = P_O + P_{LOSS} = 7.9672 \text{ W}$$

$$\eta = \frac{P_O}{P_{IN}} = 0.9414 \quad (\text{i.e., } 94.14\%)$$

Example 2

We have a non-synchronous Buck converter operating in continuous conduction mode (CCM). Its input is 12 V and its output is 5 V. It uses a FET switch with $R_{DS} = 0.1$ Ω. The catch diode is a Schottky device with forward drop $V_D = 0.4$ V. The load current is 1.5 A. What is the duty cycle?

We set: $V_O = 5\,\text{V}$, $V_{IN} = 12\,\text{V}$, $V_D = 0.4\,\text{V}$, $I_O = 1.5\,\text{A}$, $R_{DS} = 0.1\,\Omega$.

For a BJT, to a first approximation, we usually assume that its forward voltage drop is almost constant with respect to the current through it, which is the main reason why the BJT (along with its FET-driven cousin, the IGBT) is still often used in high-power applications. For a FET, the forward drop varies significantly, being considered virtually proportional to the current through it. In the simple duty cycle equation however, we need to plug in a certain *fixed* number "V_{SW}." So for a FET, we need to average the forward switch drop over the ON-time (*note*: here we do not average over the *entire* switching cycle). This is equivalent to taking the voltage drop corresponding to the *average* current through the switch during the on-time, which is simply the center-of-ramp (of the inductor current). Further, in a Buck topology, the center-of-ramp is equal to the load current I_O. Hence, denoting I_{SW} as the average current in the switch during the on-time (corresponding to the average drop V_{SW}), we get

$$I_{SW} = I_O$$
$$V_{SW} = I_O \times R_{DS}$$
$$= 1.5 \times 0.1 = 0.15 \ \text{V}$$

$$D = \frac{V_O + V_D}{V_{IN} + V_D - V_{SW}} = \frac{5 + 0.4}{12 + 0.4 - 0.15} = 0.4408$$

Example 3

What is the efficiency of the Buck converter in Example 2, if we disregard both the switch and diode drops?

Now we set: $V_O = 5\,\text{V}$, $V_{IN} = 12\,\text{V}$, $V_D = 0\,\text{V}$, $V_{SW} = 0\,\text{V}$, $I_O = 1.5\,\text{A}$.

This leads to the "ideal" duty cycle equation for a Buck. We will also confirm that, in effect, it assumes 100% efficiency.

$$D_{IDEAL} = \frac{V_O + V_D}{V_{IN} + V_D - V_{SW}} = \frac{V_O}{V_{IN}} = \frac{5}{12} = 0.4167$$

The input current of a Buck is the switch current averaged over the entire on-time. So, the input current corresponding to this duty cycle is

$$I_{IN\_IDEAL} = I_{SW} \times D_{IDEAL} = I_O \times D_{IDEAL} = 0.625 \ \text{A}$$

The corresponding input power is thus

$$P_{IN\_IDEAL} = V_{IN} \times I_{IN\_IDEAL} = 12 \times 0.625 = 7.5 \ \text{W}$$

The output power is

$$P_O = I_O \times V_O = 7.5 \text{ W}$$

Therefore, the efficiency is 7.5 W/7.5 W = 1 (i.e., 100%) as expected, validating our statement that if the switch and diode drops are set to zero, we get an "ideal" situation, with no losses.

> *Note: Of course, the only losses that were allowed in the first place, by the duty cycle equation currently in use in previous examples, are the losses related to the forward drops in the switch and diode, that is, the conduction losses in the semiconductors, no more. This indicates that since, quite obviously, not all switcher losses have been accounted for, the duty cycle equation in use so far is itself limited, and clearly just an approximation.*

Example 4

What is the efficiency of the Buck converter in Example 2, if we disregard (only) the switch drop that is, we assume only a diode drop is present. Also, what is the loss in this diode?

We set: $V_O = 5 \text{ V}$, $V_{IN} = 12 \text{ V}$, $V_D = 0.4 \text{ V}$, $V_{SW} = 0 \text{ V}$, $I_O = 1.5 \text{ A}$.

$$D_{IDEAL} = \frac{V_O + V_D}{V_{IN} + V_D} = \frac{5 + 0.4}{12 + 0.4} = 0.4355$$

The input current of a Buck is the switch current averaged over the entire on-time. So, the input current corresponding to this duty cycle is

$$I_{IN} = I_{SW} \times D = I_O \times D = 1.5 \times 0.4355 = 0.6532 \text{ A}$$

The corresponding input power is thus

$$P_{IN} = V_{IN} \times I_{IN} = 12 \times 0.6532 = 7.8387 \text{ W}$$

The output power is clearly

$$P_O = I_O \times V_O = 7.5 \text{ W}$$

Therefore, the efficiency is 7.5 W/7.8387 W = 0.9568.

Now, the average diode current in a Buck is $I_O \times (1 - D)$. So, we get $I_{D\_AVG} = 1.5 \times (1 - 0.4355) = 0.8468 \text{ A}$. The loss in the diode is therefore

$$P_D = I_{D\_AVG} \times V_D = 0.8468 \times 0.4 = 0.3387 \text{ W}$$

We can see that this is exactly equal to the difference in input power and output power: $P_{IN} - P_O = 7.8387 - 7.5 = 0.3387 \text{ W}$, as expected. So, the balance sheet of losses is complete and accurate.

In terms of input power

In terms of output power

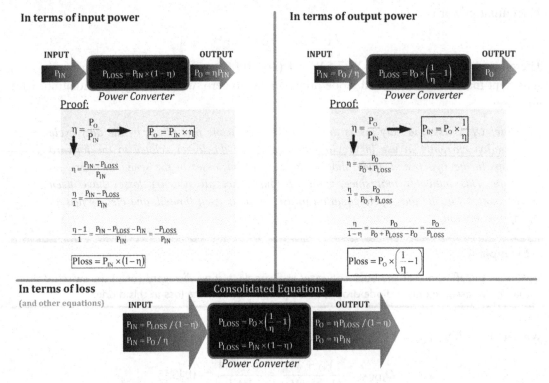

Figure 19.1: Power in, power out, power loss, and efficiency relations.

Example 5

In a Buck converter, we assume as above, that the switch is "ideal" (very low R_{DS}), and the catch diode has a voltage drop of 0.4 V. If the efficiency of the converter is 95.679%, and the diode loss is 0.3387 W, what is the input power? What is the output power?

Here we are just working backwards. Further, we are not assuming any specific input and output voltages, or even a certain load current. We are just talking in terms of power. Looking at Figure 19.1, we see all the possible relationships between input and output power, versus loss and efficiency. Keep in mind these are valid equations for any power converter in general, not necessarily just switchers. We focus our attention on the lowermost diagram in the figure (under "In terms of loss"). To use the equations here, we need to know the loss, which in this example is the loss in the diode.

We set: $P_{LOSS} = 0.3387\,\text{W}$, $\eta = 0.95679$.

So,

$$P_{\text{IN}} = \frac{P_{\text{LOSS}}}{1 - \eta} = \frac{0.3387}{1 - 0.95679} = 7.8387 \text{ W}$$

$$P_O = \frac{P_{\text{LOSS}} \times \eta}{1 - \eta} = 7.5 \text{ W}$$

This agrees with Example 4. We have thus validated the relevant equations in Figure 19.1 and also our previous calculations.

Example 6

In Example 4, correlate the diode dissipation to the additional energy drawn from the input and the increase in input current, as compared to the ideal case.

The diode loss was $P_D = 0.3387$ W. This must correspond to the additional energy per unit time drawn from the input. We recall from Example 3, that the ideal duty cycle was $D_{\text{IDEAL}} = 0.4167$. Now, with diode loss included, the duty cycle is $D = 0.4355$. The general equation for the (average) input current of a Buck is $I_O \times D$. Note that in a Buck, the input current is the switch current averaged first over the on-time, that is, $I_{\text{SW}} \equiv I_O$, and then further averaged over the entire cycle (by multiplying it with D). So, for the ideal case, we get

$$I_{\text{IN\_IDEAL}} = I_O \times D_{\text{IDEAL}} = 1.5 \times 0.4167 = 0.625 \text{ A}$$

However, for the non-ideal case (using the value of D calculated in Example 4)

$$I_{\text{IN}} = I_O \times D = 1.5 \times 0.4355 = 0.6532 \text{ A}$$

The additional energy per unit time inputted when the duty cycle stretches out from its ideal value (in turn leading to the observed increase in input current) is

$$V_{\text{IN}} \times (I_{\text{IN}} - I_{\text{IN\_IDEAL}}) = 12 \times (0.6532 - 0.625) = 0.3387 \text{ W}$$

This is equal to the diode loss in Example 4. This thus validates the following general statement:

I_{IN_IDEAL} is the baseline current level for a given P_O and input/output, corresponding to all the incoming energy being fully converted into useful energy (i.e., no losses). As explained in Chapter 1 too, any increase above and beyond this baseline level, coincides exactly with the losses in the converter.

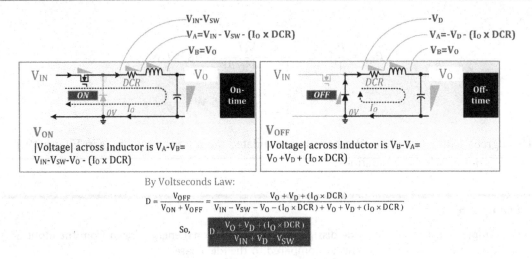

By Voltseconds Law:

$$D = \frac{V_{OFF}}{V_{ON} + V_{OFF}} = \frac{V_O + V_D + (I_O \times DCR)}{V_{IN} - V_{SW} - V_O - (I_O \times DCR) + V_O + V_D + (I_O \times DCR)}$$

So,

$$D = \frac{V_O + V_D + (I_O \times DCR)}{V_{IN} + V_D - V_{SW}}$$

Figure 19.2: Buck duty cycle equations with DCR included.

Example 7

Suppose we have a 12 to 5 V synchronous Buck, using a control FET with R_{DS} of 1 Ω, and a synchronous FET with R_{DS} of 0.8 Ω. The output current is 1.5 A. The inductor has a DC resistance (DCR) of 0.1 Ω. What is the duty cycle, the breakup of the losses, and the efficiency? Continue to ignore switching losses, as we have been doing so far.

We set: $V_O = 5$ V, $V_{IN} = 12$ V, $R_{DS\_1} = 1\,\Omega$, $R_{DS\_2} = 0.8\,\Omega$, DCR $= 0.1\,\Omega$.

Let us call the average current in the two FETs *during the on-time* (not averaged over the whole cycle) as "I_{SW_1}" and "I_{SW_2}". So, since in a Buck, that is equal to the center-of-ramp I_O, we get

$$I_{SW\_1} = I_{SW\_2} = I_O = 1.5 \text{ A}$$

The corresponding switch drops (i.e., their average values over the on-time) are

$$V_{SW\_1} = I_O \times R_{DS\_1} = 1.5 \times 1 = 1.5 \text{ V}$$
$$V_{SW\_2} = I_O \times R_{DS\_2} = 1.5 \times 0.8 = 1.2 \text{ V}$$

So, from the general duty cycle equation in Figure 19.2, with $V_D = V_{SW\_2}$

$$D = \frac{V_O + V_{SW\_2} + I_O \times DCR}{V_{IN} + V_{SW\_2} - V_{SW\_1}} = 0.5427$$

The remaining calculations are

$$I_{\text{IN}} = I_{\text{O}} \times D = 0.8141 \text{ A}$$
$$P_{\text{IN}} = I_{\text{IN}} \times V_{\text{IN}} = 9.7692 \text{ W}$$
$$P_{\text{O}} = I_{\text{O}} \times V_{\text{O}} = 7.5 \text{ W}$$

The computed efficiency is therefore $\eta = P_{\text{O}}/P_{\text{IN}} = 7.5/9.7692 = 0.7677$ (i.e., 76.8%). The losses are $P_{\text{IN}} - P_{\text{O}} = 2.2692 \text{ W}$. Let us confirm where this heat went.

We get the FET and inductor losses as

$$P_{\text{FET\_1}} = (I_{\text{O}}^2 \times R_{\text{DS\_1}}) \times D = 1.2212 \text{ W}$$

$$P_{\text{FET\_2}} = (I_{\text{O}}^2 \times R_{\text{DS\_2}}) \times (1 - D) = 0.8231 \text{ W}$$

$$P_{\text{DCR}} = (I_{\text{O}}^2 \times R_{\text{DCR}}) = 0.225 \text{ W}$$

Summing up all loss terms:

$$P_{\text{LOSS}} = 1.2212 + 0.8231 + 0.225 = 2.2692 \text{ W}$$

This agrees with the difference in input and output power $P_{\text{IN}} - P_{\text{O}}$, thus validating our equations above.

Example 8

Design a wide-input 5 V output, at 5 A, DC–DC synchronous Buck with a switching frequency of 1 MHz. The target efficiency is greater than 80% at max load over the entire input voltage range of 9–57 V.

Having understood the underlying concepts in power conversion, we now do a complete top-down design of a typical wide-input Buck converter. After going through it, the average reader should also be able to complete a similar top-down design for the Boost and Buck-Boost topologies.

Start by assuming zero switch drops (ideal case). Call that duty cycle "D_{IDEAL}." The worst-case design point for a Buck inductor (max peak currents) is V_{INMAX} (see *Chapter 7*). So, that is where we start our Buck design too.

$$D_{\text{IDEAL\_VINMAX}} = \frac{V_{\text{O}}}{V_{\text{INMAX}}} = \frac{5}{57} = 0.0877$$

As per the equations provided in the appendix

$$L = \frac{V_{\text{O}} \times (1 - D_{\text{IDEAL\_VINMAX}})}{I_{\text{O}} \times r_{\text{INITIAL\_VINMAX}} \times f} = \frac{5 \times (1 - 0.0877)}{1.5 \times 0.4 \times 1 \times 10^6} = 2.2807 \times 10^{-6} \text{ H}$$

We have used the initial estimate for r at high line as $r_{\text{INITIAL\_VINMAX}} = 0.4$. But now we need to pick a standard value of L and then recalculate the actual r at high line (and at low). Pick an inductor of standard value $2.2\,\mu\text{H}$. We thus get

$$r_{\text{VINMAX}} = \frac{V_O \times (1 - D_{\text{IDEAL\_VINMAX}})}{I_O \times L \times f} = \frac{5 \times (1 - 0.0877)}{1.5 \times 2.2 \times 10^{-6} \times 1 \times 10^6} = 0.4147$$

Note that this is still an initial estimate, since it is based on the *ideal* duty cycle.

At minimum input voltage we have the following ideal duty cycle and corresponding r

$$D_{\text{IDEAL\_VINMIN}} = \frac{V_O}{V_{\text{INMIN}}} = \frac{5}{9} = 0.5556$$

$$r_{\text{VINMIN}} = \frac{V_O \times (1 - D_{\text{IDEAL\_VINMIN}})}{I_O \times L \times f} = \frac{5 \times (1 - 0.5556)}{1.5 \times 2.2 \times 10^{-6} \times 1 \times 10^6} = 0.202$$

Part 1: FET Selection

In the top (control) FET position, we prefer a P-channel FET so that we can avoid a bootstrap rail as discussed in *Chapter 1*. Of course, the controller IC selected must be commensurate with that desire.

The output power is $5\,\text{V} \times 5\,\text{A} = 25\,\text{W}$. If we target an efficiency of over 80% over the entire operating input voltage range (not over the entire load range!), the input power is a max of $25/0.8 = 31.25\,\text{W}$. In other words, we are allowed a total loss of $6.25\,\text{W}$. To achieve a cost-effective design, we keep in mind that at low line, the top FET conducts for a longer time, whereas at high line, it is the bottom (synchronous) FET that conducts for a longer time. Since *both FETs do not see a worst-case dissipation at the same input voltage extreme*, we now set their individual dissipation targets to about $4\,\text{W}$ (for their respective conduction loss components). The rest is for switching/crossover losses (in the top FET) and other miscellaneous losses like DCR-related losses, capacitor losses, and so on. Note that our FET selection here is based on *efficiency targets*, and is not directly based on current or temperature stresses. But those should be checked out too as discussed in *Chapters 6 and 11*. Note also that with this efficiency target, if the output voltage was not $5\,\text{V}$, but say, $1\,\text{V}$ (as in many VRM applications nowadays), we would need to really choose FETs of much lower R_{DS}. If the output voltage is low, for the same power, the currents are much higher. Further, since heating goes as I^2R, to keep to within a certain power dissipation budget, the resistance "R" (R_{DS} in our case) will need to be reduced significantly (~ 25 times in this example), and/or we will need to choose a much better package too, especially for the lower FET, since the thermal resistance of the currently selected one is rather high.

Further, if we want to do 25 W with just 1 V output, that is at 25 A load current, we will actually need to use interleaved converters as discussed in *Chapter 13*.

For the current application, our first choices for FETs are:

(a) *Top position, control FET, designated #1*: P-channel FET, SUD08P06-155L from Vishay. This is rated 60 V and 6 A at 100 °C case temperature; see also the discussion in *Chapter 6* on FET current ratings. Its maximum ("hot") R_{DS} is 0.28 Ω.

(b) *Bottom position, synchronous FET, designated #2*: N-channel FET: IRFZ34S from Vishay or International Rectifier. This is rated 60 V and about 21 A at 100 °C case temperature. Its max (room temperature) R_{DS} is 50 mΩ, increasing by a factor × 1.6 at high temperatures. So, we take its max ("hot") R_{DS} as 0.05 × 1.6 = 0.08 Ω.

We thus set R_{DS_1} = 0.28 Ω and R_{DS_2} = 0.08 Ω.

> *Note: We will see that the selected Drain-to-Source resistances of the two FETs are commensurate with the target dissipation. Observe that the load current is only 1.5 A, yet we have selected a 6 A FET for the top position, and a 21 A FET for the bottom position. However, the big increase in rating (i.e., lower R_{DS}) of the bottom FET in particular is not only on account of the very low duty cycle at high line (its long conduction time) but also on the significantly worse thermal resistance characteristics it possesses as per its datasheet.*

> *Note: It seems that the voltage stress factor is not sufficient since we are using a 60 V FET in a 57 V max application. However, keep in mind that both the selected FETs have guaranteed avalanche ratings, so they can absorb narrow spikes higher than 60 V. But the application must be evaluated thoroughly to ensure that this FET selection is acceptable.*

Part 2: Conduction Losses in the FETs

For the top FET, we expect the conduction loss to be worst at low line (because of the higher duty cycle). From *Chapter 7* and the appendix, the switch RMS equation is

$$I_{FET\_1\_RMS\_VINMIN} = I_O \times \sqrt{D_{IDEAL\_VINMIN} \times \left(1 + \frac{r^2_{VINMIN}}{12}\right)}$$

$$= 5 \times \sqrt{0.5556 \times \left(1 + \frac{0.202^2}{12}\right)} = 3.7331 \text{ A}$$

Its conduction loss at low line is therefore $P_{COND\_1\_VINMIN} = 3.7331^2 \times 0.28 = \underline{3.9021 \text{ W}}$ (using R_{DS_1} = 0.28 Ω).

At high line, for the same FET, we get

$$I_{FET\_1\_RMS\_VINMAX} = I_O \times \sqrt{D_{IDEAL\_VINMAX} \times \left(1 + \frac{r_{VINMAX}^2}{12}\right)}$$

$$= 5 \times \sqrt{0.0877 \times \left(1 + \frac{0.4147^2}{12}\right)} = 1.4914 \text{ A}$$

Its conduction loss at high line is therefore $P_{COND\_1\_VINMAX} = 1.4914^2 \times 0.28 = \underline{0.6228 \text{ W}}$ (using $R_{DS\_1} = 0.28 \, \Omega$).

We now calculate the dissipation in the bottom FET, over both line extremes. We proceed as above, but with the appropriate R_{DS} and with $1 - D$ instead of D. For this FET we expect worst-case dissipation at high line because of the smaller converter duty cycle that translates into a larger on-time for this FET. At low line, the switch RMS current is

$$I_{FET\_2\_RMS\_VINMIN} = I_O \times \sqrt{(1 - D_{IDEAL\_VINMIN}) \times \left(1 + \frac{r_{VINMIN}^2}{12}\right)}$$

$$= 5 \times \sqrt{(1 - 0.5556) \times \left(1 + \frac{0.202^2}{12}\right)} = 3.339 \text{ A}$$

Its conduction loss is therefore $P_{COND\_2\_VINMIN} = 3.339^2 \times 0.08 = \underline{0.8919 \text{ W}}$ (using $R_{DS\_2} = 0.08 \, \Omega$). At high line, its RMS current is

$$I_{FET\_2\_RMS\_VINMAX} = I_O \times \sqrt{(1 - D_{IDEAL\_VINMAX}) \times \left(1 + \frac{r_{VINMAX}^2}{12}\right)}$$

$$= 5 \times \sqrt{(1 - 0.0877) \times \left(1 + \frac{0.4147^2}{12}\right)} = 4.8098 \text{ A}$$

Its conduction loss is therefore $P_{COND\_2\_VINMAX} = 4.8098^2 \times 0.08 = \underline{1.8507 \text{ W}}$ (using $R_{DS\_2} = 0.08 \, \Omega$).

Finally, we get the total FET dissipation due to conduction losses only, at low and high line as (using the numbers highlighted below)

$$P_{COND\_VINMIN} = P_{COND\_1\_VINMIN} + P_{COND\_2\_VINMIN} = 3.9021 + 0.8919 = 4.794 \text{ W}$$

$$P_{COND\_VINMAX} = P_{COND\_1\_VINMAX} + P_{COND\_2\_VINMAX} = 0.6228 + 1.8507 = 2.4735 \text{ W}$$

Note that even at low line, the RMS current in the bottom FET is comparable to that of the top FET, but because we have picked such a low R_{DS} for the bottom FET, its dissipation is relatively lower at both line extremes. Remember that especially at high line, we still need to account for crossover losses in the top FET, as calculated further below.

Part 3: FET Switching Losses

We are following the steps in *Chapter 8* (Figure 8.16 in particular). We are also trying to maintain most of that chapter's terminology here.

We are assuming that crossover losses occur in the top FET only, as explained in *Chapter 8*.

As per datasheet of the top FET, we set Qgs = 2.3 nC, Vt = 2 V, g = 8 S (Siemens, i.e., mhos).

$$\text{Ciss} = \frac{\text{Qgs}}{\text{Vt} + (I_O/g)} = 0.8762 \text{ nF}$$

Reading from the curves provided in the FET datasheet, we get a much lower value of Ciss = 0.45 nF. Therefore, as explained in *Chapter 8*, we need to apply the following scaling (correction) factor (to account for the voltage coefficient of capacitance)

$$\text{Scaling} = \frac{\text{Ciss from equations}}{\text{Ciss from curves}} = \frac{0.8762}{0.45} = 1.9471$$

From the datasheet curves, we get Coss = 0.06 nF and Crss = 0.04 nF. We need to apply the same scaling factor to these numbers too. We have also expressed all these capacitances in pF rather than nF for the equations further below. So the final values used are

$$\text{Ciss} = 876.2 \text{ pF (value from equation above, expressed in pF)}$$

$$\text{Coss} = (\text{Coss from curves, in nF}) \times 10^3 \times \text{scaling} = 0.06 \times 10^3 \times 1.9471 = 116.8254 \text{ pF}$$

$$\text{Crss} = (\text{Crss from curves, in nF}) \times 10^3 \times \text{scaling} = 0.04 \times 10^3 \times 1.9471 = 77.8836 \text{ pF}$$

Finally, the capacitances we need, to calculate the intervals are

$$\text{Cgd} = \text{Crss} = 77.8836 \text{ pF}$$
$$\text{Cgs} = \text{Ciss} - \text{Cgd} = 798.3069 \text{ pF}$$
$$\text{Cds} = \text{Coss} - \text{Cgd} = 38.9418 \text{ pF}$$

We also set a Gate drive voltage of 9 V. Assume this is very close to the minimum of the input voltage range. Keep in mind that, typically, we need a Gate drive voltage of about 10 V to drive the selected FETs properly (turn them ON fully), but 9 V will also suffice. We also assume the pull-up drive resistor is 2 Ω.

We set: Vdrive = 9 V, Rdrive = 2 Ω.

From Figures 8.7 to 8.10, we get the required durations/time constant as

$$\text{Tg} = \frac{\text{Rdrive} \times \text{Ciss}}{10^3} = \frac{2 \times 876.2}{10^3} = 1.7524 \text{ ns}$$

$$t2 = -\mathrm{Tg} \times \left[ln\left(1 - \frac{I_O}{g \times (\mathrm{Vdrive} - \mathrm{Vt})} \right) \right] = 0.1639 \text{ ns}$$

$$t3_{\mathrm{VINMAX}} = \frac{V_{\mathrm{INMAX}} \times \mathrm{Rdrive} \times \mathrm{Cgd}}{\mathrm{Vdrive} - (\mathrm{Vt} + (I_O/g))} \times 10^{-3} = 1.3927 \text{ ns (at max input)}$$

$$t3_{\mathrm{VINMIN}} = \frac{V_{\mathrm{INMIN}} \times \mathrm{Rdrive} \times \mathrm{Cgd}}{\mathrm{Vdrive} - (\mathrm{Vt} + (I_O/g))} \times 10^{-3} = 0.2199 \text{ ns (at min input)}$$

Therefore, the crossover time during turn-on is (at either voltage extreme):

$$\mathrm{tcross\_turnon}_{\mathrm{VINMAX}} = t2 + t3_{\mathrm{VINMAX}} = 0.1639 + 1.3927 = 1.5566 \text{ ns}$$

$$\mathrm{tcross\_turnon}_{\mathrm{VINMIN}} = t2 + t3_{\mathrm{VINMIN}} = 0.1639 + 0.2199 = 0.3838 \text{ ns}$$

So, the crossover losses associated with turn-on, at both input extremes are

$$\mathrm{Pcross\_turnon}_{\mathrm{VINMAX}} = \frac{1}{2} \times V_{\mathrm{INMAX}} \times I_O \times \mathrm{tcross\_turnon}_{\mathrm{VINMAX}} \times f \times 10^{-9} = 0.2218 \text{ W}$$

$$\mathrm{Pcross\_turnon}_{\mathrm{VINMIN}} = \frac{1}{2} \times V_{\mathrm{INMIN}} \times I_O \times \mathrm{tcross\_turnon}_{\mathrm{VINMIN}} \times f \times 10^{-9} = 8.6355 \times 10^{-3} \text{ W}$$

As expected, the crossover losses are higher at high line because of the higher voltage during crossover (combined with the fact that since this is a Buck, the center-of-ramp is fixed, irrespective of input voltage).

Next we calculate the losses associated with turn-off. We set the pull-down as Rdrive = 1 Ω.

From Figures 8.11 to 8.14, we get the required time constant and durations as

$$\mathrm{Tg} = \frac{\mathrm{Rdrive} \times \mathrm{Ciss}}{10^3} = \frac{1 \times 876.2}{10^3} = 0.8762 \text{ ns}$$

$$T3 = \mathrm{Tg} \times \left[ln\left(\frac{\mathrm{Vt} + (I_O/g)}{\mathrm{Vt}} \right) \right] = 0.2383 \text{ ns (ignoring Vsat in Figure 8.13)}$$

$$T2_{\mathrm{VINMAX}} = \frac{V_{\mathrm{INMAX}} \times \mathrm{Rdrive} \times \mathrm{Cgd}}{(\mathrm{Vt} + (I_O/g))} \times 10^{-3} = 1.6912 \text{ ns (at max input)}$$

$$T2_{\mathrm{VINMIN}} = \frac{V_{\mathrm{INMIN}} \times \mathrm{Rdrive} \times \mathrm{Cgd}}{(\mathrm{Vt} + (I_O/g))} \times 10^{-3} = 0.267 \text{ ns (at min input)}$$

Therefore, the crossover time during turn-off (at either voltage extreme) is

$$\mathrm{tcross\_turnoff}_{\mathrm{VINMAX}} = T3 + T2_{\mathrm{VINMAX}} = 0.2383 + 1.6912 = 1.9295 \text{ ns}$$

$$\text{tcross\_turnoff}_{\text{VINMIN}} = T3 + T2_{\text{VINMIN}} = 0.2383 + 0.267 = 0.5053 \text{ ns}$$

So, the crossover losses associated with turn-off are

$$\text{Pcross\_turnoff}_{\text{VINMAX}} = \frac{1}{2} \times V_{\text{INMAX}} \times I_O \times \text{tcross\_turnoff}_{\text{VINMAX}} \times f \times 10^{-9} = 0.2749 \text{ W}$$

$$\text{Pcross\_turnoff}_{\text{VINMIN}} = \frac{1}{2} \times V_{\text{INMIN}} \times I_O \times \text{tcross\_turnoff}_{\text{VINMIN}} \times f \times 10^{-9} = 0.0114 \text{ W}$$

The total crossover loss, therefore, is

$$\text{Pcross}_{\text{VINMAX}} = \text{Pcross\_turnon}_{\text{VINMAX}} + \text{Pcross\_turnoff}_{\text{VINMAX}} = 0.2218 + 0.2749 = 0.4968 \text{ W}$$

$$\text{Pcross}_{\text{VINMIN}} = \text{Pcross\_turnon}_{\text{VINMIN}} + \text{Pcross\_turnoff}_{\text{VINMIN}} = 0.008636 + 0.0114 = 0.02 \text{ W}$$

However, to complete the switching loss term, we also have to add one more term to the crossover loss above. This is related to the regular dumping of the energy of Cds into the FET at every turn-on. We get, using Cds = 38.94 pF (as calculated earlier)

$$\text{Pcds}_{\text{VINMAX}} = \frac{1}{2} \times \frac{\text{Cds}}{10^{12}} \times V_{\text{INMAX}}^2 \times f = 0.0633 \text{ W}$$

$$\text{Pcds}_{\text{VINMIN}} = \frac{1}{2} \times \frac{\text{Cds}}{10^{12}} \times V_{\text{INMIN}}^2 \times f = 1.5771 \times 10^{-3} \text{ W}$$

So finally, the total switching loss terms at either input voltage extreme (occurring only in the top FET) are

$$\text{Pswitching}_{\text{VINMAX}} = \text{Pcross}_{\text{VINMAX}} + \text{Pcds}_{\text{VINMAX}} = 0.56 \text{ W}$$

$$\text{Pswitching}_{\text{VINMIN}} = \text{Pcross}_{\text{VINMIN}} + \text{Pcds}_{\text{VINMIN}} = 0.0216 \text{ W}$$

Note that we have neglected the Gate driver losses calculated in *Chapter 8*, since they are relatively small, especially when doing dissipation calculations at max load. Driver losses do become significant at light loads, but we have not concerned ourselves with light-load efficiency in this example.

Adding the switching loss to the conduction loss of the top FET, we get the total loss in the top FET

$$P_{1\_\text{VINMAX}} = P_{\text{COND}\_1\_\text{VINMAX}} + \text{Pswitching}_{\text{VINMAX}} = 1.1829 \text{ W}$$

$$P_{1\_\text{VINMIN}} = P_{\text{COND}\_1\_\text{VINMIN}} + \text{Pswitching}_{\text{VINMIN}} = 3.9237 \text{ W}$$

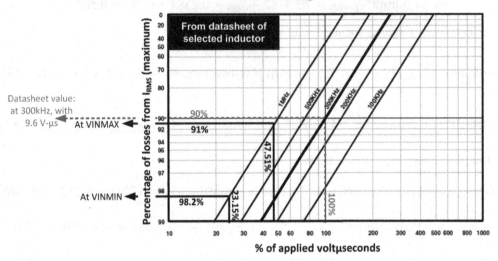

Figure 19.3: Estimating core losses in selected Coiltronics inductor.

The typical assumption is that the bottom FET has negligible switching losses. So,

$$P_{2\_\text{VINMAX}} = P_{\text{COND}\_2\_\text{VINMAX}} = 1.8507 \text{ W}$$

$$P_{2\_\text{VINMIN}} = P_{\text{COND}\_2\_\text{VINMIN}} = 0.8919 \text{ W}$$

Therefore, we get the combined dissipation in both FETs at both input extremes

$$P_{\text{FETs}\_\text{VINMAX}} = P_{1\_\text{VINMAX}} + P_{2\_\text{VINMAX}} = 1.1829 + 1.8507 = 3.0336 \text{ W}$$

$$P_{\text{FETs}\_\text{VINMIN}} = P_{1\_\text{VINMIN}} + P_{2\_\text{VINMIN}} = 3.9237 + 0.8919 = 4.8156 \text{ W}$$

Both these are well within the 6.25 W planned budget. But we still need to calculate and add a few more, relatively minor, loss terms.

Part 4: Inductor Loss

Our choice of inductance was 2.2 μH. We also know that its rating must be greater than 5 A. A suitable off-the-shelf inductor is therefore the 2.2 μH inductor "UP2C-2R2-R" from Coiltronics, rated 7.5 A. Its DCR is 6.6 mΩ and its maximum volt μ seconds is 9.6 at 300 kHz, corresponding to core losses that are 10% of the total losses for a 40 °C rise in temperature.

We need to understand that the inductor has a I_{SAT} rating of 8.67 A (which is sufficient in our application), and that the stated 7.5 A rating corresponds to the amount of *DC current* (no ripple and therefore no core loss) that can be passed through the inductor such that it produces a temperature rise of $40\,°C$. In *Chapter 2*, we discussed ways of validating an off-the-shelf inductor in a particular application, and also estimating its core loss term. This particular vendor, however, provides an easy look-up chart for the purpose, as presented in Figure 19.3. The procedure to use it will become clear soon.

First quantify the total permissible dissipation in the inductor. That is

$$P_{L\_MAX} = I_{DC\_RATING}^2 \times DCR = 7.5^2 \times 6.6 \times 10^{-3} = 0.371 \text{ W}$$

To evaluate the core loss and I_{RMS} derating in our application, we need to find out the applied voltμseconds (which we are hereby calling "*Et*" as we did in *Chapter 2*).

$$Et_{VINMAX} = V_O \frac{1 - D_{IDEAL\_VINMAX}}{f} \times 10^6 = 4.5614 \text{ V-μs}$$

$$Et_{VINMIN} = V_O \frac{1 - D_{IDEAL\_VINMIN}}{f} \times 10^6 = 2.2222 \text{ V-μs}$$

From the datasheet of the inductor, the rated voltμseconds is given by

$$Et_{RATED} = 9.6 \text{ V-μs}$$

So, the ratio of the applied voltμseconds to the rated value (at both lines extremes) is

$$Ratio\_Et_{VINMAX} = \frac{Et_{VINMAX}}{Et_{RATED}} = \frac{4.5614}{9.6} = 0.4751 \Rightarrow 47.51\%$$

$$Ratio\_Et_{VINMIN} = \frac{Et_{VINMIN}}{Et_{RATED}} = \frac{2.2222}{9.6} = 0.2315 \Rightarrow 23.15\%$$

Now looking at Figure 19.3, we see that with 47.15% of the rated voltμseconds and at 1 MHz switching frequency, the percentage of loss from the RMS heating is to be kept to about 91% of the total loss when both are combined to produce a $40\,°C$ rise. So basically, this means the total maximum (allowed) copper loss is now only $0.91 \times 0.371 \text{ W} = 0.338 \text{ W}$. This corresponds to a certain reduction in the rated RMS to a value below 7.5 A, one that we can work out easily as presented further below. But right here, we don't need that value. Because with the given information we can find out the core loss *in our application*. The logic is as follows: the reason for the required RMS derating is that the core loss makes up for the difference. So, it is easy to realize that the estimated core loss in our application (at high line) is $0.371 \text{ W} - 0.338 \text{ W} = 0.033 \text{ W}$. That makes it 9% of the total rated allowed dissipation (0.371 W) as expected. Similarly, at V_{INMIN}, the applied voltμseconds is 23.15% of the rated voltμseconds. From Figure 19.3, we see that it intersects the y-axis at 98.2%. So, with the same logic, the core loss at low input voltage is 1.8% the total rated wattage, that is, $0.018 \times 0.371 = 0.00668 \text{ W}$. We consolidate the results on core loss at both line extremes

$$P_{\text{CORE\_VINMAX}} = 33 \times 10^{-3} \text{ W}$$
$$P_{\text{CORE\_VINMIN}} = 6.7 \times 10^{-3} \text{ W}$$

Next, we calculate the actual copper loss in our application. But before we do that we should confirm that we are operating within the derated RMS rating of the inductor (with the above-mentioned core losses). If it turns out that the continuous max inductor current in our application is less than the derated rating, we can logically conclude that the inductor's temperature rise will be less than 40 °C; otherwise, it will be greater than 40 °C (but that may also be acceptable depending on our worst-case ambient temperature). The derated RMS ratings at both line extremes (in our application) are obtained by the following steps.

(a)   At V_{INMAX}, the maximum allowed copper loss is 91% of the total allowed loss (0.371 W). Now, since the DCR is 6.6 mΩ, we get the budgeted copper loss as

$$P_{\text{CU\_VINMAX}} = 0.91 \times 0.371 = I^2_{\text{RMS\_MAX\_VINMAX}} \times 6.6 \times 10^{-3}$$

Solving we get the maximum RMS rating

$$I_{\text{RMS\_MAX\_VINMAX}} = \sqrt{\frac{0.91 \times 0.371}{6.6 \times 10^{-3}}} = 7.1521 \text{ A}$$

Since the applied RMS in our application is around 5 A (see below), it is well within the max derated rating and the temperature rise will therefore be less than 40 °C.

(b)   At V_{INMIN}, similarly

$$I_{\text{RMS\_MAX\_VINMIN}} = \sqrt{\frac{0.982 \times 0.371}{6.6 \times 10^{-3}}} = 7.4297 \text{ A}$$

The applied RMS is around 5 A as we see below, so it is well within the max derated rating.

The amount of RMS current in the inductor, as per the equations in *Chapter 7* and the Appendix is

$$I_{\text{L\_RMS\_VINMAX}} = I_O \times \sqrt{1 + \frac{r^2_{\text{VINMAX}}}{12}} = 5 \times \sqrt{1 + \frac{0.4147^2}{12}} = 5.0357 \text{ A}$$

$$I_{\text{L\_RMS\_VINMIN}} = I_O \times \sqrt{1 + \frac{r^2_{\text{VINMIN}}}{12}} = 5 \times \sqrt{1 + \frac{0.202^2}{12}} = 5.0085 \text{ A}$$

Note that as expected in a typical CCM inductor design, the RMS current is very close to the DC value (center-of-ramp), so it is not really necessary to use the RMS equation above. We could just use the DC value. The copper loss is

$$P_{\text{CU\_VINMAX}} = I^2_{\text{L\_RMS\_VINMAX}} \times \text{DCR} = 5.0357^2 \times 6.6 \times 10^{-3} = 0.1674 \text{ W}$$

$$P_{\text{CU\_VINMIN}} = I^2_{\text{L\_RMS\_VINMIN}} \times \text{DCR} = 5.0085^2 \times 6.6 \times 10^{-3} = 0.1656 \text{ W}$$

The total inductor loss is finally

$$P_{\text{L\_VINMAX}} = P_{\text{CU\_VINMAX}} + P_{\text{CORE\_VINMAX}} = 0.1674 + 0.033 = 0.200 \text{ W}$$

$$P_{\text{L\_VINMIN}} = P_{\text{CU\_VINMIN}} + P_{\text{CORE\_VINMIN}} = 0.1656 + 0.0067 = 0.172 \text{ W}$$

Part 5: Input Capacitor Selection and Loss

We start by selecting the capacitor based on a target input voltage ripple of $\pm 0.5\%$ (typical of DC−DC integrated switchers and controller ICs). So, at an input of 57 V, that allows for a peak-to-peak input ripple of $1\% \times 57 = 0.57$ V. We are assuming that the maximum input ripple occurs at maximum input voltage (that statement is true for all topologies). From *Chapter 13* we have seen that the actual ripple is a mixture of an ESR-based ripple component (ignoring ESL here) and a capacitance-based ripple component. Let us assume we want to pick a ceramic input capacitor with a typical (low) ESR of 50 mΩ (including lead and trace resistances). From Figure 13.4, we can solve for C_{IN}:

$$C_{\text{IN}} = \frac{I_O \times D_{\text{IDEAL\_VINMAX}} \times (1 - D_{\text{IDEAL\_VINMAX}})}{f \times [V_{\text{RIPP\_PP\_MAX}} - \text{ESR} \times I_O \times (1 + (r_{\text{VINMAX}}/2))]}$$

$$= \frac{5 \times 0.0877 \times (1 - 0.0877)}{1 \times 10^6 \times [0.57 - 0.05 \times 5 \times (1 + (0.4147/2))]} = 1.492 \times 10^{-6} \text{ F}$$

We pick a standard cap of value of 2.2 μF, rated 63 V or higher. The RMS current in the input cap is (see *Chapter 7* and Appendix)

$$I_{\text{CIN\_RMS\_VINMAX}} = I_O \times \sqrt{D_{\text{IDEAL\_VINMAX}} \times \left(1 - D_{\text{IDEAL\_VINMAX}} + \frac{r^2_{\text{VINMAX}}}{12}\right)} = 1.4255 \text{ A}$$

$$I_{\text{CIN\_RMS\_VINMIN}} = I_O \times \sqrt{D_{\text{IDEAL\_VINMIN}} \times \left(1 - D_{\text{IDEAL\_VINMIN}} + \frac{r^2_{\text{VINMIN}}}{12}\right)} = 2.494 \text{ A}$$

Observe that this is consistent with Curve #4 in Figure 7.7, since the duty cycle closest to $D = 0.5$ is at V_{INMIN} in our application. The dissipation, therefore, is

$$P_{\text{CIN\_VINMAX}} = I^2_{\text{CIN\_RMS\_VINMAX}} \times \text{ESR} = 0.1016 \text{ W}$$

$$P_{\text{CIN\_VINMIN}} = I^2_{\text{CIN\_RMS\_VINMIN}} \times \text{ESR} = 0.311 \text{ W}$$

Maximum ESR: based on maximum output ripple

Ignoring ESR and ESL, purely based on capacitance, the maximum allowed output ripple determines a minimum output capacitance.

$$C_O = \frac{r \times I_O}{8 \times f \times V_{RIPPLE}} \qquad \text{(see Figure 13.5)}$$

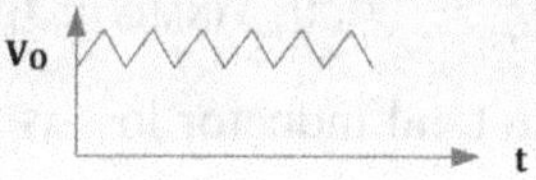

$$C_O \geq \frac{r \times I_O}{8 \times f \times V_{RIPPLE\_MAX}} \qquad \blacktriangleleft$$

Including ESR, but assuming C is large and ESL is negligible. The maximum allowed voltage ripple determines a maximum ESR

$$V_{RIPPLE} = ESR \times I_O \times r$$

$$ESR \leq \frac{V_{RIPPLE\_MAX}}{I_O \times r}$$

Minimum Capacitance: based on maximum droop

Typically, with a well-designed loop, it takes about three switching cycles for the loop to react and start correcting the output to meet a sudden load demand. During that time we do not want the output cap to fall more than a certain value Vdroop. Thus, using I = C dV/dt, we get

$$I = C\frac{\Delta V}{\Delta t} \quad \Rightarrow \quad C \geq \frac{I \times \Delta t}{\Delta V} = \frac{I \times 3T}{\Delta V_{droop}} = \frac{I \times 3}{\Delta V_{droop} \times f}$$

Here the droop is actually related to the <u>extra</u> load demand, since the normal load requirement is being met every cycle without any droop. So the current here is actually the load increase.

$$C_O \geq \frac{3 \times \Delta I_O}{\Delta V_{droop} \times f} \qquad \blacktriangleleft$$

Minimum Capacitance: based on maximum overshoot

There is another criterion. In case of a sudden release of load demand, say from max load Io to zero, the inductor energy will all get dumped into the output cap. If we do not want too much of an overshoot (to a new value Vx):

$$\frac{1}{2} \times C\left(V_X{}^2 - V_O{}^2\right) = \frac{1}{2} \times L\left(I_O{}^2\right) \quad \Rightarrow \quad C \geq \frac{L\left(I_O{}^2\right)}{(V_X + V_O) \times (V_X - V_O)} \approx \frac{L\left(I_O{}^2\right)}{(2V_O) \times (\Delta V_{overshoot})}$$

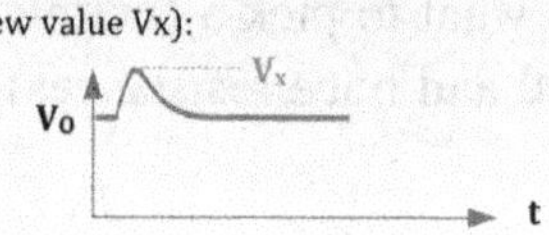

$$C_O \geq \frac{L\left(I_O{}^2\right)}{(2V_O) \times (\Delta V_{overshoot})} \qquad \blacktriangleleft$$

(where we have used the approximation
Vx + Vo ≈ 2 × Vo. Also, Vx - Vo = ΔV)

Figure 19.4: Criteria for output capacitor of a Buck converter.

Part 6: Output Capacitor Selection and Loss

We set ourselves the task of satisfying three constraints simultaneously, as shown below. The corresponding rationale and design equations are provided in Figure 19.4.

(1) Maximum peak-to-peak output ripple to be within 1% (i.e., $\pm 0.5\%$) of output rail, that is, $V_{O\_RIPPLE\_MAX} = 0.05$ V.
(2) Maximum acceptable droop during a sudden increase in load: $\Delta V_{DROOP} = 0.25$ V.
(3) Maximum acceptable overshoot during a sudden decrease in load: $\Delta V_{OVERSHOOT} = 0.25$ V.

We have minimum output capacitances based on (1), (2), and (3) above, as follows:

$$C_{O\_MIN\_1} = \frac{r_{VINMAX} \times I_O}{8 \times f \times V_{O\_RIPPLE\_MAX}} = \frac{0.4147 \times 5}{8 \times 10^6 \times 0.05} = 5.1834 \times 10^{-6} \text{ F}$$

$$C_{\text{O\_MIN\_2}} = \frac{3 \times (I_O/2)}{\Delta V_{\text{DROOP}} \times f} = \frac{3 \times (5/2)}{0.25 \times 10^6} = 3 \times 10^{-5}$$

$$C_{\text{O\_MIN\_3}} = \frac{L \times I_O^2}{2 \times V_O \times \Delta V_{\text{OVERSHOOT}}} = \frac{2.2 \times 10^{-6} \times 5^2}{2 \times 5 \times 0.25} = 2.2 \times 10^{-5}$$

Let us therefore pick a standard output capacitance of $33\,\mu\text{F}$ (rated 6.3 V or higher).

$$C_O = 33 \times 10^{-6}\ \text{F}$$

We should also double check that the ESR of the selected cap is small enough. As per Figure 7.7, we check the ESR-based portion of the ripple at high line. Refer to Figure 13.5 for the output ripple equations. The ESR should be less than

$$\text{ESR}_{\text{Co\_MAX}} = \frac{V_{\text{O\_RIPPLE\_MAX}}}{I_O \times r_{\text{VINMAX}}} = \frac{0.05}{5 \times 0.4147} = 0.0241\ \Omega \ (\text{i.e.,}\ \ 24\ \text{m}\Omega)$$

Most ceramic capacitors will have no trouble complying with this. Suppose we pick a capacitor with ESR = 20 mΩ. To calculate dissipation, we need to find out the RMS current in the output cap.

$$I_{\text{Co\_RMS\_VINMAX}} = I_O \times \frac{r_{\text{VINMAX}}}{\sqrt{12}} = 0.5985\ \text{A}$$

$$I_{\text{Co\_RMS\_VINMIN}} = I_O \times \frac{r_{\text{VINMIN}}}{\sqrt{12}} = 0.2916\ \text{A}$$

$$P_{\text{Co\_VINMAX}} = I_{\text{Co\_RMS\_VINMAX}}^2 \times \text{ESR} = 0.5985^2 \times 0.02 = 7.1647 \times 10^{-3}\ \text{W}$$

$$P_{\text{Co\_VINMIN}} = I_{\text{Co\_RMS\_VINMIN}}^2 \times \text{ESR} = 0.2916^2 \times 0.02 = 1.7005 \times 10^{-3}\ \text{W}$$

Part 7: Total Losses and Efficiency Estimate

We can now sum over the losses in the FETs, inductor, and input/output capacitors.

$$P_{\text{LOSS\_VINMAX}} = P_{\text{FETs\_VINMAX}} + P_{\text{L\_VINMAX}} + P_{\text{CIN\_VINMAX}} + P_{\text{Co\_VINMAX}}$$
$$= 3.0336 + 0.2008 + 0.1016 + 0.007 = 3.3431\ \text{W}$$

$$P_{\text{LOSS\_VINMIN}} = P_{\text{FETs\_VINMIN}} + P_{\text{L\_VINMIN}} + P_{\text{CIN\_VINMIN}} + P_{\text{Co\_VINMIN}}$$
$$= 4.8156 + 0.1722 + 0.311 + 0.0017 = 5.3006\ \text{W}$$

Estimated efficiency is

$$\eta_{\text{VINMAX}} = \frac{P_O}{P_O + P_{\text{LOSS\_VINMAX}}} = \frac{25}{25 + 3.3431} = 0.882 \text{ (i.e., } 88.2\%)$$

$$\eta_{\text{VINMIN}} = \frac{P_O}{P_O + P_{\text{LOSS\_VINMIN}}} = \frac{25}{25 + 5.3006} = 0.8251 \text{ (i.e., } 82.51\%)$$

We see that we have achieved our target of greater than 80% efficiency over the entire input range.

Note that so far we have done the calculations based on the ideal duty cycle equation. Actually, we are now in a position to calculate duty cycle more accurately — based on efficiency as explained in *Chapter 5*. We can then use the new values of duty cycle to run through the above equations once again if we want. Repeated iterations will yield progressively more accurate predictions of efficiency. The improved equations for duty cycle are

$$D_{\text{VINMAX}} = \frac{V_O}{\eta_{\text{VINMAX}} \times V_{\text{INMAX}}} = \frac{5}{0.882 \times 57} = 0.0994$$

$$D_{\text{VINMIN}} = \frac{V_O}{\eta_{\text{VINMIN}} \times V_{\text{INMIN}}} = \frac{5}{0.8251 \times 9} = 0.6733$$

Compare the values so obtained to the ideal values of 0.088 and 0.556, respectively. We realize based on the preceding examples, *that this increase in duty cycle corresponds with the total loss occurring in the converter.*

Part 8: Junction Temperature Estimates

Let us assume that the FETs are mounted reasonably far away from each other, so there are no hot-spots (thermal constriction effects). We also assume that the copper area is large enough (say, a two-sided board, with thermal vias passing through to the copper side below the FETs, and so on).

As per the datasheet of the top FET, we can assume a typical junction to ambient thermal resistance of 25 °C/W, and for the bottom FET we assume 40 °C/W. We also assume that the local ambient (in the vicinity of the FETs) is 15 °C higher than the max room ambient of 40 °C. So for the FETs, the effective max ambient is 55 °C. We also know that the worst-case for the top FET is at low line, and for the bottom FET at high line. We thus estimate the following worst-case junction temperatures for the top and bottom FETs, respectively.

$$T_{J\_1} = P_{1\_\text{VINMIN}} \times Rth_{\text{ja}} + T_{\text{amb}} = 3.9237 \times 25 + 55 = 153.1 \text{ °C}$$

$$T_{J\_2} = P_{2\_\text{VINMIN}} \times Rth_{\text{ja}} + T_{\text{amb}} = 1.8507 \times 40 + 55 = 129.0 \text{ °C}$$

Both the FETs are rated for a max junction temperature of 175 °C, so the above temperatures are considered quite acceptable, though the top FET is running a little too hot (80% temperature derating is a max junction temperature of $0.8 \times 175 = 140\,°C$).

Part 9: Control Loop Design

We need to refer to *Chapter 12* here, in particular Figures 12.25 and 12.26. We assume that the controller is a current mode IC and that it uses a "gm" error amplifier (OTA). Since the switching frequency is 1 MHz, we target a crossover frequency of $f_{sw}/3 = 333\,kHz$ in keeping with the generic guidelines of current mode control. The max current is 5 A and the time period (of switching) is 1 µs.

(1) *Slope Compensation*

Let us assume the IC has a slope compensation of 1.5 A/µs. We want to first rule out subharmonic instability at maximum D and also at $D = 50\%$ (because, remember that three conditions are required for enabling subharmonic instability: current mode control, CCM, and duty cycle exceeding 50%). As per Figure 12.24, the minimum inductances required to avoid this instability at max D and $D = 50\%$, respectively, are

$$\text{Lmin}_1 = V_{\text{INMIN}} \times \frac{D_{\text{VINMIN}} - 0.34}{\text{SlopeComp}_{A/\mu s}} = 9 \times \frac{0.6733 - 0.34}{1.5} = 2\,\mu H$$

$$\text{Lmin}_2 = 2V_O \times \frac{0.5 - 0.34}{\text{SlopeComp}_{A/\mu s}} = 10 \times \frac{0.5 - 0.34}{1.5} = 1.0677\,\mu H$$

In the latter equation we have used the fact that at $D = 50\%$, the input equals twice the output. Note also that we have used the freshly computed value of max duty cycle above, based on the preceding power and efficiency estimates.

Our selected inductance is 2.2 µH, so we will not suffer from subharmonic instability with the amount of slope compensation provided. But if that were not true, we would have needed to *increase the inductance and/or the slope compensation.*

(2) *Load Pole and Plant Transfer Function*

We are looking at Figure 12.25 here. The load pole "fp" is approximately $1/(2\pi RC_O)$, but we are going to do a more exact calculation based on $1/(2\pi AC_O)$. "A" involves either the up-slope or the down-slope of the inductor current (expressed in Amps/µs). We work it all out here.

The down-slope is

$$\text{DownSlope}_{A/\mu s} = \frac{V_O}{L \times 10^6} = \frac{5}{2.2 \times 10^{-6} \times 10^6} = 2.2727\,A/\mu s$$

Let us find "*m*" as defined in Figure 12.25, at both voltage extremes.

$$m_{\text{VINMAX}} = 1 + \left(\frac{\text{SlopeComp}}{\text{DownSlope}} \times \frac{D_{\text{VINMAX}}}{1 - D_{\text{VINMAX}}} \right) = 1.0729$$

$$m_{\text{VINMIN}} = 1 + \left(\frac{\text{SlopeComp}}{\text{DownSlope}} \times \frac{D_{\text{VINMIN}}}{1 - D_{\text{VINMIN}}} \right) = 2.3605$$

So "*A*" as defined in Figure 12.25 is (at both voltage extremes)

$$A_{\text{VINMAX}} = \frac{1}{\dfrac{1}{R} + \dfrac{m_{\text{VINMAX}} - 0.5 - (m_{\text{VINMAX}} \times D_{\text{VINMAX}})}{L \times f}}$$

$$= \frac{1}{\dfrac{1}{1} + \dfrac{1.0729 - 0.5 - (1.0729 \times 0.0994)}{2.2 \times 10^{-6} \times 1 \times 10^{6}}} = 0.8251 \ \Omega$$

$$A_{\text{VINMIN}} = \frac{1}{\dfrac{1}{R} + \dfrac{m_{\text{VINMIN}} - 0.5 - (m_{\text{VINMIN}} \times D_{\text{VINMIN}})}{L \times f}}$$

$$= \frac{1}{\dfrac{1}{1} + \dfrac{2.3605 - 0.5 - (2.3605 \times 0.6733)}{2.2 \times 10^{-6} \times 1 \times 10^{6}}} = 0.8903 \ \Omega$$

where we have used the following value for the load resistor $R = V_O/I_O = 5/5 = 1 \ \Omega$. Note that the use of "*A*" instead of *R* for finding fp, leads to a more accurate estimate of the location of the load pole. In our application, we see that "*A*" has a value between 0.8 and 0.9 Ω, that is, slightly less than $R = 1 \ \Omega$. So, in effect, it pushes out the load pole to a frequency slightly higher than that expected on the basis of the simplified (and more commonly used) value of $fp \approx 1/(2\pi RC_O)$. We thus get

$$fp_{\text{VINMAX}} = \frac{1}{2\pi \times A_{\text{VINMAX}} \times C_O} = \frac{1}{2\pi \times 0.8251 \times 33 \times 10^{-6}} = 5.8449 \times 10^{3} \ \text{Hz}$$

$$fp_{\text{VINMIN}} = \frac{1}{2\pi \times A_{\text{VINMIN}} \times C_O} = \frac{1}{2\pi \times 0.8903 \times 33 \times 10^{-6}} = 5.4171 \times 10^{3} \ \text{Hz}$$

Even though the load pole varies somewhat due to line voltage (because "*A*" varies too), the line rejection is still very good, and that is a key property of current mode control. With voltage mode control we would need to validate the control loop design at both the worst-case input extremes, even though the selection of components would obviously have to be based on one input voltage.

In Figure 12.23, we have defined the mapping resistor "Rmap." This is the characteristic resistor of current mode control, relating the sensed current to the corresponding sensed voltage. Let us assume the total control voltage range (swing) is 1 V, and that it occurs in response to a change in sensed FET current ranging from 0 A to 5 A (no load to full load). In effect, Rmap is therefore 1 V/5 A = 0.2 Ω. Now, we will use this information to calculate "B" in Figure 12.25. For a Buck, "B" equals Rmap. So we write

$$B = 0.2 \ \Omega$$

We can then calculate G_O, the DC gain of the plant at both line extremes

$$G_{O\_VINMAX} = \frac{A_{VINMAX}}{B} = \frac{0.8251}{0.2} = 4.1257$$

$$G_{O\_VINMIN} = \frac{A_{VINMIN}}{B} = \frac{0.8903}{0.2} = 4.4515$$

In terms of decibels, using $20 \times \log(G_O)$, we get 12.31 dB and 12.97 dB, respectively.

We have all the information to plot the plant transfer function if required.

(3) *Compensation Using an OTA*

Now we will complete the feedback section, using a gm-amp. First, we find the attenuation ratio "y" in Figure 12.26. This is the step-down ratio provided by the gm-amp. Note that this step is different from a standard error amplifier as explained in Figure 12.12.

$$y = \frac{V_{REF}}{V_O} = \frac{1}{5} = 0.2$$

We have assumed the reference voltage is 1 V. So, for example, in the voltage divider we could be using 4k as the upper resistor and 1k as the lower one (or 10k and 2.5k, and so on).

We also have an important math relationship in Figure 19.5. Using that, at both voltage extremes, we get

$$fp0_{VINMAX} = \frac{fcross}{A_{VINMAX}/B} \equiv \frac{fcross}{G_{O\_VINMAX}} = \frac{333 \times 10^3}{4.1257} = 80.713 \times 10^3 \ \text{Hz}$$

$$fp0_{VINMIN} = \frac{fcross}{A_{VINMIN}/B} \equiv \frac{fcross}{G_{O\_VINMIN}} = \frac{333 \times 10^3}{4.4515} = 74.806 \times 10^3 \ \text{Hz}$$

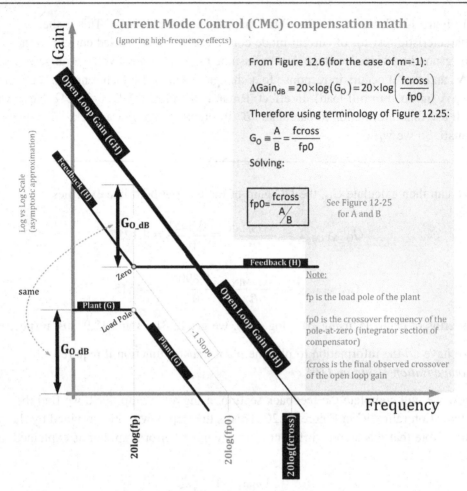

Current Mode Control (CMC) compensation math
(Ignoring high-frequency effects)

From Figure 12.6 (for the case of m=-1):

$$\Delta Gain_{dB} = 20 \times \log(G_O) = 20 \times \log\left(\frac{fcross}{fp0}\right)$$

Therefore using terminology of Figure 12.25:

$$G_O \equiv \frac{A}{B} = \frac{fcross}{fp0}$$

Solving:

$$fp0 = \frac{fcross}{A/B}$$

See Figure 12-25 for A and B

Note:

fp is the load pole of the plant

fp0 is the crossover frequency of the pole-at-zero (integrator section of compensator)

fcross is the final observed crossover of the open loop gain

Figure 19.5: A math relationship for current mode control compensation.

Using the relationships in Figure 12.26 we get two recommendations for each voltage extreme, based on a selected gm = 0.2 (but note that typically, OTAs have much smaller gm values!)

$$C1_{VINMAX} = \frac{y \times gm}{2 \times \pi \times fp0_{VINMAX}} = \frac{0.2 \times 0.2}{2 \times \pi \times 80.713k} = 7.8875 \times 10^{-8} \text{ F (i.e., 79 nF)}$$

$$C1_{VINMIN} = \frac{y \times gm}{2 \times \pi \times fp0_{VINMIN}} = \frac{0.2 \times 0.2}{2 \times \pi \times 74.806k} = 8.5103 \times 10^{-8} \text{ F (i.e., 85 nF)}$$

We see good line rejection at wok as expected. We just pick a close standard value to both.

Set C1 = 82 nF.

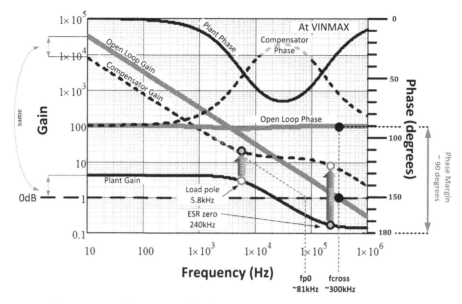

Figure 19.6: Plotting out the final results of the loop compensation.

In Figure 12.26, we see than the location of the zero "fz1" is $1/(2\pi R1C1)$. As per Figure 19.5, we set it at the location of the load pole. So we are setting fp $= 1/(2\pi R1C1)$. Since we know $C1$ from above, we can solve for $R1$ as follows

$$R1_{\text{VINMAX}} = \frac{1}{2\pi \times \text{fp}_{\text{VINMAX}} \times C_1} = 332.0719 \text{ Hz}$$

$$R1_{\text{VINMIN}} = \frac{1}{2\pi \times \text{fp}_{\text{VINMIN}} \times C_1} = 358.2936 \text{ Hz}$$

Once again, we just pick a close standard value.

Set R1 = 333Ω.

The pole fp1 in Figure 12.26 is set to cancel the ESR zero of the output cap (ESR $= 20\,\text{m}\Omega$). The ESR zero frequency is $f_{\text{ESR}} = 1/(2\pi \times \text{ESR} \times C_O) = 241.14\,\text{kHz}$. We thus get

$$C2 = \frac{1}{2\pi \times R_1 \times f_{\text{ESR}}} = 1.982 \times 10^{-9}\ \text{F}$$

We pick a standard value.

Set C2 = 2 nF.

We can plot the final results out, based on Figure 12.22 and the selected compensation components. We thus arrive at Figure 12.6, validating our selection and procedure (Figure 19.6).

Appendix

Chart 1:   DC–DC Design Chart

CCM Assumed	Buck	Boost	Buck-Boost
Voltage across inductor during ON-time "V_{ON}"	$\approx V_{IN} - V_O$	$\approx V_{IN}$	
Voltage across inductor during OFF-time "V_{OFF}"	$\approx V_O$	$\approx V_O - V_{IN}$	$\approx V_O$
Duty cycle "D"	$= \dfrac{V_{OFF}}{V_{ON} + V_{OFF}}$		
	(See Figure 2.1) V_{ON} and V_{OFF} are the magnitudes of the voltages appearing across the inductor during the ON-time and the OFF-time respectively.		
	$\approx \dfrac{V_O + V_D}{V_{IN} - V_{SW} + V_D}$	$\approx \dfrac{V_O - V_{IN} + V_D}{V_O - V_{SW} + V_D}$	$\approx \dfrac{V_O + V_D}{V_{IN} + V_O + V_D - V_{SW}}$
	In the above equation, if using synchronous topologies, consider V_{SW} as the drop across the control FET and V_D the drop across the synchronous FET.		
	$\approx \dfrac{V_O}{V_{IN}}$	$\approx \dfrac{V_O - V_{IN}}{V_O}$	$\approx \dfrac{V_O}{V_{IN} + V_O}$
	$= \dfrac{V_O}{\eta V_{IN}}$	$= \dfrac{V_O - \eta V_{IN}}{V_O}$	$= \dfrac{V_O}{\eta V_{IN} + V_O}$
	$= \dfrac{(V_O/\eta)}{V_{IN}}$	$= \dfrac{(V_O/\eta) - V_{IN}}{(V_O/\eta)}$	$= \dfrac{(V_O/\eta)}{V_{IN} + (V_O/\eta)}$
	(See Figures 5.1 and 5.19) Note that some equations are exact as indicated and some approximate, as explained in Chapter 5. The two ways above of writing D, that is, as ηV_{IN} and as V_O/η give the same duty cycle and seem equivalent in other respects too, but lead to entirely different core sizes as also explained in Chapter 5.		

(Continued)

Chart 1: (Continued)

CCM Assumed	Buck	Boost	Buck-Boost
Ideal duty cycle "D_{IDEAL}"	$= \dfrac{V_O}{V_{IN}}$	$= \dfrac{V_O - V_{IN}}{V_O}$	$= \dfrac{V_O}{V_{IN} + V_O}$
	This ignores all losses, that is, corresponds to an efficiency of 100% ($\eta = 1$). See solved examples in Chapter 19.		
DC transfer function "V_O/V_{IN}"	$= D_{IDEAL}$	$= \dfrac{1}{1 - D_{IDEAL}}$	$= \dfrac{D_{IDEAL}}{1 - D_{IDEAL}}$
	The above equations, in effect, define the ideal duty cycle.		
	$\approx D$	$\approx \dfrac{1}{1 - D}$	$\approx \dfrac{D}{1 - D}$
	$= \eta D$	$= \dfrac{\eta}{1 - D}$	$= \dfrac{\eta D}{1 - D}$
	η is the efficiency of the converter $= P_O/P_{IN}$, and D is the actual/measured duty cycle. Note which equations above are exact and which are approximate as indicated and read Chapter 5.		
Input voltage at $D = 50\%$ "V_{IN_50}"	$\approx (2V_O) + V_{SW} + V_D \approx 2V_O$	$\approx \dfrac{1}{2} \times [V_O + V_{SW} + V_D] \approx \dfrac{V_O}{2}$	$\approx V_O + V_{SW} + V_D \approx V_O$
Output voltage "V_O"	$\approx V_{IN}D - V_{SW}D - V_D(1 - D)$	$\approx \dfrac{V_{IN} - V_{SW}D - V_D(1 - D)}{1 - D}$	$\approx \dfrac{V_{IN}D - V_{SW}D - V_D(1 - D)}{1 - D}$
	$\approx V_{IN}D$	$\approx \dfrac{V_{IN}}{1 - D}$	$\approx \dfrac{V_{IN}D}{1 - D}$
Volt μ seconds "$V\mu s$" ("Et") during ON-time or OFF-time	$\approx \dfrac{V_O + V_D}{f} \times (1 - D) \times 10^6$	$\approx \dfrac{V_O - V_{SW} + V_D}{f} \times D(1 - D) \times 10^6$	$\approx \dfrac{V_O + V_D}{f} \times (1 - D) \times 10^6$
	$\approx \dfrac{V_O}{f} \times (1 - D) \times 10^6$	$\approx \dfrac{V_O}{f} \times D(1 - D) \times 10^6$	$\approx \dfrac{V_O}{f} \times (1 - D) \times 10^6$
	$\equiv \dfrac{(V_{IN} - V_O)}{f} \times D \times 10^6$	$\equiv \dfrac{V_{IN}}{f} \times D \times 10^6$	$\equiv \dfrac{V_{IN}}{f} \times D \times 10^6$
	During steady state		
Inductance "L" (μH)	$\approx \dfrac{V_O + V_D}{I_O \times r \times f} \times (1 - D) \times 10^6$	$\approx \dfrac{V_O - V_{SW} + V_D}{I_O \times r \times f} \times D(1 - D)^2 \times 10^6$	$\approx \dfrac{V_O + V_D}{I_O \times r \times f} \times (1 - D)^2 \times 10^6$
	(See Figures 5.8 and 5.9) f is the switching frequency in Hz and r is the current ripple ratio $= \Delta I_L/I_L$, where I_L is the average inductor current (center of ramp). See equations for r below. Typically, choose L such that $r = 0.4$ (that is, inductor current swing is $\pm 20\%$ of its DC value I_L); also, set r to this value at the highest input voltage for Buck and at the lowest input voltage for Boost and Buck-Boost. See last solved example in Chapter 19 for a full sample design procedure.		

Chart 1: (Continued)

CCM Assumed	Buck	Boost	Buck-Boost
Current ripple ratio "r"	$= \dfrac{\Delta I_L}{I_L} \equiv \dfrac{2 \times I_{AC}}{I_{DC}}$ (spanning)		
	$\dfrac{(V\mu s/L)}{I_O}$	$\dfrac{(V\mu s/L)}{(I_O/(1-D))}$ (spanning)	
	$\approx \dfrac{V_O + V_D}{I_O \times L \times f} \times (1-D) \times 10^6$	$\approx \dfrac{V_O - V_{SW} + V_D}{I_O \times L \times f} \times D(1-D)^2 \times 10^6$	$\approx \dfrac{V_O + V_D}{I_O \times L \times f} \times (1-D)^2 \times 10^6$
	(See Figure 2.2) L is in μH, and f in Hz. Typically, $r = 0.4$ (set at V_{INMAX} for Buck and at V_{INMIN} for Boost and Buck-Boost). (spanning)		
Average current in inductor "I_L"	$= I_O$	$= \dfrac{I_O}{1 - D}$	$= \dfrac{I_O}{1 - D}$
Peak-to-Peak current in inductor "ΔI_L"	$\equiv \Delta I_L \equiv 2 \times I_{AC} = r \times I_L$ (spanning)		
	(See equation for I_L above)		
Peak-to-Peak current in input capacitor	$= I_O \left[1 + \dfrac{r}{2} \right]$	$= \dfrac{I_O}{1-D} \times r$	$= \dfrac{I_O}{1-D} \times \left[1 + \dfrac{r}{2} \right]$
Peak-to-Peak current in output capacitor	$= I_O \times r$	$= \dfrac{I_O}{1-D} \times \left[1 + \dfrac{r}{2} \right]$ (spanning)	
Input voltage ripple (p–p) component (ESR-related)	$= I_O \left[1 + \dfrac{r}{2} \right] \times ESR_{C_{IN}}$	$= \dfrac{I_O}{1-D} \times r \times ESR_{C_{IN}}$	$= \dfrac{I_O}{1-D} \times \left[1 + \dfrac{r}{2} \right] \times ESR_{C_{IN}}$
Output voltage ripple (p–p) component (ESR-related)	$= I_O \times r \times ESR_{C_O}$	$= \dfrac{I_O}{1-D} \times \left[1 + \dfrac{r}{2} \right] \times ESR_{C_O}$ (spanning)	
Input voltage ripple (p–p) component (capacitance-related)	$= \dfrac{I_O \times D\,(1-D)}{f \times C_{IN}}$	$= \dfrac{I_O \times r}{8 \times f \times C_{IN} \times (1-D)}$	$= \dfrac{I_O \times D}{f \times C_{IN}}$
	(See Figures 13.1–13.3)		

(Continued)

Chart 1:　(Continued)

CCM Assumed	Buck	Boost	Buck-Boost
Output voltage ripple (p–p) component (capacitance-related)	$= \dfrac{I_O \times r}{8 \times f \times C_O}$	$= \dfrac{I_O \times (1-D)}{f \times C_O}$	
	(See Figures 13.1–13.3)		
RMS current in input cap	$= I_O \sqrt{D\left[1-D+\dfrac{r^2}{12}\right]} \approx \dfrac{I_O}{2}$	$= \dfrac{I_O}{1-D} \times \dfrac{r}{\sqrt{12}} \approx 0$	$= \dfrac{I_O}{1-D}\sqrt{D\left[1-D+\dfrac{r^2}{12}\right]}$
	(See Figures 7.6 and 7.7)		
	For a Buck, max RMS current in C_{IN} occurs at $D=0.5$ (i.e., when $V_{IN}=2V_O$).		
RMS current in output cap	$= I_O \times \dfrac{r}{\sqrt{12}} \approx 0$	$= I_O \times \sqrt{\dfrac{D+(r^2/12)}{1-D}}$	
RMS current in inductor	$= I_O \times \sqrt{1+\dfrac{r^2}{12}}$	$= \dfrac{I_O}{1-D} \times \sqrt{1+\dfrac{r^2}{12}}$	
RMS current in switch	$= I_O \times \sqrt{D \times \left[1+\dfrac{r^2}{12}\right]}$	$= \dfrac{I_O}{1-D} \times \sqrt{D \times \left[1+\dfrac{r^2}{12}\right]}$	
RMS current in diode (or sync FET)	$= I_O \times \sqrt{(1-D) \times \left[1+\dfrac{r^2}{12}\right]}$	$= I_O \times \sqrt{\dfrac{[1+(r^2/12)]}{(1-D)}}$	
	(See Figures 7.3 and 7.6)		
Peak current in switch and diode and inductor "I_{PEAK}"	$= I_O \times \left[1+\dfrac{r}{2}\right]$	$= \dfrac{I_O}{1-D} \times \left[1+\dfrac{r}{2}\right]$	
Average current in switch	$= I_O \times D$	$= I_O \times \dfrac{D}{1-D}$	$= I_O \times \dfrac{D}{1-D}$
Average current in diode	$= I_O \times (1-D)$	$= I_O$	$= I_O$
Average current in inductor "I_L"	$= I_O$	$= \dfrac{I_O}{1-D}$	$= \dfrac{I_O}{1-D}$
Average input current "I_{IN}"	Same as average switch current	Same as average inductor current	Same as average switch current
	$= I_O \times D$	$= \dfrac{I_O}{1-D}$	$= I_O \times \dfrac{D}{1-D}$

(Continued)

Chart 1: (Continued)

CCM Assumed	Buck	Boost	Buck-Boost
Energy cycled through inductor every cycle (μJ) "$\Delta\varepsilon$"	$= V\mu s \times I_L$		
	$= \dfrac{P_O}{\eta \times f} \times (1 - D)$	$= \dfrac{P_O}{\eta \times f} \times D$	$= \dfrac{P_O}{\eta \times f}$
	(See Figure 5.5)		
Peak energy handling capability of core "ε" (μJ)	$= \dfrac{1}{2} \times L \times I_{PEAK}{}^2 = \dfrac{\Delta\varepsilon}{8} \times \left[r \times \left(\dfrac{2}{r} + 1 \right)^2 \right]$		
	$= \dfrac{I_O \times V\mu s}{8} \times \left[r \times \left(\dfrac{2}{r} + 1 \right)^2 \right]$	$= \dfrac{I_O \times V\mu s}{8 \times (1 - D)} \times \left[r \times \left(\dfrac{2}{r} + 1 \right)^2 \right]$	
	(See Figure 5.6)		

Chart 2:   AC–DC Design Chart

CCM Assumed	Single-Ended Forward (like a Buck)	Flyback (like a Buck-Boost)
Transformer turns ratio "n"	$= \dfrac{N_P}{N_S}$	
Reflected output voltage "V_{OR}"	$\approx n \times V_O$	
	$\approx n \times (V_O + V_D)$	
	$= n \times (V_O/\eta)$	
Reflected input voltage "V_{INR}"	$\approx \dfrac{V_{IN}}{n}$	
	$\approx \dfrac{V_{IN} - V_{SW}}{n}$	
	$= \dfrac{\eta V_{IN}}{n}$	
Reflected output current "I_{OR}"	$= \dfrac{I_O}{n}$	
Reflected input current "I_{INR}"	$= n \times I_{IN}$	
	(See Figure 3.2) See also, the equation for I_{IN} further below	
Duty cycle	$= \dfrac{V_O}{V_{INR}}$	$= \dfrac{V_O}{V_{INR} + V_O}$
	$= \dfrac{V_{OR}}{V_{IN}}$	$= \dfrac{V_{OR}}{V_{IN} + V_{OR}}$
	$= \dfrac{V_O}{(\eta/n) \times V_{IN}}$	$= \dfrac{V_O}{((\eta/n) \times V_{IN}) + V_O}$
	$= \dfrac{V_O \times (n/\eta)}{V_{IN}}$	$= \dfrac{V_O \times (n/\eta)}{V_{IN} + (V_O \times (n/\eta))}$
	(See Figure 5.1)	
Ideal duty cycle "D_{IDEAL}"	$= \dfrac{\eta V_O}{V_{IN}}$	$= \dfrac{\eta V_O}{V_{IN} + n V_O}$
DC transfer function "V_O/V_{IN}"	$= (D_{IDEAL}/\eta)$	$= \dfrac{(D_{IDEAL}/\eta)}{1 - D_{IDEAL}}$
	$= D \times (\eta/n) \approx \dfrac{D}{n}$	$= \dfrac{D \times (\eta/n)}{1 - D} \approx \dfrac{1}{n} \times \dfrac{D}{1 - D}$
	η is the efficiency of the converter $= P_O/P_{IN}$, D is the actual/measured duty cycle , and n is the turns ratio.	

(Continued)

Chart 2: (Continued)

CCM Assumed	Single-Ended Forward (like a Buck)	Flyback (like a Buck-Boost)
Inductance "L" (μH)	$\approx \dfrac{V_O}{I_O \times r \times f} \times (1-D) \times 10^6$	$\approx \dfrac{V_O}{I_O \times r \times f} \times (1-D)^2 \times 10^6$
	(See Figures 5.8, 5.9, and 14.4) This refers to the inductance of the output choke of a Forward converter and the Primary side of the transformer in a flyback (measured with the Secondary windings open). f is the switching frequency in Hz, and r is the current ripple ratio; see below. Typically, choose L such that $r = 0.4$ (i.e., inductor current swing is $\pm 20\%$ of its DC or center of ramp value I_L); also, set r to this value at the highest input voltage for Forward and at the lowest input voltage for flyback.	
Average current in inductor "I_L" (or center of ramp)	$= I_O$	**Primary side:** $= \dfrac{I_{OR}}{1-D} \equiv \dfrac{(I_O/n)}{1-D}$ **Secondary side:** $= \dfrac{I_O}{1-D}$ (See Figure 3.2)
Current ripple ratio "r"	\multicolumn{2}{c}{$= \dfrac{\Delta I_L}{I_L} \equiv \dfrac{2 \times I_{AC}}{I_{DC}}$}	
	$\approx \dfrac{V_O}{I_O \times L \times f} \times (1-D) \times 10^6$	$\approx \dfrac{V_O}{I_O \times L \times f} \times (1-D)^2 \times 10^6$
	(See Figures 2.2 and 3.2) r is the current ripple ratio $= \Delta I_L/I_L \equiv 2 \times I_{AC}/I_{DC}$, where I_L is the average inductor current (the center of ramp, i.e., I_{DC}) and I_{AC} is its AC component $\equiv \Delta I_L/2$; L is in μH and f is in Hz. Typically, set $r = 0.4$, that is, inductor current swing is then $\pm 20\%$ of its DC (center of ramp) value "I_L." For a flyback, r is the same on either side of the transformer, though currents and current swings on either side of the transformer are scaled as per the turns ratio.	
Peak-to-Peak current in inductor "ΔI_L"	\multicolumn{2}{c}{$\equiv \Delta I_L \equiv 2 \times I_{AC} = r \times I_L$}	
	\multicolumn{2}{c}{(See equations for I_L above)}	

(Continued)

Chart 2:   (Continued)

CCM Assumed	Single-Ended Forward (like a Buck)	Flyback (like a Buck-Boost)
Peak-to-Peak current in input capacitor	$\approx I_{OR}\left[1 + \dfrac{r}{2}\right]$ Ignoring transformer magnetization current	$\approx \dfrac{I_{OR}}{1 - D} \times \left[1 + \dfrac{r}{2}\right]$
Peak-to-Peak current in output capacitor	$= I_O \times r$	$= \dfrac{I_O}{1 - D} \times \left[1 + \dfrac{r}{2}\right]$
Input voltage ripple (p–p) component (ESR-related)	$\approx I_{OR}\left[1 + \dfrac{r}{2}\right] \times ESR_{C_{IN}}$ Ignoring transformer magnetization current	$= \dfrac{I_{OR}}{1 - D} \times \left[1 + \dfrac{r}{2}\right] \times ESR_{C_{IN}}$
Output voltage ripple (p–p) component (ESR-related)	$= I_O \times r \times ESR_{C_O}$	$= \dfrac{I_O}{1 - D} \times \left[1 + \dfrac{r}{2}\right] \times ESR_{C_O}$
Input voltage ripple (p–p) component (capacitance-related)	$\approx \dfrac{I_{OR} \times D\,(1 - D)}{f \times C_{IN}}$ Ignoring transformer magnetization current (See Figures 13.1–13.3)	$= \dfrac{I_{OR} \times D}{f \times C_{IN}}$
Output voltage ripple (p–p) component (capacitance-related)	$= \dfrac{I_O \times r}{8 \times f \times C_O}$ (See Figures 13.1–13.3)	$= \dfrac{I_O \times (1 - D)}{f \times C_O}$
RMS current in input cap	$\approx I_{OR}\sqrt{D\left[1 - D + \dfrac{r^2}{12}\right]} \approx \dfrac{I_{OR}}{2}$ Ignoring transformer magnetization current (See Figures 7.6 and 7.7) For a Forward converter, max RMS current in C_{IN} occurs at $D = 0.5$, that is, at which $V_{IN}/\eta = 2 \times V_O$.	$= \dfrac{I_{OR}}{1 - D}\sqrt{D\left[1 - D + \dfrac{r^2}{12}\right]}$
RMS current in output cap	$= I_O \times \dfrac{r}{\sqrt{12}} \approx 0$	$= I_O \times \sqrt{\dfrac{D + (r^2/12)}{1 - D}}$

(Continued)

Chart 2: (Continued)

CCM Assumed	Single-Ended Forward (like a Buck)	Flyback (like a Buck-Boost)
RMS current in inductor and windings	**Primary side:** $$\approx I_{OR} \times \sqrt{D \times \left[1 + \frac{r^2}{12}\right]}$$ Ignoring transformer magnetization current	**Primary side:** $$= \frac{I_{OR}}{1-D} \times \sqrt{D \times \left[1 + \frac{r^2}{12}\right]}$$
	Secondary side: $$= I_O \times \sqrt{D \times \left[1 + \frac{r^2}{12}\right]}$$ **Output choke:** $$= I_O \times \sqrt{1 + \frac{r^2}{12}}$$	**Secondary side:** $$= I_O \times \sqrt{\frac{[1 + (r^2/12)]}{1-D}}$$
	(See Figure 7.3)	
RMS current in switch	$$\approx I_{OR} \times \sqrt{D \times \left[1 + \frac{r^2}{12}\right]}$$ Ignoring transformer magnetization current	$$= \frac{I_{OR}}{1-D} \times \sqrt{D \times \left[1 + \frac{r^2}{12}\right]}$$
RMS current in diode (or sync FET)	**Output diode (to transformer):** $$= I_O \times \sqrt{D \times \left[1 + \frac{r^2}{12}\right]}$$ **Freewheeling diode (to ground):** $$= I_O \times \sqrt{(1-D) \times \left[1 + \frac{r^2}{12}\right]}$$	$$= I_O \times \sqrt{\frac{[1 + (r^2/12)]}{(1-D)}}$$
	(See Figures 7.3 and 7.6)	
Average current in switch	$$\approx I_{OR} \times D$$ Ignoring transformer magnetization current	$$= I_{OR} \times \frac{D}{1-D}$$

(Continued)

Chart 2:   (Continued)

CCM Assumed	Single-Ended Forward (like a Buck)	Flyback (like a Buck-Boost)
Average current in diode	**Output diode (to transformer):** $= I_O \times D$	$= I_O$
	Freewheeling diode (to ground): $= I_O \times (1 - D)$	
Average input current "I_{IN}"	Same as average switch current	
	$= I_{OR} \times D$	$= I_{OR} \times \dfrac{D}{1 - D}$
Peak energy handling capability of core "ε" (μJ)	$= \dfrac{I_O \times V\mu s}{8} \times \left[r \times \left(\dfrac{2}{r} + 1 \right)^2 \right]$	$= \dfrac{I_{OR} \times V\mu s}{8 \times (1 - D)} \times \left[r \times \left(\dfrac{2}{r} + 1 \right)^2 \right]$
	(See Figures 5.5 and 5.6) This peak energy refers to the output choke of a Forward converter and to the transformer of a flyback. For flyback, use the $V\mu s$ appearing across the Primary winding, that is, $V_{IN} \times D/f \times 10^6$ or equivalently $V_{OR} \times (1 - D)/f \times 10^6$.	

Chart 3:   Multi-topology Voltage Stresses Design Chart

$n = N_P/N_S$ $V_{INR} = V_{IN}/n$ $V_{OR} = nV_O$		Switch	Catch Diode	Output Diode	Coupling/ Clamp Cap	Ideal Transfer Function
Buck	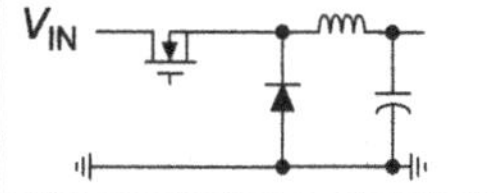	V_{INMAX}	V_{INMAX}		NA	$\dfrac{V_O}{V_{IN}} = D$
Boost	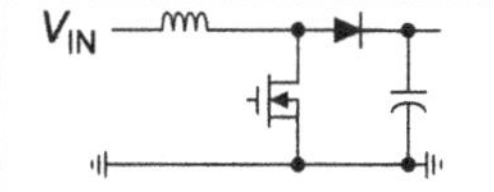	V_O	V_O		NA	$\dfrac{V_O}{V_{IN}} = \dfrac{1}{1-D}$
Buck-Boost	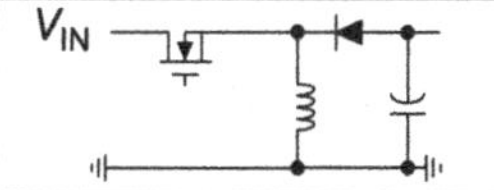	$V_{INMAX} + V_O$	$V_{INMAX} + V_O$		NA	$\dfrac{V_O}{V_{IN}} = \dfrac{D}{1-D}$
Flyback	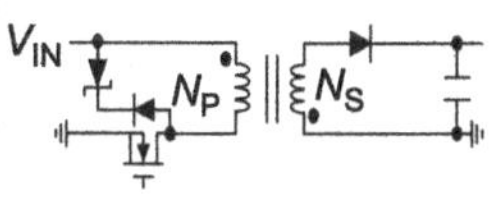	$V_{INMAX} + V_Z$	$V_{INRMAX} + V_O$		NA	$\dfrac{V_O}{V_{INR}} = \dfrac{D}{1-D}$
Forward	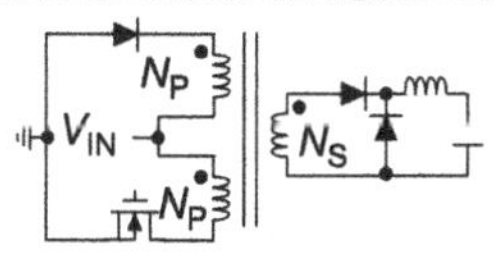	$2 \times V_{INMAX}$	V_{INRMAX}	$V_{INRMAX} + V_O$	NA	$\dfrac{V_O}{V_{INR}} = D$
2-switch Forward	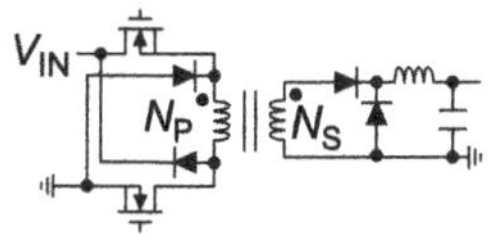	V_{INMAX}	V_{INRMAX}	$V_{INRMAX} + V_O$	NA	$\dfrac{V_O}{V_{INR}} = D$
Active Clamp	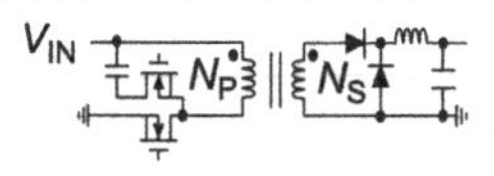	$\dfrac{V_{INMAX}}{1 - D_{MAX}}$	V_{INRMAX}	$V_{INRMAX} \times \dfrac{D_{MAX}}{1 - D_{MAX}} + V_O$	$\dfrac{V_{IN} D_{MAX}}{1 - D_{MAX}}$	$\dfrac{V_O}{V_{INR}} = D$
Half Bridge	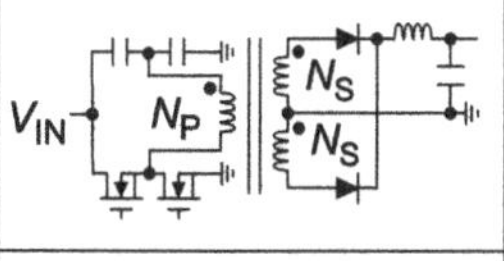	V_{INMAX}	V_{INRMAX}	V_{INRMAX}	NA	$\dfrac{V_O}{V_{INR}} = D$
Full Bridge	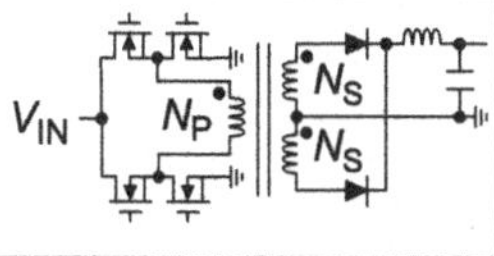	V_{INMAX}	$2 \times V_{INRMAX}$	$2 \times V_{INRMAX}$	NA	$\dfrac{V_O}{V_{INR}} = 2D$
Push-Pull	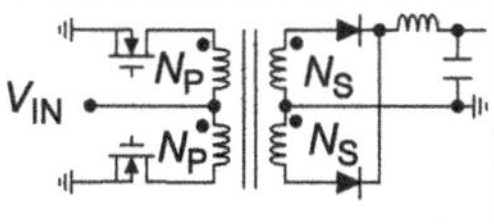	$2 \times V_{INMAX}$	$2 \times V_{INRMAX}$	$2 \times V_{INRMAX}$	NA	$\dfrac{V_O}{V_{INR}} = 2D$
Cuk	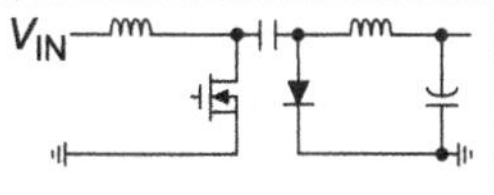	$V_{INMAX} + V_O$	$V_{INMAX} + V_O$		$V_{INMAX} + V_O$	$\dfrac{V_O}{V_{IN}} = \dfrac{D}{1-D}$

(Continued)

Chart 3:   (Continued)

	$n = N_P/N_S$ $V_{INR} = V_{IN}/n$ $V_{OR} = nV_O$	Switch	Catch Diode	Output Diode	Coupling/ Clamp Cap	Ideal Transfer Function
Sepic	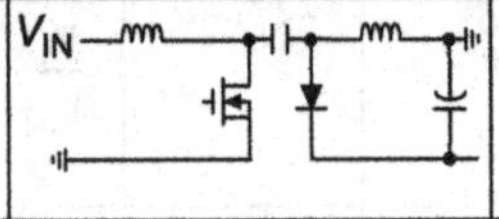	V_{INMAX} $+ V_O$		$V_{INMAX} + V_O$	V_{INMAX}	$\dfrac{V_O}{V_{IN}} = \dfrac{D}{1-D}$
Zeta	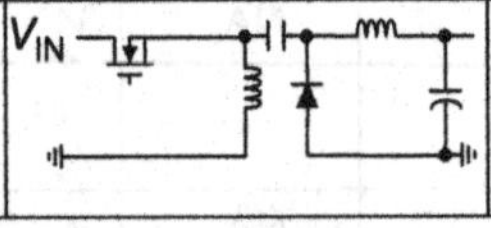	V_{INMAX} $+ V_O$		$V_{INMAX} + V_O$	V_O	$\dfrac{V_O}{V_{IN}} = \dfrac{D}{1-D}$

See additional comments in Table 7.1.

Further Reading

[1] P.C.E. Collett, Investigations into aspects affecting the design of mains filters for frequencies in the range 10 kHz–30 MHz, ERA Report No.82-145 R, ERA Technology, Surrey, UK, 1983.

[2] Capacitors for RFI suppression of the AC line: basic facts, Evox-Rifa Application Notes, fourth ed., Evox-Rifa, Lincolnshire, IL.

[3] E.C. Snelling, Soft Ferrites, Properties and Applications, second ed., Butterworths, 1988, ISBN 0408027606.

[4] Power Factor Corrector, Application Manual, first ed., SGS-Thomson Microelectronics, 1995.

[5] Data Handbook, Aluminum Electrolytic Capacitors, PA01-A, N.A. ed., Philips Components, 1993.

[6] United Chemi-Con Inc., Understanding Aluminum Electrolytic Capacitors, second ed., United Chemi-Con, Rosemont, IL, 1995.

[7] Micro Linear Corporation Data Book, Micro Linear Corporation, San Jose, CA, 1995.

[8] Fair-Rite Soft Ferrites, Databook, thirteen ed., Fair-Rite Products, Wallkill, NY.

[9] Magnetics Designer, Supplementary Information, Intusoft, 1997.

[10] UC3842/3/4/5 Provides Low-cost Current-mode Control, Application Note, U-100 A, Unitrode Integrated Circuits.

[11] K. Billings, Switchmode Power Supply Handbook, McGraw-Hill, New York, 1989, 0-07-005330-8.

[12] A.I. Pressman, Switching Power Supply Design, McGraw-Hill, New York, 1991, 0-07-050806-2.

[13] W.T. McLyman, Transformer and Inductor Design Handbook, second ed., Marcel Dekker, New York, 19880-8247-7828-6.

[14] Unitrode Power Supply Design Seminar, SEM-500, Unitrode Integrated Circuits.

[15] 3C85 Handbook, Ordering Code 9398 345 90011, Philips Electronic Components and Materials, 1987.

[16] K.K. Sum, Intuitive magnetic design, in: Electronic Design Workshops, November 15–16, 2000, Penton Media, New York.

[17] G.E. Bloom, DC-DC switchmode power converters, circuits and converters, in: National Semiconductor Corporation Seminar Presentation, April 25, 2002, Bloom Associates, CA.

[18] S.A. Mulder, Application Note on the design of Low Profile High Frequency Transformers, A New Tool in SMPS Design, Ordering Code 9398 074 80011, Philips Components Corporate Innovation Materials, Eindhoven, MD, 1990.

[19] H. Ahmadi, Calculating Creepage and Clearance Early Avoids Design Problems Later, Compliance Engineering Magazine, March/April 2001.

[20] R. Redl, Low-cost line-harmonics reduction, in: Power Quality Conference, Bremen, 1995.

[21] B. Carsten, Calculating skin and proximity Effect, conductor losses in switchmode magnetics, in: PCIM Conference, 1995.

[22] Magnetics® Ferrites, Databook, Magnetics Inc., Division of Spang and Company, 1999.

[23] S. Lee, Thermal management of electronic equipment, in: PCIM Conference, 1996.

[24] R.D. Middlebrook, S. Cuk, Advances in Switched-Mode Power Conversion, vols. I–III, TESLAco, Irvine, CA, 1983.

[25] C. Likely, Guide to selecting inductors for switching regulators, Power Electronics Technology, USA, 2003.

[26] Output Capacitor Selection for the AAT115X Series Buck Converter, AN-106, Analogic Tech.

[27] P. Wong, P. Xu, P. Yang, F.C. Lee, Performance improvements of interleaving VRMs with coupling inductors, IEEE Trans. Power Electron. 16 (July (4)) (2001) Page 499−507.

[28] G. Zhu, B.A. McDonald, K. Wang, Modeling and analysis of coupled inductors in power converters, IEEE Trans. Power Electron 26(May (5)) (2011) Page 1355−1363.

[29] G. Zhu, B. McDonald, K. Wang, Modeling and analysis of coupled inductors in power converters, Paper presented at APEC 2009, February 2009, Washington, DC.

[30] S.A. Wibowa, Z. Ting, M. Kono, T. Taura, Y. Kobori, H. Kobayshi, Analysis of coupled inductors for low-ripple fast-response buck converter, IEICE Transactions February (2009) 5.

[31] J.A. Ferriera, Improved analytical modeling of conductive losses in magnetic components, IEEE Trans. Power Electron. 9(January (1)) (1994) 127−131.

[32] X. Nan, C.R. Sullivan, An improved calculation of proximity-effect loss in high-frequency windings of round conductors, in: IEEE Power Electronics Specialists Conference, June 2003.

[33] P. Wong, P. Xu, B. Yang, Q. Wu, F.C. Lee, Investigating coupling inductors in the interleaving QSW VRM, in: CPES, Virginia Polytechnic and State University, Blacksburg, VA, March 2000.

[34] S. Maniktala, Switching power supply design & optimization, McGraw-Hill, New York, 2004, 0071434836.

[35] S. Maniktala, Troubleshooting switching power converters, Newnes, Wolgan Valley, 2007, 0750684216.

[36] C. Basso, Switch-mode power supplies spice simulations and practical designs, McGraw-Hill, New York, 2008, 0071508589.

[37] All unitrode application notes. < http://www.smps.us/Unitrode.html > .

[38] S. Maniktala, Switching Power Supplies A to Z, first ed., Newnes, Wolgan Valley, 2006, 0750679700.

Index